Power Systems

More information about this series at http://www.springer.com/series/4622

Jan A. Melkebeek

Electrical Machines and Drives

Fundamentals and Advanced Modelling

Springer

Jan A. Melkebeek
Faculty of Engineering and Architecture
Ghent University
Zwijnaarde, Ghent
Belgium

ISSN 1612-1287 ISSN 1860-4676 (electronic)
Power Systems
ISBN 978-3-319-72729-5 ISBN 978-3-319-72730-1 (eBook)
https://doi.org/10.1007/978-3-319-72730-1

Library of Congress Control Number: 2017962069

Printed on acid-free paper

This Springer imprint is published by Springer Nature
The registered company is Springer International Publishing AG
The registered company address is: Gewerbestrasse 11, 6330 Cham, Switzerland

To my late parents and brother

Foreword

It is remarkable to me that well over half of the subject matter in this book simply did not exist when I taught my first electric machines course in 1959. This is testimony to the expanding role of electromechanical conversion systems, fueled by demands for improved energy management and enabled by developments in power electronic devices and systems. It also speaks to the depth and breadth of the topical coverage in the book.

The organisation into four parts, Electric Machines, Power Electronics, Electric Drives and Drive Dynamics, provides substantial flexibility. Viewed as a textbook, the four parts together provide comprehensive content suitable for a multi-semester course sequence in electromechanical energy conversion. Alternatively, Part 1 can be used alone for a one-semester course in Electric Machines and portions of Parts 2–4 for a one-semester Electric Drives course and a more advanced course emphasising dynamics using Part 4. The book can also be viewed as a valuable reference book because of its comprehensive coverage of the subject area including many special topics such as stepping motors, switched reluctance motors, small electric motor drives and voltage surges in electrical machines.

I spent a semester in Gent as a Fulbright Lecturer and had the opportunity to collaborate with the author on the influence of magnetic saturation on electric machine dynamic behaviour. This experience left me with a deep appreciation of the author' s dedication to accurate but clear description of technical matters that has carried over to this text.

Whether as a text or a reference, the content of this book provides a comprehensive treatment of electric machines and drives spiced with a generous collection of special topics not usually included in contemporary books. It is a worthy addition to any collection of electric machines books.

Donald Novotny
Emeritus Professor, Department of Electrical and
Computer Engineering University of Wisconsin—
Madison, Madison USA

Preface

This work can be used as a comprehensive study and reference textbook on the most common electrical machines and drives. In contrast with many textbooks on drives, this book goes back to the fundamentals of electrical machines and drives, following in the footsteps of the traditional textbooks written by Richter and Bödefeld & Sequenz in German.

The basic idea is to start from the pure electromagnetic principles to derive both the equivalent circuits and the steady-state equations of these electrical machines (e.g. in Part 1) as well as its dynamic equations in Part 4. In my view, only this approach leads to a full understanding of the machine, of the steady-state behaviour of a drive and its dynamics. Much attention is paid to the electromagnetic basis and to analytical modelling. Intentionally, computer simulation is not addressed, although the students are required to use computer models in the exercises and projects, for example, for the section on power electronics or that on dynamic modelling and behaviour. I have successfully used this approach for more than 30 years, and I often receive mails and requests from former students working abroad, who would like my course texts in electronic format. Indeed, few (if any) books offer a similar in-depth approach to the study of the dynamics of drives.

The textbook is used as the course text for the Bachelor's and Master's programme in electrical and mechanical engineering at the Faculty of Engineering and Architecture of Ghent University. Parts 1 and 2 are taught in the basic course 'Fundamentals of Electric Drives' in the third bachelor. Part 3 is used for the course 'Controlled Electrical Drives' in the first master, while Part 4 is used in the specialised master on electrical energy.

Part 1 focuses mainly on the steady-state operation of rotating field machines. Nevertheless, the first two chapters are devoted to transformers and DC commutator machines: the chapter on transformers is included as an introduction to induction and synchronous machines, their electromagnetics and equivalent circuits, while that on DC commutator machines concludes with the interesting motor and generator characteristics of these machines, mainly as a reference. Chapters 3 and 4 offer an in-depth study of induction and synchronous machines, respectively. Starting from their electromagnetics, steady-state equations and equivalent circuits are derived, from which their properties can be deduced. In addition to the poly-phase machines, also special types such as capacitor motors and shaded-pole motors are discussed.

The second part of this book discusses the main power electronic supplies for electrical drives, for example, rectifiers, choppers, cycloconverters and inverters. This part is not at all intended as a fundamental course text on power electronics and its design. For the design of power electronic circuits, much more in-depth textbooks are available. The only aim is to provide the basics required for their application in electrical machine drives. After an overview of power electronic components, the following chapters provide a rather thorough analysis of rectifiers, DC and AC choppers, cycloconverters and inverters. Much attention is paid to PWM techniques for inverters and the resulting harmonic content in the output waveform.

In the third part, electrical drives are discussed, combining the traditional (rotating field and DC commutator) electrical machines treated in Part 1 and the power electronics of Part 2. Part 3 begins with a chapter on DC commutator machines and their characteristics. Next, the traditional constant frequency operation of rotating field machines is treated in detail, including its (limited) starting and variable speed operation possibilities. In the same chapter, the effect of voltage variations is also discussed, as is voltage adaptation to the load and power electronic starting of induction machines. The next chapter analyses ideal sinusoidal current supply of rotating field machines, with a special focus on main field saturation. After ideal variable frequency supply of rotating field machines is treated, the useful fundamental frequency equivalent circuits for inverters (originally presented by the colleagues of UW-Madison) are discussed. With these equivalent circuits, the main properties of rotating field machines with variable frequency inverter supply are straightforwardly derived. Next, the basics of controlled drives are presented, including field orientation of induction and synchronous machines, as well as direct torque control. The two subsequent chapters are devoted to power electronic control of small electric machines and to AC commutator machines, respectively. To end, small synchronous machines are described (i.e. permanent magnet synchronous machines, reluctance machines and hysteresis motors), as are stepping motors and switched reluctance machines.

Finally, Part 4 is devoted to the dynamics of traditional electrical machines. For the dynamics of induction and synchronous machine drives as well, the electromagnetics are used as the starting point to derive the dynamic models. Throughout Part 4, much attention is paid to the derivation of analytical models. Naturally, the basic dynamic properties and probable causes of instability of induction and synchronous machine drives are discussed in detail as well, with the derived models for stability in the small as the starting point. In addition to the study of the stability in the small, one chapter is devoted to large-scale dynamics (e.g. sudden short circuit of synchronous machines). Another chapter is dedicated to the dynamics in vector- and field-oriented control, while the last chapter discusses voltage surge phenomena in electrical machines and transformers.

In the appendices, additional background is provided on terminal markings of machines and transformers (Appendix A), static stability of a drive (Appendix B) and on phasors and space vectors (Appendix C). Some basic knowledge of terminal markings is of course required for the practical exercises. The notion of static stability is explained in Appendix B, and it is not repeated for each machine type. With regard to the appendix on space vectors and phasors, the first section is required for Parts 1 and 3, while the second section is required for Part 4.

Ghent, Belgium

Jan A. Melkebeek

Acknowledgements

This book is the result of 4 years of intensive writing and rewriting. What is more, it is the result of a 42-year career in academia, with more than 31 years as a full professor in this field. It is also the result of extensive feedback provided by assistants and students in all these years.

I would like to express my gratitude to M. Christiaen Vervust (now retired), who designed the figures in this book that were originally made for the Dutch textbook. In most cases, I only had to adapt the texts to English. I would also like to thank M. Tony Boone for drawing many of the new figures.

I am also indebted to my colleague (and successor) Prof.dr.ir. Frederik De Belie for his numerous suggestions and corrections, as well as all my colleagues for giving me the opportunity to spend so much time on this textbook and for encouraging me in the process.

Many thanks of course to Ms. Kristin Van den Eede of the Language Centre of our university, who most carefully checked and corrected the language in my original text, and even tried to make a linguistic masterpiece of it.

Last but not least, I am grateful to my wife and friends for their patience, when I was not available for them.

Contents

Symbols and Conventions

General

In most cases, lowercase letters are used for variables which are a function of time, e.g. v, i, φ . Phasors are indicated by underlined symbol, e.g. $\underline{v}$, $\underline{i}$, $\underline{\varphi}$. If required, space vectors are distinguished from phasors by an arrow under the symbol, e.g. $\underset{\rightarrow}{v}, \underset{\rightarrow}{i}, \underset{\rightarrow}{\varphi}$.

Capital letters are used for variables in sinusoidal steady state, e.g. $\underline{V}$, $\underline{I}$, $\underline{\Phi}$.

For the study of sinusoidal steady state, effective values are supposed for voltages and currents. However, for fluxes and flux density (induction), amplitude values are most common, and this is indicated by the hat symbol, e.g. $\hat{B}$, $\hat{\Phi}$.

Symbols

B	Induction (flux density) (T)
C	Capacitance (F)
E, e	Emf (induced voltage) (V)
f	Frequency (Hz)
F	Force (N)
F	mmf (A)
f	pu mmf
H	Magnetic field strength (A/m)
I, i	Current (A)
J	Inertia (kgm^2)
J	Current density (A/m^2)
j	pu inertia
j	Imaginary unit
L	Inductance (H)

l	pu inductance
M	Magnetic potential (mmf)
N_n	Speed (rpm) (1/min)
N_p	Number of pole pairs
P	Power (W)
p	pu power
p	Laplace operator
Q	Reactive power (V A)
R	Reluctance (A/Wb)
R	Resistance
r	pu resistance
s	Slip
s	Laplace operator
T	Torque (Nm)
t	pu torque
t	Time (s)
U	Voltage (V)
υ	pu voltage
V	Voltage (V)
v	pu voltage
W	Energy (J)
X	Reactance (Ω)
x	pu reactance
Z	Impedance (Ω)
z	pu impedance

Greek symbols

β	Steady-state load angle
δ	Load angle (instantaneous) (rad)
δ	Dirac
δ, ∂	Incremental value
Λ	Permeance (Wb/A)
λ	pu Laplace operator
μ	Permeability (H/m)
ν	pu speed or frequency
ρ	Resistivity (Ωm)
σ	Leakage coefficient
T	pu time ($\omega_n t$)
T	pu time constant
Φ	Flux (Wb)

φ pu flux
Ψ Flux linkage (Wb)
ψ pu flux linkage
Ω Mechanical (shaft) speed (rad/s)
ω Electrical (shaft) speed; angular frequency (rad/s)

Subscripts

o Steady state (value)
1 Primary—
2 Secondary—
d d-axis—
l Load—(e.g. Tl for load torque)
m Magnetising—
m Motor—(e.g. Tm for motor torque)
n Nominal (rated) —
q q-axis—
r Rotor—
s Stator—
s Slip—
α α-axis—
β β-axis—
σ Leakage—

Part I
Transformers and Electrical Machines

Chapter 1
Transformers

Abstract Undoubtedly, transformers are omnipresent in our society. In this chapter, the electromagnetic principles of transformers are explained in detail. Equivalent circuits are derived starting from the basic laws of Maxwell and discussed in detail. As such, the construction of transformers is also treated in some detail. Attention is also paid to numerous applications.

1.1 Introduction

Classical power stations generate electrical energy on the medium voltage (MV) level, i.e. around or lower than 10 kV (mainly because of restrictions on the insulation of the generators). Renewable energy is produced at even lower voltage levels.

However, for efficient power transport over large distances, much higher voltage levels are required, i.e. 400 kV to even more than 800 kV (HV level).

On the other hand, local distribution requires somewhat lower voltage levels, hence the MV level, e.g. 13 kV. These voltage levels can also be used directly for industrial plants. Yet, in many industrial applications and for home appliances, in particular, much lower voltages are required (usually 400 V/230 V, called the LV level).

Whereas an easy or lossless transfer of DC power to other voltage levels is not possible, the main advantage of AC power is that it can be transformed to other voltage levels rather easily and without hardly any loss. This efficient transformation of AC power to other voltages levels is the task of electromagnetic transformers. In such a classical transformer power is transformed to other voltage levels using magnetic fields as the intermediating medium.

Figure 1.1 shows some symbols of transformers as frequently used in electrical schemes, both for single-phase and three-phase transformers.

The simplest transformer is a single-phase one, consisting of an iron yoke with two windings, commonly called the primary and secondary windings. To start with, suppose that the primary winding is fed by a sinusoidal voltage supply and the secondary is open-circuited. The primary winding will draw a sinusoidal current from the supply, causing an alternating electromagnetic field in the yoke. As the

J. A. Melkebeek, *Electrical Machines and Drives*, Power Systems,
https://doi.org/10.1007/978-3-319-72730-1_1

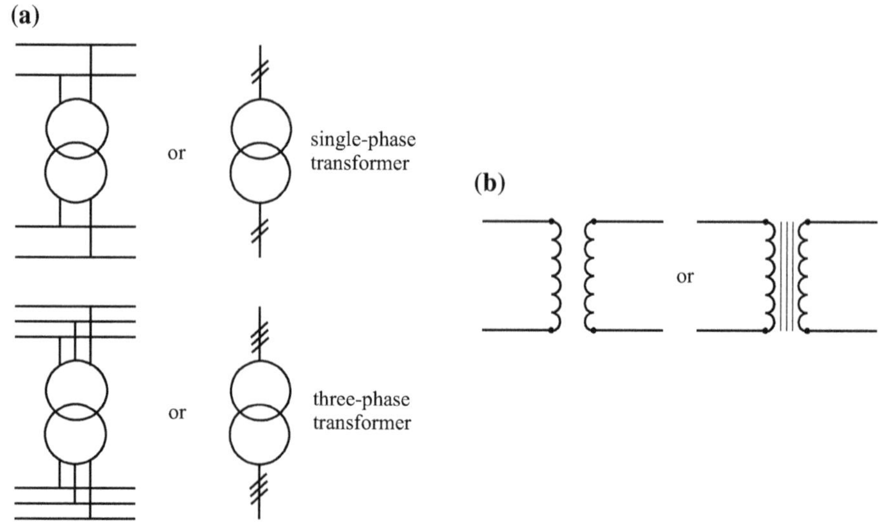

Fig. 1.1 Transformer symbols

secondary winding is linked with the field in the yoke, an alternating (sinusoidal) voltage will be induced in the secondary winding. It is easy to see that the ratio of the primary to secondary voltage is equal to the ratio of the number of turns of the primary to secondary windings. If the secondary winding is then connected to a load, a current will be drawn from the secondary winding. As this current will try to reduce its cause (the magnetic field in the yoke), this secondary current will attempt to reduce the field (Lenz's law). However this means that the primary current will increase in order to annihilate the effect of the secondary current (assuming that the primary voltage imposes the flux level). The power transmitted to the secondary load will then be drawn from the primary supply (at the voltage level of the primary).

1.2 Transformer Equations

1.2.1 Basic Electromagnetic Description and Equations

Consider the principal electromagnetic scheme in Fig. 1.2. On the (soft) iron yoke we see two windings, in this figure on two separate limbs.[1] The permeability of the iron yoke is assumed to be very large such that nearly all field lines remain within the yoke. The two windings are therefore closely magnetically coupled. As we assume that the primary winding has right turns and the secondary left turns, the usual polarity

[1] In practice, these windings are mostly placed concentric on one limb, to reduce leakage.

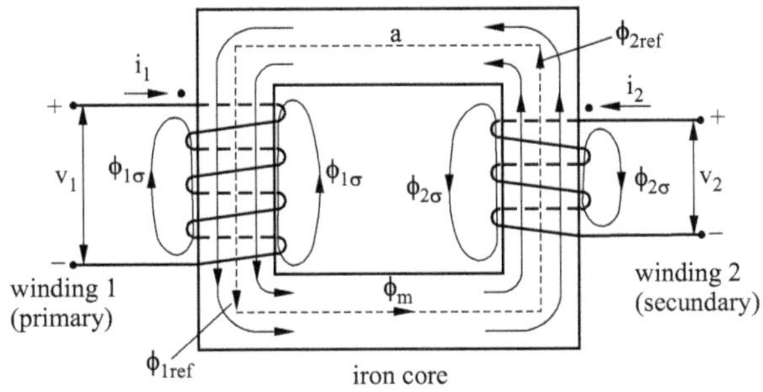

Fig. 1.2 Transformer core and windings

marks "•" may be positioned as in the figure. We will choose the reference directions of the voltages in accordance with these marks, i.e. with the voltage "+" sign at the marked ends. For both windings we use the Users Reference System (URS). With the chosen voltage polarities the arrows of the current reference must then point to the marks (positive currents will enter the windings at the marks). Conventionally, the reference directions of the fluxes coupled with the windings are chosen in accordance with the current reference directions. Therefore positive currents in either primary or secondary windings will magnetise the core in the same sense (φ_{1ref} and φ_{2ref} point into the same sense with respect to the core).

Firstly suppose that the primary winding is supplied by an AC voltage source with the secondary winding open-circuited. The primary winding will then draw a current $i_1(t)$ which will magnetise the core. The flux coupled with the primary winding will be denoted by $\Psi_1(t) = w_1\varphi_1(t)$. Here, $\varphi_1(t)$ represents the physical flux, i.e. the flux over a section of the core while $w_1$is the number of primary turns.

The flux $\Psi_1(t) = w_1\varphi_1(t)$ can be subdivided into:

- a main or magnetising flux $\Psi_{m1}(t) = w_1\varphi_m(t)$ with field lines staying completely within the iron core and thus also linking the secondary winding
- a primary leakage flux $\Psi_{1\sigma}(t) = w_1\varphi_{1\sigma}(t)$ linked only with the primary winding

Similarly one may also split up the physical flux as $\varphi_1(t) = \varphi_{1\sigma}(t) + \varphi_m(t)$.

The main or magnetising flux will therefore induce an emf in the secondary winding given by $e_2(t) = d\Psi_{m2}(t)/dt = w_2 d\varphi_m(t)/dt$.

Secondly, suppose that the secondary winding is supplied by an AC voltage source, the primary open-circuited. A secondary current will then be drawn. The secondary current will result in a flux $\Psi_2(t) = w_2\varphi_2(t)$ coupled with the secondary winding. This flux can also be split up into a main flux $\Psi_{m2}(t) = w_2\varphi_m(t)$ coupled with the secondary and a secondary leakage flux $\Psi_{2\sigma}(t) = w_2\varphi_{2\sigma}(t)$.

Thirdly, in the general case that both windings carry currents, we may write

$$\Psi_1(t) = \Psi_{1\sigma}(t) + \Psi_m(t) = w_1\varphi_1(t) = w_1\varphi_{1\sigma}(t) + w_1\varphi_m(t) \qquad (1.1)$$

$$\Psi_2(t) = \Psi_{2\sigma}(t) + \Psi_m(t) = w_2\varphi_2(t) = w_2\varphi_{2\sigma}(t) + w_2\varphi_m(t) \qquad (1.2)$$

Ampère's law along a field line in the core yields for the magnetising flux mmf

$$\oint \underline{H}_m d\underline{l} = w_1 i_1(t) + w_2 i_2(t) \tag{1.3}$$

The physical magnetising flux is the integral of the magnetising induction over a cross-section of the core

$$\varphi_m = \iint_{\Sigma} \underline{B}_m \cdot \underline{n} d\sigma \tag{1.4}$$

with $\underline{n}$ the vertical unit vector on the elementary surface element $d\sigma$ and with $\underline{B}_m = \mu \underline{H}_m$ where μ is the permeability of the core. Therefore

$$\varphi_m(t) \cdot \Re_m = \varphi_m(t) \cdot \Lambda_m^{-1} = w_1 i_1(t) + w_2 i_2(t) \tag{1.5}$$

where $\Re_m = \Lambda_m^{-1} = l_m/\mu\Sigma_m = l_m/\mu_0\mu_r\Sigma_m$ is the reluctance of the core and $\mu = \mu_0\mu_r$ the permeance of the iron (supposing the iron linear[2]). l_m and Σ_m represent the core mean length and cross-section respectively.

Similarly one may write for the leakage fluxes

$$\varphi_{1\sigma}(t) \cdot \Re_{1\sigma} = \varphi_{1\sigma}(t) \cdot \Lambda_{1\sigma}^{-1} = w_1 i_1(t) \tag{1.6}$$

$$\varphi_{2\sigma}(t) \cdot \Re_{2\sigma} = \varphi_{2\sigma}(t) \cdot \Lambda_{2\sigma}^{-1} = w_2 i_2(t) \tag{1.7}$$

The fluxes $\Psi_1(t)$ and $\Psi_2(t)$ induce in the primary and secondary windings the voltages $d\Psi_1(t)/dt$ and $d\Psi_2(t)/dt$ respectively. Taking into account also the resistive voltage drops, we may write for these voltages

$$v_1(t) = R_1 i_1(t) + \frac{d\Psi_1(t)}{dt} \tag{1.8}$$

$$v_2(t) = R_2 i_2(t) + \frac{d\Psi_2(t)}{dt} \tag{1.9}$$

or

$$v_1(t) = R_1 i_1(t) + w_1 \frac{d\varphi_{1\sigma}(t)}{dt} + w_1 \frac{d\varphi_m(t)}{dt} = R_1 i_1(t) + w_1 \frac{d\varphi_{1\sigma}(t)}{dt} + e_1(t) \tag{1.10}$$

[2]In reality the iron in the yoke is almost always saturated; in that case $\mu = \mu_0\mu_r$ represents the equivalent chord slope of the saturation characteristic in the operating point.

$$v_2(t) = R_2 i_2(t) + w_2 \frac{d\varphi_{2\sigma}(t)}{dt} + w_2 \frac{d\varphi_m(t)}{dt} = R_2 i_2(t) + w_2 \frac{d\varphi_{2\sigma}(t)}{dt} + e_2(t) \tag{1.11}$$

The voltages induced by the magnetising flux are called the (magnetising) emfs. For these magnetising emfs one may write

$$\begin{aligned} e_1(t) &= w_1 \tfrac{d}{dt} \left[\Lambda_m \left(w_1 i_1(t) + w_2 i_2(t)\right)\right] = w_1^2 \tfrac{d}{dt} \left[\Lambda_m \left(i_1(t) + \tfrac{w_2}{w_1} i_2(t)\right)\right] \\ e_2(t) &= w_2 \tfrac{d}{dt} \left[\Lambda_m \left(w_1 i_1(t) + w_2 i_2(t)\right)\right] = w_2^2 \tfrac{d}{dt} \left[\Lambda_m \left(\tfrac{w_1}{w_2} i_1(t) + i_2(t)\right)\right] \end{aligned} \tag{1.12}$$

Note that

$$\frac{e_1(t)}{e_2(t)} = \frac{w_1}{w_2} \tag{1.13}$$

As the leakage flux lines remain for the greatest part in air, the constant leakage inductances $L_{1\sigma} = w_1^2 \cdot \Lambda_{1\sigma}$ and $L_{2\sigma} = w_2^2 \cdot \Lambda_{2\sigma}$ may be introduced:

$$\begin{aligned} v_1(t) &= R_1 i_1(t) + L_{1\sigma} \tfrac{di_1(t)}{dt} + e_1(t) \\ v_2(t) &= R_2 i_2(t) + L_{2\sigma} \tfrac{di_2(t)}{dt} + e_2(t) \end{aligned} \tag{1.14}$$

If the saturation of the iron yoke can be neglected, we may introduce constant magnetising inductances $L_{m1} = w_1^2 \cdot \Lambda_m$ and $L_{m2} = w_2^2 \cdot \Lambda_m$:

$$\begin{aligned} e_1(t) &= L_{m1} \tfrac{d}{dt} \left[i_1(t) + \tfrac{w_2}{w_1} i_2(t)\right] \\ e_2(t) &= L_{m2} \tfrac{d}{dt} \left[\tfrac{w_1}{w_2} i_1(t) + i_2(t)\right] \end{aligned} \tag{1.15}$$

Therefore

$$\begin{aligned} v_1(t) &= R_1 i_1(t) + L_{1\sigma} \tfrac{di_1(t)}{dt} + L_{m1} \tfrac{d}{dt} \left[i_1(t) + \tfrac{w_2}{w_1} i_2(t)\right] \\ v_2(t) &= R_2 i_2(t) + L_{2\sigma} \tfrac{di_2(t)}{dt} + L_{m2} \tfrac{d}{dt} \left[\tfrac{w_1}{w_2} i_1(t) + i_2(t)\right] \end{aligned} \tag{1.16}$$

If the magnetic circuit is saturated, linearised chord slope inductances for the operating point considered may be used, see Fig. 1.12.

Note that Eq. 1.16 are equivalent to the classical equations for two magnetically coupled coils:

$$\begin{aligned} v_1(t) &= R_1 i_1(t) + L_1 \tfrac{di_1(t)}{dt} + M \tfrac{di_2(t)}{dt} \\ v_2(t) &= R_2 i_2(t) + L_2 \tfrac{di_2(t)}{dt} + M \tfrac{di_1(t)}{dt} \end{aligned} \tag{1.17}$$

with $L_1 = L_{1\sigma} + L_{m1}$, $L_2 = L_{2\sigma} + L_{m2}$, $M = L_{m1} \cdot \frac{w_2}{w_1} = L_{m2} \cdot \frac{w_1}{w_2}$.

From now on, we will use mostly the Eq. 1.14 with

$$\begin{aligned} e_1(t) &= L_{m1}\tfrac{d}{dt}\left[i_1(t) + \tfrac{w_2}{w_1} i_2(t)\right] \\ e_2(t) &= e_1(t)\cdot \tfrac{w_2}{w_1} \end{aligned} \tag{1.18}$$

The equivalent current

$$i_{m1} = i_1(t) + \frac{w_2}{w_1} i_2(t) \tag{1.19}$$

is called the primary magnetising current (in fact: the magnetising current as observed from the primary). It is the current which would be required in the primary winding for the magnetising field in the operating point considered (in other words, if only the primary winding current was present). Accordingly, L_{m1} is the corresponding magnetising inductance (as seen from the primary).

One may of course also use similar equations based on the secondary emf and the magnetising inductance referred to the secondary.

1.2.2 *Phasor Equations and Equivalent Circuit for Sinusoidal Supply*

For a sinusoidal supply with frequency $f = 2\pi/\omega$ Eq. 1.14 may be rewritten using the complex phasor representation:

$$\begin{aligned} \underline{V}_1 &= R_1\underline{I}_1 + j\omega L_{1\sigma}\underline{I}_1 + \underline{E}_1 \\ \underline{V}_2 &= R_2\underline{I}_2 + j\omega L_{2\sigma}\underline{I}_2 + \underline{E}_2 \end{aligned} \tag{1.20}$$

with

$$\begin{aligned} \underline{E}_1 &= j\omega L_{m1}\left[\underline{I}_1 + \tfrac{w_2}{w_1}\underline{I}_2\right] \\ \underline{E}_2 &= \tfrac{w_2}{w_1}\cdot \underline{E}_1 \end{aligned} \tag{1.21}$$

In these equations, reactances $X_{1\sigma} = \omega L_{1\sigma}$, $X_{2\sigma} = \omega L_{2\sigma}$ and $X_{m1} = \omega L_{m1}$ are commonly used instead of inductances (at least for constant frequency supply). For voltages and currents, effective values instead of amplitude values are used (whereas for fluxes and magnetic fields, amplitude values are more common: $\underline{E}_1 = j\omega w_1(\hat{\underline{\Phi}}/\sqrt{2}) = j\omega w_1\underline{\Phi}$).

For power transformers the resistances are usually relatively small as are the leakage inductances. In contrast the magnetising inductances are quite large because of the large permeance of the magnetic circuit. Therefore, taking into account $|R_1\underline{I}_1|, |X_{1\sigma}\underline{I}_1| \ll |\underline{E}_1|$ and $|R_2\underline{I}_2|, |X_{2\sigma}\underline{I}_2| \ll |\underline{E}_2|$ it follows that $\underline{V}_1 \approx \underline{E}_1$ and

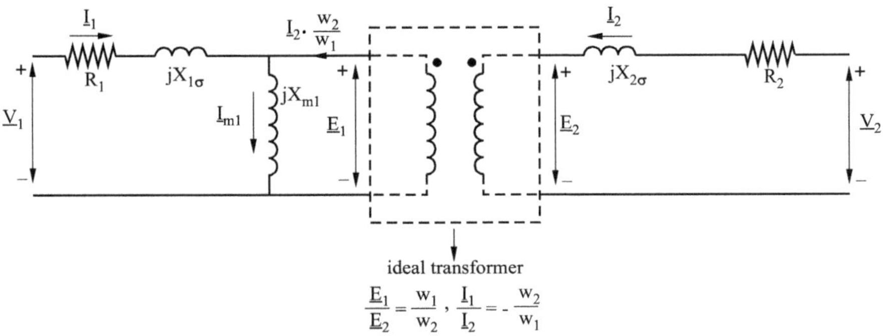

Fig. 1.3 Basic equivalent circuit

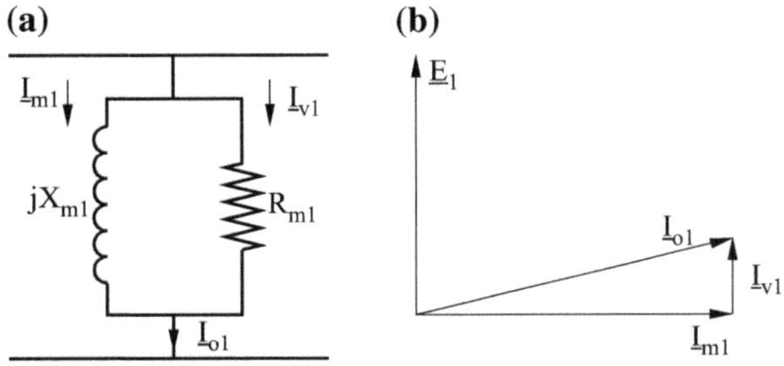

Fig. 1.4 No-load branch with iron losses

$\underline{V}_2 \approx \underline{E}_2$ and thus also $\underline{V}_2 \approx \underline{V}_1 \cdot (w_2/w_1)$. Further, $\underline{I}_{m1} = \underline{I}_1 + \underline{I}_2(w_2/w_1) \approx 0$ and thus $\underline{I}_2 \approx -\underline{I}_1(w_1/w_2)$.

Equations 1.20 through 1.21 may be represented by the equivalent circuit in Fig. 1.3. This equivalent circuit can be interpreted as replacing the real transformer by an ideal transformer with winding ratio w_1/w_2 and adding the parasitic elements (e.g. resistances, leakage inductances) externally.

It should be remarked that in reality also iron losses (eddy current losses and hysteresis losses) are present. Eddy current losses (also called Foucault losses) are caused by the electrical conductivity of the iron. To reduce these eddy currents, the core will always be laminated. These eddy current losses are proportional to the square of the frequency and the square of the magnetic induction: $P_{df} \sim \hat{\Phi}^2 \cdot f^2 \sim E^2$. The hysteresis losses are proportional to the frequency and to an iron-dependent power β of the magnetic induction: $P_{df} \sim \hat{\Phi}^{\beta} \cdot f$ with $\beta \approx 1.6 \cdots 2$.

The iron losses are often modelled as (at least for a fixed frequency): $P_{dm} = E_1^2/R_{m1}$. In the equivalent circuit the magnetising branch is then to be replaced by a resistance R_{m1} parallel to the magnetising inductance (see (a) in Fig. 1.4). The sum of the magnetising current and the iron-loss current is called the no-load current. The no-load current lags the emf by less than $\pi/2$ (see (b) in Fig. 1.4).

1.3 Referred Values: Equations and Equivalent Circuit

The equations and equivalent circuit may be simplified considerably by using *referred* (also called *reduced*) values, referring either the secondary to the primary or the primary to the secondary.

The first option, referring the secondary quantities to the primary, implies replacing $\underline{V}_2$, $\underline{E}_2$, $\underline{I}_2$, $\underline{Z}_2$ by $\underline{V}_2'$, $\underline{E}_2'$, $\underline{I}_2'$, $\underline{Z}_2'$ with $\underline{V}_2' = \frac{w_1}{w_2}\underline{V}_2$, $\underline{E}_2' = \frac{w_1}{w_2} \cdot \underline{E}_2 = \underline{E}_1$, $\underline{I}_2' = \frac{w_2}{w_1} \cdot \underline{I}_2$, $\underline{Z}_2' = \left(\frac{w_1}{w_2}\right)^2 \underline{Z}_2$. The second option, referring the primary quantities to the secondary, implies replacing $\underline{V}_1$, $\underline{E}_1$, $\underline{I}_1$, $\underline{Z}_1$ by $\underline{V}_1'$, $\underline{E}_1'$, $\underline{I}_1'$, $\underline{Z}_1'$ with $\underline{V}_1' = \frac{w_2}{w_1}\underline{V}_1$, $\underline{E}_1' = \frac{w_2}{w_1} \cdot \underline{E}_1 = \underline{E}_2$, $\underline{I}_1' = \frac{w_1}{w_2} \cdot \underline{I}_1$, $\underline{Z}_1' = \left(\frac{w_2}{w_1}\right)^2 \underline{Z}_2$.

The former option, referring the secondary to the primary, yields the equivalent circuit in Fig. 1.5 and the equations

$$\begin{aligned} \underline{V}_1 &= R_1\underline{I}_1 + jX_{1\sigma}\underline{I}_1 + \underline{E}_1 \\ \underline{V}_2' &= R_2'\underline{I}_2' + jX_{2\sigma}'\underline{I}_2' + \underline{E}_2' \end{aligned} \tag{1.22}$$

with

$$\begin{aligned} \underline{E}_2' &= \underline{E}_1 = jX_{m1}\underline{I}_{m1} \\ \underline{I}_{m1} &= \underline{I}_1 + \underline{I}_2' - \underline{I}_{d1} \end{aligned} \tag{1.23}$$

If the galvanic potential difference between primary and secondary can be neglected (or is unimportant), then the ideal transformer can be omitted and the equivalent circuit in Fig. 1.6 is obtained.

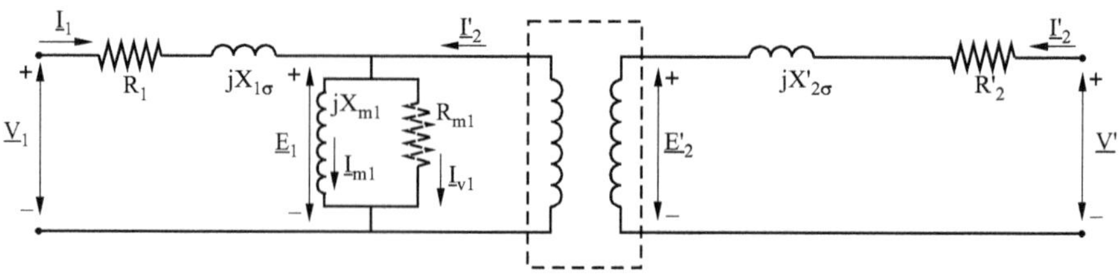

Fig. 1.5 Equivalent circuit with referred quantities and ideal transformer

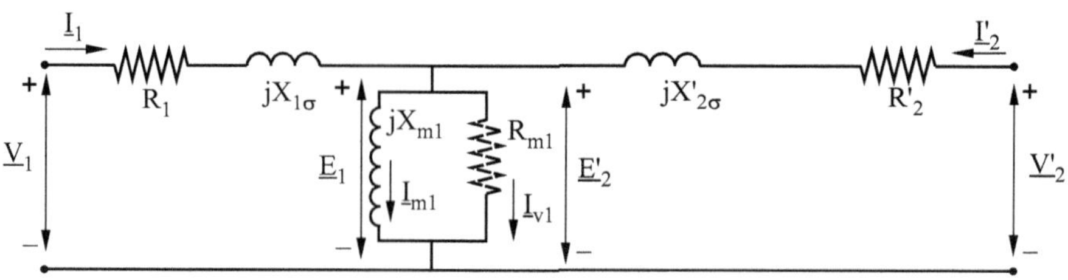

Fig. 1.6 Equivalent circuit for referred quantities

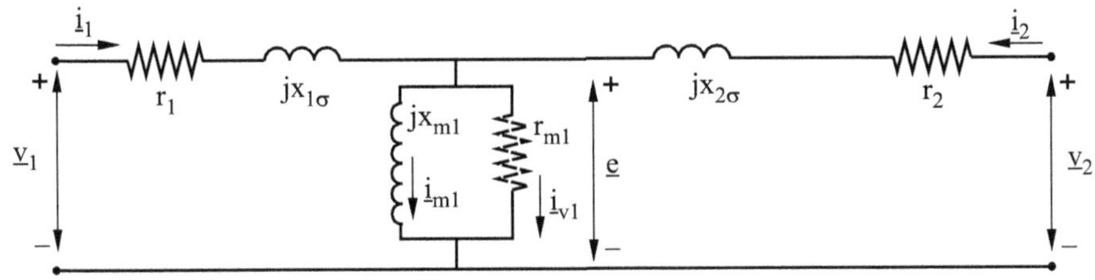

Fig. 1.7 Equivalent circuit in per unit

1.4 Per-Unit Description

Instead of absolute values, per-unit values are sometimes used for modelling a transformer. Per-unit modelling implies that all values are referred to their rated (nominal) values. The rated values for voltages and current are denoted as V_{1n}, $\grave{\imath}_{1n}$, V_{2n}, I_{2n} (where primary and secondary voltages and currents obey the transformer winding ratio, i.e. $V_{1n}/V_{2n} = w_1/w_2$ and $I_{1n}/I_{2n} = w_2/w_1$). The reference value for power must obey $S_n = V_{1n}I_{1n} = V_{2n}I_{2n}$ and those for the impedances $Z_{1n} = V_{1n}/I_{1n}$, $Z_{2n} = V_{2n}/I_{2n} = Z_{1n} \cdot (w_2/w_1)^2$.

The following per-unit equations are then obtained:

$$\begin{aligned} \underline{v}_1 &= r_1\underline{i}_1 + jx_{1\sigma}\underline{i}_1 + \underline{e} \\ \underline{v}_2 &= r_2\underline{i}_2 + jx_{2\sigma}\underline{i}_2 + \underline{e} \end{aligned} \tag{1.24}$$

$$\underline{e} = jx_{m1}\underline{i}_{m1} = jx_{m1}(\underline{i}_1 + \underline{i}_2) \tag{1.25}$$

which corresponds to the equivalent circuit in Fig. 1.7.

The advantage of a per-unit description is that it allows us to compare transformers (or machines) with different power ratings, transformer ratios or voltages. For power transformers, for example, the resistances vary within quite a broad range, depending of course on the power rating but also on other factors such as design and construction quality. Per unit description makes it possible to distinguish between the main properties and side aspects. However, for studying the behaviour of a transformer in a given grid (or, more general, an electric machine in a system) absolute values will normally be used.

1.5 Construction and Scaling Laws

Obviously, the size of a transformer is quite narrowly related with the rated power of the transformer, but this is also the case for many other parameters of a transformer, such as (p.u.) resistances or leakage inductances or the magnetising inductance. These relations between characteristic properties and dimensions are called scaling laws.

Usually, one of the dimensions is used as a reference and the others will usually vary proportionally with it. For a transformer, the average diameter D of the windings is an appropriate reference dimension. In this section, we will discuss the scaling laws and explain the background and causes of the relation with the dimensions and construction of the transformer.

1.5.1 Specific Rated Quantities

Rewriting the apparent power of a transformer as

$$S = (V/w) \cdot (I \cdot w) \tag{1.26}$$

shows that the apparent power can be expressed as the product of two factors that are (almost) equal for both primary and secondary windings, i.e. the *voltage per turn* and the *bundle current.* The rated voltage, current and apparent power related to one square metre of the winding (i.e. for one metre of conductor length and one metre of coil height, called *rated specific values)*, can be expressed as follows:

$$K_n = (V_{1n}/w_1)/\pi D = (V_{2n}/w_2)/\pi D \tag{1.27}$$

$$A_n = (w_1 I_{1n})/h = (w_2 I_{2n})/h \tag{1.28}$$

$$S_n^{\square} = A_n \cdot K_n = S_n/\pi D h \tag{1.29}$$

with D being the average winding diameter and h the coil height. K_n, A_n and $S_n^{\square}$ represent the rated voltage per metre conductor length, rated current per metre coil height and rated apparent power per square meter respectively. Like the scaling laws (see below) these specific quantities make it possible to compare transformers of quite different power ratings.

For a transformer it may be assumed as a first approximation that all dimensions change proportionally with the average winding diameter D (which will be taken as a reference). If the current density in the conductors and the flux density (induction) in the core can be considered as constant, then

$$A_n = (w_1 I_{1n})/h = (w_1 \Sigma_{cu} J_n)/h \sim D \tag{1.30}$$

$$K_n = (V_{1n}/w_1)/\pi D = (\Sigma_{Fe} \cdot B_n)/\pi D \sim D^2/D \sim D \tag{1.31}$$

$$S_n^{\square} = A_n \cdot K_n \sim D^2 \tag{1.32}$$

The rated apparent power of a transformer therefore changes as D^4.

Average values for the specific rated quantities are: $A_n = 5 \cdot 10^4 \cdot D\,[\mathrm{A/m}]$; $K_n = 25 \cdot D\,[\mathrm{V/m}]$; $S_n^{\square} = 1.25 \cdot 10^6 \cdot D^2\,[\mathrm{VA/m^2}]$.

1.5.2 *Rated Per-Unit Impedances*

1.5.2.1 Winding Resistance

From

$$r = R/Z_n = R \cdot I_n/V_n = R \cdot I_n^2/S_n \tag{1.33}$$

we see that the per-unit winding resistance is equal to the relative ohmic loss. The rated ohmic loss varies with size as D^3, indeed:

$$R \cdot I_n^2 \simeq \varrho(l/\Sigma_{Cu}) \cdot I_n^2 \simeq \varrho(l/\Sigma_{Cu}) \cdot \Sigma_{Cu}^2 J_n^2 \sim D^3 \tag{1.34}$$

If $J_n = constant$ and $S_n \sim D^4$ we can derive that $r \sim D^{-1}$. The higher efficiency of large transformers is for the greater part the result of these smaller relative winding resistances.

Inserting $\rho = 2.5{\cdot}10^{-8}[\Omega\mathrm{m}]$, $J_n = 4 \cdot 10^6[\mathrm{A/m^2}]$, $K_n = 25 \cdot D\,[\mathrm{V/m}]$ in

$$r = R \cdot I_n/V_n = \varrho(l/\Sigma_{Cu}) \cdot (\Sigma_{Cu} \cdot J_n/K_n \cdot \pi D w) = \varrho \cdot J_n/K_n \tag{1.35}$$

yields $r \simeq 0.004/D$ and thus $r = 0.02 \cdots 0.002$ for $D = 0.2 \cdots 2\,\mathrm{m}$.

1.5.2.2 Leakage Inductance

If a transformer is fed by a constant voltage, then the secondary voltage will decrease with increasing load, mainly due to the leakage inductance. To limit the voltage drop it is therefore important to aim for a small leakage inductance. However, as the leakage inductance also limits the current in case of a short circuit, some leakage inductance is required.The choice of the leakage inductance will thus be a compromise. Moreover, the effect of the scaling laws will also lead to an increasing leakage inductance with increasing size of the transformer if all aspect ratios are kept the same.

The effect of the size can be elaborated as follows. Figure 1.8 shows a longitudinal cross section of one limb of a transformer with concentric cylindrical windings (which is the usual configuration for transformers of small and not too large power ratings). The magnetic energy in the space between primary and secondary windings is

$$W_\sigma = \frac{1}{2} H_\sigma^2 \cdot (volume)$$

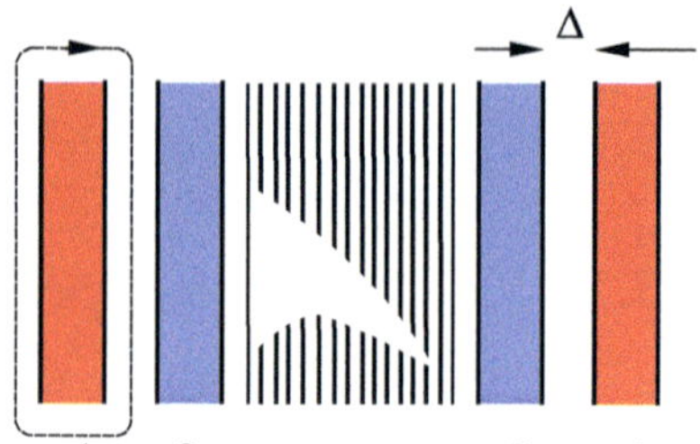

Fig. 1.8 Leakage path for cylindrical windings

For the field line (left in Fig. 1.8), we get $H_\sigma \cdot 2h = A \cdot h$ and therefore $H_\sigma \sim A$. For rated current $H_{\sigma n} \sim A_n \sim D$ and $W_{\sigma n} = \frac{1}{2} H_{\sigma n}^2 \cdot (volume) \sim D^2 \cdot D^3 \sim D^5$. For the per-unit leakage reactance, this results in

$$x_\sigma = \frac{X_\sigma}{Z_n} = \frac{X_\sigma I_n^2}{Z_n I_n^2} = \frac{2\omega W_{\sigma n}}{V_n I_n} = \frac{2\omega W_{\sigma n}}{S_n} \sim \frac{D^5}{D^4} \sim D \tag{1.36}$$

Keeping the aspect ratio of the transformer constant leads to an increasing per-unit leakage reactance. In order to limit the leakage for larger transformers, the distance between primary and secondary windings (Δ in the figure) can be varied less than proportionally with D. For very large transformers (with often high voltage ratings), however, disk windings will normally be used (see below in this chapter).

For the per-unit leakage values between 2×0.01 and 2×0.1 are found in actual transformers. The division between primary and secondary leakage is mostly unknown but not that important.[3]

1.5.2.3 Magnetic Core: Magnetising Reactance and Magnetising Current

Magnetic Core

To suppress eddy currents in the iron the magnetic core is always laminated: in fact a stack of approximately 0.3 mm thick insulated sheets is typically used.

In the case of circular coils an approximate round cross section is obtained by using for the (then stepped) stack a limited number of different sheet widths (see (a) in Fig. 1.9). For rectangular coils, a simple rectangular cross section with sheets of equal widths can be used.

To limit further eddy currents, the iron is normally alloyed with Silicium and/or Aluminium.

Oriented steel is typically used for the sheets. Oriented steel has a preferred orientation for the flux with a higher permeability in that direction. To this end (but

[3]In fact, a T of inductances can always be replaced by an equivalent L of inductances (see Sect. 1.6 below).

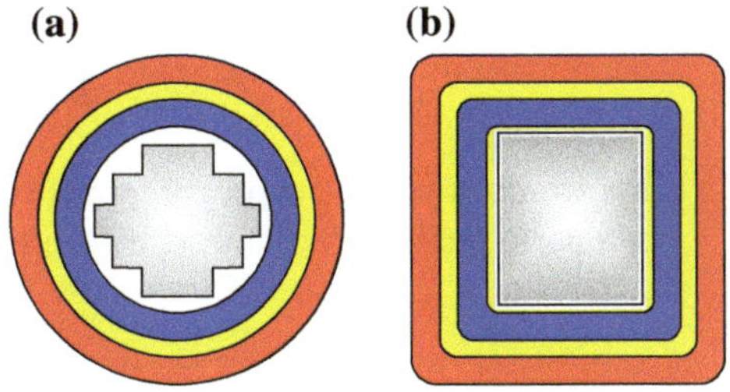

Fig. 1.9 Magnetic core cross section with windings

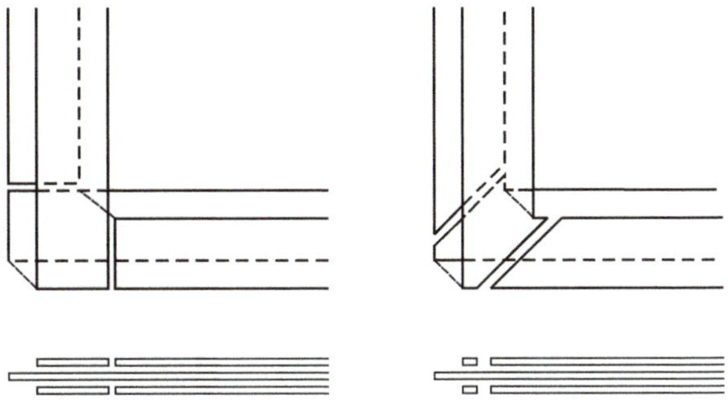

Fig. 1.10 Step or lap joints

also to enable the use of preformed coils) the core consists of separate limbs and closing yokes, all of them with oriented sheets in the flux direction. At the corners, where limbs and yokes join, the effect of parasitic air gaps is reduced by interleaving sheets of different lengths (see Fig. 1.10). These are called step or lap joints. In this way, the unreliable (and mostly larger) air gaps between sheets in the same layer are avoided. The flux will now shift to the two adjacent sheets and the associated air gap corresponds to the small (and well-defined) insulation layers of the sheets. This will normally result in a local higher saturation in these sheets but this effect is minor compared with a larger and undefined air-gap.

Magnetising current and inductance - saturation characteristic

From

$$v_1(t) = R_1 i_1(t) + w_1 \frac{d\varphi_{1\sigma}(t)}{dt} + w_1 \frac{d\varphi_m(t)}{dt} = R_1 i_1(t) + L_{1\sigma} \frac{di_1(t)}{dt} + w_1 \frac{d\varphi_m(t)}{dt} \tag{1.37}$$

where usually

$$|R_1 i_1(t)|,\ |L_{1\sigma} \frac{di_1(t)}{dt}| \ll |w_1 \frac{d\varphi_m(t)}{dt}| = |e_1(t)|$$

it follows that

$$v_1(t) \approx w_1 \frac{d\varphi_m(t)}{dt} = e_1(t) \tag{1.38}$$

or

$$\varphi_m(t) \simeq \int \frac{1}{w_1} \cdot v_1(t) \cdot dt \tag{1.39}$$

In other words, a voltage supply imposes the flux.

For a periodical (but not necessarily sinusoidal) voltage, it follows that

$$\hat{\Phi}_m \sim v_{1av} = \frac{2}{T} \int_0^{T/2} v_1(t) \cdot dt = v_{1eff}/k_f \tag{1.40}$$

with v_{1eff} the effective value of the voltage and k_f the form factor of the voltage ($k_f = \pi/2\sqrt{2}$ for a sinusoidal voltage). In case of a non-sinusoidal voltage, the higher harmonics are damped in the flux (because of the integral). It is also worth mentioning that harmonics $4k - 1$ (e.g. third harmonics) change sign in the flux, causing a flatted voltage curve to result in a pointed flux curve. Further, for a given effective value of the voltage, the hysteresis losses will be more affected by the form factor of the voltage. Indeed, the eddy current losses are proportional with $E^2_{eff} \sim (\hat{B} \cdot k_f \cdot f)^2$, whereas the hysteresis losses depend on $(\hat{B}^2 \cdot f)$.

In what follows, we will consider a purely sinusoidal voltage $v_1(t)$ with effective value V_1. As voltage and flux are imposed, also the magnetic induction is approximately sinusoidal with maximum value

$$\hat{B}_m = \sqrt{2} \cdot E_1 \cdot (\omega w_1 \Sigma_m)^{-1} \approx \sqrt{2} \cdot V_1 \cdot (\omega w_1 \Sigma_m)^{-1} \tag{1.41}$$

where Σ_m is the cross-section of the core.

The corresponding magnetic field strength $H_m(t)$ follows from the $B - H-$characteristic (or saturation characteristic) of the iron in the core. The required magnetising current can be derived from Ampère's law:

$$\oint H_m(t) \cdot dl = \sum w \cdot i(t) \tag{1.42}$$

The $B - H-$characteristic is not linear and shows hysteresis, as illustrated by the green loop in Fig. 1.11. On the one hand the non-linearity results in a non-sinusoidal and pointed curve as a function of time. The hysteresis, on the other hand, results in a leading (fundamental harmonic component of the) magnetic field strength, as is shown by the graphical derivation in Fig. 1.11. The hysteresis loss is in fact proportional to the surface area of the hysteresis loop. The leading $H_m(t)$ (or its fundamental) also corresponds to a no-load current which leads the emf $e(t)$ over less than $\pi/2$ (and which thus corresponds to a power loss).

The fundamental component of $H_m(t)$ which is in phase with $B_m(t)$ could also be obtained with a fictitious $B - H-$characteristic without hysteresis (in some way the average of the rising and descending branches of the real characteristic, cf. the red

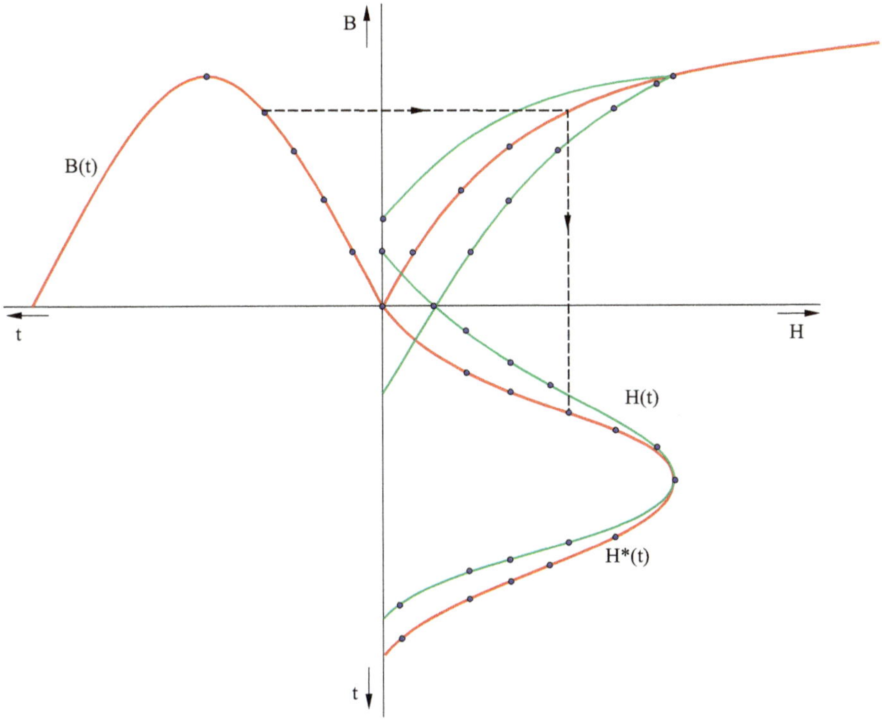

Fig. 1.11 Non-linear saturation characteristic

curve in Fig. 1.11). For the relation between the amplitude $\hat{H}_m$ and the fundamental magnetising current,

$$\hat{H}_m \cdot k_m \cdot h = \sqrt{2} \cdot w_m I_m \tag{1.43}$$

holds, where h represents the height of the winding and k_m (>1) is a factor which takes into account the remaining parts of the core as well as the joints in the corners. $w_m I_m$ can either be the primary mmf $w_1 I_{m1}$ or the secondary mmf $w_2 I_{m2}$.

For the per-unit value of the magnetising current we may write that

$$i_m = \frac{I_{m1}}{I_{n1}} = \frac{I_{m2}}{I_{n2}} = \frac{\hat{H}_m \cdot k_m \cdot h}{\sqrt{2} \cdot w_1 I_{n1}} = \frac{\hat{H}_m \cdot k_m \cdot h}{\sqrt{2} \cdot w_2 I_{n2}} = \frac{\hat{H}_m \cdot k_m}{\sqrt{2} \cdot A_n} \tag{1.44}$$

For power transformers, $\hat{H}_m \cdot k_m/\sqrt{2}$ varies in rated conditions between 10^2 and 10^3 A/m dependent on the magnetic material and transformer construction.[4] As A_n may vary between 10^4 and 10^5 A/m dependent on the power rating (see Sect. 1.5) we may

[4]Rated induction is normally $\hat{B}_m = 1 \cdots 1.5\,T$, for distribution transformers sometimes to $1.8\,T$.

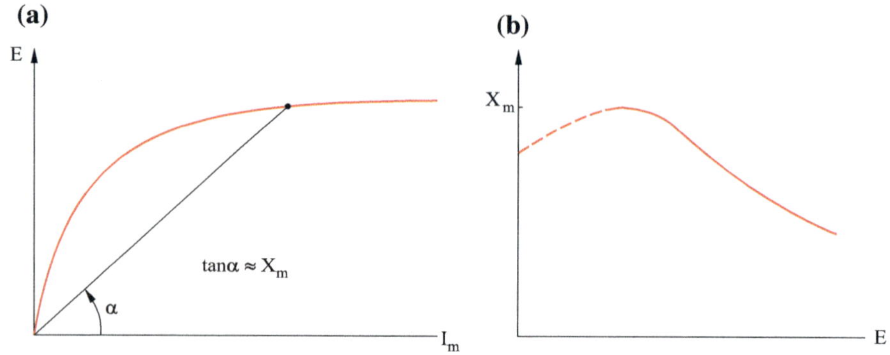

Fig. 1.12 Chord-slope magnetising inductance

expect a rather large variation for the per-unit magnetising current ($D = 0.2 \cdots 2\,\mathrm{m}$):

$$i_{mn} = (2 \cdot 10^{-2} \cdots 2 \cdot 10^{-3})/D = 0.1 \cdots 0.001 \tag{1.45}$$

Since the per-unit magnetising inductance is the inverse of the per-unit magnetising current, $x_{mn} = 10 \cdots 1000$ for $D = 0.2 \cdots 2\,\mathrm{m}$.

As mentioned above, this large variation is the result of both the normal scaling laws (smaller per-unit magnetising current for larger transformers) and the large variation in magnetic material and construction of the core. If the operating conditions differ from the rated ones (e.g. if the voltage is higher or lower) the magnetising inductance will differ from the inductance for rated conditions. Indeed, the magnetising inductance is in fact proportional to the chord slope of the operating point on the saturation characteristic (see Fig. 1.12). A higher voltage level will lead to a higher saturation level and thus a smaller magnetising inductance. Conversely, a lower-than-rated voltage will lead to larger magnetising inductances. However, loading variations will also result in some variation of the emf and thus also the magnetising inductance (due to the voltage drop over resistance and the leakage inductance in particular).

The active component of the no-load current, which corresponds to the eddy current and hysteresis losses[5] in the core, is commonly modelled by an iron loss resistance parallel to the magnetising inductance (e.g. for an equivalent circuit referred to the primary $P_{dm} = E_1^2/R_{m1}$). In per unit the iron dissipation resistance r_m varies with size approximately as $r_m \approx 5 \cdot x_m \approx (250 \cdots 2500)D$ or $r_m \approx 50 \cdots 5000$ for $D = 0.2 \cdots 2\,\mathrm{m}$.

[5]Usually manufacturers of magnetic iron will specify the iron loss in terms of the loss per kg for a given induction value and given frequency, e.g. P_{dm} = iron loss per kg for an induction of 1.5 T.

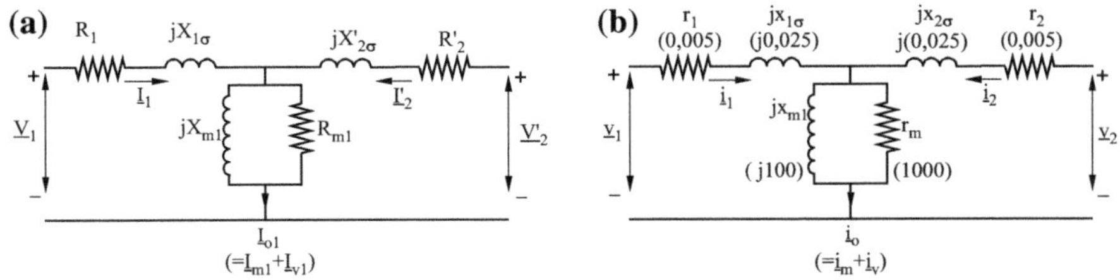

Fig. 1.13 T-equivalent circuits

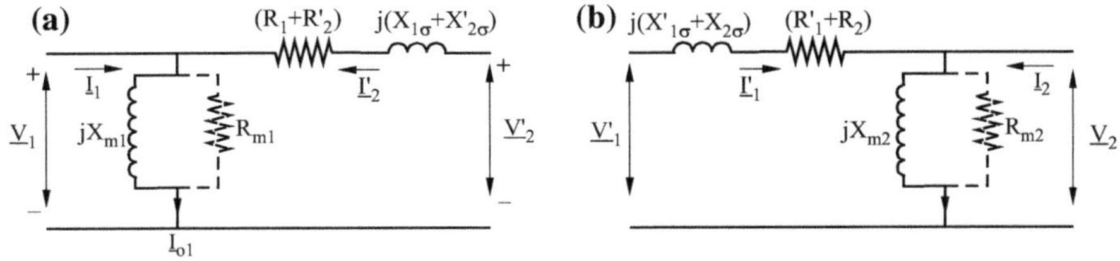

Fig. 1.14 L-equivalent circuits

1.6 Alternative and Simplified Equivalent Circuits

The equivalent circuit derived in the previous paragraphs is depicted in Fig. 1.13 (in (a) for the circuit in absolute values and (b) for the circuit in per unit). Often, the iron dissipation resistance is omitted. Such equivalent circuits (with or without the iron resistance) are commonly called "T"-circuits.

To simplify calculations, the magnetising branch can be shifted completely to the left (for a transformer supplied from a primary voltage) or to the right (for a transformer fed from a secondary voltage), as is shown in Fig. 1.14. In this way, the magnetising (or no-load) current can be directly calculated. The error is usually negligible because of the very high relative values of the magnetising branch impedances compared to the series impedances.[6] Such schemes are called "L"- schemes.

Except for studying the no-load conditions, the magnetising branch may also be completely omitted (see (a) in Fig. 1.15). For a transformer of high power rating the resistances are small and in most cases negligible (see (b) in Fig. 1.15). The sum of the leakage inductances $X_{1\sigma} + X'_{2\sigma}$ is sometimes denoted as $X_{\sigma 1}$, the total leakage as seen from the primary. The sum of these separate leakage inductances is, however, only an approximation of the total leakage as seen from the primary. A "T"-scheme of inductances can indeed be transformed correctly to an "L"-scheme, resulting in $X_{\sigma 1} = \sigma X_1$ with $\sigma = 1 - X_{m1}^2/(X_1 X'_2)$ the total leakage coefficient and $X_1 = X_{1\sigma} + X_{m1}$, $X'_2 = X'_{2\sigma} + X_{m1}$ the primary and secondary total inductances (here with the secondary inductance reduced to the primary). As the magnetising

[6]It is also possible to correctly transform a "T" of inductances to a "L"of inductances, but then the iron resistance has to be neglected.

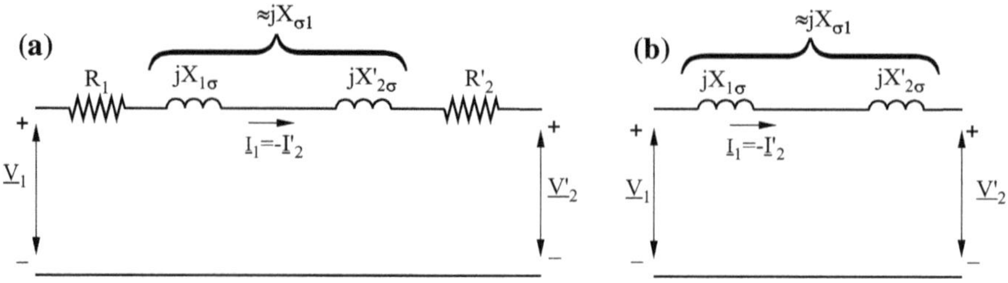

Fig. 1.15 Equivalent circuits neglecting the magnetising branch

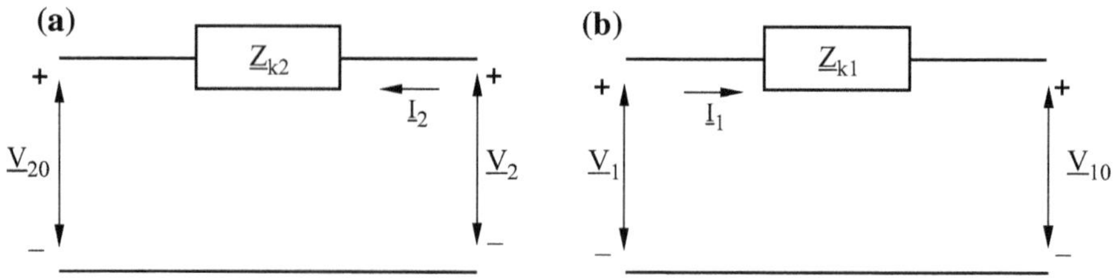

Fig. 1.16 Thévenin equivalent circuits

inductance of transformers is quite large compared with the leakage inductances, the error is small, however.

Another approach, now without approximations, is to use a Thévenin equivalent circuit, as is demonstrated in Fig. 1.16. For studying the secondary load side (when the primary is fed by a voltage source), the circuit in (a) is the preferred one. The voltage $\underline{V}_{20}$ is the no-load voltage measured at the secondary (thus with the secondary open). The secondary short-circuit (or Thévenin) impedance $\underline{Z}_{k2}$ is the impedance measured at the secondary with the primary short-circuited. Both $\underline{V}_{20}$ and $\underline{Z}_{k2}$ can be expressed in terms of the impedances of the equivalent circuit in Fig. 1.13:

$$\underline{V}_{20} = \frac{w_2}{w_1}\underline{V}_1 \frac{(jX_{m1} \| R_{m1})}{R_1 + jX_{1\sigma} + (jX_{m1} \| R_{m1})} \tag{1.46}$$

$$\underline{Z}_{k2} = R_2 + jX_{2\sigma} + jX_{m2} \| R_{m2} \| (R_1^{'} + jX_{1\sigma}^{'}) \tag{1.47}$$

with $\|$ denoting the parallel connection.

Similar expressions hold for the circuit (b) in Fig. 1.16.

Although we have derived the equations and the equivalent circuit in a symmetrical way (i.e. with the URS on both sides), in practice a transformer is in most cases intended to transfer power from one side (the supply) to the other (the load). Therefore the URS is typically chosen on the supply side and the GRS on the load side, as is shown in Fig. 1.17 in which the arrow for the secondary current is reversed. As a result, the active (as well as reactive and apparent power) does not change signs:

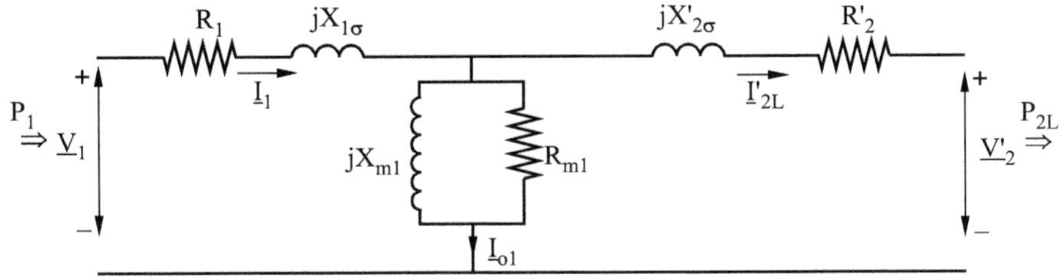

Fig. 1.17 Equivalent circuit with GRS at the secondary side

$$P_1 = P_{2L} + R_1 I_1^2 + R_2^{'} I_{2L}^{'2} + E_1^2/R_{m1} \tag{1.48}$$

where $\underline{S}_1 = P_1 + jQ_1 = \underline{V}_1 \underline{I}_1^*$ and $\underline{S}_{2L} = P_{2L} + jQ_{2L} = \underline{V}_2 \underline{I}_{2L}^* = \underline{V}_2^{'} \underline{I}_{2L}^{'*}$. Remark that the no-load current is now given by $\underline{I}_{01} = \underline{I}_1 - \underline{I}_{2L}^{'}$.

1.7 No-Load Operation

When the primary is connected to the grid with the secondary open-circuited, the primary current of the transformer is the no-load current. The following equations hold:

$$\begin{aligned} \underline{V}_1 &= \underline{I}_{10}\left[R_1 + jX_{1\sigma} + \tfrac{jX_{m1}R_{m1}}{R_{m1}+jX_{m1}}\right] \\ \underline{V}_2^{'} &= \underline{V}_{20}^{'} = \underline{E}_1 = \underline{I}_{10}\left[\tfrac{jX_{m1}R_{m1}}{R_{m1}+jX_{m1}}\right] \end{aligned} \tag{1.49}$$

In most cases, the primary resistance can be omitted as it is negligible compared with the leakage reactance. Further, as the iron dissipation resistance is much larger than the magnetising reactance, some approximations may be used, resulting in

$$\underline{V}_1 \approx \underline{I}_{10}\left[jX_{1\sigma} + jX_{m1} + \frac{X_{m1}^2}{R_{m1}}\right] = \underline{I}_{10}\left[jX_1 + \frac{X_{m1}^2}{R_{m1}}\right] \tag{1.50}$$

or

$$\underline{I}_{10} \approx \frac{\underline{V}_1}{jX_1} + \frac{\underline{V}_1}{R_{m1}} \tag{1.51}$$

If the iron loss resistance can be neglected, then $\underline{I}_{10} \approx \underline{V}_1/jX_1$. With similar approximations we find for the no-load secondary voltage that

$$\underline{V}_{20}^{'} = \underline{V}_1 - \underline{I}_{10}(R_1 + jX_{1\sigma}) \approx \underline{V}_1\left[1 - \frac{X_{1\sigma}}{X_1}\right] = \underline{V}_1\frac{X_{m1}}{X_1} = \underline{V}_1\frac{1}{1 + X_{1\sigma}/X_{m1}} = \underline{V}_1\frac{1}{1+\sigma_1} \tag{1.52}$$

Similar relations hold when the secondary is fed by a voltage source and the primary is open-circuited. Per-unit expressions can also be derived easily.

1.8 Short-Circuit Operation

1.8.1 Short-Circuit Impedance

With a short-circuited secondary, we have

$$\underline{V}_1 = \underline{I}_{1k}\left[R_1 + jX_{1\sigma} + \left(\frac{1}{R_{m1}} + \frac{1}{jX_{m1}} + \frac{1}{R_2^{'} + jX_{2\sigma}^{'}}\right)^{-1}\right] \tag{1.53}$$

The impedance between square brackets is the primary short-circuit impedance $\underline{Z}_{k1}$ (see also the Thévenin equivalent circuits in Sect. 1.6). Taking into account that R_{m1} is much larger than X_{m1} and that $R_2^{'2} \ll X_{2\sigma}^{'2} \ll X_2^{'2}$, the expression for $\underline{Z}_{k1}$ may be approximated as follows:

$$\underline{Z}_{k1} \approx R_1 + jX_{1\sigma} + \frac{jX_{m1}X_{2\sigma}^{'}X_2^{'} + jX_{m1}R_2^{'2} + jX_{m1}^2R_2^{'}}{R_2^{'2} + jX_2^{'2}} \tag{1.54}$$

or

$$\begin{aligned} \underline{Z}_{k1} &\approx \left(R_1 + R_2^{'}\tfrac{X_{m1}^2}{X_2^{'2}}\right) + j\left(X_{1\sigma} + \tfrac{X_{m1}X_{2\sigma}^{'}X_2^{'}}{X_2^{'2}}\right) \\ &\approx \left(R_1 + R_2^{'}\tfrac{X_{m1}^2}{X_2^{'2}}\right) + j\left(X_1 - \tfrac{X_{m1}^2}{X_2^{'}}\right) \\ &\approx \left(R_1 + R_2^{'}\tfrac{X_{m1}^2}{X_2^{'2}}\right) + j\sigma X_1 \end{aligned} \tag{1.55}$$

The short-circuit reactance σX_1 obtained is exactly the series reactance when the "T"-circuit of inductances in the classical transformer equivalent circuit is replaced by the equivalent "L"-circuit of inductances with the main inductance at the secondary side (and the series inductance at the primary side). This short-circuit reactance σX_1 is the total leakage seen from the primary side and is approximately equal to the sum of the primary and secondary leakage inductances: $X_{k1} = \sigma X_1 \approx X_{1\sigma} + X_{2\sigma}^{'}$. The leakage coefficient $\sigma = 1 - X_{m1}^2/(X_1X_2^{'})$ is in fact a measure for the magnetic coupling between primary and secondary windings. A leakage coefficient $\sigma = 0$ means a perfect coupling whereas a leakage coefficient $\sigma = 1$ implies no coupling at all.

The short-circuit resistance is approximately equal to the sum of the primary and (reduced) secondary resistances: $R_{k1} \approx R_1 + R_2^{'}\frac{X_{m1}^2}{X_2^{'2}} \approx R_1 + R_2^{'}$.

The short-circuit phase angle, φ_{k1} with $\tan\varphi_{k1} = X_{k1}/R_{k1}$ is another concept that is sometimes used. The secondary current can be calculated from

$$\underline{I}_2' = -\underline{I}_{1k}\left(\frac{1}{R_{m1}} + \frac{1}{jX_{m1}} + \frac{1}{R_2' + jX_{2\sigma}'}\right)^{-1} \cdot \frac{1}{R_2' + jX_{2\sigma}'} \approx -\underline{I}_{1k} \cdot \frac{X_{m1}}{X_2'} = -\underline{I}_{1k} \cdot \frac{1}{1+\sigma_2} \tag{1.56}$$

Similarly a primary short circuit with supply at the secondary can be studied.[7]

1.8.2 Procentual Short-Circuit Voltage

An important notion for the study of the short-circuit behaviour and parallel operation of transformers is the *nominal short-circuit voltage*. This is the supply voltage required to obtain the rated current in the supplied winding with the other winding shorted. For a secondary short circuit and primary supply, this short-circuit voltage is simply $V_{k1n} = Z_{k1}I_{n1}$. Mostly the procentual value of the short-circuit voltage is specified (i.e. $v_{k1n}\% = 100 \cdot Z_{k1}I_{n1}/V_{1n} = 100 \cdot z_{k1}$), thus equal to the per-unit short-circuit impedance multiplied by 100. As z_{k1} and z_{k2} are almost equal, no distinction should be made between primary or secondary values.

From the discussion in the previous sections it follows that for large transformers the procentual nominal short-circuit voltage almost completely results from the reactive part, i.e. the short-circuit reactance (equal to the total leakage reactance). It varies between 3 and 5% for transformers with power ratings between 5 and some 100 kVA. For transformers with power ratings larger than 1 MVA, $v_{kn}\% \simeq 5 \cdots 15\%$ may be observed.

For small power ratings, on the other hand, the resistive part of the short-circuit impedance becomes more important and $v_{kn}\%$ increases for lower power ratings ($5 \cdots 15\%$ for $S_n = 500 \cdots 50$ VA).

1.8.3 Remarks

In Sect. 1.2.1 we already noted that the equations for a transformer correspond to those of two magnetically coupled coils. Magnetically, such a set of two coupled coils is completely defined by the self inductances L_1 and L_2, and the mutual inductance M.

In our model, however, we introduced four magnetic parameters: the leakages of primary and secondary, the magnetising inductance and the turns ratio. It should therefore be clear that we can never obtain unique values for these four parameters.

[7] As an exercise, derive the secondary short-circuit impedance and the approximative expressions for resistive and reactive parts of it.

Indeed, to experimentally characterise a transformer, we may use: DC measurements to obtain the resistances of primary and secondary and AC no-load and short-circuit measurements.

From the short-circuit experiments, we can obtain:

$$\begin{aligned} & R_1 + R_2\left(\frac{w_1}{w_2}\right)^2 \cdot \left(\frac{1}{1+\sigma_2}\right)^2, \ \sigma X_1 \ and \ \frac{1}{1+\sigma_2} \cdot \frac{w_1}{w_2} = \frac{I_2}{I_{k1}} \\ & R_2 + R_1\left(\frac{w_2}{w_1}\right)^2 \cdot \left(\frac{1}{1+\sigma_1}\right)^2, \ \sigma X_2 \ and \ \frac{1}{1+\sigma_1} \cdot \frac{w_2}{w_1} = \frac{I_1}{I_{k2}} \end{aligned} \tag{1.57}$$

From the no-load experiments, on the other hand, we can deduce:

$$\begin{aligned} & X_1, \ R_{m1} \ and \ \frac{1}{1+\sigma_1} \cdot \frac{w_2}{w_1} = \frac{V_{20}}{V_1} \\ & X_2, \ R_{m2} \ and \ \frac{1}{1+\sigma_1} \cdot \frac{w_1}{w_2} = \frac{V_{10}}{V_2} \end{aligned} \tag{1.58}$$

The results obtained from these four experiments are thus not completely independent. As to the magnetic parameters we may derive σX_1, σX_2, X_1 and $X_2$and thus σ, X_1, X_2. However, we cannot derive the separate leakages $X_{1\sigma}$, $X_{2\sigma}$ (or the leakage coefficients $\sigma_1$and σ_2) and the winding ratio. We may *choose* a value for the turns ratio which will then correspond to some specific (but not necessarily physically realistic) values for the primary and secondary leakages. In other words, the total leakage can be measured but the separate leakages of primary and secondary cannot. This division of the leakage is not important however. As observed from the terminals the behaviour of the transformer is completely determined. As seen from the terminals the transformer is as a black box; what is inside as to turns ratio and magnetic coupling is not unique and not important as long as the behaviour at the terminals is the same.

In practice, only one no-load and one short-circuit experiment will be used and the leakage will normally be assumed to be equally divided between primary and secondary.

1.9 Voltage Variation with Load

In this section we suppose the transformer is supplied at the primary side by a voltage source and the secondary is connected to a load. As a result of the series impedances R_1, R_2, $X_{1\sigma}$ and $X_{2\sigma}$ the secondary voltage of a transformer will change as to amplitude and phase with the load. Figure 1.18 shows the phasor diagrams for an inductive load and a capacitive load (with the URS at the primary side and the GRS at the secondary side).

For the most common case of an inductive load the amplitude of the secondary voltage will decrease with increasing load: $|V_2| < |V_1^{'}| \approx |V_{20}|$. In contrast, a sufficiently capacitive load may result in an increasing secondary voltage amplitude $|V_2| > |V_1^{'}| \approx |V_{20}|$.

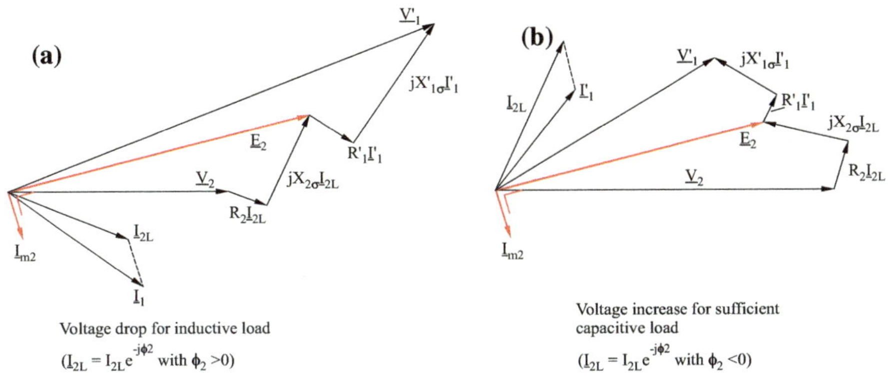

Fig. 1.18 Phasor diagrams for inductive and capacitive load

The voltage variation may be calculated as follows, using the Thévenin equivalent circuit as seen from the secondary. We choose the real axis along the voltage $\underline{V}_2$, as a result $\underline{V}_2 = V_2$. The load current is then represented as $\underline{I}_{2L} = I_{2L}\exp(-j\phi_2)$ with $\phi_2 > 0$ for an inductive load and $\phi_2 < 0$ for a capacitive load. We define the active and reactive components of $\underline{I}_{2L}$ as

$$\underline{I}_{2L} = I_{2L}\exp(-j\phi_2) = I_w + jI_b \tag{1.59}$$

With $\underline{V}_2 = \underline{V}_{20} - \underline{Z}_{k2}\underline{I}_2$ and $\underline{Z}_{k2} = Z_{k2}\exp(-j\phi_{k2}) = R_{k2} + jX_{k2}$ we obtain

$$\begin{aligned}\underline{V}_2 &= \underline{V}_{20} - Z_{k2}I_{2L}\left[\cos(\phi_k - \phi_2) + j\sin(\phi_k - \phi_2)\right]\\ &= \underline{V}_{20} - \left[(R_{k2}I_w - X_{k2}I_b) + j(X_{k2}I_w + R_{k2}I_b)\right]\end{aligned} \tag{1.60}$$

If $\phi_2 = \phi_k$ then $\underline{V}_2 = \underline{V}_{20} - (R_{k2}I_w - X_{k2}I_b)$. For load angles ϕ_2 around ϕ_k, $V_2 = V_{20} - Z_{k2}I_{2L}\cos(\phi_k - \phi_2)$ may still be used as an approximation. For larger differences the following approach may be used. We will use the following abbreviations: $R_{k2}I_w - X_{k2}I_b = |AB'| = a$ and $X_{k2}I_w + R_{k2}I_b = |B'B| = b$ (see Fig. 1.19). As $\underline{V}_{20} = V_2 + (a + jb)$ and thus $V_{20} = \sqrt{(V_2 + a)^2 + b^2}$ we easily find after some approximations (i.e. using $\sqrt{1+x} \approx 1 + x/2$) that

$$\frac{V_{20} - V_2}{V_{20}} = \frac{a}{V_{20}} + \frac{1}{2}\cdot\frac{b^2}{V_{20}^2} = \frac{R_{k2}I_w - X_{k2}I_b}{V_{20}} + \frac{1}{2}\cdot\frac{(X_{k2}I_w + R_{k2}I_b)^2}{V_{20}^2} \tag{1.61}$$

This result may also be obtained geometrically making use of the rectangular triangle BB'B" and the isosceles triangle OBB" in Fig. 1.19.

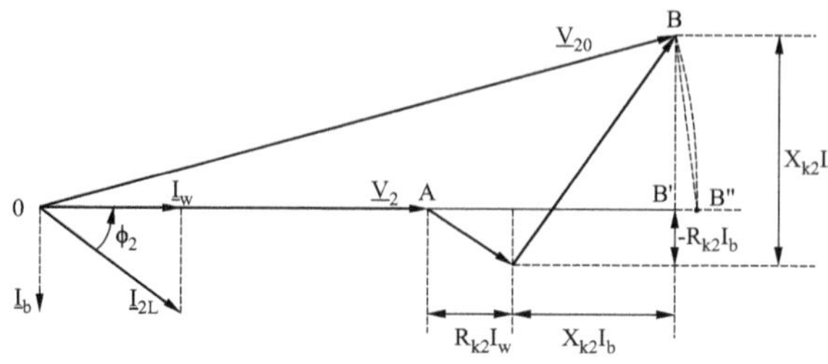

Fig. 1.19 Calculation of voltage regulation

In per-unit, with $V_{20} = V_{2n}$, Eq. 1.61 reads

$$\Delta v_2 = (r_k i_w - x_k i_b) + \frac{1}{2}(x_k i_w + r_k i_b)^2 \tag{1.62}$$

The quadratic term can be neglected for $z_k < 0.05$ (or $v_k\% < 5\%$).

Please note that it follows from Eq. 1.60 that for $\phi_2 \simeq \phi_k$ and $I_{2L} = I_{2n}$ the p.u. voltage drop is equal to the p.u. short-circuit impedance, $\Delta v_2 = z_k$.

1.10 Parallel Operation of Transformers

Parallel connected transformers supply the same load and are supplied from the same voltage source, see Fig. 1.20. In practical applications, additional impedances due to the grid conductors (which might ease the parallel operation) should be taken into account. These will not be considered here.

In Fig. 1.21, the two transformers ("A" and "B") are represented by their equivalent circuits referred to the secondary. The GRS is used for the secondary windings. By replacing the equivalent circuits by their Thévenin equivalents, we obtain the scheme in Fig. 1.22. The following equations hold:

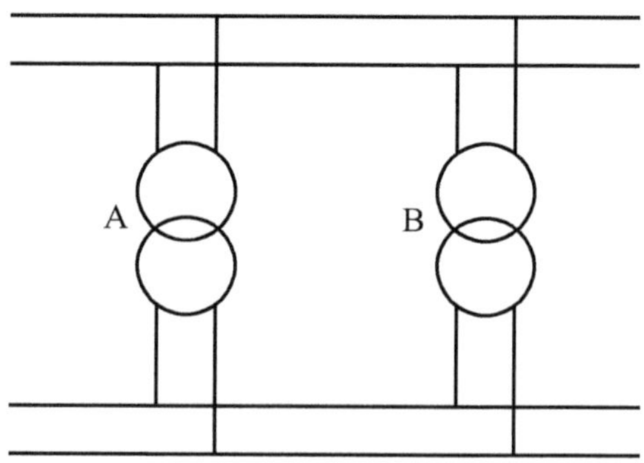

Fig. 1.20 Parallel operation of transformers

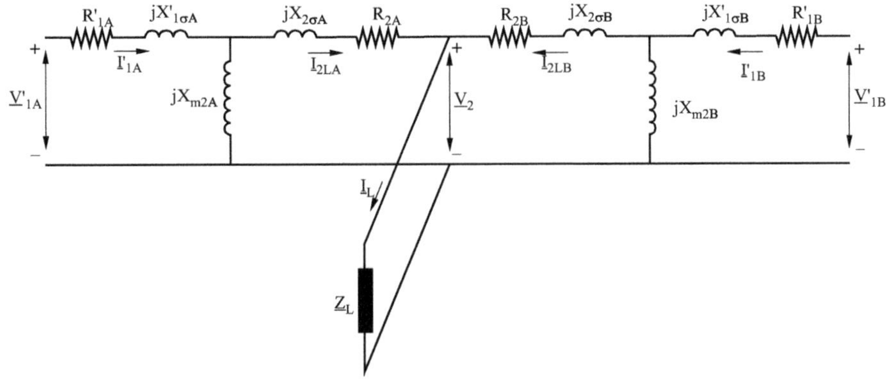

Fig. 1.21 Parallel operation: equivalent circuits

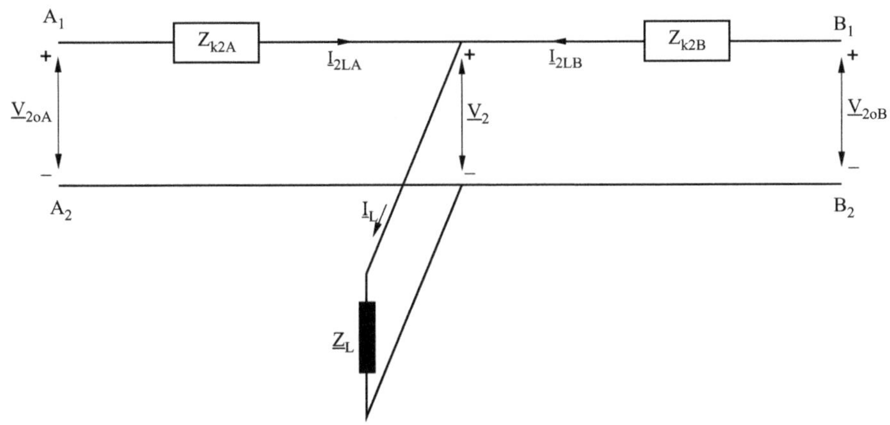

Fig. 1.22 Parallel operation study using Thévenin equivalent circuits

$$
\begin{aligned}
\underline{V}_2 &= \underline{Z}_L \underline{I}_L \\
\underline{I}_L &= \underline{I}_{2LA} + \underline{I}_{2LB} \\
\underline{V}_{20A} &= \underline{Z}_{K2A}\underline{I}_{2LA} + \underline{V}_2 \\
\underline{V}_{20B} &= \underline{Z}_{K2B}\underline{I}_{2LB} + \underline{V}_2
\end{aligned}
\tag{1.63}
$$

Elimination of $\underline{V}_2$ yields

$$
\begin{aligned}
\underline{I}_{2LA} &= \frac{\underline{V}_{20A} - \underline{V}_{20B}}{\underline{Z}_{K2A} + \underline{Z}_{K2B}} + \underline{I}_L \cdot \frac{\underline{Z}_{K2B}}{\underline{Z}_{K2A} + \underline{Z}_{K2B}} \\
\underline{I}_{2LB} &= \frac{\underline{V}_{20B} - \underline{V}_{20A}}{\underline{Z}_{K2A} + \underline{Z}_{K2B}} + \underline{I}_L \cdot \frac{\underline{Z}_{K2A}}{\underline{Z}_{K2A} + \underline{Z}_{K2B}}
\end{aligned}
\tag{1.64}
$$

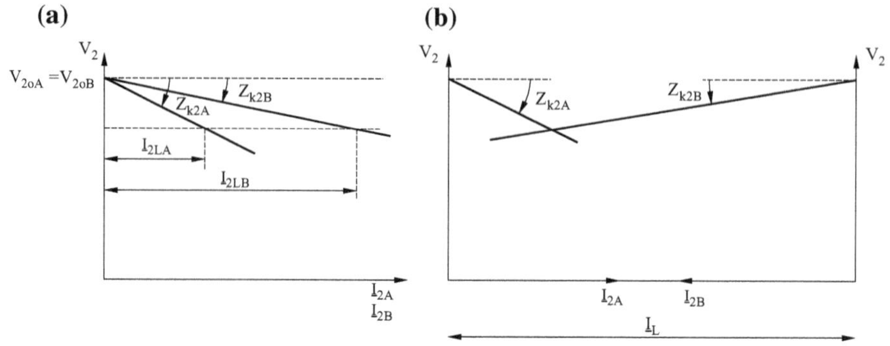

Fig. 1.23 Load sharing

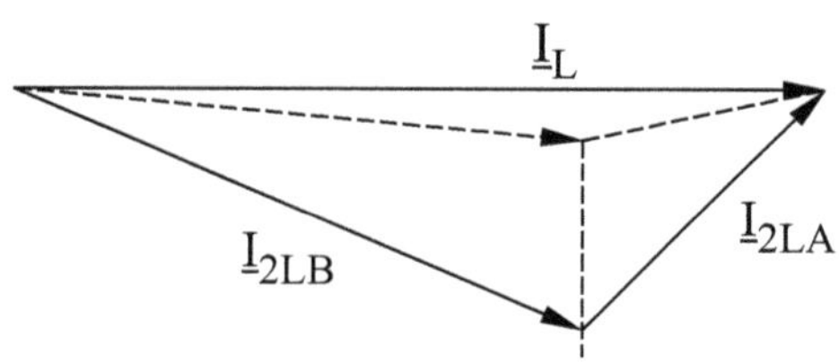

Fig. 1.24 Unequal short-circuit phase angles

In addition to a component proportional to the load current, the transformer currents therefore also contain a circulating component.

It should be clear that the circulating component is a disturbance. To avoid it, the no-load voltages of the two transformers should be equal, implying that the turns ratios should be equal.

Between the two transformers the load current is shared inversely proportional to their short-circuit impedances. In order to have an acceptable load sharing, the short-circuit impedances should be such that the load current is shared in proportion to their power ratings. An acceptable load sharing in terms of the amplitude of the currents is achieved if the moduli of the p.u. short-circuit impedances are equal.[8] If this condition is not fulfilled, the transformer with the smallest short-circuit impedance takes the largest share of the load current, as is illustrated in Fig. 1.23.

However, it is also important that the phase angles of the short-circuit impedances are equal. If the short-circuit angles are not equal, then the phase angle difference between the two secondary currents $\underline{I}_{2LA}$ and $\underline{I}_{2LB}$ results in larger-than-necessary amplitudes of these currents (for the given load current $\underline{I}_L$) and therefore extra losses (see Fig. 1.24).

The discussion above shows that parallel operation of identical transformers does not pose any problems. Given the evolution of the p.u. short-circuit impedances with

[8]Prove this using the equality of the rated secondary voltages V_{2nA} and V_{2nB}.

size (scaling laws), parallel operation of transformers with rather different power ratings may be difficult to achieve.[9] For that reason, the standards limit the ratio of the power ratios of parallel connected transformers.

1.11 Construction of Single-Phase and Three-Phase Transformers

1.11.1 Single-Phase Transformers

Single-phase transformers are either core-type or shell-type transformers. In a core-type transformer (see (a) in Fig. 1.25), the windings surround the core. Primary and secondary windings may be put on different legs or on the same leg. Usually, both windings are split in half, with one half on each leg (and with primary and secondary halves concentric). In a shell-type transformer, the core has three legs, with both windings concentric on the middle leg (see (b) in Fig. 1.25). In a shell-type transformer the outer legs (as well as the yokes) need only half the cross section of the middle leg.

The core of small transformers is in many cases assembled using E-shaped laminations with a closing lamination above or below (alternately stacked to reduce the air gaps). Another construction type stacks L-shaped laminations, also stacked alternately. The disadvantage of these construction types is that the flux is not everywhere in the preferred magnetic direction for oriented steel. Larger transformers therefore use (basically rectangular) lamination sheets that are cut with the preferred magnetic direction along the length of the sheets. To avoid large air gaps, the laminations are cut at an angle in the corners and somewhat different lengths are used for subsequent layers so that the laminations overlap (see Fig. 1.10).

1.11.2 Three-Phase Transformers

1.11.2.1 Construction Types: Core

Transformer Bank

The most logical (albeit not very common) way to obtain a three-phase transformer is to use three separate single-phase transformers, one for each phase. An advantage of such a *transformer bank* is that the three magnetic flux paths are independent of each other, which may be advantageous for non-symmetrical operation (see also

[9] Minor differences in short-circuit impedances may, however, be compensated by using inductors in series connection.

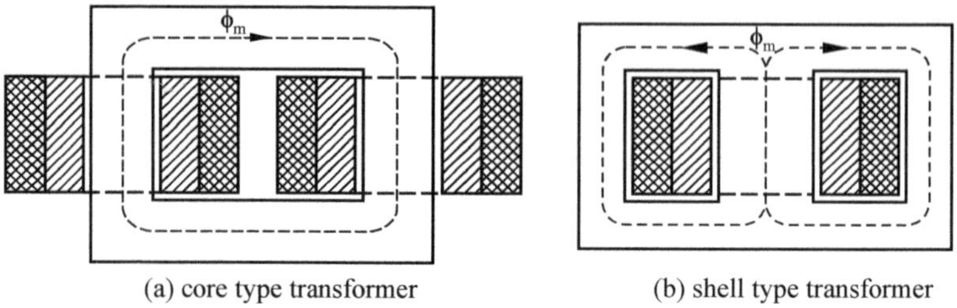

Fig. 1.25 Single-phase transformers: core and shell types

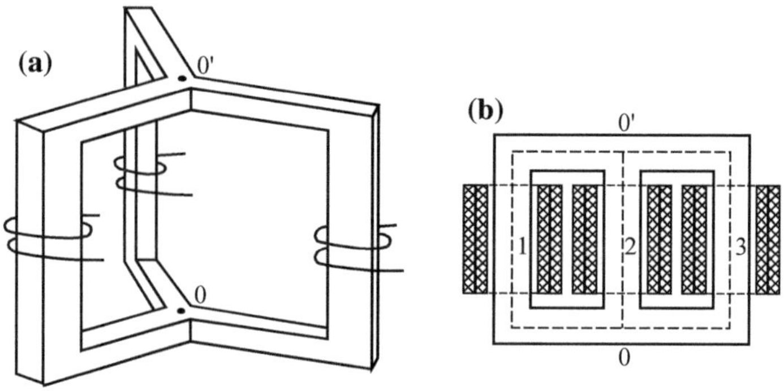

Fig. 1.26 Three-phase temple and core transformers

Sect. 1.12.4 below). However, such a transformer bank is generally quite expensive compared with the three-phase transformer units discussed below.

Three-Phase Core-Type Transformer

Consider three single-phase core-type transformers, each with both primary and secondary windings on one leg. If the three legs without windings are merged (see the 3-D sketch (a) in Fig. 1.26), then for a symmetrical sinusoidal supply (and loading) the sum of the three fluxes is zero and the combined leg (return yoke) can be completely removed, resulting in a structure as (b) in Fig. 1.26.

Such a 3-D structure is not very practical, however. If we put the three legs for each phase in the same plane as in (b) in Fig. 1.26, we obtain an almost equivalent magnetic structure that can be used for symmetrical operation. The only difference (and slight disadvantage) is that the magnetic paths for the outer legs are not exactly equal to that of the inner leg - a slight unsymmetry that is of minor importance. This structure is called a three-phase core-type.

Three-Phase Five-Leg Core-Type Transformer

For the sake of symmetry, two outer return legs may also be added (see (a) in Fig. 1.27). This structure, called a five-leg (core-type) 3-phase transformer, offers a return through iron if the three fluxes do not form a three-phase symmetrical set and have a non-zero sum.

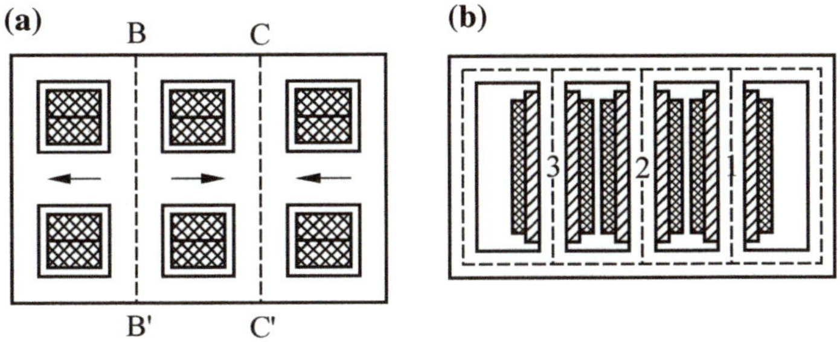

Fig. 1.27 Three-phase 5-leg and shell transformers

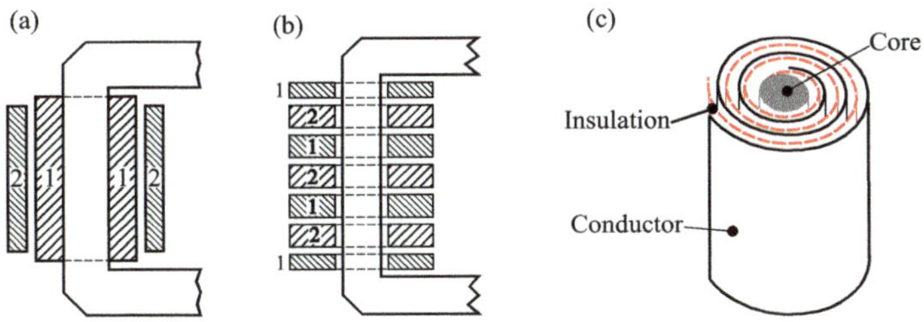

Fig. 1.28 Winding types

Three-Phase Shell-Type Transformer

We may also broaden the idea of a single-phase shell-type transformer to a three-phase structure by combining three single-phase shell-type transformers as illustrated in (b) in Fig. 1.27. This structure, called a shell-type three-phase transformer, also offers a return through iron if the three fluxes do not form a three-phase symmetrical set and have a non-zero sum. There is a slight asymmetry for the transformer leg in the centre compared to the two outer transformer legs, but this is irrelevant. This core type also offers a return path through the iron core for the flux if the sum of the three fluxes is non-zero.

1.11.2.2 Winding Types

In the discussion of the single-phase transformer in Sect. 1.5, we considered only the *cylindrical winding,* i.e. a winding structure in cylindrical layers with the subsequent turns in each layer below or above the previous turn (see (a) in Fig. 1.28). Between the layers an extra insulation layer is usually required. The primary and secondary cylinder windings are concentric with a phase-insulation layer in between. Two disadvantages of cylindrical windings are: turns with high voltage difference (belonging to different layers) may be rather close to each other and that the p.u. leakage increases linearly with size, as the power rating goes up. Because of the insulation problems, this winding type is used only for low- to medium-voltage applications. However, single-phase transformers are mostly used for low-voltage (LV) applications only, i.e. grid voltages below 600 V.

For three-phase transformers that are used in electrical energy distribution or transport in the grid the voltage levels are medium voltage (MV) or high voltage (HV). These transformers are often also subject to lightning disturbances. For these higher voltage levels and higher power ratings, another type of winding is mostly used, known as *disk winding* or *pancake* winding. In this type of winding, primary and secondary windings consist of a series connection of flat coils (disks) in which the turns form a spiral. The disks (or groups of disks) of primary and secondary are placed alternating on the core (see (b) in Fig. 1.28). The leakage can be influenced by changing the distance between the disks, modifying the interleaving of primary and secondary disks (including more or less grouping disks belonging to primary or secondary) or even by interleaving the turns inside one single disk (although this last interleaving is mainly intended to modify the capacitance between the turns for better behaviour in case of over-voltage as caused by lightning, for example).

A third winding type is the *foil winding*, shown in (c) in Fig. 1.28. In this type of winding, primary and secondary windings consist of insulated aluminium or copper

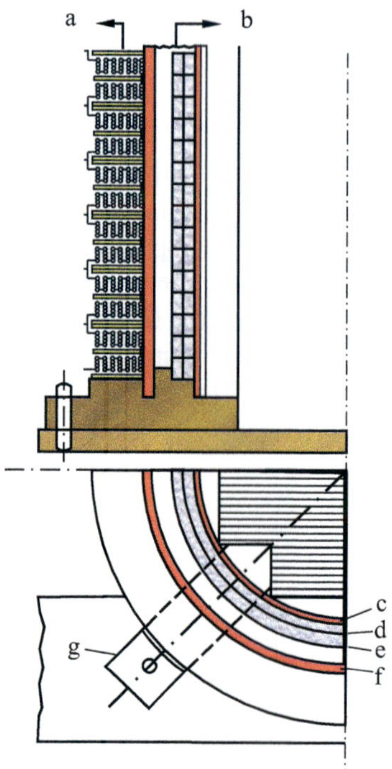

a) disc winding, HV
b) cylinder winding, LV
c) support and insulation cylinder
d) layer insulation
e) insulation cylinder
f) support and insulation cylinder
g) support - fixation

Fig. 1.29 Transformer assembly

foils that form a spiral around the core. This winding type is mostly used for LV applications.

Figure 1.29 provides an example of the assembly of core and windings. Here a cylindrical winding is used for the LV, while a disk-type HV winding is used.

1.12 Connection and Vector Group of a Three-Phase Transformer

1.12.1 Winding and Terminal Markings

The three windings and the corresponding core legs of a three-phase transformer are marked by the letters U, V, W. The primary and secondary windings are indicated by a prefix (e.g. 1 or 2) and their terminals by a suffix consisting of 1 or 2 numbers[10] (e.g. 1U01 or 1U1, 1U2, etc.). For windings on the same core leg, the order of the markings must correspond to the signs of the induced voltage for a common main flux direction. The markings of the terminals in the connection box or on the external connection plate correspond to the winding terminals connected to them. Figure 1.30 provides some examples together with a commonly used representation of the terminals in electrical schemes.[11] The schemes in (a) and (b) in the figure show the connections for a $Y - \Delta$ and a $\Delta - \Delta$ transformer, while (c) does it for a $Y - Z$ transformer. A $Z-$connection consist of the series connection of two equal winding sections belonging to two different core legs so that the phases of their voltages differ by 30° (see also the next paragraph).

1.12.2 Modelling of a Three-Phase Transformer

For modelling a three-phase transformer with symmetrical supply and symmetrical load, the single-phase equivalent circuit discussed above can be used for each phase. This holds for both positive-sequence and negative-sequence situations. The single-phase equivalent circuit can relate corresponding phase windings. For a $\Delta-$connection, the equivalent circuit may relate the fictitious phase windings of an equivalent Y.

For a supply or load containing zero-sequence components, the behaviour (and equivalent circuit to be used) will depend on the construction of the transformer and the connections, as discussed below.

---

[10]Suffixes 1 and 2 are used for the end points while 3, 4, etc. are used for tap connections.

[11]Note the polarity dots "**.**", indicating the common flux direction: one must imagine their physical position on their leg corresponding to these windings pivoted vertically below the plane around the terminal plate.

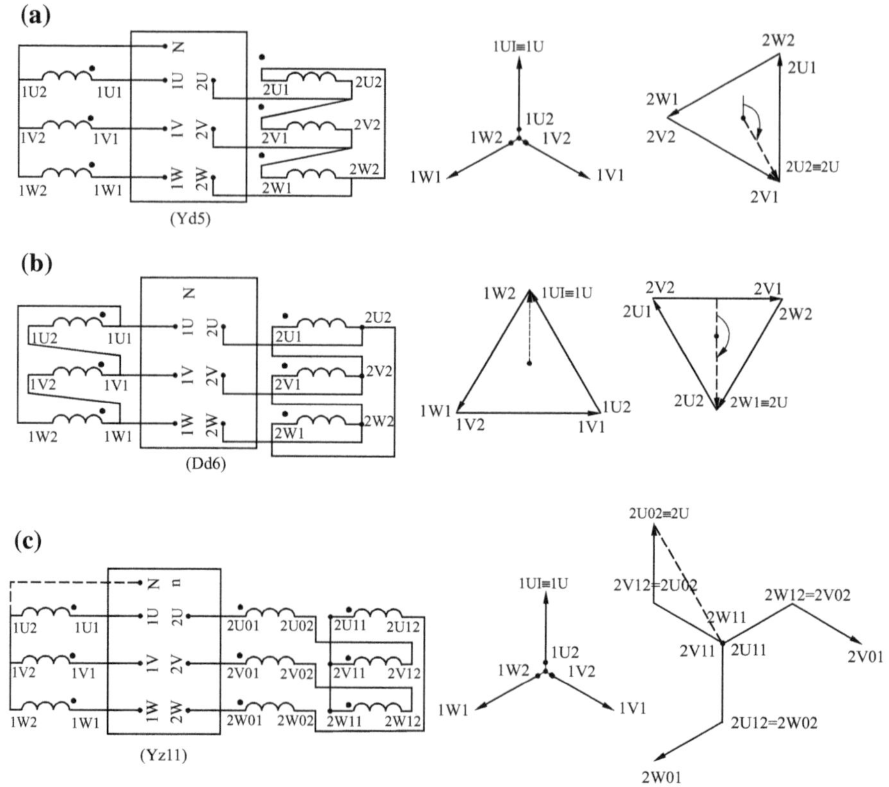

Fig. 1.30 Connection Yd5, Dd6 and Yz11

1.12.3 *Connections and Vector Groups*

The choice of the primary and secondary winding connections in a three-phase transformer determines not only the (secondary) voltage levels but also the phase difference between primary and secondary voltages and the behaviour of the transformer with respect to asymmetrical supply and loading conditions. The phase difference between primary and secondary is always a multiple of 30°. It is always considered as the lagging of the secondary voltage $2N - 2U$ with respect to the primary voltage $1N - 1U$. The neutral point can be the actual neutral point (in case of a $Y-$connection) or the fictitious or equivalent neutral point for a $\Delta-$connection. This lagging of the secondary voltage with respect to the primary voltage is expressed by the *vector group,* which is the multiple of 30° by which the secondary voltage lags the primary. When specifying or determining the phase shift or vector group the phase sequence is supposed to be $U \rightarrow V \rightarrow W$ (i.e. phase U leads phase V, which in turn leads phase W).

The connection of a transformer is indicated by three symbols:

- the primary connection (Y, D or Z for a $Y-$ or $\Delta-$ or $Z-$connection)
- the secondary connection (y, d or z for a $Y-$ or $\Delta-$ or $Z-$connection)
- the vector group or phase shift in multiples of 30°

Figure 1.30 illustrates the connections Yd5, Dd6 and Yz11. For a $Z-$connection of a (secondary) winding, each phase winding must contain two equal halves. In the $Z-$connection, half a phase winding is connected in opposite series with half a winding on another core leg. The secondary voltage is then the phasor sum of these two voltages (with the appropriate sign). In the next paragraph, we will discuss the advantages of this connection with regard to asymmetrical operation.

To determine the vector group the primary voltage vectors first need to be sketched. Taking into account the polarity marks ("$\cdot$") and the connections of the winding terminals, we may then derive the secondary voltage vectors. For the examples in Fig. 1.30, this results in phase shifts of 150°, 180° and 330°, corresponding with the vector groups 5, 6 and 11.

For the voltage ratios, in the given examples $\sqrt{3}w_1/w_2$, w_1/w_2 and $2w_1/(\sqrt{3}w_2)$ are obtained, where w_1 and w_2 represent the total primary and secondary turns, respectively.

Figure 1.31 shows the most common connections of three-phase transformers used in Europe. Remark that all these connections generate phase shifts of ±180°, ±30°, or combinations of these. A vector group of 1 is obtained by a combination of Y and Δ, or a combination of a Y or Δ with a $Z-$connection. For a vector group of 6 (or phase shift of ±180°), the transformer terminals are to be connected to the other end of the windings.

Further, a phase shift of ±120° can be achieved if the transformer terminals U, V, W are connected with winding terminals V, W, U or W, U, V. A phase shift of ±60° is obtained if the connections in a Δ are switched (e.g. connecting $2U2$ with $2V1$ instead of $2W1$yields Dd8 instead of Dd6).

When three-phase transformers are connected in parallel, it is clear that the phase shift (vector group) must not be ignored, and neither must the phase sequence and the equality of voltages (and, as much as possible, the equality of the p.u. short-circuit impedances).

1.12.4 Asymmetrical Operation of 3-Phase Transformers

A general three-phase voltage or current system can be split up into a positive sequence, a negative sequence and a zero-sequence system, using the symmetrical components transformation. The behaviour of a transformer is fundamentally equivalent for positive-sequence and negative-sequence components. This is, however, not the case for zero-sequence components. For these components, the core construction has a fundamental effect on the resulting zero-sequence fluxes. Also the electrical connection comes into play.

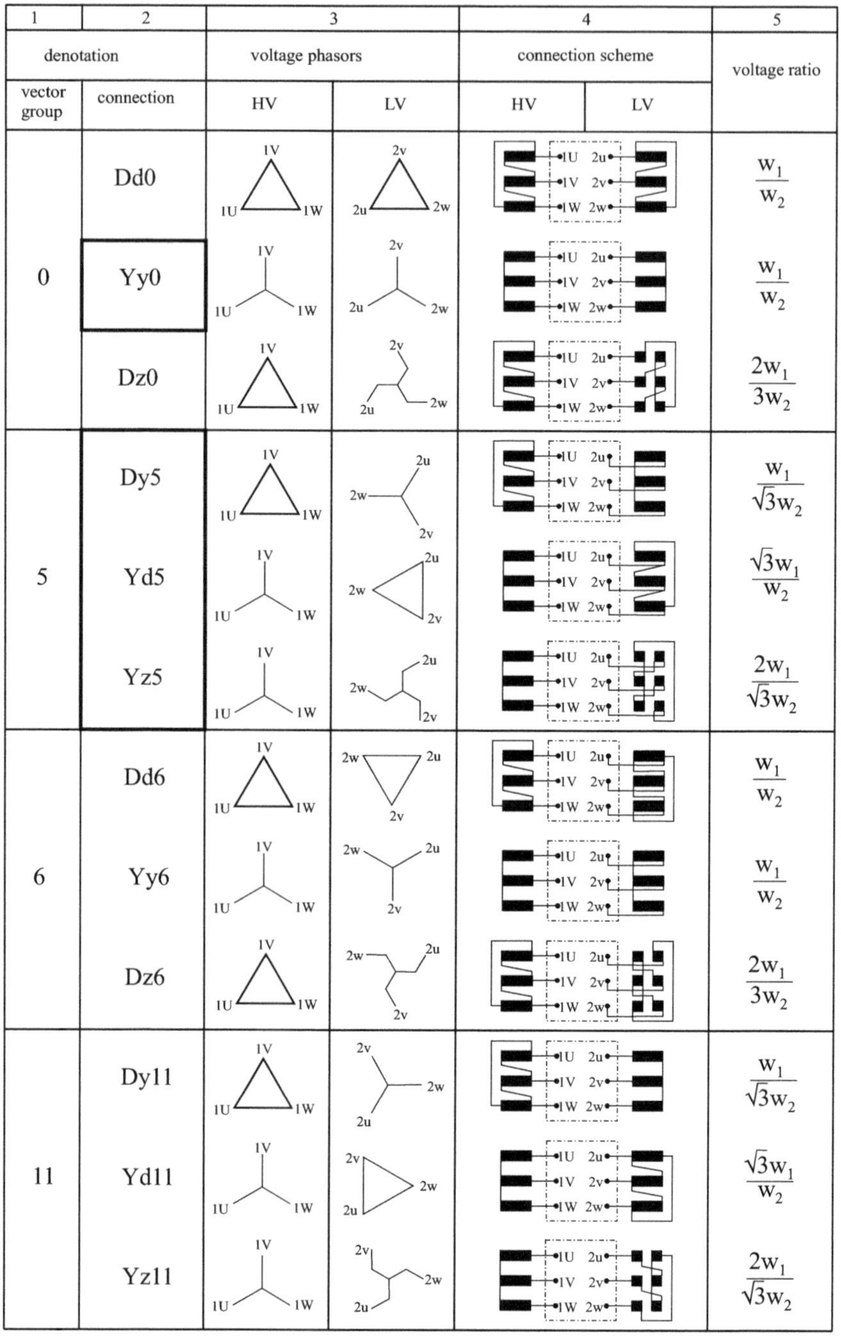

Fig. 1.31 Commonly used connections and vector groups

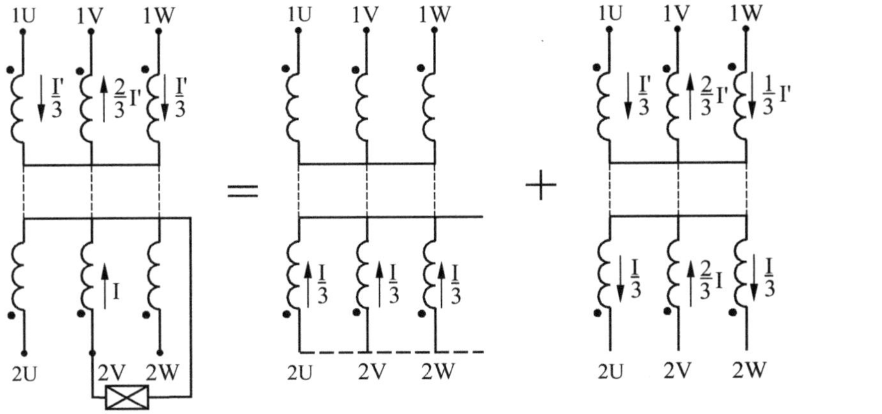

Fig. 1.32 Secondary load with zero-sequence current

As the transformer bank, the five-leg transformer and the three-phase shell-type transformer offer a magnetic return for zero-sequence fluxes, the corresponding zero-sequence magnetising inductance is large (and the zero-sequence magnetising current small for an imposed zero-sequence flux). For the three-phase core-type transformer, the zero-sequence flux has to return through the air and the corresponding magnetising inductance is small (and thus the zero-sequence magnetising current is quite large for an imposed zero-sequence flux).

However, whether or not zero-sequence fluxes are present or pose problems also depends on the electrical connection.

Here we will restrict ourselves to the example of a symmetrical primary supply, but with an asymmetrical secondary load imposing an asymmetrical current loading.

In order to impose a zero-sequence current load, the secondary must be $Y-$connected with neutral connection. Consider for example a single-phase load between phase V and the neutral. The symmetrical components transformation of $\underline{I}_{2U} = 0,\ \underline{I}_{2V} = \underline{I},\ \underline{I}_{2W} = 0$ yields $\underline{I}_{20} = \underline{I}/3,\ \underline{I}_{2+} = a\underline{I}/3,,\ \underline{I}_{2-} = a^2\underline{I}/3$.

The primary currents corresponding with these secondary currents depend on the connection of the primary.

If the primary is $Y-$connected without neutral connection, zero-sequence currents cannot flow in the primary. The primary positive- and negative-sequence currents will compensate the secondary positive- and negative-sequence currents (neglecting the magnetising currents) but the zero-sequence component of the secondary will not be compensated. This corresponds with the situation in Fig. 1.32 where $\underline{I}'$ represents the current $\underline{I}$ referred to the number of windings of the primary, i.e. $\underline{I} \cdot (w_2/w_1)$. The primary currents are $\underline{I}_{1U} = \underline{I}'/3,\ \underline{I}_{1V} = -2\underline{I}'/3,\ \underline{I}_{1W} = \underline{I}'/3$.

Whereas the bi-symmetrical components (i.e. the sum of positive- and negative–sequence components) are compensated (neglecting the magnetising current), the zero-sequence component is not compensated. This zero-sequence-component,

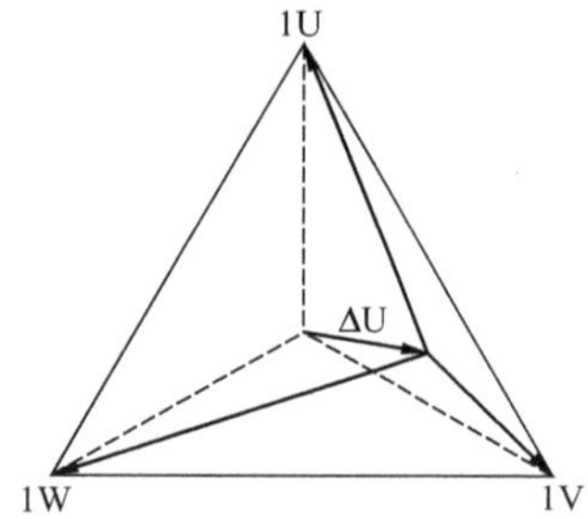

Fig. 1.33 Neutral voltage displacement

whose order of magnitude is that of a load current, results in a zero-sequence mmf in each of the three phase legs of the core. This zero-sequence mmf will result in a zero-sequence voltage induced in each phase, both in the primary and secondary. As a consequence, the potential of the neutral will be displaced with regard to the potentials of the lines, both in primary and secondary (see Fig. 1.33). If a core-type (three-leg) transformer is used, the zero-sequence mmf will see a very low magnetising inductance and the displacement will be limited. In contrast, in a five-leg core-type or a shell-type transformer, the neutral potential displacement will be unacceptably large. In that case it may result in an important decrease of the secondary V to N voltage. In addition, the higher flux in the other phase legs may increase saturation to an unacceptable level. On the other hand, for a core-type transformer the zero-sequence flux will try to close through the transformer casing, resulting in high eddy-current losses and heating of the shell.

We conclude that a zero-sequence load for a $Y - Y$ transformer without neutral connection in the primary:

- should be avoided for shell-type transformers, five-leg transformers and transformer banks
- should be limited for core-type transformers; according to the standards no more than 10% of the rated current.

Nevertheless, a single-phase load is acceptable if:

- the primary is $\Delta-$connected
- the secondary is $Z-$connected
- a tertiary compensation winding is provided

With a primary winding in Δ or a compensation winding in Δ, currents in the delta will compensate the zero-sequence fluxes. A secondary in Z will avoid any zero-sequence fluxes.

Remarks:

1. The transformer connection may also affect the magnetising. When we discussed the magnetising of a single-phase transformer we saw that the saturation of the magnetic circuit leads to a sharped current shape and, as a result, third harmonic

currents if the flux is imposed sinusoidally (by the supply voltage). In a three-phase transformer such third harmonics are zero-sequence currents as they are the same in all three phases. If the transformer is $Y - Y$–connected without neutral connection, then these third harmonic currents cannot flow. The flux curve will then have to be flattened. The flux and the voltages induced in the primary and secondary phase windings will contain third harmonics. In the line-to-line voltages these third harmonics will not appear, however.
As to the magnitude of these third harmonic fluxes and voltages there is, however, a difference dependent on the construction of the transformer. For transformers with a closing path for the zero-sequence fluxes through the iron (i.e. five-leg, shell-type or transformer bank), the induced third harmonic voltages will be much larger than for transformers without closing path through the iron yoke (i.e. the core-type transformer with three legs).
If the primary (or secondary) is connected in Δ or if there is a tertiary compensation winding in Δ, the third harmonic currents can flow and the zero-sequence fluxes will then be weakened.
2. In a winding with neutral connection, the supply voltage (grid) might also impose zero-sequence voltages. These may result in zero-sequence fluxes and magnetising currents. Here too, the construction of the transformer (whether or not a return path for the zero-sequence fluxes through the core) and the transformer connection will be determinant for the behaviour of the transformer.

1.13 Autotransformer

In an autotransformer, primary and secondary windings share a common winding part and are also physically connected (i.e. there is no galvanic separation). Figure 1.34a shows an autotransformer with the secondary voltage higher than the primary. Here, the primary winding is taken from a tap on the secondary. The primary has w_1 turns, while the secondary has $w_1 + w_{2+}$ turns.

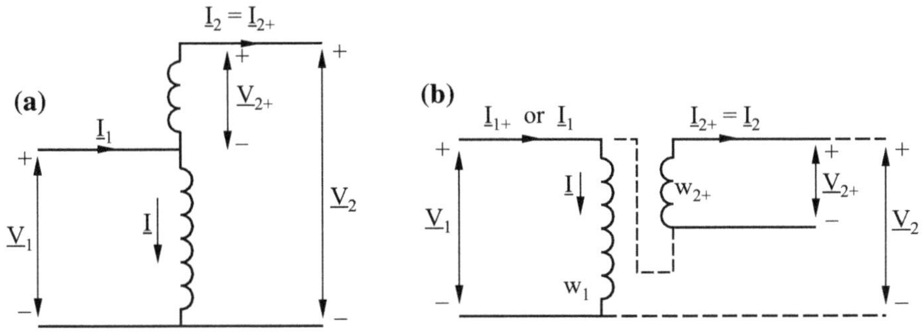

Fig. 1.34 Autotransformer

To study the autotransformer, we may first consider the transformer in (b) (without the dashed connection lines). The number of turns of primary and secondary is w_1 and w_{2+}, respectively. Disregarding the series voltage drop and the magnetising current we have, denoting $\alpha = w_{2+}/w_1$,

$$\begin{aligned} V_{2+} &= \alpha V_1 \\ I_{2+} &= I_{1+}/\alpha \\ I &= I_{1+} \end{aligned} \tag{1.65}$$

With the dashed connection lines, we obtain $V_2 = V_1 + V_{2+}$ as secondary voltage. Supposing the same output current $I_2 = I_{2+}$, we will see the same primary winding current as before, $I{=}\alpha I_2$. For the terminal quantities we obtain now

$$\begin{aligned} V_2 &= (1+\alpha)V_1 \\ I_1 &= I + I_2 = (1+\alpha)I_2 \end{aligned} \tag{1.66}$$

This transformer, which is identical to the autotransformer in (a) in Fig. 1.34, can thus be regarded as a transformer with turns ratio $w_1/w_2 = 1/(1+\alpha) = w_1/(w_1 + w_{2+})$.

The (apparent) power transferred by the autotransformer is $S = V_1 I_1 = V_2 I_2$. This power is, however, not completely transferred by transformer action, $S = V_1 I + V_1 I_2 = V_1 I_2 + V_{2+} I_2$. Only $V_1 I = V_{2+} I_2$ is actual transformer power, corresponding to the actual voltages over and currents through the winding parts and therefore determining the design of the transformer.

The autotransformer has the following advantages:

- for the same dimensions and weight as a normal transformer, the transferred power is larger by the factor $(1+\alpha)/\alpha$
- as the losses are proportional to the dimensions to the third power (i.e. the volume of iron and copper), the relative losses are lower and decrease by a factor $[\alpha/(1+\alpha)]^{3/4}$
- the relative magnetising current is lower: the p.u. magnetising current decreases by a factor $\alpha/(1+\alpha)$
- there is a lower leakage and therefore also a lower voltage drop (also by a factor $\alpha/(1+\alpha)$)

The main disadvantage of an autotransformer is of course the lack of galvanic isolation between primary and secondary. Further, as the lower leakage results in higher short-circuit currents, its design should take this into account (e.g. increase the leakage; rugged construction to better withstand short-circuit forces on the windings). Figure 1.35 illustrates that the short-circuit impedance is in absolute values the same as in the transformer without interconnection; as the reference power with the autotransformer connection is larger, the p.u. value is lower. Clearly, a normal transformer *as such* cannot be used in an autotransformer connection.

An autotransformer is merely used in applications where the voltages in the two network parts are not very different. In many cases taps are provided to provide

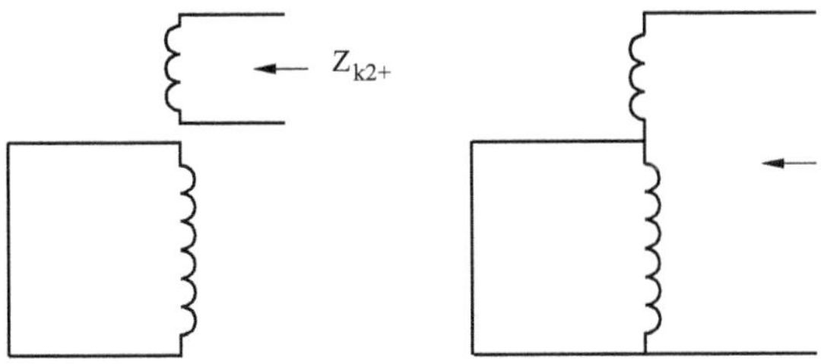

Fig. 1.35 Short-circuit impedance of an autotransformer

voltage regulation. Autotransformers are also typically used to start an induction motor (but in this case this is done to lower the voltage, thus in Fig. 1.34 the supply is at the secondary side and the induction motor load is at the primary side).

1.14 Phase-Number Transformation

1.14.1 Three to Six or Twelve Phases

For the transformation from three into six phases, the configuration in Fig. 1.36 is used. This transformer has two identical sets of three-phase windings, connected as shown in the figure. In applications like rectifiers the neutral point may be connected using a current limiting inductor.

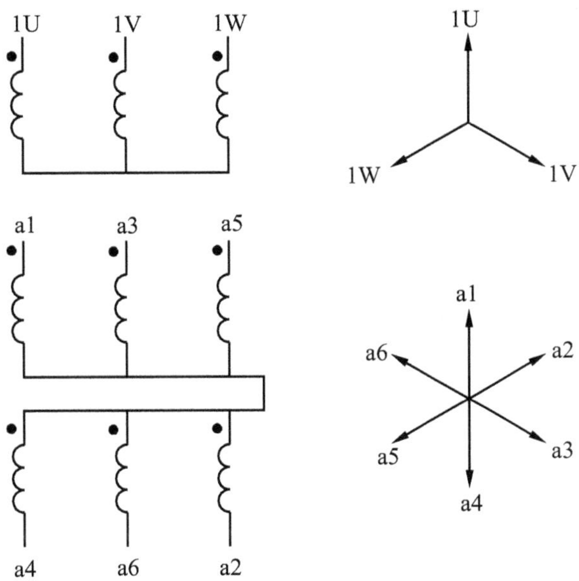

Fig. 1.36 Three to six phases

To transform three into twelve phases an additional 30° phase shift is obtained by means of a Δ−connection for the primary or secondary.

1.14.2 Three to Two Phases

For the transformation from three into two phases (or vice versa), the Scott connection in Fig. 1.37 can be used. The Scott connection makes use of two single phase transformers with the same number of secondary turns but a primary number of turns in a proportion of $1 : \sqrt{3}/2$. The coil with the higher number of turns has a centre tap connected to one end of the other primary winding. The other end of this primary winding is connected to phase U of the primary three-phase supply, while the end point of the winding with centre tap is connected to phases V and W. The end points of the two secondaries are connected to the four conductors of the two-phase grid. Both transformers, but in particular the one with centre tap) should have a tight magnetic coupling, implying large magnetising reactances and small leakage.

The proof of the resulting transformation consists of three parts.

First, we need to prove that (under no load) a symmetrical three-phase voltage system is transformed into a symmetrical two-phase voltage (or vice versa). The phase-to-neutral voltage $\underline{V}_U = \underline{V}_{Un} = \underline{V}$ is used as reference (n is the actual or fictitious neutral point of the grid). The potential of the midpoint "o" can be assumed equal to the average potentials of b ($\underline{V}_V$) and c ($\underline{V}_W$). Indeed, the emfs in the two

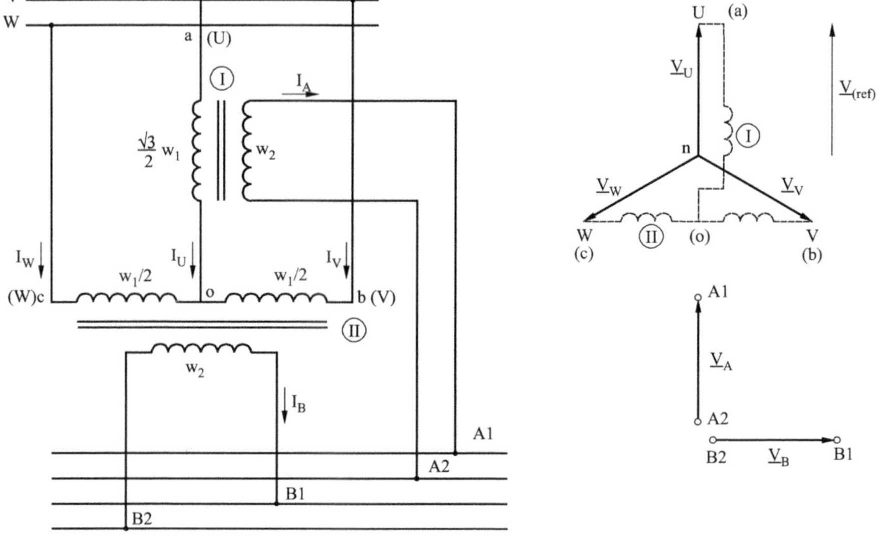

Fig. 1.37 Scott transformer

halves are the same and we assume that the magnetising current is negligible.[12] Then:

$$\begin{aligned} \underline{V}_A &= \underline{V}_{A1} - \underline{V}_{A2} = \left(w_2 \cdot 2/w_1\sqrt{3}\right)\left(\underline{V}_a - \underline{V}_o\right) = (w_2/w_1)\sqrt{3} \cdot \underline{V} \\ \underline{V}_B &= \underline{V}_{B1} - \underline{V}_{B2} = (w_2/w_1)\left(\underline{V}_b - \underline{V}_c\right) = -j(w_2/w_1)\sqrt{3} \cdot \underline{V} \end{aligned} \tag{1.67}$$

or: $\underline{V}_B = -j\underline{V}_A$.

Next, we show that a symmetrical two-phase current system is transformed in a symmetrical three-phase current system. Again disregarding the magnetising current, we may deduce from the equilibrium of the mmfs in both transformers:

$$\begin{aligned} \underline{I}_U \cdot \tfrac{\sqrt{3}}{2} w_1 &= \underline{I}_A \cdot w_2 \\ \underline{I}_V \cdot \tfrac{w_1}{2} - \underline{I}_W \cdot \tfrac{w_1}{2} &= \underline{I}_B \cdot w_2 \end{aligned} \tag{1.68}$$

With $\underline{I}_U + \underline{I}_V + \underline{I}_W = 0$ and supposing that $\underline{I}_B = -j\underline{I}_A \equiv -j\underline{I}$, we find ($a = \exp(j2\pi/3) = -1/2 + j\sqrt{3}/2$)

$$\begin{aligned} \underline{I}_U &= \tfrac{w_2}{w_1} \cdot \tfrac{2}{\sqrt{3}} \cdot \underline{I} \\ \underline{I}_V &= \tfrac{w_2}{w_1} \cdot \tfrac{2}{\sqrt{3}} \cdot a^2\underline{I} \\ \underline{I}_W &= \tfrac{w_2}{w_1} \cdot \tfrac{2}{\sqrt{3}} \cdot a\underline{I} \end{aligned} \tag{1.69}$$

Finally, the assumed equal p.u. short-circuit impedances will guarantee that a symmetrical load impedance results in still symmetrical currents and voltages.

1.15 Voltage Regulation Transformers

Voltage regulation transformers (also called load tap changing transformers) are mainly used to compensate voltage drops in grids with variable loads. They have taps on at least one of the windings (primary and/or secondary). In most cases, these transformers make it possible to compensate voltage drops in steps of 2.5%.

The most important problems for regulating transformers is to guarantee that the induction remains constant (as much as possible) and that the forces on the windings remain balanced for whatever tap position. Variations of the core induction should be limited to $5 \cdots 10\%$. If the induction values are too high, this will lead to high no-load currents and losses. If the induction values are too low, which is rare, they point to a non-economic usage of the core. Balanced forces on the windings are quite important as under load and particularly in short-circuit conditions these forces can be rather high. The radial forces should not pose a problem, especially not for round coils. Asymmetrical axial forces, however, may cause a displacement of the coils in

---

[12]Under load as well, we will assume that the voltage drops are small because of the small leakage inductances and small resistances for these transformers.

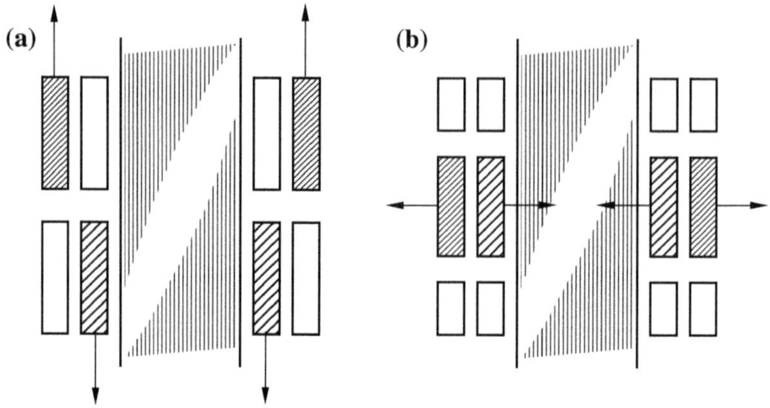

Fig. 1.38 Forces on active winding parts

axial direction on the core. Tap placement should therefore ensure that for every tap position, the equilibrium in axial direction of active winding sections is maintained (see Fig. 1.38 in which active winding sections are shaded).

1.16 Measurement Transformers

Measurement transformers are used to measure high voltages (voltage transformers) or high currents (current transformers) on a much lower measurement level, in many cases also isolated from the dangerously high voltage level of the grid. Measurement transformers are always single-phase transformers and have a very low rated power (less than 1 kVA and usually less than 100 VA). As to construction and properties, low magnetising currents, low resistances and low leakage will be aimed for. To obtain a low magnetising current, special core materials will normally be used together with special core construction (to keep the induction level and saturation levels low).

1.16.1 Current Transformers

Current transformers are high-precision transformers where the ratio of primary to secondary current is a known constant that changes very little with the load. They have only a few turns (albeit with large cross section) in the primary winding, but a large number of turns in the secondary. The primary is put in series with the (grid) conductor (e.g. high voltage line). The secondary is connected to a low-impedance ammeter. In this way, the primary current is imposed while the secondary winding is

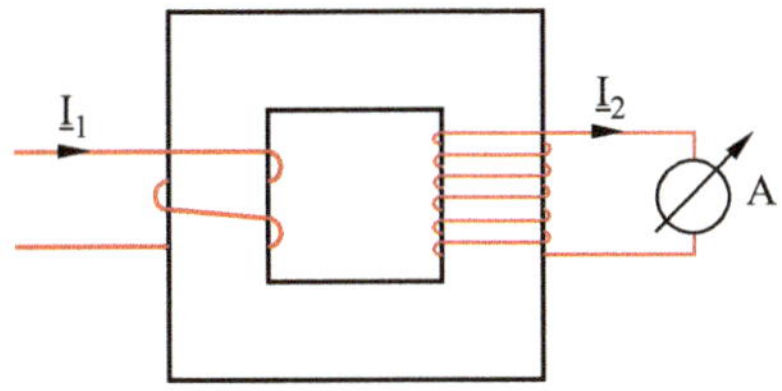

Fig. 1.39 Current transformer

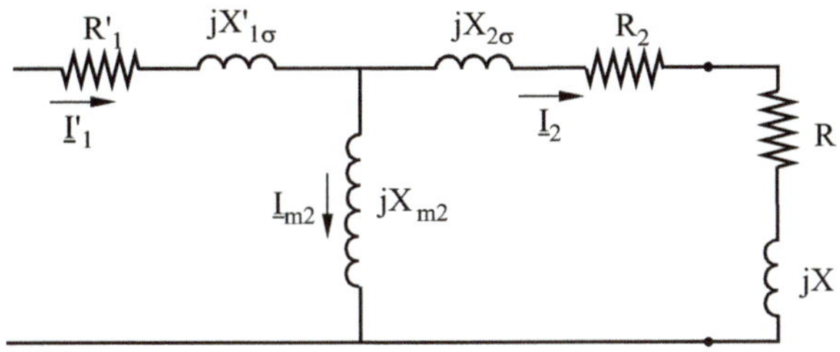

Fig. 1.40 Equivalent circuit for measuring transformer

Table 1.1 Current transformer classes

Class	ϵ_i (% for I_n)	δ_i (arc minutes for I_n)
0.1	0.1	5
0.5	0.5	30
1	1	60

almost short-circuited. A current transformer also isolates the measuring apparatus from the possibly high voltage of the line (Fig. 1.39).

If the transformer were ideal, then

$$\underline{I}_2 = \underline{I}_1 \cdot \frac{w_1}{w_2} = \underline{I}_1^{'}$$

However, because of the magnetising current a small magnitude and phase error will exist (see also the equivalent circuit in Fig. 1.40):

$$\underline{I}_2 = \underline{I}_1^{'} \cdot \frac{jX_{m2}}{R_2 + R + jX_2 + jX} \simeq \underline{I}_1^{'} \cdot \frac{\exp(j\delta_i)}{1 + \frac{X+X_{2\sigma}}{X_{m2}}} \tag{1.70}$$

δ_i is the angle error with $\tan\delta_i = (R_2 + R)/(X_2 + X)$. The amplitude error is $\epsilon_i = (I_2 - I_1^{'})/I_1^{'} \approx -(X + X_{2\sigma})/X_{m2}$ which illustrates the importance of a low magnetising current (or large magnetising inductance).

Specifications of current transformers include: the accuracy class, the maximum voltage protection, rated primary current and over-current possibility, rated apparent power (or rated load impedance, in most cases for $\cos\varphi = 0.8$), own losses and short-circuit durability (see for example Table 1.1).

Please note that the secondary of a current transformer may never be opened when the primary is carrying current. If it was opened, the imposed primary current would act as magnetising current and cause a dangerously high voltage at the secondary, a high induction level and losses in the core.

1.16.2 Voltage Transformers

Voltage transformers are high-precision transformers that allow us to measure high voltages at an acceptable lower level (e.g. 100 or 200 V). The number of primary turns is thus much higher than the number of secondary turns. In normal conditions, the secondary impedance (voltmeter) is very large, so that the voltage transformer works under practically no load. A voltage transformer also isolates the measuring apparatus from the high voltage of the line (Fig. 1.41).

If the no-load current is negligible and the load impedance infinitely large, then

$$\underline{V}_2 = \underline{V}_1 \cdot \frac{w_2}{w_1} = \underline{V}_1^{'}$$

In reality, the non-zero no-load current and the load current cause an amplitude and a phase error. The error can be calculated as follows (see Fig. 1.40 and the corresponding Thévenin equivalent):

$$\underline{V}_{20} = \underline{V}_1^{'} \cdot \frac{jX_{m2}}{R_1^{'} + jX_1^{'}} \tag{1.71}$$

$$\underline{V}_2 = \underline{V}_{20} \cdot \frac{R + jX}{\underline{Z}_{k2}^{'} + R + jX} \tag{1.72}$$

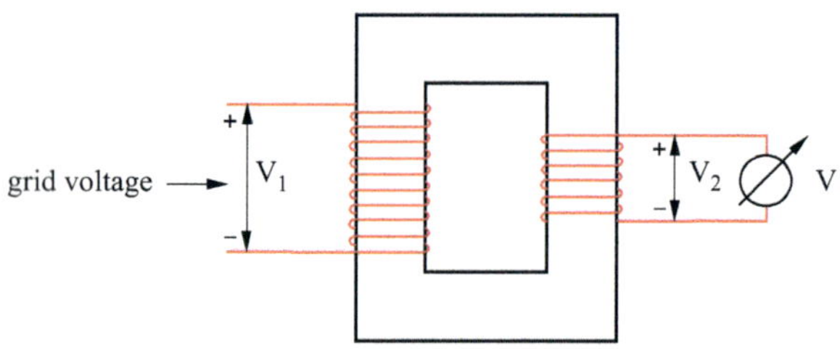

Fig. 1.41 Voltage transformer

Table 1.2 Voltage transformer classes

Class	ϵ_i (% for I_n)	δ_i (arc minutes for I_n)	Operation range
0.1	0.1	5	$0.8 \cdots 1.2\ V_{1n}$
0.5	0.5	20	$0.8 \cdots 1.2\ V_{1n}$
1	1	40	$0.8 \cdots 1.2\ V_{1n}$

For voltage transformers there are also accuracy classes, as illustrated in Table 1.2. Other characteristic data are: the rated voltages, the rated apparent power (the secondary rated power for V_{1n}), the maximal power (maximal secondary power in order not to exceed the maximum allowed heating), and the own power loss (for V_{1n} and open secondary).

Chapter 2
Direct Current Commutator Machines

Abstract Although DC commutator machines are nowadays largely being replaced by rotating field machines, they remain an interesting study object. The basic principles for energy conversion are similar to those of other machines. Moreover, their excellent control properties (e.g. speed control, torque control) are the inspiration for modern drive control schemes for rotating field machines. Starting from the basic electromagnetic laws, the electromagnetic energy conversion in DC machines is explained. Attention is paid also to commutation and armature reaction. Motoring and generating characteristics are discussed in detail.

2.1 Introduction

DC commutator machines (abbreviated as 'DC machines') are (some of) the oldest electrical machines. Thanks to the simplicity of direct current, DC machines were able to maintain a leading position for many years, even after the introduction of AC and induction and synchronous machines in the beginning of the 20th century.

Later on, this position was threatened by the advantage of AC as to lossless transforming voltage and current levels, at least for generators and constant speed drives. For variable speed drives, however, DC machines were still the preferred drives for variable speed drives (i.e. controlled drives). This changed completely with the introduction of power electronics in the second half of the 20th century, which facilitated controlled drives using induction and synchronous machines.

Nevertheless, DC machines are still used for low power applications (automotive). However, this may not last as they are less reliable and generate higher maintenance costs compared to pm-synchronous machines with low-cost power-electronic supplies.

J. A. Melkebeek, *Electrical Machines and Drives*, Power Systems,
https://doi.org/10.1007/978-3-319-72730-1_2

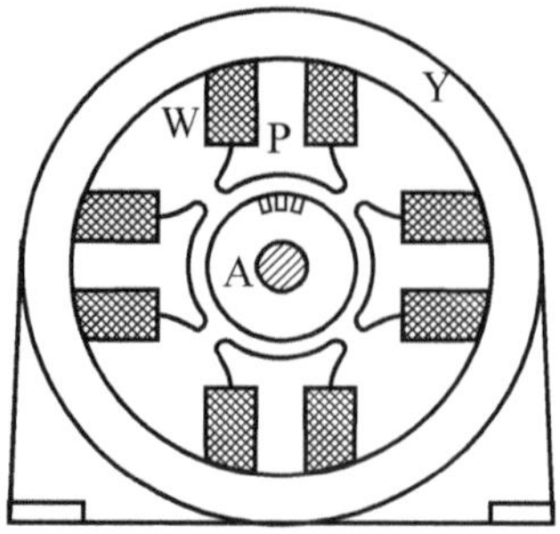

Fig. 2.1 Main electromagnetic parts of a DC machine

Fig. 2.2 Pictures of the armature

2.2 Construction of the DC Machine

2.2.1 Basic Construction - Operating Principle

Construction

The most important electromagnetic parts of a DC machine are the following (see also Fig. 2.1):

- The standstill excitation system mainly consisting of
  - the main poles P
  - the excitation winding W (or permanent magnets for small machines)
  - the closing yoke Y
- The rotating armature A: the core of the armature is always laminated, with iron sheets of less than 0.5 mm. The armature has slots for the armature conductors. The armature winding is a closed winding and is connected to the commutator segments in several places. Brushes make contact with the segments to provide current to the armature.

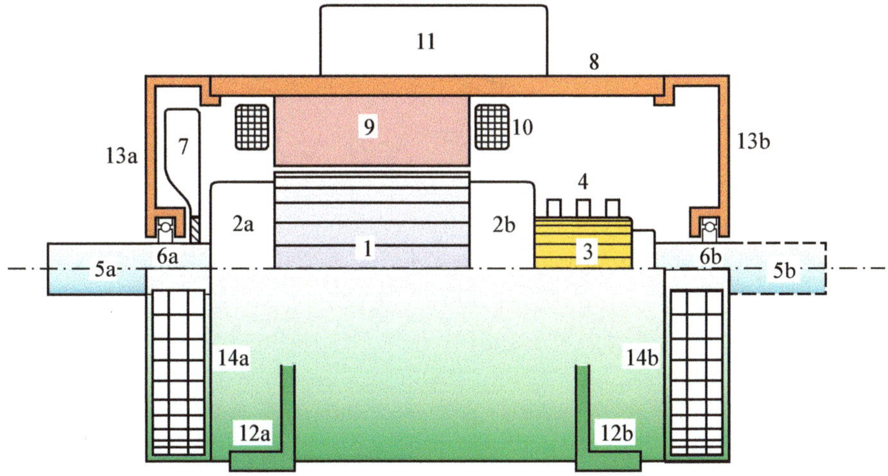

1	armature core (segmented) with conductors in slots
2	end coils of the armature winding
3	commutator (copper segments)
4	brush holders with brushes
5a, 5b	shaft ends
6a, 6b	bearings
7	ventilator
8	housing (with yoke, segmented or massive)
9	main field pole (in most cases segmented)
10	main field coil
11	electrical terminals
12a, 12b	feet (supports)
13a, 13b	bearing houses
14a, 14b	ventilator in- and outlets

Fig. 2.3 Construction details of DC machine

Figure 2.2 shows some pictures of both sides of the armature, while Fig. 2.3 provides more details of the construction of a DC machine. Auxiliary poles and compensation windings (see further on) have not been depicted.

Operating Principle

The operating principle of a DC commutator machine is explained in Fig. 2.4. The excitation poles (either with a DC fed excitation winding or permanent magnets) produce a magnetic field in the air gap. Because of the high permeability of the iron in the poles and the armature core, the field lines are almost orthogonal to the armature and field pole boundaries. As a result, the induction in the air gap as a function of the linear coordinate x (or the angular coordinate θ) along the air gap is as depicted in (c) in Fig. 2.4. When the armature rotates, emfs are induced in the armature conductors: $e = v \cdot B \cdot l_a$ with $v = \Omega_a \cdot r_a$(where r_a is the armature radius) and with l_a the armature length. The curve of this emf e as a function of time is

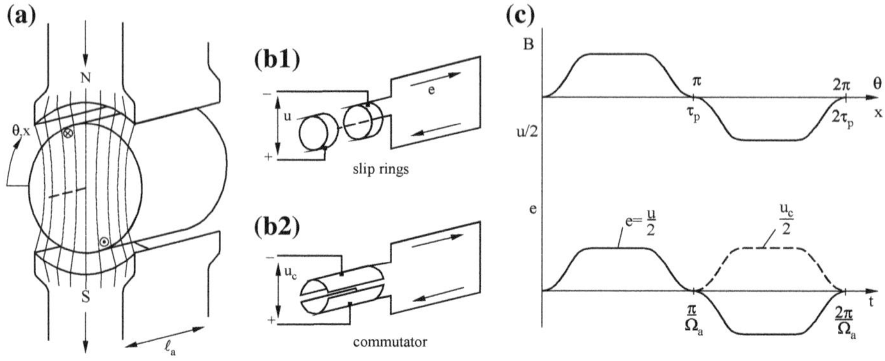

Fig. 2.4 Operating principle

similar to the curve of the air-gap induction as a function of the linear coordinate of the circumference.

Now consider two conductors that are positioned diametrically on the armature. These are series connected on one end to form a coil, while the other ends are connected to slip rings (see (b1) in the figure). The voltage u that can be measured using two brushes that make contact with the slip rings will be an alternating voltage with the same shape as e but with twice the amplitude (see the lower curve in the full line in (c)). However, if the free ends of the coil are connected to two segments of a commutator (see b2 in the figure), then the voltage u_c measured using two brushes will be a rectified voltage (see the lower dashed curve in (c)). A smoother curve for this rectified voltage is obtained using more commutator segments, each pair connected to similar coils with conductors distributed along the circumference of the armature.

In contrast, if a DC supply is connected to the armature conductors via the brushes, then a current will flow in the armature conductors. When the armature rotates, the current direction *in a given conductor* will be an AC current. Indeed, the commutator operates as a frequency converter. When a conductor is under the north pole, the current will always flow in the same direction but it will be in an opposite direction when the conductor is under the south pole. The current direction in the conductor changes when it passes the neutral position between north and south poles. This holds for all conductors and coils, i.e. the current direction for all conductors under a north pole will be the same while the current direction for all conductors under a south pole will be the opposite.

This may also be formulated as follows. An observer at standstill with respect to the poles will see a constant current image while an observer rotating with the armature (or conductors) will see alternating currents in the conductors. Conversely, the current in the coils or conductors is an alternating current which is rectified by the commutator.

As a result, the tangential force on the conductors due to the interaction of the currents with the magnetic field of the poles under the north and south poles will act in the same direction and cause a net torque on the armature.

The operating principle of the DC commutator machine as an electromechanical power converter should therefore be clear. Rotation of the armature in the presence of an excitation will cause emfs in the armature conductors. The connection of the armature conductors with the commutator causes the emfs to add to a DC voltage at the brushes. If the brushes are connected to a load resistance (or a lower DC voltage source in series with a resistance), a current will result in the load, converting the mechanical energy into electrical energy in the resistances and/or the voltage source. To maintain the speed, a mechanical driving torque must be exerted on the shaft of the armature, as the currents in the conductors will cause a braking torque. In other words, the DC machine works as a generator.

In contrast, if the brushes are connected to a voltage source which is higher than the combined armature emfs at the brushes (which occurs at a sufficiently low speed), the voltage source connected at the brushes will bring about a current in the armature conductors, producing a driving torque that will try to accelerate the machine and load. Here, the machine works as a motor.

In both generating and motoring, a constant equilibrium speed will be achieved when the driving and braking torques are equal in magnitude. Remark that power reversal is an inherent property of all electromagnetic power converters (transformers, rotating machines).

2.2.2 *Excitation*

Construction

The stator of a DC machine contains an even ($2N_p$) number of identical salient poles. These *main poles* are distributed evenly around the circumference. The distance between subsequent poles is called the pole pitch τ_p, measured along the median circle (in the middle of the air gap). The width of the pole shoes is referred to as the pole width. These poles generate a heteropolar field around the circumference, i.e. a succession of north and south poles. The excitation is either a DC-fed concentrated winding around the poles or a permanent-magnet excitation. The field lines for a two-pole machine and for the two-pole part of a multipole machine are illustrated in (a) and (b) in Fig. 2.5, respectively.

Such a pair of poles ((a) in Fig. 2.6) is usually representative of the whole machine. For further analysis, we may represent it by a flat representation as in (b) in Fig. 2.6. Because of the high permeability of the iron of poles and armature, most field lines cross the air gap orthogonally to the iron-air boundary. Only at the corners of the poles does some flux fringing occur. This flux fringing results in a smoother induction curve (see the full lines in the figure) versus the ideal rectangular curve (in dashed lines). In addition, the air gap is often widened at the corners, and the armature slots also cause some deformation of the flux curve. However, later on we will see that

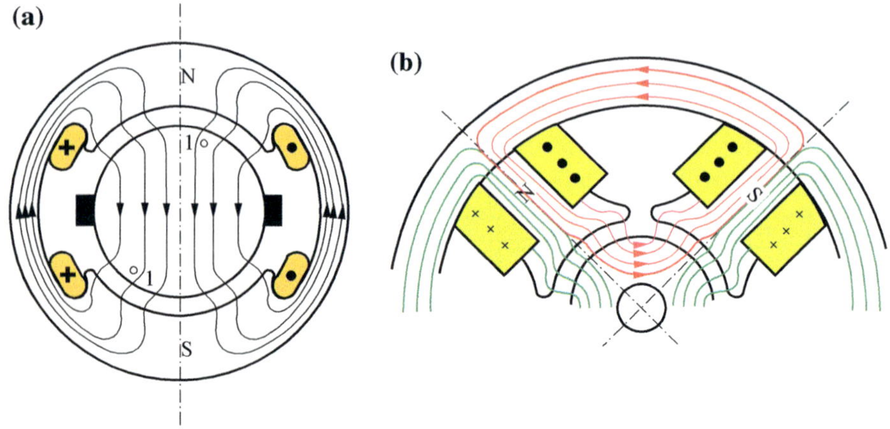

Fig. 2.5 **a** Field lines; **b** real and idealised flux curves

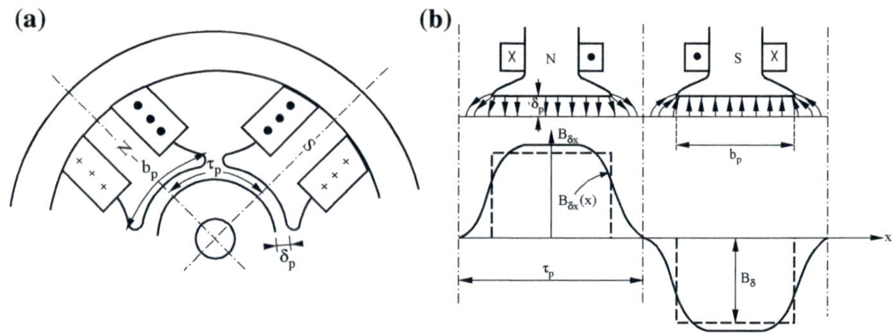

Fig. 2.6 **a** Field lines; **b** real and idealised flux curves

the real shape of the flux curve is not significant. Only the average flux over a pole pitch comes into play for the energy conversion (emf, torque).

Note that the magnetic field is zero in the middle between two subsequent poles (i.e. the geometrical neutral zone. When also armature currents are present, the zone with zero induction may shift away from this geometrical neutral zone (see the armature reaction in Sect. 2.4).

Saturation Characteristic

The air-gap induction in a point x under a pole can be calculated as follows. We denote by w_m the number of turns per pole, I_m the excitation current in the excitation winding conductors, δ_x the air-gap length in a point x, $B_{\delta x}$ the air-gap induction in x and l_a the armature length. Applying Ampère's law to a closed field line yields

$$\oint \underline{H} \cdot d\underline{l} = 2w_m I_m \tag{2.1}$$

The left hand part of this equation contains the magnetic field drops over the stator yoke, the poles, the air gaps, the armature slot regions and the armature yoke:

$$H_{yx}\delta_{yx} + 2H_{px}\delta_{ap} + 2H_{\delta x}\delta_x + 2H_{sx}\delta_{sx} + H_{ax}\delta_{ax} = 2w_m I_m \tag{2.2}$$

If the permeability of the iron parts is infinite, then the left hand part reduces to the mmf drop over the two air gaps and thus

$$2H_{\delta x}\delta_x = 2w_m I_m \tag{2.3}$$

The physical flux per pole is given by

$$\Phi_m = \int_{-\tau_p/2}^{\tau_p/2} B_{\delta x} \cdot l_a \cdot dx = \int_{-\tau_p/2}^{\tau_p/2} \mu_0 H_{\delta x} \cdot l_a \cdot dx \tag{2.4}$$

If only the mmf drop over the air gap comes into play, then, using Eq. 2.3

$$\Phi_m = \int_{-\tau_p/2}^{\tau_p/2} \mu_0 \frac{w_m I_m}{\delta_x} \cdot l_a \cdot dx = w_m I_m \cdot \Delta_\delta \tag{2.5}$$

with Δ_δ the permeance of the air gap. If the air-gap length is constant and equal to δ_p under the poles ($-b_p/2 \le x \le b_p/2$) and almost infinite in between the poles (such that the induction is about zero in between the poles), then $\Delta_\delta = \mu_0 l_a b_p/\delta_p = \mu_0 \Sigma_\delta/\delta_p$.

In reality, the permeability of the iron is finite. Moreover, there is always saturation (reducing the iron permeability), especially in the teeth of the armature because of the reduced iron cross section. The expression for permeance is therefore often written as $\Delta_m = \mu_0 l_a b_p/\delta_p'' = \mu_0 \Sigma_\delta/\delta_p''$ with $\delta_p'' = k_c k_i \delta_p$. The factor k_c (>1) is Carter's factor, which takes into account the effect of the slots, and $k_i > 1$ is a factor that considers the mmf drops in the iron parts (in particular the armature teeth).

The relation between the pole flux and the excitation current is called the saturation characteristic. Because of the iron, this is a non-linear characteristic. Moreover it shows hysteresis (see (b) in Fig. 2.7). Although the hysteresis is important for the self-excitation of generators (see below) we will usually neglect the hysteresis effect and replace it with an average curve as in (a) in Fig. 2.7.

To obtain a given flux, the required excitation mmf F_m consists of the mmf F_δ for the air gap, the mmf F_s for the armature slots and the mmf F_i for the iron. The latter two are represented by the mentioned factors k_c and k_i, respectively.

Remarks:

- with increasing saturation, the leakage flux of the excitation winding (i.e. the flux linking the excitation winding but not the armature) also increases.

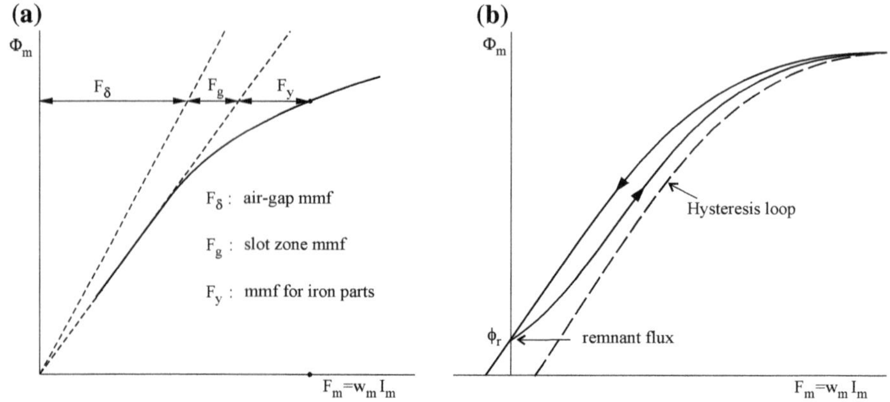

Fig. 2.7 Saturation characteristic

- the iron in the machine is required to concentrate the flux under the pole shoes linked with the armature.
- the actual induction curve is sometimes represented by a fictitious square curve with amplitude B_δ under the pole width so that the pole flux remains the same, which means that $\Phi_m = B_\delta \cdot b_p \cdot l_a$ (see the dashed line in (b) in Fig. 2.7).

2.2.3 *Armature*

The main parts of the armature are the core with the windings and the commutator. In this section, we will mainly discuss the core with the windings. The commutator will be examined in more detail in Sect. 2.5.

The oldest version of the armature winding is the ring winding (or Gramme[1] winding). Figure 2.8a depicts the cross section of a 2-pole Gramme-winding machine. Although this kind of winding does not have any industrial or practical applications nowadays, it is a nice tool to explain the windings and the connections to the commutator. The Gramme winding is in fact a coil closed on itself around a ring (or cylinder). The outer conductors are exposed to the magnetic field of the excitation (when under the poles), while the inner conductors are connected to the commutator segments. The brushes make contact with the segments in between the poles.

In Fig. 2.8b, c, schematic layouts of a 2-pole and a 4-pole machine are shown (the commutator has now been omitted and simplified by brush contacts with the outer conductors). If the armature rotates in the magnetic field by the poles, voltages will be induced in the outer conductors, with opposite signs under north and south poles as marked in the figure. In the 2-pole version, the voltages in the conductors 2, 3, 4,

[1]Zenobe Gramme was a Belgian inventor; there is a statue of him on the bridge over the Meuse in Liège.

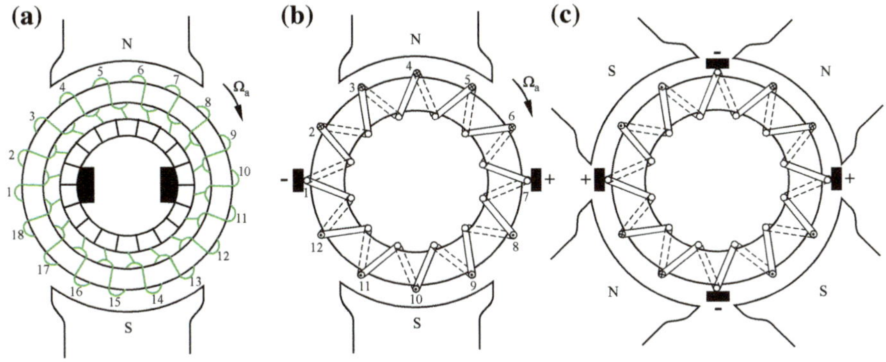

Fig. 2.8 Gramme winding

5 and 6 will add up. Similarly, the voltages in conductors 8, 9, 10, 11 and 12 will add up. These two resulting voltages have the same magnitude because of symmetry. Seen from the brush contacts, these voltages have the same sign. At the brushes, we therefore see the winding as two parallel branches. A similar reasoning holds for a current injected at the brushes, for example. Half of this current will flow in the upper conductors and the other half in the lower conductors.

In the 4-pole depicted, it is assumed that the brushes with the same polarity are connected externally. So there are four parallel branches in the armature winding as observed from the external armature terminals.

In the Gramme winding, less than half of the conductor copper is used actively as the inner conductors serve only as connections and do not take part in the energy conversion. Nowadays, drum windings are always used. The drum winding may be derived from the Gramme winding if the inner conductors are moved to the outside under the opposite poles (for the 2-pole version, this is the diametrically opposed pole). In this way, a winding is obtained with two conductors near each other but belonging to two different coils and to different parallel branches (the two coil sides are normally placed below each other in slots). The width of such coils is then equal to the pole pitch τ_p. Figure 2.9 shows the flat representation of a 4-pole drum winding with 23 slots (and also 23 commutator segments). In each slot, two coil sides can be observed. One is called the entrance coil side (full lines) and the other the exit coil side (dashed). In this example, the coil width (i.e. the distance between entrance and exit coil sides of a single coil) is $y = \tau_p$ (although this is not always the case). Remark that this drum winding is also closed on itself. Brushes with the same polarity are connected externally. In this way, again four parallel branches are obtained.

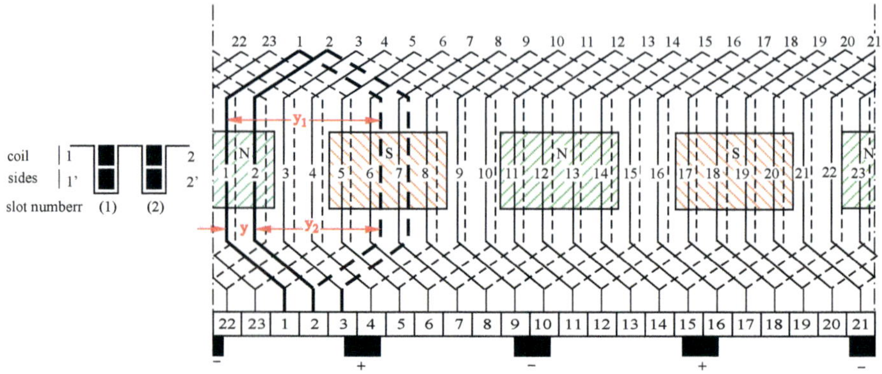

Fig. 2.9 Actual armature drum winding

2.3 Electrical Power Conversion in a DC Machine

2.3.1 Voltage Induction (emf)

When the armature rotates in the magnetic field of the excitation poles, electromotive forces (emfs) are induced in the armature conductors according to $e_c(x) = v_a \cdot b(x) \cdot l_a$ with $v_a = (N_p \tau_p / \pi) \cdot \Omega_a$ the linear speed of the conductor in x, $b(x)$ the induction in x and l_a the armature (or conductor) length. In order to calculate the total emf induced between two brushes, the following notations are used: z is the number of conductors of the armature, $2a$ the number of parallel branches, Φ_m the flux per pole, Ω_a the rotation speed of the armature (rad/s) and r_a the average radius of the armature.

When the conductor moves over a pole pitch τ_p, the evolution in time of the emf is similar to the curve of the air-gap induction along the pole pitch. The average conductor emf can thus be calculated from

$$e_{c,av} = \frac{1}{\tau_p} \int_{-\tau_p/2}^{\tau_p/2} e_c(x) \cdot dx = \frac{v_a}{\tau_p} \int_{-\tau_p/2}^{\tau_p/2} B_{\delta x}(x) \cdot l_a dx = \frac{v_a}{\tau_p} \Phi_m = \frac{N_p \Omega_a \Phi_m}{\pi} \tag{2.6}$$

The voltage measured at the armature terminals is the sum of the emfs of the $z/2a$ series connected conductors in one of the $2a$ parallel branches (see Fig. 2.10), which means that

$$E_a = \frac{z}{2\pi a} N_p \cdot \Omega_a \cdot \Phi_m \tag{2.7}$$

This result can also be obtained by the summation of the emfs in the $z/2a$ conductors in a branch, assuming that the number of conductors is very large so that the summation can be replaced by a surface integral over a pole pitch. However, in reality the

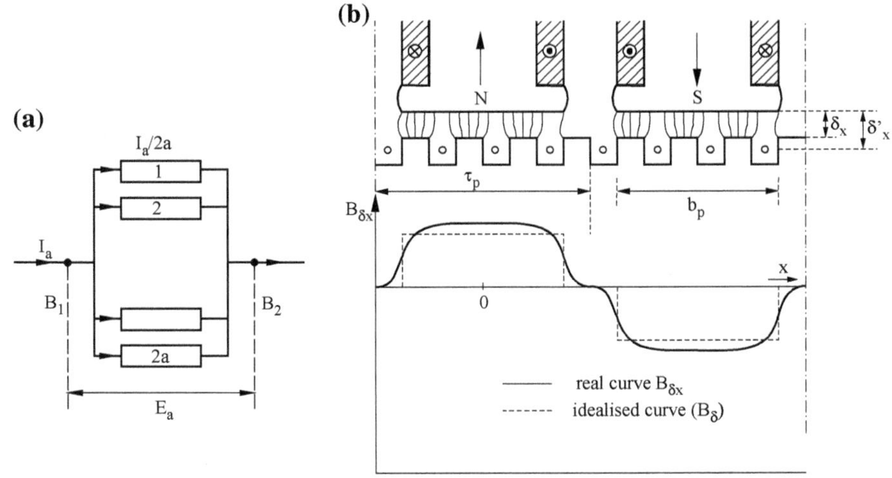

Fig. 2.10 **a** Parallel branches, **b** flat representation of pole flux

number of conductors is finite and the actual voltage shows some undulation. This undulation is usually of no practical importance and only the time-average voltage of equation is interesting (2.7).

Please note that this average voltage is only dependent on the average flux per pole. The actual shape of the induction curve over the pole pitch does not come into play.

From Eq. 2.7 it also follows that the saturation characteristic $\Phi_m(i_m)$ is similar to the emf curve as a function of the excitation current $E_a(i_m)$ (for a constant speed), which offers a practical and easy means to experimentally record this characteristic.

2.3.2 *Torque*

If an armature current I_a is flowing into the armature terminals, this current will be divided into $2a$ equal parts in the $2a$ parallel branches. In each armature conductor, we will thus see a current $I_a/2a$. As a result, on a conductor in point x of the armature a tangential force $f_c(x) = \frac{I_a}{2a} \cdot l_a \cdot B_{\delta x}(x)$ will be exerted (see Fig. 2.11). This force corresponds to a torque on the conductor $t_c(x) = \frac{N_p \tau_p}{\pi} \cdot \frac{I_a}{2a} \cdot l_a \cdot B_{\delta x}(x)$. The average values of this force and torque when the conductor moves through a pole pitch are then

$$f_{c,av} = \frac{1}{\tau_p} \int_{-\tau_p/2}^{\tau_p/2} f_c(x) \cdot dx = \frac{1}{\tau_p} \cdot \frac{I_a}{2a} \cdot \Phi_m \tag{2.8}$$

Fig. 2.11 Tangential force and torque on a conductor

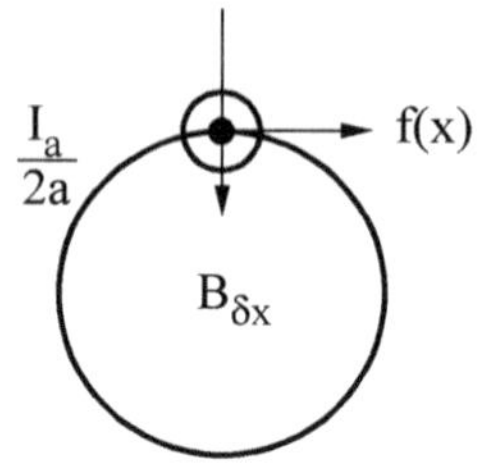

$$t_{c,av} = \frac{1}{\tau_p} \int_{-\tau_p/2}^{\tau_p/2} t_c(x) \cdot dx = \frac{N_p}{\pi} \cdot \frac{I_a}{2a} \cdot \Phi_m \tag{2.9}$$

For the total electromagnetic torque on the armature, this average conductor torque must be multiplied by the total number of conductors z:

$$T_{em} = \frac{z}{2\pi a} N_p \cdot I_a \cdot \Phi_m \tag{2.10}$$

The torque is independent of the actual shape of the induction over a pole pitch, as long as the flux per pole remains the same.

Note that, as expected, the product of the electromotive force and the current equals the product of the electromagnetic torque and the armature speed (i.e. conservation of energy):

$$E_a \cdot I_a = T_{em} \cdot \Omega_a \tag{2.11}$$

This product is called electromagnetic power P_{em}.

2.3.3 *Electrical Power Conversion*

When the machine is loaded, the armature terminal voltage differs from the emf by the armature resistance drop and the transition voltage drop between commutator and brushes. In the URS and GRS (see also Fig. 2.12):

$$V_a = E_a + R_a I_a + \Delta V_b \tag{2.12}$$

$$V_a = E_a - R_a I_a - \Delta V_b \tag{2.13}$$

respectively, where $\Delta V_b = |\Delta V_b| \cdot \text{sign}(I_a)$. In other words, this transition voltage always has the sign of the armature current. Its magnitude is approximately $2V$.

To simplify the equations, the transition voltage is sometimes lumped together with the armature resistance, as $V_a = E_a + R_a^{'} I_a$and $V_a = E_a - R_a^{'} I_a$, respectively.

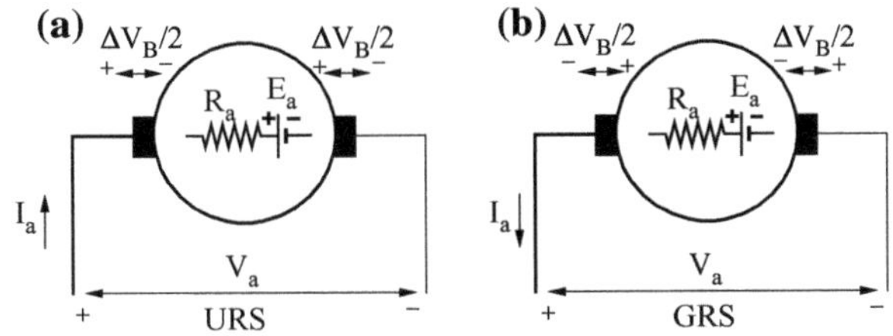

Fig. 2.12 Armature electrical circuit: **a** URS, **b** GRS

As to the torque, the shaft torque differs from the electromagnetic torque by the friction torque, the ventilation torque and a small load torque due to the iron losses in the armature:

$$T_{sh} = T_{em} - T_w \tag{2.14}$$

$$T_{sh} = T_{em} + T_w \tag{2.15}$$

for the GRS and the URS at the mechanical side, respectively (Fig. 2.13).

The power conversion in a DC machine can thus be described by the following sets of equations.

- for the M-convention (URS at the electrical side and GRS at the mechanical side):

$$\begin{aligned} V_a &= E_a + R_a I_a \\ T_{sh} &= T_{em} - T_w \\ P_{el} &= V_a I_a = P_{em} + R_a I_a^2 \\ P_{sh} &= T_{sh} \Omega_a = P_{em} - T_w \Omega_a \end{aligned} \tag{2.16}$$

or also $P_{el} = P_{sh} + T_w \Omega_a + R_a I_a^2$.

- for the G-convention (URS at the mechanical side and GRS at the electrical side):

$$\begin{aligned} V_a &= E_a - R_a I_a \\ T_{sh} &= T_{em} + T_w \\ P_{el} &= V_a I_a = P_{em} - R_a I_a^2 \\ P_{sh} &= T_{sh} \Omega_a = P_{em} + T_w \Omega_a \end{aligned} \tag{2.17}$$

or $P_{sh} = P_{el} + R_a I_a^2 + T_w \Omega_a$.

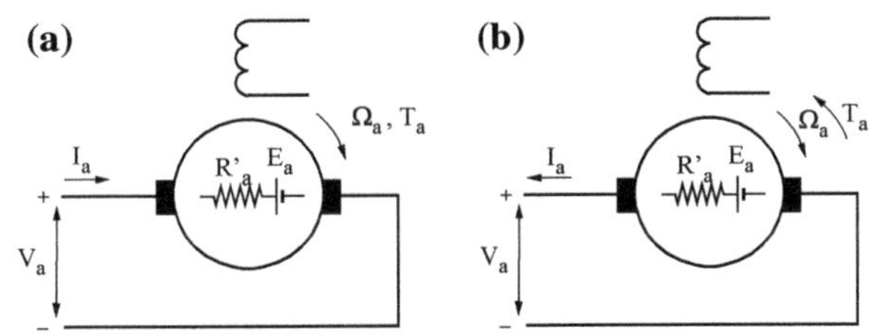

Fig. 2.13 **a** M-convention and **b** G-convention

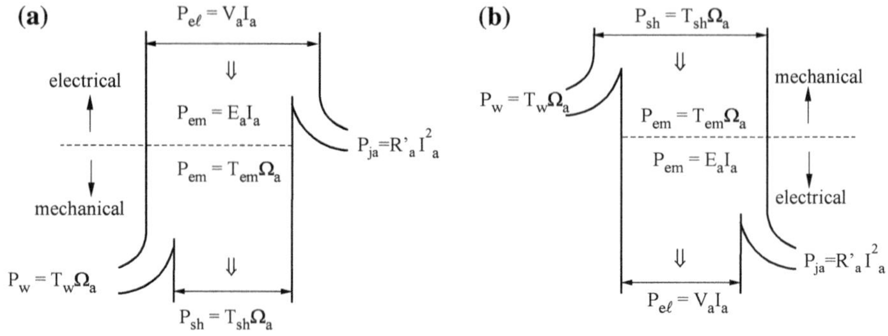

Fig. 2.14 Sankey diagrams

These equations are represented by the Sankey diagrams in Fig. 2.14.

Please note that in the preceding equations the electric power for the excitation is not taken into account. In case of a series of shunt excitation, the energy loss in the excitation winding must be provided at the armature connections as well.

Remarks:

1. The electromagnetic power may also be expressed as a function of the linear force and the linear speed (e.g. for a linear motor):

$$\begin{aligned} F_{em} &= \tfrac{z}{2\tau_p a} \cdot I_a \cdot \Phi_m \\ v_a &= \Omega_a \cdot \tfrac{N_p \tau_p}{\pi} \\ P_{em} = E_a \cdot I_a = F_{em} \cdot v_a &= \tfrac{z}{2\pi a} N_p \cdot \Omega_a \cdot I_a \cdot \Phi_m \end{aligned} \tag{2.18}$$

2. Force, torque and power may also be expressed per square meter of active air-gap surface

$$\begin{aligned} F_{em}^{\nabla} &= \tfrac{F_{em}}{2N_p b_p l_a} \\ T_{em}^{\nabla} &= \tfrac{T_{em}}{2N_p b_p l_a} \\ P_{em}^{\nabla} = \tfrac{P_{em}}{2N_p b_p l_a} = F_{em}^{\nabla} \cdot v_a &= T_{em}^{\nabla} \cdot \Omega_a = K_a \cdot A_a \end{aligned} \tag{2.19}$$

   with

$$\begin{aligned} K_a &= \tfrac{E_a}{(z/2a)\cdot(b_p/\tau_p))\cdot l_a} = v_a \cdot B_\delta \\ A_a &= \tfrac{z \cdot I_a/2a}{2N_p \tau_p} \end{aligned} \tag{2.20}$$

   where K_a represents the emf per meter of active armature length and A_a the current layer amplitude (or current per meter of armature circumference).

3. The efficiency of a DC machine is determined by the following three types of losses:

   a. no-load losses (e.g. mechanical losses and iron losses)

b. load losses (i.e. Joule losses in the armature and brush contacts and extra losses in the armature iron due to the slot effects; the latter two are in most cases also included in the term $R_a^{'} I_a^2$)
c. excitation losses (unless permanent magnets are used for the excitation).

2.4 Armature Reaction and the Compensation Winding

Armature Reaction

From the discussion in Sect. 2.2.2, we learned that the excitation results in a magnetic field along the pole axes (and an induction curve which is symmetrical around the pole axes). However, this changes when armature currents are also present. In that case, the curve is not symmetrical any more and a flux reduction may occur.

Figure 2.15 shows the field lines due to an armature current (in dash-dot lines). For example, to the right of the north pole the armature mmfs add to the mmf of the excitation, while to the left they will subtract from the mmf of the excitation. Figure 2.16 illustrates this in more detail by means of a flat representation. Sketch (b) in this figure shows the excitation mmf and induction without armature current. In (c), on the other hand, the mmf and induction curves are depicted for armature current without excitation (the induction in between the poles is low due to the large air gap). The armature mmf varies with the coordinate x according to

$$F_a(x) = H_a(x) \cdot \delta_x = \int_0^x A_a(x) \cdot dx \tag{2.21}$$

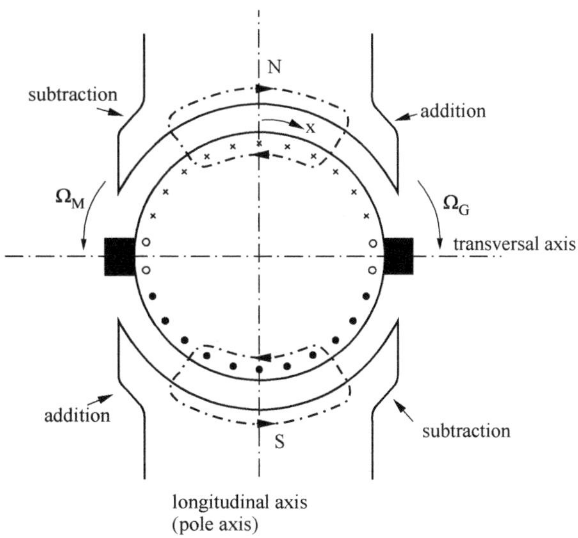

Fig. 2.15 Field lines of armature reaction

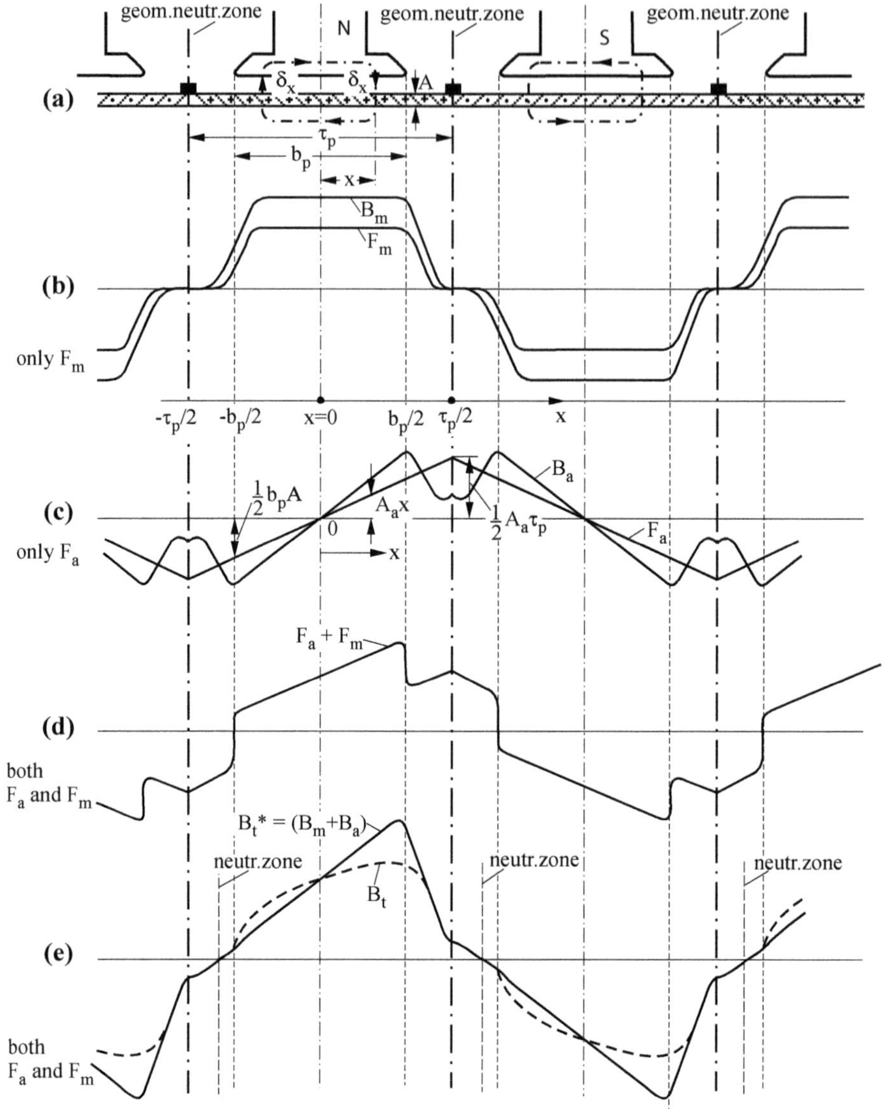

Fig. 2.16 Flat representation of armature reaction

supposing that the field lines are orthogonal to the iron boundaries.

For a sufficiently fine distribution of the armature conductors, A_a may be considered as a constant (the average current layer amplitude), and the armature mmf then varies linearly with x.

When both excitation and armature mmfs are present, the resulting mmf curve is not symmetrical any more around the pole axes, as shown in (d). The flux over a pole pitch will not change if the magnetic material is linear, as the induction curve will then always remain proportional to the total local mmf. Indeed, if the magnetic circuit is linear, the separate induction values of excitation mmf and armature mmf may also be added. As the integral of the armature mmf over a pole pitch - symmetrical with respect to the pole axes, thus from $-\tau_p/2$ to $\tau_p/2$ - is equal to zero, the pole flux will not change.The only result will be a shift of the magnetically neutral zone (away from the geometrically neutral zone). This transversal armature reaction is illustrated by the induction curve B^* in full lines in (e).

If, however, saturation of the iron occurs, the pole (longitudinal) flux will decrease, as the relative induction increase for lower mmfs will not compensate the relative induction decrease for higher mmfs (see the dashed induction curve B_t in (e)). The outcome is therefore a flux reduction along the pole axes, resulting in a lower emf and a lower torque. This effect is called the longitudinal armature reaction.

Compensation Winding

To mitigate the longitudinal armature reaction, larger DC machines often contain a compensation winding. This is a winding in slots of the pole shoes with ampère-windings that oppose the underlying armature currents (see Fig. 2.17). As this winding should compensate the armature mmf in all operating conditions, this compensation winding must be connected in series with the armature winding. Because this winding is not fixed to the armature, no torque effects will occur. The only negative effect is that the slots will reduce the permeability in the longitudinal axis.

However, note that the compensation winding will only compensate the armature mmf under the pole shoes (and thus the longitudinal armature reaction) but will not affect the armature mmf in the inter-polar space. As a result, the transversal armature reaction and the shift of the neutral zone will not be eliminated.

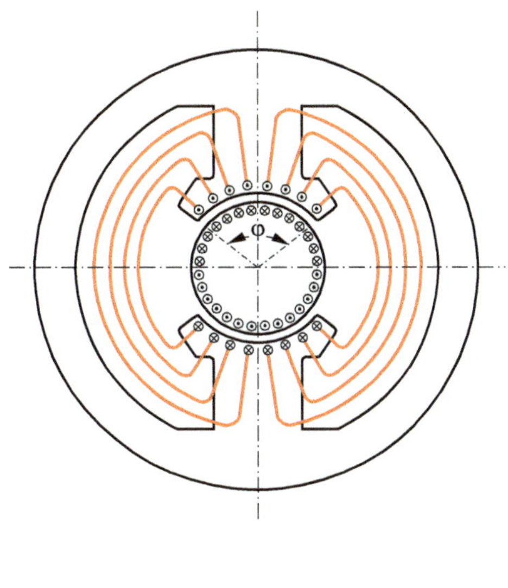

Fig. 2.17 Compensation winding

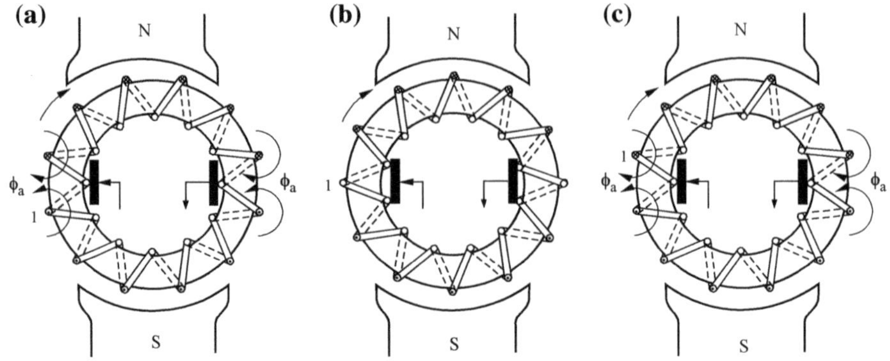

Fig. 2.18 Commutation for a Gramme winding

2.5 Commutation and the Commutation Poles

Commutation

As discussed above, the armature winding is a winding closed in itself, with $2a$ parallel branches. Current is supplied to (or from) the armature by the brushes. These brushes make contact with the commutator segments that are connected to the connection points of subsequent coils. As explained above, the currents in the armature conductors are AC currents. The polarity of the current in a coil changes when the segments connected to the coil end points pass the brushes (see Fig. 2.18). For conductor 1, the current direction changes from upwards (a) to downwards (c) when the armature rotates clockwise. In between, the brushes short-circuit the coil to allow the current to reverse. This is called commutation.

The current should reverse smoothly and on time during the short-circuit interval. If not, sparking will occur, causing wear of commutator and brushes. The commutation is affected by both mechanical and electrical issues. The former, mechanical issues, relate to vibrations, eccentricity and roundness of the commutator, insufficient or variable contact pressure of the brushes, among other things. The latter, electrical issues, are far more common causes of commutation problems, which is why they will be further discussed below.

A schematic representation of the commutation is shown in Fig. 2.19. $I = I_a/a$ is called the brush current. Before commutation, the current $I/2$ in the coil connected to segments 1 and 2 flows from left to right (see (a)). After commutation, it will flow from right to left (see (c)). During the commutation interval, the coil is short-circuited by the brushes. In the coil, we can then observe a short-circuit current $i_k(t)$. During this commutation, the currents in the connections to segments 1 and 2 are $i_1(t) = I/2+i_k(t)$ and $i_2(t) = I/2-i_k(t)$, respectively. Indeed, we may suppose that the brush current I does not change during the commutation. An important criterion for a good commutation is that the current densities in the contact surfaces between the brushes and the segments remain limited and, if possible, constant. These current

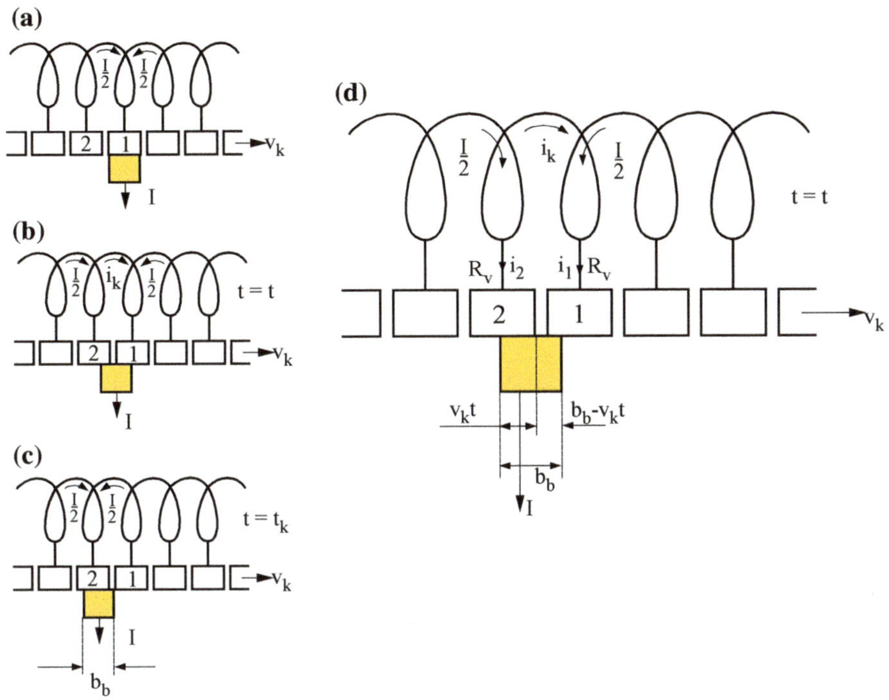

Fig. 2.19 Schematic representation of commutation

densities in the contacts of segments 1 and 2 can easily be calculated taking into account the contact surfaces of the brush with segments 1 and 2: $b_b l_b (1 - t/t_k)$ and $b_b l_b (t/t_k)$, respectively, where b_b is the brush width, l_b the brush length and $t_k = b_b/v_k$ the commutation time (v_k represents the linear circumferential speed of the commutator segments).

Figure 2.20a gives a graphical representation in which the current densities can be derived as $J_1 = J_0 \tan \alpha_1$ and $J_2 = J_0 \tan \alpha_2$ with $J_0 = I/b_b l_b$. The ideal situation is a linear commutation, as shown in (b). In particular, if the current density J_2 is too high, this can be problematic and lead to sparking. However, many electrical and electromagnetic factors affect the commutation:

1. The contact resistance of the brushes
   The contact resistance with the segments is position-dependent (and thus time-dependent). It is easily demonstrated that the larger the ratio $\lambda = (R_s + 2R_v)/R_b$ (where R_s, R_v and R_b represent the resistances of the short-circuited coil, the resistance of the connection between the coil and the segment and the maximum brush-segment contact resistance, respectively), the more linear the commutation, as illustrated in (c) in Fig. 2.20.

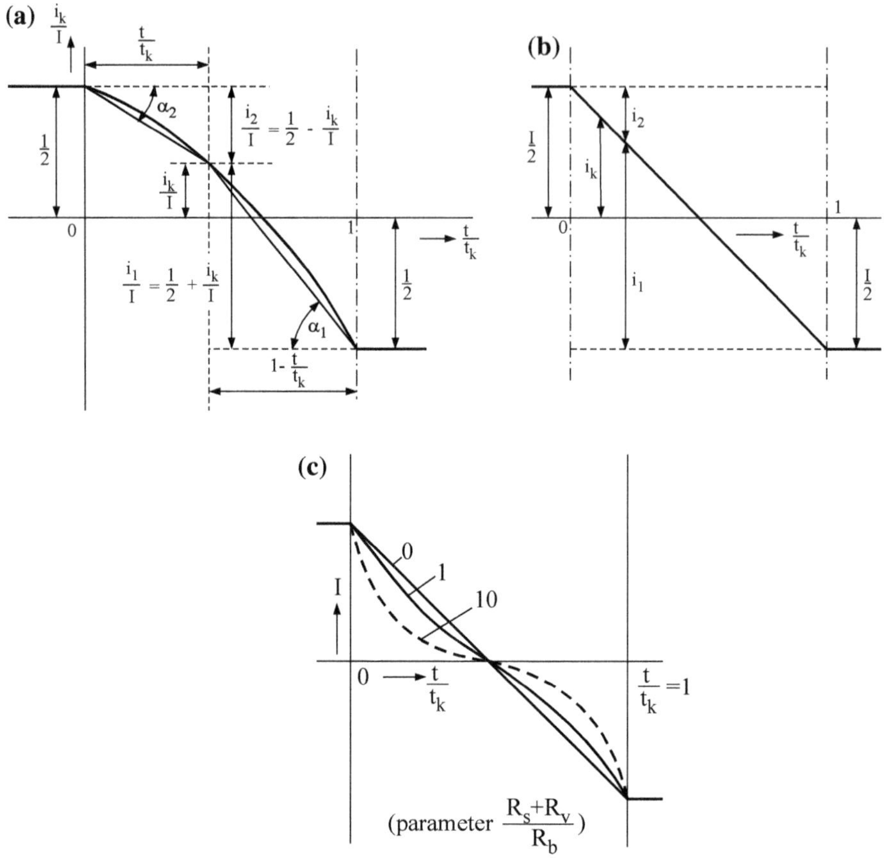

Fig. 2.20 Commutation current: **a** general; **b** linear commutation; **c** resistive commutation

2. The leakage inductance of the short-circuited coils
   The leakage inductance of the short-circuited coils will obviously oppose a current change, according to $v_\sigma = L_\sigma(di_k/dt)$.
3. The transversal armature reaction
   As discussed above, he transversal armature reaction results in a shift of the neutral zone, away from the geometrically neutral zone. As a result, the commutating armature conductors are now subjected to a magnetic field that opposes the current change, i.e. emfs are induced that counteract the current reversal.[2] This is also clear from (a) in Fig. 2.21, where we can see that the field lines of the transversal armature reaction $B_a$and those of the leakage field B_σ have the same direction and therefore induce emfs with the same polarity in the commutating coils.

[2]Can you prove this?

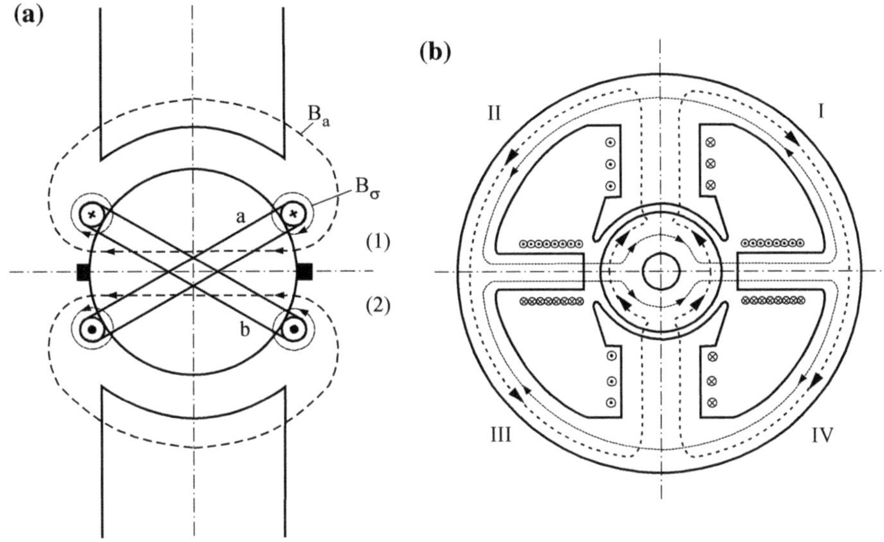

Fig. 2.21 **a** relative signs of leakage and armature reaction fields; **b** commutation poles

To mitigate the negative factors and improve the commutation, there are two possible solutions:

1. The brush axes may be shifted into the real neutral zone or even somewhat further. The main disadvantage is that the direction of the shift depends on the current direction: for a given voltage polarity, the direction of the shift depends on motoring or generating. This brush shift is usually only applied for small machines that are to be operated either as motor or as generator (not both)
2. Auxiliary or commutation poles (see (b) in Fig. 2.21) may be provided, inducing a local[3] field for the commutating coil sides, which will improve the commutation. This field will have to counteract the leakage field and the transversal armature reaction, both of which are proportional to the current. Thus the auxiliary pole winding must be connected in series with the armature. It will work for all speeds, as both the voltages induced by the leakage and the transversal armature reaction are proportional to the speed. Because of the more complicated machine construction, this solution will only be used for somewhat larger power ratings.

[3]The auxiliary poles act only on the commutating conductors in the inter-polar space while the compensation windings act only on the air gap between pole shoes and armature.

2.6 Steady-State Characteristics

2.6.1 Introduction - Per-Unit

The variables determining an operating point of a DC machine are the terminal voltage V_a, the armature current I_a, the excitation current I_m (or the excitation mmf F_m or the excitation flux for a permanent magnet excitation), the speed Ω_a and the torque T_a. These five variables are not independent of each other because of the electrical armature relation (Eqs. 2.12 and 2.7) and the torque equation 2.10.

Generator characteristics include the no-load characteristic, the load characteristic, the control characteristic and the external characteristic. The main motor characteristic is the torque-speed characteristic.

Before discussing these characteristics in more detail, we will briefly describe the per-unit system usually applied for DC machines.

For the five variables, only three reference values can be chosen independently. For a generator, the reference values to start with will typically be the rated armature voltage, the rated armature current and the rated speed. The reference values for flux and torque then follow from the relations between emf, torque, flux, current and speed (and the saturation characteristic for the excitation mmf or excitation current reference values). For a motor, the rated values for speed, torque and armature current will usually be taken as a starting point.

2.6.2 Basic Characteristics and Derivation Methods

2.6.2.1 No-Load Characteristic

The no-load characteristic is the basis for all other generator characteristics. It can be measured by driving the DC machine at rated speed and measuring the armature voltage as a function of the magnetising current (or mmf if the number of windings of the excitation is known). The no-load characteristic is similar to the saturation characteristic:

$$V_{a0}(I_m, \Omega_{an}) = E_a(I_m, \Omega_{an}) = k\Phi_m(I_m) \cdot \Omega_{an} \tag{2.22}$$

In general, the relation between the magnetising flux and the magnetising current or mmf is indeed non-linear because the permeance of the magnetic circuit is saturation-dependent:

$$\Phi_m(I_m) = w_m I_m \cdot \Lambda_m(I_m) \tag{2.23}$$

The no-load characteristic may also be measured at a speed different from the rated speed. Multiplying the emfs by Ω_{an}/Ω_a then yields the characteristic for rated speed.

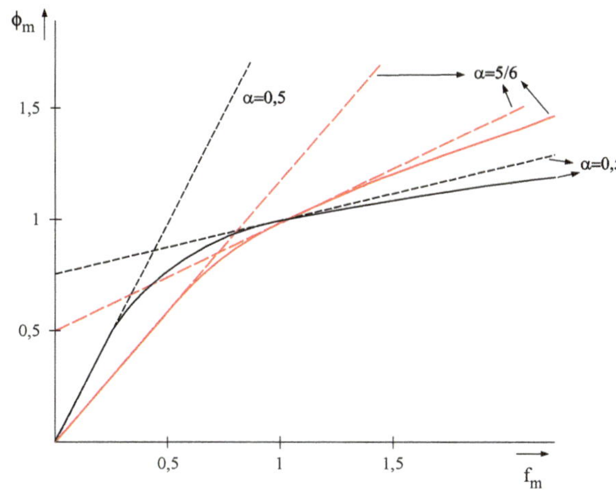

Fig. 2.22 Parametric p.u.-characteristic

The per-unit no-load or saturation characteristic is easily derived from the no-load characteristic at rated speed:

$$e_a(i_m) = \varphi_m(i_m) = \frac{E_a(I_m, \Omega_{an})}{V_{an}} \tag{2.24}$$

The per-unit magnetising current i_m is also equal to the per-unit mmf f_m. Therefore, the per-unit characteristic $\varphi_m(f_m) \equiv e_a(i_m)$ is sometimes called the generalised saturation characteristic. The ordinate represents both the p.u. flux and p.u. emf, the abscissa both the p.u. magnetising current and p.u. mmf.

A sometimes useful approximation for a saturation characteristic (neglecting hysteresis) is of the form:

$$f_m = \alpha \cdot \varphi_m + (1 - \alpha) \cdot \varphi_m^z \tag{2.25}$$

where the parameter $z \approx 5 \ldots 7$ mainly depends on the magnetic material. The parameter α determines the relative importance of the linear and the non-linear parts, see Fig. 2.22.

Although hysteresis is not considered in the parametric equation 2.25, hysteresis **is** important for self-excitation of DC machines.

2.6.2.2 General Derivation Method for Generator Characteristics

There are several excitation types for DC machines, as illustrated in Fig. 2.23. The basic types are: separate (or independent) excitation, shunt excitation and series excitation. DC machines may also have more than one excitation winding in order to obtain more suitable characteristics. When a shunt excitation is combined with a series excitation this is called a compound generator (or motor). The series excitation

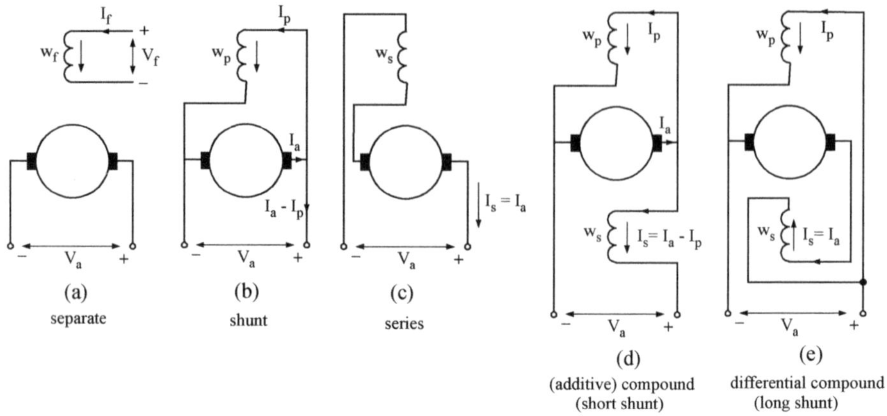

Fig. 2.23 Excitation types of DC machines

may either reinforce the independent (or shunt) excitation or counteract it, see (d) and (e) in the figure. These are called compound[4] and differential compound excitation, respectively.

Consider a DC generator with an independent excitation winding as well as a series excitation winding. The separate (independent) excitation provides an excitation mmf F_{f0}. In no-load, the resulting emf at a given speed is E_0. When the generator is loaded, the armature current I_a results in a series mmf $\pm K I_a$, with the plus sign representing an (over-)compound and the minus sign a differential compound.

For a differential compound the actual mmf $F_{f0} - K I_a$ will be lower when loaded, which in turn leads to a lower emf. In addition, the armature resistance drop will also cause a voltage reduction at the terminals. The load characteristic (for constant armature current I_a) may be derived graphically from the no-load characteristic, as illustrated in Fig. 2.24. The mmf, when loaded, will be reduced to $OP' = F_{res} = F_{f0} - K I_a$. The resulting emf (as derived from the saturation characteristic) will now be $P'N = E_{res}$. The terminal voltage, however, will be somewhat lower: $V_a = E_{res} - R_a I_a$. In the rectangular triangle *NMP*, two sides are proportional to the current. For a constant armature current, this triangle will remain congruent with itself. The load characteristic $V_a = G(F_f)$ for a constant armature current I_a is thus the no-load characteristic shifted by the oriented line *NM* or the vector $\overrightarrow{NM}$. The triangle *NMP* is called Potier's triangle or characteristic triangle or short-circuit triangle. The latter name is explained by the triangle *QRS* which shows the required mmf OR to obtain this particular armature current I_a when the armature is shorted ($V_a = 0$).

It can be shown that the longitudinal armature reaction (which also brings about a flux reduction when the magnetic circuit is saturated) may be modelled approximately

[4]Which is sometimes also called additive compound or over-compound excitation.

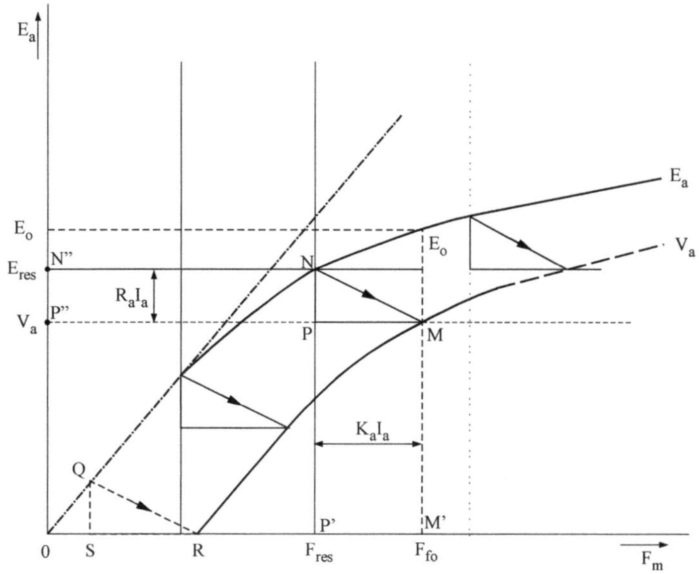

Fig. 2.24 General derivation method of DC machine characteristics

as a differential compound proportional to the armature current. If represented as an mmf reduction $-K_a I_a$, the factor K_a is in reality saturation-dependent.

2.6.3 *Generator Characteristics*

2.6.3.1 Load Characteristic of a Separately Excited Generator

In the previous section, we derived the load characteristic $V_a = G(F_f)$ of a separately excited generator for a constant armature current, taking into account a differential compound winding. The same sort of characteristic and derivation also holds for a pure armature reaction ($K = K_a$). If there is no armature reaction, the load characteristic simply follows from a vertical shift of the no-load characteristic

Another useful load characteristic is the characteristic $V_a = G(F_f)$ for a given load resistance R_L and a given speed (usually the rated speed). The characteristic and its derivation are illustrated in Fig. 2.25. Using the construction on the left (i.e. $V_a = R_L I_a$ and $E_a = (R_a + R_L)I_a$) the shift of the no-load characteristic by the vector $\overrightarrow{NM}$ of the triangle *NMP*, which now only remains similar and not congruent, should be clear. The sides of the triangle are proportional to I_a (and E_a).

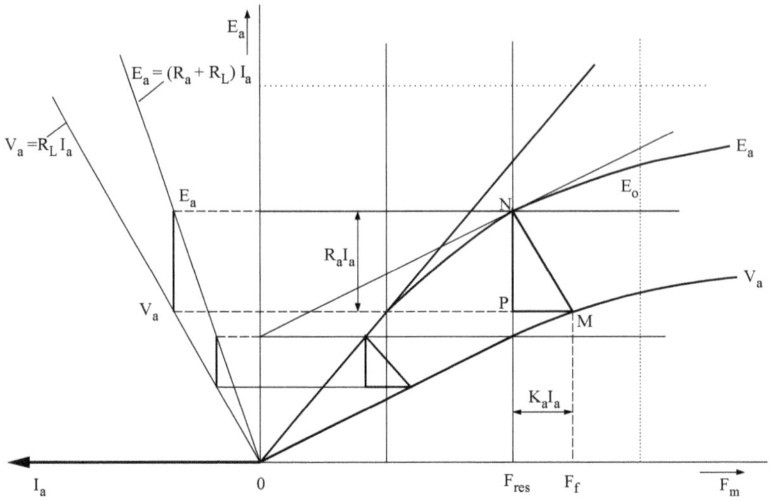

Fig. 2.25 Load characteristic of a separately excited DC generator for constant load resistance

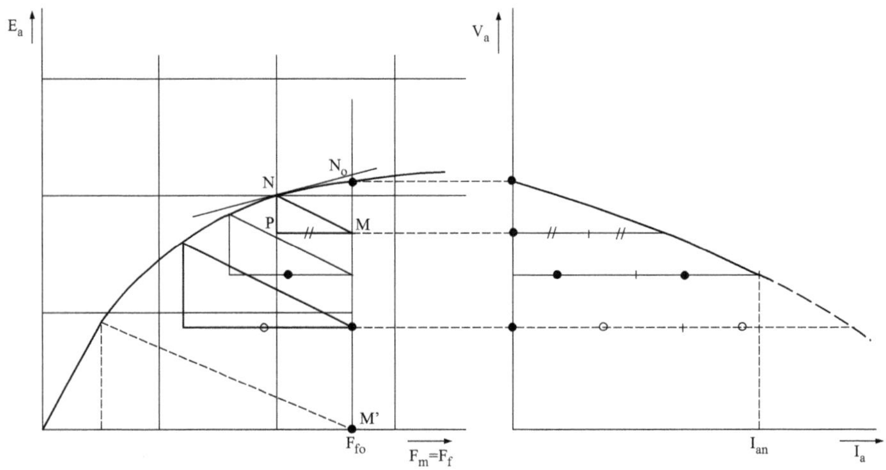

Fig. 2.26 External characteristic of a separately excited generator

2.6.3.2 External Characteristic of a Separately Excited Generator

The external characteristic, $V_a = G(I_a)$ for constant excitation mmf F_{f0} (or excitation current I_{f0}) can be derived using the construction on the left in Fig. 2.26. The triangle *NMP* remains similar, while point *N* moves on the no-load characteristic and *M* remains on the vertical line corresponding with the excitation F_{f0}. The armature currents are proportional to the lengths of the sides of the triangles.

The voltage drop with increasing armature current results from both the armature reaction and the resistive voltage drop in the armature resistance. Nevertheless, short-circuiting of such a generator will usually lead to excessive armature currents.

2.6.3.3 Control Characteristic of a Separately Excited Generator

The control characteristic $F_f = G(I_a)$ for constant armature voltage V_a and speed of a separately excited generator, Fig. 2.27, is derived in a similar way. Now the points P and M move on the horizontal line $V_a = V_{a0}$ while N is on the saturation characteristic. The armature currents are proportional to the lengths of the sides of the triangles.

2.6.3.4 Characteristics of a Shunt Excited Generator

The external characteristic $V_a = G(I_a)$, for a given resistance of the shunt winding and given speed, is the only important characteristic of a shunt generator. It is illustrated in Fig. 2.28, along with its derivation. The sides of the triangles are again proportional to the armature currents.

If hysteresis is taken into account, the dash-dot lines are obtained. Hysteresis is in fact required for self-excitation if no external source is available. For example, when the load is a pure resistance, the remanent magnetism will result in an emf and an armature voltage, which in turn yields an excitation current in the parallel connected excitation winding. The point N_0 is called the no-load point. Neglecting the excitation current $I_a = I_p \approx 0$, it corresponds with the operating condition $V_a = V_{a0}$, $\mathrm{I}_a \approx 0$. V_{a0} is referred to as the no-load voltage.

2.6.3.5 Characteristics of a Series Excited Generator

For a series excited generator, load and external characteristics coincide. The characteristic is called the self-excitation characteristic. As the no-load characteristic is measured without armature current (i.e. the excitation winding is then fed from a separate source) the armature reaction must be subtracted form the series ampère-turns. In other words, the horizontal side of the triangle now represents $(w_s - K_a)I_a$. Remark that in addition to the triangle NMP, also the triangle NPP' remains similar (with the point P' on the $y-$axis, see Fig. 2.29).

The voltage of a series excited generator is clearly very variable with variable armature current, which is one of the reasons why such a generator is hardly used. Yet, thanks to the remnant flux, self-excitation on a resistive load is possible without an external DC source.

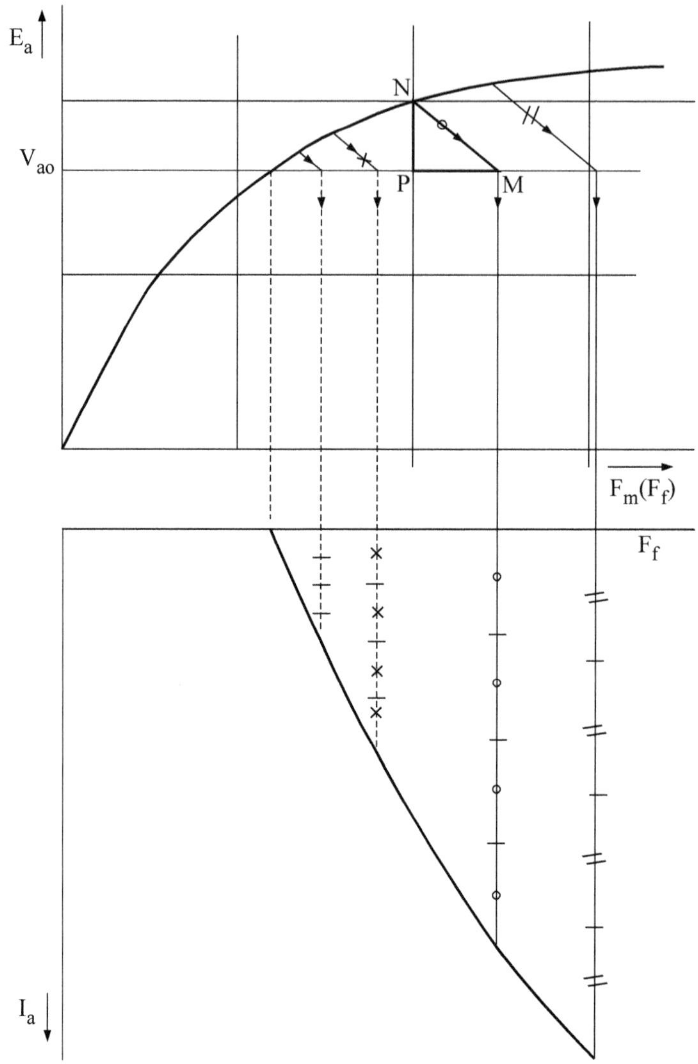

Fig. 2.27 Control characteristic of a separately excited generator

2.6.3.6 Characteristics of Compound and Differential Compound Generators

In addition to an independent and/or shunt excitation, a (differential) compound generator also has a series excitation. When the series excitation reinforces the independent or shunt excitation, it is called a compound (generator). When it opposes the other excitation(s), the term'differential compound' is used. A generator with

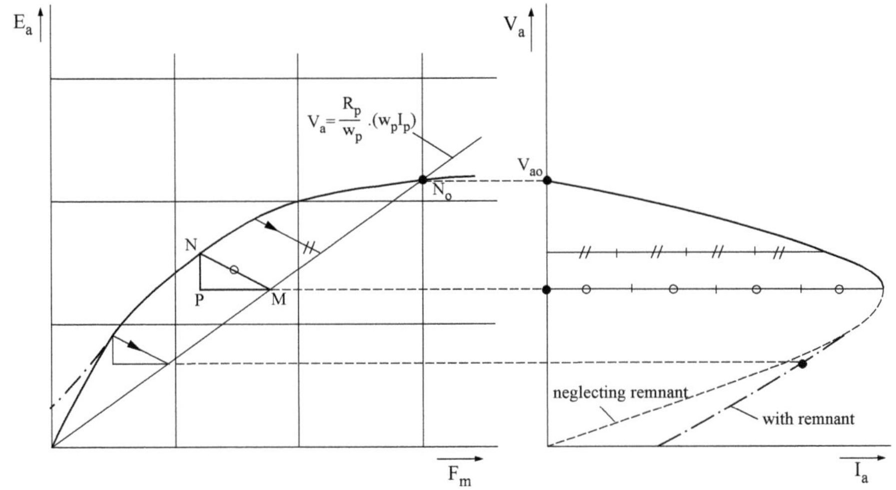

Fig. 2.28 Self-excitation characteristic of a shunt generator

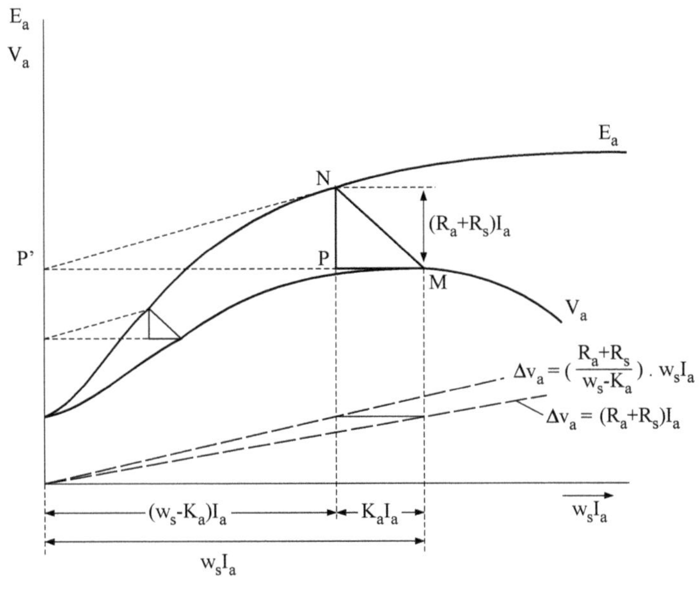

Fig. 2.29 Self-excitation characteristic of a series generator

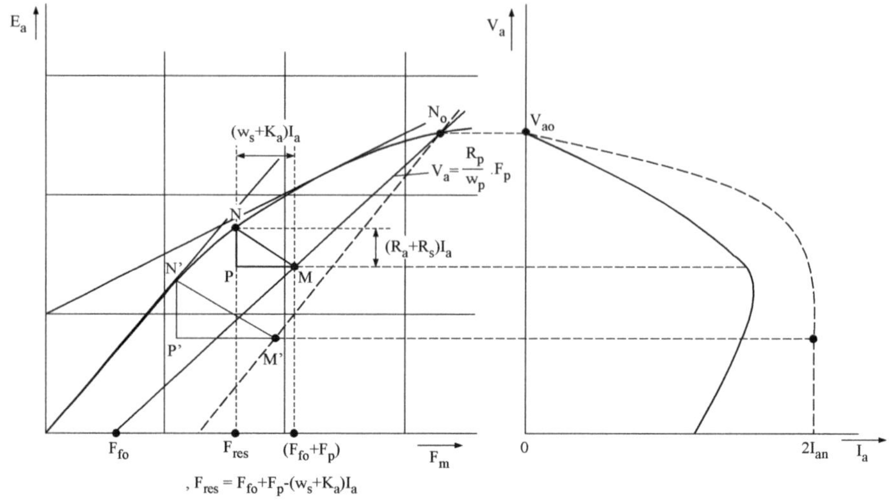

Fig. 2.30 Differential compound generator characteristic

separate, shunt and series excitation is referred to as a three-field generator, and when the series excitation opposes the other excitations, it is called a Krämer dynamo.

Figure 2.30 illustrates some external characteristics of a Krämer dynamo. By combining the separate and shunt excitation windings with a differential compound winding, a current limitation or even a constant current source is obtained in a fairly large range.

In contrast, a series excitation which reinforces the separate and shunt excitation (compound) results in an output voltage that may initially increase (for smaller current output) (see Fig. 2.31). Saturation will limit the output voltage for larger currents.

2.6.4 Motor Characteristics

The derivation of the torque-speed characteristics is based on the equations

$$V_a = k \cdot \Phi_m \cdot \Omega_a + R_a I_a + R_u I_a$$
$$T_a = k \cdot \Phi_m \cdot I_a$$

where R_u is an external series resistance (if present). Most characteristics can be derived analytically if saturation and armature reaction may be neglected.

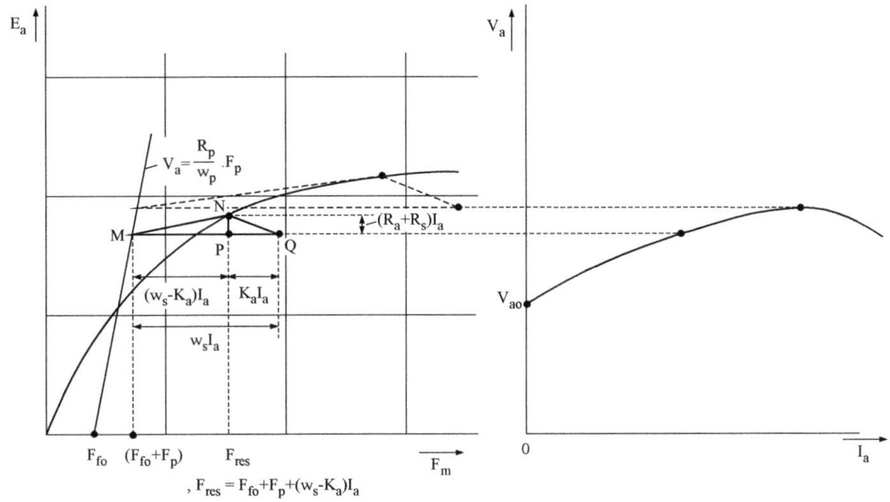

Fig. 2.31 Additive compound generator characteristics

2.6.4.1 Separately Excited DC Motor

If armature reaction can be neglected, the torque-speed characteristic is a straight line with a slope determined by the sum of armature and external resistances:

$$\Omega_a = \frac{V_a}{k \cdot \Phi_m} - \frac{(R_a + R_u)}{(k \cdot \Phi_m)^2} T_a$$

The no-load speed is

$$\Omega_{a0} = \frac{V_a}{k \cdot \Phi_m}$$

The characteristic for rated voltage and flux and without external resistance is called the natural characteristic.

With increasing load speed also decreases, which is in most cases undesirable. However, in the (distant) past, this property was also used for speed control and starting, by adding a sufficiently high external resistance.

Nowadays, the preferred method for speed control (and starting) is to vary the armature voltage (at least below the rated speed and thus voltage[5]) as this introduces no extra losses. Power-electronic converters indeed permit an almost lossless variation of the DC voltage. The characteristics for variable armature voltage (and rated flux) are lines parallel to the natural characteristic (see the full drawn lines in Fig. 2.32). To obtain higher-than-rated speeds the flux can be reduced, although

[5]The insulation of the windings is usually designed for rated voltage; higher than rated voltages will reduce the machine's life expectancy.

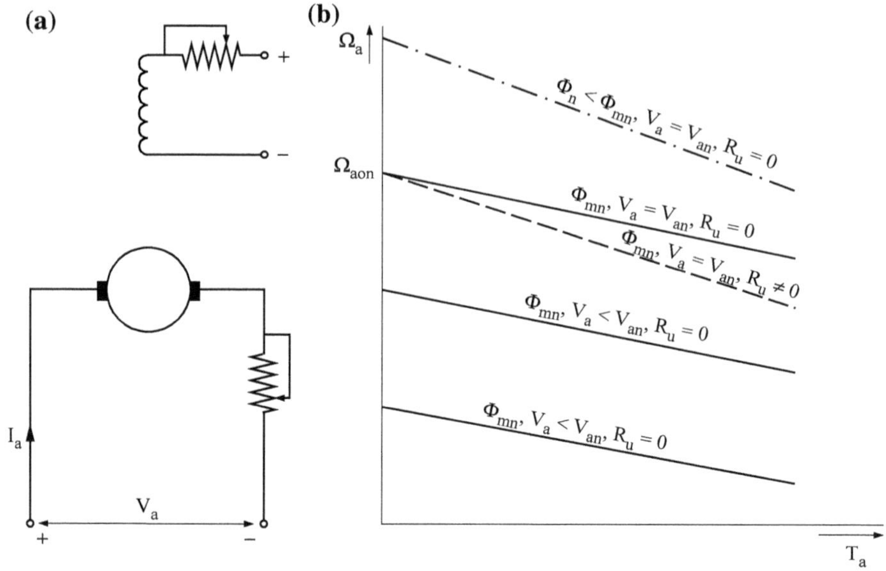

Fig. 2.32 Separately excited DC motor and characteristics

this leads to a steeper slope of the characteristics (see the dash-dot line). This is called *field weakening*. Remark, however, that in the field weakening range, a higher armature current is required for the same torque value.

Armature reaction results in a lower flux with increasing armature current (or torque). As a result, the speed will rise with increasing load compared to the linear characteristics derived above (see (b) in Fig. 2.33). A minimum in the speed characteristic may also occur. In most cases, only the range before the minimum is useful because of stability of the operating point (e.g. for a constant or almost speed-independent load torque).

The torque characteristic including armature reaction can be calculated as follows. Suppose that $E_a(F_m, \Omega_{an})$ represents the no-load characteristic at rated speed. Without armature reaction, the speed $\Omega_a^{'}$ corresponding with an armature current I_a would be given by

$$\frac{\Omega_a^{'}}{\Omega_{an}} \cdot E_a(F_f, \Omega_{an}) = V_a - R_a I_a$$

Armature reaction results in a lower mmf and therefore also a lower emf and thus a higher speed Ω_a for the same armature current

$$\frac{\Omega_a}{\Omega_{an}} \cdot E_a(F_f - K_a I_a, \Omega_{an}) = V_a - R_a I_a$$

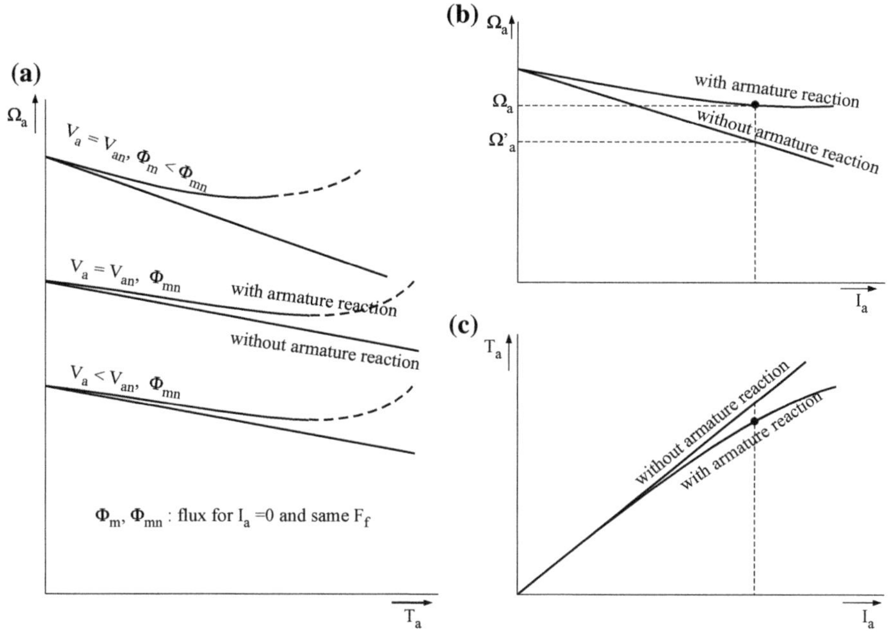

Fig. 2.33 Armature reaction effect on a separately excited DC motor

The resulting characteristics for the speed versus the current are shown in (b) in Fig. 2.33.

The relations between torque and armature can be derived in a similar way (see (c) in the figure).

From the two characteristics including armature reaction, the relation between torque and speed can be derived by eliminating I_a (see (a) in Fig. 2.33). In general, field weakening may lead to a more pronounced effect of the armature reaction, as is illustrated in the figure. A compound winding may be used to compensate the armature reaction (see Sect. 2.6.4.4).

2.6.4.2 Shunt DC Motor

The torque-speed characteristic of a shunt motor is analogous to the one for the separately excited motor. However, voltage lowering (as such) cannot be used for speed control as the excitation current would decrease as well. Of course, adding resistances in series with only the armature remains an option. Field weakening for higher speeds is possible if an additional resistance is put in series with the excitation winding.

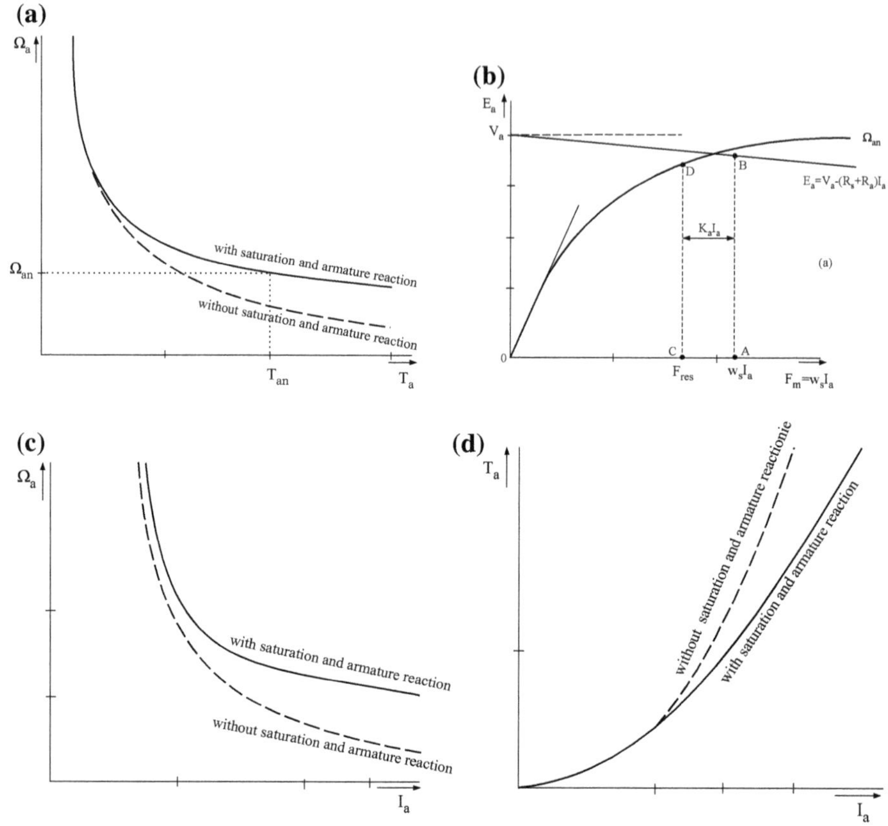

Fig. 2.34 Series excited DC motor characteristics

2.6.4.3 Series DC Motor

If saturation (and armature reaction) can be neglected, the torque-speed characteristic is hyperbolic, as can easily be demonstrated. Indeed, without saturation, the flux is proportional with the excitation current (i.e. $k \cdot \Phi_m(I_a) = k' \cdot I_a$). From $I_a = V_a / \left(R_a + R_u + k'\Omega_a\right)$ and $T_a = k' \cdot I_a^2$, it follows that

$$T_a = \frac{k' V_a^2}{(R_a + R_u + k'\Omega_a)^2}$$

Figure 2.34a illustrates this characteristic (dashed line). Armature reaction will create a flatter characteristic for higher torque values (see the full line).

The derivation of the effect of armature reaction is somewhat analogous to the derivation for the separately excited DC motor. With now $F_{res} = w_s I_a - K_a I_a$ and

Fig. 2.35 Series motor: speed control

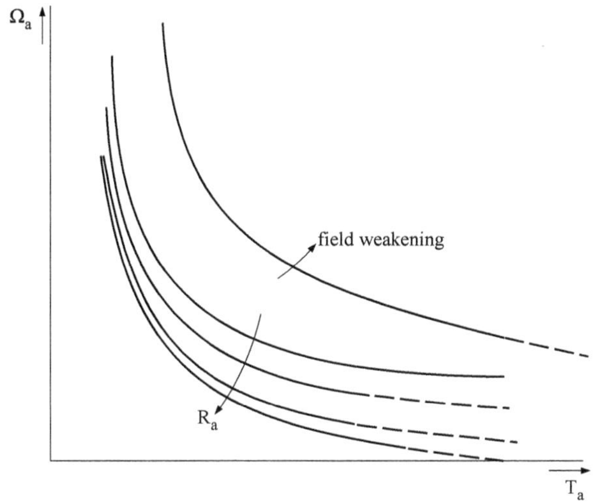

$$\frac{\Omega_a}{\Omega_{an}} \cdot E_a(w_s I_a - K_a I_a, \Omega_{an}) = V_a - (R_a + R_s) I_a$$

we derive the characteristic Ω_a versus I_a (see (c) in Fig. 2.34). A similar derivation yields the characteristic T_a vs. I_a (see (d) in Fig. 2.34):

$$T_a = k\Phi_m(F_{res}) \cdot I_a = \frac{1}{\Omega_{an}} \cdot E_a(w_s I_a - K_a I_a, \Omega_{an}) \cdot I_a$$

Eliminating I_a then yields the characteristic T_a versus Ω_a. As Φ_m becomes more constant (instead of proportional to the current), the torque tends to become more linear with the current for higher armature currents.

The series DC motor is an ideal motor for traction. It offers a very high torque at low speed (starting) and the shaft power varies little with speed. The characteristic also exhibits very stable behaviour for various load torque characteristics. The main disadvantage is that it must never be operated in no-load as the speed would rise to intolerably high values (Fig. 2.35).

Speed regulation can be obtained by:

- voltage control
- inserting series resistances R_u between source and motor
- field weakening by inserting resistances in parallel with the series excitation winding (for higher speeds).[6]

[6]Prove that the characteristics for field weakening are approximately obtained by shifting the characteristic to higher torque values for the same speeds (consider operating points for the same flux, with and without parallel resistances; neglect the resistances in armature and series excitation winding).

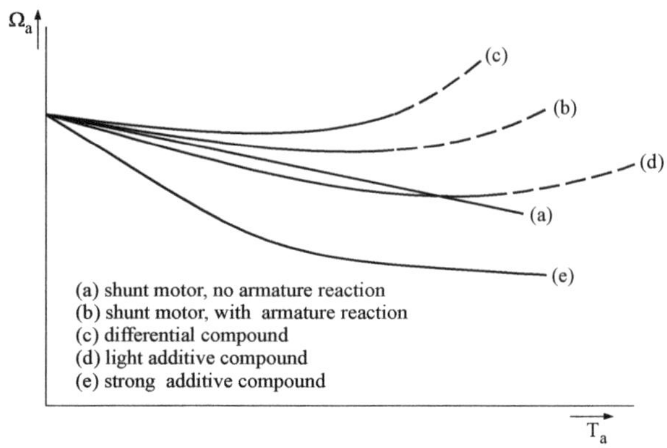

Fig. 2.36 Compound DC motor characteristics

2.6.4.4 Compound DC Motor

As derived above, the armature reaction causes the initially slightly decreasing speed of a shunt motor to show a minimum value and even an increase for higher torque values. In some cases, such an almost constant speed is desirable and this behaviour can be reinforced by adding a differential compound winding (see (c) in Fig. 2.36). On the other hand, if unstable behaviour is to be avoided and/or a more decreasing speed with higher torque values is desired, an additive compound winding may help (see (d) and (e)).

Chapter 3
Rotating Field Machines: mmf, emf and Torque

Abstract Rotating fields are the basis for most electric drives (induction and synchronous machines). First, the generation of a rotating field is discussed. As in the previous chapters, we start again from the basic electromagnetic laws. Both using a graphical depiction and a more mathematical method the rotating field generation is explained. Next, the emf is discussed. Finally, the torque on a (rotating) current layer in a (rotating) field is discussed.

3.1 Generation of a Rotating Field

Nowadays (and since long), induction and synchronous machines are by far the most important electrical machines. Their operation principle is based upon rotating fields (or travelling fields for linear machines). In this section we will see two ways to realise such a rotating magnetic field. At the same time it is explained how it can be ensured that this rotating field has an almost sinusoidal shape in space.

3.1.1 Magnetic Field by (stator) Salient Poles with Concentrated Windings

Consider the pair of poles in Fig. 3.1 If we feed the excitation windings with a DC current i, then a magnetic field originates with field lines as shown in (a). If we call w_p the number of windings per pole pair (i.e. $w_p = w/N_p$ with w the total number of windings for the machine and N_p the number of pole *pairs*), then we may write for each of the field lines:

$$\oint \underline{h}.d\underline{l} = w_p.i \tag{3.1}$$

If the magnetic voltage drop in the iron parts can be neglected, the left hand side of this equation reduces to the magnetic voltage drop over the two air-gaps. If we call b_x the induction in the air-gap in x and δ_x the air-gap length in x, we get

J. A. Melkebeek, *Electrical Machines and Drives*, Power Systems,
https://doi.org/10.1007/978-3-319-72730-1_3

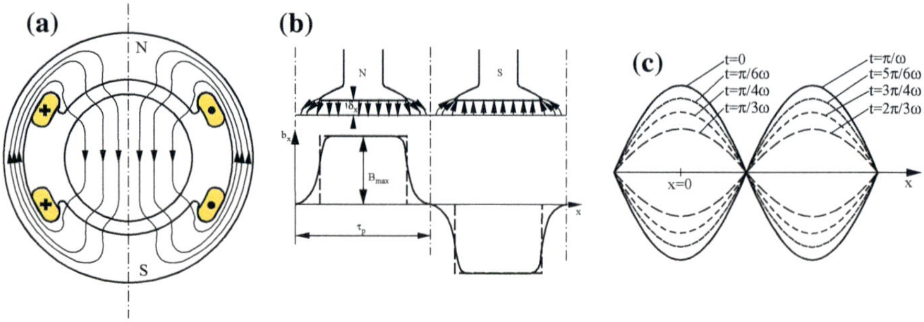

Fig. 3.1 Salient poles with concentrated windings

$$b_x = \mu_o . \frac{w_p i}{2\delta_x} = \mu_o h_x \tag{3.2}$$

where h_x is called the air-gap mmf in x.

When we draw the magnitude of the magnetic field along the circumference, we get a wave shape like (b) in Fig. 3.1. The dashed line is the ideal (theoretical) curve, neglecting the fringing at the edges of the poles (i.e. when the permeability of the iron were infinite).

The field curve can be developed in a *(space)* Fourier series as follows (with $x = 0$ along the north pole axis):

$$b_x = b(x) = \hat{B}_1 \cdot \cos \frac{x\pi}{\tau_p} + \hat{B}_3 \cdot \cos \frac{3x\pi}{\tau_p} + \hat{B}_5 \cdot \cos \frac{5x\pi}{\tau_p} + \cdots \tag{3.3}$$

where τ_p represents the pole pitch.

As we will see below, for rotating field machines only the *space* sinusoidal and fundamental component (with respect to $2\tau_p$) is important:

$$b_1(x) = \hat{B}_1 \cdot \cos \frac{x\pi}{\tau_p} \tag{3.4}$$

Until now, the current i has not been specified. It could be a DC current, in which case the field is constant *in time* (but with an approximately sinusoidal space distribution). If the excitation current i is an AC current $i(t)$, then the field curve (b) in Fig. 3.1 represents the (sinusoidal) field distribution over the circumference at a given instant. Neglecting saturation the *(space)* shape of this field curve will remain the same, but the local induction values will vary periodically and proportionally with the current $i(t)$. This alternating field will therefore resemble the field of a transformer.

If the current $i(t)$ is sinusoidal in time:

$$i(t) = \hat{I} \cdot \cos \omega t \tag{3.5}$$

then we obtain a field distribution

$$b_1(x,t) = \hat{B}_1 \cdot \cos\frac{x\pi}{\tau_p} \cdot \cos\omega t \tag{3.6}$$

This is the equation of a standing wave, see (c) in Fig. 3.1. The radial direction of the maximum of the field, called the field axis, is coincident with the axis of the winding (the north-south axis). This alternating field with a *sinusoidal distribution in space* can be represented by a *space vector* with as direction the field axis:

$$\underset{\rightarrow}{b}_1(t) = \underline{\hat{B}}_1 \cdot \cos\omega t \tag{3.7}$$

with $\underline{\hat{B}}_1 = \hat{B}_1 e^{j0}$. More general, if the maximum of the field is displaced over an electrical angle γ in the direction of positive x, i.e.

$$b_1(x,t) = \hat{B}_1 \cdot \cos\left(\frac{x\pi}{\tau_p} - \gamma\right) \cdot \cos\omega t$$

the space vector is

$$\underset{\rightarrow}{b}_1(t) = \underline{\hat{B}}_1 \cdot \cos\omega t = \hat{B}_1 e^{j\gamma}\cos\omega t \tag{3.8}$$

The instantaneous value of this field (with sinusoidal space distribution) in a point x along the circumference is obtained by projecting the vector onto the radial through this point x. For example at $x\pi/\tau_p = \gamma$, we get the maximum value $\hat{B}_1 \cdot \cos\omega t$ or at $x\pi/\tau_p = \gamma + \pi/3$ it's $(\hat{B}_1/2) \cdot \cos\omega t$. Back transformation to the real time domain is obtained by multiplying the space vector (3.8) with $e^{-j\frac{x\pi}{\tau_p}}$ and taking the real part.[1]

3.1.2 *Magnetic Field by Rotating Salient Poles with Concentrated Windings*

The same field distribution as in Fig. 3.2 is also obtained when the salient poles are on the rotor of a machine. Or formulated more correctly: the same distribution but now with respect to this rotor.

Suppose now that the excitation current is a DC current. Then this sinusoidal field distribution is constant in time (*and space as well, with respect to the rotor*). We will now use x' as the coordinate with respect to the rotor, i.e. for an observer on the rotor the field looks like

[1]Remark that the sign of $j\gamma$ in Eq. 3.8 is opposite to that used in time phasors; more on that convention later on.

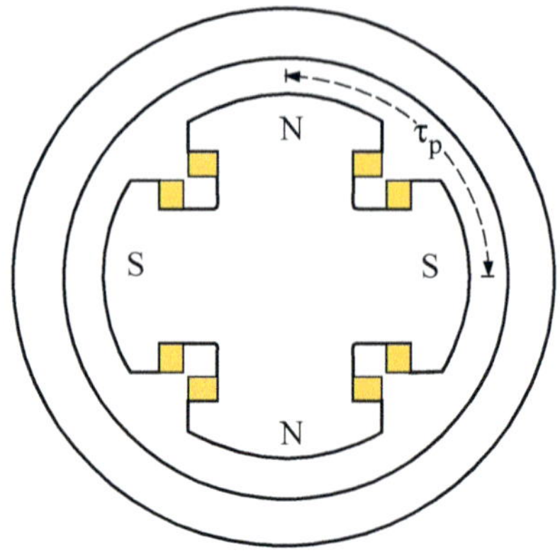

Fig. 3.2 Salient poles with excitation windings on the rotor

$$b_1(x', t) = \hat{B}_1 \cdot \cos \frac{x'\pi}{\tau_p} \tag{3.9}$$

If the rotor is stationary, an observer on the stator sees the same field distribution. For the stator observer we will use a coordinate x in the same direction as x' (and with $x = x'$ at $t = 0$). Suppose now that the rotor rotates in the positive x direction with a rotational speed N [1/s]. The linear speed of a point on the rotor is then $v = 2\pi r \cdot N = 2N_p\tau_p \cdot N$. An observer on the stator sees now a field distribution (substitute $x = x' + vt$ or $x' = x - vt$ in Eq. 3.9)

$$b_1(x, t) = \hat{B}_1 \cdot \cos\left(\frac{x\pi}{\tau_p} - v\frac{\pi}{\tau_p}t\right) = \hat{B}_1 \cdot \cos\left(\frac{x\pi}{\tau_p} - \omega t\right) \tag{3.10}$$

Equation 3.10 represents a rotating wave or travelling wave, see (a) in Fig. 3.3. Its shape remains constant and sinusoidal (with a wave length of $\lambda = 2\tau_p$), but the wave moves at a constant linear speed $dx/dt = v = \omega \cdot \tau/\pi = 2\tau_p \cdot f = 2\tau_p \cdot N_pN$ or a constant rotational speed ω.

The rotational speed $\omega = 2\pi \cdot N_p.N = v\pi/\tau_p$ is the *electrical* rotational speed: for an observer on the armature (stator), N_pN north poles and N_pN south poles are passing each second. The frequency $f = N_pN{=}\omega/2\pi$ also is the frequency of the voltages induced by this field in a conductor on the armature; the amplitude of these voltages also is proportional to f (in the next section all this will be discussed in more depth).

Remark that the *mechanical*[2] rotational speed [1/rad] is equal to $\Omega = 2\pi \cdot N = \omega/N_p$.

---

[2]One revolution of the rotor corresponds to 2π mechanical radians and to $2\pi.N_p$ electrical radians where N_p is the number of pole pairs; otherwise said the circumference of the rotor corresponds to $2N_p$ pole pitches, i.e. $2\pi r = 2N_p\tau_p$ while each pole pitch corresponds to πelectrical radians.

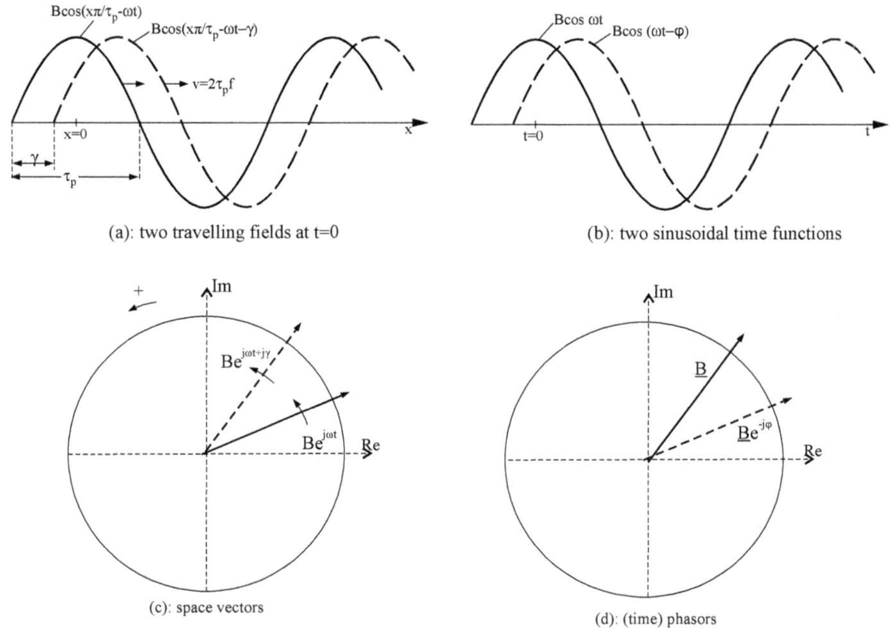

Fig. 3.3 Travelling wave, space vectors and time phasors

A rotating field can also be represented by a ***space vector*** (see (c) in Fig. 3.3):

$$\underset{\rightarrow}{b}_1 = \underline{\hat{B}}_1 e^{j\omega t} = \hat{B}_1 e^{j\omega t} \tag{3.11}$$

This is clearly a vector rotating *in space* at an *electrical* rotational speed ω. The length of the space vector is equal to the maximum value of the rotating field. The space vector also shows the instantaneous position of the maximum of the field at each instant: the instantaneous value of the field at a given point on the armature is obtained by projection of the space vector on the radial through this point.

A rotation in the negative x direction will result in a sign change in Eqs. 3.10 and 3.11: $\cos\left(\frac{x\pi}{\tau_p} + \omega t\right)$ and $\hat{B}_1 e^{-j\omega t}$ respectively.

More general, for a field that is shifted by an angle γ

$$b_1(x,t) = \hat{B}_1 \cdot \cos\left(\frac{x\pi}{\tau_p} - \gamma \pm \omega t\right) \tag{3.12}$$

the space vector is

$$\underset{\rightarrow}{b}_1 = \underline{\hat{B}}_1 e^{\mp j\omega t} = \hat{B}_1 e^{+j\gamma} e^{\mp j\omega t} \tag{3.13}$$

The maximum of this field is found at $x = \gamma \cdot \tau_p/\pi$ at $t = 0$ (or at $x = 0$ at $t = \gamma/\omega$).

As mentioned before, returning from the space vector domain to the real time domain is achieved by multiplying with $e^{-j\frac{x\pi}{\tau_p}}$ and taking the real part.

This sign difference compared to the time domain (phasors) - where returning to the real time domain is obtained by multiplying with $e^{+j\omega t}$ - can be explained as follows. Consider the two travelling waves in (a) in Fig. 3.3 (i.e. in the *space* domain). The travelling field in dashed line is clearly *leading* the travelling field in solid line: an observer in a point x_0 will indeed see passing the former *before* the latter. In contrast, (b) in Fig. 3.3 shows two sinusoidal *time* waveforms. In this time domain, the dashed waveform is *later* than the solid waveform and is thus *lagging* by φ! In this figure, (d) shows the corresponding phasor representation in the *complex time plane* ,while (c) shows the *complex space plane corresponding with (a).*

Remark: An alternating field can be decomposed into two counter-rotating fields, each with half the amplitude of the alternating field:

$$b_1(x,t) = \hat{B}_1 \cdot \cos\frac{x\pi}{\tau_p} \cdot \cos\omega t = (\hat{B}_1/2) \cdot \cos\left(\frac{x\pi}{\tau_p} - \omega t\right) + (\hat{B}_1/2) \cdot \cos\left(\frac{x\pi}{\tau_p} + \omega t\right) \tag{3.14}$$

or in space vector form

$$\underset{\rightarrow}{b}_1(t) = \underline{\hat{B}}_1 \cdot \cos\omega t = (\underline{\hat{B}}_1/2) \cdot e^{j\omega t} + (\underline{\hat{B}}_1/2) \cdot e^{-j\omega t}. \tag{3.15}$$

3.1.3 Magnetic Field by a Distributed AC Winding

In Sect. 3.1.1 we have seen how a concentrated winding can be used to obtain a magnetic field. In most cases the space distribution of such a field is not at all sinusoidal, unless the pole faces are somewhat adapted. However, for rotating field AC machines, a sinusoidal distribution in space is almost a must (in order to avoid losses). To this end, distributed windings are preferred. In this section we will show how a distributed winding may result in a fair approximation for a sinusoidal space distribution.

We consider therefore a single-phase AC-fed winding and look for making the field distribution as sinusoidal as possible, or at least more sinusoidal than that of a concentrated winding.

First we consider a single-phase concentrated winding, Fig. 3.4. This is a winding with N_p coils (thus $2N_p$ coil sides) on the circumference. All conductors carry the same current and are in most cases series-connected. The diameter of the coils is thus equal to the pole pitch τ_p. Each coil has $w_p = w/N_p$ windings, where w is the total number of phase windings and w_p denotes the number of windings per pole pair (or coil). Normally the conductors are located in slots[3] (as in Fig. 3.1), but here we suppose for simplicity that they are located 'on the surface'.

---

[3]Slots protect the windings but they also result in a useful spread of the forces on the iron.

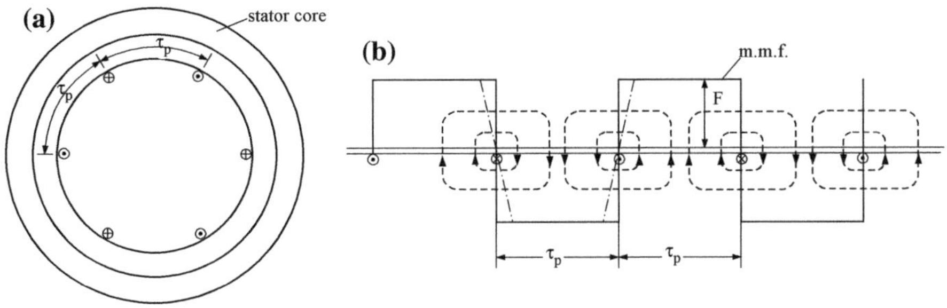

Fig. 3.4 Concentrated surface winding

A DC current in this winding results in a magnetic field with field lines as the dashed lines in (b) in Fig. 3.4. For each of these field lines, the mmf is $w_p i$ but this mmf is divided equally between the two (equal because constant) air-gaps δ_p (neglecting the mmf drop in the iron, i.e. we suppose $\mu_{ir} \approx \infty$). This magnetic voltage drop over the air-gap (i.e. the magnetic potential difference) we will call the *(air-gap) mmf*[4], and denote it by F. When we draw the value of this mmf F over the circumference, we get an (almost) rectangular curve with amplitude $F = w_p i/2$. In reality, there is a slight deviation from the rectangular shape due to the finite width of the conductors and the effect of the slots (see the dash-dot sides).

A Fourier analysis of this curve yields:

$$f_x = \frac{4}{\pi} F \left[\cos \frac{x\pi}{\tau_p} - \frac{1}{3} \cos \frac{3x\pi}{\tau_p} + \frac{1}{5} \cos \frac{5x\pi}{\tau_p} \cdots \right] \tag{3.16}$$

where x is the distance measured from a north pole axis of the winding (also called winding axis).

Next, we consider a single-phase distributed diameter[5] winding, see (a) in Fig. 3.5. In each double pole pitch there are q *diameter coils*, displaced with respect to each other over a distance λ [m]. λ is the distance between successive coils of the same polarity, either in or out. The subsequent coils may span a complete pole pitch, or only partly as in this figure. In Fig. 3.6a, b the field lines are shown for a 2-pole and a 4-pole machine respectively. The curve of the mmf values over the circumference is shown in (b) in Fig. 3.5.

At each coil side, there is an mmf jump equal to $\pm(w_p/q)i$. This staircase curve can also be derived from superposition of the q rectangular and, with respect to each other, over λ delayed step curves in Fig. 3.4.

A superposition of the q Fourier series, delayed with respect to each other, according to Eq. 3.16 yields:

[4]Actually, the notion mmf is, strictly spoken, always related to a closed field line but in machine theory it is also used in lieu of magnetic potential difference.

[5]In a diameter winding the coil width, i.e. the distance between entrance and exit conductors of a turn, is exactly one pole pitch.

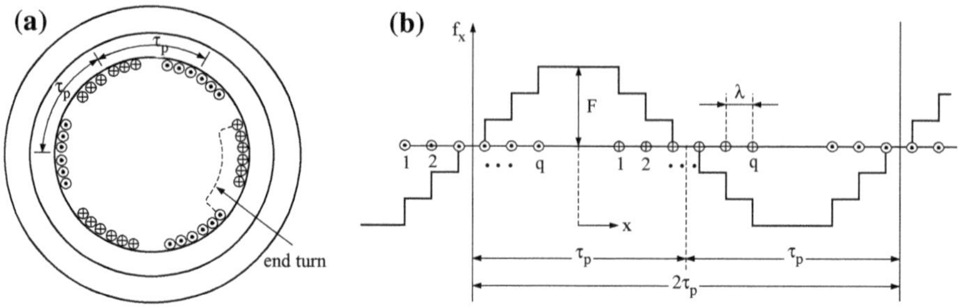

Fig. 3.5 Distributed winding

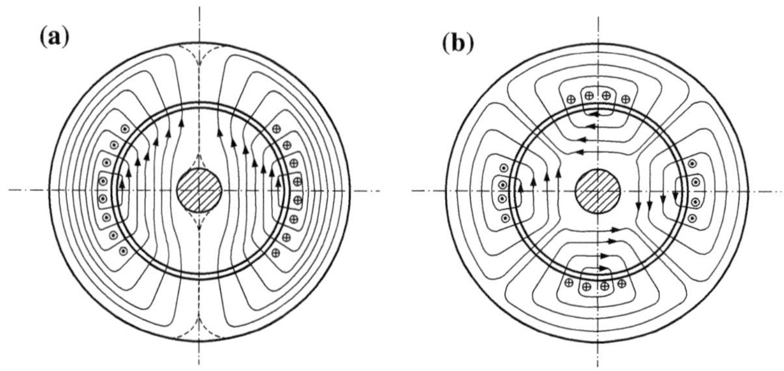

Fig. 3.6 Field lines for distributed windings

$$f_x = \hat{F}_1 \cos \frac{x\pi}{\tau_p} + \hat{F}_3 \cos \frac{3x\pi}{\tau_p} + \cdots + \hat{F}_\nu \cos \frac{\nu x\pi}{\tau_p} + \cdots \tag{3.17}$$

when $x = 0$ is chosen along the winding axis. For the amplitudes $\hat{F}_\nu$ a straightforward calculation yields

$$\hat{F}_\nu = \frac{4}{\nu\pi}\xi_\nu \cdot (-1)^{\frac{\nu-1}{2}} \cdot \frac{w_p i}{2} \tag{3.18}$$

where w_p is the number of turns per pole pair (i.e. in the q coils) and ξ_ν represents the *distribution factor* or *zone factor* (which is equal to the *winding factor* for a *diameter*[6] winding):

$$\xi_\nu = \frac{\sin q \frac{\nu\alpha}{2}}{q \cdot \sin \frac{\nu\alpha}{2}} \tag{3.19}$$

[6]As mentioned above, in a diameter winding the coil diameter is equal to a pole pitch.

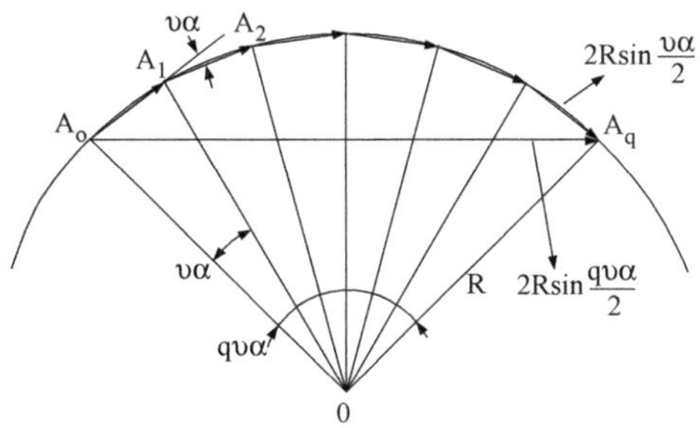

Fig. 3.7 Winding factor

Table 3.1 Winding factors for some q-values

q	ξ_1	ξ_5	ξ_7	ξ_{11}	ξ_{13}
1	1	1	1	1	1
2	0.9659	0.2588	0.2588	0.9659	0.9659
3	0.9598	0.21756	0.17736	0.17736	0.21756
4	0.95766	0.2053	0.15756	0.12608	0.12608
...	...	...	...	...	...
∞	0.955	0.1909	0.1364	0.0868	0.0735

Herein $\alpha = \lambda \cdot \pi/\tau_p$ is the displacement between successive coil sides converted to electrical radians (with respect to the period of the fundamental harmonic or $2\tau_p$).

The amplitudes of Eq. 3.18 and the winding factors may also be obtained graphically, see Fig. 3.7. For each harmonic the resultant amplitude is the *vectorial* sum of the contributions of each of the q coils in a pole pair (for a higher harmonic ν, the delay angle is $\nu\alpha$, as a double pole pitch $2\tau_p$ corresponds to $2\nu\pi$ radians for this harmonic).

The distribution factor is thus the ratio of the length of the vector sum A_oA_q to the sum of the lengths of the vectors A_oA_1, A_1A_2,... Compared to a concentrated winding with a same number of turns w (or same w_p), the amplitude of the fundamental is thus reduced (i.e. $\xi_1 < 1$). However the distribution factors for the higher harmonics are much smaller $\xi_\nu \leq \xi_1$ (see Table 3.1) and thus the resultant waveform is much more sinusoidal than that for a concentrated winding. From Eq. 3.18, it follows that for a distributed winding not the number of turns w or w_p but the product $\xi_1 w$ (or $\xi_1 w_p$) determines the (fundamental) mmf. The product $\xi_1 w$ (or $\xi_1 w_p$) is called the *effective number of turns*.

Note that for further reducing the harmonic content, in multiphase machines a double (or multi-) layer winding can be used. In such a winding the layers are displaced over one or more slots so that one (or more) slots contain conductors of two phases.

Remark:

Figure 3.8 shows a (simplified) depiction of a distributed winding configuration. The winding is a single-layer diameter winding (i.e. a winding where the entry and exit slots of a slot pair are a distance of τ_p apart, as measured on the mean circumference)

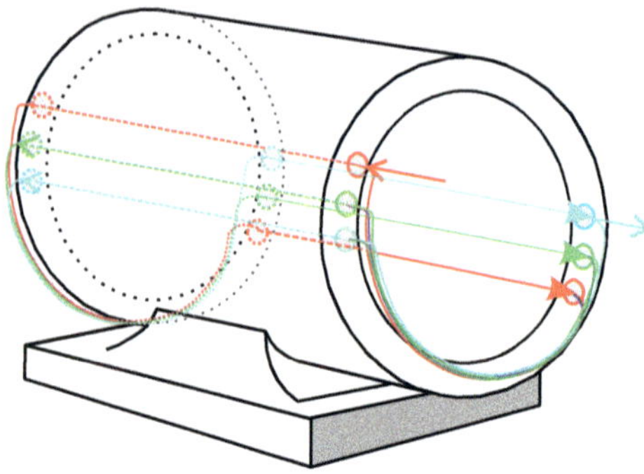
Fig. 3.8 Winding configuration

with $q = 3$. The red arrow indicates the entry conductor (and slot). The blue arrow represents the exit conductor (and slot). Each pair of slots (red, green, blue) may contain more than one turn. After the conductors in the red pair of slots, the conductors in the green pair of slots are connected in series, thereafter the conductors in the blue slots.

3.1.4 Magnetic Field by a Multiphase AC Winding

The previous section explained how a distributed winding may yield a (more) sinusoidal field distribution in space. What we need however is a rotating (or travelling) field with a sinusoidal space distribution. Section 3.1.2 presented a first method for obtaining a rotating field. However, it requires a rotating member (DC excited rotor, rotated by external means). Although it was presented using a concentrated winding, a DC excited distributed winding as in the previous section could be used as well (and will result in a more sinusoidal space distribution).

In this section we will show how an AC fed multiphase (distributed) winding can be used to obtain a rotating field (without *a priori* necessitating a rotating member).

As a starting point, the distributed winding of the previous section is supplied with an AC current $i = \hat{I}.\cos\omega t$. The air-gap mmf then becomes a standing wave (along the winding axis):

$$f_1(x,t) = \hat{F}_1 \cos\frac{x\pi}{\tau_p}.\cos\omega t \tag{3.20}$$

with

$$\hat{F}_1 = \frac{2}{\pi} w_p \xi_1.\hat{I} = \frac{2}{\pi}\frac{w\xi_1}{N_p}.\hat{I} \tag{3.21}$$

(we recall that w is the total number of turns of the phase; each pole pair has $w_p = w/N_p$ turns which are evenly distributed in the q coils of the pole pair; the complete phase winding contains $N_p \cdot q$ coils in $2N_p \cdot q$ slots).

The alternating mmf of Eq. 3.20 can be decomposed in the sum of two counter-rotating fields, each with half the amplitude of the alternating field:

$$f_1(x,t) = \hat{F}_1 \cos\frac{x\pi}{\tau_p} \cdot \cos\omega t = (\hat{F}_1/2) \cdot \cos\left(\frac{x\pi}{\tau_p} - \omega t\right) + (\hat{F}_1/2) \cdot \cos\left(\frac{x\pi}{\tau_p} + \omega t\right) \tag{3.22}$$

The rotation speed of each of these mmfs is $\pm\omega$ (in electrical radians per second) and the linear speed is $v = \pm\omega\tau_p/\pi = \pm 2\tau_p f$ [m/s]. Note also that at the instant the current in the phase is maximum, the (alternating) phase mmf is maximum as well. At the same instant, both rotating mmfs are along the winding axis.

If saturation of the magnetic circuit can be neglected, the air-gap induction in each point of the armature circumference follows directly from the local mmf:

$$b_x = \mu_o h_x = \mu_o \frac{f_x}{\delta_x} \tag{3.23}$$

If, moreover, the air gap length can be considered as constant (which implies that the effect of the slots is neglected) then $\delta_x = \delta_p$ and the distribution of the induction over the circumference is identical in shape to that of the mmf. Saturation will result in some deformation and thus space harmonics, but the periodicity and zero crossings remain the same. Even if there are space harmonics (either due to the mmf harmonics or to saturation), the space fundamental with $2N_p$ poles determines the basic properties of the machine; the higher harmonics result in fields with $2\nu N_p$ poles which cause some undesirable effects (e.g. losses).

Now we consider a multiphase (armature) winding. An m-phase (distributed) winding normally contains 2 m phase belts[7] in a double pole pitch. Figure 3.9a, b show schematically the winding configurations of a three-phase armature with 2m = 6 phase belts and q = 3, for four and two poles respectively. In (c) in Fig. 3.9a linear representation of a double pole pitch (2π electrical radians) is depicted.

The belts without prime contain the entry wires, those with prime the corresponding exit wires of the coils for these phases. All coils for a phase carry the same current, e.g. for phase U the currents in the coils of belt U are $\underline{I}_u$ and those for belt U' are $-\underline{I}_u$.

Observe that the currents in subsequent phase belts are displaced <u>*in time*</u> with respect to each other over $2\pi/2m = \pi/m$ (thus $\pi/3$ for a 3-phase winding), whereas the subsequent phase belts are also displaced <u>*in space*</u> with respect to each other over the same angle (thus with respect to 2π electrical radians or one double pole pitch, over π/m or $\pi/3$ for a 3-phase winding). It is easily shown that under these conditions, a rotating field results with a sinusoidal distribution in space. This field contains $2N_p$ poles and rotates at a constant speed in the direction of the lagging phases (thus $U \to V \to W \to U \ldots$ for a 3-phase winding).

---

[7]In some winding arrangements m.

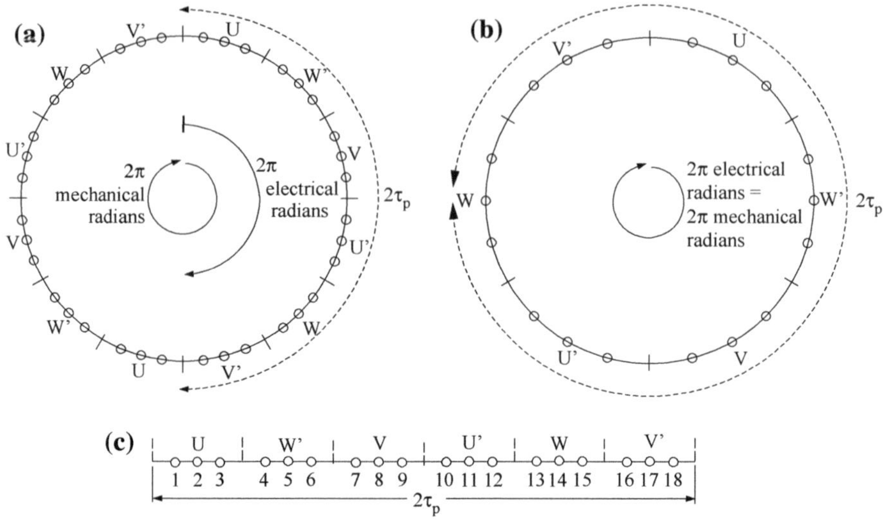

Fig. 3.9 Multiphase winding

This rotating mmf of a multiphase winding can be derived by superposition, similarly as we have done before for a distributed winding.

A purely mathematical way is as follows. We consider the alternating mmfs of the three phases, taking into account their phase lag in both space and time:

$$f_{u1}(x,t) = \hat{F}_1 \cos\frac{x\pi}{\tau_p}\cdot\cos\omega t = (\hat{F}_1/2)\cdot\cos\left(\frac{x\pi}{\tau_p}-\omega t\right)+(\hat{F}_1/2)\cdot\cos\left(\frac{x\pi}{\tau_p}+\omega t\right) \tag{3.24}$$

$$f_{v1}(x,t) = \hat{F}_1 \cos\left(\frac{x\pi}{\tau_p}-\frac{2\pi}{3}\right)\cdot\cos\left(\omega t-\frac{2\pi}{3}\right) = (\hat{F}_1/2)\cdot\cos\left(\frac{x\pi}{\tau_p}-\omega t\right) + (\hat{F}_1/2)\cdot\cos\left(\frac{x\pi}{\tau_p}+\omega t-\frac{4\pi}{3}\right) \tag{3.25}$$

$$f_{w1}(x,t) = \hat{F}_1 \cos\left(\frac{x\pi}{\tau_p}-\frac{4\pi}{3}\right)\cdot\cos\left(\omega t-\frac{4\pi}{3}\right) = (\hat{F}_1/2)\cdot\cos\left(\frac{x\pi}{\tau_p}-\omega t\right) + (\hat{F}_1/2)\cdot\cos\left(\frac{x\pi}{\tau_p}+\omega t-\frac{8\pi}{3}\right) \tag{3.26}$$

Adding these three mmfs results in a purely rotating mmf in the direction of the lagging phases with an amplitude equal to 3/2 times the amplitude of the separate alternating mmfs:

$$f_1(x,t) = \frac{3}{2}\hat{F}_1 \cos\left(\frac{x\pi}{\tau_p} - \omega t\right) \tag{3.27}$$

The speed of this rotating field is $v_1 = dx/dt = \omega\tau_p/\pi = 2\tau_p f$.

In general, an m-phase symmetrical winding results in a rotating wave, rotating in the direction of the lagging phases with an amplitude equal to m/2 times the amplitudes of the separate alternating mmfs.

If the single-phase mmfs contain space harmonics ν then a similar approach leads - for a 3-phase winding - to

- absence of 3rd (i.e. $\nu = 3k$) harmonics (for a 3-phase winding only)
- 5th, 11th,... (i.e. $\nu = 6k - 1$ for $k = 1, 2, \ldots$) harmonics that rotate in negative direction (compared to the fundamental)
- 7th, 13th,... (i.e. $\nu = 6k + 1$ for $k = 1, 2, \ldots$) harmonics that rotate in positive direction (compared to the fundamental)

The rotation speed of these harmonics is $(1/\nu)^{th}$ of that of the fundamental: $v = dx/dt = \omega\tau_p/\nu\pi = 2\tau_p f/\nu = v_1/\nu$ (note however that these space harmonic fields have $2\nu N_p$ poles; thus, for each harmonic, an observer on the stator sees passing the same number of poles per second).

If the air-gap is constant (i.e. the effect of the slots can be neglected) and saturation can be neglected, the fundamental mmf and the harmonics result in corresponding rotating induction fields. Saturation and the slotting may add some additional field harmonics.

There exist another, graphical, way to derive and visualise the rotating field of a symmetrical m-phase winding. This graphical method starts from the ampere-turns in the slots. In each slot there are $w_p/q = w/N_p q$ turns carrying a current $i(t)$ - for a 3-phase winding either $\pm i_u(t)$ or $\pm i_v(t)$ or $\pm i_w(t)$). We denote the ampere-turns in a slot k by $a_k = a_k(t) = i(t). \cdot w/q$. We denote now the mmf between slot k and slot $k+1$ by $f_k = f_k(t)$. According to Ampère's law at each slot k the mmf jumps by an amount a_k. Thus, for the mmf difference of the teeth behind the slots n and m we have

$$f_n - f_m = \sum_{k=m+1}^{n} a_k \tag{3.28}$$

Let us now represent the ampere-turns and the mmfs by their (time) phasors $\underline{A}_k = \hat{\underline{I}}_k.w_p/q$ and $\underline{F}_k$ respectively (their phase and length correspond to the phase and amplitude of the time functions $a_k(t)$ and $f_k(t)$ respectively). If we draw the successive current phasors $\underline{A}_k$, they form the sides of a polygon and the mmf phasors are the radii, see Fig. 3.10. For the time being, an arbitrary phasor has been drawn for the *start* mmf phasor F_o.

For the winding configuration in Fig. 3.9, we then get a regular closed polygon (as $\sum \underline{A}_k = 0$) with three (indeed, $q = 3$) current phasors on each side. Moreover, the star of the mmf phasors must be symmetrical as the sum of the mmfs over the

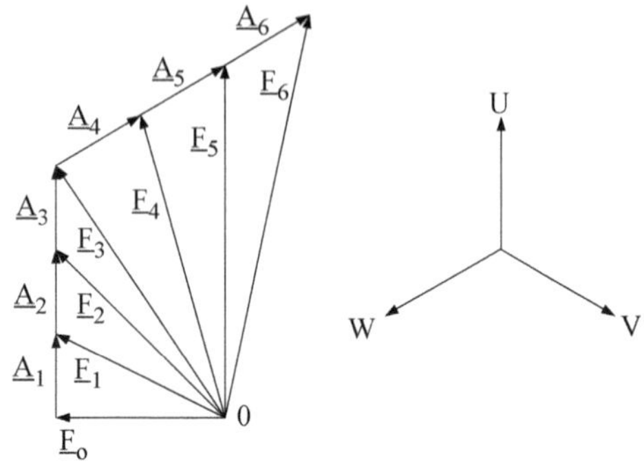

Fig. 3.10 Current layer and mmf phasors

machine is zero at each instant (the field lines are closed over the circumference, i.e. there is no axial flux). Thus the origin of the mmf star is the midpoint of the polygon and the start mmf phasor F_o is determined by this symmetry.

In Fig. 3.11 (left side), the polygon - known in literature as the polygon of Görges - is drawn for three instants in time, i.e. $t = 0, t = \pi/6\omega$ and $t = \pi/3\omega$ (corresponding to maximum current in phase U, zero current in phase V and maximum negative current in phase W respectively). The mmf time values between two slots can be derived by projection of the mmf phasors on the vertical time axis. As is clear from the figures on the right, the mmf curves shift synchronously with the frequency of the currents *in the direction of lagging phases*. The maximum of the mmf curves are always *along the axis of the phase winding where the current is maximum at that instant*. It is true that the shape of the mmf curves varies periodically (with 6 times the mains frequency), but this is due to the mmf harmonics that have different propagation speeds (and which differ from the speed of the fundamental).

3.1.5 Current Layer - Linear Current Density

In the previous sections we considered the real winding and slot configuration (although we neglected the reluctance variation due to the slots and their finite width). For the study of the basic properties of a machine, one may also suppose that the winding is a surface winding distributed evenly over the circumference (i.e. with $q = \infty$ but preserving the same total number of ampere-turns in a phase belt as the real winding with finite q). The result is a (fictitious) stepwise linearly constant current layer with a linear current density expressed in A/m. This comes down to neglecting the mmf harmonics due to the slots.

First we consider a ***single-phase*** $2N_p$ pole winding with in total w turns. In each pole pitch the conductors are distributed over a width S ('phase width'). A DC current i in such a winding results in the total ampere-turns of wi/N_p in a pole pitch.

Now we suppose the conductors to be distributed evenly over the same width S, keeping the total number of ampere-turns constant. This results in a linear current

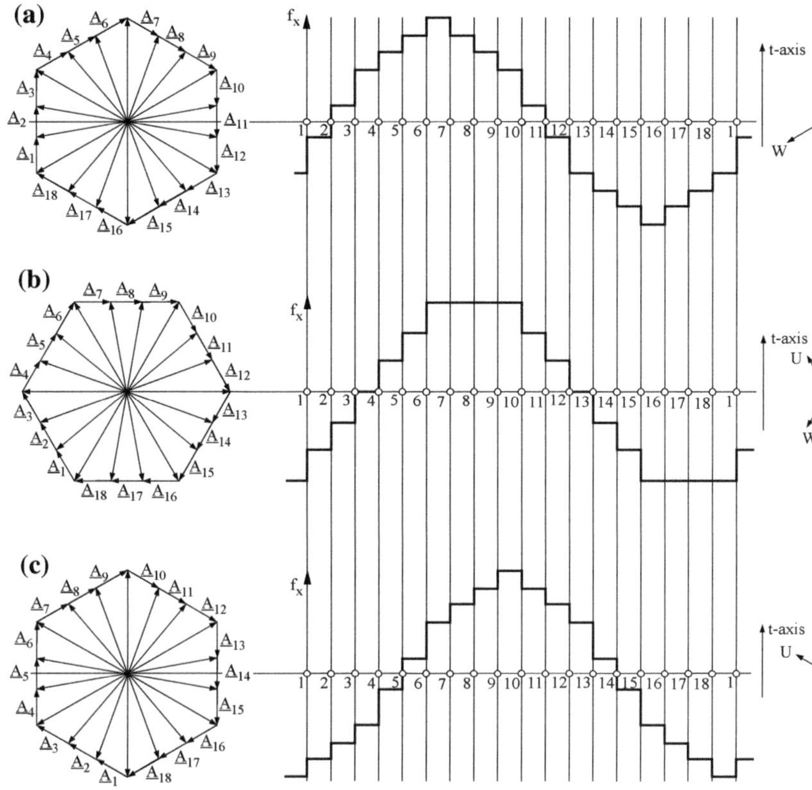

Fig. 3.11 Görges polygons

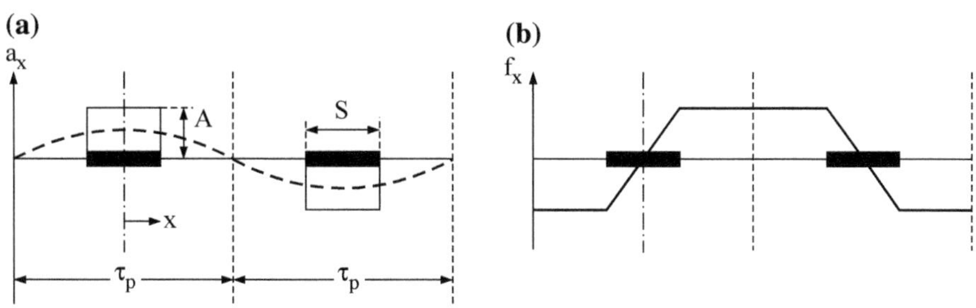

Fig. 3.12 Single-phase current layer

density (see (a) in the outstretched representation in Fig. 3.12), which is constant over the phase width and equal to [A/m]:

$$a = \frac{wi}{N_p S} \tag{3.29}$$

If the current is an AC current, then the height of this block changes with supply frequency and its maximum amplitude is equal to

$$\hat{A} = \frac{w\hat{I}}{N_p S} \tag{3.30}$$

A Fourier analysis of this alternating current layer yields

$$a(x,t) = \sum_{\nu=2k\pm1} \hat{A}_\nu \cdot \cos\frac{\nu x \pi}{\tau_p} \cdot \cos\omega t \tag{3.31}$$

with

$$\hat{A}_\nu = \frac{4}{\nu\pi}\hat{A}.\sin\frac{\nu\beta}{2} = \frac{4}{\nu\pi}\frac{w\hat{I}}{N_p S}\sin\frac{\nu\beta}{2} \tag{3.32}$$

where $\beta = \pi.S/\tau_p$ the phase width S [m] expressed in electrical radians.

The alternating current layer can thus be decomposed in a series of standing waves with wave lengths $\lambda = 2\tau_p/\nu$, all pulsating with mains frequency.

For the fundamental wave, the wave length is $\lambda = 2\tau_p$ and its amplitude is

$$\hat{A}_1 = \frac{4}{\pi}\hat{A}.\sin\frac{\beta}{2} = \frac{4}{\pi}\frac{w\hat{I}}{N_p S}\sin\frac{\beta}{2} = 2\frac{w\xi_1}{N_p\tau_p}\hat{I} \tag{3.33}$$

with the fundamental distribution factor

$$\xi_1 = \frac{\sin\beta/2}{\beta/2} = \lim_{q\to\infty q\alpha=\beta}\frac{\sin q\alpha/2}{q\sin\alpha/2} \tag{3.34}$$

Similar relations hold for the harmonics.

Next we consider a multiphase winding, see Fig. 3.13 for a double pole pitch (outstretched). The winding considered has $S/\tau_p = 1/3$. A similar approach as for a single-phase winding can be adopted. However, we must take into account the time shift between the phase currents and the space shift of the zones of the different phases. This leads to the a-curves of (a) in Fig. 3.13, drawn for $t = 0$, $t = T/12$ and $t = T/6$. It is clear that the curves (and in particular the space fundamental harmonic) move synchronously with the supply frequency in the direction of the lagging phases. The shape is varying somewhat (with 6 times the supply frequency), which is due to the harmonic fields that have different propagation speeds. Note also that the *maximum of the current layer* is always in the *centre of the phase belt* where the current is maximum at that instant.

A Fourier analysis of the current layer results in ($x = 0$ located in the middle of phase belt U):

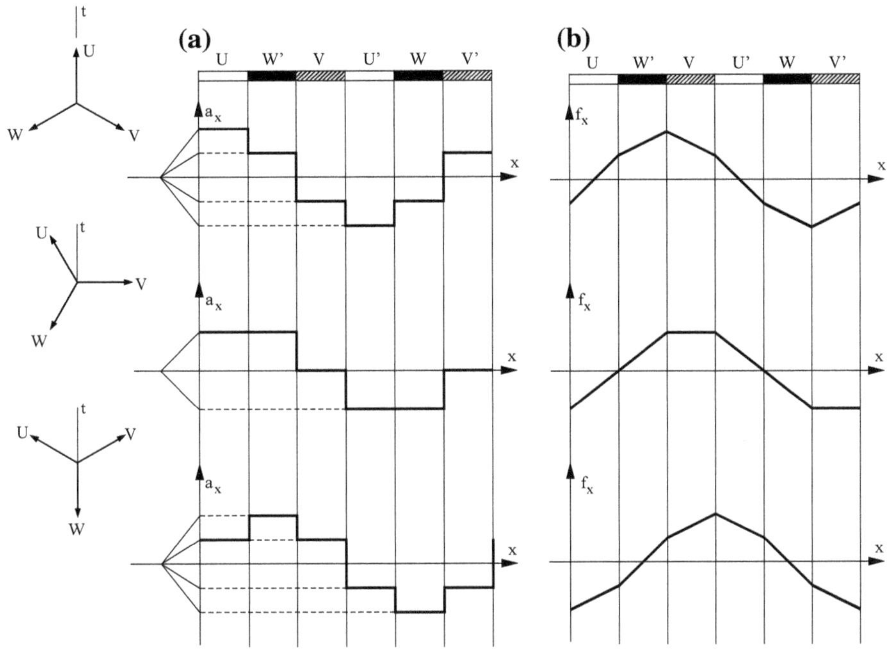

Fig. 3.13 3-phase current layer

$$a_m(x,t) = \sum_{\nu=2k\pm1} a_{\nu,m}(x,t) = \sum_{\nu=2k\pm1} \hat{A}_{\nu,m} \cdot \cos\left(\frac{\nu x \pi}{\tau_p} \pm \omega t\right) \tag{3.35}$$

with

$$\hat{A}_{\nu,m} = \frac{m}{2}.\hat{A}_{\nu} = \frac{m w \xi_{\nu}}{N_p \tau_p} \hat{I} \tag{3.36}$$

For the fundamental we have

$$a_{1,m}(x,t) = \hat{A}_{1,m} \cdot \cos\left(\frac{x\pi}{\tau_p} - \omega t\right) \tag{3.37}$$

with

$$\hat{A}_{1,m} = \frac{m}{2} \cdot A = \frac{m w \xi_1}{N_p \tau_p} \hat{I} \tag{3.38}$$

From the current distribution, the mmf distribution can easily be derived by integration, according to Ampère's law (see also the discrete form in Eq. 3.28):

$$f_x = \int_{x_o}^{x} a.dx \tag{3.39}$$

Normally, symmetry requirements will allow to locate the neutral point where $f_x(x_o) = 0$.

Integration of the current distributions results in the mmf distributions (b) in Figs. 3.12 and 3.13 for the single-phase and three-phase windings respectively. Instead of the stepwise mmf curves in Figs. 3.5 and 3.11, the mmf curves are now piecewise continuous lines.[8]

Fourier analysis yields:

$$f_m(x,t) = \sum_{\nu=2k\pm1} f_{\nu,m}(x,t) = \sum_{\nu=2k\pm1} \hat{F}_{\nu,m} \cdot \cos\left(\frac{\nu x \pi}{\tau_p} \pm \omega t - \frac{\pi}{2}\right) \tag{3.40}$$

with

$$\hat{F}_{\nu,m} = \frac{m w \xi_\nu}{\nu \pi \tau_p} \hat{I} \tag{3.41}$$

For the fundamental we have

$$f_{1,m}(x,t) = \hat{F}_{1,m} \cdot \cos\left(\frac{x\pi}{\tau_p} - \omega t - \frac{\pi}{2}\right) \tag{3.42}$$

with

$$\hat{F}_{1,m} = \frac{m w \xi_1}{\pi \tau_p} \hat{I} \tag{3.43}$$

Note that $\hat{F}_{1,m}$ can be derived directly from the expression for $\hat{A}_1$ by multiplying with m/2 (m-phase instead of single-phase) and with $\tau_p/\nu\pi$ (corresponding with the integration w.r.t. x).

3.1.6 Discussion and Conclusions

Common conclusions for both the real winding in slots and the idealised evenly distributed (surface) winding are:

1. An AC fed single-phase winding gives a standing wave AC current layer with maximum value at the centre of the phase belt. The corresponding mmf curve as well is a standing wave with maximum value at the axis of the winding.
2. A symmetrical multiphase winding fed by a symmetrical multiphase current results in a current layer which is a travelling or rotating wave. The maximum of the current layer is always at the centre of the phase belt where the current is

[8] Indeed, in Figs. 3.5 and 3.11 the jumps of the mmf curve at each slot correspond to the Dirac functions for the ampere-turns in the slots.

maximal *at that instant*. The corresponding travelling or rotating mmf wave has its maximum always at the axis of the phase belt where the current is maximal *at that instant*.
Current layer and mmf move in the direction of the phases that are lagging in time. During each period of the supply a double pole pitch is covered, corresponding with 2π electrical radians. The complete circumference of a $2N_p$ pole machine (corresponding with 2π mechanical degrees or $2N_p\pi$ electrical radians) is passed through in N_p periods of the supply. Thus the mechanical synchronous speed (in rad/s) is equal to $1/N_p$ times the supply pulsation.

3. The amplitude of the fundamental (or of a space harmonic) rotating current layer or mmf is equal to m/2 times the amplitude of the single-phase alternating current layer or mmf respectively (Ferraris' theorem).
4. Compared to a concentrated winding with the same number of turns per phase - and for the same current - the amplitudes of fundamental and harmonics are reduced with the winding factor ξ_ν (cf. effective number of turns $w\xi_\nu \leq w$).
A distributed winding is therefore less *effective* for the fundamental than a concentrated winding. However, with a distributed winding the harmonics are, in general, much more reduced than the fundamental, as shown in Table 3.1 for $S/\tau_p = 1/3$ or $\beta = \pi/3$ (exceptions are the harmonics that correspond to the slot harmonics, i.e. $\nu = k.2mq \pm 1$ or $\nu = k.6q \pm 1$ for m = 3 for which $\xi_\nu = \xi_1$). The higher the rank ν of the harmonics, the more they are reduced. For $q = \infty$ (an evenly distributed surface winding), $\xi_\nu = \xi_1/\nu$ and the mmf harmonics are reduced by $\xi_\nu/\nu = \xi_1/\nu^2$.

Remark: *For a diameter winding (coil width equal to pole pitch) the winding factor is equal to the distribution factor (or zone factor), as was explained above. Normally such a winding will be a so-called single-layer winding, i.e. in each slot there is one coil side. Most multiphase windings (and especially those for larger machine ratings) will use multilayer windings, mostly dual-layer. Such a winding contains two (for a dual layer winding) coil sides per slot. Then the coil width y will be chosen different from the pole pitch, e.g. the pole pitch minus or plus one slot pitch. As a result, at the ends of a phase belt, one will see one or more slots containing coils with currents belonging to different phases. This results in a 'softer' transition to the next phase belt. Mathematically an additional reduction of the harmonics is obtained, i.e. by the chord factor:*

$$\xi_\nu'' = \pm \sin(\pi\nu y/2\tau_p) \tag{3.44}$$

The winding factor is then the product of the previously derived zone factor ξ' (Eq. 3.19 or 3.34) and the chord factor, $\xi = \xi'\xi''$. The chord factor can be used to eliminate specific harmonics.

3.2 Induced Voltage (Electromagnetic Force or emf)

In the previous section, we have seen how a rotating field can be obtained: either by a rotating DC excited rotor or by a multiphase symmetrical winding fed by a symmetrical multiphase AC current. In this section we discuss how a rotating field can induce a (symmetrical) voltage in a (symmetrical) multiphase winding. Where this rotating field originates from, either from the same or another multiphase winding or from a rotating DC-fed rotor, is not important.

3.2.1 *Sinusoidal Rotating Field*

Consider a pure sinusoidal rotating field

$$b_1(x,t) = \hat{B}_1 \cdot \cos\left(\frac{x\pi}{\tau_p} - \omega t\right) \tag{3.45}$$

and a stationary multiphase AC winding (armature). We suppose $x = 0$ along the axis of the reference phase (phase U) of the winding. The origin of the field is unimportant, either by a stationary multiphase armature winding (the same winding or another winding) or by a DC-fed rotating rotor winding but the number of poles of the field and the AC winding should be equal.

The relative motion of the rotating field and the armature winding results in an induced voltage in the conductors of the armature winding. For the one in x:

$$e_{co}(x,t) = k(x,t).l = b_1(x,t) \cdot v \cdot l \tag{3.46}$$

with $e_{co}(x,t)$ and $k(x,t)$ the induced voltage in a single conductor and the voltage induced per meter of conductor length respectively and $b_1(x,t)$ the magnetic field (all in the conductor at x). v is the linear speed of the field with respect to the armature and l the armature length.

Remark that the superposition of a stationary armature and a rotating field is not essential. The same voltage would be induced in a rotating armature and a stationary field (as long as their relative speed is the same); in that case x is of course a rotating coordinate.

The conductor voltage in x is thus a sinusoidal time function and corresponds with the value of the induction in x at time t. The maximum value of the conductor voltage is

$$\hat{E}_{co} = \hat{B}_1 \cdot v \cdot l = \hat{B}_1 \cdot 2\tau_p \cdot f \cdot l = \hat{B}_1 \cdot \frac{\omega\tau_p}{\pi} \cdot l \tag{3.47}$$

or with effective value

$$E_{co} = \frac{1}{\sqrt{2}}\hat{B}_1 \cdot v \cdot l = \frac{1}{\sqrt{2}}\hat{B}_1 \cdot 2\tau_p \cdot f \cdot l = \frac{1}{\sqrt{2}}\hat{B}_1 \cdot \frac{\omega\tau_p}{\pi} \cdot l \tag{3.48}$$

If this conductor belongs to a **concentrated** diameter winding with w turns in series then the voltages in all conductors are in phase (entry and exit conductors are located in opposing field values, but connected in opposite series). The total winding voltage is therefore (in effective value):

$$E = 2w \cdot E_{co} = \frac{1}{\sqrt{2}}\hat{B}_1 \cdot v \cdot l = \frac{4\tau_p l}{\sqrt{2}} \cdot \hat{B}_1 \cdot wf = \frac{2\tau_p l}{\pi\sqrt{2}} \cdot \hat{B}_1 \cdot w \cdot \omega \tag{3.49}$$

In Eq. 3.49 it is supposed that the windings of all pole pairs are series connected and that all pole pairs are identical (as is normally the case). If the windings of different pole pairs are connected in parallel, in Eq. 3.49 w should be replaced by w_p.

We may also calculate the fundamental flux per pole (with $x = 0$ along the axis of the winding):

$$\phi(t) = \phi_1(t) = \int_{-\tau_p/2}^{\tau_p/2} b_1(x, t) \cdot l \cdot dx = \frac{2}{\pi}\hat{B}_1 l \tau_p \cos\omega t \equiv \hat{\Phi}_1 \cos\omega t \tag{3.50}$$

Therefore, we can now rewrite the winding voltage as:

$$E = \frac{1}{\sqrt{2}} \cdot w \cdot \omega \cdot \hat{\Phi}_1 = \frac{2\pi}{\sqrt{2}} \cdot w \cdot f \cdot \hat{\Phi}_1 \ (\approx 4,44 \cdot w \cdot f \cdot \hat{\Phi}_1) \tag{3.51}$$

Note that, although this is a rotational emf, this expression resembles the common expression for a transformer voltage: $e(t) = +w \cdot \frac{d\phi(t)}{dt} = -\hat{E}\sin\omega t$. As we adopt here the URS (users reference system), see the + sign in the second term of the previous equation, the voltage is leading the flux. When we take a look at the space diagram in Fig. 3.14, we observe that the induction at the conductors is maximum $\pi/2$ radians *before* the flux linked with the winding is maximum. Thus, in the time domain, the motional voltage is indeed leading the flux linked with the winding.

Next, we consider a **distributed** diameter winding with q slots per pole and per phase. The voltages induced in the conductors of a phase belonging to different slots (of the same pole pitch) are now not in phase as the field is distributed sinusoidally around the circumference. These voltages are phase-delayed sinusoidal time functions and they must be added vectorially. In case the windings of the N_p pole pairs are series connected, the sum of all these voltages is equal to $2N_p$ times the vectorial sum of the q voltages of *one pole pitch,* as entry and exit conductors of each turn are located under exactly opposite induction values (diameter winding):

Fig. 3.14 emf sign

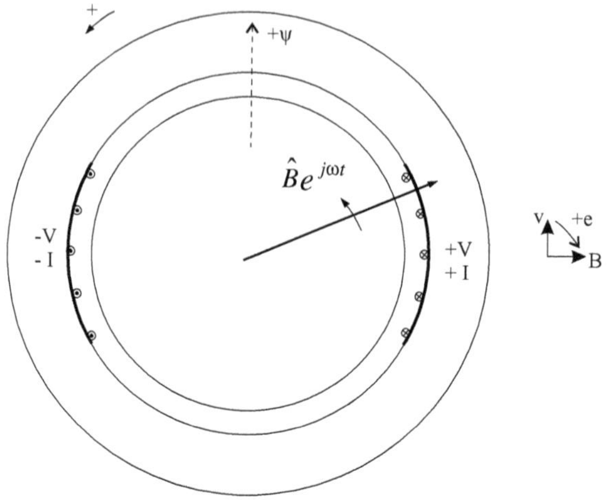

Positive direction of flux follows positive direction of current
In URS positive voltage and current directions in winding are the same
Induced emf sign results from right hand rule and is positive in winding 90 degrees before coupled flux is maximal

$$e_f(t) = 2N_p \cdot \sum \hat{B}_1 \cdot \frac{\omega\tau_p}{\pi} \cdot l \cdot \frac{w}{N_p q} \cdot \cos\left(\omega t - x_j \frac{\pi}{\tau_p}\right) \tag{3.52}$$

Herein w is the number of series connected turns of one phase; $w/N_p q$ is thus the number of series connected conductors of one slot. The phase delay between the slot voltages is $\alpha = \lambda\pi/\tau_p$, corresponding to the displacement of the subsequent slots expressed in electrical radians (see Sect. 3.1.3). The sum in Eq. 3.52 thus leads to the same reduction factor (winding factor ξ_1) as in Sect. 3.1.3:

$$\hat{E}_f = 2\hat{B}_1 \cdot \frac{\omega\tau_p}{\pi} \cdot l \cdot w \cdot \xi_1 \tag{3.53}$$

The time function of this voltage is $e_f(t) = -\hat{E}_f \cdot \sin\omega t$ if $x = 0$ is chosen along the axis of the winding.

Equation 3.53 can also be written as

$$\hat{E}_f = (w \cdot \xi_1) \cdot \omega \cdot \hat{\Phi}_1 \tag{3.54}$$

or

$$E_f = \frac{1}{\sqrt{2}} \cdot (w\xi_1) \cdot \omega \cdot \hat{\Phi}_1 = \frac{2\pi}{\sqrt{2}} \cdot (w\xi_1) \cdot f \cdot \hat{\Phi}_1 \; (\approx 4,44 \cdot w\xi_1 \cdot f \cdot \hat{\Phi}_1) \tag{3.55}$$

with $\hat{\Phi}_1$ the flux of one pole pitch

$$\hat{\Phi}_1 = \frac{2}{\pi}\hat{B}_1 l \tau_p$$

These expressions are similar to those of a concentrated winding, Eq. 3.51, but the number of windings w must be replaced by the number of effective windings $w\xi_1$ (for a chord winding ξ_1 is the product of zone factor and chord factor, see above).

3.2.2 Alternating Field

The voltage induced in an armature winding by an alternating field can be derived from Eqs. 3.51 and 3.55. As seen before an alternating field can be decomposed into two counter-rotating fields with half the amplitude as the original alternating field, see (a) in Fig. 3.15. Each of these alternating fields induces a sinusoidal voltage in the armature winding with amplitude given by

$$\hat{E}' = (w\xi) \cdot \omega \cdot \hat{\Phi}/2 \tag{3.56}$$

If the alternating field is directed along the winding axis, these two voltages are in phase and the amplitude of the total induced voltage is given by

$$\hat{E} = (w\xi) \cdot \omega \cdot \hat{\Phi} = \omega \cdot \hat{\Psi} \tag{3.57}$$

This voltage can also be interpreted as a transformer voltage, where $\hat{\Psi} = (w \cdot \xi) \cdot \hat{\Phi}$ is the amplitude of the alternating flux coupled with the armature winding. Here, the winding factor takes into account that for a distributed winding the axes of the q coils

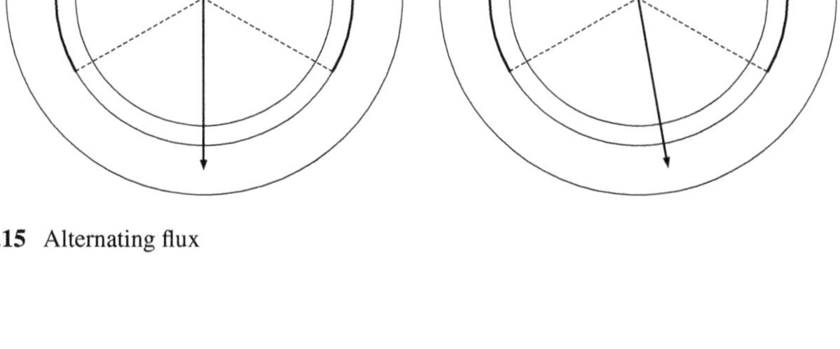

Fig. 3.15 Alternating flux

are not all aligned with the flux axis. The alternating flux coupled with them is then lower by the cosine of the inclination angle.

If the winding axis is inclined over an angle γ with respect to the flux axis, see (b) in Fig. 3.15, the induced voltage of Eq. 3.57 (or the coupled flux) must be multiplied by $\cos\gamma$.

3.2.3 Non-sinusoidal Field

For a non-sinusoidal field, the induced voltages are calculated from the Fourier components of the field. The calculation should then take into account the wave length and the speed of these harmonic fields.

First, we consider the mmf (or field) harmonics from an AC-fed multiphase winding (the same winding or another). In this case the νth harmonic has a wave length of $\lambda_\nu = 2\tau_p/\nu$ (i.e. N_p pole pairs), but a speed of $1/\nu$th of the fundamental speed. Thus all these harmonics induce voltages in the armature with the *same frequency* as the fundamental

$$\hat{E}_{f\nu} = (w \cdot \xi_\nu) \cdot \omega \cdot \hat{\Phi}_\nu \tag{3.58}$$

or

$$E_{f\nu} = \frac{1}{\sqrt{2}}(w \cdot \xi_\nu) \cdot \omega \cdot \hat{\Phi}_\nu \tag{3.59}$$

with $\xi_\nu = \xi_\nu^{'} \cdot \xi_\nu^{"}$ (see Eqs. 3.19 and 3.44) and $\hat{\Phi}_\nu = \frac{2}{\pi} \cdot \frac{\tau_p}{\nu} \cdot l \cdot \hat{B}_\nu$

Thus for the ratio of the harmonic emf to the fundamental emf we obtain

$$\frac{E_{f\nu}}{E_{f1}} = \frac{\xi_\nu}{\xi_1} \cdot \frac{\hat{\Phi}_\nu}{\hat{\Phi}_1} = \frac{\xi_\nu}{\xi_1} \cdot \frac{\hat{B}_\nu}{\nu \hat{B}_1} \tag{3.60}$$

The factor ν in the denominator stems from the flux calculation where $\nu - 1$ half pulses annihilate (ν is always odd).

We remark that the already small field harmonics from a distributed winding (reduced by $1/\nu\xi_\nu^\star \approx 1/\nu^2$ for large q, where $\xi_\nu^\star$ is the winding factor of the field winding) are again reduced by a factor $1/\nu\xi_\nu \approx 1/\nu^2$. Thus the total reduction is about $(1/\nu)^4$. For large q (in both field winding and armature), the emf harmonics are therefore negligible.

For the field harmonics stemming from a rotating DC-fed excitation winding the result is much less rosy. First of all, the amplitudes of these field harmonics are only $(1/\nu)$th of the fundamental amplitude. Secondly, these field harmonics move with the same speed as the fundamental, i.e. the speed of the excitation winding. Thus they induce emfs with a frequency ω_ν which is ν times the fundamental emf frequency. Therefore

$$\hat{E}_{f\nu} = (w \cdot \xi_\nu) \cdot \omega_\nu \cdot \hat{\Phi}_\nu \tag{3.61}$$

or

$$E_{f\nu} = \frac{1}{\sqrt{2}}(w \cdot \xi_\nu) \cdot \omega_\nu \cdot \hat{\Phi}_\nu \tag{3.62}$$

with $\omega_\nu = \nu \cdot \omega$ and $\hat{\Phi}_\nu = \frac{2}{\pi} \cdot \frac{\tau_p}{\nu} \cdot l \cdot \hat{B}_\nu$.

Thus we obtain for the ratio of the harmonic emf to the fundamental emf induced in the armature winding:

$$\frac{E_{f\nu}}{E_{f1}} = \frac{\xi_\nu}{\xi_1} \cdot \omega \cdot \nu \cdot \frac{\hat{\Phi}_\nu}{\hat{\Phi}_1} = \frac{\xi_\nu}{\xi_1} \cdot \frac{\hat{B}_\nu}{\hat{B}_1} \tag{3.63}$$

For an armature winding with large q the reduction is thus only $(1/\nu)^2$.

Thus the higher the number of slots per pole and per phase (q), the more the higher harmonics in the induced voltage are suppressed. There is one exception: for the slot harmonics the winding factor is equal to the fundamental winding factor. We note also that, if third harmonics are present in the field curve (e.g. with a concentrated field winding), third harmonic voltages may be induced in the armature winding. However, in the line voltages these third harmonic voltages are not present as they cancel out. For an armature winding with $S/\tau p = 2/3$ (a rather special two-layer winding), the third harmonics are never present, as in that case $\xi'_\nu = 0$ for $\nu = 3$k.

3.3 Magnetising Inductance of an Armature Winding

The magnetising inductance of a winding can be calculated from the ratio of coupled air-gap flux and excitation current, or from the ratio of induced emf and current (multiplied by the angular frequency).

3.3.1 Single-Phase Winding

Consider a single-phase $2N_p$-pole winding with w in series connected turns. An AC current, e.g. $i = \hat{I} \cos \omega t$, in this winding gives an air-gap magnetic field with an amplitude of the fundamental equal to (see also Eq. 3.21):

$$\hat{B}_1 = \mu_o \cdot \frac{2}{\pi} \cdot \frac{w\xi_1}{\delta_p N_p} . \hat{I} \tag{3.64}$$

The flux coupled with the winding is equal to

$$\hat{\Psi} = (w\xi_1) \cdot \hat{\Phi} = (w\xi_1) \cdot \frac{2}{\pi} \cdot \tau_p \cdot l \cdot \hat{B}_1 \tag{3.65}$$

The magnetising inductance of the winding for the fundamental field is therefore (dimensions [H]):

$$L_1 = \frac{\hat{\Psi}}{\hat{I}} = \frac{8}{\pi^2} \cdot \mu_o \cdot \frac{\tau_p \cdot l}{2N_p \cdot \delta_p} \cdot (w\xi_1)^2 \tag{3.66}$$

In this as well as the next section, we may denote this magnetising inductance also as the self-inductance of the single-phase winding, because the leakage field is not considered.

3.3.2 *Multiphase Winding*

We consider now a multiphase symmetrical winding fed by a multiphase symmetrical current. The resulting rotating magnetic field has an amplitude equal to m/2 times the amplitude of each composing alternating field (of the m single-phase windings). Of course this holds also for the corresponding fluxes. The (so-called cyclic) magnetising inductance for a multiphase winding, which is the ratio between the *m-phase* rotating *flux amplitude* and the *(single-phase)* current amplitude in one of the phase windings is therefore m/2 times the single-phase magnetising inductance of the previous section:

$$L_{m1} = \frac{\hat{\Psi}}{\hat{I}} = \frac{m}{2} \cdot \frac{8}{\pi^2} \cdot \mu_o \cdot \frac{\tau_p \cdot l}{2N_p \cdot \delta_p} \cdot (w\xi_1)^2 \tag{3.67}$$

This cyclic magnetising inductance, is by a factor of m/2 larger than the single-phase magnetising inductance (self-inductance). This is caused by the other phases which contribute to the field and thus to the coupled flux, while the current in the phase remains the same as in the single-phase case.

This result can also be derived by considering the inductance matrix for the set of coupled three-phase windings. For example, for a symmetrical three-phase winding the mutual inductance M_1 between the phases is equal to minus half the single-phase self-inductance L_1, i.e. $M_1 = -L_1/2$ (because of the space shift of $2\pi/3$). Taking into account the symmetric three-phase currents (the sum of the three currents is always zero), it is easily shown that the cyclic inductance is 3/2 times the self-inductance.

Remark: One may also derive a magnetising inductance or reactance per square meter of air-gap. For an air-gap winding (uniformly spread, $q = \infty$) we may write for the ratio of the emf per meter armature length to the current per meter of armature circumference

$$X_m^{\square} = \frac{\hat{K}_1}{\hat{A}_1} = \frac{\frac{\omega\tau_p}{\pi}\cdot\hat{B}_1}{\hat{A}_1} = \frac{\frac{\omega\tau_p}{\pi}\cdot\frac{\mu_o}{\delta_p}\cdot\frac{\tau_p}{\pi}\hat{A}_1}{\hat{A}_1} = \omega\cdot\mu_o\cdot\left(\frac{\tau_p}{\pi}\right)^2\cdot\frac{1}{\delta_p} \tag{3.68}$$

An actual slot winding with a finite number of slots may, for the fundamental flux, also be treated as an air-gap winding. Indeed, we have (with $\xi_1$the actual winding factor for finite q)[9]

$$\hat{A}_1 = \frac{m}{2}\cdot\frac{(2w\xi_1)\cdot\hat{I}_f}{N_p\tau_p} = \frac{(2w\xi_1)\cdot\hat{I}_f}{2N_p(\tau_p/m)}$$

3.4 Torque

In the previous sections we have seen that a symmetrical multiphase current in a symmetrical multiphase winding generates a rotating field. Further, we derived that a symmetrical emf is induced in a symmetrical multiphase winding when it is located in a rotating magnetic field. Note that current and voltage (emf) occurring at the same time in the same winding will (or rather may) correspond to electrical power.

In this section we will analyse the torque that will (or may) come into being when a current carrying (symmetrical multiphase) winding is located in a (rotating) magnetic field. This torque will (or may) give rise to mechanical power).

3.4.1 General

We start by considering a general air-gap field $b(x,t)$ (oriented orthogonally on the stator or rotor surface) and a current layer $a(x,t)$ (in a surface or slot winding on stator or rotor). The magnetic field exerts a tangential force on the current and therefore a torque will result. This torque can be calculated as follows.

The elementary tangential force, and thus torque, on an elementary segment dx can be calculated according to Lorentz' law:

$$dF = b(x,t)\cdot a(x,t)\cdot l\cdot dx \tag{3.69}$$

$$dT = \frac{N_p\tau_p}{\pi}\cdot b(x,t)\cdot a(x,t)\cdot l\cdot dx \tag{3.70}$$

The total equivalent tangential force and the torque can then be calculated by integration over the circumference, or, because of symmetry, as $2N_p$ times the integral over one pole pitch:

---

[9]Give an interpretation of the last term?

$$F = 2N_p \int_{-\tau_p/2}^{\tau_p/2} b(x,t) \cdot a(x,t) \cdot l \cdot dx \tag{3.71}$$

$$T = \frac{N_p \tau_p}{\pi} \cdot F = \frac{N_p \tau_p}{\pi} \cdot 2N_p \int_{-\tau_p/2}^{\tau_p/2} b(x,t) \cdot a(x,t) \cdot l \cdot dx \tag{3.72}$$

In these equations $b(x,t)$ and $a(x,t)$ can be replaced by their Fourier expansion in space harmonics. Force and torque are then obtained as the sum of integrals of products of harmonics of induction and current layer. Because of the integration only the products of harmonics with the same order (in space) may give a non-negative result.

In the next sections we will apply this to alternating and rotating field and current layers.

3.4.2 Alternating Field and Alternating Current Layer

Consider the space-fundamental harmonics of an alternating field distribution and alternating current layer (with the same number of poles). A constant space angle ψ between the axes of field and current layer is assumed. For the time being the time functions of field and current are left unspecified.

$$b_1(x,t) = B_1(t) \cdot \cos \frac{x\pi}{\tau_p} \tag{3.73}$$

$$a_1(x,t) = A_1(t) \cdot \cos \left(\frac{x\pi}{\tau_p} - \psi \right) \tag{3.74}$$

The resulting torque (and tangential force) is then:

$$T = \frac{N_p \tau_p}{\pi} \cdot F = \frac{N_p \tau_p}{\pi} \cdot A_1(t) \cdot B_1(t) \cdot 2N_p \int_{-\tau_p/2}^{\tau_p/2} \cos \frac{x\pi}{\tau_p} \cdot \cos \left(\frac{x\pi}{\tau_p} - \psi \right) \cdot l \cdot dx \tag{3.75}$$

or

$$T = \frac{N_p \tau_p}{\pi} \cdot F = \frac{N_p \tau_p}{\pi} \cdot N_p \tau_p \cdot l \cdot A_1(t) \cdot B_1(t) \cdot \cos \psi \tag{3.76}$$

Equation 3.76 can also be rewritten in terms of the pole flux $\Phi_1(t)$:

$$T = \frac{1}{2} N_p^2 \tau_p \cdot A_1(t) \cdot \Phi_1(t) \cdot \cos \psi \tag{3.77}$$

Consider now the case that the time functions of induction and current are sinusoidal with a time shift α

$$B_1(t) = \hat{B}_1 \cdot \cos \omega t \tag{3.78}$$

$$A_1(t) = \hat{A}_1 \cdot \cos(\omega t - \alpha) \tag{3.79}$$

This yields

$$T = \frac{1}{4} N_p^2 \tau_p \cdot \hat{A}_1 \cdot \hat{\Phi}_1 \cdot \cos \psi \cdot \cos \alpha + \frac{1}{4} N_p^2 \tau_p \cdot \hat{A}_1 \cdot \hat{\Phi}_1 \cdot \cos \psi \cdot \cos(2\omega t - \alpha) \tag{3.80}$$

An alternating field and alternating (i.e. single-phase) current layer thus results in an average torque (first term), proportional to the cosine of the space shift angle and the cosine of the time shift angle. In addition there is a pulsating torque with twice the supply (mains) frequency (second term). The latter is of course an unwanted disturbance (giving rise to vibrations and noise).

For a space shift $\psi = \pi/2$, i.e. the axes of current layer (i.e. in the centre of the phase belt) and flux are orthogonal, the total torque is identically zero (*both* average torque and pulsating torque). In fact, in this case the maximum of the field occurs where the current is zero (and *vice versa*).

For a time shift $\alpha = \pi/2$, the average torque is zero but here remains nonetheless a non-zero pulsating torque (unless $\psi = \pi/2$).

For the space harmonics of field and current layer (same number of poles), a similar discussion can be held (but only the harmonics with the same rank will or may give a non-zero result).

3.4.3 Rotating Field and Rotating Current Layer

Next, we consider a multiphase winding carrying a symmetrical multiphase current and a rotating field (with the same number of poles). For the space fundamentals of these we may write for example

$$b_1(x,t) = \hat{B}_1 \cdot \cos\left(\frac{x\pi}{\tau_p} - \omega t\right) \tag{3.81}$$

$$a_1(x,t) = \hat{A}_1 \cdot \cos\left(\frac{x\pi}{\tau_p} - \omega t - \psi\right) \tag{3.82}$$

For the torque we obtain now

$$T = \frac{2N_p^2\tau_p}{\pi} \cdot l \cdot \hat{A}_1 \cdot \hat{B}_1 \cdot \int_{-\tau_p/2}^{\tau_p/2} \frac{1}{2}\left[\cos\psi + \cos\left(\frac{2x\pi}{\tau_p} - 2\omega t - \psi\right)\right] dx$$
$$= \frac{1}{2} N_p^2 \tau_p \cdot \hat{A}_1 \cdot \hat{\Phi}_1 \cdot \cos\psi \qquad (3.83)$$

In contrast with the case of alternating field and current layer, the torque for rotating field and current layer is constant and does not contain a pulsating component.

It is quite interesting to investigate the relation between the mechanical quantities torque-speed on the one hand and the electrical quantities current-emf on the other hand.

The rotating field induces an emf in the winding given by (see Eq. 3.54)

$$\hat{E}_{f1} = (w \cdot \xi_1) \cdot \omega \cdot \hat{\Phi}_1 \qquad (3.84)$$

The current layer corresponds to a phase current in the (same) winding (see Eq. 3.38)

$$\hat{A}_1 = \frac{m w \xi_1}{N_p \tau_p} \hat{I}_f \qquad (3.85)$$

Substitution in Eq. 3.83 gives

$$T \cdot (\omega/Np) = \frac{m}{2} \cdot \hat{E}_{f1} \cdot \hat{I}_f \cdot \cos\psi \qquad (3.86)$$

or

$$P_{em} = T \cdot \Omega = m E_{f1} \cdot I_f \cdot \cos\psi \qquad (3.87)$$

The left hand side of Eq. 3.87 is the product of the ***electromagnetic torque on the winding*** and the ***speed of the rotating field with respect to this winding*** (in mechanical radians per second). The right hand side resembles an electrical power. However, the angle ψ came into play as the space angle between the magnetic field and the current layer. It is easily shown, however, that ψ is also the time delay angle between emf and current. Indeed, at the instant the field axis is at the centre of the (positive) phase belt of the reference winding, the emf induced in this reference winding is maximal (as the conductors are then located in maximum field strength). Further, at the instant the axis of the current layer is along the centre of the (positive) phase belt of this reference winding, the current in this reference winding is maximal (see Sect. 3.1.4). Thus, if there is a space angle ψ between field and current layer axes, the corresponding emf and current in the reference winding will have a time delay of ψ as well, as both field and current layer rotate synchronously with synchronous speed (Fig. 3.16).

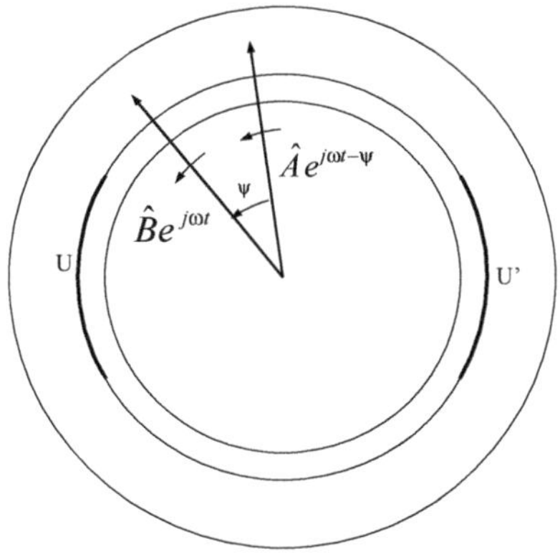

Fig. 3.16 Rotating field power

The right hand side of Eq. 3.87 is therefore the electrical power of the voltage induced by the field with the current in this reference winding. It is called "rotating field power" or "electromagnetic power" and it is basic to the operation of all rotating field machines (induction machines, synchronous machines and the now obsolete multiphase commutator machines). It is the electrical power that is converted by the rotating field in the current-carrying winding. But it is not necessarily a mechanical power (left hand side) as the synchronous speed of the field is not necessarily the speed of a mechanical part (but it can be).

Remark:

Equation 3.72 can also be applied to derive the torque of a commutator machine, both DC and AC (although this is not a rotating field machine).

For a commutator machine the current layer is fixed in space, i.e. between subsequent brushes the current layer has the same instantaneous value: $a(x, t) = A(t)$. Thus

$$T = \frac{2}{\pi} N_p^2 \tau_p \cdot \int_0^{\tau_p} A(t) \cdot b(x, t) \cdot l \cdot dx = \frac{2}{\pi} N_p^2 \tau_p \cdot A(t) \cdot \Phi(t) \tag{3.88}$$

with $\Phi(t)$ the physical flux entering the armature between two subsequent brushes. This flux is independent of the space shape of the field (just like in the DC machine: i.e. a decomposition into space harmonics is not required, it's only the total flux that comes into play). For a DC commutator machine, Eq. 3.88 leads to a constant torque, but for AC commutator machines a pulsating torque component will be present.

Chapter 4
The Induction Machine

Abstract Nowadays, the induction machine is by far the most commonly used electrical machine in electrical drives (for applications that require a highly dynamic behaviour or where energy efficiency or compactness is primordial the permanent magnet synchronous motor may be preferred, however). The main reasons for this are its straightforward and robust construction and its quite efficient energy conversion. Moreover, the last 30 or so years variable speed operation of induction machines using power electronic converters has ousted almost completely the DC commutator machine in variable speed applications. This chapter starts from the traditional transformer properties of an induction motor at standstill. Then, the operating principle of an induction machine is explained both intuitively and from a more mathematical point of view. In the subsequent sections we treat the energy conversion and torque, equivalent circuits and equations for an induction machine, the current locus and single-phase induction machines. Much attention is paid to per-unit values and scaling laws as these determine the behaviour of the machine to a great extent.

4.1 Construction

The main components of a (multiphase) induction machine are

- the primary, mostly the stator, which contains a multiphase symmetrical winding (m_1 phases, N_p pole pairs)
- the secondary, mostly on the rotor, containing either a multiphase symmetrical winding (m_2 phases, N_p pole pairs) or a squirrel cage.

Other construction details of the electromagnetic and mechanical parts and components are illustrated in Fig. 4.1.

J. A. Melkebeek, *Electrical Machines and Drives*, Power Systems,
https://doi.org/10.1007/978-3-319-72730-1_4

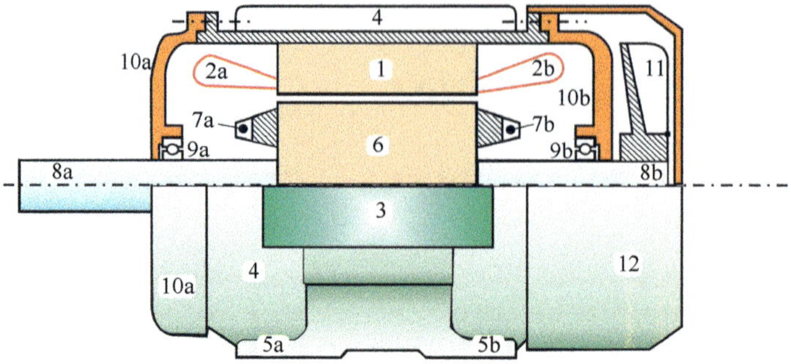

1	laminated stator core with 3-phase winding in slots
2a, 2b	end-turns of stator winding
3	connection box
4	case with fins
5a, 5b	feet
6	laminated rotor core with bars in slots (squirrel-cage rotor)
7a, 7b	(short circuit) end-rings with cooling blades
8a, 8b	shaft with shaft end
9a, 9b	bearings
10a, 10b	bearing caps
11	ventilator
12	ventilator cap

Fig. 4.1 Induction motor

4.2 Transformer Properties of the Induction Machine at Standstill

The operating principle of an induction machine can be explained most conveniently with the rotating field transformer. A rotating field transformer is in fact an induction machine with on both stator and rotor a symmetrical multiphase winding (with of course the same number of pole pairs and, in most cases, also the same number of phases), but where the rotor is at standstill (blocked). The rotor can be rotated in different positions using worm gear and pinion.

4.2.1 The Axes of Stator and Rotor Windings Are Co-linear

First, we consider the case that the magnetic axes of the reference windings (1U and 2U) of stator and rotor are co-linear, Fig. 4.2.

The stator is fed by a symmetrical AC voltage source with frequency $f_1 = \omega_1/2\pi$, while the rotor windings are for the moment open-circuited. The stator windings will then draw a symmetrical m_1-phase current $i_1(t)$ from the supply; for the 3-phase case thus

$$i_u(t) = \hat{I}_1 \cos\omega_1 t \; ; \; i_v(t) = \hat{I}_1 \cos(\omega_1 t - 2\pi/3) \; ; \; i_w(t) = \hat{I}_1 \cos(\omega_1 t - 4\pi/3) \quad (4.1)$$

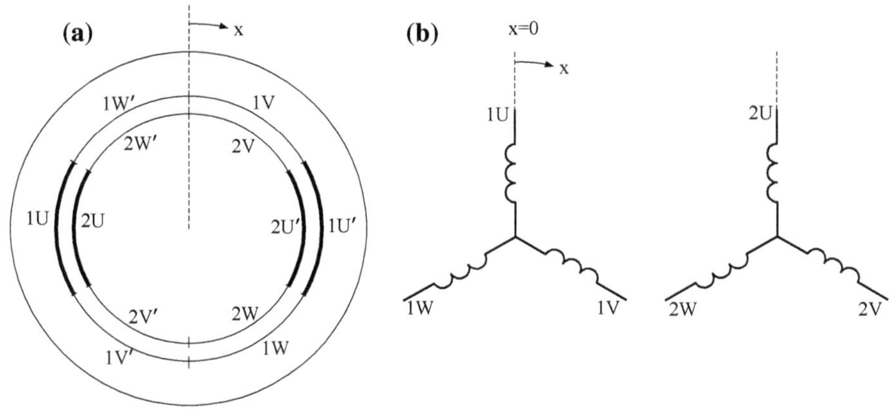

Fig. 4.2 Windings axes co-linear

This current results in a fundamental current layer (where $x = 0$ is chosen along the winding axis of phase 1U)

$$a_1(x,t) = \hat{A}_1 \sin\left(\omega_1 t - \frac{x\pi}{\tau_p}\right) \tag{4.2}$$

and therefore also a fundamental rotating field

$$b_1(x,t) = \hat{B}_1 \cos\left(\omega_1 t - \frac{x\pi}{\tau_p}\right) \tag{4.3}$$

with

$$\hat{B}_1 = \frac{\mu_o}{\delta_p} \cdot \hat{F}_1 = \frac{\mu_o}{\delta_p} \cdot \frac{\tau_p}{\pi} \cdot \hat{A}_1 = \frac{\mu_o}{\delta_p} \cdot \frac{\tau_p}{\pi} \cdot \frac{m_1 w_1 \xi_1}{N_p \tau_p} \cdot \hat{I}_1 \tag{4.4}$$

Herein we neglected the reluctance (and saturation) of the iron core; w_1 is the number of series connected turns per phase of the stator (primary), which corresponds to w in Chap. 3.

The rotating field induces a symmetrical emf in the primary winding which, for the reference winding 1U, can be written as[1] (see Sect. 3.2.1)

$$e_1(t) = e_{11}(t) = \frac{d\psi_{11}(t)}{dt} = w_1 \xi_1 \cdot \frac{d\phi_{11}(t)}{dt} \tag{4.5}$$

with ψ_{11} the air-gap flux coupled with phase 1U (due to current in the primary) and ϕ_{11} the physical air-gap flux over one pole pitch along the axis of phase 1U

[1]This is the emf in the reference winding 1U; the first index 1 indicates the winding where it is induced while the second index 1 indicates where the flux originates.

$$\phi_{11}(t) = \phi_{m1}(t) = \int_{-\tau_p/2}^{\tau_p/2} b_1(x,t) \cdot l \cdot dx = \frac{2}{\pi} \hat{B}_1 l \tau_p \cos\omega_1 t \equiv \hat{\Phi}_{m1} \cos\omega_1 t \quad (4.6)$$

The index m is used to indicate that this flux is the magnetising flux (later on we will introduce another flux, the leakage flux).

This coupled magnetising flux is an alternating flux which can also be written as (see Sect. 3.2.1)

$$\psi_{11}(t) = L_{m1} \cdot i_1(t) \quad (4.7)$$

with the (cyclic) magnetising inductance given by

$$L_{m1} = \frac{m_1}{2} \cdot \frac{8}{\pi^2} \cdot \mu_o \cdot \frac{\tau_p \cdot l}{2N_p \cdot \delta_p} \cdot (w_1 \xi_1)^2 \quad (4.8)$$

The field of the primary current also induces a voltage in the secondary winding, given by (for the reference phase 2U)

$$e_2(t) = e_{21}(t) = \frac{d\psi_{21}(t)}{dt} = w_2 \xi_2 \cdot \frac{d\phi_{21}(t)}{dt} \quad (4.9)$$

As the winding axes of primary and secondary are aligned, the flux $\phi_{21}(t)$ is equal to the flux $\phi_{m1}(t)$. For the flux coupled with the secondary $\psi_{21}(t) = w_2\xi_2 \cdot \phi_{m1}(t)$ we may also write

$$\psi_{21}(t) = M_{21} \cdot i_1(t) \quad (4.10)$$

with the (cyclic) mutual inductance M_{21} between primary and secondary reference phases given by

$$M_{21} = L_{m1} \cdot \frac{w_2 \xi_2}{w_1 \xi_1} \quad (4.11)$$

Remark that the ratio of the induced voltages is equal to the (effective) turns ratio, similarly as in a transformer

$$\frac{e_{11}(t)}{e_{21}(t)} = \frac{w_1 \xi_1}{w_2 \xi_2} \quad (4.12)$$

When the rotating field results from a symmetrical multiphase (m_2 phases) current $i_2(t)$ in the secondary winding instead of the primary, similar equations result, i.e.

$$e_2(t) = e_{22}(t) = \frac{d\psi_{22}(t)}{dt} = w_2 \xi_2 \cdot \frac{d\phi_{m2}(t)}{dt} \quad (4.13)$$

$$e_1(t) = e_{12}(t) = \frac{d\psi_{12}(t)}{dt} = w_1 \xi_1 \cdot \frac{d\phi_{m2}(t)}{dt} \quad (4.14)$$

with (assuming the current reference directions are so that positive currents in both windings give a field with the same polarity)

$$\psi_{22}(t) = L_{m2} \cdot i_2(t) \tag{4.15}$$

$$L_{m2} = \frac{m_2}{2} \cdot \frac{8}{\pi^2} \cdot \mu_o \cdot \frac{\tau_p \cdot l}{2N_p \cdot \delta_p} \cdot (w_2\xi_2)^2 \tag{4.16}$$

$$\psi_{12}(t) = M_{12} \cdot i_2(t) \tag{4.17}$$

$$M_{12} = L_{m2} \cdot \frac{w_1\xi_1}{w_2\xi_2} \tag{4.18}$$

In complex (time phasor) notation the preceding equations can be summarised as

$$\underline{E}_{11} = j\omega_1 L_{m1}\underline{I}_1 \tag{4.19}$$

$$\underline{E}_{21} = j\omega_1 M_{21}\underline{I}_1 \tag{4.20}$$

$$\underline{E}_{22} = j\omega_1 L_{m2}\underline{I}_2 \tag{4.21}$$

$$\underline{E}_{12} = j\omega_1 M_{12}\underline{I}_2 \tag{4.22}$$

with[2] $\frac{L_{m1}}{M_{21}} = \frac{w_1\xi_1}{w_2\xi_2}, \frac{L_{m2}}{M_{12}} = \frac{w_2\xi_2}{w_1\xi_1}$.

Apparently both windings behave as the primary and secondary of a multiphase transformer, i.e. the ratio of the induced voltages is equal to the (effective) turns ratio. The inductances as well are similar if the number of phases is equal.

Next, suppose that both windings carry symmetrical multiphase currents (with the same frequency and phase sequence). This could be the case when for example the primary is fed from the mains and the secondary is connected to a symmetrical load. In that case each of both windings produces a rotating mmf. These two mmfs $f_1(x,t)$ and $f_2(x,t)$ (or in space vector notation $\underset{\rightarrow}{F}_1$ and $\underset{\rightarrow}{F}_2$) rotate in the same direction with the same speed ω_1/N_p and they can therefore be added vectorially to a resulting mmf $\underset{\rightarrow}{F}_m$. This resulting mmf $\underset{\rightarrow}{F}_m$ produces a resulting rotating magnetic field $\underset{\rightarrow}{B}_{res}$. This resulting rotating magnetic field induces the multiphase symmetrical emfs $e_1(t)$ and $e_2(t)$ (or in phasor form $\underline{E}_1$ and $\underline{E}_2$) in the primary and secondary windings respectively.

When the magnetic circuit is linear the resulting magnetic field $b_{res}(x,t)$ can be obtained as the (vectorial) sum of the magnetic fields due to the separate currents $i_1(t)$ and $i_1(t)$ in both windings. Under the same simplifying assumption, this also holds for the emfs $e_1(t)$ and $e_2(t)$ and we may write

[2]Calculate the ratio of the (cyclic) magnetising field inductances. If the number of phases of stator and rotor is the same, would you indeed expect this result?

$$\underline{E}_1 = \underline{E}_{11} + \underline{E}_{12} = j\omega_1 L_{m1}\underline{I}_1 + j\omega_1 M_{12}\underline{I}_2 \tag{4.23}$$

$$\underline{E}_2 = \underline{E}_{22} + \underline{E}_{21} = j\omega_1 L_{m2}\underline{I}_2 + j\omega_1 M_{21}\underline{I}_1 \tag{4.24}$$

or

$$\underline{E}_1 = j\omega_1 L_{m1}\left[\underline{I}_1 + \frac{m_2 w_2 \xi_2}{m_1 w_1 \xi_1}\underline{I}_2\right] \tag{4.25}$$

$$\underline{E}_2 = j\omega_1 L_{m2}\left[\underline{I}_2 + \frac{m_1 w_1 \xi_1}{m_2 w_2 \xi_2}\underline{I}_1\right] = j\omega_1 L_{m1}\cdot\frac{w_2\xi_2}{w_1\xi_1}\left[\underline{I}_1 + \frac{m_2 w_2 \xi_2}{m_1 w_1 \xi_1}\underline{I}_2\right] \tag{4.26}$$

or

$$\underline{E}_1 = j\omega_1 L_{m1}\underline{I}_{m1} \tag{4.27}$$

$$\underline{E}_2 = \frac{w_2\xi_2}{w_1\xi_1}\cdot j\omega_1 L_{m1}\underline{I}_{m1} = \frac{w_2\xi_2}{w_1\xi_1}\cdot\underline{E}_1 = \underline{E}_1^{'} \tag{4.28}$$

with the prime indicating quantities referred to the other winding and $\underline{I}_{m1}$ denoting the magnetising current referred to the primary winding:

$$\underline{I}_{m1} = \left[\underline{I}_1 + \frac{m_2 w_2 \xi_2}{m_1 w_1 \xi_1}\underline{I}_2\right] = \underline{I}_1 + \underline{I}_2^{'} \tag{4.29}$$

We conclude that the rotating field transformer behaves like a normal transformer. For the voltages the conversion ratio from secondary to primary is $\frac{w_1\xi_1}{w_2\xi_2}$ while for the currents the conversion ratio also depends on the number of phases, $\frac{m_2 w_2 \xi_2}{m_1 w_1 \xi_1}$. If the primary and secondary have the same number of phases then we obtain almost identical conversion ratios as for a transformer (replacing only the number of turns by the effective number of turns). At the end of the next section a set of equations for the rotating field transformer will be derived.

4.2.2 The Axes of Stator and Rotor Windings Are Displaced

Next, we study the case where the reference winding axis of the rotor winding is rotated by an electrical angle $\alpha = N_p \cdot \theta$ with respect to the reference winding axis of the stator, see Fig. 4.3. $\alpha = N_p \cdot \theta$ is considered positive when in the direction of the rotating field.

A symmetrical m_1-phase primary current $i_1(t) = \hat{I}_1 \cos\omega_1 t$ results in a rotating field, just as in the previous section:

$$b_1(x,t) = \hat{B}_1 \cos\left(\omega_1 t - \frac{x\pi}{\tau_p}\right) \tag{4.30}$$

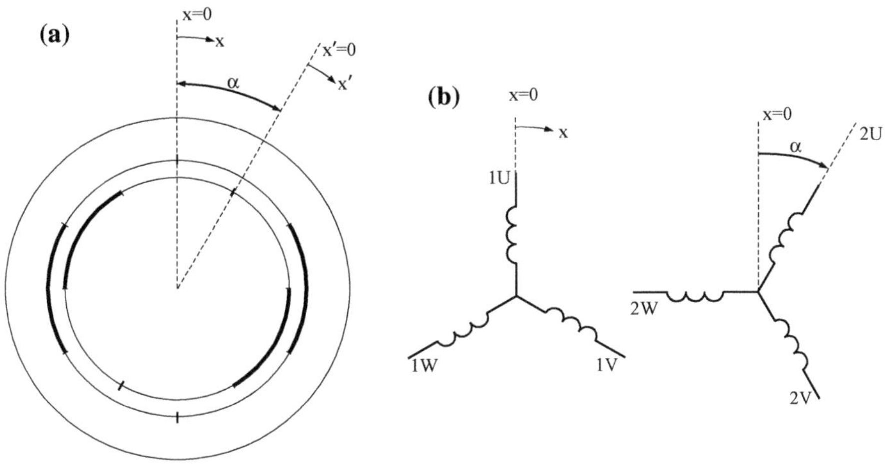

Fig. 4.3 Winding axes not co-linear

For the flux coupled with the primary winding, $\psi_{11,\alpha} = w_1\xi_1 \cdot \phi_{11,\alpha}$ we obtain the same result as in the previous section, i.e. the physical flux along the primary winding axis is also given by

$$\phi_{11,\alpha}(t) = \int_{-\tau_p/2}^{\tau_p/2} b_1(x,t) \cdot l \cdot dx = \frac{2}{\pi}\hat{B}_1 l \tau_p \cos\omega_1 t \equiv \hat{\Phi}_{m1}\cos\omega_1 t \tag{4.31}$$

As to the flux coupled with the secondary winding, $\psi_{21,\alpha} = w_2\xi_2 \cdot \phi_{21,\alpha}$ the physical flux along the secondary winding axis is delayed by α

$$\phi_{21,\alpha}(t) = \int_{-\tau_p/2+\alpha\tau_p/\pi}^{\tau_p/2+\alpha\tau_p/\pi} b_1(x,t) \cdot l \cdot dx = \frac{2}{\pi}\hat{B}_1 l \tau_p \cos(\omega_1 t - \alpha) \equiv \hat{\Phi}_{m1}\cos(\omega_1 t - \alpha) \tag{4.32}$$

Thus, for the induced voltages we see that also the secondary induced voltage is delayed by α

$$e_1(t) = e_{11}(t) = \frac{d\psi_{11,\alpha}(t)}{dt} = -w_1\xi_1 \cdot \hat{\Phi}_{m1}\sin\omega_1 t \tag{4.33}$$

$$e_2(t) = e_{21}(t) = \frac{d\psi_{21,\alpha}(t)}{dt} = -w_2\xi_2 \cdot \omega_1 \cdot \hat{\Phi}_{m1}\sin(\omega_1 t - \alpha) \tag{4.34}$$

or in phasor form

$$\underline{E}_{11} = j\omega_1 L_{m1}\underline{I}_1 \tag{4.35}$$

$$\underline{E}_{21} = j\omega_1 M_{21} \underline{I}_1 \cdot e^{-j\alpha} \tag{4.36}$$

On the other hand, a symmetrical m_2-phase secondary current $i_2(t) = \hat{I}_2 \cos\omega_1 t$ results in a rotating field (with $x' = 0$ along the secondary axis):

$$b_2(x', t) = \hat{B}_2 \cos\left(\omega_1 t - \frac{x'\pi}{\tau_p}\right) = \hat{B}_2 e^{j\alpha} \cos\left(\omega_1 t - \frac{x\pi}{\tau_p} + \alpha\right) \tag{4.37}$$

as $x' = x - \alpha\tau_p/\pi$. Therefore we obtain for the physical fluxes along the axes of primary and secondary

$$\phi_{12,\alpha}(t) = \int_{-\tau_p/2}^{\tau_p/2} b_2(x, t) \cdot l \cdot dx = \hat{\Phi}_{m2} \cos(\omega_1 t + \alpha) \tag{4.38}$$

$$\phi_{22,\alpha}(t) = \int_{-\tau_p/2}^{\tau_p/2} b_2(x', t) \cdot l \cdot dx' = \hat{\Phi}_{m2} \cos\omega_1 t \tag{4.39}$$

For the induced voltages we see therefore that the primary induced voltage is leading the secondary voltage by α

$$e_1(t) = e_{12}(t) = -w_1\xi_1 \cdot \omega_1 \cdot \hat{\Phi}_{m2} \sin(\omega_1 t + \alpha) \tag{4.40}$$

$$e_2(t) = e_{22}(t) = -w_2\xi_2 \cdot \omega_1 \cdot \hat{\Phi}_{m2} \sin\omega_1 t \tag{4.41}$$

or in phasor form

$$\underline{E}_{12} = j\omega_1 M_{12} \underline{I}_2 \cdot e^{j\alpha} \tag{4.42}$$

$$\underline{E}_{22} = j\omega_1 L_{m2} \underline{I}_2 \tag{4.43}$$

When both currents are present, in principle one should vectorially add the mmfs of both currents to calculate the resulting rotating magnetic field from the resulting mmf. If the magnetic circuit is (or can be supposed) linear, one may also add the magnetic fields due to both mmfs considered separately and thus also add the corresponding emfs, i.e.

$$\underline{E}_1 = \underline{E}_{11} + \underline{E}_{12} = j\omega_1 L_{m1} \underline{I}_1 + j\omega_1 M_{12} \underline{I}_2 \cdot e^{j\alpha} = j\omega L_{m1} \left[\underline{I}_1 + \frac{m_2 w_2 \xi_2}{m_1 w_1 \xi_1} \underline{I}_2 \cdot e^{j\alpha}\right] \tag{4.44}$$

$$\underline{E}_2=\underline{E}_{22}+\underline{E}_{21}=j\omega_1 L_{m2}\underline{I}_2+j\omega_1 M_{21}\underline{I}_1\cdot e^{-j\alpha}=j\omega_1 L_{m2}\left[\underline{I}_2+\frac{m_1 w_1\xi_1}{m_2 w_2\xi_2}\underline{I}_1\cdot e^{-j\alpha}\right] \tag{4.45}$$

or

$$\underline{E}_1=j\omega L_{m1}\left[\underline{I}_1+\underline{I}_2^{'}\cdot e^{j\alpha}\right]=j\omega L_{m1}\left[\underline{I}_1+\underline{I}_2^{''}\right]=j\omega_1 L_{m1}\underline{I}_{m1} \tag{4.46}$$

$$\underline{E}_2=\frac{w_2\xi_2}{w_1\xi_1}\cdot e^{-j\alpha}\cdot\underline{E}_1=\underline{E}_1^{'}\cdot e^{-j\alpha}=\underline{E}_1^{''} \tag{4.47}$$

with the magnetising current seen from the primary

$$\underline{I}_{m1}=\underline{I}_1+\frac{m_2 w_2\xi_2}{m_1 w_1\xi_1}\underline{I}_2\cdot e^{j\alpha}=\left[\underline{I}_1+\underline{I}_2^{'}\cdot e^{j\alpha}\right]=\left[\underline{I}_1+\underline{I}_2^{''}\right] \tag{4.48}$$

Alternatively, one may rewrite the secondary emf referred to the primary

$$\underline{E}_2^{''}=\frac{w_1\xi_1}{w_2\xi_2}\cdot e^{j\alpha}\cdot\underline{E}_2=\underline{E}_2^{'}\cdot e^{j\alpha}=\underline{E}_1 \tag{4.49}$$

The double prime is used here to indicate quantities referred to the other side when the axes are rotated over an electrical angle α.

The complete equations of the rotating transformer are then obtained by adding the resistive and leakage voltage drops (as in a normal transformer):

$$\underline{V}_1=(R_1+jX_{1\sigma})\underline{I}_1+\underline{E}_1=\underline{Z}_1\underline{I}_1+\underline{E}_1 \tag{4.50}$$

$$\underline{V}_2=(R_2+jX_{2\sigma})\underline{I}_2+\underline{E}_2=\underline{Z}_2\underline{I}_2+\underline{E}_2 \tag{4.51}$$

with $\underline{E}_1$ and $\underline{E}_2$ given by Eqs. 4.46 and 4.47. Usually the equations are being referred to one or the other side. When referred to the primary the second voltage equation reads

$$\underline{V}_2^{''}=(R_2^{''}+jX_{2\sigma}^{''})\underline{I}_2^{''}+\underline{E}_1=\underline{Z}_2^{''}\underline{I}_2^{''}+\underline{E}_1 \tag{4.52}$$

where

$$\underline{V}_2^{''}=\underline{V}_2\cdot\frac{w_1\xi_1}{w_2\xi_2}\cdot e^{j\alpha} \tag{4.53}$$

$$\underline{I}_2^{''}=\underline{I}_2\cdot\frac{m_2 w_2\xi_2}{m_1 w_1\xi_1}\cdot e^{j\alpha} \tag{4.54}$$

$$\underline{Z}_2^{''}=\underline{Z}_2\cdot\frac{m_1(w_1\xi_1)^2}{m_2(w_2\xi_2)^2} \tag{4.55}$$

These equations are thus quite similar to the equations of a normal transformer. They may also be represented by a similar equivalent circuit. There is however a quantitative difference compared to the normal transformer as here the main (or magnetising) field has to cross an air-gap. Therefore the magnetising inductance is much smaller, in per-unit $x_m \approx 2$ (compared to $x_m \gtrapprox 10$ for a transformer). In addition, the leakages are somewhat larger, i.e. in per-unit $x_{1\sigma} \approx x_{2\sigma} \approx 0.1$, compared to $x_\sigma \approx 0.025$ for a transformer. However, there are also other reasons for designing induction machines with a higher leakage, see Sect. 4.9.

4.2.3 *Energy Conversion and Forces for an Induction Machine at Standstill*

We suppose here that the rotor is not necessarily short-circuited (as in the case of a cage rotor) but that the machine has a slip-ring rotor. From the (fundamental) air-gap magnetic field $b(x, t)$ and the current layer $a(x, t)$ in one of the windings, we may calculate the tangential force and torque on *this* winding, see Sect. 3.4.

The energy converted by the rotating field can be calculated from Eq. 3.87

$$T \cdot \Omega = m \cdot E \cdot I \cdot \cos \psi \tag{4.56}$$

where Ω is the speed of the field with respect to the winding and ψ is the space angle between $b(x, t)$ and the current layer $a(x, t)$ in *this* winding. However, as shown in Sect. 3.4 ψ also is the time shift between the emf and the current in *this* winding. Thus only the active current in the winding is responsible for the torque on *this* winding.

Applying the right hand side of Eq. 4.56 on both primary and secondary windings and recalculating using Eqs. 4.46 and 4.47 for the primary and secondary emf, yields

$$m_1 \cdot E_1 \cdot I_1 \cdot \cos \psi_1 = -m_2 \cdot E_2 \cdot I_2 \cdot \cos \psi_2 \tag{4.57}$$

This equation shows that the **rotating field power** or **electromagnetic power** is the same for stator and rotor when the induction motor rotor is at standstill. At standstill the rotating field power is thus converted between primary and secondary in a purely electrical way, like in a transformer. Indeed, as the rotor is at standstill (Ω is the synchronous speed of the *field* with respect to both *windings*) no mechanical energy is converted.

Equation 4.57 also illustrates that the torque on stator and rotor is the same in absolute value (action = reaction); the minus sign results from the application of the user reference system in both windings.

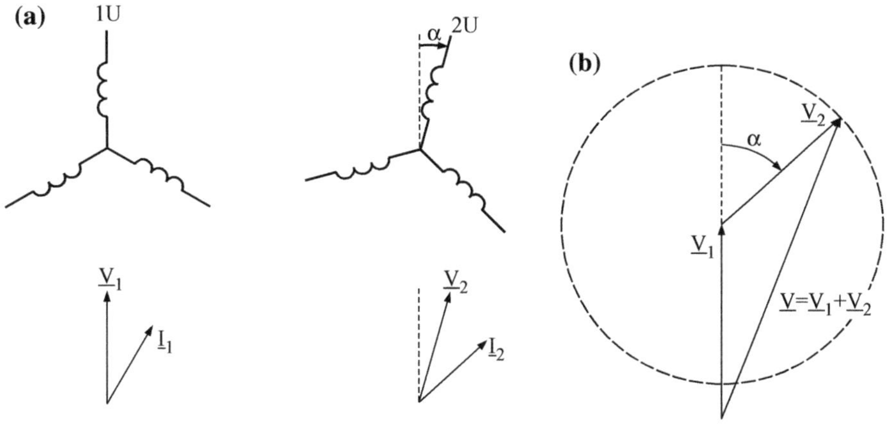

Fig. 4.4 Rotating field transformer

4.2.4 Applications of the Rotating Field Transformer

Some applications require a transformer with a variable phase angle between primary and secondary voltages or one with a variable transformation ratio.

Figure 4.4a illustrates the use of a rotating field transformer as a phase shifter. It is sometimes used in ring connections of a power grid.

By a connection as a (multiphase) auto-transformer one gets a variable transformation ratio, see (b) in Fig. 4.4. A disadvantage is that together with the transformation ratio also the phase angle varies.[3]

4.3 The Rotating Induction Machine: Operating Principle

The induction machine utilised as a (rotating field) transformer is a rarity. In by far the majority of its applications, the induction machine is used in drives, thus as a motor, sometimes also as a generator. Of course the transformer action remains inherent to its operating principle.

Suppose the multiphase (three-phase) winding of the stator is connected to a multiphase (three-phase) voltage source (e.g. the electrical grid) with frequency $f_1 = \omega_1/2\pi$. The stator winding will then draw a multiphase (three-phase) symmetrical current which will give rise to a rotating current layer with mechanical speed $\Omega_{sy} = \omega_1/N_p$. If this is the only current-carrying winding, this current layer will result in a corresponding rotating field with the same speed.

[3]One obtains the same transformation ratio for a rotation in the rotation direction of the field or opposite to it; it can be (e.g. graphically) shown, however, that dependent on the direction of the active current one or the other is slightly advantageous.

The rotor winding of an induction machine is either a normal multiphase (three-phase) winding connected to slip-rings, or a squirrel-cage winding.[4]

4.3.1 Motoring

For motoring, the rotor winding is short-circuited, either by the short-circuit rings for a squirrel cage, or through the slip-rings (direct short-circuit or on external resistances).

Suppose for example that the rotor is initially at standstill. Switching on the supply of the stator results in a multi-phase current ($\underline{I}_1$) and thus a rotating current layer and a rotating field. This field, also rotating with respect to the rotor, will result in a symmetrical multiphase emf in the rotor. Thus, as the rotor windings are short-circuited, a multiphase symmetrical current ($\underline{I}_2$) will start to flow in the rotor. From Lenz' law it follows that the current will result in a torque that will try to oppose its cause, i.e. the relative speed of the rotating field with respect to the rotor. Thus the resulting torque exerted by the field on the rotor will try to rotate the rotor in the direction of the field. If the rotor is free to rotate, the rotor will try to reach synchronism with the field. Finally, at synchronous speed, $\Omega_m = \Omega_{sy} = \omega_1/2N_p$, the induced emf in the rotor, the rotor current and the torque will become zero.

Note that the frequency of the induced emf (and current) in the rotor is equal to the stator frequency when the rotor is at standstill. When the rotor rotates with a speed Ω_m the relative speed of the field with respect to the rotor becomes $\Omega_{sy} - \Omega_m$ (in mechanical radians per second) or $\omega_1 - N_p\Omega_m$ (in electrical radians per second). The induced emf and currents in the rotor then have as frequency the **slip frequency** $f_2 = f_1 - N_p\Omega_m/2\pi$ [Hz] or **slip pulsation** $\omega_2 = \omega_1 - N_p\Omega_m$ [rad/s]. The ratio of the secondary and primary frequency is called the **slip**

$$s = \frac{\omega_2}{\omega_1} = \frac{\omega_1 - N_p\Omega_m}{\omega_1} = \frac{\Omega_{sy} - \Omega_m}{\Omega_{sy}} \tag{4.58}$$

However, when a current flows in the rotor windings, the resulting rotor mmf will affect the field. Indeed, the rotor mmf rotates with the slip speed $\omega_2 = s\omega_1$ (electrical rad/s) with respect to the rotor, which itself rotates with a speed of $N_p\Omega_m = (1-s)\omega_1$. Thus this mmf also rotates with the synchronous (electrical) speed ω_1 with respect to the stator, thus synchronously with the stator mmf. Thus both mmfs will add to form the resulting mmf and magnetic field.[5]

[4]A squirrel cage winding consists of (aluminium or copper) bars cast inside slots in the rotor core and that are short-circuited by short-circuit rings cast at the same time; it can be proved that a cage winding behaves as a normal multiphase symmetrical winding as to the fundamental.

[5]In fact, this rotor current will, in accordance with Lenz' law, try to reduce the field and thus oppose the mmf by the stator current (which will have to increase again).

4.3.2 Generating

Suppose now that the rotor is (externally) driven faster than the synchronous speed. In this case as well there is a non-zero relative speed of the field with respect to the rotor windings and thus an emf will be induced (although with the opposite phase order, as can be seen intuitively). The torque resulting from the current in the short-circuited rotor winding will try to oppose its cause, i.e. the relative motion of the field with respect to the winding. This torque will thus exert a braking action, i.e. opposite to the field rotation direction. The corresponding input mechanical power will then be converted into electrical power that is delivered to the mains via the primary.

The final remark in the previous subsection regarding the contribution of the rotor mmf to the resulting field also holds true here.

4.3.3 Frequency Converter

4.3.3.1 Rotor Winding Open

Suppose the rotor is driven at a speed Ω_m, while the stator is fed by the AC supply with frequency ω_1. The relative speed of the field with respect to the rotor is then $\Omega_{sy} - \Omega_m$ (mechanical rad/s) or $\omega_2 = \omega_1 - N_p\Omega_m = s\omega_1$ (electrical rad/s). This is also the frequency of the voltages induced in the rotor. The magnitude as well of these voltages is proportional to it. This is easily derived form Eq. 4.32 where now $\alpha = N_p\Omega_m t + \alpha_o$

$$\phi_{21,\alpha}(t) = \hat{\Phi}_{m1}\cos(\omega_1 t - \alpha) = \hat{\Phi}_{m1}\cos(\omega_1 t - N_p\Omega_m t - \alpha_o) = \hat{\Phi}_{m1}\cos(\omega_2 t - \alpha_o) \tag{4.59}$$

For the rotor emf we obtain therefore

$$e_2(t) = e_{21}(t) = \frac{d\psi_{21,\alpha}(t)}{dt} = -w_2\xi_2 \cdot \omega_2 \cdot \hat{\Phi}_{m1}\sin(\omega_2 t - \alpha_o) \tag{4.60}$$

or as <u>time vector</u> (i.e. phasor with explicit exponential time function)

$$\underline{E}_{21} = j\omega_2 M_{21}\underline{I}_1 \cdot e^{j\omega_2 t - j\alpha_o} \tag{4.61}$$

At standstill, $s = 1$, the rotor frequency equals the stator frequency. With the rotor rotating in the direction of the field ($0 < s < 1$) the frequency in the rotor decreases with increasing speed until synchronism ($s = 0$). At synchronism the induced voltage in the rotor becomes zero. When driving the rotor faster than the synchronous speed the phase order (and polarity) of the rotor voltages reverses; both amplitude and frequency increase with increasing speed (above synchronous speed).

In contrast, driving the rotor against the direction of the field ($s > 1$) results in increasing voltage amplitudes and increasing rotor frequencies.

4.3.3.2 Rotor Winding Connected to a Symmetrical Load

When the rotor winding is connected to a symmetrical load, the induced rotor voltages give rise to symmetrical rotor currents with the slip frequency $\omega_2 = s\omega_1$. The resulting rotor mmf $\vec{F}_2$ rotates with the (mechanical) speed $\omega_2/N_p = s\omega_1/N_p = \Omega_{sy} - \Omega_m$ with respect to the rotor, but as the rotor itself rotates with the speed Ω_m, the speed of $\vec{F}_2$ with respect to the stator is Ω_{sy}. Stator mmf $\vec{F}_1$ and rotor mmf $\vec{F}_2$ are therefore synchronous and add vectorially to form the resulting mmf $\vec{F}_m$. It is this resulting mmf that determines the air-gap field (inducing the stator and rotor emfs). Whatever the speed of the rotor, for an observer on the stator everything happens with stator frequency, while for an observer on the rotor it's like everything has the slip or rotor frequency. For a machine with slip rings the rotor voltage with slip frequency can be used for an external load connected to the slip rings. Obviously all these considerations are also valid when the machine is working as a motor or generator.

4.4 Equations and Equivalent Circuit of an Induction Machine

From the discussion in the previous section, it is clear that a rotating induction machine also performs as a rotating field transformer, but with a rotor frequency depending on the rotor speed. Suppose that the stator is connected to a primary supply with frequency f_1 (angular frequency ω_1) while the rotor carries currents with frequency f_2 (angular frequency ω_2). Hereby the rotor rotates with a speed which corresponds to the difference of these frequencies: $N_p\Omega_m = \omega_m = \omega_1 - \omega_2$. The rotor winding might be short-circuited or connected to an external load or an external supply with frequency f_2.

The currents in stator and rotor are symmetrical multi-phase, of the form

$$i_1(t) = \hat{I}_1 \cos(\omega_1 t - \beta_1) \tag{4.62}$$

$$i_2(t) = \hat{I}_2 \cos(\omega_2 t - \beta_2) \tag{4.63}$$

in the reference phases of stator and rotor respectively.

The stator current results in an mmf $\underline{F}_1$ that rotates with the speed Ω_{sy} with respect to the stator (and thus $s\Omega_{sy}$ with respect to the rotor). The rotor current yields an mmf $\vec{F}_2$ which rotates with the speed $s\Omega_{sy}$ with respect to the rotor (and thus $s\Omega_{sy} + \Omega_m = \Omega_{sy}$ with respect to the stator). Both mmfs rotate therefore synchronously (with a speed Ω_{sy} with respect to the stator or $s\Omega_{sy}$ with respect to the rotor). The resulting magnetic field should be calculated from the resulting mmf, i.e. the vectorial sum of $\vec{F}_m = \vec{F}_1 + \vec{F}_2$. However if the magnetic circuit is linear (or can be approximated as such) the magnetic fields, the fluxes and the emfs from the stator and rotor currents can be added. For the coupled fluxes we may write (by analogy with Sects. 4.2.2 and 4.3.3.1, substituting now $\alpha = \omega_m t = (1 - s)\omega_1 t$):

$$\psi_{11}(t) = L_{m1} \cdot \hat{I}_1 \cos(\omega_1 t - \beta_1) \tag{4.64}$$

$$\psi_{12}(t) = M_{12} \cdot \hat{I}_2 \cos(\omega_2 t - \beta_2 + \alpha) = M_{12} \cdot \hat{I}_2 \cos(\omega_1 t - \beta_2) \tag{4.65}$$

$$\psi_{21}(t) = M_{21} \cdot \hat{I}_1 \cos(\omega_1 t - \beta_1 - \alpha) = M_{21} \cdot \hat{I}_1 \cos(\omega_2 t - \beta_1) \tag{4.66}$$

$$\psi_{22}(t) = L_{m2} \cdot \hat{I}_2 \cos(\omega_2 t - \beta_2) \tag{4.67}$$

As already mentioned above, from the stator side everything appears as if everywhere the frequency is ω_1; from the rotor side it appears as everything has the frequency ω_2.

We may write the emfs corresponding to these fluxes directly in complex time vector form (see also Sect. 4.2.2):

$$\underline{E}_1(t) = \underline{E}_{11}(t) + \underline{E}_{12}(t) = j\omega L_{m1}\left[\underline{I}_1 e^{j\omega_1 t} + \frac{m_2 w_2 \xi_2}{m_1 w_1 \xi_1}\underline{I}_2 \cdot e^{j\omega_2 t} \cdot e^{j\omega_m t}\right] \tag{4.68}$$

$$\underline{E}_2(t) = \underline{E}_{22}(t) + \underline{E}_{21}(t) = j\omega_2 L_{m2}\left[\underline{I}_2 e^{j\omega_2 t} + \frac{m_1 w_1 \xi_1}{m_2 w_2 \xi_2}\underline{I}_1 \cdot e^{j\omega_1 t} e^{-j\omega_m t}\right] \tag{4.69}$$

or

$$\underline{E}_1(t) = j\omega L_{m1}\left[\underline{I}_1 + \frac{m_2 w_2 \xi_2}{m_1 w_1 \xi_1}\underline{I}_2\right] e^{j\omega_1 t} = j\omega L_{m1}\left[\underline{I}_1 + \underline{I}_2^{'}\right] e^{j\omega_1 t} \tag{4.70}$$

$$\underline{E}_2(t) = j\omega_2 L_{m2}\left[\underline{I}_2 + \frac{m_1 w_1 \xi_1}{m_2 w_2 \xi_2}\underline{I}_1\right] e^{j\omega_2 t} = j\omega_2 L_{m2}\left[\underline{I}_2 + \underline{I}_1^{'}\right] e^{j\omega_2 t} \tag{4.71}$$

Until now, we retained the explicit (exponential) time dependence in the equations to emphasise the different frequencies in stator and rotor. In the sequel we will omit these however. By introducing the magnetising current referred to the stator $\underline{I}_{m1} = \underline{I}_1 + \underline{I}_2^{'}$ and referring the magnitude of the rotor emf to the number of turns of the stator we obtain

$$\underline{E}_1 = j\omega L_{m1}\left[\underline{I}_1 + \underline{I}_2^{'}\right] = j\omega L_{m1}\underline{I}_{m1} \tag{4.72}$$

$$\underline{E}_2 = s\frac{w_2 \xi_2}{w_1 \xi_1}\underline{E}_1 = js\omega_1 L_{m1}\frac{w_2 \xi_2}{w_1 \xi_1}\underline{I}_{m1} \tag{4.73}$$

or

$$\underline{E}_2^{'} = \frac{w_1 \xi_1}{w_2 \xi_2}\underline{E}_2 = js\omega_1 L_{m1}\underline{I}_{m1} = s\underline{E}_1 \tag{4.74}$$

Fig. 4.5 Basic equivalent circuit

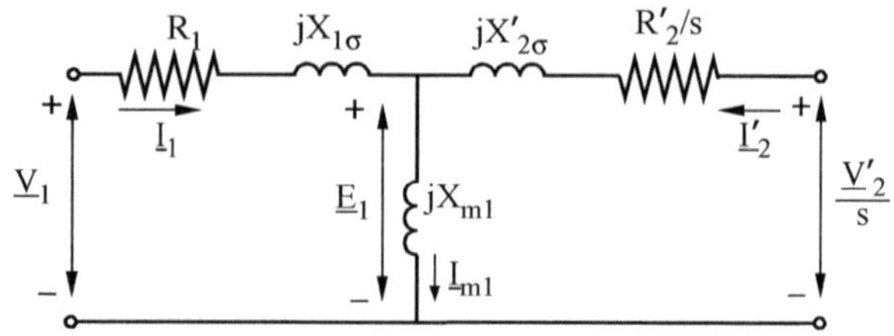

For the relation between voltages and currents we have to add the resistive and inductive leakage voltage drops in the windings:

$$\underline{V}_1 = (R_1 + jX_{1\sigma})\underline{I}_1 + \underline{E}_1 = \underline{Z}_1\underline{I}_1 + \underline{E}_1 \tag{4.75}$$

$$\underline{V}_2 = (R_2 + jsX_{2\sigma})\underline{I}_2 + \underline{E}_2 = \underline{Z}_2\underline{I}_2 + \underline{E}_2 \tag{4.76}$$

with[6] $\underline{E}_1 = j\omega L_{m1}\underline{I}_{m1}$, $\underline{E}_2 = s\frac{w_2\xi_2}{w_1\xi_1}\underline{E}_1$, $\underline{I}_{m1} = \underline{I}_1 + \frac{m_2 w_2\xi_2}{m_1 w_1\xi_1}\underline{I}_2$.

We may rewrite these equations with all voltages and currents referred to the number of primary turns

$$\underline{V}_1 = \underline{Z}_1\underline{I}_1 + \underline{E}_1 \tag{4.77}$$

$$\underline{V}_2^{'} = \underline{Z}_2^{'}\underline{I}_2^{'} + s\underline{E}_1 \tag{4.78}$$

$$\underline{E}_1 = j\omega L_{m1}\underline{I}_{m1} \tag{4.79}$$

$$\underline{I}_{m1} = \underline{I}_1 + \underline{I}_2^{'} \tag{4.80}$$

$$\underline{I}_2^{'} = \frac{m_2 w_2\xi_2}{m_1 w_1\xi_1}\underline{I}_2 \tag{4.81}$$

$$\underline{V}_2^{'} = \frac{w_1\xi_1}{w_2\xi_2}\underline{V}_2 \tag{4.82}$$

$$\underline{Z}_2^{'} = \frac{m_1(w_1\xi_1)^2}{m_2(w_2\xi_2)^2}\underline{Z}_2 \tag{4.83}$$

Dividing Eq. 4.78 by the slip yields

$$\underline{V}_2^{'}/s = (\underline{Z}_2^{'}/s)\underline{I}_2^{'} + \underline{E}_1 = (R_2^{'}/s + jX_{2\sigma}^{'})\underline{I}_2^{'} + \underline{E}_1 \tag{4.84}$$

With Eqs. 4.77, 4.84, 4.79 and 4.80 corresponds the equivalent circuit in Fig. 4.5.

[6]Remark that we will use the notation X for a reactance referred to the primary frequency. $X_{2\sigma}$ is thus the rotor leakage reactance referred to the primary frequency.

Table 4.1 Per-unit values

	Induction machine	Transformer
$r_1 \approx r_2$	$0.005/\tau_p$	$0.005/D$
$x_{1\sigma} \approx x_{2\sigma}$	0.1	0.025
x_m	$2 \cdots 10$	$10 \cdots 1000$

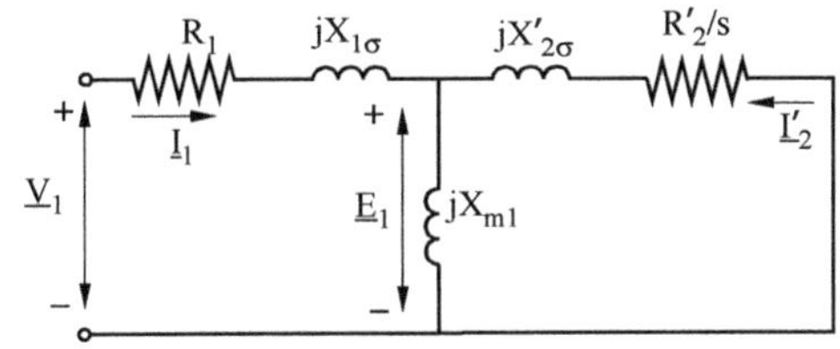

Fig. 4.6 Rotor winding short-circuited

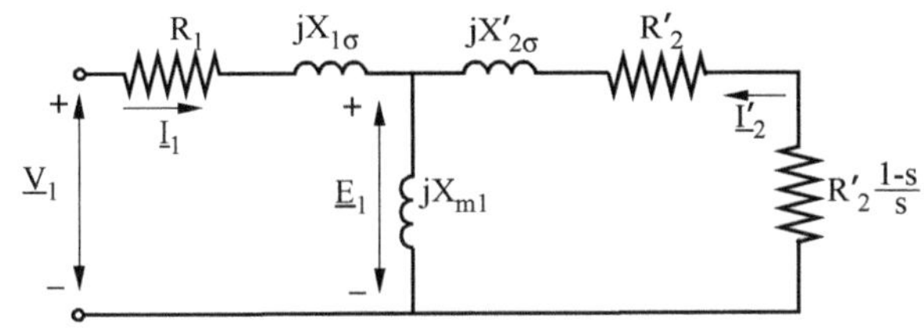

Fig. 4.7 Equivalent rotor resistance split up

This equivalent scheme is about the same as that of a transformer, except for the rotor resistance and rotor voltage that are divided here by the slip s.

In a similar way as for a transformer, simplified equivalent circuits can be derived. However, there are some differences as the magnetising inductance of an induction machine is (relatively) much smaller while the leakage reactances are much higher, see the Table 4.1 below. Therefore, the magnetising inductance cannot be neglected as is frequently done for large transformers.[7]

In most applications the rotor winding is short-circuited, either directly or on an external resistance (in case of slip-rings). The equivalent circuit is that in Fig. 4.6 (where the rotor resistance R_2' should be replaced by the sum of the rotor resistance and the external resistance $R_2' + R_{2u}'$ if an external resistance is connected to the slip rings, if any).

The equivalent resistance R_2'/s can be split up into the real resistance R_2' and a resistance $R_2'(1-s)/s$, see Fig. 4.7. The resistance R_2' represents the real rotor joule losses while the resistance $R_2'(1-s)/s$ corresponds to the conversion of electrical energy into mechanical energy (or vice versa), see also the next section.

[7] Also the number of poles has a large effect on the magnetising inductance: with increasing number of pole pairs the magnetising inductance decreases as it varies approximately as $1/\sqrt{N_p}$. As a result the leakage coefficient σ is also much higher for induction machines, the more so as the number of pole pairs is larger.

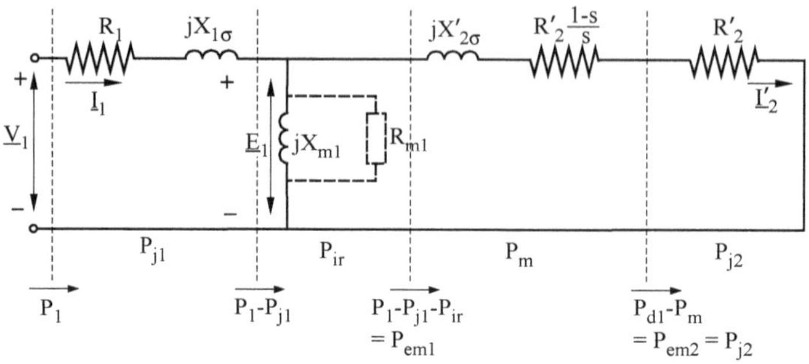

Fig. 4.8 Equivalent circuit with power flows

4.5 Energy Conversion and Torque

The most common use of an induction machine is in motoring where the stator is supplied by a three-phase voltage source and the rotor is short-circuited (on itself or on an external resistance in case of a slip-ring motor). From the supply, a symmetrical stator current is drawn which results in a rotating mmf with speed $\Omega_{sy} = \omega_1/N_p$ with respect to the stator. With respect to the rotor (with a mechanical speed Ω_m) the field rotates with a speed $\Omega_{sy} - \Omega_m$ and induces in the rotor an emf and thus a rotor current with the slip frequency $\omega_2 = s\omega_1 = N_p(\Omega_{sy} - \Omega_m)$. The resulting rotor mmf together with the stator mmf yield the resulting mmf responsible for the actual air-gap field,[8] see above.

In such a steady-state condition, we thus normally have both (non-zero) emfs E_1, E_2 and both (non-zero) currents I_1, I_2. From Chap. 3, Sect. 3.4 we know that the electromagnetic or rotating field power of a winding is equal to the product of the torque on the winding and the speed of the field with respect to the winding, thus for stator and rotor respectively

$$P_{em1} = T\Omega_{sy} = m_1 E_1 I_1 \cos\psi_1 \tag{4.85}$$

$$P_{em2} = Ts\Omega_{sy} = m_2 E_2 I_2 \cos\psi_2 \tag{4.86}$$

Splitting up the first equation yields

$$P_{em1} = T(1-s)\Omega_{sy} + sT\Omega_{sy} = T\Omega_m + P_{em2} \tag{4.87}$$

The equality 4.87 (and $P_{em2} = sP_{em1}$) also follows from the equivalent circuit[9] in Fig. 4.8 (but without the equivalent iron loss resistance). In fact, substituting $E_2^{'} =$

[8]If there was only a stator current and mmf, the field would be that resulting from the stator mmf; actually both stator and rotor mmfs contribute to the resulting field, except in no-load.

[9]In accordance with the M-convention we reversed the positive direction of $I_2^{'}$.

Table 4.2 Speed and power relations

	Primary (stator)		Secondary (rotor)		Mechanical
Speed	$\Omega_{sy} = \omega_1/N_p$	$=$	$s\Omega_{sy}$	$+$	$(1-s)\Omega_{sy}$
Frequency	ω_1	$=$	$s\omega$	$+$	$(1-s)\omega_1$
Power	$P_{em1} = T\Omega_{sy}$	$=$	$sP_{em1} = sT\Omega_{sy}$	$+$	$(1-s)P_{em1} = (1-s)T\Omega_{sy}$
Power for shorted rotor	$m_2 \frac{R_2}{s} I_2^2$	$=$	$m_2 R_2 I_2^2$	$+$	$m_2 \frac{(1-s)}{s} R_2 I_2^2$

sE_1 yields:

$$m_2 E_2 I_2 \cos\psi_2 = m_1 E_2^{'} I_2^{'} \cos\psi_2 = s \cdot [m_1 E_1 I_1 \cos\psi_1] \tag{4.88}$$

Equation 4.87 shows that the primary (stator) electromagnetic power is transferred *by the rotating field* partly into mechanical power $P_m = T\Omega_m$ and partly by transformer action into electrical power in the rotor, the secondary (rotor) electromagnetic power P_{em2}. For a short-circuited rotor this secondary electromagnetic power is equal to the joule losses in the rotor resistance, $P_{em2} = m_2 E_2 I_2 \cos\psi_2 = m_2 R_2 I_2^2 = m_1 R_2^{'} I_2^{'2}$. If an external circuit or power source (with slip frequency) is connected to the slip rings the secondary electrical power is partly (P_2) transferred to this circuit or power source (and partly into rotor joule losses).

The power P_1 taken from the supply consists of the primary electromagnetic power P_{em1}, the primary joule losses $m_1 R_1 I_1^2$ and the iron losses $m_1 E_1^2/R_{ir}$. Of the primary electromagnetic power P_{em1} the fraction $(1-s)$ is converted into mechanical energy

$$P_m = (1-s)P_{em1} = T\Omega_m \tag{4.89}$$

The available shaft power is somewhat smaller due to the friction and ventilation losses P_{vf}.

The foregoing is valid for either motoring or generating or braking, of course with the proper signs for P_{em1}, P_{em2}, P_m and the slip s. The losses are however always positive (thus also P_{em2} in the case of a shorted rotor).

Table 4.2 summarises the relation between speed and power for the M-convention. In the M-convention motoring implies P_m positive while generating corresponds to P_{em1} negative.[10]

The last row is valid only for a short-circuited rotor.

For a short-circuited rotor (in the M-convention) we have $P_{em2} = Ts\Omega_{sy} = m_2 R_2 I_2^2$. As $P_{em2} = Ts\Omega_{sy} = m_2 R_2 I_2^2 \geq 0$, T and s then always have the same

[10]Remark however that a negative mechanical power does not necessarily implies generating! Conversely, however, generating always implies input of mechanical power.

sign. A positive slip thus implies a positive torque, a negative slip implies a negative torque.

The different operating conditions for a short-circuited rotor are summarised in Table 4.3.

The Sankey diagram in (a) in Fig. 4.9a illustrates the power flow for motoring ($0 < s < 1$) while the one in (b) in Fig. 4.9 shows the power flow for generating ($s < 0$). The dotted line corresponds with the power transfer by the rotating field. For a shorted rotor the secondary supply power P_2 is zero. As will be shown in Part 3, Chap. 13, a secondary supply can be used for enlarging the possible operating states, while at the same time also increasing the efficiency.

For $s > 1$ as well the torque is positive, but as the speed is negative, the mechanical power is negative (i.e. input of mechanical power). In this braking operating condition the input mechanical power is dissipated in the (short-circuited) rotor together with the positive input power from the supply.

Note also that, disregarding iron and stator joule losses, the efficiency for motoring is given by $\eta = 1 - s$; for generating the efficiency is $\eta = 1/(1 - s)$. It is therefore important to keep the slip magnitude ($|s|$) as small as possible.

Table 4.3 Speed, slip, torque and power signs for shorted rotor

	Braking	Motoring	Generating
Speed	$\Omega_m < 0$	$0 < \Omega_m < \Omega_{sy}$	$\Omega_{sy} < \Omega_m$
Slip	$1 < s$	$0 < s < 1$	$s < 0$
Torque	$T > 0$	$T > 0$	$0 > T$
Primary electromagnetic power	$P_{em1} > 0$	$P_{em1} > 0$	$P_{em1} < 0$
Mechanical power	$P_m < 0$	$P_m > 0$	$P_m < 0$

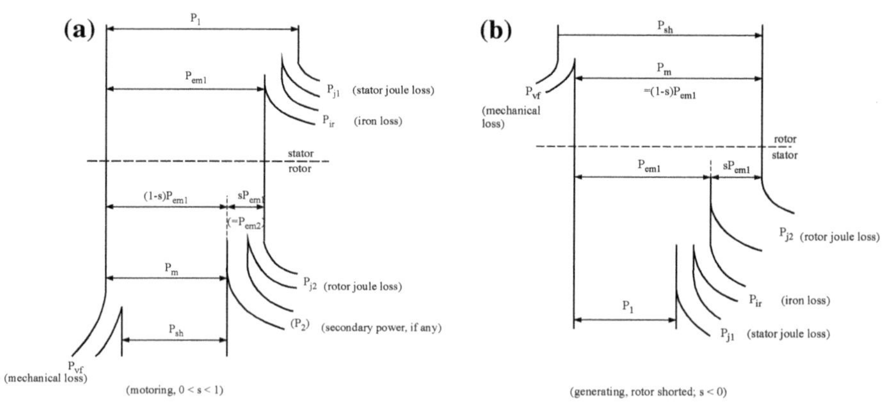

Fig. 4.9 Sankey diagrams

4.6 Torque and Torque-Slip Characteristic

The torque expression can be derived from the rotating field power. For a short-circuited rotor we may easily obtain the torque from the secondary electromagnetic power $P_{em2} = Ts\Omega_{sy} = m_2 R_2 I_2^2 = m_1 R_2^{'} I_2^{'2}$.

To derive the expression for $\underline{I}_2^{'}$ we use the Thévenin equivalent circuit as observed from the rotor, Fig. 4.10 (which can be derived from Fig. 4.8, omitting the iron loss resistance, or from Fig. 4.6 by reversing the sign of $\underline{I}_2$):

$$\underline{I}_2^{'} = \frac{\underline{V}_{2o}^{'}}{R_{k2}^{'} + jX_{k2}^{'}} \tag{4.90}$$

with

$$\underline{V}_{2o}^{'} = \underline{V}_1 \frac{jX_{m1}}{R_1 + jX_{1\sigma} + jX_{m1}} \approx \frac{\underline{V}_1}{1+\sigma_1} \tag{4.91}$$

$$\underline{Z}_{k2}^{'} = R_{k2}^{'} + jX_{k2}^{'} \approx \left[\frac{R_1}{(1+\sigma_1)^2} + R_2^{'}/s\right] + j\left[X_{\sigma 2}^{'} + \frac{X_{1\sigma}}{1+\sigma_1}\right] \tag{4.92}$$

where the primary leakage coefficient is defined by $\sigma_1 = X_{1\sigma}/X_{m1}$, and with $\left[X_{\sigma 2}^{'} + \frac{X_{1\sigma}}{1+\sigma_1}\right] = \sigma X_2^{'} = X_{2\sigma}^{'}$ the total leakage as observed from the secondary.

This results in

$$T \simeq \frac{m_1}{\Omega_{sy}} \left[\frac{\underline{V}_1}{1+\sigma_1}\right]^2 \frac{R_2^{'}/s}{\left[\frac{R_1}{(1+\sigma_1)^2} + R_2^{'}/s\right]^2 + \left[X_{\sigma 2}^{'} + \frac{X_{1\sigma}}{1+\sigma_1}\right]^2} \tag{4.93}$$

If the stator resistance can be neglected, this simplifies to

$$T \simeq \frac{m_1}{\Omega_{sy}} \left[\frac{\underline{V}_1}{1+\sigma_1}\right]^2 \frac{R_2^{'}/s}{\left[R_2^{'}/s\right]^2 + X_{2\sigma}^{'2}} \tag{4.94}$$

For small slip values ($s \ll R_2^{'}/X_{2\sigma}^{'}$), on the one hand, Eq. 4.94 can be approximated to

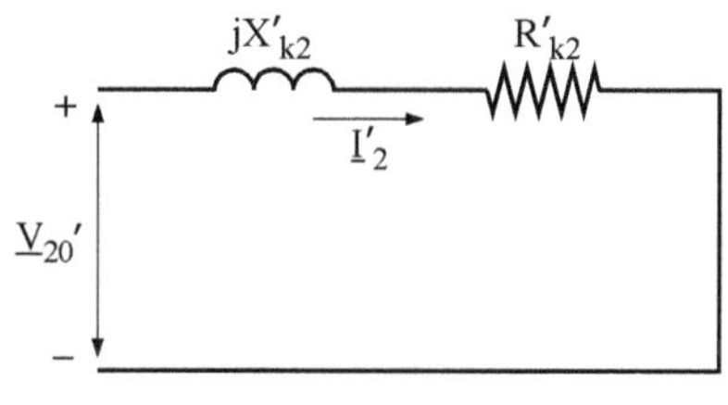

Fig. 4.10 Thévenin equivalent circuit

$$T \simeq \frac{m_1}{\Omega_{sy}} \left[\frac{V_1}{1+\sigma_1} \right]^2 \frac{s}{R_2'} \tag{4.95}$$

i.e. the torque varies about linearly with s.

For large values of the slip ($s \gg R_2'/X_{2\sigma}'$), on the other hand, the torque varies hyperbolically with the slip:

$$T \simeq \frac{m_1}{\Omega_{sy}} \left[\frac{V_1}{1+\sigma_1} \right]^2 \frac{R_2'/s}{X_{2\sigma}'^2} \tag{4.96}$$

In between the torque reaches a maximum value, the pull-out (or breakdown) torque:

$$T_{po} \simeq \pm \frac{m_1}{\Omega_{sy}} \left[\frac{V_1}{1+\sigma_1} \right]^2 \frac{1}{2X_{2\sigma}'} \tag{4.97}$$

The corresponding slip value is called the pull-out slip:

$$s_{po} = \pm R_2'/X_{2\sigma}' \tag{4.98}$$

The plus sign is for motoring, the minus sign for generating. The value of the pull-out slip varies between 0.25 (for small power ratings, e.g. 5 kW) and 0.025 (for larger power ratings, e.g. 10 MW).

The linear part of the torque characteristic resembles the speed-torque characteristic of an independently excited DC motor. The cause for the maximum torque and the further decreasing torque for slip values larger than the pull-out slip is that with increasing slip, the rotor current increases considerably. As a result, also the primary current will have to increase, attempting to keep the air-gap flux constant. However, the voltage drop over the primary leakage (and the stator resistance) causes the air-gap flux to decrease for these large slip values. The torque will therefore decrease considerably for large slip values.

Figure 4.11 illustrates the torque as a function of the speed (or slip). The dot-dash curve shows the effect of the stator resistance (in essence a slight decrease of the torque for motoring and a slight increase for generating[11]).

For small stator resistance, the torque-slip characteristic can also be written as

$$\frac{T}{T_{po}} = \frac{2}{s/s_{po} + s_{po}/s} \tag{4.99}$$

where pull-out torque and slip are given by Eqs. 4.97 and 4.98 respectively. This formula is known as Blondel's formula.

In addition to the linearity of the torque characteristic for small slip values, two other properties are of uttermost importance:

---

[11]Prove that with non-zero stator resistance the pull-out slip slightly decreases and the pull-out torque decreases for motoring and increases for generating.

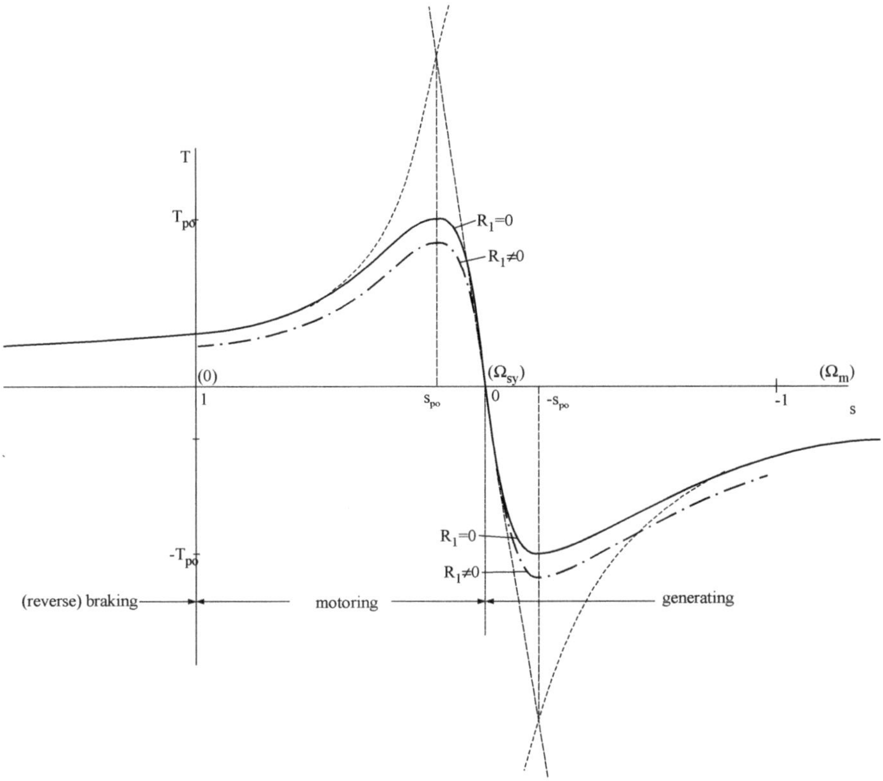

Fig. 4.11 Torque-slip characteristic

- the quadratic dependence on the voltage (a voltage dip of e.g. 20% results in a torque decrease of 36%!)
- the torque depends on the ratio s/s_{po} or s/R_2' and not simply on s, while the pull-out torque is independent of the rotor resistance.
  Adding an additional resistance R_{2e} in the rotor circuit (for a slip-ring motor) permits to obtain the same torque for a larger slip (or smaller speed) while the maximum torque does not change. The torque characteristics for some rotor resistance values are illustrated in Fig. 4.12. The pull-out slip changes proportionally to the total rotor resistance but the pull-out torque remains unchanged, i.e. the torque characteristic just shifts to higher slip values, or $T(s^*) = T(s)$ and $T(s_{po}^*) = T(s_{po}) = T_{po}$ with

$$\frac{s_{po}^*}{s_{po}} = \frac{s^*}{s} = \frac{R_2 + R_{2e}}{R_2} \tag{4.100}$$

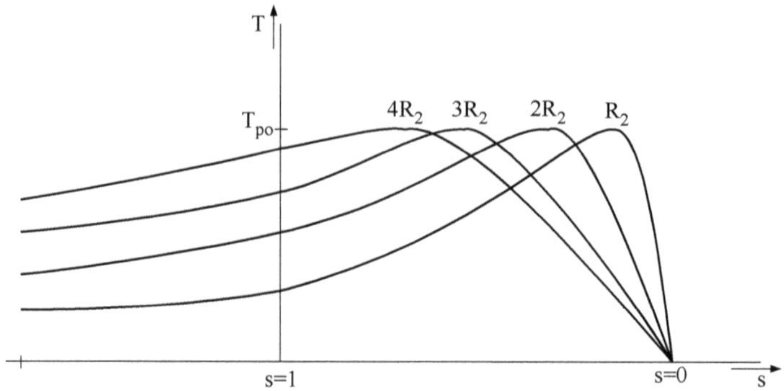

Fig. 4.12 Effect of additional rotor resistance

This property can be used for speed control and for starting purposes. Nowadays, however, adding rotor resistance is rarely used for speed adaptation as the efficiency ($\eta \approx (1 - s)$) is rather low for higher slip values. Instead, power electronic means are now used for variable speed operation.

4.7 The Current Locus of an Induction Machine

To derive the current vector locus, we will neglect the stator resistance (and the iron loss resistance).

A quite handy equivalent circuit is obtained if the conventional T-scheme of inductances in Fig. 4.8 is replaced by an L-scheme with all leakage transferred to the secondary. This is possible without any loss of information or accuracy.[12] This results in the equivalent scheme in Fig. 4.13. Stator voltage and current remain unchanged, but the secondary current and the impedances are converted as follows:

$$X_m = X_{1\sigma} + X_{m1} = (1 + \sigma_1) X_{m1} \equiv X_1 \tag{4.101}$$

$$X_\sigma = \sigma X_2^{'} (1 + \sigma_1)^2 = \frac{\sigma}{1 - \sigma} X_1 \tag{4.102}$$

$$\frac{R}{s} = (1 + \sigma_1)^2 \frac{R_2^{'}}{s} \tag{4.103}$$

$$\underline{I} = \underline{I}_2^{'} / (1 + \sigma_1) \tag{4.104}$$

[12]In fact, the separate leakages of stator and rotor cannot be measured, see the remarks at the end of this section.

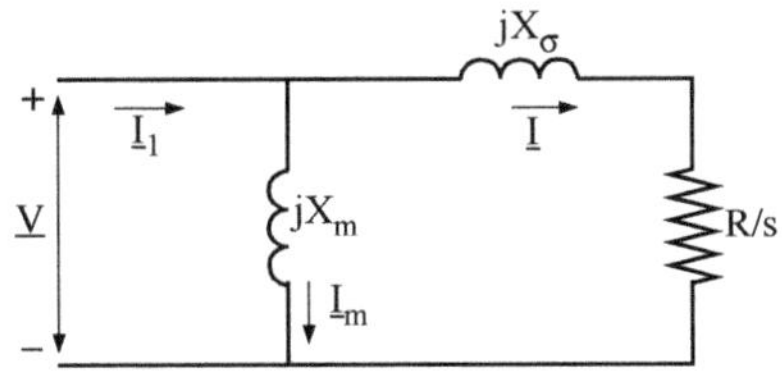

Fig. 4.13 L-scheme

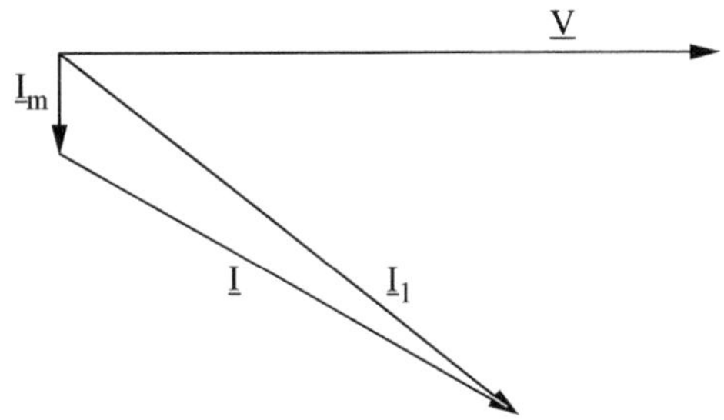

Fig. 4.14 Current vectors

From this equivalent circuit we get

$$\underline{I}_m = \frac{\underline{V}_1}{jX_m} \tag{4.105}$$

$$\underline{I} = \frac{\underline{V}_1}{R/s + jX_\sigma} = \frac{\underline{V}_1}{jX_\sigma} \cdot \frac{\underline{V}_1}{1 - js_{po}/s} \tag{4.106}$$

$$\underline{I}_1 = \underline{I}_m + \underline{I} \tag{4.107}$$

where the pull-out slip is given by $s_{po} = R/X_\sigma$.

The torque (and electromagnetic power) can be written as

$$T = \frac{P_{em1}}{\Omega_{sy}} = \frac{3}{\Omega_{sy}} V_1^2 \cdot \frac{R/s}{(R/s)^2 + X_\sigma^2} \tag{4.108}$$

and the pull-out torque (for $s = \pm s_{po}$) as[13]

$$T_{po} = \pm \frac{3}{\Omega_{sy}} \cdot \frac{V_1^2}{2X_\sigma} \tag{4.109}$$

The phasor diagram corresponding to Eq. 4.107 *(choosing the voltage phasor along the real axis)* is illustrated in Fig. 4.14.

[13] Show that all these expressions are equivalent to those derived above for the original circuit.

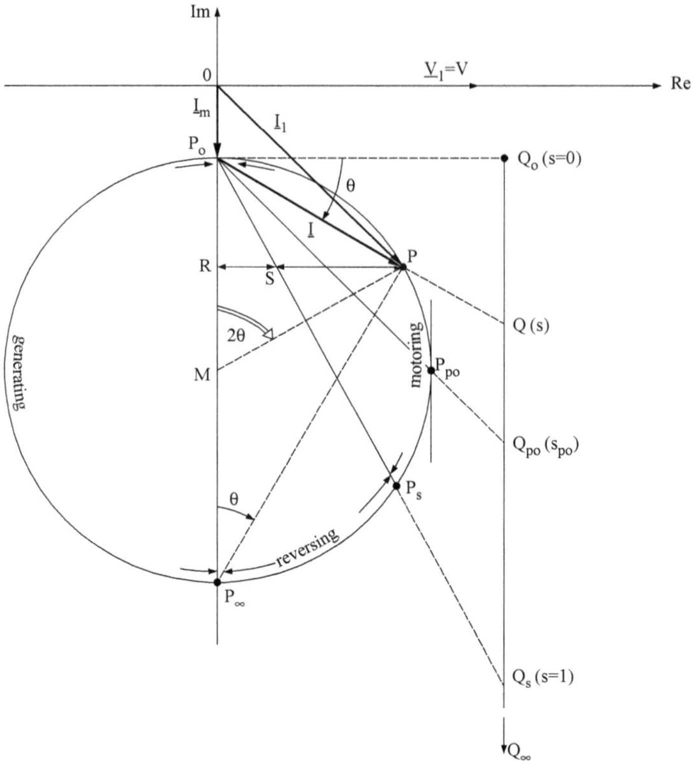

Fig. 4.15 Current locus

With varying slip the locus of the phasor $\underline{I}$ (and thus of $\underline{I}_1$) is a circle,[14] see Fig. 4.15. For zero slip, $\underline{I} = 0$ and $\underline{I}_1 = \underline{I}_m$ (point P_o). For infinite slip ($s = \pm\infty$), the current vector endpoint is at P_∞ with $\underline{I}_\infty = \underline{P_o P_\infty} = V_1/jX_\sigma$, from which we derive that the diameter of the circle is V_1/X_σ.

Other important endpoints are:

- the pull-out condition at P_{po} for $s_{po} = R/X_\sigma$ and
- standstill or starting condition at P_s for $s = 1$.

Note that the position of the standstill point on the circle depends heavily on the rotor resistance (in contrast to the other mentioned points). This has important consequences, as for large machines (relatively small resistances) this standstill point moves closer to P_∞, while for small machines (relatively large resistances) the standstill point is located nearer to P_{po}, or for extremely high resistances even between P_{po} and P_o.

A linear slip scale can be obtained on a line parallel with $P_o P_\infty$. The extension of the vector $\underline{P_o P}$ cuts on this line an oriented segment $\underline{Q_o Q}$ that is proportional to

[14]This is easily shown by introducing θ and substituting $s/s_{po} = \tan\theta$ in Eq. 4.106.

the slip s ($Q_oQ = P_oQ_o \cdot \tan\vartheta = P_oQ_o \cdot (s/s_{po})$; remark that Q_o is on the tangent from P_o).

On this current locus the torque and power values can also be found. For a given operating point P, call R the projection on the line P_oP_∞ and S the intersection of RP - or its extension - with P_oP_s. As RP (with sign corresponding to the oriented segment $\underline{RP}$) is equal to the active current component $\Re e(\underline{I})$ it is proportional to the primary electromagnetic power P_{em1} and, because of the zero stator resistance, also proportional to the supply power P_1: $RP = I_w = P_{em1}/3V = P_1/3V$. As the primary rotating field power is proportional to the torque, $RP = T\Omega_{sy}/3V$.

The line P_oP_s cuts the line RP in S. From the pairs of similar triangles P_oPR, P_oQQ_o and P_oSR, $P_oQ_sQ_o$ it is easily shown that the (oriented) segment $\underline{RS} = s\underline{RP}$ is proportional to the slip losses while the remaining segment $\underline{SP} = (1-s)\underline{RP}$ represents the mechanical power. The line P_oP_s is thus the reference line for mechanical power while P_oP_∞ is the reference line for primary power and torque.[15]

In addition to the special points (P_o, P_{po}, P_s, P_∞) mentioned above, other important points are e.g. the rated operating point P_n or the pull-out condition in generating P'_{po}. This will be discussed in the next sections.

Remarks:

1. It is feasible to take into account the stator resistance (and the iron loss resistance) for the current locus. The locus is still a circle but the no-load point is now not any more located on the imaginary axis, the diameter of the locus is rotated with respect to the imaginary axis, and the P_∞ point is not on this diameter, see Fig. 4.16.
2. An L-scheme can also be used without neglecting the stator resistance, Fig. 4.17. However, the main advantage of the scheme in Fig. 4.13 is now lost, i.e. the voltage over the magnetising inductance is not any more equal to the supply voltage and depends on the load (current). Also an equivalent iron resistance can be added in this scheme (although its meaning and value are not the same as the one in Fig. 4.13).
   Whereas neglecting the stator resistance is quite acceptable for large machines (>100 kW), for machines of smaller power ratings the effect of the stator resistance can be important to calculate the characteristics (e.g. torque characteristic). For assessing the losses, the stator resistance should always be taken into account of course (as well as the iron resistance).
3. In principle, the L-schemes are not less accurate than the *'general'* T-scheme if saturation is neglected (as we did). Indeed, the no-load and short-circuit tests provide only two independent complex equations. These four real equations allow to estimate the stator resistance (which can also be measured by a DC test if skin-effect is neglected), the magnetising resistance for the scheme in Fig. 4.6 (or the sum of primary leakage and magnetising inductance for the scheme in Fig. 4.17) and the real and imaginary parts of the short-circuit impedance. However, this

[15] Work out that the figure also gives the correct sign for generating and braking (all segments are oriented).

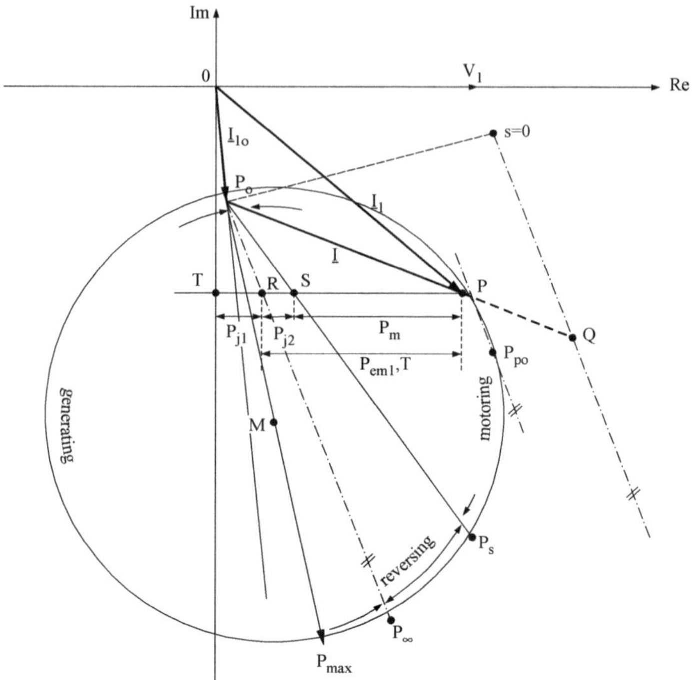

Fig. 4.16 Current locus for $R_s \neq 0$

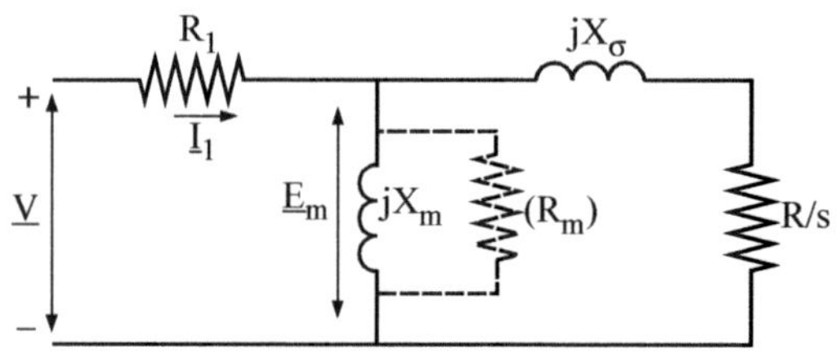

$$X_m = X_1$$

$$X_\sigma = \frac{\sigma}{1-\sigma} X_1 = (1+\sigma_1)^2 . \sigma X'_2$$

$$R = (1+\sigma_1)^2 R'_2$$

Fig. 4.17 L-scheme with Rs

does not permit to estimate the leakage division between stator and rotor, nor thus the separate leakages (neither the turns ratio).[16]

For a cage induction machine it is even more extreme: disregarding the real parts of the impedances, for a cage machine we only have two equations (from no-load and short-circuit tests) for three unknowns, i.e. the two self-inductances and the mutual inductance of the inductance matrix. Indeed, for a cage induction machine the rotor windings are not accessible.

[16]As we know, this also is the case for a transformer.

Note also that this becomes even more complex if saturation is taken into account, as the magnetising inductance and all leakages vary with the supply voltage and the load. Then for each value of the supply voltage and load, another scheme is obtained. Then load tests are also required to estimate the equivalent parameters of the schemes.

4.8 Per-Unit Description

A per-unit description of a machine implies choosing proper reference values for electrical and mechanical quantities, i.e. voltage, current, torque and speed (also frequency if variable frequency operation is considered). However these reference values cannot be chosen completely arbitrarily, as the basic laws like conservation of energy have to be fulfilled. Thus, only three reference values can be chosen independently. For electrical machines that are mainly used in motoring, usually one starts from the rated (nominal) values of supply voltage, torque and speed. For machines mainly used as generator the starting values are the rated values of supply voltage, current and speed.

As induction machines are mainly used as motor, the per-unit reference system for an induction machine is based on the rated supply voltage, rated (electromagnetic) torque and synchronous speed at rated frequency. Using the synchronous speed (at rated frequency) as reference instead of rated speed, results in the reference power being equal to the rated electromagnetic power (rotating field power) and thus somewhat larger than the rated electromagnetic shaft power, i.e. greater by a factor $1/(1-s_n)$. Nevertheless, there also are some advantages. For example, at rated frequency the per-unit torque is also equal to the per-unit power. Note that with this reference system, the reference current is about equal to the rated active current (exactly equal when neglecting the effect of the stator resistance and iron loss).

This reference system results in the per-unit equivalent circuit in Fig. 4.18. The per-unit values of the impedances are for most machines of the order of $x_\sigma \approx 0.15\cdots 0.25$; $x_m \approx 2\cdots 10$; $r \approx 0.005/\tau_p$.

The corresponding equations for current, power and torque are quite similar to those for absolute values:

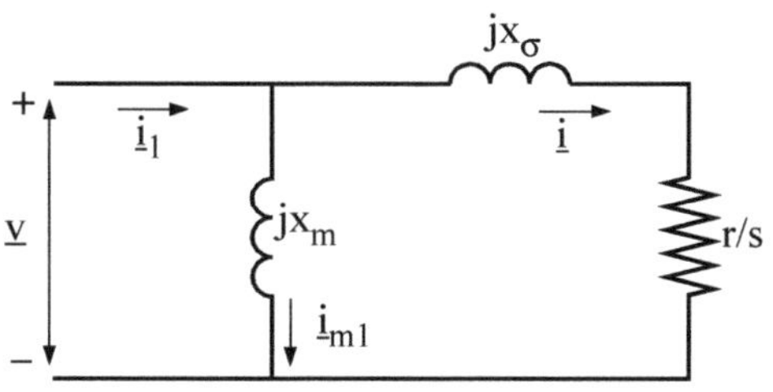

($x_\sigma \simeq$ 0,15...0,25, typically 0,20)

($x_m \simeq$ 1,2...10, typically 2)

($r \simeq 0{,}005/\tau_p = 0{,}05...0{,}005$ for $\tau_p = 0{,}1...1$ m)

Fig. 4.18 p.u. equivalent circuit

$$\underline{i}_1 = \underline{i}_m + \underline{i} = \frac{\underline{v}}{jx_m} + \frac{\underline{v}}{r/s + jx_\sigma} \tag{4.110}$$

$$p_1 = p_{em1} = t \cdot 1 = i^2 \cdot (r/s) = v^2 \frac{r/s}{(r/s)^2 + x_\sigma^2} \tag{4.111}$$

Mechanical power and slip losses are obtained by splitting up the electromagnetic power according to

$$p_{em1} = s \cdot p_{em1} + (1 - s) \cdot p_{em1} \equiv p_{jr} + p_m \tag{4.112}$$

As the reference power is equal to the rated[17] electromagnetic or rotating field power ($P_{em,n} = T_n \cdot \Omega_{sy}$), in rated conditions of voltage, frequency and speed (or slip) we then have

$$p_{em1,n} = 1 = 1^2 \frac{r/s_n}{(r/s_n)^2 + x_\sigma^2} \tag{4.113}$$

from which we get

$$r/s_n = \frac{1}{2}\left[1 \pm \sqrt{1 - 4x_\sigma^2}\right] \approx \frac{1}{2}\left[1 \pm (1 - 2x_\sigma^2)\right] \tag{4.114}$$

From these two solutions, only the largest one (corresponding with the smallest slip) is here of interest to us:

$$r/s_n \approx 1 - x_\sigma^2 \approx 1 \tag{4.115}$$

With a fair approximation, the rated slip is thus equal to the per-unit rotor resistance, $s_n \approx r$. For this slip value we have $p_{em1} = t \approx 1$ and $i_w = 1$. This rated slip is on the (approximately) linear part of the torque characteristic.

The other slip value corresponding with unity torque and electromagnetic power is rather large, $s \approx r/x_\sigma^2$, i.e. on the other side (the hyperbolic part) of the pull-out torque on the torque characteristic. In per-unit the pull-out torque is given by $t_{po} = v^2/2x_\sigma$ for $s_{po} = r/x_\sigma$. Blondel's torque formula is in per-unit

$$t = t_{po} \cdot \frac{2}{s/s_{po} + s_{po}/s} \tag{4.116}$$

Figure 4.19 shows the current vector locus in per-unit (drawn for $x_m = 2$ and $x_\sigma = 0.2$). In the rated operating point (for motoring), we have for the current $\underline{i}_1 = v(1 - jx_\sigma - j/x_m)$. The operating point with maximum power factor is found at the tangent to the circle from the origin. This point is, generally, not the rated point.[18]

[17]The notation n (for nominal) is used for rated values.

[18]Calculate the optimal power factor from Fig. 4.19 and compare to the power factor at rated slip.

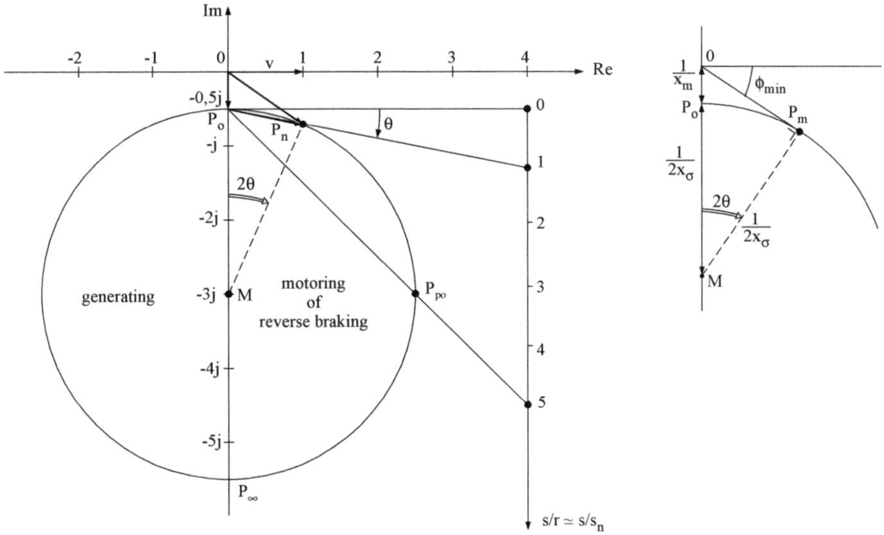

Current locus for $x_\sigma = 0,2$; $x_m = 2$ and $v = 1$

Fig. 4.19 Current locus in p.u.

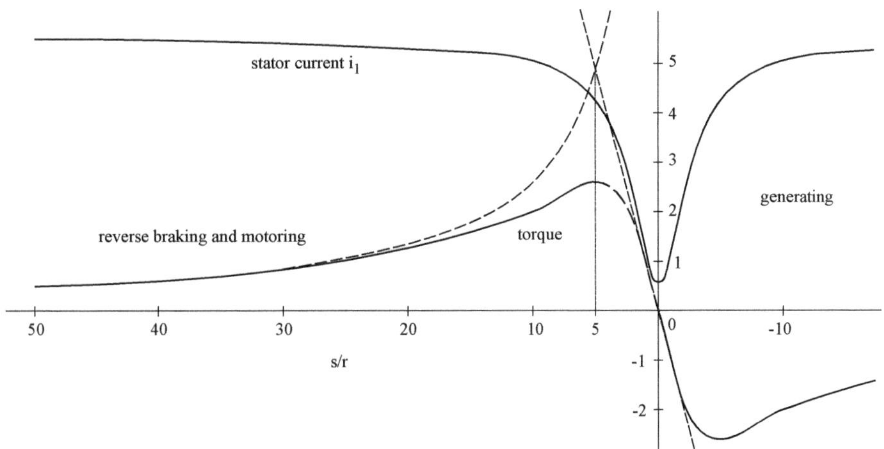

Current and torque as a function of s/r for x_σ=0.2 ; x_m = 2 at v = 1

Fig. 4.20 Current and torque as a function of s/r

4.9 Effect of s/r, x_σ and x_m on Current and Torque

In Fig. 4.20, we have drawn the evolution of the current and the torque as a function of the generalised slip parameter s/r (for $x_m = 2$ and $x_\sigma = 0.2$).

The current is minimal for zero slip and increases monotonously for increasing slip parameter. The torque, however, increases from zero to the pull-out torque and then decreases again (to zero) for infinite slip. For continuous operation, we want of course a sufficiently high torque with a low or modest current (and high efficiency). For transient purposes, higher currents (and/or with lower torque) can be acceptable. Therefore, we may distinguish three ranges for the slip parameter:

1. s/r between 0 and 2: this is the normal continuous operating range, centred around $s/r = 1$, i.e. the rated speed. Torque production is efficient and the power factor is high.
2. s/r between 2 and 5: in this range the torque production is less efficient because of large lagging currents. This range might be necessary to overcome sudden and unwanted load torque increases without the risk to stalling the motor.
3. s/r higher than 5: here the currents are rather large and torque low. But brief operation in this range might be required for starting purposes (cage motors).

For a further discussion, the value of the **rotor resistance** itself must be considered as well. Also the distinction between cage induction machines and slip-ring machines is important.

Figure 4.21 shows the torque characteristics for some resistance values ($v = 1$ and $x_\sigma = 0.2$ have been assumed). A small rotor resistance is preferable as it results in a good efficiency in the normal operating range (small slip). However, in this case the starting torque is rather limited resulting in large start-up times. And in Fig. 4.20 we have seen that for $s/r > 1$ the current is rather large as well (resulting also in high stator joule losses). However, remember that the slip range between $s = 1$ and $s = s_n$ must always be passed through when starting.

While machines of small power ratings mostly have rather large resistances because of economic (and constructional) reasons, machines of large power ratings always have low (per-unit) resistances because of efficiency reasons (efficiency $\eta \approx 1 - s$). In fact, it turns out (from an analysis of data of a large number of machines) that the per-unit resistance of induction machines varies with the pole pitch approximately as $r \approx 0.005/\tau_p$. This results in a rather low starting torque for

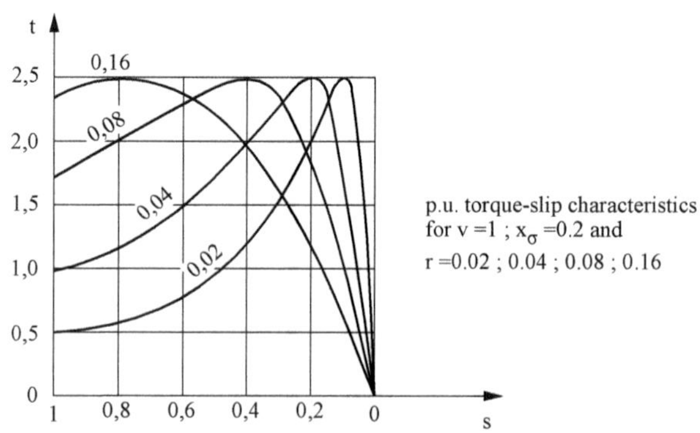

Fig. 4.21 Effect of the rotor resistance on torque characteristic

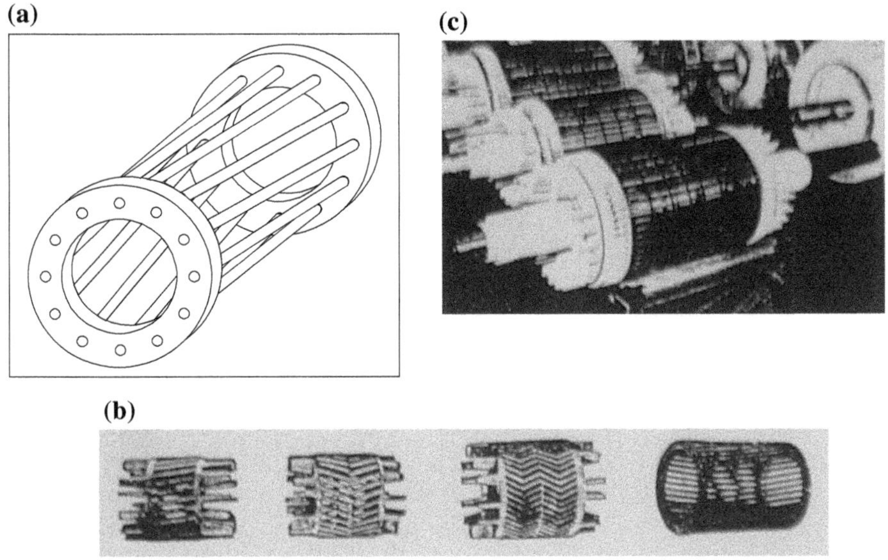

Fig. 4.22 Rotor cage constructions

large machines. Indeed, with $s_{po} = r/x_\sigma \approx 0.025/\tau_p$, the starting torque (for $s = 1$) can be approximated as $t \approx t_{po} \cdot 2s_{po} \approx t_{po} \cdot (0.05/\tau_p)$, which is very small for large pole pitches (nevertheless the corresponding starting current is more than 5 times the rated current).

If the machine is a slip-ring machine, additional rotor resistances may be connected at the slip-rings for starting, while these resistances are afterwards removed, for continuous operation.[19]

For large squirrel cage machines, however, starting may pose a problem without additional measures. However, special construction of the cage may relieve the starting problem (see below).

Figures 4.22 and 4.23 illustrate the basic construction of a cage rotor. Electromagnetic sheets with punched slot openings are stapled on the shaft. Then, an alloy containing mainly aluminium (or, more recently, also copper for high-efficiency motors) is cast into these slots. Together with the cage slot fillings, the end-rings (short-circuit rings) are cast, sometimes provided with ventilation blades as well. By a special shape of the slot and slot openings, either in the form of narrow deep slots (a, b and c in Fig. 4.23) or as a double-cage (d), thus both with the largest cross-section to the inside of the rotor, the skin effect in these slots is increased and with a high frequency in the rotor conductors, the rotor currents will shift to the upper part of the cage conductors where the resistance is higher.

[19]In the past additional rotor resistances were also used for speed regulation, but since the advent of power electronics this method for speed control all but disappeared.

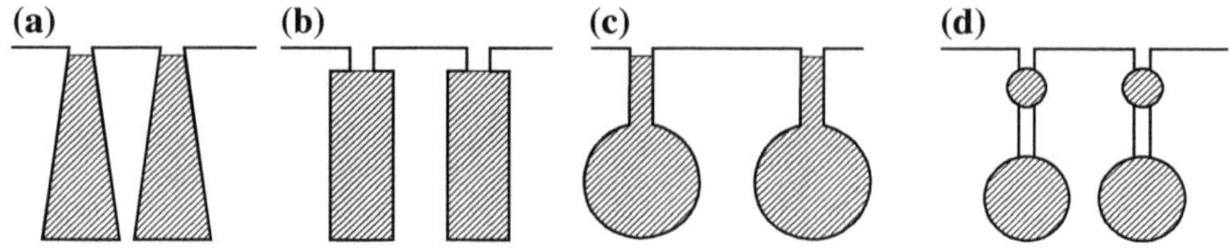

Fig. 4.23 Deep slot and double cage

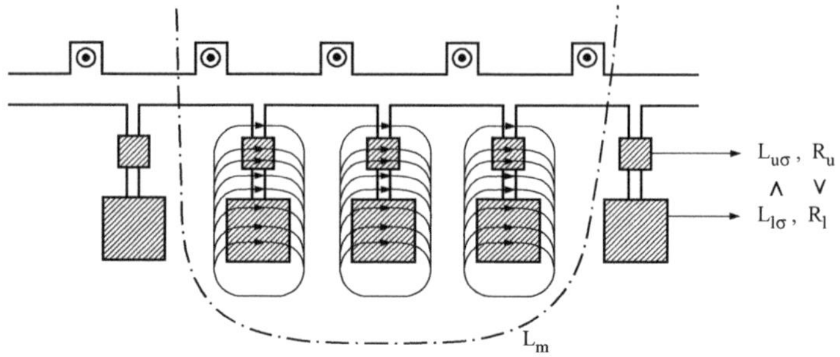

Fig. 4.24 Field lines in double cage rotor

For explaining this operation principle for starting, we refer to Fig. 4.24, illustrating the field lines of the main air-gap field (crossing the air-gap and linking the stator winding with the rotor slots, i.e. the dash-dot line) and those of the leakage field of the rotor cage (i.e. the full lines, linking only the rotor slots). Remark that nearly all field lines will close ***through the iron below*** the rotor slots because of the lower reluctance of the rotor iron compared to the aluminium or copper conductor in the slots. Thus the lower portion of the rotor cage slots is linked with nearly all field lines, in particular with all field lines of the rotor leakage field.

At low speed (high slip), the rotor currents will shift to the higher parts of the slots, near the air-gap. This can be explained in two ways. To start with, keep in mind that at low speed and thus high slip, the frequency in the rotor is fairly high (e.g. 50 Hz at $s = 1$).

The main field (dashed line) linked with the rotor induces an emf in the rotor cage. This emf gives rise to currents in the short-circuited rotor. But these currents result in a leakage field with field lines as in Fig. 4.24. In turn these alternating fields results in emfs induced in the cage that oppose their origin and thus counteract the currents. As the lower parts of the cage bars are linked with most or all leakage flux lines, the counteracting emf is the higher the lower this part of the cage bar and thus the current will be pushed to the upper parts of the cage bars. The higher the slip, the higher the frequency and the higher the counteracting emfs. Thus for high slip, almost all current will flow in the upper parts where the resistance is higher. For low slip values, the counteracting emf is lower or negligible and the current will be distributed evenly over the cross section of the cage bars, i.e. an equivalent lower resistance. This is analogous to the well known skin effect in a conductor in free space, where the current is shifted to the outer skin.

Another way to explain this skin-effect is illustrated in Fig. 4.25 (for a double-cage motor). For the upper conductor the resistance is high (small cross section) but its leakage inductance is low as this upper conductor is linked with only a small portion of the leakage field lines. For the lower conductor the resistance is low (high cross section) but the leakage inductance is high as it is linked with nearly all leakage flux lines. Both upper and lower conductors can be represented as acting in parallel with the main field as in the figure. For high slip on the one hand, the impedance of the leakage will dominate and the current will be shifted to the upper conductor, resulting in an equivalent high resistance. In contrast, for low slip values the current will be distributed over both conductors, inversely proportionally to the resistance and the equivalent resistance of the cage will be that of the parallel connection of both upper and lower conductors, i.e. a low resistance.

As a result, the torque-slip characteristic has for high slip a similar shape as for an induction machine with high rotor resistance, while for low slip values the characteristic resembles that of a machine with low resistance, Fig. 4.26.

For the **leakage inductance** we supposed above a per-unit value of 0.2. This value is in fact a compromise between two contradictory requirements:

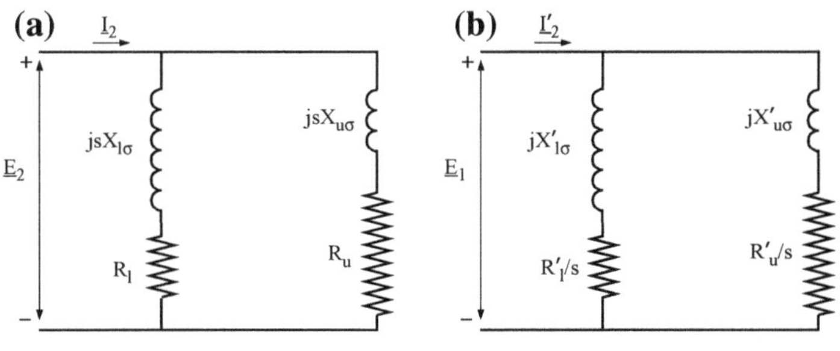

Fig. 4.25 Equivalent circuit for double cage

Fig. 4.26 Torque characteristic for double cage

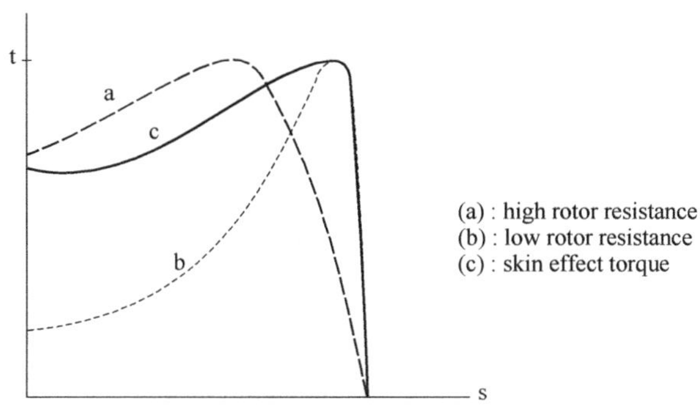

- a sufficient low value to obtain a pull-out torque that is at least twice the rated torque (as prescribed by the standards, as sudden load increases may not stall the motor); from $t_{po} = v^2/2x_\sigma$ and $v = 1$, this yields therefore $x_\sigma \leq 0.25$
- a maximum starting current of maximum 7 times the rated current; from $i_s \approx v/\sqrt{(r^2 + x_\sigma^2)} \approx v/x_\sigma$, this yields (at rated voltage) values of $x_\sigma \geq 0.15$.

A per-unit value of $x_\sigma = 0.2$ corresponds with a pull-out torque of 2.5 and a starting current of 5 and is an ideal compromise.

The (per-unit) value of the **magnetising inductance** depends on the reluctance of the main flux path. To reduce the magnetising current, the most important parameter is an as small as possible air-gap. However, some other factors impose a lower limit on the air-gap, in particular mechanical limitations (bearing slack) and electromechanical restrictions (reducing pulsating losses and forces and the resulting vibrations). This results in an approximate scaling law for the air-gap as $\delta \approx 3 \cdot 10^{-3} \cdot \tau_p \cdot \sqrt{N_p}$ [m]. Further, also the quality of the iron used in the construction plays an important role (e.g. from economic point of view, large machines allow for the use of higher quality iron).

It is clear therefore that mainly the power rating of the machine (or its size) will determine the p.u. value of the magnetising inductance[20]: $x_{mn} \approx 10 \cdot \sqrt{\tau_p/N_p}$. For example, this yields for four-pole machines $x_{mn} = 1.2 \ldots 10$ for $\tau_p = 0.03 \ldots 1$ [m].

4.10 Scaling Laws - Rated Specific Values

In the preceding sections, we frequently referred to scaling laws as a means to obtain an idea of the values of impedances of induction machines. Scaling laws reflect the effect of the machine size on the properties of a machine. Indeed, induced voltage, force and energy conversion per unit of air-gap surface area depend on three basic quantities: air-gap induction B [T], linear current density A [A/m] and linear surface speed $v \approx v_{sy}$ [m/s]. As the pole pitch length τ_p has an important effect (i.e. the scaling laws) on the value of these basic quantities in rated conditions, so it also has on the basic properties of a machine.

For an efficient energy conversion the air-gap induction should be as large as possible, but it is limited by the minimum allowable air-gap length.[21] Nevertheless, with increasing machine size the air-gap induction turns out to increase slightly, thanks to a somewhat larger useful iron section available for the flux (and possibly also better quality iron) for larger machine sizes:

$$\hat{B}_n \approx \tau_p^{1/6} \tag{4.117}$$

[20]That the magnetising inductance decreases with the number of poles can be readily understood considering that a high number of pole pairs results in multiple crossings of the air-gap over the circumference.

[21]This minimum air-gap length also depends of the size, with an absolute minimum of usually 0.25 mm.

As to the current loading, a constant current density J (say 4 A/mm^2) would correspond to a current sheet or linear current loading A [A/m] proportional to the pole pitch. As the linear speed of the circumference also increases with size ($v = 2\tau_p f$) this would seem acceptable with a cooling ventilator driven by the shaft. However, with increasing machine size and thus axial length the cooling efficiency of the airflow in the air gap decreases. Therefore the losses per square meter of air gap should decrease slightly more than linear with size or linear speed, and thus the linear current loading may increase less than linear with the pole pitch:

$$A_n \approx 10^5 \tau_p^{2/3} \tag{4.118}$$

These scaling laws for induction and current loading (together with $v = 2\tau_p f$) lead to the following scaling laws for emf per metre, force density per square metre and power per square metre (the numeric data are for $\tau_p = 0.03 \ldots 1$ m):

$$K_n = v_{sy} B_n \approx 70 \tau p^{7/6} \approx 1 \ldots 70 \quad [\text{V/m}] \tag{4.119}$$

$$F_n^{\square} = A_n B_n \approx 7 \cdot 10^4 \tau_p^{5/6} \approx 3.5 \cdot 10^3 \ldots 7 \cdot 10^4 \quad [\text{N/m}^2] \tag{4.120}$$

$$P_n^{\square} = F_n v_{sy} \approx 7 \cdot 10^6 \tau_p^{11/6} \approx 10^4 \ldots 7 \cdot 10^6 \quad [\text{W/m}^2] \tag{4.121}$$

For the relation between the axial length of the machine and the pole pitch, one may assume $l_a \approx D = 2N_p \tau_p / \pi$, which leads to the following scaling law for the rated power:

$$P_n \approx 10^7 \tau_p^{23/6} \cdot N_p^2 \tag{4.122}$$

The rated power thus increases with approximately the fourth power of the dimensions (pole pitch), slightly less than for transformers.

4.11 Single-Phase and Two-Phase Induction Machines

4.11.1 Two-Phase Induction Machines

Two-phase (squirrel cage) induction machines have two phase windings per pole-pair. These windings are orthogonal in space (in electrical degrees). When supplied by a symmetrical two-phase supply, a rotating field results and the energy conversion is identical to that of three-phase machines.

A disadvantage however is that with a lower number of phases the actual field contains more space harmonics of lower order (and thus higher amplitude).

4.11.2 Single-Phase Induction Machines

The term 'single-phase squirrel cage machines' is used to denote all squirrel cage induction machines that are supplied by a single-phase supply, i.e. true single-phase motors, capacitor motors and also multi-phase motors fed from a single-phase supply.

4.11.2.1 True Single-Phase Motors

These motors have only a single-phase stator winding, but a squirrel cage rotor winding (thus indeed a multi-phase rotor).

The alternating current in the stator winding (from the supply) results in an alternating current layer. As explained before, this alternating current layer can be decomposed into two counter-rotating current layers (with half the amplitude of the original alternating current layer).

In the absence of rotor currents, the alternating stator current sheet would result in an alternating air-gap field (which can also be decomposed into two counter-rotating fields of equal amplitude).

As the (short-circuited) rotor cage is linked with this field, a rotor current will be induced. This rotor current will contribute to the resulting field in the same way as the stator current sheet, i.e. the mmfs of both current distributions will add to form the resulting mmf, which in turn determines the resulting (induction) field.

If the rotor is at standstill, the rotor current sheet is an alternating current (which is co-linear in space with the stator current sheet). The mmfs of both stator and rotor current layers are thus also alternating and displaced 90° in space with respect to the current distributions. This holds true for the magnetic field as well, and thus no torque will be produced. As there is no starting torque, the rotor remains at standstill.[22]

Suppose now that the rotor is pushed in one or the other direction (say, in positive direction). The slip s for this positive field direction then becomes smaller than 1 while the slip $s' = 2 - s$ for the negative field direction is larger than 1. The field rotating in the positive direction will induce in the rotor a symmetrical current with frequency $sf < f$ while the field rotating in the negative direction induces a current with frequency $s'f > f$. The amplitudes of both rotor current layers differ however: as $s'f > sf$ the rotor current layer rotating in the negative direction is larger (larger induced emf) and will pull down the field rotating in the negative direction, while the field component rotating in the positive direction benefits as it is reduced to a lesser degree.[23]

---

[22]This may also be derived by considering the counter-rotating current layers and the counter-rotating fields corresponding to alternating current layers: from symmetry the torques resulting from these cancel out.

[23]This is similar to a transformer where the load current tries to decrease the main field; the primary current will of course increase trying to keep the total flux at the same level but because of the leakage inductance the main flux will decrease. For this case of an induction machine the equivalent

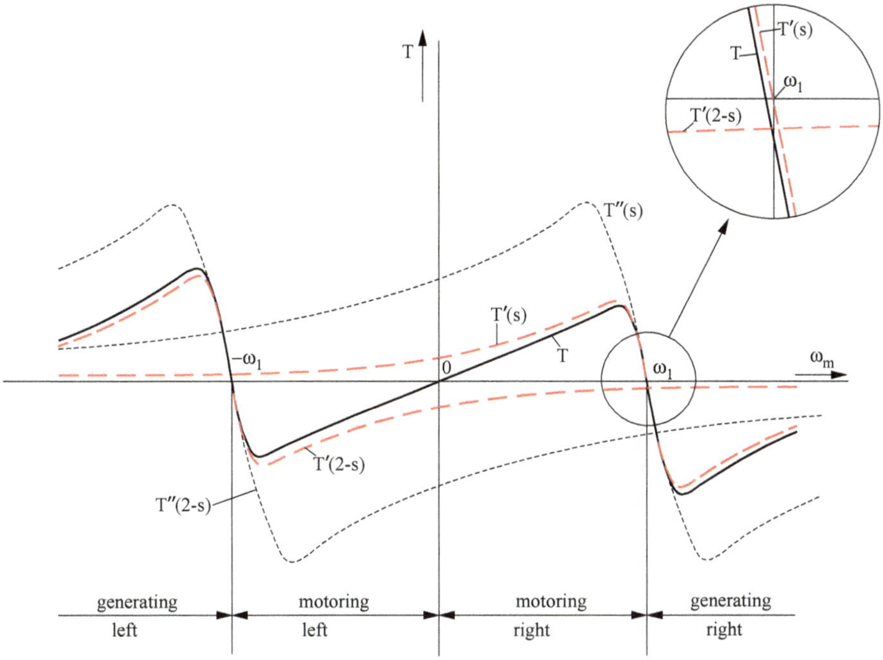

Fig. 4.27 Single-phase torque

The electromagnetic torque is, in principle, the sum of four components, i.e. the torques of the left and right rotating fields on the left and right rotating current layers in the rotor. However the torque components resulting from fields and current layers that rotate in different directions are just pulsating and do not lead to a net torque. The net torque is therefore the sum of the torque of the right rotating field on the right rotating current layer and the torque of the left rotating field on the left rotating current layer: $T = T'(s) + T'(2-s)$.

If the rotor is rotating to the right (positive direction), the latter is negative but smaller in absolute value than the former (which is positive). As a result, there remains a net torque to the right, i.e. in the actual direction of rotation of the rotor.

This is represented in Fig. 4.27. Here $T''(s)$ and $T''(2-s)$ are the torques for a normal induction motor (and full rated voltage) for the right and left rotation direction respectively. If for the single-phase machine the decomposition of the alternating field (or current layer) into two equal counter-rotating components of half the original amplitude remained valid also with a rotating rotor, then the torque components $T'(s)$ and $T'(2-s)$ would attain only 25% of the torques $T''(s)$ and $T''(2-s)$ (as he torque is proportional to the square of the voltage).

rotor resistance R/s for the positive rotating field is larger than the resistance R/s' for the negative rotating field and thus the remaining field for the former will be larger than the one for the latter.

However as explained above, for a right rotation of the rotor, the right rotating field increases while the left rotating one decreases (and vice versa for a left rotating rotor). Thus[24] for $s < 1$, $T'(s)$ may attain 50% of the corresponding torque for rated voltage (while $T'(2-s)$ decreases) for $s < 1$; for $s' < 1$ (or $s = 2 - s' > 1$) the same holds true for $T'(s')$.

Although the torque values seem acceptable, some disadvantages of the single-phase induction machine are apparent:

- there is no starting torque as $T'(s) = -T'(2-s)$ for $s = 1$. When the rotor is pushed in one or the other direction it will start rotating in this direction
- synchronism will never be attained, as the total torque attains zero for some small non-zero slip (because of the small negative torque in the other direction).
- the presence of non-negligible pulsating torques (remember that this is a single-phase load with inherent pulsating electrical power)

Therefore, true single-phase machines are seldom used. Instead capacitor motors and shaded-pole induction machines are frequently used when only a single-phase supply is available, as explained in the next sections.

4.11.2.2 Capacitor Motors

A capacitor motor contains, in addition to the main single-phase supplied winding, a second (auxiliary) winding placed orthogonally (i.e. a space shift of 90 electrical degrees) with respect to the main winding. This auxiliary winding is fed by the same supply, but connected in series with a capacitor in order to imitate a two-phase supply. Figure 4.28 shows a two-pole representation of the windings and connections. The main A-winding is directly supplied by the mains. The auxiliary B-winding is also fed by the same mains supply, but in series with a capacitor.

From the scheme in Fig. 4.28 we derive:

$$\underline{V} = \underline{V}_A = \underline{Z}(s) \cdot \underline{I}_A = [R(s) + jX(s)] \cdot \underline{I}_A \tag{4.123}$$

$$\underline{V} = \underline{V}_B - jX_C \underline{I}_B = \underline{Z}(s) \cdot \underline{I}_B - jX_C \underline{I}_B = [R(s) + jX(s)] \cdot \underline{I}_B - jX_C \underline{I}_B \tag{4.124}$$

where $\underline{Z}(s) = R(s) + jX(s)$ is the slip-dependent impedance of the induction machine (cf. equivalent circuit) and X_c is the impedance of the capacitor.

From the condition for two-phase symmetry, i.e. $\underline{V}_B = j\underline{V}_A$ and $\underline{I}_B = j\underline{I}_A$, we may derive

$$X_c = (1 - j)[R(s) + jX(s)] \tag{4.125}$$

This implies that, in order to achieve 2-phase symmetry, $R(s) = X(s)$ and $X_c = 2R(s) = 2X(s) = \sqrt{2}Z(s)$.

[24]This single-phase machine is in fact equivalent with two mechanically connected machines that are electrically series connected, one for the left and one for the right rotation direction.

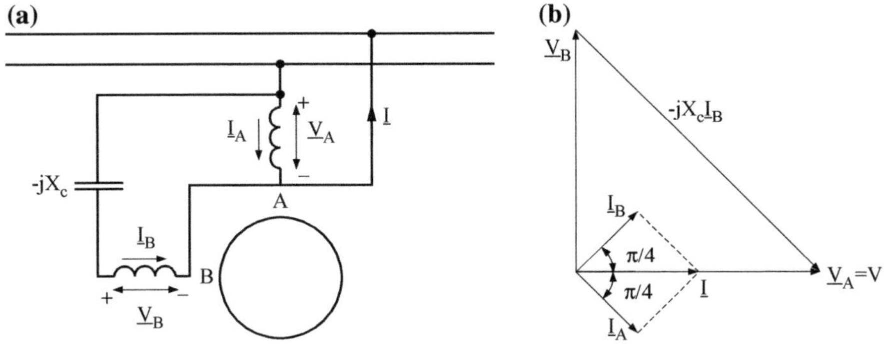

Fig. 4.28 Capacitor motor

Two-phase symmetry is therefore only possible if in the operating point (defined by the slip s), the power factor angle is 45° and the capacitor is chosen according to Eq. 4.125.

The phasor diagram in Fig. 4.28 illustrates this condition and reveals an additional advantage: the supply current is purely active. The capacitor compensates all reactive current.

Also from the machine point of view, this operating condition is ideal: completely symmetrical, with constant torque and thus without pulsating torque components (as in pure single-phase machines). Of course, the mains power still contains a pulsating component with twice the mains frequency (as always for single-phase supply), but the capacitor effectively compensates this pulsating power for the machine.

For an efficient operation of the machine, the slip s where the power factor angle is about 45° should occur at the rated operating point. This implies that the power factor

Fig. 4.29 Phase angle condition

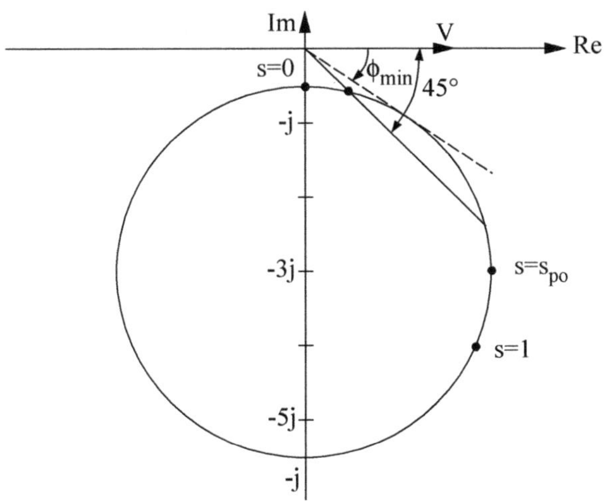

angle should be about 45° at the rated slip s_n. Figure 4.29 shows a circle diagram, where the 45° line intersects the circle in two points. One intersection point is at fairly low slip while the second one is at rather high slip. If the rated power of the machine is not too high, the p.u. rotor resistance is rather high and the standstill point (s = 1) is not far from the pull-out point. The low-slip intersection point is then near to the rated operating point and thus quite useful. Nevertheless, it should be remarked that power factor angles between 30° and 60° still result in acceptable symmetrical operation conditions.

Whereas the interval between 30° and 60° corresponds for small machines with the major part of slip values between $s = 0$ and $s = 1$, for large machines the larger value of x_m and the smaller value of the rotor resistance r_r result in intersection points that are not very useful (i.e. at small slip near no-load and at larger slip near pull-out).

Capacitor motors may exhibit a problem at no-load, not only because of bad symmetrical behaviour with a power factor angle around 90°, but also because the capacitor voltage (and current) may become extremely high. From

$$v = \underline{v}_B - jx_c\underline{i}_B \approx jx_m\underline{i}_B - jx_c\underline{i}_B \tag{4.126}$$

where the per-unit value of x_c is about $\sqrt{2}$ (see Eq. 4.125) and x_m is for small machines between 1.2 and 2, we get a phasor diagram as in Fig. 4.30.

For starting, the 'run capacitor', as described above, is not an ideal solution. Therefore sometimes two capacitors are used, one for starting (chosen corresponding to the intersection point at large slip, i.e. around s = 1) and one for continuous operation (the 'run capacitor'). The value of the start capacitor is much larger than that of the run capacitor. Whereas the run capacitor corresponds to $X_c = 1/\omega C \approx \sqrt{2}Z_r$ (or $x_c \approx \sqrt{2}$ in per unit), the start capacitor corresponds to the machine impedance at standstill, i.e. $X_c = 1/\omega C \approx \sqrt{2}X_\sigma$ (or in per-unit $0.2 \cdot \sqrt{2} \approx 0.28$). This corresponds to 133 µF/kW or 2 kVA/kW at 230 V, compared to only 30 µF/kW or 500 VA/kW for the run capacitor. The start capacitor is not continuously on, but switched by a centrifugal or a current switch. Then also a cheaper electrolytic version can be used for this start capacitor (another solution to reduce the cost is to use a higher number of turns for the auxiliary winding, resulting in a lower capacitor value).

In contrast, the run capacitor should be of high quality, i.e. with a low loss factor $\tan\delta (\approx 0.01)$. Usually metal foil (with paper insulation) or the newer plastic capacitors are used.

In the standards, capacitor motors are referred to by several names, depending on the type. For Europe they are addressed in IEC Publication 50, Chap. 411, Rotating machines (e.g. 411-03-24: Capacitor motor; 411-03-25: Capacitor start motor; 411-03-26: Capacitor start and run motor; 411-03-27: Two-value capacitor motor).

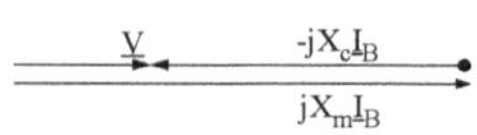

Fig. 4.30 No-load phasor diagram for capacitor motor

4.11.2.3 Single-Phase Supply of Multi-phase Motors

A capacitor can also be used to obtain a rotating field in 3-phase motors when only a single-phase supply is available, see Fig. 4.31. This connection is usually called the 'Steinmetz' scheme. The mains voltage must of course correspond to either the phase voltage (delta connection) or the line voltage (wye connection) of the machine.

Assume that a symmetrical voltage and current system is obtained. From the phasor diagrams in Fig. 4.32 (corresponding to the schemes in Fig. 4.31), it is easily derived that the power factor angle must be 60° in order to obtain 3-phase symmetry (capacitor voltage and current are orthogonal and, dependent on wye or delta connection, either the capacitor voltage or the capacitor current correspond to a machine phase value).

However, at rated slip the power factor angle of multi-phase induction machines normally differs greatly from the ideal angle of 60° and thus symmetrical operation in rated conditions will be more difficult to obtain than for single-phase machines with auxiliary winding. As a result there will be more extra losses and heating of the

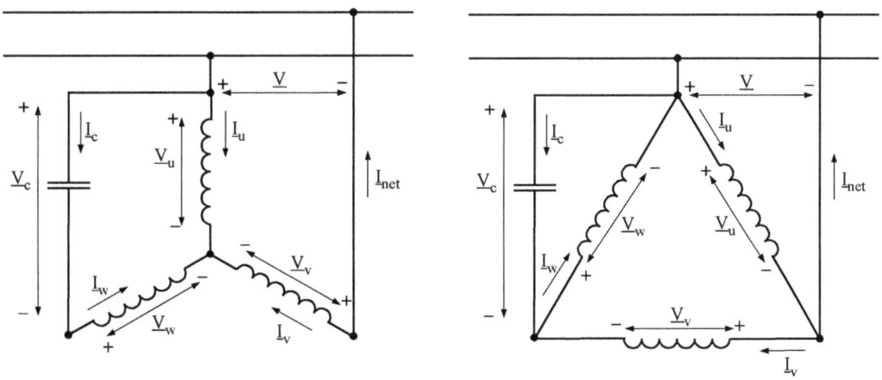

Fig. 4.31 Steinmetz connection

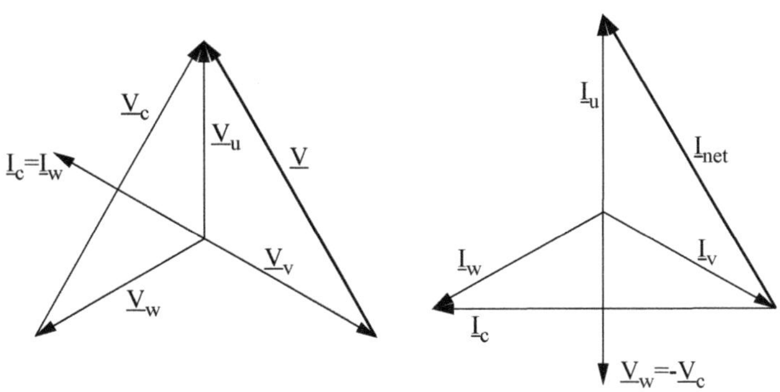

Fig. 4.32 Steinmetz: phasor diagrams

machine. Moreover, in most cases a start capacitor will not be present, leading to a low starting torque, limiting its use to starting under no-load.

4.11.2.4 Shaded-Pole Motors

A shaded-pole motor has a normal squirrel cage rotor, but a single-phase fed stator with a special electromagnetic construction to obtain a rotating field. The coils around the salient poles in the stator are fed by the mains. One side of these poles is shielded electromagnetically by a short-circuited winding (or ring) in a slot (see Fig. 4.33, for a 4-pole machine).

As the short-circuited winding acts as the secondary of a transformer, the current in it will oppose the alternating field excited by the mains-fed coils. If the coupling between the main coils and the shorted ring were ideal, the field in the shielded part would vanish completely. Because of the incomplete coupling (leakage field) and the resistance of the ring, there remains a field in the shielded part δe ((a) in Fig. 4.34). This field, however, lags the field in the non-shielded part. As the symmetry axis of the shielded part is also displaced in space with respect to the non-shielded part, a rotating field, although imperfect, is created. Mathematically this can be elaborated as follows, see also (b) in Fig. 4.34.

With an ideal coupling, we would have (the prime indicating referred to the number of turns of the short-circuit ring)

$$w_c \underline{I}_c + w_e \underline{I}_e = 0 \tag{4.127}$$

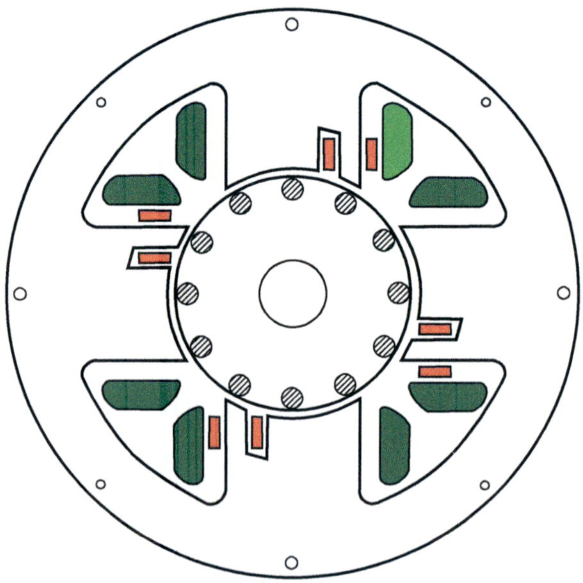

Fig. 4.33 Shaded pole motor

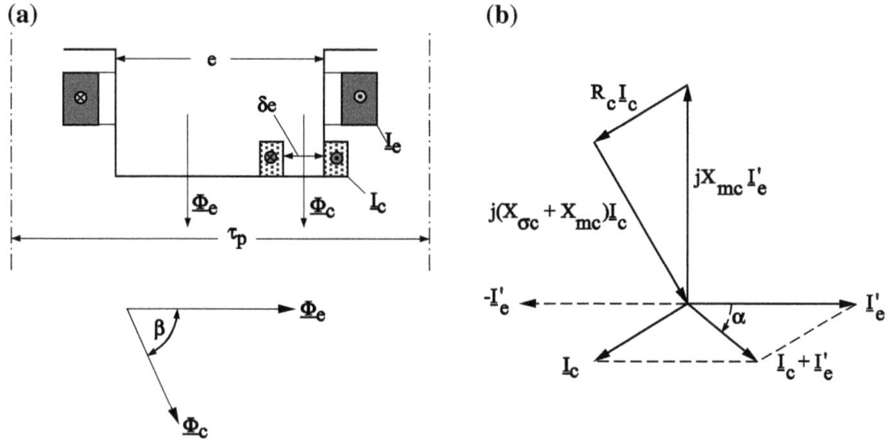

Fig. 4.34 Schematic of shaded pole motor

or

$$\underline{I}_c = -\underline{I}_e^{'} = -\underline{I}_e \cdot \frac{w_e}{w_c} \tag{4.128}$$

In this case, the field in the shielded part would vanish completely, $\underline{\phi}_{mc} = X_{mc}(\underline{I}_c + \underline{I}_e^{'}) \sim w_c \underline{I}_c + w_e \underline{I}_e = 0$.

In reality, due to the leakage field and resistance of the short-circuit ring, we have

$$R_c \underline{I}_c + jX_{\sigma c}\underline{I}_c + jX_{mc}(\underline{I}_c + \underline{I}_e^{'}) = 0 \tag{4.129}$$

or

$$R_c \underline{I}_c + j(X_{\sigma c} + X_{mc})\underline{I}_c = -jX_{mc}\underline{I}_e^{'} \tag{4.130}$$

As depicted in Fig. 4.34 the flux $\underline{\phi}_{mc} = X_{mc}(\underline{I}_c + \underline{I}_e^{'})$ lags the flux $\underline{\phi}_{me} = X_{me}\underline{I}_e$ by an angle α. But the symmetry axes of the fluxes $\underline{\phi}_{mc}$ and $\underline{\phi}_{me}$ (thus the axes of the parts δe and $(1-\delta)e$ of the pole) are also shifted in space, by an angle β. By decomposing both alternating fields into two equal and counter-rotating fields we obtain a situation as depicted in Fig. 4.35. The left rotating fields may almost completely annihilate each other while the right rotating fields add up. The result is a large right rotating field and a small relic of the left rotating field. A pure rotating field requires $\underline{\phi}_{mc} = \underline{\phi}_{me}$, $\alpha + \beta = \pi$ and $\alpha = \beta$ or thus $\alpha = \beta = \pi/2$. Although a time shift of 90° is quite feasible, a space shift of 90° is physically impossible (as this would boil down to a magnetic short-circuit of adjacent north and south poles).

Another way to obtain a rotating-field effect is to use a step in the air-gap, see (a) in Fig. 4.36. Where the air-gap is larger, the coupling of the rotor with the stator is smaller. That part of the rotor has therefore a smaller air-gap inductance and a larger leakage. Seen from the stator, the rotor may be modelled as the series connection of

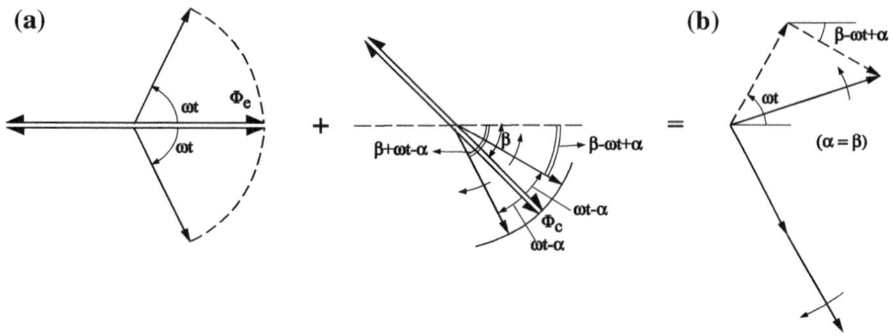

Fig. 4.35 Field of shaded-pole motor

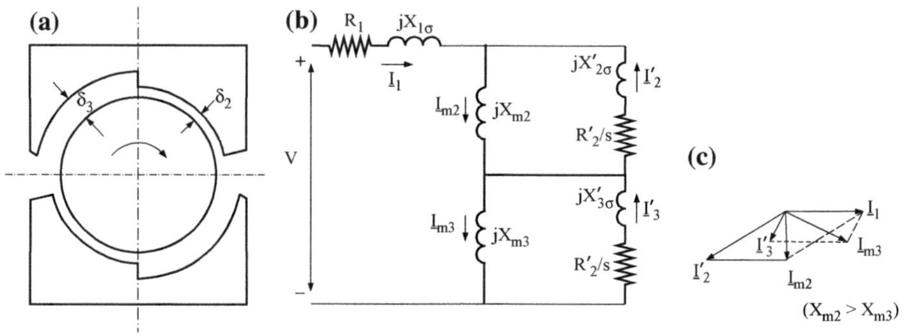

Fig. 4.36 Stepped air-gap motor

Fig. 4.37 Torque of shaded-pole motor

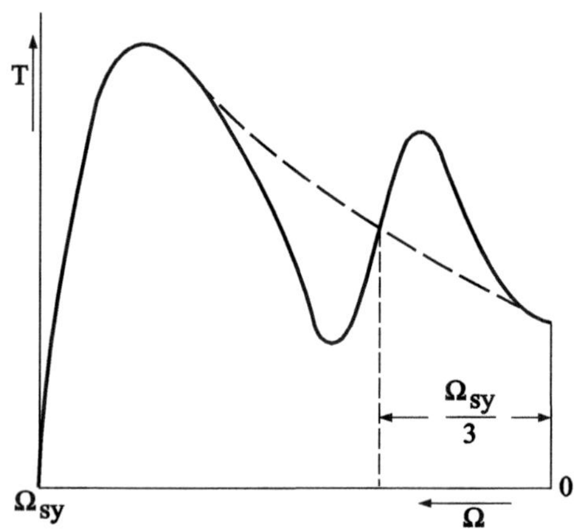

two parts, one with a larger air-gap inductance and small leakage and the other one with a smaller air-gap inductance and larger leakage inductance, see (b) in Fig. 4.36. Because of the above and with $\underline{I}_1 = \underline{I}_{m2} - \underline{I}_2' = \underline{I}_{m3} - \underline{I}_3'$ we easily obtain a phasor diagram like (c) in Fig. 4.36. The magnetising current for the smaller air-gap is lagging the magnetising current for the larger air-gap. Thus also the air-gap field for the smaller air-gap will lag the one for the larger air-gap. Because of the displacement in space we will again obtain a rotating-field effect.

A shaded pole and a step in the air-gap may also be combined to reinforce the rotating-field effect (i.e. to reduce the inverse rotating field).

A disadvantage of all single phase motors with concentrated windings (and thus also of the shaded pole motor or the stepped air-gap variant) is that the concentrated winding gives rise to mmf harmonics, in particular the third mmf harmonic. The torque characteristic will then show a distortion at 1/3rd of the synchronous speed (in addition to torque components from the inverse rotating field), Fig. 4.37.

The third harmonic may be reduced by shaping the stator pole, by slanting the rotor bars,.... It is also possible to use a distributed stator winding to eliminate the third harmonic. The combination with a step in the air-gap may also be used to reduce the third mmf harmonic.

Chapter 5
The Synchronous Machine

Abstract In this chapter we discuss the synchronous machine mainly from its traditional function as generator. Motoring, in particular with permanent magnet synchronous motors, is treated in a later chapter. The first section gives an overview of the main two rotor types and their properties. In the following section, the smooth rotor synchronous machine is treated in detail, including armature reaction, phasor diagrams, equivalent circuits, current diagram, torque but also a thorough discussion of the non-linear generator characteristics. In the next section, salient pole synchronous machines are treated. One of the last sections is devoted to the operation (mainly as generator) on an infinitely strong grid.

5.1 Introduction - Construction

Whereas the induction machine is used mainly (but not exclusively) as a motor, the synchronous machine is the obvious machine at the generating side of the grid. Indeed, thanks to the (DC or permanent-magnet) excitation on the rotor the synchronous machine is able to supply reactive power in addition to its main role for active power conversion.

The synchronous machine is, however, also used for motoring, either for very large power, or in applications where special requirements (efficiency, speed control accuracy, dynamic properties, compactness,...) justify its higher acquisition costs. The main electromagnetic parts of a synchronous machine are:

- the primary with a poly-phase (in most cases three-phase) winding, mostly on the stator
- the excitation, either a DC field winding or a permanent magnet excitation, in most cases on the rotor

For a DC excited field one may discern two designs:

- salient-pole synchronous machines (see a, b in Fig. 5.1). The field winding is a concentrated winding around poles, similar to the excitation winding of a DC machine. The number of poles $2N_p$ (of alternating polarity around the circumference) is normally quite large for these machines. The magnetic core (of the rotor

J. A. Melkebeek, *Electrical Machines and Drives*, Power Systems,
https://doi.org/10.1007/978-3-319-72730-1_5

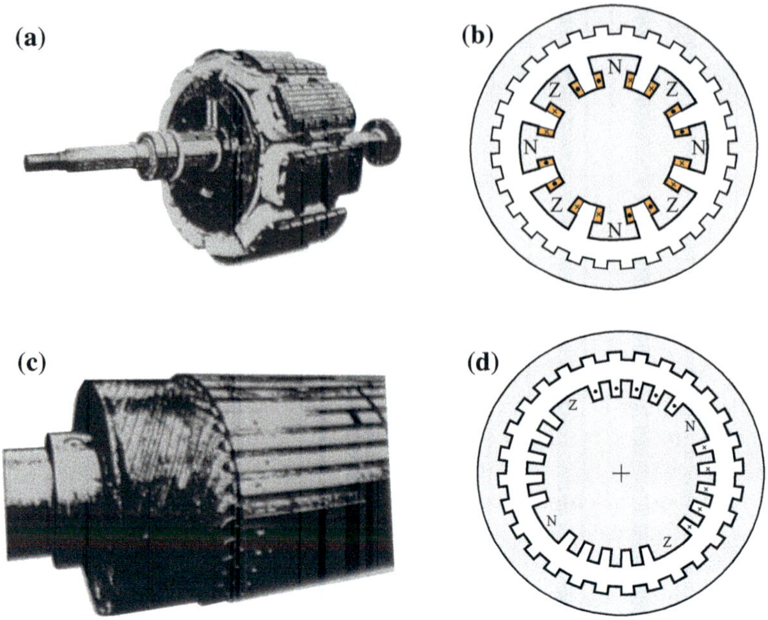

Fig. 5.1 Rotor constructions

poles) can be massive iron, but is in many cases laminated. The air gap under the pole shoes can be widened slightly at the ends in order to obtain a more sinusoidal magnetic field shape. Because of the salient poles the reluctance around the circumference varies strongly, with a minimum along a pole axis and a maximum in between the poles. These low speed machines with a large diameter and relatively short armature length are the typical generators in hydraulic or diesel power stations.

- smooth rotor synchronous machines (see c, d in Fig. 5.1). The field winding is a distributed DC winding with a low number of poles, $2N_p = 2$ or 4. In most cases the magnetic core is massive, as these synchronous machines are typical for large turbo generators. Indeed, the high power ratings, up to 2 GW, and the high speeds require long forged rotor structures. Further, because of the DC excitation, the massive poles are not much of a problem (and the eddy currents in the iron yield the required damping for oscillations). These machines are called smooth-rotor machines because of the theoretical symmetry of the two electrical axes (pole axis and intermediate axis). In reality, however, the reluctance along a pole axis is a little higher than along an intermediate axis,[1] but the difference is quite small. The armature length of these machines is mostly a multiple of their diameter.

---

[1]The slots are usually omitted at the centre of the poles, as the mmf of these slots would not add much to the field but would introduce additional reluctance.

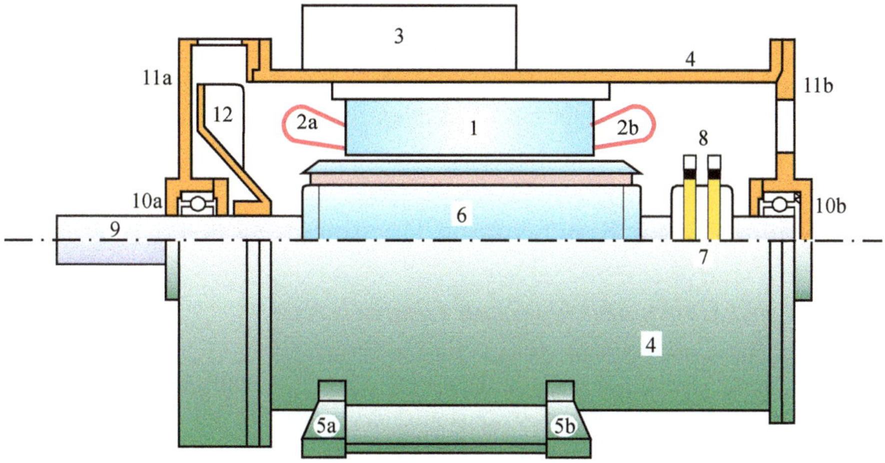

1	stator core (laminated) with poly-phase winding
2a, 2b	connection box
3	case
4	feet
5a, 5b	pole with pole shoe (massive) and field coil
6	slip rings
7	brush holder with brushes
8	shaft with shaft end
9	bearings
10a, 10b	bearing caps
11a, 11b	ventilator
12	ventilator cap

Fig. 5.2 Cross section of a synchronous machine with DC excitation and slip rings

In addition to these DC-excited synchronous machines, there also exist permanent-magnet excited synchronous machines although these machines are mainly used in motoring (e.g. robotics). For these PM machines, axial field designs gained a lot of attention in recent years, e.g. for electric vehicles where their small axial length makes them quite feasible as wheel motors. In essence, these PM machines behave as 'normal' DC-excited synchronous machines (see Part 3, Chap. 20), but their excitation is fixed and, if required, the induced emf must be influenced by other means.

Compared to induction machines, synchronous machines normally have a much higher air gap. Indeed, a small air gap is of primordial importance for induction machines as the (for the magnetic field) required reactive current from the supply should be kept at a minimum. For synchronous machines the DC excitation is about 'free' and the larger air-gap offers a better reliability (especially for large power ratings). In addition, for DC-excited machines a larger air gap yields a smaller ratio of maximum to minimum excitation which results in easier excitation control.

Other constructional details of synchronous machines (generators) are illustrated in Fig. 5.2.

In the sequel of this chapter, we will use the G-convention (with at the electrical side the GRS or Generator Reference System) for the synchronous machine as we will concentrate on its use as a generator (although in Part 3 we will use the M-convention, with the URS at the electrical side).

5.2 Smooth Rotor Synchronous Machines

5.2.1 *Field Curve and No-Load Characteristic*

The field curve of a smooth rotor synchronous machine ('turbo-rotor') with distributed DC winding has an almost sinusoidal shape, see Fig. 5.3. Denoting by x' a linear co-ordinate on the rotor (with $x' = 0$ along the north pole axis) we may write

$$b_p(x', t) = \sum_{\nu} \hat{B}_{p\nu} \cos\left(\nu \frac{x'\pi}{\tau_p}\right) \approx \hat{B}_{p1} \cos\left(\frac{x'\pi}{\tau_p}\right) \tag{5.1}$$

The higher harmonics in the field curve are indeed suppressed by the winding factors of the distributed winding, see Chap. 3:

$$\hat{F}_{p\nu} \cong \frac{2}{\pi}.(-1)^{\frac{\nu-1}{2}}.\frac{w_f \xi_{f\nu}}{\nu N_p} i_f \tag{5.2}$$

where w_f represents the total number of series connected field windings, $\xi_{f\nu}$ the winding factor for the νth space harmonic and i_f the DC excitation current.

If the reluctance of the iron core can be neglected, the almost constant air-gap results in an induction curve that is proportional to the mmf:

$$\hat{B}_{p\nu} \approx \hat{F}_{p\nu} \frac{\mu_o}{\tau_p} \tag{5.3}$$

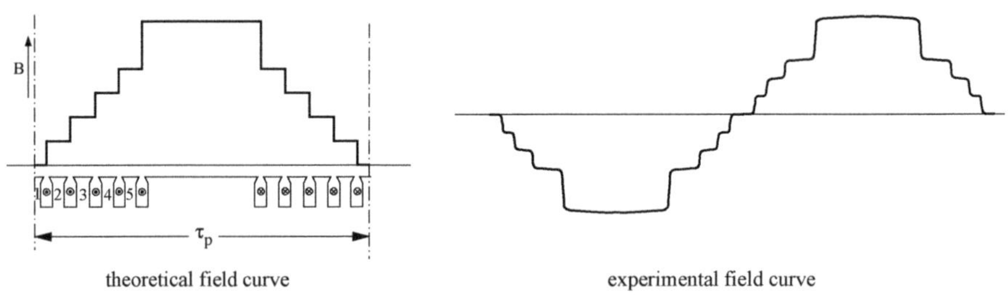

Fig. 5.3 Field curves

In the sequel of the analysis, we will neglect all higher harmonics in the field (and omit the subscript "1"). When the rotor rotates with a mechanical speed Ω, a fundamental harmonic rotating field with respect to the stator (the 'armature') is created:

$$b_p(x,t) = \hat{B}_p \cos\left(\frac{x\pi}{\tau_p} - \omega t\right) \tag{5.4}$$

with $\omega = \Omega \cdot N_p$ (see also Chap. 3).

This rotating field induces a symmetrical three-phase emf in the stator winding with angular frequency ω. The corresponding phase voltage in the reference phase 1U is given by (remark the + sign because of the GRS)[2]:

$$e_p(t) = \hat{E}_p \sin \omega t \tag{5.5}$$

where the emf amplitude is given by[3]

$$\hat{E}_p = \omega \cdot \left(w\xi_1/N_p\right) \cdot \hat{\Phi}_p \tag{5.6}$$

with the flux over one pole pitch centred around x = 0 (the axis x = 0 supposed co-incident with the rotor north pole axis $x' = 0$ at the instant t = 0) given by

$$\phi_p(t) = \hat{\Phi}_p \cos \omega t \tag{5.7}$$

$$\hat{\Phi}_p = \frac{2}{\pi} \cdot l \cdot \tau_p \cdot \hat{B}_p \tag{5.8}$$

In phasor (time vector) form this is written as:

$$\underline{E}_p = -j\omega \cdot (w\xi_1/N_p) \cdot (\underline{\hat{\Phi}}_p/\sqrt{2})$$

(in the GRS, the voltage is indeed lagging the corresponding flux).

Higher space harmonics in the field curve also induce voltages in the armature winding (although with a higher frequency, see Chap. 3), but because of the much lower winding factors these induced voltages are negligible.

At constant speed, the effective (or amplitude) value of the fundamental induced voltages is thus proportional to the fundamental flux over a pole pitch, $\hat{\Phi}_p$.

With a variable excitation current, saturation of the iron of core and poles results in a non-linear curve of the emf as a function of the excitation current, see Fig. 5.4. The ordinate is the emf amplitude (or effective value), which is proportional to the flux per pole as mentioned before. This curve is called the no-load characteristic of

---

[2]The subscript p is used whenever quantities are referred to the armature; thus although this emf results from a DC field current ('f') we use the subscript 'p'.

[3]This is the emf for all windings or conductors in series of one single pole-pair.

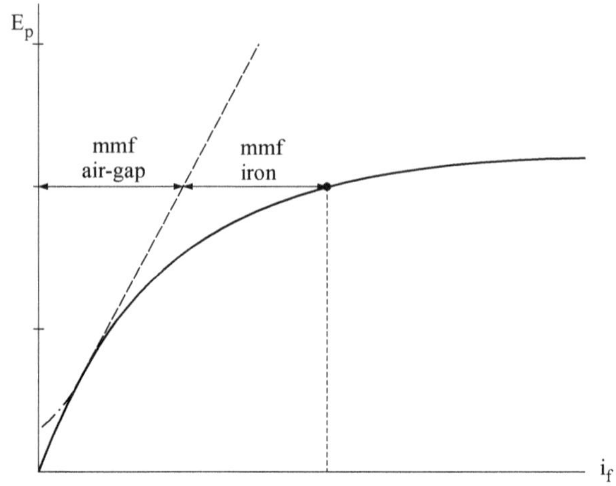

Fig. 5.4 No-load characteristic

the synchronous machine. It allows to derive the required excitation current for a given emf value. As the excitation is a DC current, the real no-load characteristic does not go through the origin but is in fact a hysteretic curve (see the dot-dash line in Fig. 5.4) with a remanent induction of about 5% of the rated voltage.

5.2.2 *Armature Reaction*

If the stator three-phase winding is connected to a symmetrical load (or to the grid or other supply with the right frequency), a symmetrical three-phase current results with the same frequency as the emf, $\omega = N_p \cdot \Omega$. The corresponding stator current sheet with $2N_p$ poles rotates also with the speed Ω and yields a rotating mmf with the same speed as the rotor. The amplitude of this armature mmf is given by (see Chap. 3):

$$\hat{F}_a = \frac{3}{2} \cdot \frac{4}{\pi} \cdot \frac{w\xi_1}{2N_p} \cdot \hat{I}_1 = \frac{3}{\pi} \cdot \frac{w\xi_1}{N_p} \cdot \hat{I}_1 \tag{5.9}$$

As a result, there are now two rotating mmfs in the air-gap, thus two sine waves with the same (mechanical) speed Ω and the same wavelength, i.e. the rotor mmf $f_p(x, t)$ and the armature mmf $f_a(x, t)$. These two mmfs add *vectorially* to the resulting mmf $f_r(x, t)$ which determines the resulting rotating magnetic field $b_r(x, t)$ in the air-gap.

Indeed, the mmfs must be added vectorially as, generally, these mmfs are not in phase and their phase difference depends on the load.

The maximum of the rotor mmf is always coincident with the north pole of the rotor. Thus an observer (at standstill) on the symmetry axis of the reference phase 1U (see (a) in Fig. 5.5) will see the maximum rotor mmf at the instant the rotor

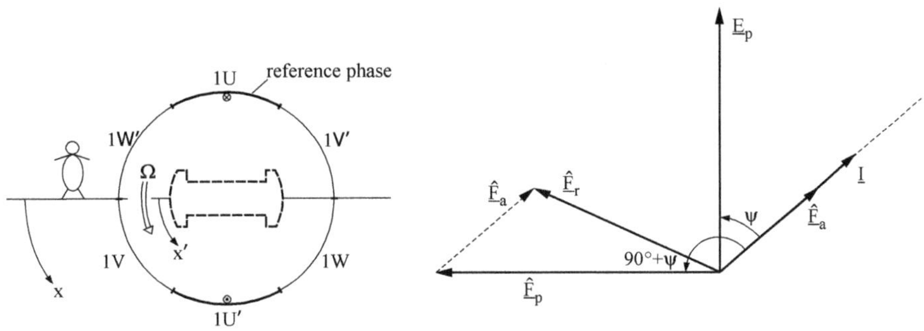

Fig. 5.5 Armature reaction and phasor diagram

north pole passes along his position. This is also the instant when the rotor emf $e_p(t)$ induced in 'his' reference phase 1U goes through zero (i.e. this is the position of the maximum flux coupled with this reference winding). The rotor emf $e_p(t)$ (i.e. the voltage induced in the reference phase if only this rotor mmf were present) will attain its maximum value 90° later (see Sect. 3.2.1 but now in the GRS), thus 90° after the north pole of the rotor is along the axis of the reference phase.

Connecting now a symmetrical load to the stator winding, a symmetrical stator current will result. Suppose this current lags the (original) emf $e_p(t)$ by an angle ψ. The armature mmf $f_a(x,t)$ resulting from this symmetrical current will lie along the axis of the reference phase 1U at the instant the current in the phase 1U is maximal. Note also that the maximum of the corresponding rotating armature current layer lies along the centre of the phase belt of phase 1U where this current at that instant is positive (indeed, the mmf is the integral of the current layer and will therefore lead the current layer by 90°: when the current layer is at the centre of the phase belt the corresponding mmf will lie along the axis of this phase).

The phasors of the sinusoidal time functions emf $e_p(t)$ and current $i(t)$ can be represented as $\underline{E}_p$ and $\underline{I}$ illustrated in (b) in Fig. 5.5.

The mmfs, however, are space quantities, i.e. functions of the space co-ordinate x and time t, and a phasor representation of these does not speak for itself. But for the analysis we might replace (or represent) these mmfs by the time quantities of the corresponding fluxes coupled with the reference winding 1U, i.e. $\Psi_p(t)$ for the mmf $f_p(x,t)$ and $\Psi_a(t)$ for the mmf $f_a(x,t)$. These fluxes are maximal for the reference phase when the corresponding mmfs lie along the axis of the reference phase, i.e. at the position of our observer in (a) in Fig. 5.5. When this observer *could* look at the mmfs in the air-gap he/she would see that these mmfs have their maximum value along the axis of phase 1U at the same instant the corresponding fluxes coupled with the reference winding are maximal. Therefore we may use the notation of mmf 'phasors' instead of phasors for the corresponding fluxes,[4] see (b) in Fig. 5.5.

[4]One could also consider the phasors for the time functions $f_{..}(0,t)$ with $x=0$ along the axis of the reference phase.

The phasor $\underline{\hat{F}}_a$ is in phase with the current phasor $\underline{I}$: when the current is maximal in the reference phase, the flux coupled with the reference phase due to the armature currents is also maximal at that instant.

The phasor $\underline{\hat{F}}_p$ leads the emf $\underline{E}_p$ by 90°: when the phasor $\underline{\hat{F}}_p$ lies along the reference winding axis, the flux coupled with the reference winding due to this mmf is maximal (and the induced voltage is zero at that instant); the induced voltage due to this mmf or flux is maximal 90° later (GRS) - which this same observer on the axis of the phase could see on an oscilloscope connected to the phase winding -.

The vectorial sum of the mmfs $\underline{\hat{F}}_p$ and $\underline{\hat{F}}_a$ is equal to the resulting mmf $\underline{\hat{F}}_r$. This resulting mmf determines the resulting air-gap field $\underline{\hat{B}}_r$ and thus the resulting induced voltage $\underline{E}_r$ which lags it by 90°.

Dependent on the delay angle ψ (called the *internal angle*) between the no-load emf $\underline{E}_p$ and the armature current $\underline{I}_a$ we may discern three cases:

- $\psi = \pi/2$ (or, more general, $\pi/2 \geq \psi > 0$): in this case the amplitude of the resulting mmf f_r is lower than that of the mmf f_p and thus the resulting emf E_r decreases in comparison with the no-load emf E_p (this is also called field weakening). As observed from the point of view of the no-load emf, the machine delivers reactive power to the load.
- $\psi = -\pi/2$ (or, more general, $-\pi/2 \leq \psi < 0$): in this case the amplitude of the resulting mmf f_r is higher than that of the mmf f_p and thus the resulting emf E_r increases in comparison with the no-load emf E_p. From the point of view of the no-load emf, the machine uses reactive power from the load.
- $\psi = 0$: there is merely an active power exchange with the load (at least as seen from the point of view of the no-load emf) while the emf does not change (significantly).

In the general case of $\psi \neq \pm\pi/2$ the armature current can, if desired, be split up into two components: one component in the direction of $\underline{\hat{F}}_p$ and another orthogonal to it. The component in the direction of $\underline{\hat{F}}_p$ results in a decrease or an increase of the resulting mmf and thus of the resulting emf (compared to the no-load case). This *longitudinal* effect is called *demagnetising* and *magnetising*, respectively. The orthogonal component gives a slight *transverse* magnetising effect, but the main effect is active power conversion (at least with respect to the no-load emf). It should be stressed however that the active or reactive power exchange with the load or supply should be regarded from the terminal voltage and not from the emf (see the next section). In addition, when the machine is loaded, i.e. when there is a non-zero armature current, the no-load emf does not exist any more and the only physical emf is the resulting emf.

Space Vector Representation

The mathematical basis of space vectors has already been introduced in Chap. 3 (and for a more elaborated explanation and discussion, see Appendix C):

- phasors (or time vectors) are a complex representation of sinusoidal time functions like $\hat{A}\cdot\cos(\omega t-\varphi)$; the phasor representation in the complex plane is $\hat{A}\cdot\exp(-j\varphi)$. To return to the real time representation, multiply by $\exp(j\omega t)$ and take the real part of the product. The delay angle φ is recovered in the complex plane as the phase <u>delay</u> of the phasor with respect to the real axis.[5]

- space vectors are a complex representation of sinusoidal space functions like $\hat{A} \cdot \cos(\frac{x\pi}{\tau_p} - \varphi)$; the space vector representation in the complex plane is $\underset{\rightarrow}{a} = \hat{A} \cdot \exp(j\varphi)$. To return to the real space representation, multiply by $\exp(-j\frac{x\pi}{\tau_p})$ and take the real part of the product. The 'delay' angle φ is recovered in the complex plane as the phase <u>advance</u> of the space vector with respect to the real axis. Note that φ may, for example, also be a function of time (*not* necessarily a *sinusoidal* function of time).

The background for the difference as to the handling of the sign of φ, is that in real time a positive φ (in the above time example) corresponds to an event that occurs *later in time*, thus lagging. To the contrary, in space, a positive φ in the above space function example corresponds to something that occurs *farther in space*. When we consider for example a rotating field $\hat{A} \cdot \cos(\frac{x\pi}{\tau_p} - \omega t - \varphi)$ we see that the maximum in $x = 0$ occurs at $t = -\varphi/\omega$ whereas the maximum at $t = 0$ is found in $x = \varphi \cdot \tau_p/\pi$. The field $\hat{A} \cdot \cos(\frac{x\pi}{\tau_p} - \omega t - \varphi)$ is indeed leading[6] in time <u>and</u> space a field like $\hat{A} \cdot \cos(\frac{x\pi}{\tau_p} - \omega t)$. The diagram in Fig. 5.5 is in fact a phasor diagram. The vectors are time vectors or phasors for the reference phase quantities (voltage, current,...). These phasors allow to derive their mutual phase relationships (as well as the absolute phase, if the time axis is specified). However, the diagram can also be regarded as a space vector diagram which, in particular for the mmfs, shows the instantaneous space position of these mmfs in a two-pole representation of the machine.

The direction of the mmf $\underset{\rightarrow}{\hat{F}}_p$ is coincident with the (rotating) north pole of the rotor. For an observer fixed to the rotor, the whole diagram (a) in Fig. 5.6 is at rest. For an observer fixed to the stator or stator windings (see (b) in Fig. 5.6), this diagram rotates counter-clockwise[7] with the angular frequency of the machine. The flux coupled with a phase winding of the stator is maximal when the north pole axis of the rotor is coincident with the axis of this stator phase. When the rotor rotates further over 90°, the emf induced in the reference winding is maximal, i.e. when the emf vector $\underline{E}_p$ is along the axis of the reference phase. The rotor north pole axis is usually called the d-axis (direct axis) while the axis 90° in the opposite direction of rotation (along $\underline{E}_p$) is called the q-axis (quadrature axis).[8]

---

[5]There also exists a complex time representation, which is nothing else but the phasor multiplied by $\exp(j\omega t)$.

[6]Please, compare the phasor and space vector representations of both $\hat{A} \cdot \cos(\frac{x\pi}{\tau_p} - \omega t - \varphi)$ and $\hat{A} \cdot \cos(\frac{x\pi}{\tau_p} - \omega t)$.

[7]In the complex space plane, we choose the counter-clockwise direction as the positive rotation direction, like for the complex time plane.

[8]Many authors choose the q-axis in the other direction, thus 90° leading the d-axis; however in a 1968 IEEE recommendation the 90° lagging direction is preferred.

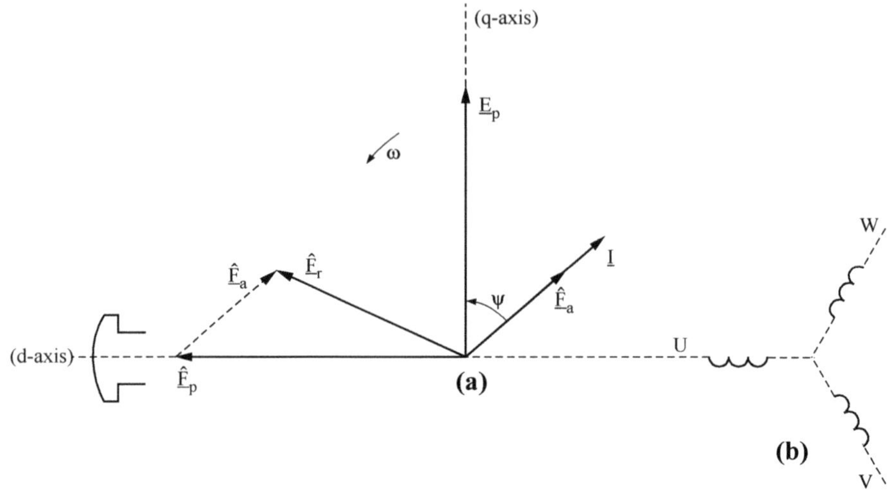

Fig. 5.6 Space vector representation

For the mmfs $\hat{\underline{F}}_p$, $\hat{\underline{F}}_a$, $\hat{\underline{F}}_r$, the interpretation as space vectors is in complete accordance with the mathematical definition of space vectors, as they are of the form $\hat{F} \cdot \cos(\frac{x\pi}{\tau_p} - \varphi)$. For the emfs or voltages and the current vectors in this diagram an interpretation as space quantities is somewhat less obvious. For the armature current, one could consider the corresponding current sheet which is in fact also a rotating space wave. For the current space vector, one should consider the vector along the <u>*axis*</u> *of the current layer (and not, as usual for a current layer, a vector along the maximum of the current layer). For emfs (or voltages), the notion of space vectors is even less obvious and may look somewhat artificial. One way to interpret emfs as space quantities is as follows: one considers a uniform distribution of conductors along the circumference of the machine; a (sinusoidal) rotating mmf will induce emfs in these conductors and the distribution of these emfs along the circumference will be sinusoidal in space (and time), i.e. of the form $\hat{E} \cdot \cos(\frac{x\pi}{\tau_p} - \varphi)$; as the corresponding space vector we consider the vector along the* <u>*axis*</u> *of this distribution (and not along the maximum of the emf), just like for the current space vector. See also Appendix C.*

5.2.3 Phasor Diagram of Voltages and Currents

For the synchronous machine with smooth rotor, the resulting air-gap mmf follows from the vectorial addition of the rotor mmf and the armature mmf:

$$\hat{\underline{F}}_r = \hat{\underline{F}}_p + \hat{\underline{F}}_a \tag{5.10}$$

This resulting mmf determines the resulting air-gap induction $\underline{\hat{B}}_r$ or flux $\underline{\hat{\Psi}}_r$ and thus the resulting emf $\underline{E}_r$. Note that an addition of the fluxes $\underline{\hat{\Psi}}_p$ and $\underline{\hat{\Psi}}_a$ - which would exist if the rotor mmf or the armature mmf existed separately - is not allowed as saturation of the magnetic circuit precludes an addition of induction or fluxes. Instead, the mmfs must be added and from the resulting mmf the air-gap induction or flux can be determined using the no-load characteristic. In order to derive the amplitude of the resulting flux $\underline{\hat{\Psi}}_r$ or emf $\underline{E}_r$ from the no-load characteristic, the armature current must be rescaled to an equivalent excitation current $i' = \alpha \cdot I$, where α is determined by the equivalence of the resulting fundamental air-gap mmf. With w_f the number of turns of the excitation winding and w the number of series connected turns of one phase of the armature, we have

$$\frac{2}{\pi} \cdot \frac{w_f \cdot \xi_{f1}}{N_p} \cdot i' = \frac{3}{\pi} \cdot \frac{w \cdot \xi_1}{N_p} \cdot \hat{I} \tag{5.11}$$

or, as $\hat{I} = \sqrt{2} I$,

$$\alpha = \frac{3}{\sqrt{2}} \cdot \frac{w \cdot \xi_1}{w_f \cdot \xi_{f1}} \tag{5.12}$$

Using α we may rewrite Eq. 5.10 as follows, introducing an equivalent magnetising current $\underline{i}_m$

$$\underline{i}_m = \underline{i}_f + \underline{i}' \tag{5.13}$$

From the no-load characteristic the amplitude or effective value of the resulting emf E_r can now be determined as the ordinate corresponding with i_m, see Fig. 5.7.

If required, the resulting mmf can be derived from $\underline{i}_m$ as $\underline{\hat{F}}_r = \frac{2}{\pi} \cdot \frac{w_f \cdot \xi_{f1}}{N_p} \cdot \underline{i}_m$

The triangle corresponding with the vector addition Eq. 5.10 can also be replaced by a similar triangle for the magnetising currents, according to Eq. 5.13, as illustrated in Fig. 5.8.

The resulting emf vector $\underline{E}_r$ lags (GRS!) the vector $\underline{i}_m$ by 90°, see Fig. 5.8.

To obtain the terminal voltage $\underline{V}$ we have to subtract from $\underline{E}_r$ the resistive voltage drop in the armature $R_s \cdot \underline{I}$ and the reactive voltage drop due to the leakage field of the armature, $jX_{s\sigma} \cdot \underline{I}$, see Fig. 5.8.

$$\underline{E}_r = \underline{V} + R\underline{I} + jX_\sigma \underline{I} \tag{5.14}$$

The phasor diagram of the synchronous machine with smooth rotor in Fig. 5.8 is also called the Potier diagram. It consists of two parts, an mmf or current part and a voltage part. Using this phasor diagram and the no-load characteristic, the excitation current required for a given terminal voltage and given load (I, $\cos\varphi$) can be determined. Indeed, from V, I and φ we derive E_r using Eq. 5.14. The no-load characteristic then yields the magnetising current i_m corresponding with E_r. The magnetising current

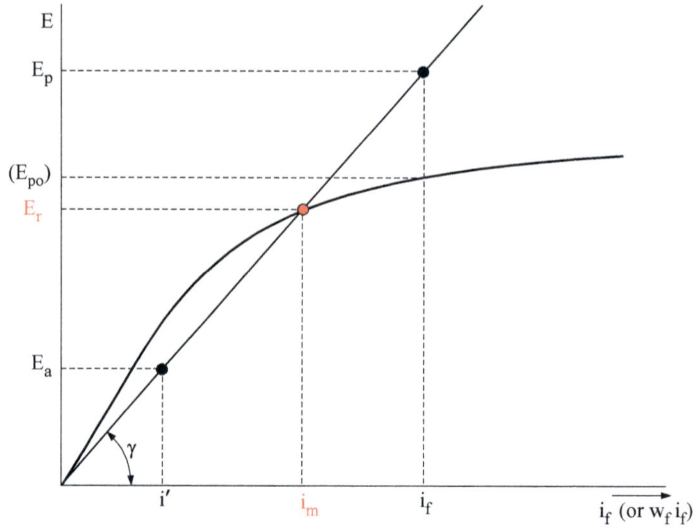

Fig. 5.7 Resulting emf from the no-load characteristic

Fig. 5.8 Basic SM phasor diagram

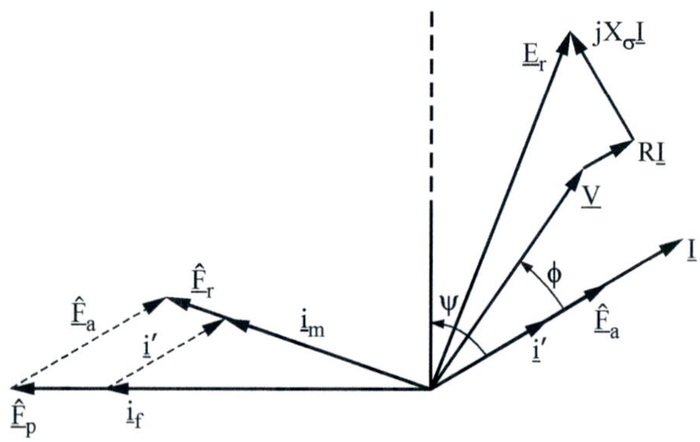

phasor $\underline{i}_m$ is therefore known in magnitude and phase (as it is leading $\underline{E}_r$ by 90°). From $\underline{i}_m$ and $\underline{i}' = \alpha \cdot \underline{I}$ we obtain $\underline{i}_f$.

Note that the greater the inductive loading of the machine, the larger the required excitation will be, as already derived before.

For the inverse problem, i.e. determining the terminal voltage corresponding with a given excitation current and load (I, $\cos\varphi$), only a iterative procedure can be followed.

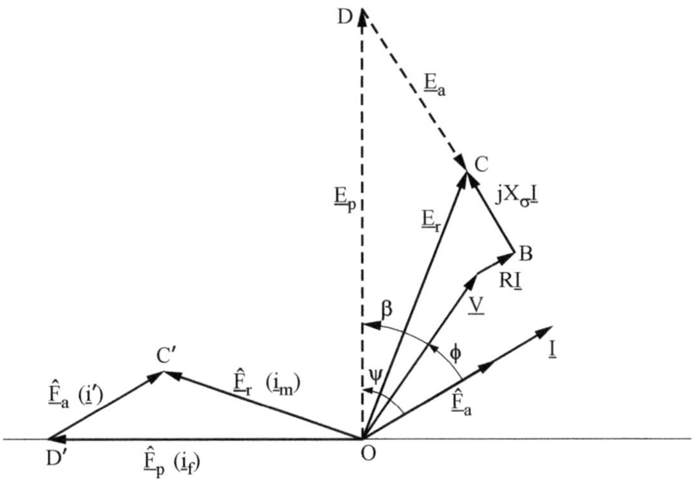

Fig. 5.9 Linearised SM phasor diagram

5.2.4 Linearised Equivalent Circuit of a Smooth Rotor Synchronous Machine

Constructing now a triangle OCD similar to the triangle OC'D' of the mmfs or the magnetising currents on the side of the emf vector $\underline{E}_r$ we may write $\underline{E}_r = \underline{OC}$ as the sum of vectors $\underline{E}_p = \underline{OD}$ and $\underline{E}_a = \underline{DC}$. Because of the similarity of the triangles, the magnitudes OD and OC can be derived from the construction in Fig. 5.7. We also have

$$\frac{OD}{i_f} = \frac{DC}{i'} = \frac{OC}{i_m} = \frac{E_r}{i_m} \tag{5.15}$$

In this way we have in fact linearised the synchronous machine around the operating point defined by $\underline{E}_r$. Indeed, the magnitude of $\underline{E}_r$ (or i_m) defines the flux level and thus the saturation level of the machine. The saturation characteristic is for this operating point linearised to the line through the origin and the point (i_m, E_r), Fig. 5.7.

It is important to stress that both $\underline{E}_p = \underline{OD}$ and $\underline{E}_a = \underline{DC}$ are **fictitious voltages** which *would* be induced by the mmfs $\underline{\hat{F}}_p$ and $\underline{\hat{F}}_a$ respectively if these *were* to act separately and at the saturation level defined by E_r or i_m. Remark also that $\underline{E}_p$ is not at all equal to the no-load emf $\underline{E}_{po}$ corresponding to the given excitation current i_f.

The phasor equations corresponding to Fig. 5.9 can be written as:

$$V = \underline{E}_r - R\underline{I} - jX_\sigma \underline{I} \tag{5.16}$$

$$\underline{E}_r = \underline{E}_p + \underline{E}_a \tag{5.17}$$

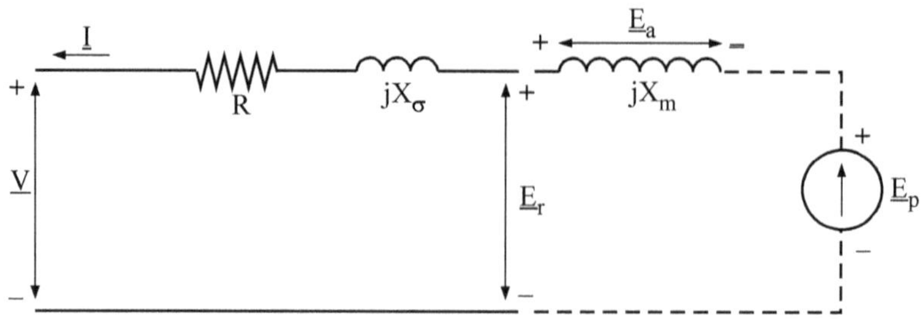

Fig. 5.10 Equivalent circuit of smooth rotor SM

where $\underline{E}_a$ is lagging $\underline{I}$ by 90°. Its algebraic value is proportional toi', thus with I. Defining the magnetising reactance $X_m = \alpha \cdot \tan\gamma$, $\underline{E}_a$ can thus be written as

$$\underline{E}_a = -jX_m\underline{I} \tag{5.18}$$

Therefore

$$V = \underline{E}_p - R\underline{I} - jX_\sigma\underline{I} - jX_m\underline{I} \tag{5.19}$$

or

$$V = \underline{E}_p - R\underline{I} - jX\underline{I} \tag{5.20}$$

The *total armature reactance* $X = X_\sigma + X_m$ is called the synchronous reactance and consists of the *magnetising reactance* X_m corresponding with that part of the armature field linked with the rotor and the *leakage reactance* X_σ corresponding with that part of the field linked only with the armature. $Z = R + jX$ is called the *synchronous impedance*. The magnetising reactance is sometimes called the *armature reaction component* of the synchronous reactance.

The equivalent circuit corresponding with these equations is shown in Fig. 5.10. The left part of the equivalent circuit (including $\underline{E}_r$) is always valid while the right part (dashed lines) is the linearised approximation valid only for the saturation level defined by E_r.

The phasor diagram in Fig. 5.9 indicates also the three important angles for a synchronous machine: the *internal angle* $\psi = \arg(\underline{E}_p, \underline{I})$, the *load angle* (also called *torque angle* or *power angle*) $\beta = \arg(\underline{E}_p, \underline{V})$ and the *phase angle* $\varphi = \arg(\underline{V}, \underline{I})$. The above definition of the load angle is valid in the GRS; in contrast, in the URS the definition of the load angle is: $\beta = \arg(V, \underline{E}_p)$.

Remarks:

1. In general, the stator leakage field of synchronous machines is not very saturation-dependent. To the contrary the main or magnetising field normally is rather

saturation-dependent, and therefore the linearised equations and the corresponding part of the equivalent circuit are only valid in the operating point defined by E_r. However, as large synchronous generators are usually connected to the grid with a nearly constant voltage, the resulting emf E_r will not vary that much in normal operation conditions (the voltage drop over stator resistance is negligible for large machines and that over the leakage reactance is relatively small).

2. The linearised equivalent circuit in Fig. 5.10 can also be transformed into an equivalent circuit that is more similar to that of induction machines, see Fig. 5.11. In this equivalent circuit the fictitious rotor emf $\underline{E}_p$ is now replaced by a current source $\underline{I}_p$, representing the excitation current as seen from the stator.
The excitation current referred to the stator, $\underline{I}_p$, is proportional to the real DC excitation current i_f, $I_p = i_f/\alpha$. However it has, as it is referred to the stator, the same phase angle as the rotor mmf $\underline{\hat{F}}_p$ (one could also interpret $\underline{I}_p$ as the rotating rotor excitation current sheet as seen from the stator).

$$\underline{I}_p = \underline{i}_f/\alpha \tag{5.21}$$

The voltage over jX_m, with as current the total magnetising current $\underline{I}_p + \underline{I}$, is equal to the resulting emf $\underline{E}_r$. Indeed

$$-jX_m\underline{I}_m = -jX_m(\underline{I}+\underline{I}_p) = -jX_m\underline{I} - jX_m\underline{i}_f/\alpha = \underline{E}_a + \underline{E}_p \equiv \underline{E}_r \tag{5.22}$$

or also

$$-jX_m\underline{I}_m = -jX_m\underline{i}_m/\alpha \equiv \underline{E}_r \tag{5.23}$$

One could argue that this equivalent circuit is more correct as mmfs or currents are added instead of emfs or fluxes. However, the magnetising reactance is still dependent on the magnetising level or total magnetising current and for a non-linear magnetic circuit (as is generally the case) the outcome is the same.

3. If we record the armature current as a function of a variable terminal voltage,[9] at rated speed but with zero excitation current, we obtain the *no-load characteristic with poly-phase AC excitation*. This saturation characteristic is similar to the usual no-load characteristic with DC excitation.
Indeed, let's indicate the no-load (saturation) characteristic at rated speed with DC excitation by $E_p = f(i_f)$.
In the general case, where there is an armature current in addition to DC excitation current, we have

$$V = \underline{E}_r - R\underline{I} - jX_\sigma\underline{I} \tag{5.24}$$

with

$$E_r = f(i_m) = f(|\underline{i}_f + \underline{i}'|) = f(|\underline{i}_f + \alpha\underline{I}|) \tag{5.25}$$

[9]Using a variable auto-transformer between the grid and the synchronous machine.

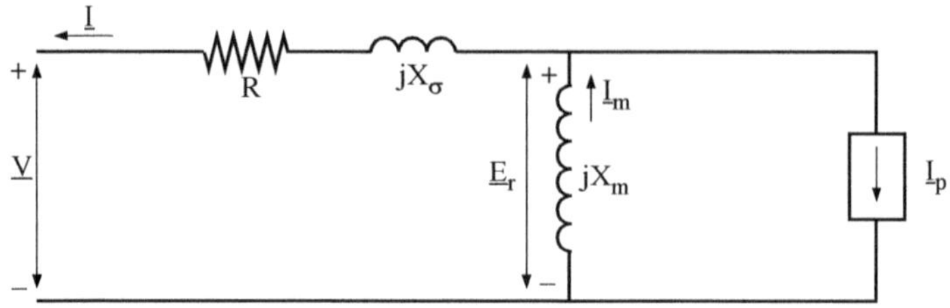

Fig. 5.11 Alternative equivalent circuit

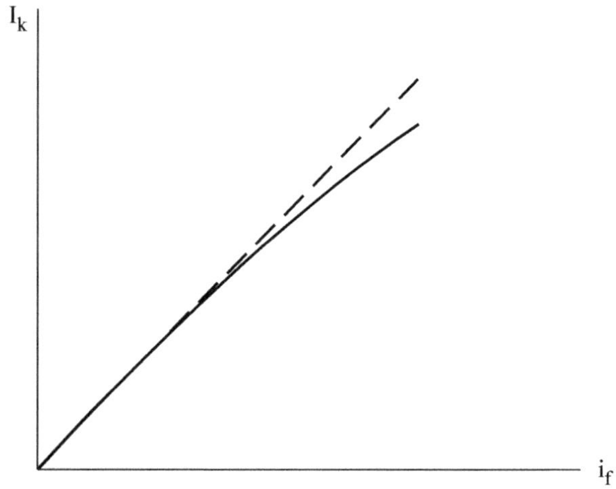

Fig. 5.12 Short-circuit characteristic

with $f(\cdot)$ the no-load characteristic. If now the DC excitation is zero, then $E_r = f(\alpha I)$ and thus $V = f(\alpha I) - R\underline{I} - jX_\sigma \underline{I}$. The recorded characteristic of the terminal voltage as a function of the armature current for zero DC excitation therefore also represents the usual no-load characteristic, yet with a slight vertical shift due to the resistive and leakage voltage drops and a rescaling of the abscissa (α). However, for not too small power ratings, the resistive voltage drop can be neglected. As $\underline{E}_r = -jX_m \underline{I}_m$, we may also write $\underline{V} = -j(X_m + X_\sigma)\underline{I}$ and thus we may also derive the synchronous reactance as a function of I (recall that X_m is not a constant but saturation-dependent!).

4. The short-circuit characteristic shows the short-circuit current as a function of the excitation current. It is measured at rated speed with the armature shorted (and reduced excitation current so as to limit the armature current). As illustrated in Fig. 5.12 the short-circuit characteristic is almost linear, except for very large armature currents.[10]

[10] The leakage reactance in per unit is about 0.2; limiting the armature current to 1 or 2 p.u. results in a resulting emf which remains on the linear part of the no-load characteristic, well below saturation.

5.2.5 Torque - Power - Energy Flow

The torque expression for the synchronous machine can be derived from the general equations for rotating field power in Sect. 3.4. From Eq. 3.87 follows:

$$T = \frac{3}{\Omega} \cdot Re(\underline{E}_r \cdot \underline{I}^*) \tag{5.26}$$

Substituting $\underline{E}_r = \underline{E}_p + \underline{E}_a = \underline{E}_p - jX_m\underline{I}$ this yields

$$T = \frac{3}{\Omega} \cdot Re(\underline{E}_p \cdot \underline{I}^*) = \frac{3}{\Omega} \cdot E_p \cdot I \cdot \cos\psi \tag{5.27}$$

Usually, the torque is expressed as a function of the terminal voltage and the rotor emf. Using (see Fig. 5.10)

$$I = \frac{\underline{E}_p - \underline{V}}{R + jX} \tag{5.28}$$

and neglecting the stator resistance we get

$$T = \frac{3}{\Omega} \cdot \frac{E_p \cdot V \cdot \sin\beta}{X} \tag{5.29}$$

If the real axis is chosen along the terminal voltage, i.e. $\underline{V} = V$, then $\underline{E}_p = E_p \cdot \exp(j\beta)$, Fig. 5.9.

For a constant voltage and constant[11] rotor emf, the torque varies sinusoidally with the load angle. The strong link between torque (or power) and the load angle is obvious: when $\beta = 0$ voltage and emf are in phase and the active current is zero; a non-zero active current component requires $\beta \neq 0$.

We recall that with the GRS chosen for the electrical side (and thus the URS for the mechanical side, i.e. the G-convention) the load angle β is positive when the rotor voltage is leading the terminal voltage,[12] see Fig. 5.9. In the G-convention a positive load angle implies generator operation ($P = 3 \cdot Re(\underline{V} \cdot \underline{I}^*) > 0$) while the mechanical power and the torque are positive as well (i.e. the torque of the machine constitutes a mechanical load for the driving engine, e.g. a turbine).

Figure 5.13 illustrates the torque as a function of the load angle for constant V and E_p. In both the generating and motoring region there is a maximum torque

$$T_{po} = \pm\frac{3}{\Omega} \cdot \frac{E_p \cdot V}{X} \tag{5.30}$$

[11] I.e. constant excitation current for a constant saturation level.

[12] With the URS at the electrical side, the load angle is positive when the terminal voltage is leading the rotor voltage.

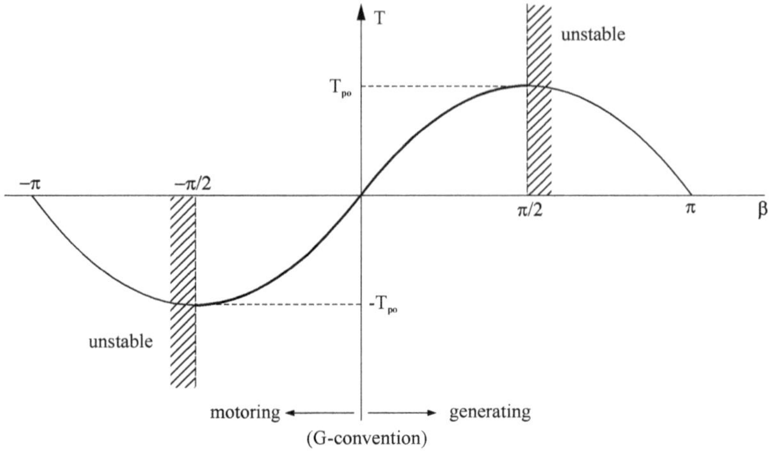

Fig. 5.13 Torque versus load angle characteristic of a smooth rotor synchronous machine

for $\beta = \pm\pi/2$. For a constant driving torque (or load torque in the motoring region), only the range of load angles between 0 and $\pm\pi/2$ corresponds to a stable behaviour.[13]

The mechanical power delivered to the machine is mainly converted into electrical power via the rotating field (electromagnetic power, see Sect. 3.4). Some additional shaft power is also required for the friction and ventilation losses (and also for a portion of the iron losses). If we neglect these friction and ventilation losses we may write

$$P_{shaft} \approx P_{em} = T \cdot \Omega = 3 \cdot Re(\underline{E}_p \cdot \underline{I}^*) = 3 \cdot E_p \cdot I \cdot \cos\psi \tag{5.31}$$

The power delivered to the grid (or more general, the supply) is given by

$$P_1 = 3 \cdot Re(V \cdot \underline{I}^*) = 3 \cdot V \cdot I \cdot \cos\varphi = 3 \cdot V \cdot I_w = 3 \cdot E_p \cdot I \cdot \cos\psi - 3 \cdot RI^2 \tag{5.32}$$

If we can neglect the stator joule losses, we may write

$$P_1 \approx P_{em} = T \cdot \Omega = 3 \cdot V \cdot I_w = 3 \cdot V \cdot E_p \sin\beta/X \tag{5.33}$$

If the friction and ventilation losses are taken into account, we must write

$$P_{shaft} = P_{em} + P_{fv} \tag{5.34}$$

[13]Prove this by considering the equation of motion; you will also have to relate a change of the load angle with an increasing or decreasing speed, e.g. for a constant supply frequency.

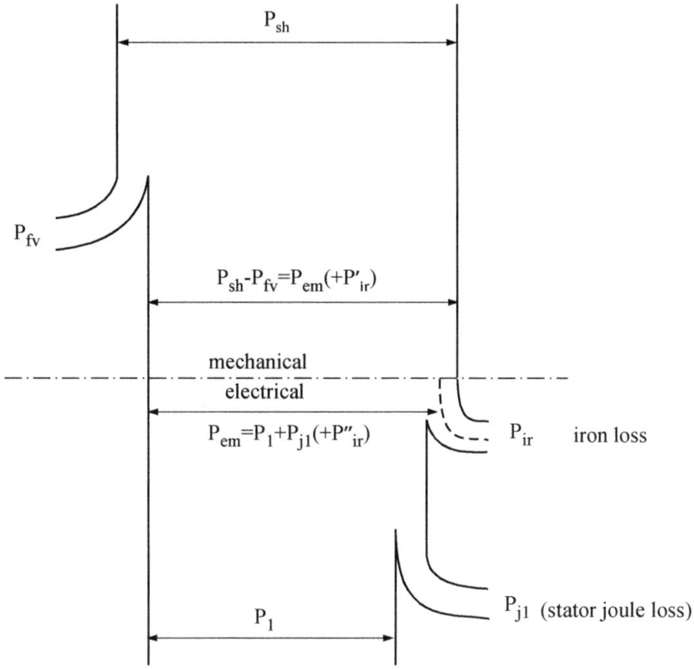

Fig. 5.14 Sankey diagram for generating

The power that can be converted into electrical power is the difference of the shaft power and the mechanical loss. This electromagnetic power $P_{em} = T \cdot \Omega$ is converted into electrical form and is supplied to the grid, after some reduction for the stator joule losses and (a portion of) the iron losses:

$$P_1 = P_{em} - P_{j1} - P_{ir} = P_{em} - 3RI^2 - 3E_r^2/R_m \tag{5.35}$$

Figure 5.14 shows the corresponding Sankey diagram. If required, the excitation DC power can be accounted for as well.

For motoring, the grid or electrical power supply must compensate all these losses. Then only the difference $P_1 - P_{j1} - P_{ir} - P_{fv} = P_{em} - P_{fv}$ is available as mechanical shaft power.

Remark:
If the stator resistance is taken into account, we obtain the following torque expression

$$T = \frac{3}{\Omega} \cdot \left\{ \frac{E_p^2 R}{R^2 + X^2} + \frac{VE_pX \sin\beta - VE_pR\cos\beta}{R^2 + X^2} \right\} = \frac{3}{\Omega} \cdot \left\{ \frac{E_p^2 R}{R^2 + X^2} + \frac{VE_pX \sin(\beta - \rho)}{\sqrt{R^2 + X^2}} \right\} \tag{5.36}$$

where $\varrho = \arctan R/X$.

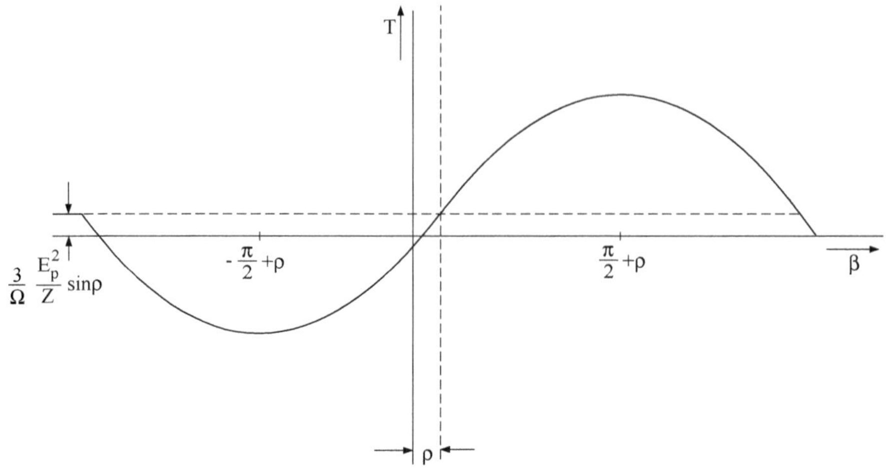

Fig. 5.15 Torque characteristic for non-zero stator resistance

The pull-out torque occurs now at $\beta = \varrho \pm \pi/2$ and is somewhat smaller for motoring and larger for generating, see Fig. 5.15.

5.2.6 *Per-Unit Values*

The per-unit values for a synchronous generator are based upon the reference values for the rated voltage V_n, the rated armature current I_n and the rated speed $\Omega_n = \omega_n/N_p$. From these basic reference values, the reference values for power $P_{ref} = 3V_nI_n$, for torque $T_{ref} = 3V_nI_n/\Omega_n$ and for the armature impedances $Z_{ref} = V_n/I_n$ are derived. Note that the reference value for the power is not the rated value for the active power $P_n = 3V_nI_n\cos\varphi_n$, neither is the reference value for the torque equal to the rated torque $T_n \approx Pn/\Omega_n$.

With these reference values, the p.u values for voltages and emfs (e.g. v, e_p), current (i), impedances (z, r, x), active, reactive and apparent power (p, q, s) and torque (t) can be derived.

The speed of a synchronous machine has a one-to-one relation with the frequency. A speed different from the rated speed thus implies also a frequency different from the rated frequency. When variable speed (and thus frequency) operation is considered it is good practice to write reactances as ωL instead of X. For per-unit, the reference value for the frequency should correspond with the reference value for the speed ($\Omega_n = \omega_n/N_p$). Per unit values for reactances are then of the form νl with $\nu = \omega/\omega_n$ the p.u frequency (equal to the p.u synchronous speed). Note however that also the magnitude of the emf of motion is proportional to the speed or frequency, i.e. $E_p(i_f, \omega) = \nu \cdot E_p(i_f, \omega_n)$.

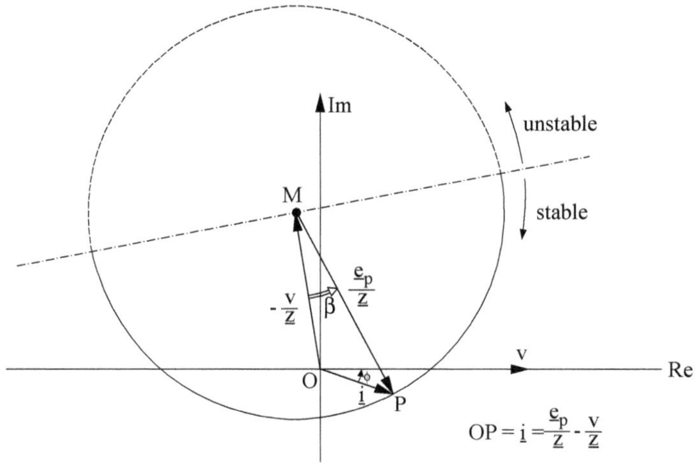

Fig. 5.16 Current locus of a smooth rotor synchronous machine

5.2.7 The Current Locus for Constant Excitation

From the equivalent circuit in Fig. 5.10 and choosing the real axis along the terminal voltage ($\underline{V} = V$) it follows that

$$\underline{I} = \frac{\underline{E}_p - V}{R + jX} = \frac{E_p \exp(j\beta) - V}{R + jX} \tag{5.37}$$

For constant rotor emf (or constant excitation if the saturation condition is constant), constant voltage and variable load angle, the locus of the current vector is a circle with radius $E_p/Z = E_p/\sqrt{R^2 + X^2}$. The centre of the circle is at $-V/(R + jX)$.

In per-unit we obtain

$$\underline{i} = \frac{\underline{e}_p - v}{r + jx} = \frac{e_p \exp(j\beta) - v}{r + jx} \tag{5.38}$$

This locus is drawn in Fig. 5.16. The dashed line marks the boundary of stable operation for a constant driving or load torque.

If the stator resistance can be neglected then

$$\underline{i} = \frac{\underline{e}_p - v}{jx} = \frac{e_p \exp(j\beta) - v}{jx} \tag{5.39}$$

Figure 5.17 illustrates the current loci (supposing $r = 0$) for three values of the ratio e_p/v. For under-excitation $e_p/v < 1$ the synchronous machine will always absorb reactive power from the grid or supply. In order to deliver reactive power to the grid, the machine must be over-excited, i.e. $e_p/v > 1$. Reactive power delivery (both for

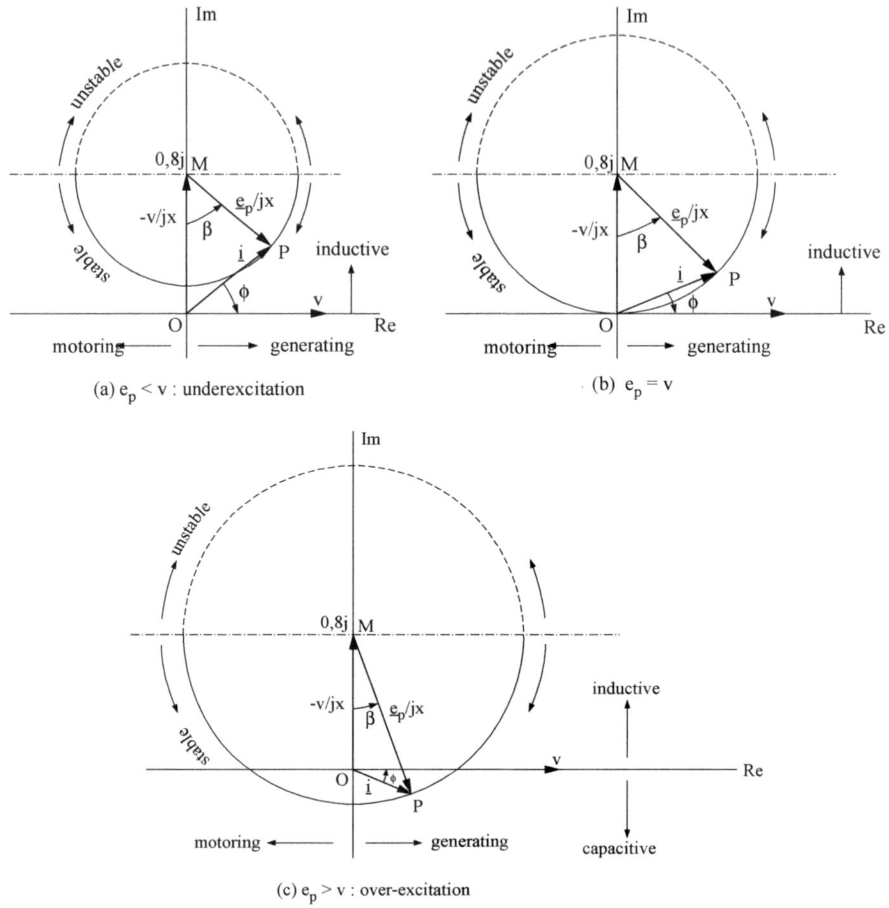

Fig. 5.17 Circle diagrams for zero stator resistances

motoring or generating) is, however, only possible for sufficiently small load angles, see (c) in Fig. 5.17. This clearly illustrates the compromise that the operation of a synchronous machine presents. For an appreciable active power delivery (either generating or motoring) the load angle should be sufficiently large, but as in most cases also reactive power delivery is required, the load angle should be limited as well.

For an idea about the numerical values for the load angle and rotor emf, suppose $x = 1.25$, $r \approx 0$ and rated values for voltage $v = 1$, for apparent power $s = 1$, reactive power $q = s \cdot \sin \varphi_n = -0.6$ and active power $p = s \cdot \cos \varphi_n = \pm 0.8$. A straightforward calculation leads to $e_{pn} \approx 2$ and $\beta \approx \pm\pi/6$. This (rated) operating point lies then below the real axis (meaning that reactive power is delivered to the supply, $q = -0.6$) while the active power is 0.8 (or -0.8 for motoring). It should be stressed again that this $e_{pn} \approx 2$ is the value according to the linearised model and not the no-load emf e_{po} for the same excitation current.

For very large power ratings (e.g. turbo-generators in nuclear power stations), the load angle can be increased to e.g. $\pi/4$ so as to increase the active power. However, when large variations in reactive power of the grid can be expected, a load angle smaller than $\pi/6$ might be chosen.

5.2.8 Characteristics of Synchronous Machines

Characteristics of synchronous generators are quite similar to those of DC generators. Important characteristics are the load characteristics, the control characteristics and the external characteristics. These characteristics are typically non-linear and can be derived from the no-load characteristic as shown below.

5.2.8.1 Load Characteristics

The load characteristics are similar to the no-load characteristic, but instead of the no-load voltage as a function of the excitation current, the load characteristics show the evolution of the terminal voltage as a function of the excitation current for a given load (I, $\cos\varphi$). For pure inductive or capacitive loads, the load characteristics can be derived graphically from the no-load characteristic.

Load Characteristic for Pure Inductive Load

Disregarding the armature resistance, the equivalent circuit of the synchronous machine with its inductive load is shown in Fig. 5.18 (only the left part will be used here, as the right part in dashed line depends on the saturation level, which is inherently variable). For a purely inductive load the phasor relations $\underline{E}_r = \underline{V} + jX_\sigma \underline{I}$ and $\underline{i}_m = \underline{i}_f + \underline{i}'$ can be written as scalar equations $E_r = V + X_\sigma I$ and $i_m = i_f - i' = i_f - \alpha I$, see the phasor diagram in Fig. 5.18.

Suppose $M(V, i_f)$ is a point on this load characteristic for an inductive current I (Fig. 5.19). Then the corresponding point $P(E_r, i_m)$ will be on the no-load charac-

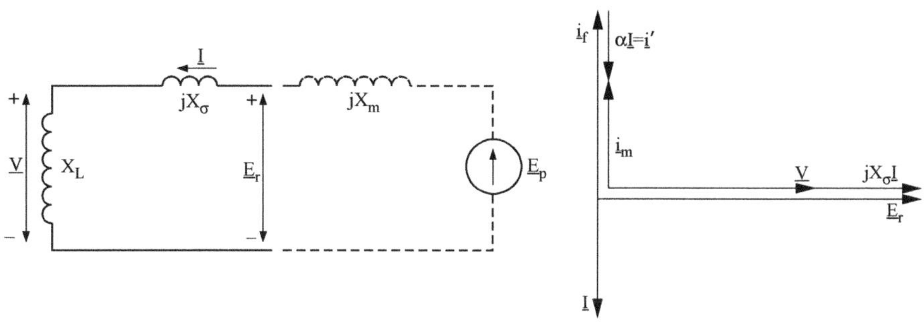

Fig. 5.18 Equivalent circuit and phasors for inductive load

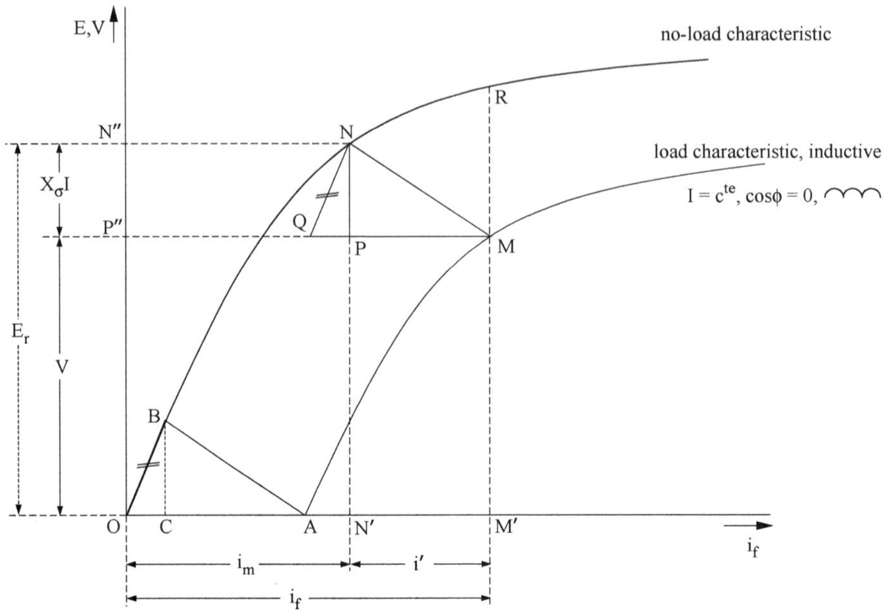

Fig. 5.19 Inductive load characteristic

teristic. The vertical distance between N and M is equal to $X_\sigma I$, while the horizontal distance between these two points is equal to αI. Both points are therefore connected by a rectangular triangle[14] with all sides proportional to I. For a constant I, this triangle remains congruent while point P stays on the no-load characteristic. Thus the load characteristic for this constant I is found by shifting the no-load characteristic in the direction of the vector $\overrightarrow{NM}$.

From Fig. 5.19 it is clear therefore that an inductive load will reduce the terminal voltage. This is due to two causes: the armature reaction on the one hand (without armature reaction the emf would be given by the ordinate of point R) and the leakage voltage drop NP on the other hand.

From an experimentally recorded load characteristic and the no-load characteristic, the vector NM can be reconstructed and therefore also the triangle MNP for a given armature current. Thus α and X_σ can be determined as well. Moreover, if also the short-circuit characteristic is known, one single measured point M on the load characteristic suffices to reconstruct the triangle and thus the complete load characteristic.[15]

---

[14]This triangle is called the Potier triangle; for a given machine (X_σ and α) these triangles are always similar; the triangle is also called the short-circuit triangle, see the triangle ABC.

[15]From the short-circuit characteristic the excitation current corresponding with a short-circuit current I can be derived and thus the point A on the short-circuit triangle can be found. From the point Q on the horizontal through M with QM = OA a line parallel with the linear initial part of the no-load characteristic the point N is found. This is known as the method of Fisher-Hinnen.

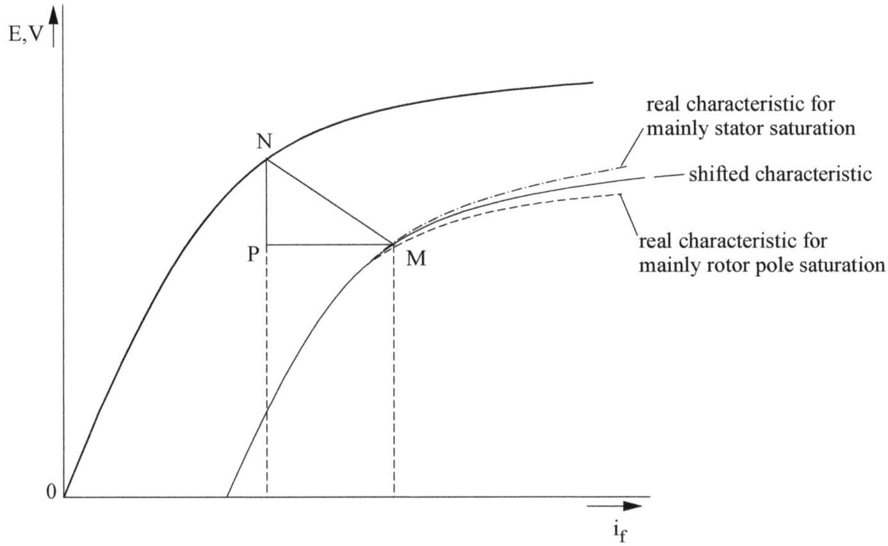

Fig. 5.20 Measured inductive load characteristics

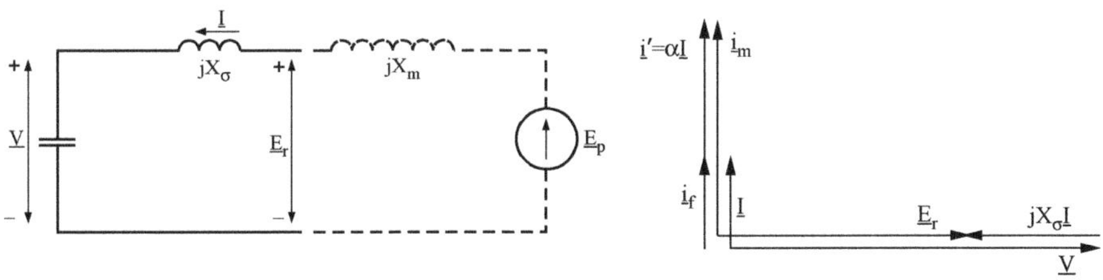

Fig. 5.21 Equivalent circuit and phasors for capacitive load

However, it should be noted that the real measured load characteristics may differ somewhat from the shifted no-load curves, especially for high saturation levels, see Fig. 5.20. The cause is the not always uniform saturation of the core.

Load Characteristic for a Pure Capacitive Load

For a purely capacitive load, the phasor relations $\underline{E}_r = \underline{V} + jX_\sigma \underline{I}$ and $\underline{i}_m = \underline{i}_f + \underline{i}'$ can be written as scalar equations $E_r = V - X_\sigma I$ and $i_m = i_f + i' = i_f + \alpha I$, see the phasor diagram in Fig. 5.21.

The load characteristic for a pure capacitive load is therefore found by shifting the no-load characteristic upwards in the direction of the vector $\overrightarrow{NM}*$ (see Fig. 5.22).

For a capacitive load, the armature reaction is magnetising, i.e. the terminal voltage increases compared to the no-load case. For a given maximum terminal voltage there is a maximum capacitive load as the excitation current may not be made negative.[16]

[16] For a smooth rotor synchronous machine this would result in a reversal of the torque as the polarity of the rotor poles switches.

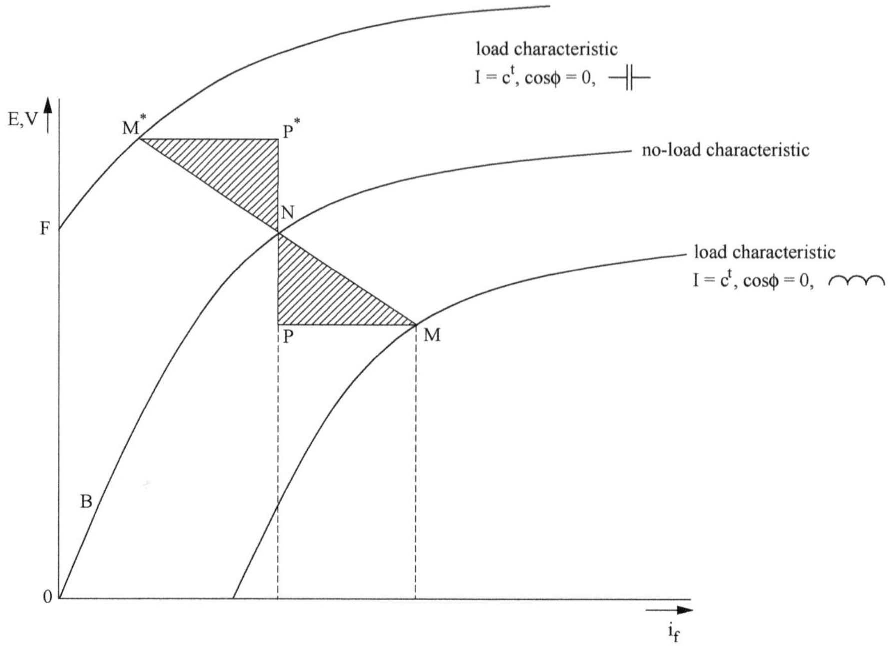

Fig. 5.22 Capacitive load characteristics

This is illustrated in Fig. 5.23, where the length of the vector $\overrightarrow{BC}$ represents the maximal capacitive armature current for the given voltage V_n. In B in Fig. 5.23 (or F in Fig. 5.22) all magnetising current is drawn from the supply (grid), just like for an induction machine.

Load Characteristics for a General Load

For a general load the load characteristics cannot easily be derived graphically. For a given armature current amplitude, the corresponding load characteristics for different $\cos\varphi$ lie between the pure capacitive and pure inductive load characteristics for the same current amplitude, see Fig. 5.24. However, all these load characteristics pass through the same short-circuit point for this same current, whatever the $\cos\varphi$.[17]

5.2.8.2 Control Characteristics

The control characteristics show the evolution of the required excitation current as a function of the armature current for a given terminal voltage and a given $\cos\varphi$, see Fig. 5.25.

[17]Please, try to explain! Hint: when a complex impedance $\underline{Z}$ tends to zero, it doesn't matter what the phase angle is.

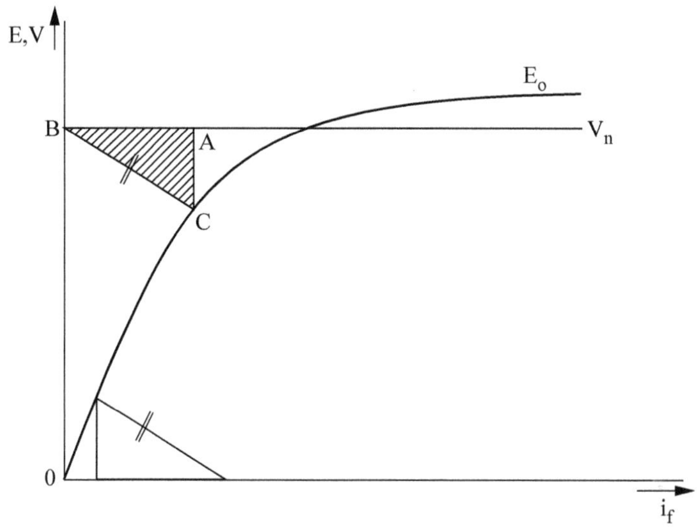

Fig. 5.23 Maximum capacitive load

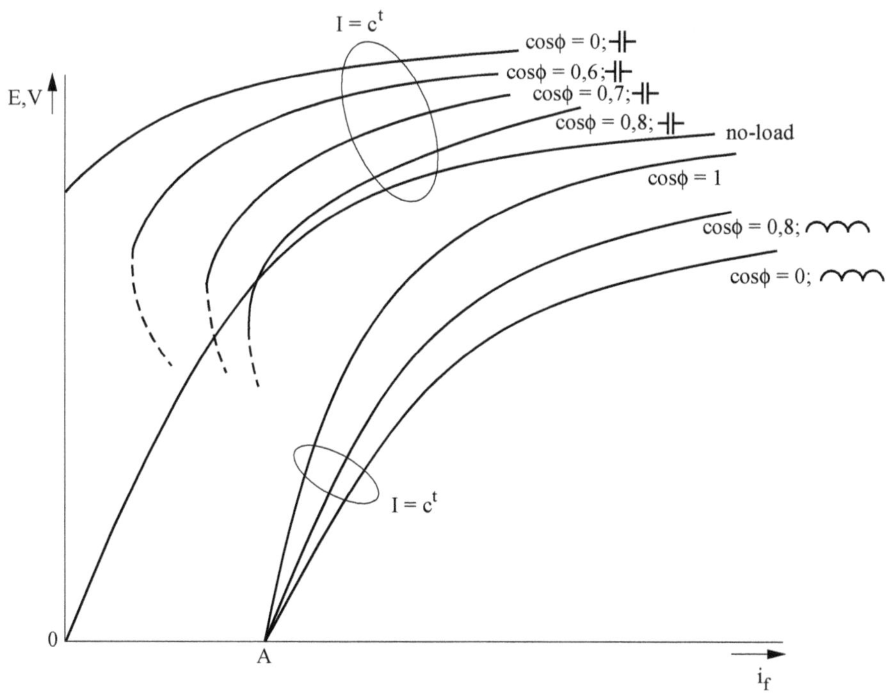

Fig. 5.24 Load characteristics for a general load

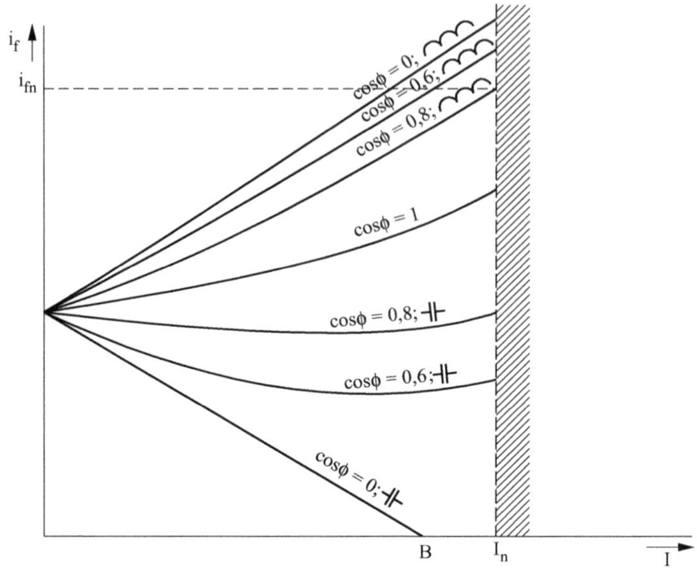

Fig. 5.25 Control characteristics

Figure 5.25 illustrates again that an inductive load requires a larger excitation current in order to maintain the same terminal voltage. With a capacitive load, the excitation current must be reduced. The characteristics also show the maximum capacitive current for a given terminal voltage (point B).

These control characteristics can be derived graphically[18] starting from the load characteristics.

5.2.8.3 External Characteristics

The external characteristics show the relation between voltage and armature current for a given excitation current and a given $\cos\varphi$. These characteristics also illustrate the magnetising effect of capacitive loads versus the demagnetising effect of inductive loads.

These external characteristics can also be derived graphically[19] starting from the load characteristics (Fig. 5.26).

---

[18]In the load characteristics, draw a horizontal line corresponding to the given terminal voltage. Consider then the Potier triangles with point P on the no-load characteristic and the sides QM on this horizontal line.

[19]In the load characteristics, draw a vertical line corresponding to the given excitation current. Consider then the Potier triangles with point P on the no-load characteristic and the sides QP on this vertical line.

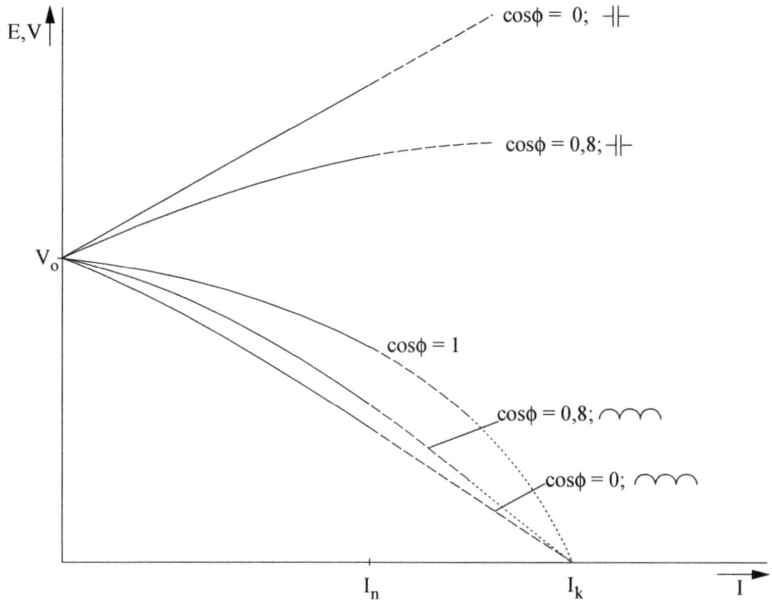

Fig. 5.26 External characteristics

5.3 Salient-Pole Synchronous Machines

5.3.1 Emf Induced by a Salient-Pole Rotor with Concentrated DC Winding

As mentioned before, the field curve of a salient-pole rotor with concentrated field winding is not at all a pure sine. By shaping the pole shoes the field curve can be made somewhat more like a sine (instead of rectangular) but important space harmonics remain present.

Rotation of the rotor will induce emfs or fluxes in the stator *conductors* with the same non-sinusoidal shape in time. However, as discussed in Chap. 3, the distributed armature *winding* will filter[20] out the harmonics, and the winding voltage will be fairly sinusoidal. Therefore, we will concentrate on the fundamentals of the rotor mmf $f_p(t)$ and emf $e_p(t)$. As we have done for the smooth rotor synchronous machine, we will represent these fundamentals by the (time) phasors $\underline{\hat{F}}_{p1}$ and $\underline{E}_p$. To recall, these phasors stand for the rotor flux (or rotor mmf) measured along the axis of the

[20]Because of the lower winding factors for higher harmonics.

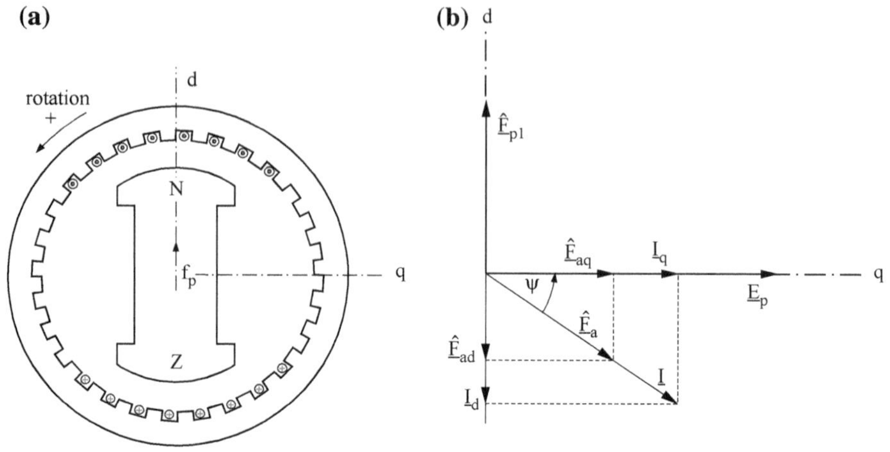

Fig. 5.27 Mmf and emf phasor diagram for salient pole synchronous machines

reference phase and the emf induced by the rotor mmf or flux in this same reference phase. In the chosen G-convention (GRS at the electrical side), the mmf phasor $\hat{\underline{F}}_{p1}$ leads the emf phasor $\underline{E}_p$, see (b) in Fig. 5.27.

In space, see (a) in Fig. 5.27, the maximum of the rotor mmf is always along the north-south axis of the rotor. This axis, usually the axis with the smallest reluctance, is called the d-axis (direct axis or longitudinal axis). The axis which is lagging $\pi/2$ *electrical* radians (lagging is referred to as lagging with respect to the positive rotation direction), is called the q-axis (quadrature axis or transverse axis).

Just like for the smooth rotor synchronous machine, we may also interpret the (time) phasors $\hat{\underline{F}}_{p1}$ and $\underline{E}_p$ as (rotating) space vectors. Indeed, the flux (time function) coupled with a winding is maximal when the maximum of the mmf (space quantity) lies along the axis of this phase, i.e. when the d-axis is along the reference phase axis. The maximal voltage induced in this reference winding occurs $\pi/2$ radians later (in time), i.e. when the q-axis of the rotor is along the reference phase axis (the reference winding in the figure is the one with the conductors shown; its axis is horizontal and the positive direction is to the right). Imagine the vector diagram on the right rotating counter-clockwise (synchronously with the rotor in the left figure). In the position shown, the flux coupled with the reference winding is zero and the induced emf is maximal. When the north pole or d-axis is rotated $\pi/2$ (electrical) radians to the left, the flux coupled with the reference winding becomes maximal and then the induced emf is zero (cf. $e(t) \sim \frac{d\Psi}{dt}$).

The time values of mmf or emf therefore correspond with the projection at that instant of these vectors on the axis of the reference phase. Here as well the no-load characteristic gives the relation between the emf amplitude values and the excitation current (or mmf).

5.3.2 Armature Reaction

In the previous section we analysed the no-load condition. When the armature winding is connected to a symmetrical electrical load (e.g. the grid), generally an armature current will flow. This poly-phase armature current results in an armature mmf which rotates synchronously with the rotor mmf. In the air gap there are thus two rotating mmfs which are synchronous, but in general not in phase. The resulting mmf, which determines the resulting rotating magnetic field and the resulting emf, is the vectorial sum of both.

In contrast with the smooth rotor synchronous machine, one cannot derive the resulting field directly from the resulting mmf. Indeed, as the air-gap is not uniform - for a salient pole machine the reluctance in the d-axis is smaller than in the q-axis - an mmf (or mmf component) acting in the d-axis (i.e. with the symmetry axis of the mmf in the d-axis) will result in a larger magnetic field than a similar mmf (or mmf component) acting in the q-axis.[21]

Therefore an approximate method is used, as explained below.

In principle, the vectorial addition of the rotor mmf and the armature mmf phasors yields the resulting mmf phasor $\underline{\hat{F}}_r = \underline{\hat{F}}_{p1} + \underline{\hat{F}}_a$. This resulting mmf determines the resulting air-gap induction $\underline{\hat{B}}_r$ or flux $\underline{\hat{\Psi}}_r$ and thus the resulting emf $\underline{E}_r$. However, to obtain $\underline{\hat{B}}_r$ we must now use the permeances (or saturation characteristics) along the d- and q-axes for the corresponding components of the resulting mmf.

The resulting mmf component along the d-axis is the sum of the rotor mmf $\underline{\hat{F}}_{p1}$ and the d-axis component of the armature mmf $\underline{\hat{F}}_{ad}$ with $\hat{F}_{ad} = -\hat{F}_a \sin\psi$. The mmf component along the q-axis stems from the armature mmf only: $\underline{\hat{F}}_{aq}$ with $\hat{F}_{aq} = \hat{F}_a \cos\psi$. Instead of the armature mmf components one may also consider the d- and q-axis components $\underline{I}_d$ and $\underline{I}_q$ of the armature current $\underline{I}$ for which $I_d = -I \sin\psi$ and $I_q = I \cos\psi$ (and calculate then the corresponding mmf components from these). To calculate the emfs corresponding with these mmf components along the d- and q-axes, we must use the corresponding permeances (or saturation characteristics).

For the mmf along the d-axis - resulting in an emf along the q-axis - we can use the no-load characteristic. However this no-load characteristic has been measured for the quasi-rectangular mmf shape of a salient pole (with DC excitation in a concentrated winding), see Fig. 5.28. The (no-load) emf considered in the no-load characteristic is the *fundamental* which is in fact the emf induced by the *fundamental* of the rotor mmf. So we must compare the amplitude of the (nearly) sinusoidal armature mmf with that of the fundamental of the quasi-rectangular rotor mmf (and not with the maximum value of this mmf). When using the no-load characteristic to calculate the q-axis emf resulting from the resulting d-axis mmf $\hat{F}_d$, we should correct for this difference by reducing the d-axis armature mmf with a factor k_d ($\approx \pi/4$ if the rotor mmf was a pure rectangle). Thus, we should use $\hat{F}_d = \hat{F}_{p1} + k_d \cdot \hat{F}_{ad} = \hat{F}_{p1} - k_d \cdot \hat{F}_a \sin\psi$ (with as fundamental $\hat{F}_{d1} = \frac{4}{\pi}\hat{F}_d$). As the no-load characteristic is measured with the

[21] An additional problem is that, in general, the saturation characteristic in the q-axis is not known and cannot easily be measured.

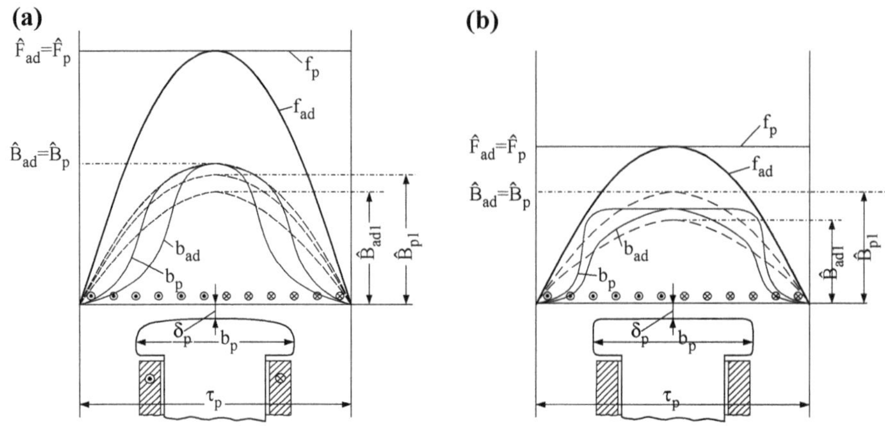

Fig. 5.28 Mmf d-axis waveshapes for salient pole machines

DC current as abscissa, these mmfs can preferably be transformed into equivalent excitation currents: $i_{md} = i_f + i_d^{'} = i_f - k_d \cdot \alpha \cdot I \sin\psi$. For a salient pole rotor we have $\alpha = \frac{6\sqrt{2}}{\pi} \cdot \frac{w\xi_1}{w_f}$, as the rotor winding is a concentrated winding (see also Chap. 3).

These equations may also be written in phasor form as follows:

$$\underline{\hat{F}}_d = \underline{\hat{F}}_{p1} + k_d \cdot \underline{\hat{F}}_{ad} \tag{5.40}$$

and

$$\underline{i}_{md} = \underline{i}_f + \underline{i}_d^{'} \tag{5.41}$$

Corresponding with i_{md}, we then find on the no-load characteristic the q-axis component of the resulting emf, E_q, see Fig. 5.29.

Similar to the smooth rotor synchronous machine, we may linearise this q-axis emf as

$$\underline{E}_q = \underline{E}_p + \underline{E}_{aq} = \underline{E}_p - jX_{ad}\underline{I}_d \tag{5.42}$$

where the d-axis armature (or magnetising) reactance $X_{ad} = k_d \cdot \alpha \cdot \tan\gamma$ depends on the saturation level determined by $\hat{F}_d$ or i_{md}. Remark again that $\underline{E}_p$ is not the no-load emf $\underline{E}_{po}$ corresponding with i_f, but a fictitious linearised value. To determine the d-axis emf $\underline{E}_d$ corresponding to the q-axis mmf $\underline{\hat{F}}_{aq}$, with $\hat{F}_{aq} = \hat{F}_a \cos\psi$, ideally we should use a q-axis saturation characteristic. Unfortunately, such a q-axis saturation characteristic is seldom available.[22] As an approximation one will use the d-axis saturation characteristic, with the q-axis mmf corrected for the (much)

[22] Normally, a q-axis excitation winding is not present; the only way to record a q-axis saturation characteristic is to use AC excitation with all AC mmf in the q-axis. However, this is problematic as well because of stability issues (please explain).

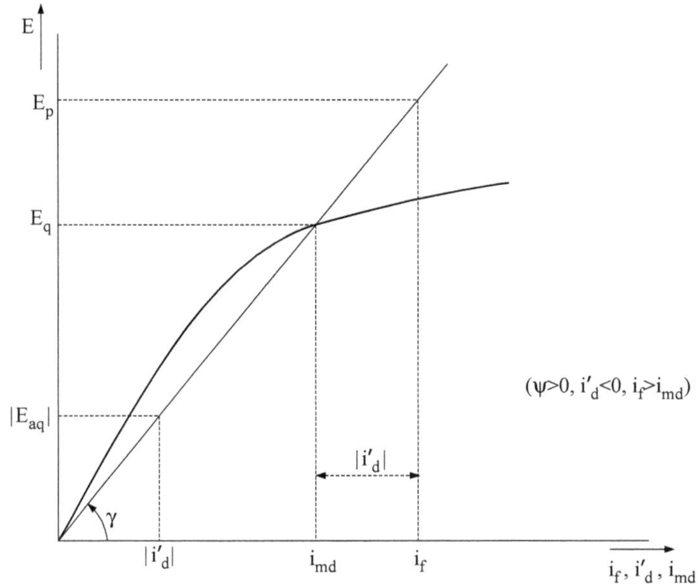

Fig. 5.29 Saturation characteristic in the d-axis

lower fundamental[23] in the q-axis compared to the rectangular rotor mmf in the d-axis, see Fig. 5.30: $\hat{F}_q = k_q \cdot \hat{F}_{aq} = k_q \cdot \hat{F}_a \cos\psi$. Instead, one may also use the corresponding armature current component $i_{mq} = k_q \cdot \alpha \cdot I \cos\psi$. Whereas in the d-axis $k_d \approx 0.7 \ldots 0.8$, the reduction in the q-axis is much more important, $k_q \approx 0.5 \cdot k_d$.

Often, these equations are written in phasor form as:

$$\underline{\hat{F}}_q = k_q \cdot \underline{\hat{F}}_{aq} \tag{5.43}$$

$$\underline{i}_{mq} = \underline{i}'_q = k_q \cdot \alpha \cdot I \cos\psi \tag{5.44}$$

The emf induced in the d-axis by the q-axis mmf can also be written in linearised (phasor) form as

$$\underline{E}_d = \underline{E}_{ad} = -jX_{aq}\underline{I}_q \tag{5.45}$$

with $X_{aq} \sim 0.5 \cdot X_{ad}$.

[23] Which is caused by the large air-gap in the inter-polar space of the q-axis.

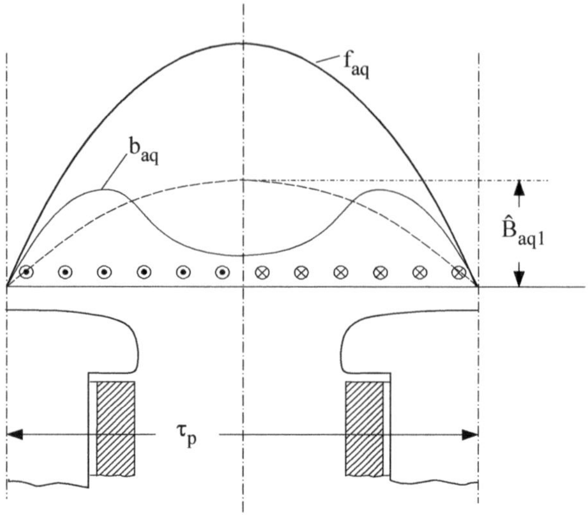

Fig. 5.30 Mmf waveshapes in the q-axis

Like X_{ad}, the reactance X_{aq} as well is saturation dependent. In fact, the flux lines of d- and q-axis follow many common saturated parts of the magnetic circuit, like the armature teeth and the pole tips. For that reason X_{aq} (or E_{ad}), is often determined on the line set by the saturation level of the d-axis mmf, see Fig. 5.31. Then $X_{aq} = k_q \cdot \alpha \cdot \tan\gamma$, where $\tan\gamma$ is determined by the d-axis mmf or magnetising current i_{md} or thus by E_q. Some authors argue that the q-axis is much less saturated (because of the large reluctance of the inter-polar space), and that rather the tangent line in the origin of the saturation characteristic should be used, thus leading to E^*_{ad} (with $X_{aq} = k_q \cdot \alpha \cdot \tan\gamma_o$) instead of E_{ad}.

5.3.3 *Equations and Phasor Diagram of the Salient Pole Synchronous Machine*

The total induced voltage[24] can be obtained as the vectorial sum of the emfs induced in d-axis and q-axis:

$$\underline{E}_r = \underline{E}_q + \underline{E}_d = \underline{E}_p + \underline{E}_{aq} + \underline{E}_{ad} = \underline{E}_p - jX_{ad}\underline{I}_d - jX_{aq}\underline{I}_q \tag{5.46}$$

[24]This is in fact also an approximation or linearisation as adding these emfs is equivalent to adding induction values which is strictly not allowed in case of saturation.

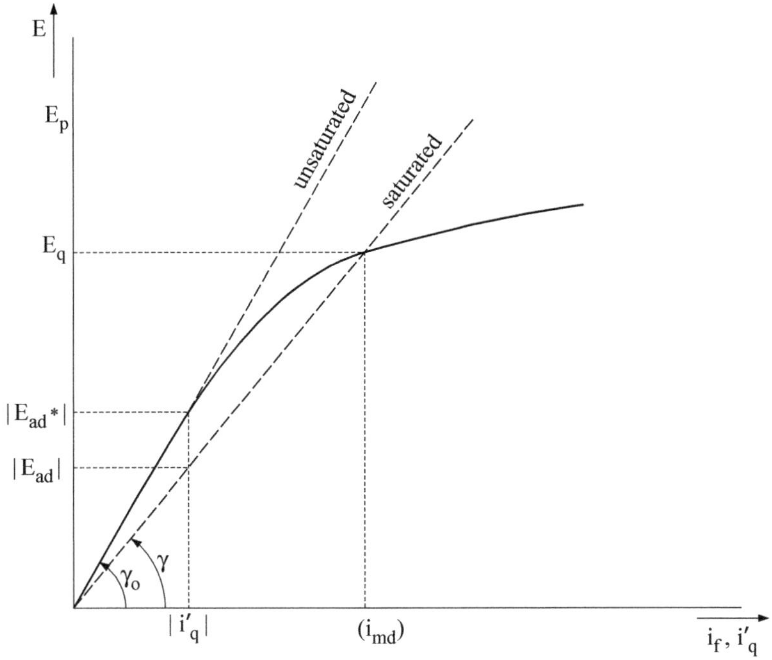

Fig. 5.31 Saturation characteristic in the q-axis

If the real axis is chosen along the q-axis, we may write[25] $\underline{I}_d = jI_d = -jI\sin\psi$; $\underline{I}_q = I_q = I\cos\psi$; $\underline{E}_{aq} = E_{aq} = -X_{ad}I\sin\psi$; $\underline{E}_{ad} = jE_{ad} = -jX_{aq}I\cos\psi$.

The terminal voltage of the machine differs from the emf by the resistive voltage drop and the leakage voltage drop:

$$\underline{V} = \underline{E}_r - R\underline{I} - jX_\sigma\underline{I} = \underline{E}_p - jX_{ad}\underline{I}_d - jX_{aq}\underline{I}_q - jX_\sigma\underline{I} - R\underline{I} \qquad (5.47)$$

Defining the total or 'synchronous' d- and q-axis reactances $X_d = X_{ad} + X_\sigma$ and $X_q = X_{aq} + X_\sigma$, this can be simplified to

$$\underline{V} = \underline{E}_p - jX_d\underline{I}_d - jX_q\underline{I}_q - R\underline{I} \qquad (5.48)$$

The phasor diagram corresponding with Eq. 5.46 is shown in Fig. 5.32. To draw this diagram starting from (for example) the terminal voltage and armature current, the armature mmf $\underline{\hat{F}}_a$ or the armature current $\underline{I}$ must first be split up into its d- and q-axis components. However, this requires that the position of the d- and q-axes is known. Note that only he portion of the diagram with the terminal voltage $\underline{V} = \underline{OA}$ and the resulting emf $\underline{E}_r = \underline{OG}$ (which differ by the resistive and leakage voltage drops)

[25]Alternatively, one may use the scalar form of these equations for d- and q-axes but then the risk for errors is greater.

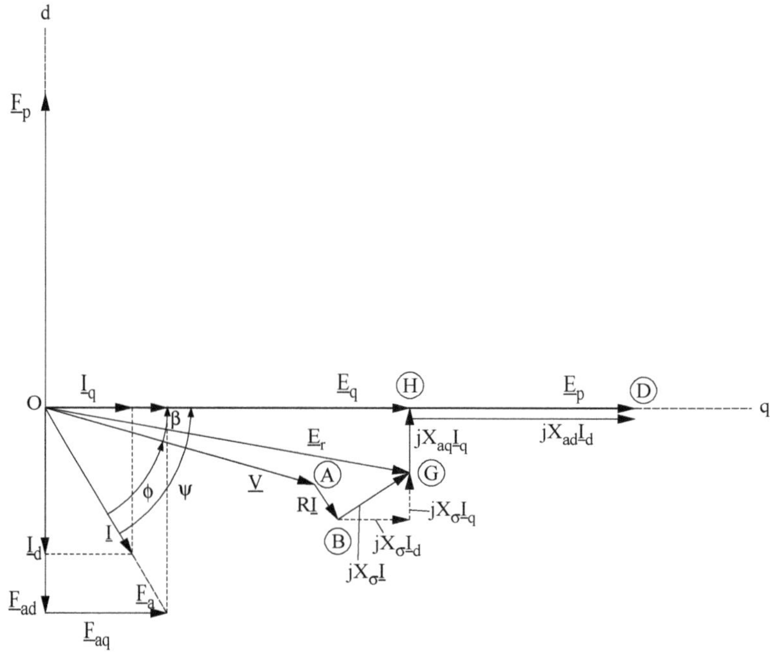

Fig. 5.32 Phasor diagram of a salient rotor synchronous machine

represent the real physical quantities. The remaining voltage vectors $\underline{OH} = \underline{E}_q$, $\underline{OD} = \underline{E}_p$, $\underline{GH} = jX_{aq}\underline{I}_q$ and $\underline{HD} = jX_{ad}\underline{I}_d$ represent imaginary vectors from the linearised model.

By introducing the synchronous reactances X_d and X_q, the diagram can be simplified, see Fig. 5.33. Starting from the knowledge of the terminal voltage as well as the current and its phase angle, the d- and q-axes can be reconstructed if the synchronous reactances are known, see the construction[26] in Fig. 5.33.

For the inverse problem, i.e. calculating the terminal voltage for a given excitation current and load, only an iterative solution method exists.

5.3.4 *Equivalent Circuits for a Salient Pole Synchronous Machine*

From Eq. 5.47 or 5.48, it is clear that a single equivalent circuit for the phasors $\underline{V}$ and $\underline{I}$ does not exist[27] if $X_d \neq X_q$.

[26]In B, draw the vector $\underline{BF} = jX_d\underline{I}$; the endpoint E of the vector $\underline{BE} = jX_q\underline{I}$ is on the q-axis and the projection D of F on the line OE yields the emf $\underline{OD} = E_p$.

[27]Indeed, verify that $\underline{I}_d$ and $\underline{I}_q$ cannot be replaced by $\underline{I}$ when $X_d \neq X_q$.

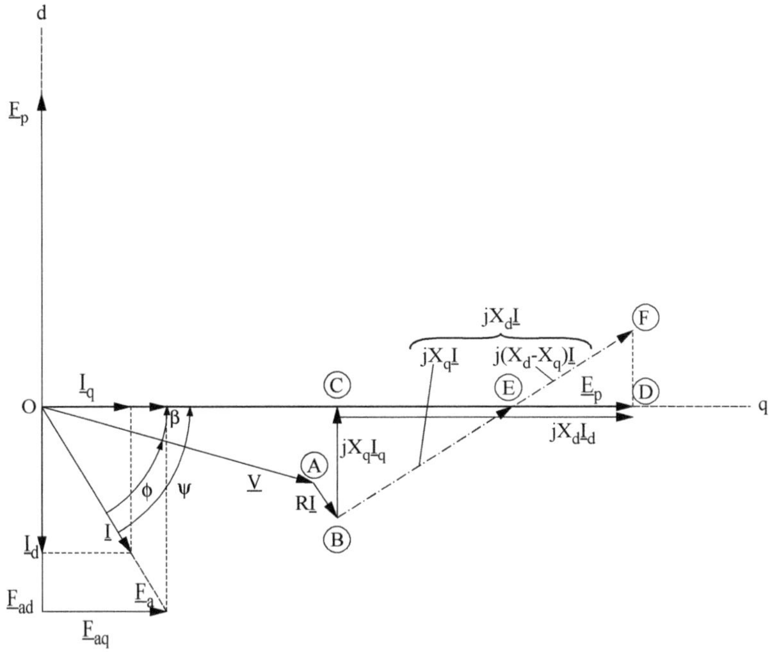

Fig. 5.33 Simplified phasor diagram of a salient pole synchronous machine

If, however, the stator resistance can be neglected, then a system of two 'scalar' equivalent circuits can be devised, one for the q-axis current and one for the d-axis current. Indeed, projecting Eq. 5.48 on the d- and q-axes (for $R = 0$) yields the following set of two equations:

$$\underline{V}_q = \underline{V} \cdot \exp j\beta \cdot \cos\beta = \underline{E}_p - jX_d\underline{I}_d \tag{5.49}$$

$$\underline{V}_d = \underline{V} \cdot \exp(j\beta - \pi/2) \cdot \sin\beta = -jX_q\underline{I}_q \tag{5.50}$$

(these equations may also be written in real form). Figure 5.34 shows the two corresponding equivalent circuits.

When choosing the real axis along $\underline{V}$ we get

$$\underline{V}_q = (V \cdot \cos\beta) \cdot \exp j\beta = E_p \cdot \exp j\beta - jX_d\underline{I}_d \tag{5.51}$$

$$\underline{V}_d = (-jV \cdot \sin\beta) \cdot \exp j\beta = -jX_q\underline{I}_q \tag{5.52}$$

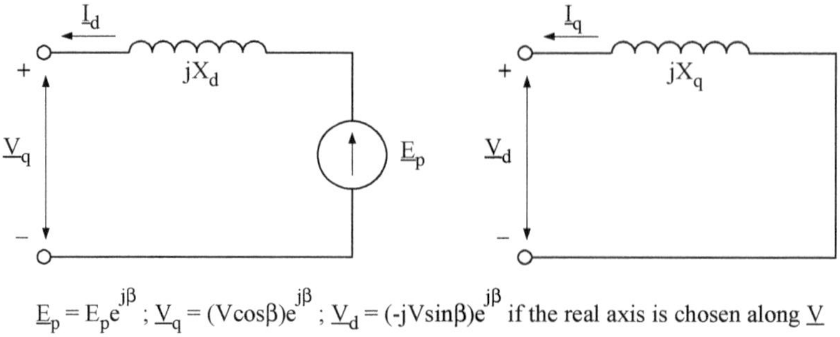

Fig. 5.34 Equivalent circuits for a salient-pole synchronous machine

with $\underline{I}_q = (I\cos\psi)\exp j\beta$ and $\underline{I}_d = (-jI\sin\psi)\exp j\beta$, see also the phasor diagram in Fig. 5.33.

5.3.5 *Torque, Power and Energy*

The torque of a salient pole synchronous machine can be derived from the basic power Eq. 3.87 for rotating field machines:

$$T = \frac{3}{\Omega} \cdot Re(\underline{E}_r \cdot \underline{I}^*) \tag{5.53}$$

with $\Omega = \omega/N_p$. Using Eq. 5.47 or 5.48, the torque can then be written as a function of $\underline{E}_p$ and the current. To eliminate the current, Eqs. 5.49 and 5.50 can be used.

An alternative approach is to start from

$$T = \frac{3}{\Omega} \cdot \left[Re(\underline{V} \cdot \underline{I}^*) + RI^2\right] \tag{5.54}$$

When the stator resistance is not negligible, the resultant torque equation is rather complicated. Here, we will study only the case of a negligible stator resistance. From Eqs. 5.50 and 5.49, or the corresponding equivalent circuits, we find (with the real axis along $\underline{V}$)

$$\underline{I}_d = (\underline{E}_p - V\cos\beta)\exp j\beta / jX_d \tag{5.55}$$

$$\underline{I}_q = (V\sin\beta)\exp j\beta / X_q \tag{5.56}$$

Projection of these current components on the real and imaginary axes then yields the real (or active) and imaginary (or reactive) current components

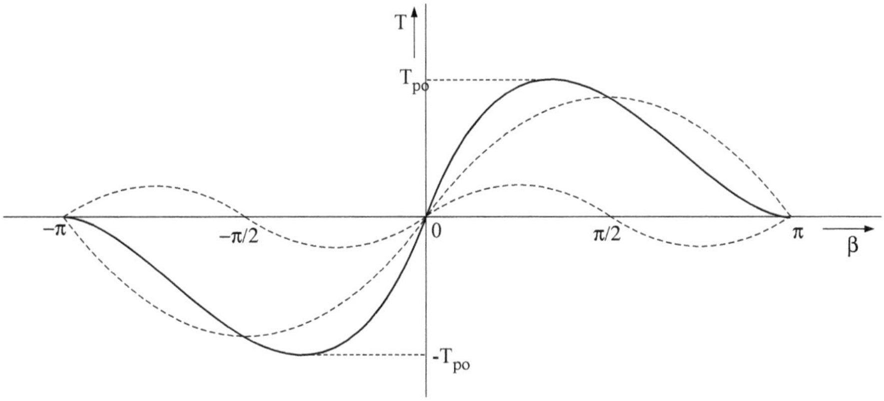

Fig. 5.35 Torque of a salient pole synchronous machine

$$I_{re} = I\cos\varphi = \frac{E_p \sin\beta}{X_d} + \frac{1}{2}V\left(\frac{1}{X_d} - \frac{1}{X_q}\right)\sin 2\beta \tag{5.57}$$

$$I_{im} = -I\sin\varphi = \left[-\frac{E_p \cos\beta}{X_d} + V\left(\frac{\sin^2\beta}{X_q} + \frac{\cos^2\beta}{X_d}\right)\right] \tag{5.58}$$

The expressions for torque, active and reactive power are therefore

$$T = \frac{3}{\Omega}\left(\frac{E_p \sin\beta}{X_d} + \frac{1}{2}V\left(\frac{1}{X_d} - \frac{1}{X_q}\right)\sin 2\beta\right) \tag{5.59}$$

$$P = 3\left(\frac{VE_p \sin\beta}{X_d} + \frac{1}{2}V^2\left(\frac{1}{X_d} - \frac{1}{X_q}\right)\sin 2\beta\right) \tag{5.60}$$

$$Q = 3\left(\left[\frac{VE_p \cos\beta}{X_d} - V^2\left(\frac{\sin^2\beta}{X_q} + \frac{\cos^2\beta}{X_d}\right)\right]\right) \tag{5.61}$$

Torque and active power consist of two parts. The first term stems from the DC excitation in the rotor (and is similar to the one for the smooth rotor synchronous machine). The second term, the reluctance torque or power, is due to the difference in saliency in d- and q-axes.

Observe that the torque component due to saliency is proportional to the square of the voltage, whereas the torque component due to excitation is proportional to the product of voltage and emf. A supply voltage reduction (e.g. a grid voltage drop) will therefore severely affect this reluctance torque (compare with the induction machine). The reluctance torque also varies with 2β, whereas the excitation component varies with β.

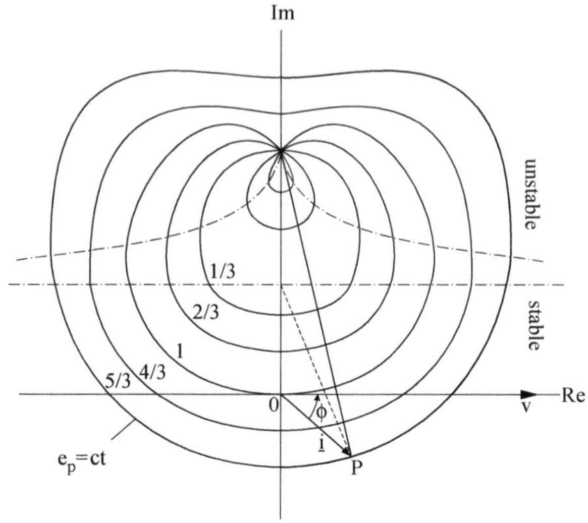

Fig. 5.36 Current locus for a salient pole synchronous machine

Figure 5.35 illustrates the evolution of the torque as a function of the load angle. When $X_q < X_d$ the reluctance torque is in the same sense as the excitation torque for load angles $-\pi/2 \leq \beta \leq \pi/2$ and thus reinforces the torque in this range. The maximum (or pull-out) torque is somewhat larger than that of only the excitation torque, and occurs for smaller values of $|\beta|$, i.e. for values $|\beta| < \pi/2$, which might be advantageous as in most cases reactive power delivery is required whilst at the same time sufficient active power should be converted. For per-unit values $x_d = 1.25$, $x_q = 0.75$, $v = 1$, $e_p = 2$ the maximum torque occurs for $\beta = 80.8°$. The value $e_p = 2$ corresponds to the rated operating condition $v = 1$, $i = 1$ and $\cos\varphi_n = 0.8$ which occurs here at $\beta_n = 22.5°$.

5.3.6 *Current Diagram*

For the salient pole synchronous machine the current diagram for constant voltage and excitation is not a circle as for the smooth rotor synchronous machine, see Fig. 5.36. This is a so-called snail's line of Pascal.

For a pure reluctance machine (i.e. $E_p = 0$ in the equations above), the current locus is again a circle which lies, however, completely above the real axis (in the G-convention), see Fig. 5.37 (for $R = 0$). A reluctance motor or generator will therefore always absorb reactive power from the supply.

Fig. 5.37 Current locus for reluctance machine

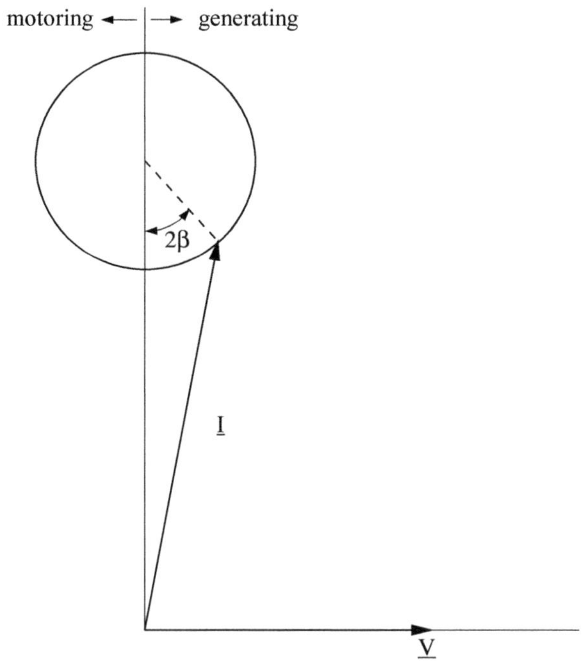

5.4 Synchronous Machines Connected to a Power Grid

In this section we study the behaviour of a synchronous machine connected to an infinitely strong grid.

To start with, we suppose the synchronous machine driven at rated speed ($\Omega = \omega/N_p$), and synchronised with an infinitely strong grid with angular frequency ω and voltage V (using an excitation current so that the no-load voltage $E_{po} = V$). To simplify matter, we consider a smooth rotor synchronous machine with negligible stator resistance.

For such an ideal synchronisation we have $\underline{V} = \underline{E}_p$ and thus $\underline{I} = 0$.

First, we increase the excitation current, keeping the phase angles of $\underline{V}$ and $\underline{E}_p$ equal to each other. With now $E_p > V$ we get the phasor diagram (a) in Fig. 5.38. The synchronous machine is now delivering pure reactive current to the grid, i.e. it acts as a capacitor.

If we decrease the excitation current so that $E_p < V$, then the SM starts drawing pure reactive current from the grid, i.e. it acts as a inductance (see (b) in Fig. 5.38).

Next, (starting again from the no-load condition) we increase the torque of the engine driving the synchronous machine. As a result, the rotor speed momentarily starts to increase and $\underline{E}_p$ starts leading $\underline{V}$, see (c) in Fig. 5.38. However, (with β positive) the SM will now develop a braking torque to make equilibrium with the

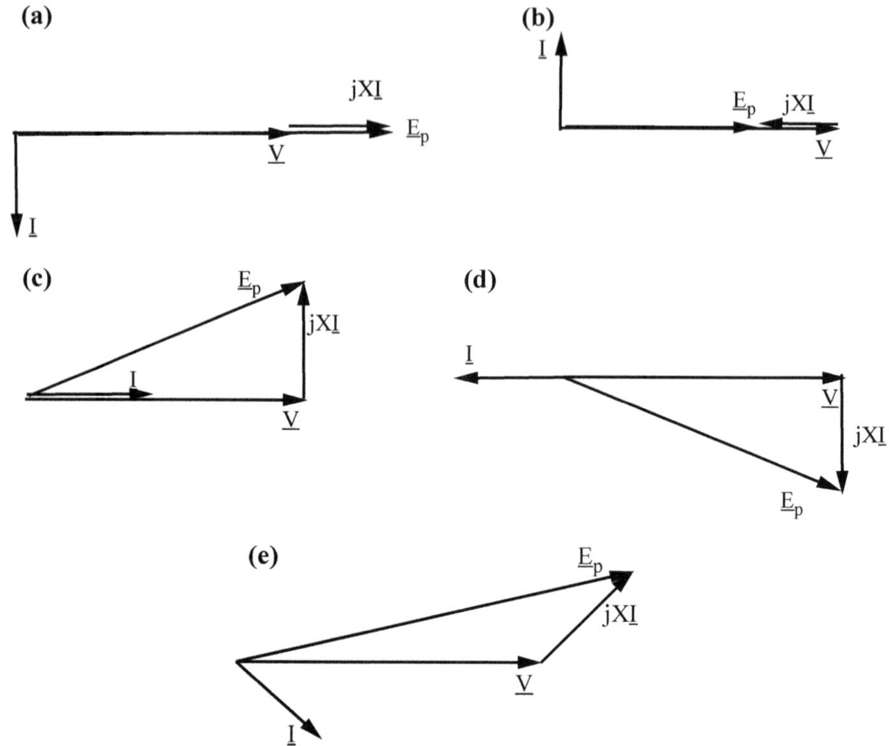

Fig. 5.38 Grid operation of a synchronous machine

increased driving torque. Thus synchronism will be retained. But at the same time an active current starts to flow and active power will be delivered to the grid.

If we decrease the torque of the driving engine (below the ventilation and friction torque), the speed of the SM would like to decrease below the (constant) rotational speed of the voltage $\underline{V}$. The load angle β becomes negative (G-convention) and the synchronous machine starts developing a driving torque until equilibrium with the engine and mechanical losses is obtained, see (d) in Fig. 5.38. The resulting negative active current represents the power delivered by the grid to the synchronous "motor".

In most cases, a synchronous machine will at the same time convert mechanical power into electrical power (or vice versa) and take care of reactive power exchange with the supply (see (e) in Fig. 5.38). In the case of a synchronous generator connected to the grid, automatic speed control will take care of the driving torque of the engine (by monitoring the frequency of the grid, in collaboration with other generators on the grid). The reactive power exchange with the grid on the other hand is controlled based on the voltage level at the point of connection with the grid.[28]

[28] Indeed, in reality the grid is not 'infinitely' strong and the generator active and reactive power exchange will affect frequency and voltage level of the grid.

5.5 Synchronous Motors

In the analysis above, we concentrated on the operation of the synchronous machine as generator. Of course, as the greater part of electrical power in the grid is delivered by (large) synchronous generators (e.g. 1 GW generators of nuclear or thermal power stations), this is a most important application of synchronous machines (as to power).

But synchronous machines are also frequently used as a motor and, although in terms of total energy converted this might seem less important, in terms of the number of applications (and machines) this may far outreach by far the number of synchronous generators.

Some of the oldest applications of synchronous motors are those where very large power output is required (for induction machines the power rating is limited to 20 . . . 40 kW, e.g. because of limitations regarding magnetising current). Another important consideration for the use of synchronous motors is that the efficiency of synchronous motors is much better than that of induction machines. Also, synchronous motors are more compact than induction motors for the same power rating.

With the high degree of development of power electronics, making variable frequency supply very common, and the advent of powerful permanent magnets instead of the DC excitation in smaller synchronous motors, synchronous motors have almost entirely replaced DC drives. In particular for servo drives or controlled drives (e.g. in robotics), the synchronous motor offers the same possibilities as to controllability and dynamics (without issues regarding maintenance or cost).

In Part 3, the permanent-magnet synchronous motor will be studied in depth, after a thorough study of power electronics in Part 2.

The basic equations for the study of motoring of synchronous machines are exactly the same as for generating, although one will in general opt for the M-convention instead of the G-convention. Moreover, permanent magnet excitation can be modelled like DC excitation as will be discussed in Part 3.

Part II
Basics of Power Electronics

Chapter 6
Power Electronic Components

Abstract This chapter aims to give an overview of the main power electronic components for application in electrical drives and electrical grids. It is not at all intended as a power electronics components course, but rather as the basis for the later chapters on the main power electronic circuits like rectifier, chopper or inverter. In addition to the traditional components (for example: diode, thyristor, IGBT) also a section is devoted to new developments regarding Si-C and GaN components.

6.1 Introduction

In most cases, if not all, power electronic components are operated as pure switches. A general representation is shown in Fig. 6.1. Anode and cathode (or source and drain) are the main terminals while the other connections (i.e. gate, basis), if present, are used to control the flow of current between the main terminals.

Power electronic switches may be classified as:

- electronic valves: the typical representative is the diode which allows current in one direction only, on condition that the voltage has the right sign. A diode will only allow current from anode to cathode (if the voltage anode-cathode is positive), but in the reverse direction (with negative voltage) only a negligibly small current can flow.
- electronic valves that can be switched on (but not off): a typical representative is the thyristor. With a positive voltage between anode and cathode a current can flow from anode to cathode if a (small) positive voltage or current is present at the gate. Once conduction has started, the gate signal is not required any more. When a positive anode-to-cathode voltage is present, but no gate signal, the thyristor remains blocked. In the reverse direction (with negative voltage) only a negligible current can flow, just like in a diode.
- power electronic elements that can be switched on and off: for example, the Gate Turn-off Thyristor (GTO), the Bipolar (Junction) Transistor (BJT), Mosfet, IGBT, or IGCT.

In the following sections these power-electronic switches will be described into more detail.

J. A. Melkebeek, *Electrical Machines and Drives*, Power Systems,
https://doi.org/10.1007/978-3-319-72730-1_6

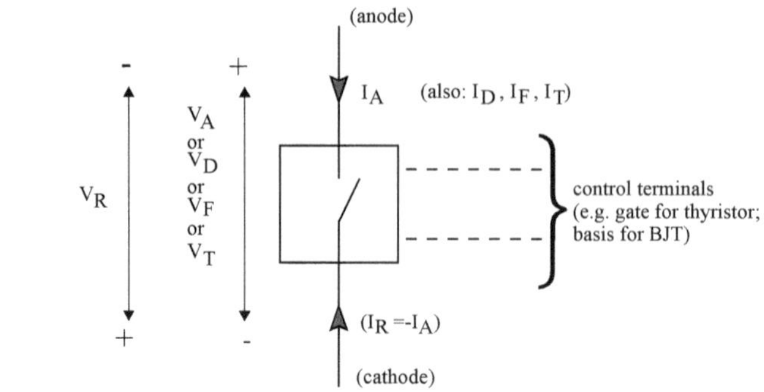

Fig. 6.1 General representation of a power electronic switch

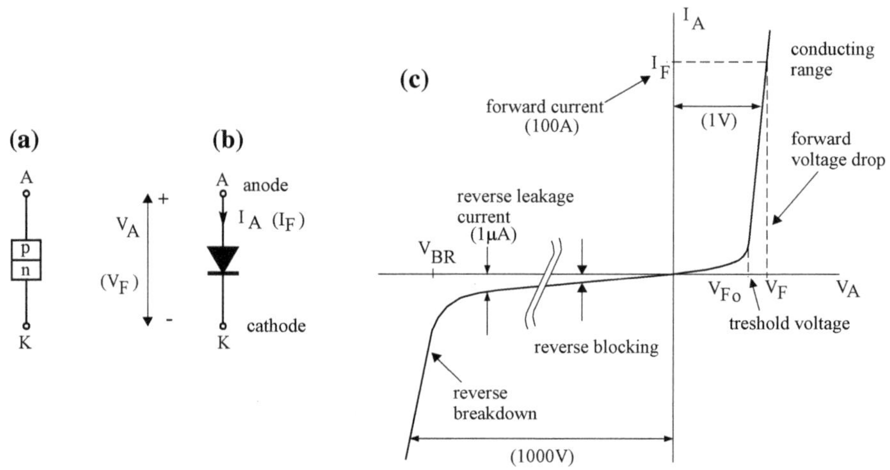

Fig. 6.2 Diode: symbol and characteristics

Tables 6.1, 6.2, 6.3, 6.4, 6.5 and 6.6 at the end of this chapter provide overviews of the main power semiconductors, their main properties and application examples in drives.

6.2 The Diode

A diode is a two-layer $p - n$ element; its symbol and schematic layout are shown in Fig. 6.2a, b. Conduction is performed by both minority and majority charge carriers. Typical characteristics are illustrated in Fig. 6.2c.

With a sufficiently positive voltage between anode and cathode, the $p - n-$ junction will be forward biased and current ($I_A = I_F > 0$) will flow with quite a small voltage drop. However, the voltage should still be higher than the voltage

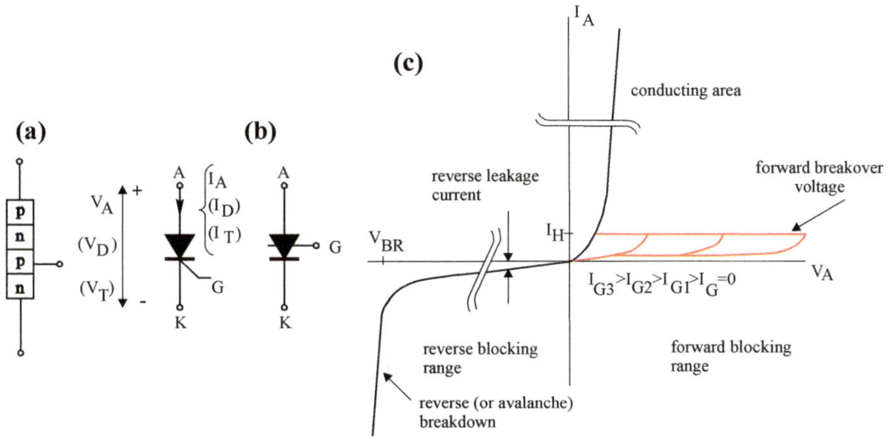

Fig. 6.3 Thyristor: symbol and characteristics

threshold V_{F0} so as to obtain a small and often negligible voltage drop. The voltage threshold is in the range of $0.7 \ldots 1.2$ V. The forward voltage drop in conduction, $V_A = V_F > 0$, depends on the type of semiconductor material and is current-dependent. For rated diode current, the voltage drop is on the order of 0.3–1 V above the threshold voltage V_{F0}.

Negative voltages $V_A = -V_R < 0$ will result in only a very small negative current ($I_A = -I_R < 0$), meaning that the diode essentially behaves as an open circuit because the $p-n-$junction is blocking the current. For very high negative voltages $V_R > V_{BR}$, breakdown may occur with a large negative current. The reverse breakdown voltage V_{BR} is normally high; depending on the semiconductor material, it may be on the order of some tens of volts to some kilovolts.

When the diode switches from the conduction state to the non-conducting state (by a reverse voltage), momentarily a higher reverse current will flow. This current will remove the charge in the junction so as to block the current flow. The time for this reverse current (i.e. the recovery time) varies from some $1 \ldots 5\,\mu$s in conventional diodes to some hundred *ns* in fast-recovery diodes.

6.3 The Thyristor

The thyristor (also called the Silicon Controlled Rectifier or SCR) is a four-layer $p-n-p-n$ semiconductor element. Its schematic layout and symbol are shown in (a) and (b) in Fig. 6.3. A thyristor has three terminals: in addition to the power terminals anode and cathode, the gate terminal serves to control the conduction.

Figure 6.3c illustrates the characteristics of a thyristor. Like the diode, it conducts current in only one direction, from anode to cathode. For negative voltages, it remains in the reverse blocking region.

To conduct, together with a positive voltage $V_A = V_F$, also a (small) gate signal is required to first turn the thyristor on. When a positive anode voltage is applied without gate current ($i_g = i_G = 0$), the anode current remains small (cf. the red line). The thyristor remains in the forward blocking region. If at the same time a sufficiently positive gate signal i_g is applied ($I_{G1} > 0$ or $I_{G2} > 0$ or $I_{G3} > 0$), the thyristor comes into conduction[1] with now a high anode current and a small voltage drop (this range is called the high conduction region).

Applying too high an anode voltage may, however, also cause the thyristor to enter into conduction without gate signal (i.e. forward break-over).[2]

Note that a thyristor will come out of conduction if the anode current drops below the holding current I_H.

For negative anode-to-cathode voltages, the thyristor will show only a small negative leakage current (called the reverse blocking region). Now junctions J_1 and J_3 are reverse biased and junction J_2 is forward biased. Too high a negative voltage may result in avalanche (i.e. reverse breakdown), which is likely to destroy the thyristor.

When a thyristor should come out of conduction, a negative anode-to-cathode voltage must be applied to allow the blocking charge for junction J_2 to build up. In practice, this can be achieved by discharging a capacitor with the right (negative) charge to the anode of the thyristor, or by injecting a negative anode current; both are realised in most cases using resonant $L - C-$circuits (called *killer circuits)*. This process is referred to as forced commutation, whereas natural or load commutation is what happens when the external (AC) load or the (AC) power source causes a reversal of the voltage, thus reducing the current below the holding current.

Note that a thyristor may come into conduction in other abnormal ways (in addition to an anode-to-cathode voltage exceeding the forward breaker voltage):

- by a very high dV_A/dt, resulting in a capacitive ignition
- when, after a conduction interval, the positive anode-to-cathode voltage returns too quickly.

The latter phenomenon, in particular, limits the switching speed of a thyristor. In order to allow the blocking charge for the junction J_2 to build up, the anode-to-cathode voltage should be negative for a minimum time span, called the recovery time t_{rr}. This recovery time is the largest part of the minimum delay before a positive anode-to-cathode voltage can be applied again. To this recovery time other (smaller) delays add. The total minimum delay between the instant the anode current becomes zero and the instant the device is capable of blocking forward voltage is called the turn-off

[1] With a positive anode voltage, junctions J_1 and J_3 are forward biased but junction J_2 is reverse biased. The function of the gate signal is to lower the breakdown voltage of this junction.

[2] Another interpretation of normal conduction is that the gate signal reduces the breakdown voltage; for high gate currents, the entire forward blocking region is removed and the thyristor behaves as a diode (see Fig. 6.3).

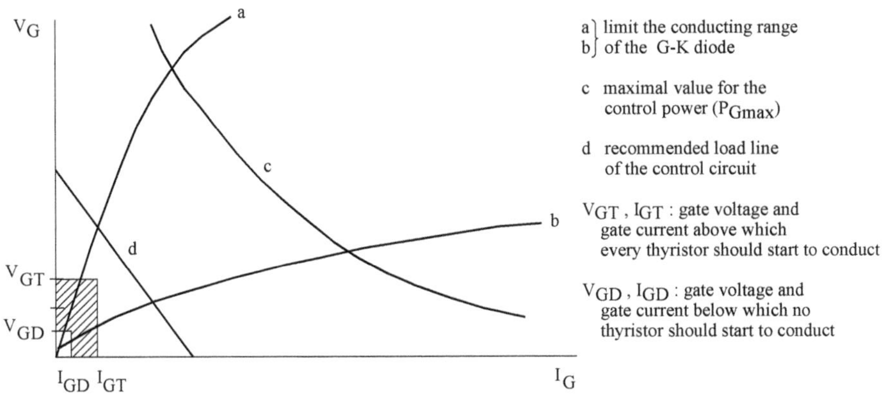

Fig. 6.4 Gate characteristics of a thyristor

or hold-off time[3] t_{off}. Whereas the turn-on time t_{on}(or t_{gt}) is very short, on the order of $1 \cdots 3\,\mu$s, the turn-off time is rather large, on the order of $50 \cdots 100\,\mu$s for normal types or $1 \cdots 50\,\mu$s for fast-switching types. For mains-commutated thyristors the turn-off time is more around $100 \cdots 200\,\mu$s, but these thyristors possess a very small forward voltage drop in conduction.

Applying a safety factor of 10, a turn-off time of $100\,\mu$s results in a maximum frequency of 1 kHz for the circuit. For mains applications (controlled rectifiers, see further chapters) this is more than sufficient but not for all other applications (e.g. inverters need higher frequencies).

Despite the limited speed of a thyristor, its important advantages are the very low conduction and switching losses as well as the low power requirements[4] for the gate circuit.

Table 6.5 (at the end of this chapter) lists the main parameters required to select a thyristor. Figure 6.4 shows an example of the gate characteristics of a typical thyristor.

In addition to fast fuses, a thyristor requires other means of protection. If the current rises too fast, a local current concentration on the substrate may destroy the device. To avoid this, some inductance must be present in series with the thyristor. Because a thyristor may turn on by high dV_A/dt, a snubber circuit in parallel with the device is required as well (a simple snubber circuit consists of a series connection of a resistance and capacitor, for example).

Maximum values for thyristors are:

- $V_{DRM} \approx 6000$ V (V_{DRM} = repetitive maximum direct (forward) off-state voltage)
- $V_{RRM} \approx 5000$ V (V_{RRM} = repetitive maximum reverse blocking voltage)
- $I_{TRM} \approx 5000$ A (I_{TRM} = repetitive maximum direct on-state current)
- $f_{max} \approx 50$ kHz

[3]The delay the circuit actually applies is the circuit turn-off time t_q and this must be larger than the device hold-off time.

[4]Gate current is quite low and the gate signal must only be applied at the start of the conduction interval and can be removed afterwards.

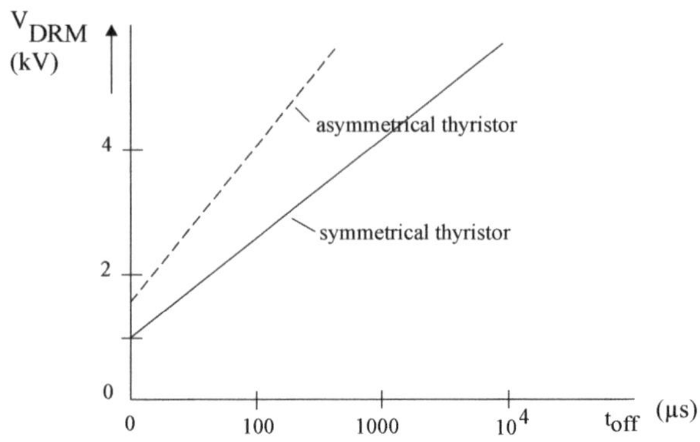

Fig. 6.5 Symmetrical and asymmetrical thyristors

Note that these maximum values will not occur at the same time. There is a trade-off between high forward blocking voltage, high reverse blocking voltage, low voltage drop in conduction, high current, high di_A/dt and high frequency (low turn-off time). For example, Fig. 6.5 shows the relation between the forward blocking voltage and the turn-off time for common types. The full line applies to symmetrical types with $V_{DRM} = V_{RRM}$, while the dashed line is for asymmetrical thyristors (with $V_{DRM} > V_{RRM}$). Asymmetrical thyristors have a lower forward voltage drop and shorter turn-off times but also lower reverse blocking voltages.

6.4 The Triac

A triac can be considered as two thyristors connected in anti-parallel. Figure 6.6 shows the symbol and characteristics. Triacs have somewhat less performant characteristics than thyristors, such as lower forward and reverse blocking voltages and higher turn-off time. They are not available with high voltage and current ratings, but are more economical than two thyristors in anti-parallel and are used in many low-power applications (e.g. blenders, vacuum cleaners).

6.5 The GTO

A Gate Turn Off thyristor (GTO) is essentially a thyristor-like device, but in contrast with the thyristor it can be turned off by a pulse of negative gate current. Figure 6.7 shows symbols and turn-off characteristics.

When a negative gate current pulse is applied for turn-off, the anode current first falls abruptly, while the anode-to-cathode voltage starts to rise. Subsequently, the anode current falls rather slowly to zero. This tail current results in relatively high losses, as the anode-to cathode voltage is already high (see also (b) in Fig. 6.7).

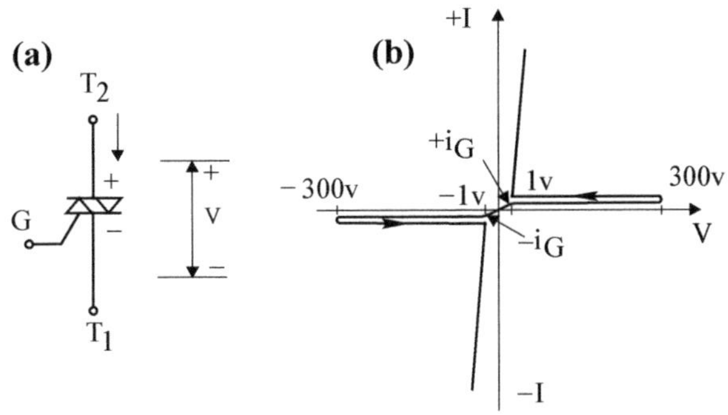

Fig. 6.6 Triac: symbol and characteristics

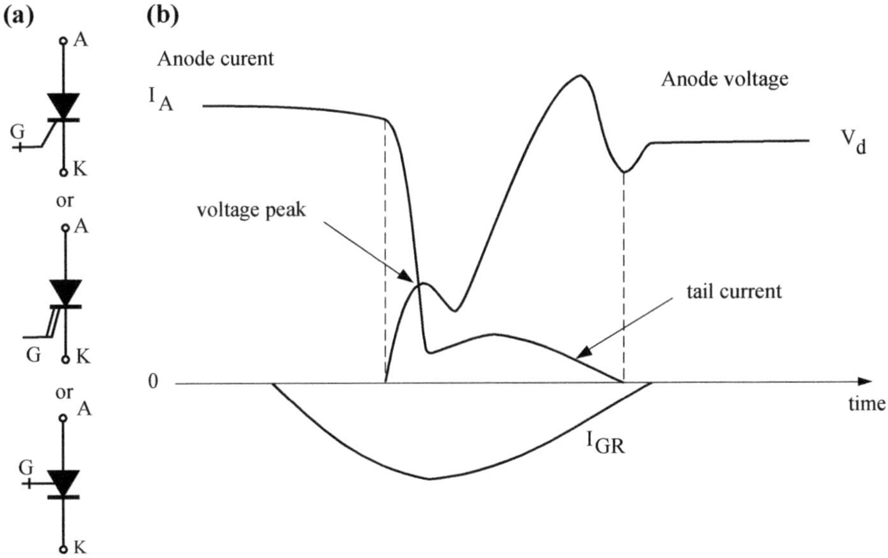

Fig. 6.7 GTO: symbol and turn-off

Also note that, whereas the (positive) gate current to turn it on is quite low, the negative gate current to turn it off is rather high. Typically, the ratio of the anode current to the negative gate current for turn-off is only 3 to 5. For example, to turn off an anode current of 1000 A, a negative gate current of 250 A is required.

The maximum characteristics of GTOs are somewhat lower than those of a thyristor (e.g. $V_{DRM} \approx 4000$ V, $I_{TRM} \approx 4000$ A). Nevertheless, when forced commutation is required, GTOs are nowadays used instead of thyristors with killer circuits.[5]

Other disadvantages of GTOs include the relatively high turn-off losses (due to the tail current), and the limited or sometimes absent reverse blocking voltage.

[5] In turn, GTOs have made way for Integrated Gate Commutated Thyristors (IGCTs), which are now often preferred to GTOs (see Sect. 1.6).

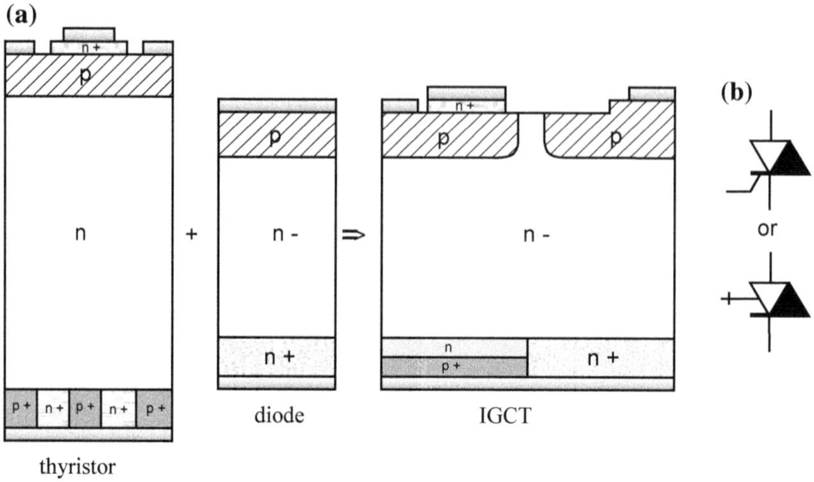

Fig. 6.8 IGCT: symbol and structure

6.6 The IGCT

The Integrated Gate Commutated Thyristor (IGCT) is a relatively new device that was invented and patented by ABB at the beginning of this century. Its symbol and basic structure can be found in Fig. 6.8.

Basically, the IGCT is a GTO-like device with several enhancements. First, the co-axial gate consists of several fingers evenly distributed along the circumference of the die. As a result, when the device is switched on (or off), the current distribution is more even, thus avoiding local high current densities. Second, the gate control circuit is integrated into the package, avoiding additional leakage inductance by gate leads. Third, most types have an integrated anti-parallel diode which eases the turn-off.

As a result, higher switching frequencies (up to 2 kHz or even higher) are possible, snubbers can be avoided because of the anti-parallel diode, the conduction and switching losses are smaller and the power requirements for the gate are reduced, especially for turn-off.

6.7 The BJT

The Bipolar Junction Transistor (BJT or BT) is a three-layer $n-p-n$ (or $p-n-p$) device. In Fig. 6.9, its structure, symbol and characteristics are shown.

The three terminals are called collector, emitter and base. In power electronic applications, the transistor is normally used as a switch by providing sufficient base current if it is switched on. When the basis current I_B is zero, the BJT essentially

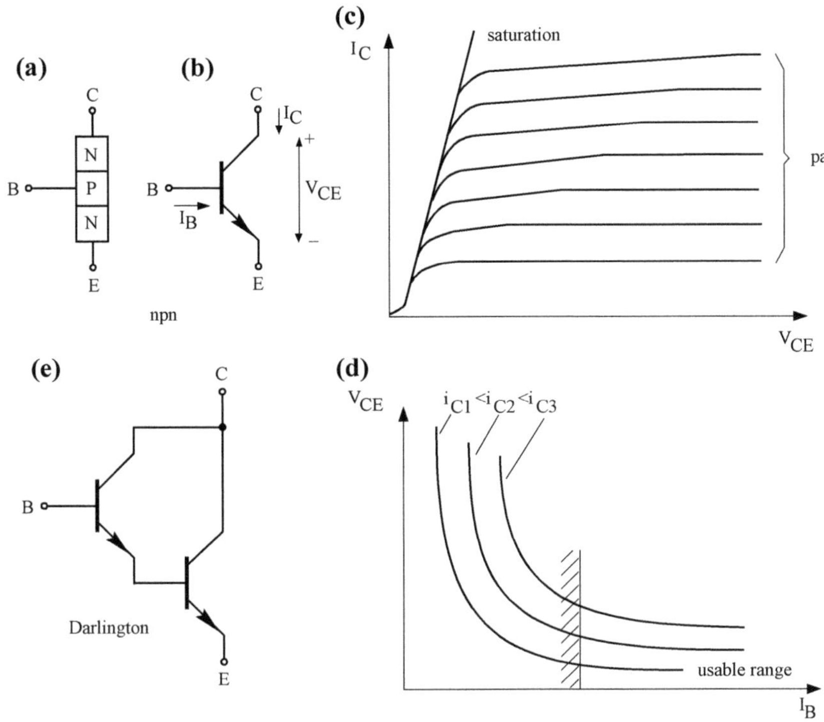

Fig. 6.9 BJT: symbol and characteristics

behaves as an open circuit, with high high collector-emitter voltage V_{CE} and low collector current I_C. For a sufficiently high basis current, the transistor enters saturation and approximately behaves as a closed switch with a relatively low voltage drop (see the characteristics $I_C - V_{CE}$ in (b) in Fig. 6.9 or the characteristics $V_{CE} - I_B$ in (d) in Fig. 6.9).

A transistor is current-controlled: the basis current determines the open or closed condition. To keep the collector-emitter voltage sufficiently low, the basis current must be high enough; for high collector currents, the gain I_C/I_B may become less than 10. A much higher gain (100 or more) can be obtained using a Darlington pair, as is illustrated in (c) in Fig. 6.9.

The operating point of a BJT must lie within the Safe Operating Area (SOA) (see Fig. 6.10). The SOA is defined by the maximum collector current $I_{C,max}$, the maximum dissipation P_{max}, the second breakdown limit[6] ($\sim V^2 I$), and the maximum collector-emitter voltage $V_{CE,max}$.

[6]A BJT does not only have a first breakdown limit (if V_{CE} is too high) but also a second breakdown or avalanche limit caused by a local current concentration in a limited area of the collector junction, resulting in a thermal runaway. Such a second breakdown can occur when switching the device on, but by switching off highly inductive loads as well.

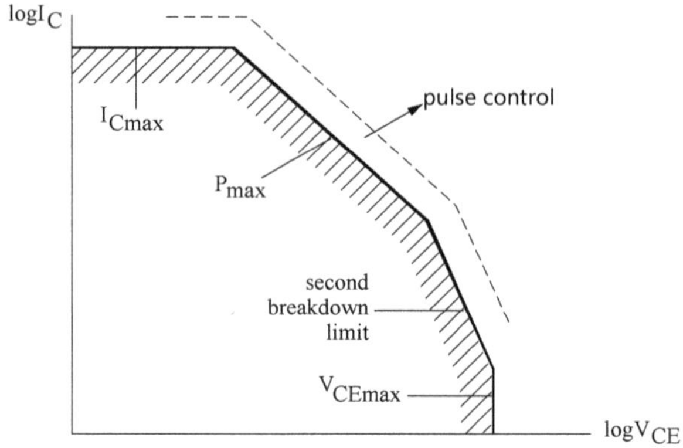

Fig. 6.10 SOA of a BJT

BJTs have much shorter switching times than thyristors (i.e. 1 μs for switching-on and 3 μs for switching-off) and can be used for frequencies up to 100 kHz. Another advantage is that BJTs can be switched off quite easily: thyristors require killer circuits, and GTOs require rather high negative gate currents. Further, protection against over-currents can be realised in quite a straightforward manner and the switching losses are lower than for thyristors or GTOs.

On the other hand, BJTs (Darlingtons) are not available for high voltages and high currents (e.g. maximum 500 V and 2000 A). Moreover, for higher voltages and currents they are rather expensive (compared to GTOs). The reverse blocking voltage is also fairly limited, and the power requirement for the control (basis) is relatively high compared with thyristors or GTOs. Nowadays, they are mostly substituted by Mosfets (for lower power) or IGBTs (for high-power applications).

6.8 The Mosfet

The power Mosfet (Metal Oxide Semiconductor Field Effect Transistor) is a unipolar device, which means that current is solely carried by majority charge carriers (in contrast with thyristors, GTOs or BJTs). Figure 6.11 shows the basic layout, symbol and characteristics of an N-channel Mosfet.

Positive and negative terminals are called drain and source, respectively, and current flows from drain to source. By applying a positive voltage signal (dependent on the type 3 . . . 4 V to 10 . . . 15 V) at the gate, we are able to form an N-type conduction channel between drain and source. Because the gate is only capacitively coupled, the gate current is minimal, and the power required to control a Mosfet is negligible. Other advantages include the absence of second breakdown (because of a positive temperature coefficient), the very short switching times (<50 ns), the

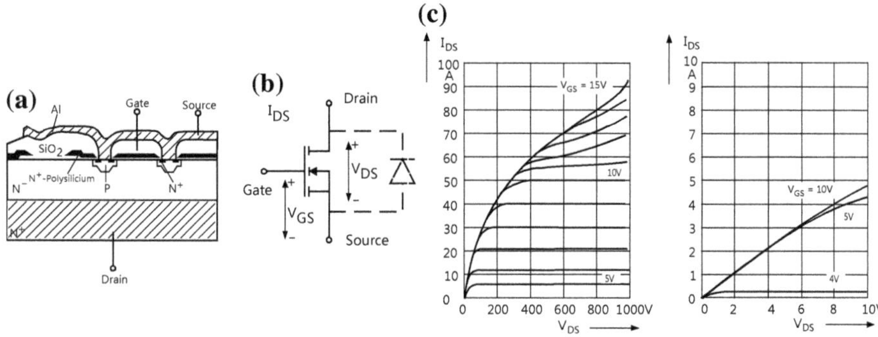

Fig. 6.11 Mosfet: structure, symbol, and characteristics

small or even negligible switching losses and the fact that the positive temperature coefficient facilitates parallel connections (for higher currents).

However, the conduction losses of Mosfets are rather significant compared to BJTs or thyristors, especially for higher power ratings. As an example, compare the characteristics in (c) in Fig. 6.11 for a BUZ54 (1000 V, 25 A). with the ON-state resistance for a BUZ15 (50 V; 45 A): 0.03 Ω for rated current. Other disadvantages include the fact that reverse blocking is almost non-existent (since Mosfets have an inherent anti-parallel diode) and their sensibility to high dV/dt, which somewhat limits the switching speed. Nevertheless, the switching speed is still very high compared to BJTs or thyristors, with frequencies up to 100 MHz being realistic.

The SOA of a Mosfet can be found in Fig. 6.12.

6.9 The IGBT

The Insulated Gate Bipolar Transistor (IGBT) can be regarded as combination of a BJT for the power part and a MOS-structure for the steering part. As such, it combines the advantages of BJTs and Mosfets, i.e. low conduction losses and low control losses. Figure 6.13 shows its structure, symbol and an example of its characteristics.

Switching speed, steering power and robustness are comparable to those of a Mosfet, while the conduction properties are similar to those of a BJT (i.e. much smaller conduction losses than a Mosfet). This is immediately clear from a comparison between the characteristics in Fig. 6.13c, d for a BUP 304 (1000 V, 25 A) with those of (c) and (d) in Fig. 6.11 for a BUZ 54 (1000 V, 25 A). Note that both semiconductors have the same semiconductor area. Table 6.6 gives an overview of the main characteristics of the IGBT BUP 304.

IGBTs are more advantageous than Mosfets for higher voltage levels (>500 V) and frequencies that are not too high. For lower voltage applications, IGBTs offer minor advantages as Mosfets also have low conduction resistance. However, in the

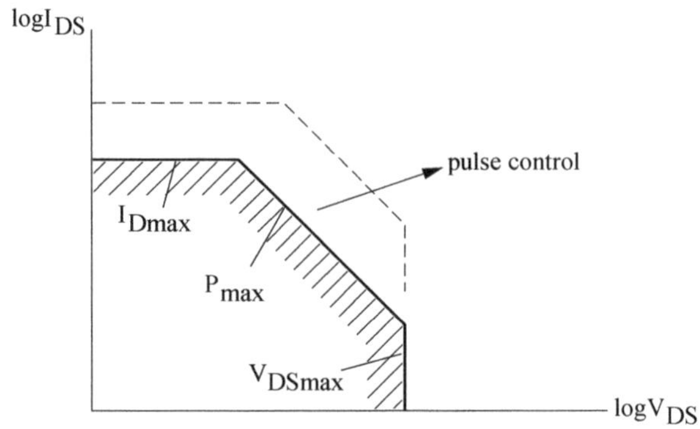

Fig. 6.12 SOA of a Mosfet

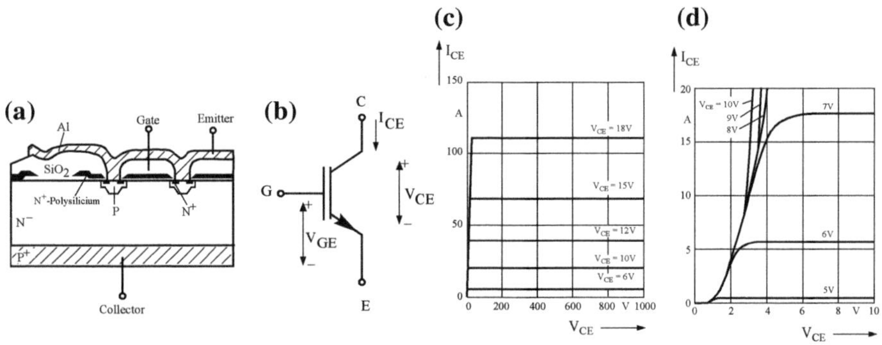

Fig. 6.13 IGBT: symbol, structure and characteristics

last few years the application range for IGBTs has been extended to both higher and lower power applications. It is important to note here, however, that the tail current when an IGBT is switched off may result in higher total losses than for a comparable Mosfet (see Fig. 6.14).

6.10 SiC and GaN Devices

Silicium Carbide (SiC) and Gallium Nitride (GaN) semiconductors are quite new devices. They entered the market a few years ago and are now further developed for higher voltages and power range.

Both SiC and GaN are wide-bandgap (WBG) devices. The band gap (between conduction and valence band) is approximately three times larger than the one for Si. As a result, WBG material allows high electric fields, with depletion regions possibly very short or narrow, so device structures can have high carrier density and can be packed very densely. In contrast with most Si devices, SiC and GaN are unipolar and only use majority charge carriers. A limitation at this time is that the SiC and GaN

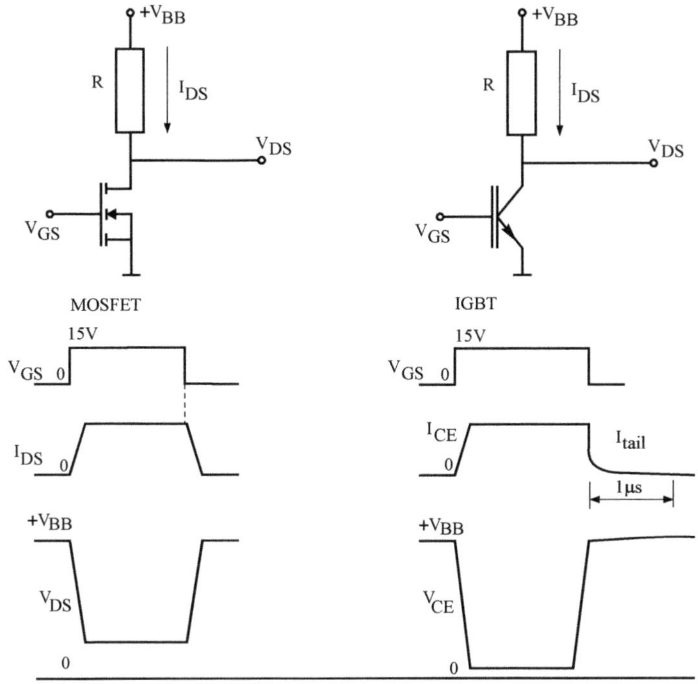

Fig. 6.14 Mosfet and IGBT switching

semiconductors are to be grown on Si substrates, which may introduce dislocations. However, it is expected that, in time, lateral GaN will replace Si FETs and vertical SiC will replace Si IGBTs.

SiC has shown excellent capability to operate at voltages well above 1 kV and even above 10 kV in some applications. Another advantage is that the operating temperature may reach 250–300 °C (versus 125 °C for Si). It is optimal for higher power applications (10 kW to 1 MW) and voltages from 1 to 7 kV, thus replacing silicon insulated-gate bipolar transistors (IGBTs). SiC semiconductors can exploit a vertical structure for lowest resistance and benefit from the greatly reduced switching losses (see Fig. 6.15). At lower voltages, SiC FETs become limited by relatively low channel mobility, high field stresses in gate dielectrics, and high substrate resistance [36]. An advantage of SiC semiconductors is the higher thermal conductivity of SiC (compared to Si), leading to better and easier cooling.

Lateral GaN may replace Si FETs for a range of applications that require breakdown voltages of 100–650 V. Currently, applications up to 10 kW and voltages up to 1200 V are envisaged. Lateral GaN offers the highest switching speed and lowest switching losses, and it will enable frequencies to increase well above 1 MHz in many applications (even up to 10 MHz) [16]. Further, generally, ON-resistance and therefore also conduction losses are smaller than those for Si devices. Figure 6.16 shows a schematic layout of a lateral GaN FET. Optimal performance is obtained in

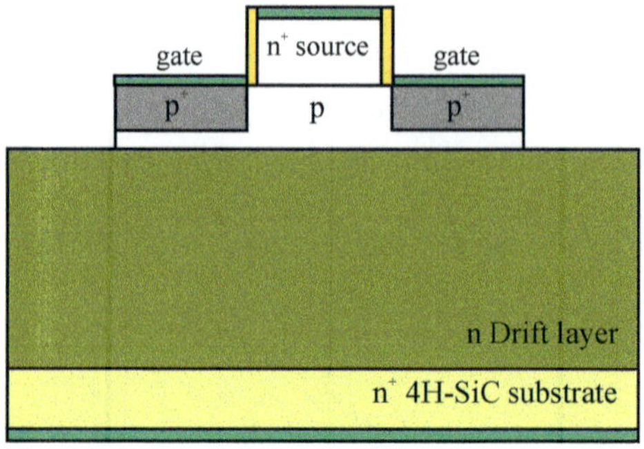

Fig. 6.15 SiC structure

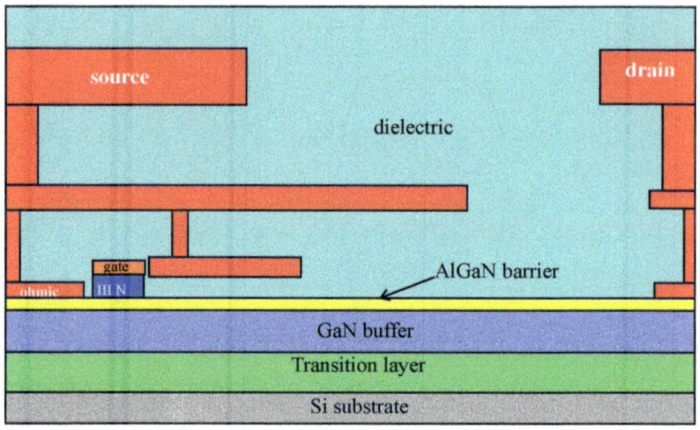

Fig. 6.16 GaN: structure

monolithic chips, meaning that logic, driver and FET are incorporated in the same package.

Currently, SiC Schottky barrier diodes are preferred for systems between 300 and 1200 V and SiC MOSFETS have already penetrated the market. Lateral high-electron mobility transistor GaN devices for power-electronic systems under 650 V have also entered the market. Future research should enable developments related to bidirectional 600 V GaN MOSFETs, 10 kV SiC bidirectional insulated-gate bipolar transistors (IGBTs), SiC MOS-FETs and IGBTs rated above 10 kV, as well as more GaN-based power application-specific integrated circuits, for example incorporating optoelectronics. Of course, it is expected that current and future research will also improve Si-based semiconductors (i.e. Mosfets, IGBTs).

6.11 Other Power Electronic Devices

In the past, numerous other semiconductor power devices have been developed. A few of these have found special applications, but others are not widely used. The devices discussed below (which are used in special applications) are the SIT (Static Induction Transistor), the SITh (Static Indiction Thyristor) and the MCT (MOS Controlled Thyristor).

Fig. 6.17 SIT: structure and symbol

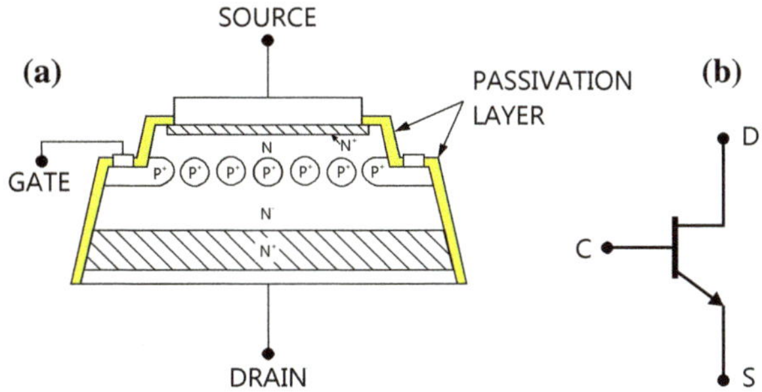

Fig. 6.18 SITh: structure and symbol

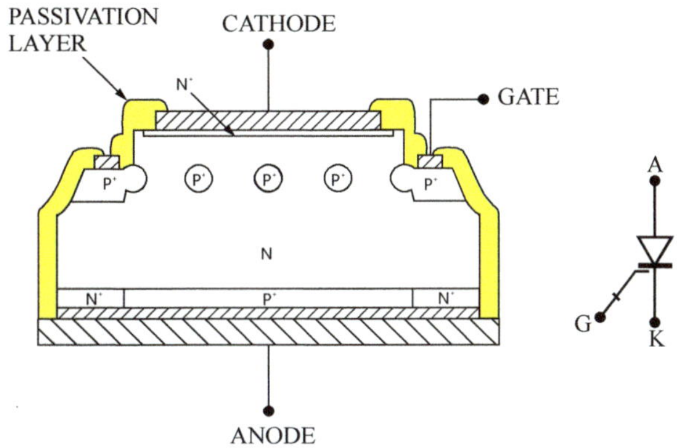

The SIT (Static Induction Transistor) is a unipolar device just like the JFET. It is also a normally-on device (as is the JFET). Figure 6.17 illustrates its structure and symbol. The SIT can be considered as the semiconductor equivalent of the vacuum triode. Its main application is for high frequency *and* high power (e.g. audio, VHF/UHF and microwave amplifiers). For general power electronic applications, the voltage drop (and thus conduction loss) is too large, unless an exceptionally high frequency justifies its use.

Similar to the SIT, the SITh (Static Induction Thyristor) is also a normally-on device, but it has a bipolar structure (a $p+nn+$ diode-like power part with a raster-like $P+$gate-structure). Figure 6.18 shows its symbol and layout. It is important to note that the SITh has no reverse blocking capacity. The turn-off requires a significant negative gate current and exhibits a tail current (along with the associated losses) like the GTO.

The MCT (MOS Controlled Thyristor) has a bipolar $pnpn-$structure for the power part and a MOS control part (see Fig. 6.19). Like a GTO, the MCT can be switched on and off by current pulses at the gate. However, the current gain for switching off is much larger. The switching speed is of the same magnitude as for an IGBT, but the voltage drop in conduction is smaller. Nevertheless, the MCT has not found wide application.

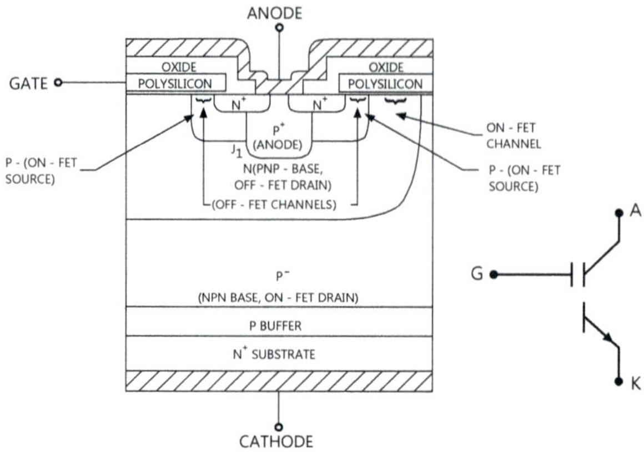

Fig. 6.19 MCT: structure and symbol

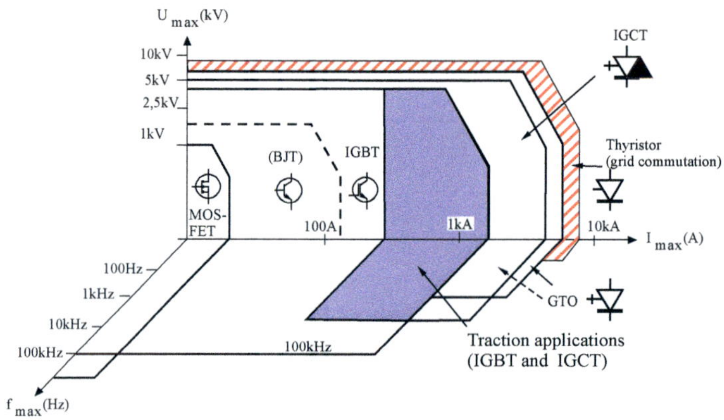

Fig. 6.20 Application range of power semiconductors

6.12 Concluding Remarks

Power electronics semiconductor technology is in continuous evolution. It is therefore rather difficult to describe the state of the art. Figure 6.20 gives a broad overview of the industrial application range of the most common power semiconductors at this moment. SiC and GaN semiconductors are not yet mentioned in this figure, but this may change in a few years.

The tables below provide some more illustrative data of common power semiconductors.

Table 6.1 Classification of power semiconductors

(a) Classification of semiconductors

	Unipolar	Bipolar
Conduction	Majority charge carriers	Majority and minority charge carriers
Types	Schottky-Diode, MOSFET, JFET, SIT	Diode, thyristor, BJT, IGBT, GTO, IGCT, SITh, MCT
Conduction modulation	None	Injection of charge carriers
Forward voltage drop	High (exc.: Schottky)	Low
Switching time	Short	Medium to long
Switching frequency	High	Low to medium
Control by	Voltage	Current (voltage for IGBT)
Control power	low	High (low for IGBT)
Driver cost	Low	High (low for IGBT)
cost/chip area	High	Low (high for SITh)

(b) Typical values

Type	V_{max} [V]	I_{max} [A]	t_{off} [μs]	P_{max} [kVA]	Frequency [kHz]
BJT/darlington (with fine structure)	1400	300	15 to 25	500	0.5 to 5
	600	500	5 to 10	150	0.5 to 5
	1000	80	1 to 3	40	2 to 50
IGBT	3000	1000	0.5 to 3	3000	to 150
	1000	2000	0.5 to 3	2000	to 150
MOSFET	1000	20	0.1 to 0.3	50	to 3000
	60	150	0.1 to 0.3	20	to 3000
SIT	1400	300	0.1 to 0.3	200	30 to 300
GTO	4500	3000	10 to 25	10000	0.2 to 1
IGCT	6000	3000	2 to 5	10000	0.5 to 2
SITh	2000	600	2 to 4	300	1 to 10

Table 6.2 Comparison of the main properties of BJT, Mosfet, IGBT

Semiconductor	Darlington BJT	MOSFET	IGBT
Symbol	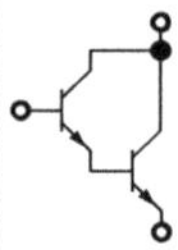	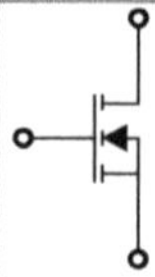	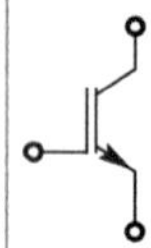
Structure	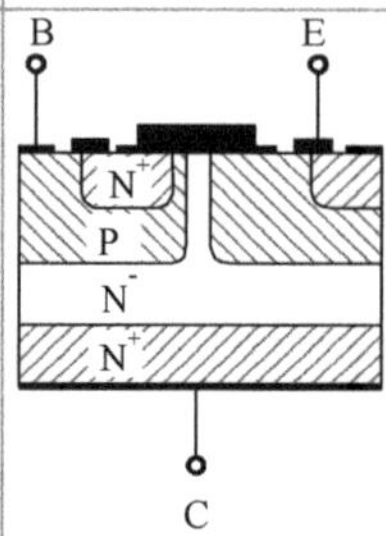	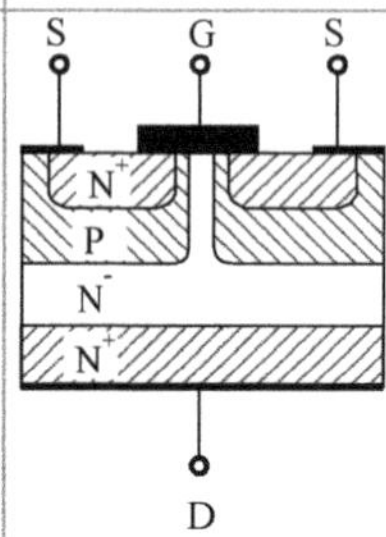	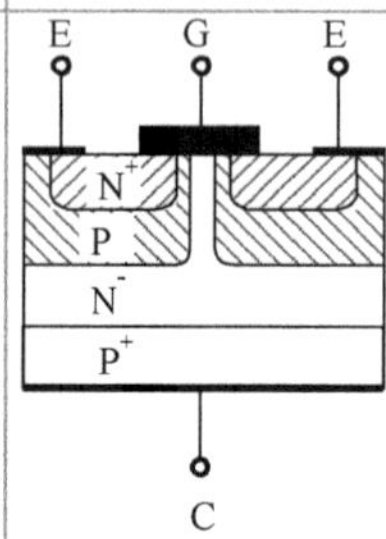
Blocking properties (upper limits)	Average	Low	High
Control circuit complexity & power	average	small	minimal
	high	low	low
Switching properties Switch-on time Switch-off time Switching loss	average	short	average
	long	short	average
	high	low	average
Conducting properties Current Power losses	quite high	low	high
	low	high	rather low
Pulse frequency limit for 0.5 I_{DC}	= 4kHz	= 250kHz	10kHz

Table 6.3 Semiconductors for drives

	Max forward voltage [V]	Max reverse voltage [V]	Max forward current [A]	Max switching frequency [kHz]	Max power (3-phase)	Properties
Thyristor (Silicon Controlled Rectifier, SCR)	600–6000	0–5000	5000	0.4	20 MW	• Low current pulse for switching-on • Switch-off requires killer circuit • High overload allowed • Low switching frequency
Gate Turn-Off Thyristor (GTO)	800–6000	200	6000	1 (500 Hz for high power)	10 MW	• Low current pulse for switching-on • High current pulse for switching-off • Lower maximum values than SCR
IGCT	4500 6000	–	3000 2000	2	10 MW	• Improved switching compared to GTO • Lower switching and conduction losses than GTO en IGBT • Snubber circuits not required
Bipolar transistor (BJT of HFBT) and darlington	50 | 1400	50	500 | 300	10 (500 Hz for high power)	800 kW (now replaced by IGBT)	• Conduction requires continuous basis current • Reverse current pulse for switching-off • High switching frequencies

(continued)

Table 6.3 (continued)

	Max forward voltage [V]	Max reverse voltage [V]	Max forward current [A]	Max switching frequency [kHz]	Max power (3-phase)	Properties
Power-Mosfet	50 | 1000	0	150 | 20	3000	40 kW (5–10 kW economical limit, for higher rather IGBT)	• Quite low pulse for switching on and off • Quite high switching frequencies • Rather high ON-state resistance • No reverse voltage allowed
Insulated-Gate Bipolar Transistor (IGBT)	500 | 3000	200 (lateral) 500 (vertical)	2000 | 1000	100	2000 kW	Quite low pulse for switching on and off

Table 6.4 Comparison of characteristics of power semiconductors for drives

	BJT		SIT		MOSFET		IGBT		GTO		SITh		IGCT	
typical (maximum) values for voltage and current	800 V 100 A		600 V 20 A		500 V 20 A		3000 V 1000 A		4 kV 2 kA		1500 V 600 A		4 kV 2 kA	
typical converter power (three-phase)	220kW		100 kW		5 kW (10 kW)		10 MW		20 MW		1 MW		20 MW	
typical gate signals														
required power for the control	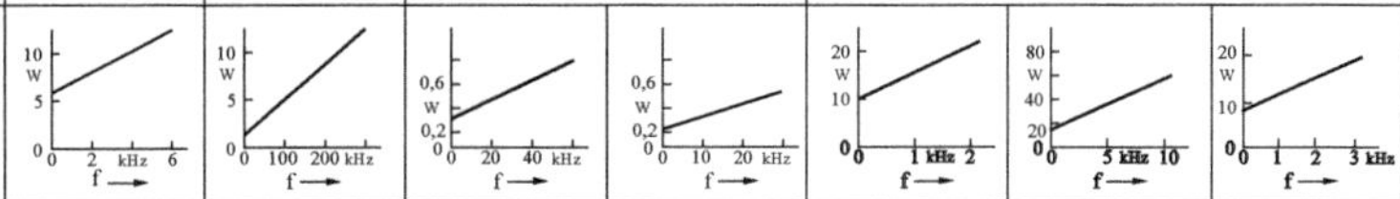													
f_p kHz	1	10	50	200	20	100	5	20	0.05	2	1	5	1	3
total control power P_S W	39	52	21	60	2.7	3.9	1.8	2.0	60	90	63	195	30	60
specific control power P_{ST}/P_{SR} %	0.1	0.13	0.35	1.0	0.05	0.08	0.007	0.008	0.03	0.05	0.035	0.1	0.02	0.04
complexity of the control circuit	average - high		average		minimal				high		average - high		minimal (integrated)	

Table 6.5 Definitions and abbreviations for power thyristors and diodes

V_R reverse blocking voltage also V_{RRM} repetitive; R_{RSM} peak (surge); V_{RNM} non-repetitive } maximal reverse blocking voltage	I_R reverse leakage current also I_{RRM} repetitive maximal reverse leakage current $I_{RRMS\,(M)}$ effective maximal reverse leakage current	(also for diodes)
V_D forward blocking voltage also V_{DRM} repetitive; V_{DSM} peak (surge); V_{DNM} non-repetitive } maximal direct (forward) blocking voltage	I_D forward leakage current also I_{DRM} repetitive maximal direct leakage current $I_{DRMS\,(M)}$ effective maximal direct leakage current	(for thyristors)
	I_L latching current I_H holding current	
V_T on-state voltage forward voltage drop	I_T direct current ($I_{TAV\,(M)}$; $I_{TRMS\,(M)}$; I_{TSM}; I_{TCM}; ...)	P_T conduction loss (for diodes: T→F)
$\left(\frac{dv}{dt}\right)_{max}$	$\left(\frac{di}{dt}\right)_{max}$	
t_{rr} reverse recovery time		(also for diodes)
t_{off} of t_q turn-off time $= t_s$ (carrier storage time) + t_f (fall time)	$t_{on} = t_{gt}$ gate controlled turn-on time $= t_{gd}$ (delay time + t_r (rise time)	(for thyristors)
t_{off} of t_{gq} gate controlled turn-off time $= t_{gs}$ (carrier storage time) + t_{gf} (fall time)	ton $= t_{gt}$ gate controlled turn-on time $= t_{gd}$ (delay time + t_r (rise time)	(for GTOs)
V_G gate voltage V_{GT} minimum gate triggered voltage V_{GD} maximum gate non-triggered voltage	I_G gate current I_{GT} minimum gate triggered current I_{GD} maximum gate non-triggered current	P_G control (gate) power

Table 6.6 Data of IGBT BUP304

Quantity	Symbol	Value			Unity	Conditions
Collector current	I_C	25			A	TC = 25 °C
Power loss	P_{tot}	200			W	TC = 25 °C
Avalanche current, periodical	I_{AR}	5			A	$T_{Jmax} = 150\,°C$
Operating temperature limits	T_J T_{Stg}	−55 to +150			°C	–
Thermal resistance chip-package	R_{thjC}	≤0.63			°K/W	–
		min.	typ.	max.		
Collector-emitter break-over voltage	$V_{(BR)CES}$	1000	–	–	V	$V_{GE} = 0$ $I_C = 0.25\,mA$
Gate voltage (min)	$V_{GE(th)}$	4.0	5.0	6	V	$V_{GE} = V_{CE}$, $I_C = 1\,mA$
Collector-emitor saturation voltage	$V_{CE(sat)}$	-	3.5	5	V	$V_{GE} = 15\,V$, $I_C = 15\,A$
Inductive load						
Turn-off time (recovery)	$t_{d(off)}$	–	200	–	ns	$T_j = 125\,°C$ $V_{CC} = 600\,V$ $V_{GE} = 15\,V$ $I_C = 15\,A$ $R_{g(on)} = 3.3\,\Omega$ $R_{g(off)} = 3.3\,\Omega$
Fall time	t_f	–	200	–	ns	
Switching-off power loss ($E_{off} = E_{roff1} + E_{off2}$)	E_{off1}	–	0.7	–	mWs	
	E_{off2}	–	1	–		

Chapter 7
Rectifier

Abstract This chapter mainly discusses the classical rectifier. Starting from the diode rectifier, next the controlled rectifier is treated (both mainly for inductive loads). A section also is devoted to operation on capacitive loads. Much attention is paid to reactive power requirements and to harmonics in the grid (and load as well).

7.1 Introduction

There are two types of classical rectifiers, uncontrolled diode rectifiers and controlled rectifiers. Both types convert the sinusoidal mains voltage (or current) into a DC voltage (or current).

The first type, the (uncontrolled) diode rectifier, is the most straightforward rectifier, as the relation between the DC voltage and the AC voltage is fixed. The energy flow always goes from the AC side to the DC side.

The second type, the controlled rectifier, however, makes it possible to adapt (i.e. reduce) the DC voltage compared to the uncontrolled voltage. It also makes it possible to reverse the energy flow and make it go from the DC side to the AC side. However, as will be explained further on, the presence of an AC grid able to provide reactive energy to switch the power electronic switches (diodes, thyristors) remains essential: a rectifier cannot convert a DC voltage into an AC voltage without the prior presence of an AC grid.

Both types will be discussed in the next section. The last section will discuss the application of rectifiers for DC machine supply.

7.2 Basic Theory of the Rectifier

7.2.1 Uncontrolled Diode Rectifier

7.2.1.1 Wye-Connected Rectifier

Figure 7.1 shows the layout of an m-phase wye-connected (also called centre-connected) rectifier. On the left, in (a), the AC voltages are represented as ideal

J. A. Melkebeek, *Electrical Machines and Drives*, Power Systems,
https://doi.org/10.1007/978-3-319-72730-1_7

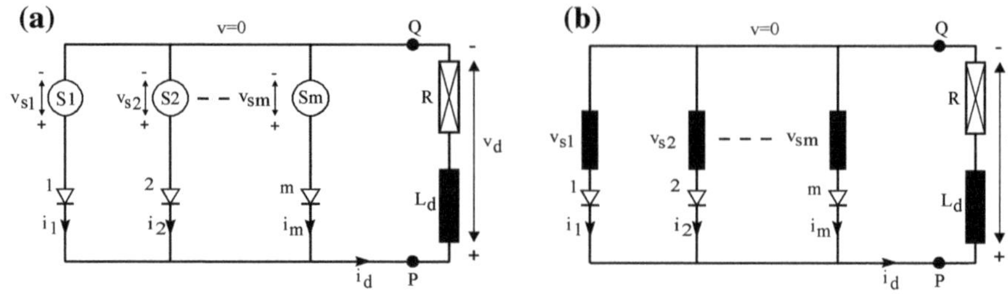

Fig. 7.1 Basic circuit of an m-phase wye-connected rectifier

Fig. 7.2 Output voltage of an m-phase wye-connected rectifier

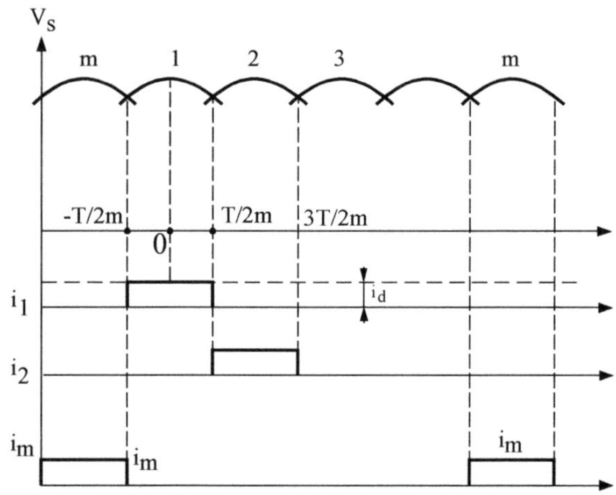

sources $v_{s1}, v_{s2}, \ldots v_{sm}$. In reality, these AC voltages are the secondary voltages of a transformer, as is illustrated on the right, in (b). In a more thorough analysis in a later section, these voltages will be replaced by their Thevenin equivalents.

Here, the m-phase voltage source is wye connected. In series with each phase voltage source, we can observe a diode. In our figure, the anodes are connected to the voltage sources while the cathodes of all diodes are connected to the minus pole of the DC side. The plus pole of the DC side is connected to the star point of the m-phase AC source.

Consider the instant $t_1 = 0$ (Fig. 7.2). At t_1, the instantaneous voltage v_{s1} is higher than the other source voltages. Thus, diode $D_1$is positively polarised and conducts current to the DC load, while all other diodes are negatively polarised and blocked. From $t_2 = T/2m$ on, the instantaneous voltage v_{s2} becomes higher than the voltage v_{s1}. Diode D_2 becomes positively polarised and starts conducting. At this instant a short-circuit current[1] starts flowing in the circuit $v_{s2} - D_2 - D_1 - v_{s1}$ and diode

[1]This infinite short-circuit current only flows during an infinitely short interval; in reality the DC inductance is finite and the short-circuit current is also finite and flows during a finite time interval.

D_1 will block. Then, from $t_2 = T/2m$ to $t_3 = 3T/2m$ (when v_{s3} takes over), only v_{s2} will deliver the DC current to the load (via diode D_2).

Suppose that the load inductance is very large, so that the DC current $i_d = I_d$ can be considered as constant. The resulting (constant) current at the DC side is therefore composed of m blocks of T/m (or in radians $2\pi/m$) wide, each one delivered by one of the m AC sources, as illustrated in Fig. 7.2.

As the DC current is constant (because of the supposition of an infinite inductance at the DC side), the AC current in each phase is also a block current with a width of T/m (or in radians $2\pi/m$), separated by intervals of $(m-1)T/m$ (or in radians $(m-1)2\pi/m$). Besides a fundamental harmonic with AC frequency, the AC current therefore also contains a DC component, as well as current harmonics with order $km \pm 1$. The effective value of the fundamental is given by:

$$I_{f1} = I_1 = \frac{\sqrt{2}}{\pi} \cdot I_d \cdot \sin\frac{\pi}{m} \tag{7.1}$$

The (total) effective value of these currents is $I_d/\sqrt{m}$. Note that the mean value of the phase currents is I_d/m, which means that for a wye connection of the AC source, each phase at the AC side of the rectifier contains a non-zero DC component. A wye-connected rectifier therefore requires a suitable[2] transformer between the grid and the rectifier in order to eliminate the DC component in the grid.

The current through each semiconductor switch also has an instantaneous (i.e. maximum) value I_d, a mean value I_d/m and an effective value (responsible for conductive power losses) of $I_d/\sqrt{m}$.

Note also that the DC voltage is not quite constant but contains harmonics of order km:

$$v_d = V_{do}\left[1 + \sum_{\nu = km} (-1)^{k+1} \frac{2}{\nu^2 - 1} \cos \nu\omega t\right] \tag{7.2}$$

with the mean DC voltage V_{d0} given by

$$V_{do} = \sqrt{2} \cdot V_s \frac{\sin \pi/m}{\pi/m} = \hat{V}_s \frac{\sin \pi/m}{\pi/m} \tag{7.3}$$

With a sufficient inductance at the DC side, the ripple in v_d is found mainly across the inductance and the DC part V_{do} mainly over the load resistance R.

Note that the harmonics in the voltage become smaller the greater the number of phases m there are. Also the order of these harmonics becomes larger with larger m. We can conclude that the current harmonics at the DC side (for a realistic finite inductance) will decrease significantly, the greater the phase-number m is.

[2] What would be a suitable connection for the multiphase transformer?

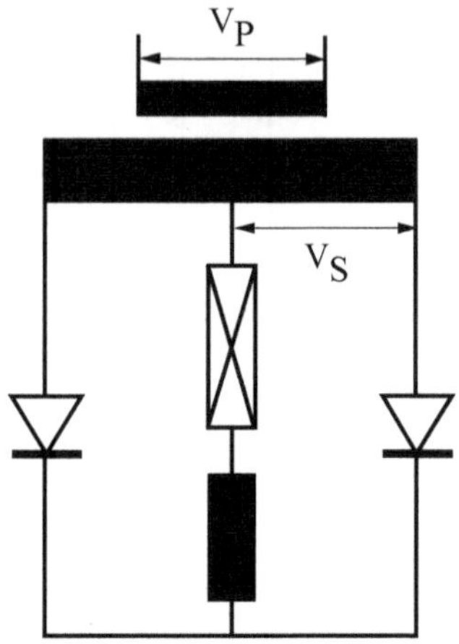

Fig. 7.3 Two-phase wye rectifier

Consider, for example, the single-phase (also called two-phase) wye-connected rectifier with mid-tapped transformer (see Fig. 7.3). The harmonics are much more significant and of a lower order than those of a three-phase rectifier.[3]

7.2.1.2 Bridge Rectifier

An important disadvantage of the above-mentioned wye (or centre-tap) rectifier is the DC component in the AC current, necessitating an oversized transformer to avoid DC currents in the grid. This is why a second type of uncontrolled diode rectifier, the bridge rectifier, is mostly preferred.

Figure 7.4a - without the dashed line - shows the schematic diagram of a three-phase bridge rectifier (also called *Graetz-bridge*). Here, the current flows at the right side from the transformer[4] pole where the voltage is at a maximum at that instant, through the diode in series to point P of the load, then through the load to point Q, and then back to the AC source through the diode that is connected with that phase where the voltage is minimal at that instant. The diodes on the right commutate among themselves, so that P is always connected to the transformer pole with the highest voltage at that instant; similarly, the diodes at the left side commutate among themselves, so that Q is always connected to the transformer pole with the lowest voltage at that instant. As a result (see (b) in Fig. 7.4), the voltage across the load follows the maximum values of the line-to-line voltages $U - V$, $V - W$, $W - U$ and their inverse.

With respect to the (real of fictitious) star point of the AC source, the voltage potential v_P of point P follows the positive tops of the line-to-neutral voltages $U - n$,

[3]Calculate from Fourier analysis the mean DC voltage and harmonic content for this two-phase rectifier. Do this directly, without using Eq. 7.3; because of the mid-tapped transformer, m = 2.

[4]For a bridge rectifier a transformer between grid and rectifier is not strictly required, however.

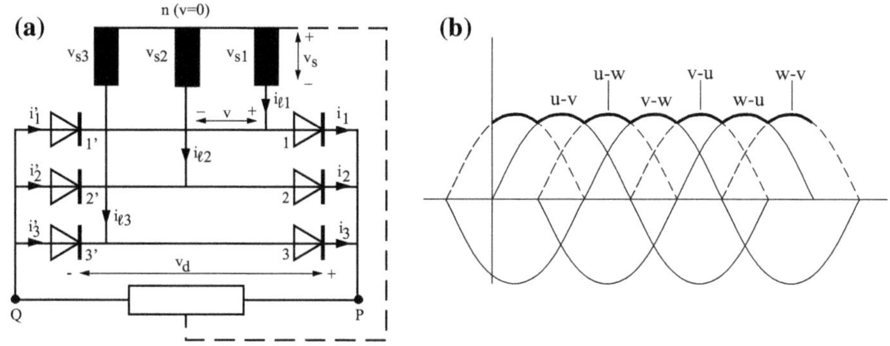

Fig. 7.4 Three-phase bridge rectifier

$V-n$, $W-n$, while the voltage potential v_Q follows the negative tops of these line-to-neutral voltages (see (a) in Fig. 7.5). The load voltage v_d is the difference between v_p and v_Q (see (b) in Fig. 7.5). Indeed, as to the voltage output, the bridge rectifier can also be considered as the series connection of two wye rectifiers (see the dashed line in (a) in Fig. 7.4). Therefore, for the DC output voltage we may write, for both wye and bridge rectifier:

$$V_{do} = s \cdot \sqrt{2} \cdot V_s \frac{\sin \pi/m}{\pi/m} = s \cdot \hat{V}_s \frac{\sin \pi/m}{\pi/m} \tag{7.4}$$

with $s=1$ for a wye and $s=2$ for a bridge connection. In Eq. 7.4, m represents the commutation number. This is the number of commutations per period *in a group commutating cells*. For the Graetze bridge, $m=3$ for both left and right diodes. In most cases, the commutation number equals the number of phases. One exception is the single-phase bridge in Fig. 7.6, where $m=2$ and where for the application of Eq. 7.4 half the secondary voltage $\hat{V}_s$ of the transformer must be inserted as voltage.

In contrast, the harmonics in the output voltage do not depend on the commutation number m but on the pulse number p. The pulse number is the *total number of non-simultaneous commutations* per period.[5] For the three-phase bridge, $p=6$ as the commutations of the upper and lower diodes are not coincident. In general, for a bridge rectifier with m odd, $p=2m$. For a bridge with m even, $p=m$ holds (e.g. the single-phase bridge in Fig. 7.6, also referred to as a two-phase bridge).

Next, the currents in the rectifier will be examined more closely. We suppose again that the load inductance is very large, so that the DC load current can be considered as constant. Then, the diode currents are again blocks of width T/m (see (c) in Fig. 7.5). For the average and effective value of the diode currents, the same equations hold as before. However, now the secondary line currents (i_{l1}, i_{l2}, i_{l3} see (d) in Fig. 7.5) do not contain a DC component. For the fundamental of the secondary line current, we obtain

[5] Explain why p determines the harmonics, referring to Eq. 7.2.

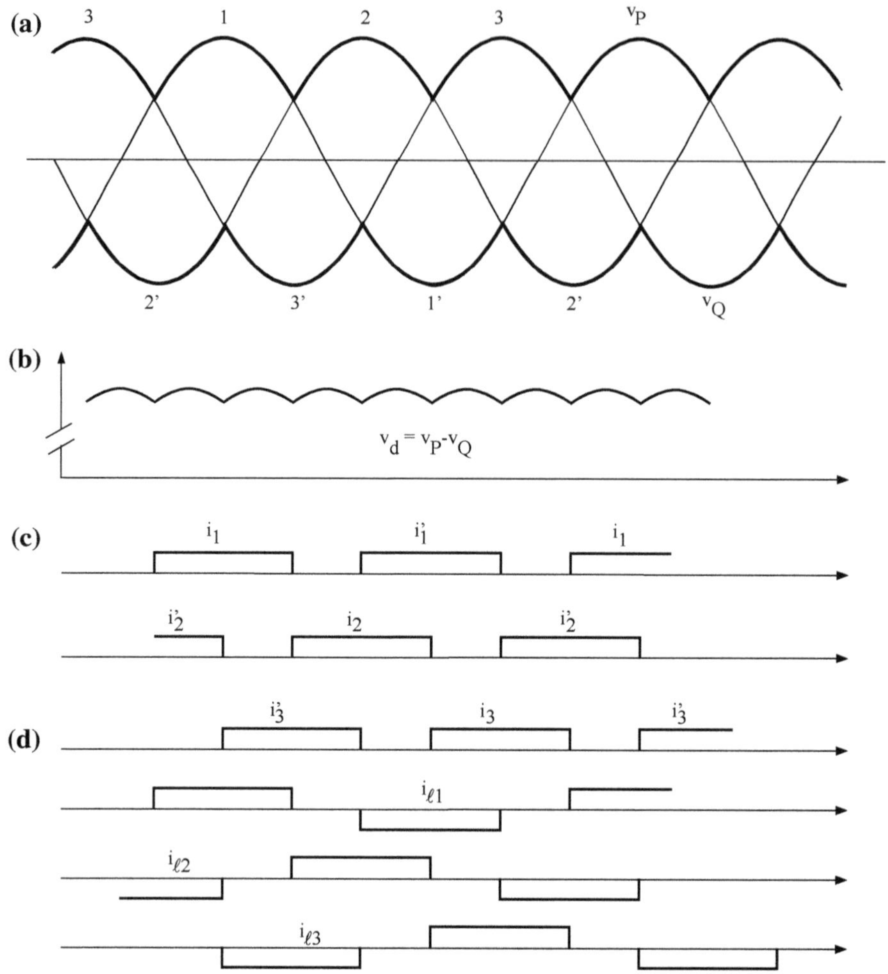

Fig. 7.5 Voltages and currents of the three-phase bridge

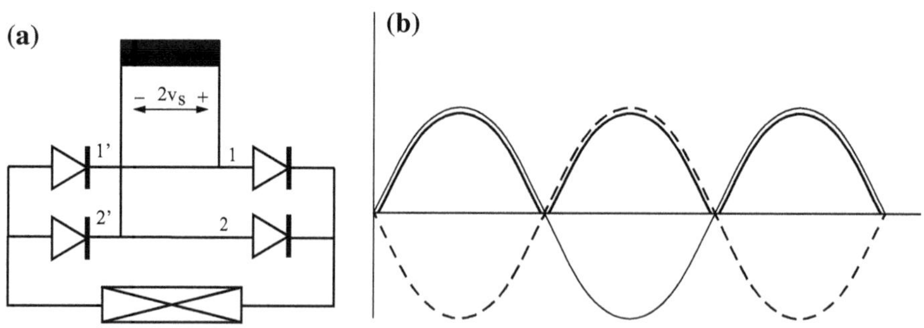

Fig. 7.6 Single-phase bridge

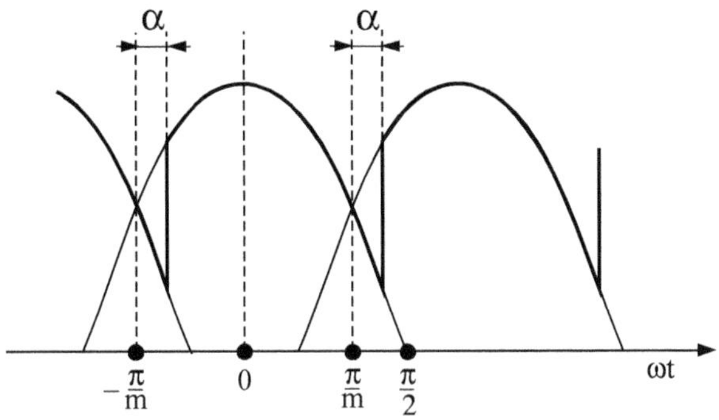

Fig. 7.7 Phase-controlled rectifier

$$I_{l1} = 2 \cdot \frac{\sqrt{2}}{\pi} \cdot I_d \cdot \sin\frac{\pi}{m} = s \cdot \frac{\sqrt{2}}{\pi} \cdot I_d \cdot \sin\frac{\pi}{m} \tag{7.5}$$

Again, the expression on the right also holds in general, with $s = 1$ or $s = 2$ for wye and bridge, respectively.[6]

7.2.2 *Phase-Controlled Rectifier*

In the diode rectifier of the previous section, within a group commutating switches, the diodes commutate automatically from one phase to the next when the voltage of the next phase becomes higher at the anode side (or lower for those connected to the cathodes). This results in a DC voltage that is fixed for a given AC voltage.

When we substitute the diodes with valves that can be switched on, like thyristors, the commutation instant can be delayed by an angle α (called the delay angle or phase angle). Such a rectifier is called a *controlled rectifier* (sometimes also a *mutator*). For a wye-connected rectifier, we therefore obtain a voltage diagram as in Fig. 7.7. The average DC output voltage now becomes

$$V_{d\alpha} = \hat{V}_s \frac{\sin \pi/m}{\pi/m} \cdot \cos\alpha = V_{do} \cdot \cos\alpha \tag{7.6}$$

For a *wye rectifier*, negative instantaneous voltage values occur for $\alpha > 30°$; similarly, for $\alpha \geq 150°$ no positive instantaneous positive voltage values remain (see (b) in Fig. 7.8).[7]

For a *bridge rectifier,* (usually) both the lower and upper switches are delayed over the same angle (see (a) in Fig. 7.8). The output voltage for the bridge can then be written as

---

[6]As an exercise, prove that for a three-phase bridge $V_{do} = (3/\pi)\hat{V}_l$ (with $\hat{V}_l$ the secondary line-to-line voltage). Calculate the line current for the three-phase bridge from Eq. 7.5 and verify whether power is indeed preserved.

[7]Nevertheless, the current must remain positive; what does this imply? (see also below).

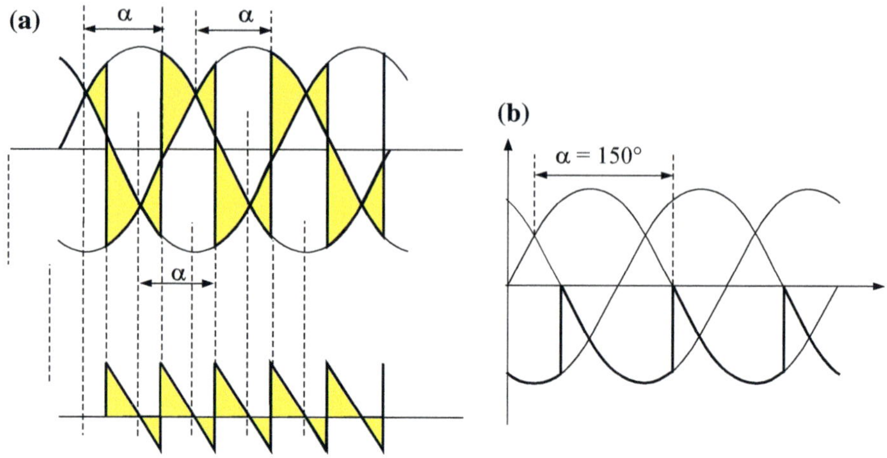

Fig. 7.8 Output voltage **a** bridge with $\alpha = 80°$ **b** wye with $\alpha = 150°$

$$V_{d\alpha} = V_{do} \cdot \cos\alpha \tag{7.7}$$

where for the value of V_{do}, Eq. 7.4 should be used with $s = 2$.

Sketch (a) in Fig. 7.8 illustrates the voltages for a *bridge rectifier* with $\alpha = 80°$.

For a bridge with $\alpha > 60°$, the *instantaneous* voltage shows intervals with negative values (see Fig. 7.9). For $\alpha > 90°$ the *average* voltage becomes negative. This implies that a suitable emf must be present in the DC circuit ("load") as the current direction (I_d or the average of i_d) is confined by the thyristors. Moreover, such an emf also is required to reverse the power flow from DC to AC.

Theoretically, the maximum value of α for a bridge rectifier is 180° but in reality the finite commutation duration (see below) limits α to approximately 150°.

The delay of the commutation over an angle α also affects the voltage waveform and thus the harmonics at the DC side. Fourier analysis of this DC voltage yields the following result for the harmonics of order ν:

$$V_\nu = \frac{2}{\nu^2 - 1} V_{d\alpha} \sqrt{1 + \nu^2 \tan^2 \alpha}$$

As for the current at the AC side, it is clear that, for an infinite or very large inductance at the DC side, the current waveform has not changed with the commutation delay, i.e. we still see current blocks of T/m wide with amplitude I_d. Neither do the harmonics. However, because of the commutation delay, these current blocks as well as the harmonics (including the fundamental) have shifted by an angle of α. As a result, the fundamental power input has also changed, corresponding with the decrease of the DC voltage on the one hand, and the decrease of the output power (for the same

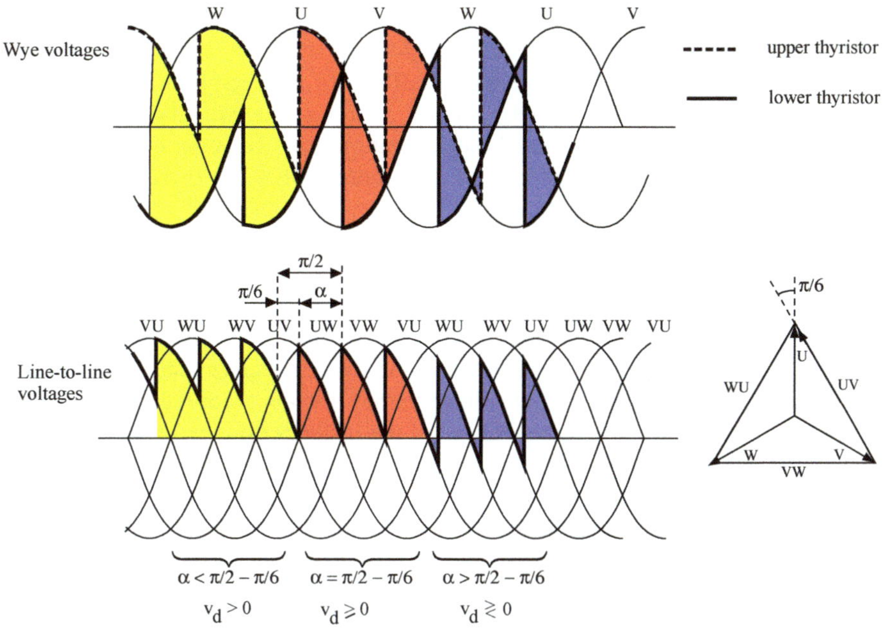

Fig. 7.9 Bridge instantaneous output voltages for several α

I_d) on the other hand. As the delay angle cannot be negative,[8] the controlled rectifier will always absorb reactive power (see also below).

7.2.3 Discontinuous Conduction Mode

In the previous sections, we assumed that the inductance at the DC side was very large, allowing us to consider the DC current as constant (and thus continuous). In reality of course, the DC side inductance is finite. If the DC side voltage shows intervals with instantaneous negative voltages, the current may decrease to zero during these intervals at which time the diodes or thyristors will cease to conduct. The diodes will start conducting again if the DC voltage becomes positive again, or in the case of thyristors if at the same time a gate signal is present. Of course, this will also affect the (average) output load voltage, as during the intervals with zero current, the output voltage will be zero (and not negative).

This phenomenon may occur with both diode and (phase controlled) thyristor rectifiers. For a diode rectifier, it can only occur if a counter-emf is present in the DC circuit.

[8]Why not?

For a purely resistive load, the rectifier diodes or thyristors will be blocked as soon as the instantaneous voltage is negative, i.e. for $\alpha > \pi/2 - \pi/m$ for a wye rectifier and $\alpha > \pi/2 - \pi/2m$ for a bridge rectifier (thus for a three-phase wye or bridge rectifier, at 30° and 60°, respectively; see Fig. 7.9 for a bridge rectifier). Then, input and output are no longer connected and the output voltage is zero. If the voltage becomes positive again, the diodes will start conducting, but thyristors will also require a new gate signal at that instant.

This discontinuous conduction may also occur for a resistive-inductive load if the inductance is insufficient to keep the current from attaining zero during negative voltage intervals.

Discontinuous conduction may even occur for small delay angles, $\alpha < \pi/2 - \pi/m$ or $\alpha < \pi/2 - \pi/2m$ (rectifier mode) if the load contains a counter-emf that is larger than the instantaneous output voltage. It may also occur for regenerating mode ($\alpha \ggg \pi/2 - \pi/m$ or $\alpha \ggg \pi/2 - \pi/2m$) if the instantaneous negative DC output voltage is more negative than the (then negative) emf.

Discontinuous operation for negative instantaneous *load* voltages ($v_d - E$) can be avoided if the load inductance is sufficiently large. Of course, in all cases, the average load voltage ($V_{d\alpha} - E$) must still be positive as the DC current direction is confined by the switches.

7.2.4 Rectifier with a Capacitive Load

Rectifiers with discontinuous conduction will result in increased mains current harmonics. Such discontinuous conduction always occurs in rectifiers with capacitive or resistive-capacitive load. Rectifiers with a capacitive load are nevertheless omnipresent, for example as a battery charger or supply for small appliances.

Consider the single-phase configuration depicted in (a) in Fig. 7.10. If the output current were continuous, we would get an output voltage as in (b) in Fig. 7.10. This could be realised by adding a sufficiently large inductance at the DC side, but such an inductance is normally avoided because of the cost and weight penalty. In the majority of such small power applications, the rectifier has a purely resistive-capacitive load.

For a pure R-C load, we get a (steady-state) output voltage as in Fig. 7.11. In the same figure, the voltage $|v_s(t)|$ is drawn in dotted lines by way of reference. The actual output voltage waveform can be explained as follows. From t_1 to t_2 diodes 2 and 1' are conducting and the output voltage follows $|v_s(t)|$. The source current flows via the diodes to the resistance on the one hand, and the capacitor on the other hand. In the meantime, the capacitor is being charged.

Once beyond $\omega t = \pi$, the current through the capacitor reverses, $i_C = C \cdot dv_d/dt < 0$. While the capacitor discharges, diodes 2 and 1' continue to conduct as long as $i_d > 0$, i.e.

$$i_d = i_C + i_R = C\frac{dv_d}{dt} + \frac{v_d}{R} = C\frac{d}{dt}|\hat{V}_s \cos\omega t| + \frac{1}{R}|\hat{V}_s \cos\omega t| > 0$$

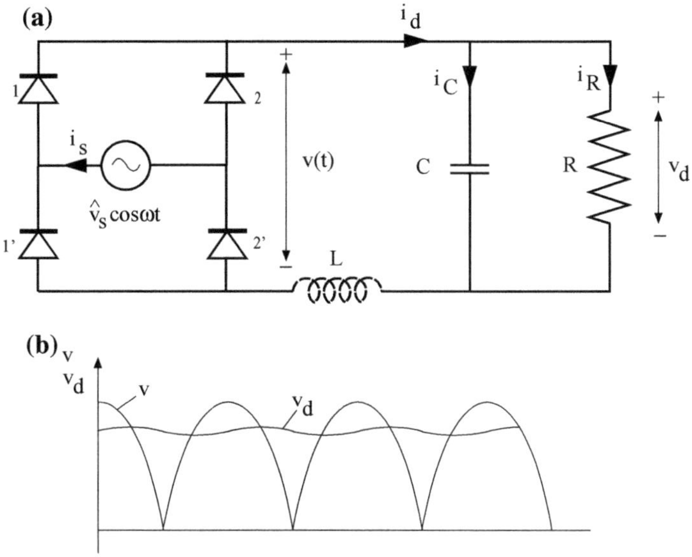

Fig. 7.10 Two-phase rectifier with capacitive load

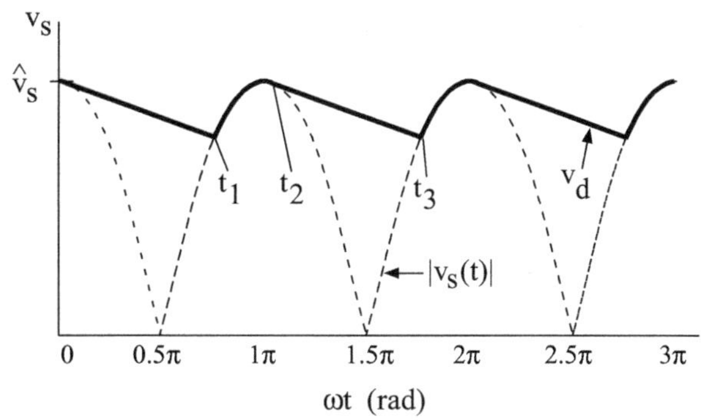

Fig. 7.11 Output voltage for capacitive-resistive load

The negative capacitor current becomes larger and larger (in absolute value), and finally i_d turns zero at $t_2 = t_{off}$. Then, diodes 2 and 1' stop conducting. This instant can be calculated as $\tan(\omega t_{off}) = 1/\omega RC$. From that moment on, the capacitor discharges in the resistance and the capacitor voltage v_d decreases exponentially with the time constant RC. When v_d becomes smaller than $\hat{V}_s \cos \omega t$, the diode bridge starts conducting again at $t_3 = t_{on}$, in the case of Fig. 7.10 via diodes 1 and 2'. The instant t_{on} can be calculated from the transcendental equation

$$\hat{V}_s |\cos \omega t_{off}| \cdot \exp[-(t_{on} - t_{off})/RC] = \hat{V}_s |\cos \omega t_{on}| \tag{7.8}$$

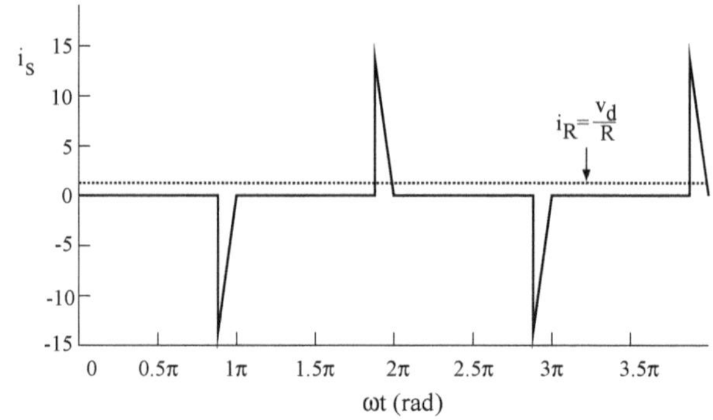

Fig. 7.12 Line current for capacitive load

To obtain a low ripple in the mains voltage, the time constant RC should be sufficiently large, i.e. $RC >> T/2$.

Despite its widespread applications, the main disadvantage of the diode bridge with a capacitor load is the far from sinusoidal mains current (see Fig. 7.12), especially for large capacitor values.

7.2.5 *Non-ideal AC Source: Finite Commutation Duration*

In the previous sections, we supposed that the AC source was an ideal voltage source and did not have any internal impedance. This results in an immediate commutation, as these ideal sources are short-circuited over the switches at the instant of commutation.

In reality, this AC source is either directly the grid or a transformer between the grid and the rectifier. The AC source can then be represented by its Thévenin equivalent. In most cases, the internal impedance of this source can be simplified to an inductance (in the case of a transformer, approximately the total leakage inductance), as is illustrated in Fig. 7.13.

The effects of this internal impedance are twofold: the commutation between the switches takes a finite time and the output DC voltage changes compared to the case of an ideal source.

7.2.5.1 Inductive Commutation

At the commutation instant between thyristors 1 to 2, the sources v_{s1} and v_{s2} are short-circuited over the two internal inductances L_{k1} and L_{k2} The DC inductance is supposed to be very large such that $i_d = I_d$ can be considered constant.

Suppose that switch 1 is conducting. When $v_{s2}(t)$ is or becomes larger than $v_{s1}(t)$ - and, in the case of thyristors, switch 2 receives a gating signal - switch 2 starts to conduct. Then both switches 1 and 2 are conducting and thus both phases 1 and 2

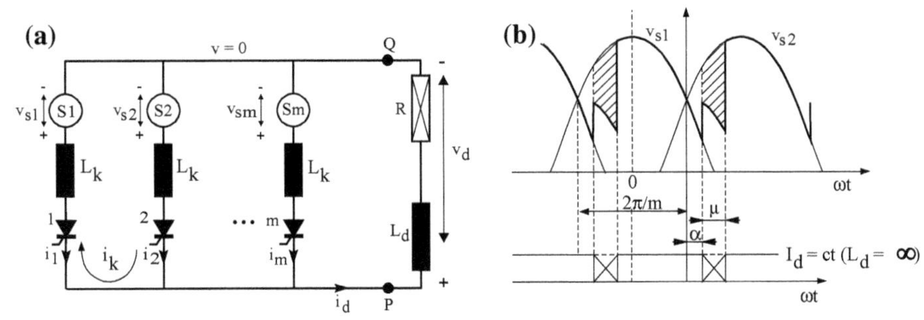

Fig. 7.13 Inductive commutation

are short-circuited over the inductances L_k (and connected to the load). This results in the following equations:

$$v_d = v_{s1} - L_k \frac{di_1}{dt} = v_{s2} - L_k \frac{di_2}{dt} \tag{7.9}$$

Adding and subtracting the Eq. 7.9 and taking into account $i_1 + i_2 = i_d = I_d = ct$ results in

$$v_d = (v_{s1} + v_{s2})/2 \tag{7.10}$$

$$\frac{di_2}{dt} = -\frac{di_1}{dt} = \frac{v_{s2}(t) - v_{s1}(t)}{2L_k} \tag{7.11}$$

During the commutation time (which we will denote by μ, in radians), the output voltage takes on the mean value of the previous and the next phase voltages. The current in the upcoming phase (2) increases while the one in the preceding phase decreases with the same rate. This rate is directly proportional to the voltage difference between the new and old phases. As soon as the current in the upcoming phase reaches I_d (and the one in the preceding phase reaches zero), the new phase takes over completely. The output voltage then becomes determined by the new phase voltage only.

During the commutation, however, the output voltage has changed compared to an instantaneous commutation. Between $\pi/m + \alpha$ and $\pi/m + \alpha + \mu$ the voltage potential of point P is not $v_{s2}(t)$, but instead equal to the average voltage given by Eq. 7.10. The average rectifier output voltage change can thus be computed from:

$$D_x = s \cdot \frac{m}{2\pi} \cdot \int_{\pi/m+\alpha}^{\pi/m+\alpha+\mu} \left[v_{s2} - \frac{v_{s1} + v_{s2}}{2} \right] d\omega t = s \cdot \frac{m}{2\pi} \cdot \int_{\pi/m+\alpha}^{\pi/m+\alpha+\mu} \left[\frac{v_{s2} - v_{s1}}{2} \right] d\omega t \tag{7.12}$$

where $s = 1$ for a wye rectifier and $s = 2$ for a bridge rectifier.

Substituting the time functions for $v_{s1}(t)$ and $v_{s2}(t)$ in Eq. 7.12 results in

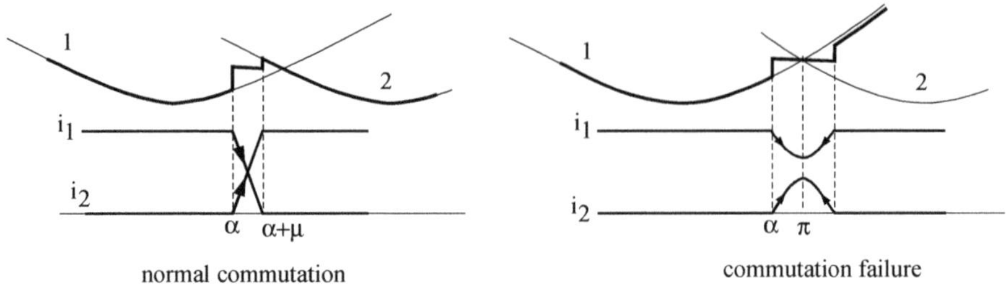

Fig. 7.14 Commutation failure

$$D_x = V_{do}\frac{[\cos\alpha - \cos(\alpha+\mu)]}{2} \tag{7.13}$$

where μ is still unknown.

On the other hand, substituting the expression for the voltage difference $v_{s2} - v_{s1}$ in terms of di_2 or di_1 according to Eq. 7.11 (and using $\int di_2 = I_d$) results in

$$D_x = s \cdot m \cdot f \cdot L_k \cdot I_d \triangleq X_d I_d \tag{7.14}$$

The average DC voltage then becomes

$$V_d = V_{d\alpha} - D_x = V_{do}\frac{[\cos\alpha + \cos(\alpha+\mu)]}{2} = V_{d\alpha} - X_d I_d \tag{7.15}$$

If the DC current is known, the commutation duration μ can be calculated from Eq. 7.15. The DC current can be calculated from Eq. 7.15 if the DC load is known (making use of the equality $V_d = R \cdot I_d$).

The commutation duration is largest when the difference $v_{s2} - v_{s1}$ is small, thus near $\alpha = 0$ and near $\alpha = \pi$. A rather long commutation may pose problems for delay angles α near π (regeneration from the DC side). Indeed, consider Fig. 7.14.

When the commutation from 1 to 2 is finished before $\alpha = \pi$, no problems arise as phase 2 will have taken over completely from phase 1 and switch 1 is then blocked. However, when the commutation lasts beyond $\alpha = \pi$, i.e. when the voltage $v_{s1}(t)$ again becomes more positive than $v_{s2}(t)$, commutation from phase 1 to phase 2 will reverse and phase 1 will take over again. This commutation failure then results in an output voltage which is larger (less negative) than expected and desired. As the rectifier works in regenerating mode at values of α larger than $\pi/2$, the DC load is a *negative* counter-emf. Combined with the now less negative - and possibly positive - rectifier output voltage, this may or will result in much too large currents.

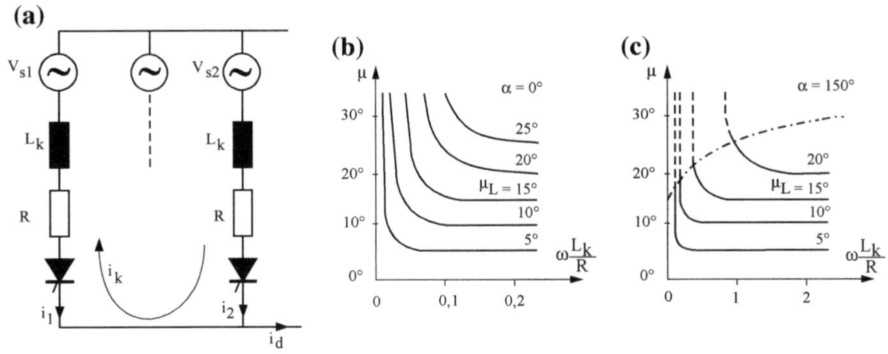

Fig. 7.15 Resistive-inductive commutation

7.2.5.2 Resistive-Inductive Commutation

In most cases, the effect of a resistive impedance in the AC source can be disregarded as to commutation. For large resistive values, or in the case of a low frequency AC source, the resistance in the AC source may significantly lengthen the commutation duration. The commutation can be calculated by introducing the short-circuit current i_k and solving the corresponding differential equation.

Figure 7.15 shows the calculated relation between the commutation duration with resistance μ and the one without the resistance effect μ_L as a function of the per unit time constant $\omega L_k/R$. For small $\omega L_k/R$ (or large resistances), the commutation duration may become very large, prohibiting effectively delay angles larger than or near $5\pi/6$.

7.2.6 Power Exchange Between Rectifier and Grid

7.2.6.1 Active, Reactive and Harmonic Power

In this section, we will consider an ideal rectifier, i.e. with negligible commutation duration, constant output DC current, without conductive losses and fed by an ideal multiphase sinusoidal AC source.

At the DC side the instantaneous output voltage is $v_d(t)$, with average value $V_{d\alpha}$. The DC current is constant, $i_d = I_d$. The instantaneous DC output power is given by

$$p_d(t) = v_d(t) \cdot I_d = V_{d\alpha} \cdot I_d + I_d \cdot \sum v_{d\nu}(t) \tag{7.16}$$

In addition to the average (real) power $P_d = V_{d\alpha} \cdot I_d$, the instantaneous power contains pulsating terms oscillating back and forth between DC and AC side.

At the <u>*AC side*</u> the supply is supposed an ideally sinusoidal voltage source. The current at the AC side consists of rectangular current blocks with amplitude I_d and width $2\pi/m$ (for a wye rectifier, only positive current blocks as well as a DC component; for a bridge rectifier, both positive and negative current blocks displaced by π radials with respect to each other).

The fundamental of the AC phase (or line) current follows from Fourier analysis of the current blocks:

$$\hat{I}_{l1} = s \cdot I_d \cdot \frac{2}{\pi} \cdot \sin\frac{\pi}{m} \tag{7.17}$$

In a diode rectifier the currents blocks are in phase with the voltage, while in a controlled rectifier these current blocks as well as the fundamental are displaced by the delay angle α. The instantaneous power at the AC side is given by

$$p(t) = \sum_{i=1}^{m} v_{si}(t) \cdot i_{li}(t) \tag{7.18}$$

with $v_{si}(t)$and $i_{li}(t)$ the instantaneous values of phase voltage and phase (line) current, respectively.

As the AC voltage is sinusoidal, only the fundamental current will contribute to the average active power:

$$P = \frac{m}{2} \cdot \hat{V}_s \cdot \hat{I}_{l1} \cdot \cos\alpha = V_{d\alpha} \cdot I_d \tag{7.19}$$

The instantaneous power Eq. 7.16 contains other (pulsating) terms in addition to the average active power. As is well known, these pulsating power components lead to additional losses in the grid and transformers.

The pulsating power terms due to higher harmonic currents are referred to as harmonic (pulsating) power. However, first we will concentrate here on the pulsating power due to the fundamental current. For a diode rectifier, this pulsating power is zero as the grid is supposed m-phase symmetric. For a controlled rectifier, however, the phase delay α leads to non-zero m-phase pulsating power, which is generally represented by the reactive power

$$Q = \frac{m}{2} \cdot \hat{V}_s \cdot \hat{I}_{l1} \cdot \sin\alpha \tag{7.20}$$

This *control* reactive power can be very large, especially for large delay angles. For a given active power P, the grid currents (and associated losses) are much higher with higher delay angles.

<u>Remarks:</u>

- In addition to the reactive power Q for a controlled rectifier, the commutation delay μ also creates a (small) reactive power component, i.e. the commutation reactive power. It can be approximated by an additional phase delay of $\mu/2$, as

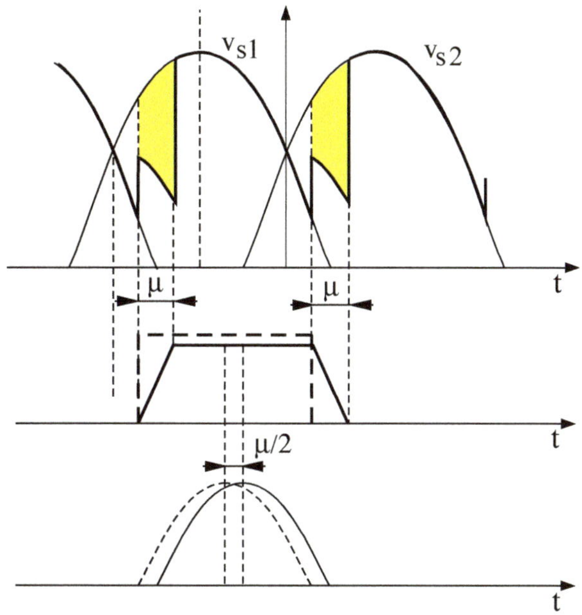

Fig. 7.16 Commutation reactive power

is illustrated in Fig. 7.16. However, the commutation reactive power is generally negligible compared to the reactive power Q for the control of the rectifier.

- The DC components in the phase current for a wye rectifier also generate additional losses in the secondary of the transformer that feeds the rectifier. These DC components are not transformed to the grid side, but the transformer does have to be oversized.

7.2.6.2 Reduction of the Reactive Power Requirements for a Controlled Rectifier

As explained above, the reactive power requirements of a controlled rectifier may be huge, especially for delay angles around $\pi/2$. Active and reactive power are given by Eqs. 7.19 and 7.20 respectively, while for a given DC current I_d, the apparent power is constant, $S = S_o = V_{do} \cdot I_d$. The locus of the complex power $\underline{S} = P + jQ$ is the circle (a) in Fig. 7.17.

There are several solutions to mitigate the reactive power requirements, but what is most suitable depends on the technical requirements and costs.

If regeneration is required and cost is not very important, a variable transformer between the grid and the rectifier could be considered, with the rectifier being operated with delay angles either around $\alpha = 0$ for rectifier operation, or around $\alpha = \pi$ for regenerative operation.

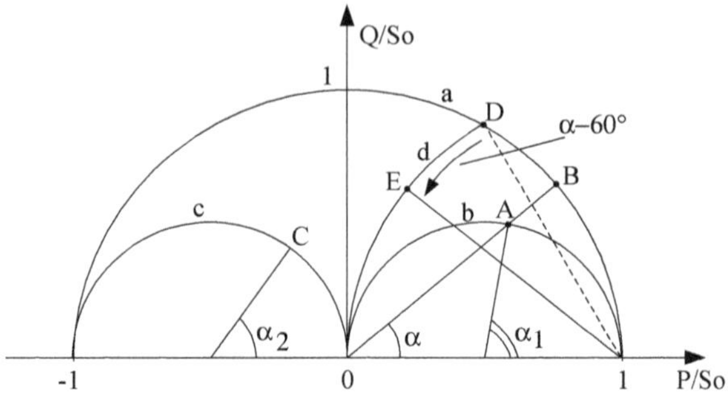

Fig. 7.17 P-Q-S-loci

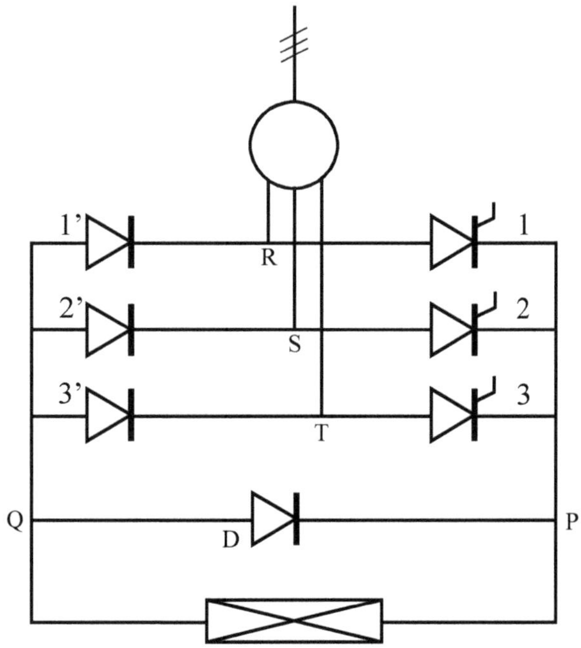

Fig. 7.18 Semi-controlled bridge rectifier

The second solution, at least if regeneration is not required, is a semi-controlled (bridge) rectifier (see Fig. 7.18) which has much lower reactive power requirements than a fully-controlled rectifier.

The lower switches are diodes, while the upper switches are controlled with a delay angle α_1. The average DC output voltage is thus given by

$$V_d = V_{do}\frac{1 + \cos\alpha_1}{2} \tag{7.21}$$

However, the positive and negative current blocks in each phase are now displaced by π radians. While the negative block is still in phase with the phase voltage, the positive block is displaced by the delay angle α_1, as is demonstrated in Fig. 7.19.

Fig. 7.19 AC current blocks for a semi-controlled bridge

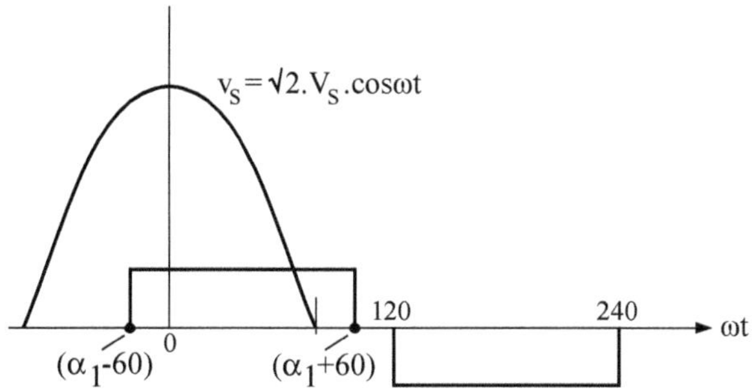

As a result, the (fundamental) harmonic current changes compared to that of the symmetrically controlled bridge. This fundamental can be calculated from a Fourier analysis but may also (and more easily) be derived from the active and reactive components. From energy conservation (for the three-phase case)

$$P = V_{do} \frac{1 + \cos \alpha_1}{2} \cdot I_d = 3 \cdot V \cdot I_w \tag{7.22}$$

it follows that the active component is given by $I_w = I_d \cdot \frac{\sqrt{6}}{2\pi}(1 + \cos \alpha_1)$. As only the current block corresponding to the upper controlled half of the bridge contributes to the reactive current, we get $I_r = -I_d \cdot \frac{\sqrt{6}}{2\pi} \sin \alpha_1$.

Therefore, defining $So = V_{do} \cdot I_d$, we may write the expressions for active, reactive and apparent power as follows:

$$P = S_o \frac{1 + \cos \alpha_1}{2} = S_o \cos^2(\alpha_1/2) \tag{7.23}$$

$$Q = S_o \cos(\alpha_1/2) \cdot \sin(\alpha_1/2) \tag{7.24}$$

$$S = S_o \cos(\alpha_1/2) \tag{7.25}$$

The locus of the corresponding complex power $\underline{S} = P + jQ$ is the circle (b) in Fig. 7.17. The advantage of this semi-controlled bridge is that the output voltage (and active power) can be regulated to (almost) zero, while at the same time also the reactive power is reduced. The fundamental phase displacement is $\cos \varphi = \cos(\alpha_1/2)$.

The anti-parallel diode is not strictly required but does offer some advantages, as will be discussed below.

For values of α larger than π/m, there are intervals with zero output voltage (see (a) in Fig. 7.20).

Without an anti-parallel diode, in these intervals the load current flows through the diode and thyristor of the same branch, e.g. $1 - 1' - P - load - Q - 1$. When the anti-parallel diode is present, the current switches from this branch to the anti-parallel diode as the voltage drop in it is lower. As a result, the thyristor (1 in our

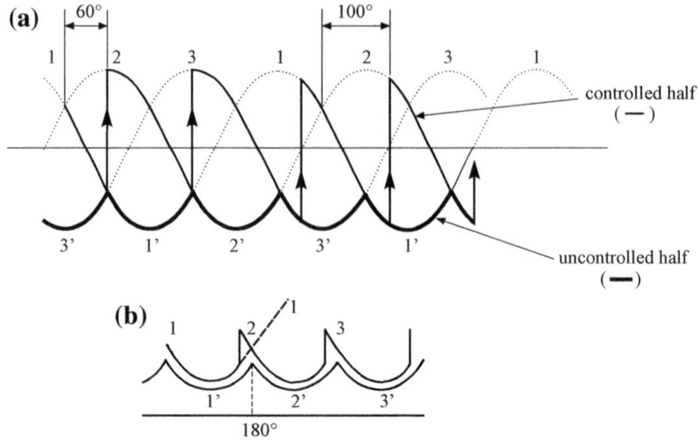

Fig. 7.20 Anti-parallel diode to avoid commutation failure

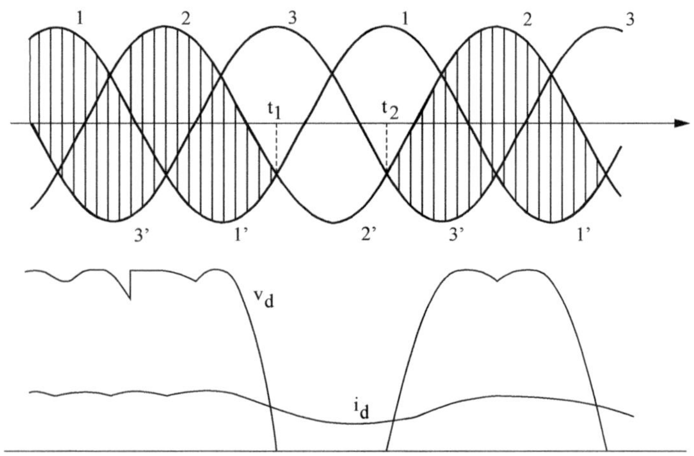

Fig. 7.21 Unwanted continued operation

example) turns out of conduction and will not start to conduct again without a gating signal.

A first advantage is that the anti-parallel diode may avoid commutation failure. Indeed, if the delay angle α_1 is large, the commutation at $\alpha_1 = \pi$ may not always be completed (e.g. from 1 to 2 in our example). Without an anti-parallel diode, thyristor 1 will continue to conduct. As illustrated in (b) in Fig. 7.20, the output voltage (resulting from the conduction of thyristor 1 and diode 2') will increase beyond the aimed value. This will be avoided by an anti-parallel diode: the current will shift to the anti-parallel diode, and thyristor 1 will be blocked. The next commutation will be from the anti-parallel diode to the next switch, at the right instant.

Another advantage of the anti-parallel diode is that it will avoid an unwanted continuation of power delivery to the load when the rectifier thyristor switches are turned off. As shown in Fig. 7.21, without an anti-parallel diode, thyristor 2 may continue to conduct even as no gating signals are delivered to the thyristors from time t_1 on. From t_1 to t_2, the load current flows through thyristor 2 - diode 2' and thyristor 2 will not be blocked if the DC inductance is large. From t_2 on, the load current will increase again as thyristor 2 and diode 3' provide a positive voltage to the load. With an anti-parallel diode, the load current will shift from the branch 2-2' to the anti-parallel diode and thyristor 2 will not conduct any more.

A disadvantage of the semi-controlled bridge is that the six-pulse character that exists for $\alpha = 0$ (or for a symmetrically controlled bridge) gradually becomes three-pulse for $\alpha \neq 0$ and becomes completely three-pulse (thus with higher harmonic content) for $\alpha \geq \pi/3$.

A third possibility to reduce the reactive power requirements is to use an anti-parallel diode in a controlled rectifier bridge (see Fig. 7.22).

For $\alpha < \pi/3$, the anti-parallel diode has no effect. The complex power follows the circle (a) until point B in Fig. 7.17. From $\alpha \geq \pi/3$ on, the anti-parallel diode will take over the load current during the intervals with otherwise zero or negative output voltage. As a result, the currents shown in Fig. 7.22 are obtained. AC phase currents are zero when the anti-parallel diode conducts. Fourier analysis of these AC currents yields the following result:

$$\begin{aligned} I_w &= I_d \cdot \tfrac{\sqrt{6}}{\pi} \cdot [1 - \sin(\alpha - \pi/6)] \\ I_r &= -I_d \cdot \tfrac{\sqrt{6}}{\pi} \cdot \cos(\alpha - \pi/6) \end{aligned} \tag{7.26}$$

and thus

$$\begin{aligned} P &= S_o \cdot [1 - \sin(\alpha - \pi/6)] \\ Q &= S_o \cdot \cos(\alpha - \pi/6) \end{aligned} \tag{7.27}$$

which corresponds with segment (d) in Fig. 7.17. Note that the active power $P = 0$ for $\alpha \geq 2\pi/3$. Two disadvantages are that regeneration is not possible with this circuit and that some harmonics may be rather large. For example, the fifth harmonic for $\alpha = 84°$ turns out to be 40% of the fundamental.

A fourth solution to mitigate the reactive power requirement is to use a series connection of two controlled rectifiers, each of these providing half of the output voltage (see Fig. 7.23).

While the delay angle for rectifier 2 is kept at 0, the delay angle of rectifier 1 may vary between 0 and π, which is demonstrated in circle (b) in Fig. 7.17. While α_1 is then kept at π, the delay angle of rectifier 2 may vary between 0 and π, as is shown in circle (c) in Fig. 7.17.

In other words, the average output voltage is

$$V_d = V_{do} \cdot \frac{\cos\alpha_1 + \cos\alpha_2}{2} \tag{7.28}$$

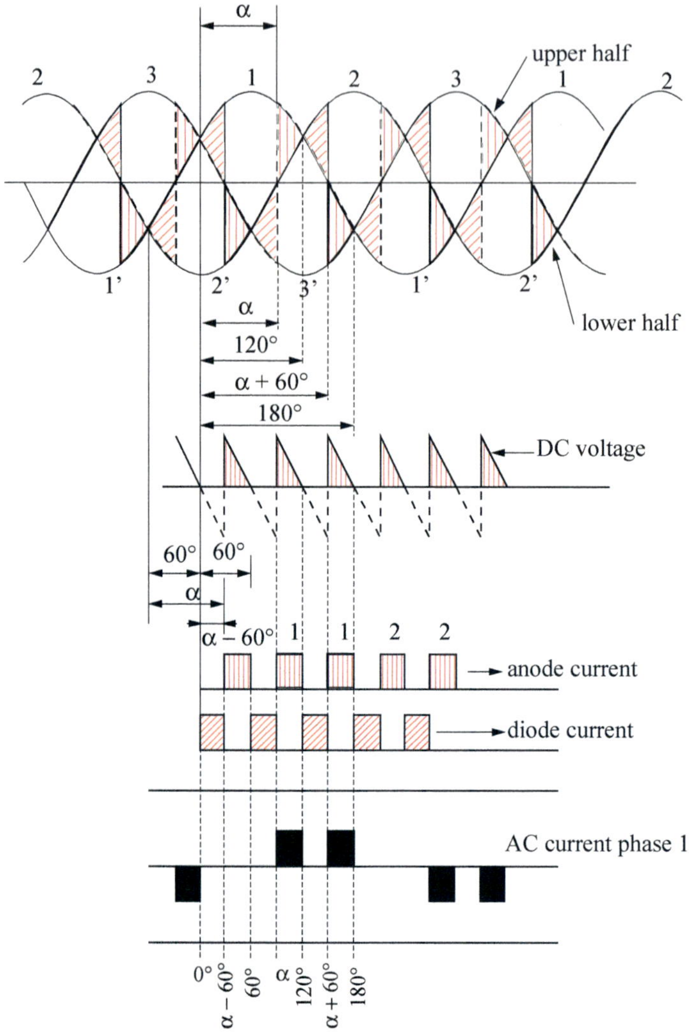

Fig. 7.22 Fully-controlled bridge with anti-parallel diode

In practice, the delay angles α_1 and α_2 are limited to approximately $5\pi/6$, to avoid commutation failures. Thus, circle (b) cannot be run through until $P = Q = 0$.[9] This solution offers both rectifier and regenerative operation, but the investment cost is higher.

[9] Draw the complete locus for the complex power in that case: how do the circles (b) and (c) transform if the delay angles are limited to $5\pi/6$.

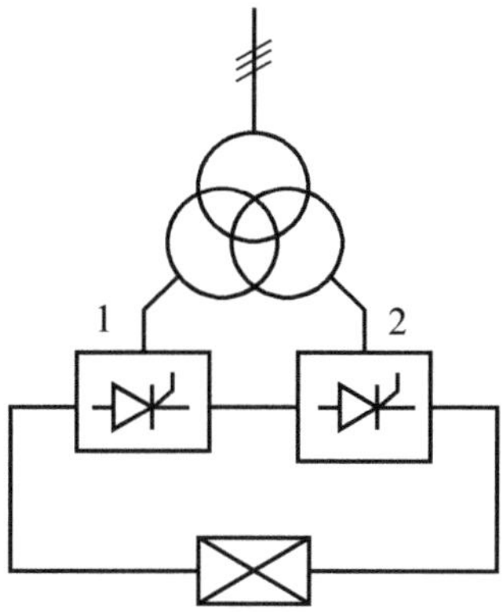

Fig. 7.23 Series connection of two controlled rectifiers

7.2.6.3 Grid Harmonics

Section 7.2.6.1 already mentioned the higher current harmonics in the grid current. Although these harmonics do not correspond with real power if the mains voltage is purely sinusoidal, these higher current harmonics may propagate in the grid and result in losses or other problems in other loads connected to the grid (generally, these problems are indicated by the term EMC or electromagnetic compatibility).

If the commutation delay is negligible, the current blocks are perfectly rectangular; for a bridge rectifier, positive and negative blocks in each phase are displaced by π radians and between similar current blocks in different phases the displacement is $2\pi/m$. For a wye rectifier, there is also a DC component I_d/m at the secondary side of the transformer, but this DC component is not transformed to the grid side.

Disregarding the transformer turns ratio, the Fourier analysis of the currents yields (with $s = 1$ for a wye rectifier and $s = 2$ for a bridge rectifier):

$$i(t) = s \cdot \frac{2}{\pi} \cdot I_d \cdot \sum_{\nu \neq km} \sin \frac{\nu\pi}{m} \cdot \frac{1}{\nu} \cdot \cos \nu\omega t \tag{7.29}$$

with ν being odd for a bridge rectifier. In these equations $t = 0$ is chosen at the centre of the current blocks for reference phase U; for a diode rectifier this is also at the maximum of the phase voltage U, but for a controlled rectifier $t = 0$ occurs α radians later than this maximum.

It can be proven that the order of the current harmonics *at the grid side* is determined by the pulse number of the rectifier, i.e. $\nu = kp \pm 1$. A general proof will not be provided here, but this will be illustrated by some examples.

The first example is a three-phase wye rectifier (see Fig. 7.24). The order of the grid harmonic currents is $1, 2, 4, 5, 7, \ldots$. As the pulse number for this rectifier is $p = m = 3$, these harmonic orders correspond to $kp \pm 1$.

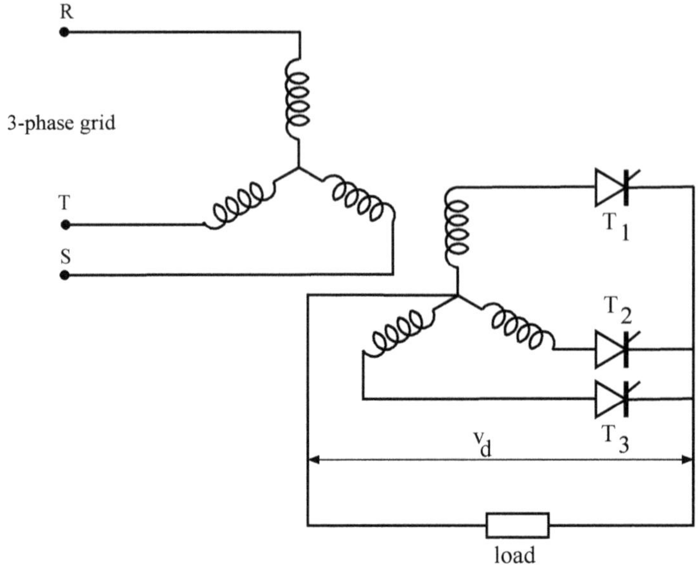

Fig. 7.24 Three-phase wye rectifier (three-pulse)

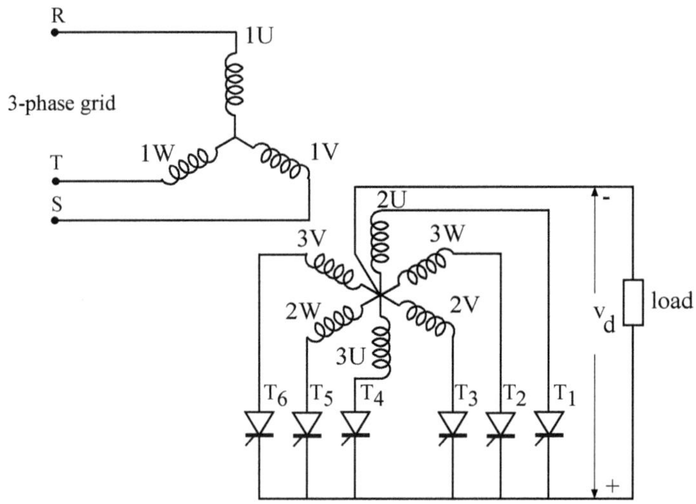

Fig. 7.25 Six-phase wye rectifier (six-pulse)

As a second example, we consider a six-phase wye-connected rectifier (see Fig. 7.25). The harmonic orders at the rectifier side (i.e. the six-phase secondary transformer side) are (according to Eq. 7.29): $1, 2, 3, 4, 5, 7, \ldots$ As the even harmonics are canceled out at the three-phase primary, at the primary transformer side we only find $1, 5, 7, \ldots$; in other words, again $kp \pm 1 = km \pm 1$. with $p = m = 6$.

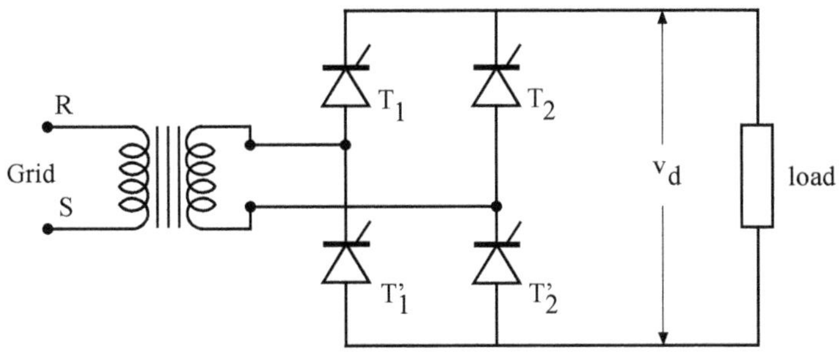

Fig. 7.26 Two-phase bridge rectifier (2-pulse)

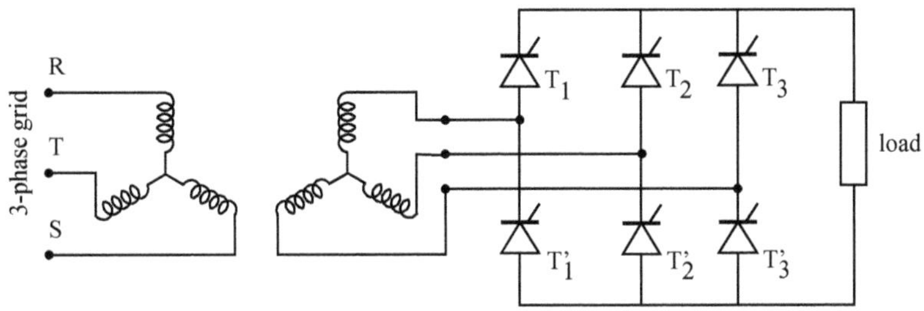

Fig. 7.27 Three-phase bridge rectifier (six-pulse)

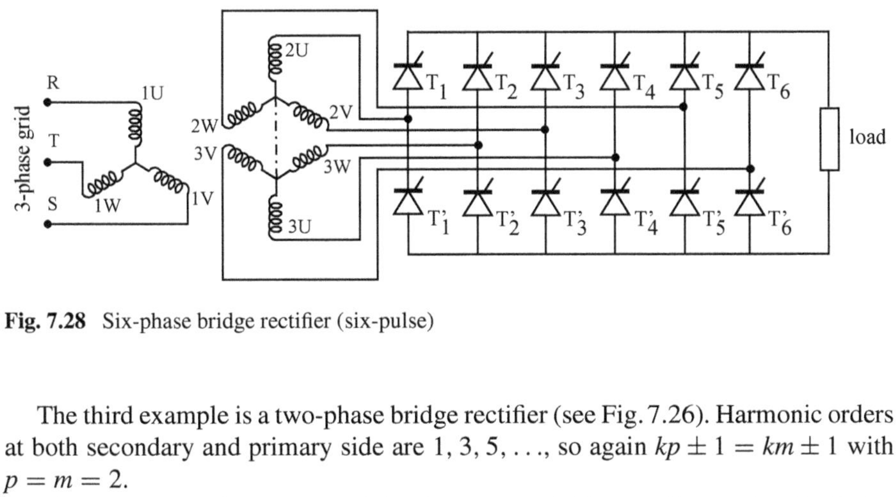

Fig. 7.28 Six-phase bridge rectifier (six-pulse)

The third example is a two-phase bridge rectifier (see Fig. 7.26). Harmonic orders at both secondary and primary side are $1, 3, 5, \ldots$, so again $kp \pm 1 = km \pm 1$ with $p = m = 2$.

The line current harmonics of the three-phase bridge rectifier in Fig. 7.27 also contain only orders $kp \pm 1$ with $p = 6$ (and $m = 3$).

For the six-phase bridge rectifier in Fig. 7.28, the rectifier side harmonic orders are $1, 3, 5, 7, 9, \ldots$ but as the third harmonics are cancelled out at the primary side of the transformer, only harmonics $1, 5, 7, 11, \ldots$ remain, which means that $\nu = kp \pm 1$ with $p = m = 6$. In all these examples (and in general), the amplitude of the $\nu - th$ harmonic is $\hat{I}_\nu = \hat{I}_1/\nu$.

Table 7.1 C_{eff} as a function of the pulse number

p	3	6	12	∞
C_{eff}	1.21	1.05	1.012	1

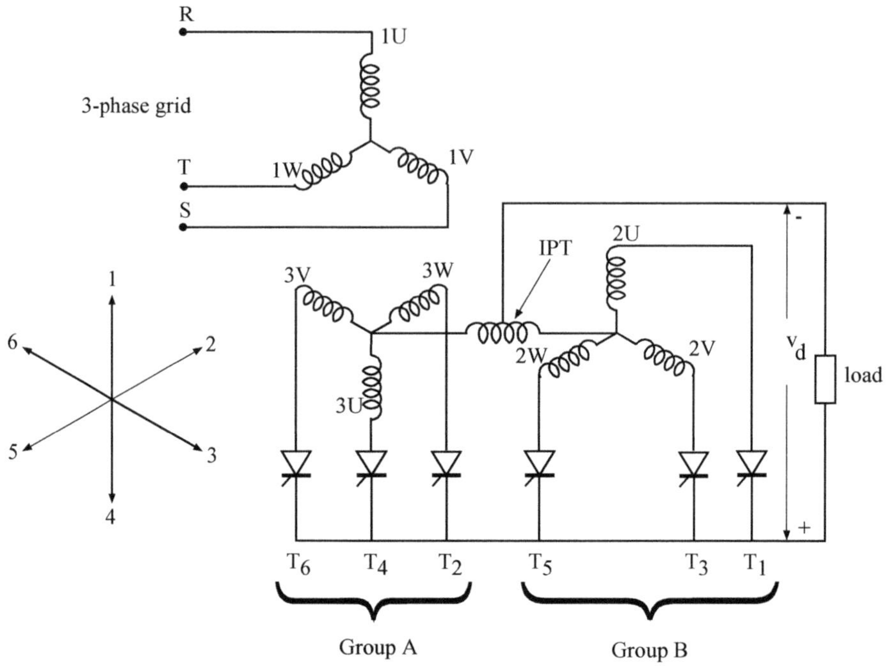

Fig. 7.29 Six-phase wye rectifier with IPT (p = 6)

To limit the grid pollution, a high order (and low amplitude) of the lowest harmonic is of utmost importance. To avoid a too oversized transformer, the ratio $C_{eff} = I_{eff}/I_1$ with $I_{eff} = \sqrt{\sum I_\nu^2}$ must be as low as possible as well. Therefore, a rectifier with a high pulse number will be regarded as ideal, in particular if the power rating is high. However, higher than 12 (or exceptionally 24) is not very common, as the additional harmonics reduction with pulse numbers higher than 12 is limited (see Table 7.1).

A pulse number of six can be obtained by a three-phase bridge rectifier, or by a six-phase wye rectifier as shown in Fig. 7.25. A disadvantage of the configuration in Fig. 7.25 is that each switch conducts only during $\pi/3$.

Figure 7.29 shows a better scheme where a large inductance, the IPT (interphase transformer), results in independent commutation of the two three-phase rectifiers, so that the conduction of each switch is now $2\pi/3$. The parallel connection of the two three-phase rectifiers with π radians phase difference elicits six-pulse behaviour.

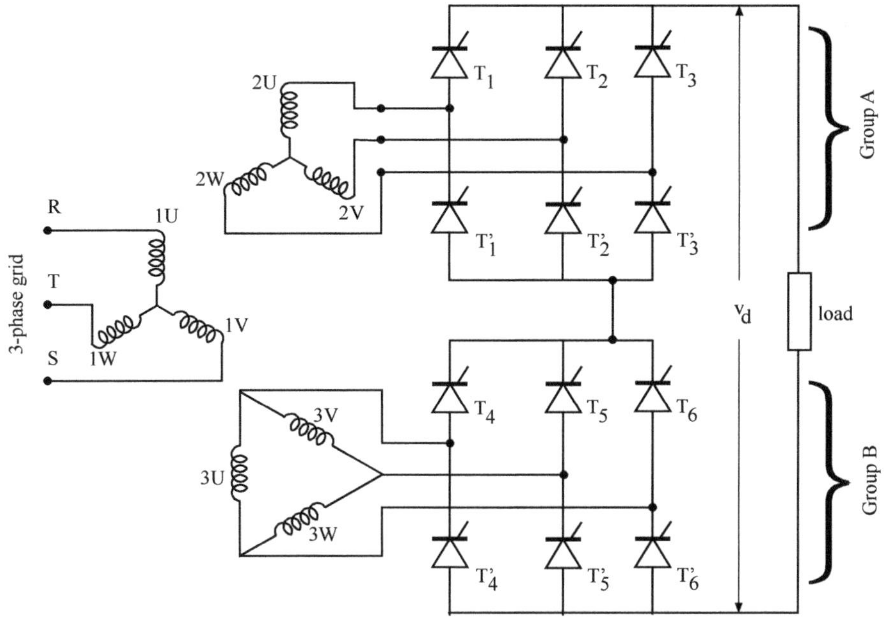

Fig. 7.30 Series connection of two three-phase bridges (p = 12)

Another connection with higher pulse number is the twelve-pulse rectifier in Fig. 7.30. The series connection of the two rectifiers with a phase difference of $\pi/6$ elicits twelve-pulse behaviour.

If necessary, the propagation in the grid of remaining harmonics can also be mitigated by filters.

7.3 Rectifier Supply of DC Machines

A controlled rectifier (mutator) allows both rectifier and regenerative operation. However, only the average output voltage may change signs, while the current direction is fixed due to the switches. Nevertheless, for the supply of DC machines, a reversal of the current direction is often required (e.g. independently excited DC machines with both motoring and generating). There are two possible (related) solutions: the anti-parallel connection of two rectifiers and the cross-connection.

7.3.1 Anti-parallel Connection

The basic idea is to use two anti-parallel connected rectifiers, one for the positive current direction and one for the negative current direction. By using the delay

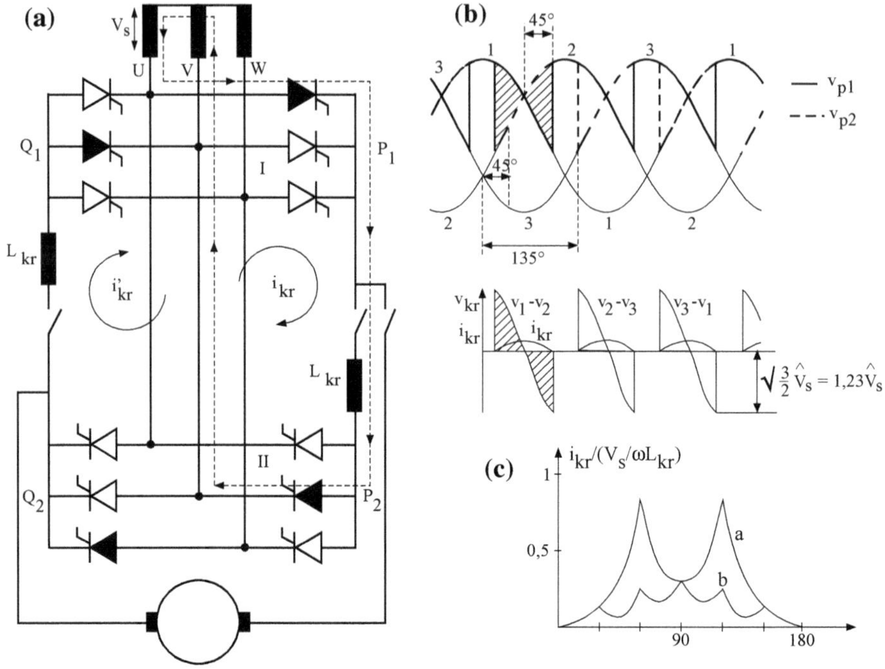

Fig. 7.31 Anti-parallel connection of two bridges

angle $\alpha$for one of these rectifiers and the $\pi-$complement $\alpha' = \pi - \alpha$ for the other, the average DC output voltages are complementary and, apparently, the two rectifiers can be connected in anti-parallel. However, although the average voltages are exactly the opposite of each other, the difference between the instantaneous voltages is quite important (see (b) in Fig. 7.31). To mitigate the resulting short-circuit currents between the two rectifiers, large inductances are inserted at both sides of the converters: indeed, in this configuration there are two independent short-circuit circuits, one between $P_1 - P_2$ and one between $Q_1 - Q_2$. These large inductances are indispensable if both converters are steered continuously at α and α', respectively. The resulting short-circuit currents are illustrated by curve a in (c) of Fig. 7.31.

Nowadays, the steering of the converters is usually adapted based on a current sensing circuit:

- either only one is steered (e.g. the one for the positive current when a positive load current is sensed) while the other is blocked completely;
- or the converter whose current direction corresponds to the sensed load current is steered completely, while the other one is steered using a current-limiting algorithm *until a load current reversal is sensed;*

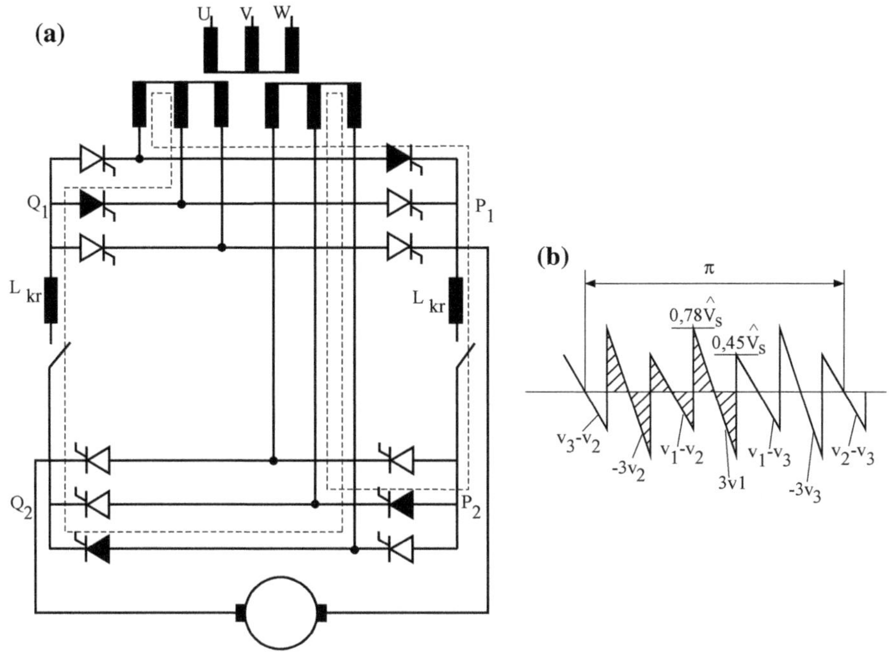

Fig. 7.32 Cross-connection of two bridges

- or the converter whose current direction corresponds to the sensed load current is steered completely, while the other one is steered using a current-limiting algorithm until some time before a load current reversal *can be expected,* after which both are steered completely during some interval around the expected current reversal.

Although most anti-parallel connected rectifiers use some kind of current-limiting algorithm, the practical realisation can sometimes be problematic. Detecting the current reversal may be plagued by uncertainty as the DC load inductance is not infinite in reality and thus the current may oscillate around zero at the current reversal.

7.3.2 *Cross Connection*

In the cross connection in Fig. 7.32, the two rectifiers are fed from two secondary windings of the transformer.

As a result, there is one single short-circuit loop (see the dashed line). In this loop, the instantaneous voltage differences $P_1 - P_2$ and $Q_1 - Q_2$ subtract, resulting in a much smaller voltage in the loop (see (b) in Fig. 7.32 and curve b in (c) in Fig. 7.31). Note that the short-circuit loop now contains the series connection of the leakage inductances of the two secondaries, also limiting the current.

Chapter 8
DC Chopper

Abstract For AC, there is a straightforward and almost lossless way to transfer electrical energy from one (voltage) level to another level, i.e. the classical electromagnetic transformer. For DC such a standard solution is not available. Energy transfer from a higher to a lower level is indeed possible using a resistance network, but this comes at the expense of high losses. The chopper is an elegant power-electronic solution that does not lead to any significant losses. Starting from a fixed DC voltage, it adapts the load voltage by periodically switching the input DC voltage (or current), and as such adapting the average load voltage. Two advantages are that some chopper circuits also allow energy transfer to a higher voltage level, and that the chopper principle eases electronic control. Another way to control the average current is by using periodic switching of a series resistor, but this method is not energy-efficient. For low-power applications (e.g. telecommunications, computer power supply), yet another alternative using resonant circuits is quite common (see Sect. 8.5). In the first section, the basic chopper circuits will be discussed. The next section shortly discusses the power-electronic switches used in choppers. In other sections, chopper applications in traction and DC drives are illustrated. The final section gives a brief overview of the principles of resonant chopper circuits.

8.1 Basic Chopper Circuits

The two basic chopper circuits are the step-down chopper (buck chopper) and the step-up chopper (boost chopper). They also can be combined as discussed in Sect. 8.1.3. Finally, an (albeit rather energy inefficient) chopper variant is discussed in Sect. 8.1.4.

8.1.1 Step-Down Chopper (Buck Chopper)

The first type of chopper is the step-down or buck converter, which allows energy transfer from a fixed voltage level to a load with *lower* voltage level. The basic circuit

J. A. Melkebeek, *Electrical Machines and Drives*, Power Systems,
https://doi.org/10.1007/978-3-319-72730-1_8

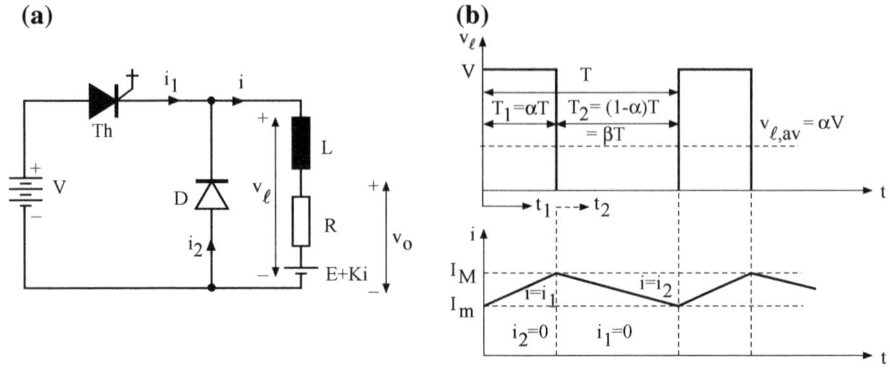

Fig. 8.1 Step-down chopper

is shown in Fig. 8.1. Here, the load is assumed to contain an L-R impedance and a counter-emf with a current-dependent part (e.g. a series-excited DC machine).

The switch Th must be able to switch on and off (e.g. a GTO or IGBT or or IGCT or Mosfet). During interval T_1, the source delivers current to the load (Th conducting). When Th is switched off at $t = T_1$, the current continues flowing through the diode as the load is inductive. The load voltage is then zero. At $t = T_1 + T_2$ the switch is turned on again, and so on periodically with period $T = T_1 + T_2$. Assuming a steady-state condition, the following equations are valid in the intervals $0 \le t \le T_1$ and $T_1 \le t \le T_1 + T_2$:

$$V - E = (K + R) \cdot i + L\tfrac{di}{dt} \;\; for\ 0 \le t \le T_1 \tag{8.1}$$

$$-E = (K + R) \cdot i + L\tfrac{di}{dt} \;\; for\ T_1 \le t \le T_1 + T_2 \tag{8.2}$$

This yields, again for intervals $0 \le t \le T_1$ and $T_1 \le t \le T_1 + T_2$

$$i = I_1 + (I_m - I_1)\exp(-t/\tau) \;\; for\ 0 \le t \le T_1 \tag{8.3}$$

$$i = I_2 + (I_M - I_2)\exp(-(t - T_1)/\tau) \;\; for\ T_1 \le t \le T_1 + T_2 \tag{8.4}$$

in which $I_1 = (V - E)/(K + R)$, $I_2 = -E/(K + R)$ and $\tau = L/(K + R)$.

Expressing the boundary conditions of steady state, i.e. continuity of i at $t = T_1$ ($i = I_M$) and at $t = T_1 + T_2$ i.e. $i(T_1 + T_2) = i(0) = I_m$, we obtain

$$I_M = I_1 + (I_m - I_1)\exp(-T_1/\tau) \tag{8.5}$$

$$I_m = I_2 + (I_M - I_2)\exp(-T_2/\tau) \tag{8.6}$$

Table 8.1 Control methods of α

Method	T	T_1	T_2
I	ct	var	var
II	var	ct	var
III	var	var	var
IV	var	var	ct

From Eqs. 8.5 and 8.6, I_M and I_m can be calculated. In most cases, $T \lll \tau$. Using $\exp(-x) \approx 1 - x$, we can approximate I_m and I_M by

$$I_m = \alpha I_1 + \beta I_2 - \alpha\beta \frac{T}{\tau} I_1 \tag{8.7}$$

$$I_M = \alpha I_1 + \beta I_2 - \alpha\beta \frac{T}{\tau} I_2 \tag{8.8}$$

where $\alpha = T_1/T$ and $\beta = 1 - \alpha = T_2/T$. A straightforward calculation yields for the average current

$$I \equiv i_{av} = \alpha I_1 + \beta I_2 = \frac{\alpha V - E}{R + K} \tag{8.9}$$

which shows that the average current corresponds to an average load voltage of αV. Note also that $i_{av} \approx (I_M + I_m)/2$.

The ripple of the current is

$$w = \frac{I_M - I_m}{2I} = \frac{\alpha(1-\alpha)}{2} \cdot \frac{T}{\tau} \cdot \frac{V}{\alpha V - E} = \frac{\alpha(1-\alpha)}{2} \cdot T \cdot \frac{V}{LI} \tag{8.10}$$

To limit the ripple of the current, the ratio T/τ should be small (e.g. smaller than 0.1) and therefore the inductance should be sufficiently high.

As mentioned before, from Eq. 8.9 it follows that, as to the average current, the load is supplied with a voltage αV. With a step-down chopper, the voltage can be regulated to a lower level without incurring any important losses.

Applications of step-down choppers include speed or torque control of DC machines and switched power supplies (thus as DC converters). For the control of α (called the *duty ratio*), there are four possibilities, as shown in Table 8.1. Methods I and II are most common. Method II might be cheaper to implement than method I, but has as a disadvantage that the variable frequency makes adequate filtering difficult and may therefore lead to EMC problems. Methods III and IV as well have this disadvantage.

It is important to note that the analysis above (in particular the calculation of the average current) is only valid if the current is continuous. If the inductance is too small, the current in the interval T_2 may become zero and the diode may block.

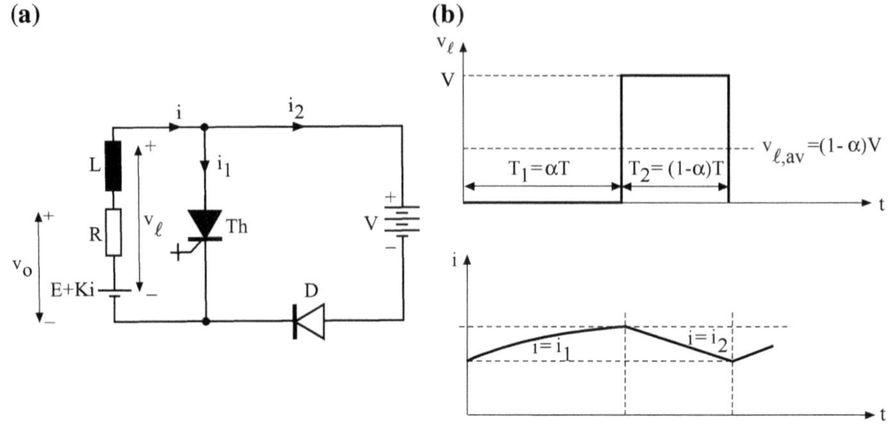

Fig. 8.2 Step-up chopper

8.1.2 *Step-Up Chopper (Boost Chopper)*

The step-down converter of the previous section can only be used for power transfer to a load with a lower voltage level. This is where the step-up chopper comes in, as this does allow transfer of DC power to a higher voltage level (see Fig. 8.2). A current-dependent emf in series with an inductance is assumed to be the energy source. The aim is to transfer the DC power to a fixed DC grid or battery (*load*) at a higher voltage level, i.e. $V > E$.

When switch Th is conducting, the current builds up in the resulting short circuit. Subsequently, the switch Th is turned off, and the current that has been built up in the inductive circuit is returned to the *load* V via the diode. The corresponding differential equations are, for the two corresponding intervals:

$$E = (R - K) \cdot i + L\frac{di}{dt} \tag{8.11}$$

$$E - V = (R - K) \cdot i + L\frac{di}{dt} \tag{8.12}$$

Note that the time constant of the circuit is $\tau = L/(R - K)$. If $R > K$, we get an unconditionally stable behaviour with negative exponentials. Limiting values ($t \to \infty$) in intervals 1 and 2 are $I_1 = E/(R - K) > 0$ and $I_2 = (V - E)/(R - K) > 0$, respectively.

However, if $R < K$, the time constant is negative and we get positive exponentials. As a result, the behaviour is only stable when at the end of interval 1 the voltage $v_l = E + Ki - Ri$ is less than the voltage V. If not, the current will continue to increase in interval 2 (until failure).

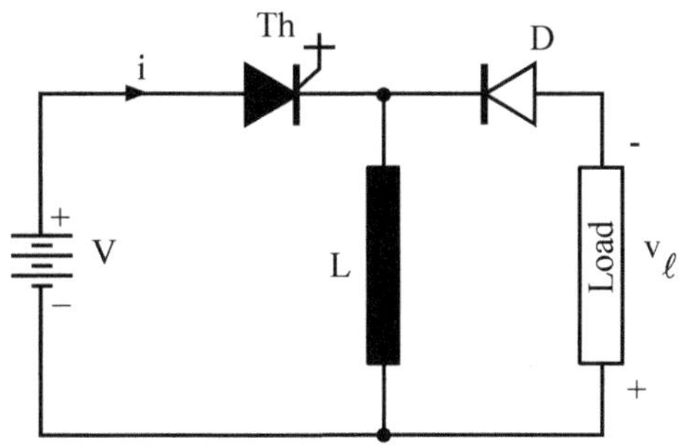

Fig. 8.3 Combined step-down and step-up chopper

Important applications include regenerative braking in DC traction, which is used in older trains with DC traction where the emf of the braking train is not sufficient to feed into the supply, but also numerous low-power appliances.

8.1.3 Mixed Step-Down and Step-Up Chopper Circuits

Chopper circuits that can handle both step-down and step-up conversion can be obtained by a combination of the circuits in Sects. 8.1.1 and 8.1.2, by parallel connection or cascade connection or in a combined circuit.

An example of a combined circuit is shown in Fig. 8.3. Importantly, in this circuit, the load must not be inductive: the polarity of *source* and *load* are reversed compared to the circuits in Figs. 8.1 and 8.2.

8.1.4 Resistance Chopping

Current regulation can also be obtained by chopping a resistance. Figure 8.4a shows a circuit where the series resistance of a DC machine is chopped between 0 and R, for current or speed control, for example. To reduce the ripple, the resistance can be split into parts that are chopped with a phase difference. Of course, this control method is not lossless.

A more interesting application of resistance chopping is in resistance braking of DC machines, as illustrated in (b) and (c) in Fig. 8.4. The disadvantage of circuit (b) is that the voltage over the switch can be rather high. Circuit (c) can only be used at sufficiently high speed.

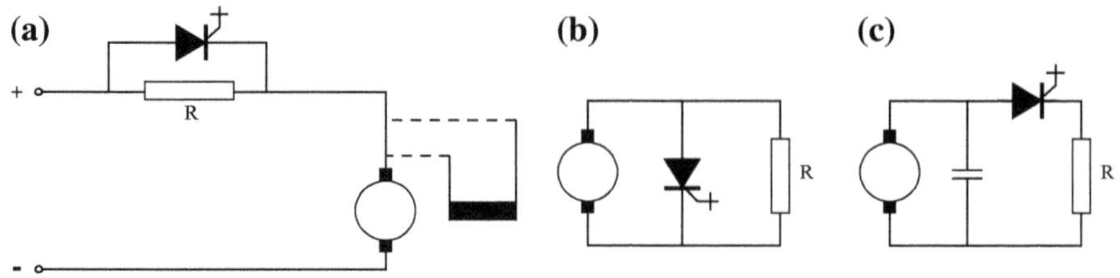

Fig. 8.4 Resistance chopping

Fig. 8.5 Chopper circuit with only thyristors

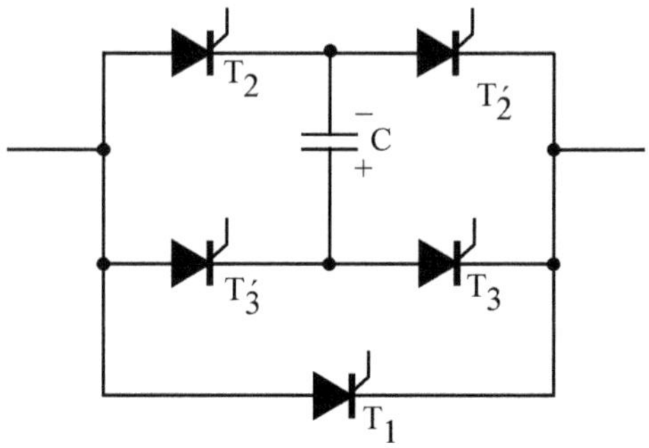

8.2 Practical Switches for Choppers

At the beginning of the power-electronic era, the only available switches were thyristors with a killer circuit for turn-off (as the switch in a chopper must be turned on and off). Later on, GTOs became available and a much less demanding turn-off circuit was possible. BJTs (Darlington) have also been used for smaller power ratings.

Nowadays, Mosfets are used for lower power ratings, while IGBTs are the normal choice for all but the very low or very high power ratings. For the latter, IGCTs or GTOs could be used, although practical applications of very high power choppers have practically disappeared nowadays.

Nevertheless, at least from a historical perspective, the circuit in Fig. 8.5 is worth mentioning. The switch is replaced by five thyristors and a capacitor. When T_1 is conducting and the capacitor is charged, as is shown in the figure, the thyristor can be switched off by switching-on T_2 and T_3. The current will temporarily continue to flow until the capacitor charge is reversed. In the next cycle, T_2' and T_3' must be used to turn off T_1 (as the charge of the capacitor will have reversed). It is possible to omit T_1 and use T_2 with T_2' and T_3 with T_3' to conduct the load current, but in that case these four thyristors have to be dimensioned for conducting continuously the load current, and not only the turn-off current.

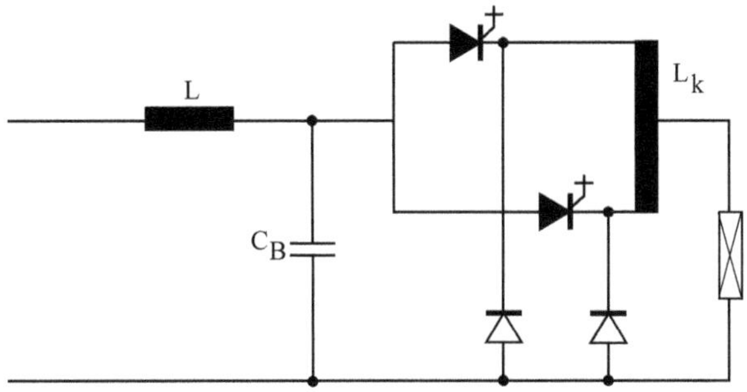

Fig. 8.6 Multiphase chopping

8.3 Buffer Capacitor and Multiphase Chopping in Traction Applications

For strongly inductive supply lines, like the long overhead lines for trams or trains, it can be necessary to use a capacitor on board of the vehicle to limit the oscillations of the supply voltage.

However, the overhead wire inductance and the capacitor also form an $L-C$-circuit with a resonance frequency of $f_o = 1/2\pi\sqrt{LC}$. For a reduction of the voltage pulsations, the frequency f of the chopper circuit should be higher than twice this resonance frequency, i.e. $f > 2f_o$. Thus either f_o should be low (but this requires a large capacitor) or the switching frequency should be high. This is problematic if GTOs are used as switches, because their switching frequency is limited.

A solution to this is the use of two or more choppers in parallel and phase displaced with respect to each other (i.e. *multiphase chopping*). The circuit in Fig. 8.6 shows two choppers in parallel, effectively doubling the line switching frequency compared to the switching frequency of each chopper. The inductance or IPT (interphase transformer) must ensure that both choppers work independently.

8.4 Chopper Supply of DC Machines

8.4.1 *Motoring*

Supplying a DC motor from a chopper is equivalent to a supply from a variable voltage source, apart from the current ripple which is to be limited by a sufficiently large inductance (for a series-excited DC machine there is already the large inductance of the excitation winding). Therefore, starting and speed control are possible without almost any additional losses. In addition, the chopper is ideal for electronic control and automation.

With series-excited DC machines, the flux is often reduced to obtain higher speeds. Figure 8.7 illustrates a circuit where field flux weakening is obtained automatically

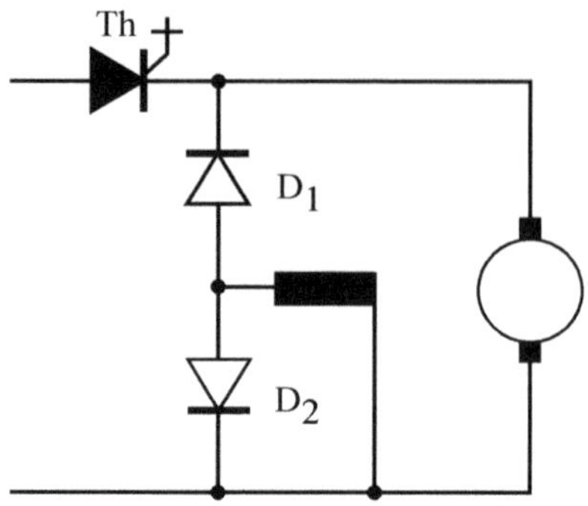

Fig. 8.7 Automatic field control

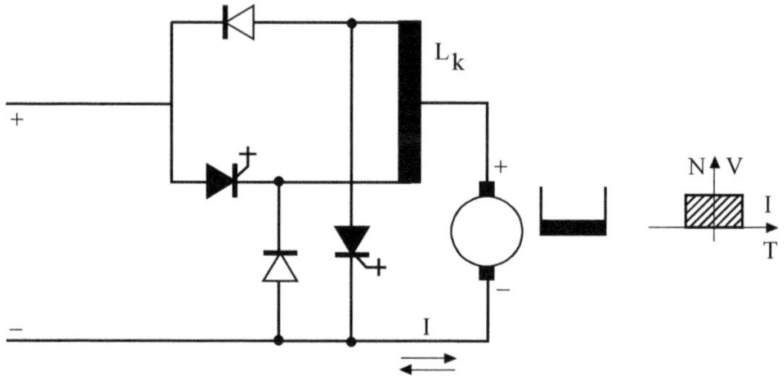

Fig. 8.8 Chopping in quadrants 1 and 2

when the voltage supplied to the motor increases. When the switch is on, current is supplied from the DC source to the armature winding. When the switch is turned off, the inductance of the armature winding results in an increasing current in the excitation winding (via diode D_1). When the switch is turned on again, the remaining excitation current will continue to flow via diode D_2.

8.4.2 *Two-Quadrant Operation*

For operation in more than one quadrant (the first, for example) a combination of the step-up and step-down choppers can be used.

On the one hand, Fig. 8.8 shows a configuration which allows operation in quadrants I and II. This is a combination of the circuits in Figs. 8.1 and 8.2.

On the other hand, Fig. 8.9 demonstrates a circuit for operation in quadrant I and IV. With T_1 and T_2 conducting together, the supply is connected to the motor. When one of the switches is switched on and off periodically (the other switch continuously on), we get an operation as in Fig. 8.1. If one of the switches is turned off continuously

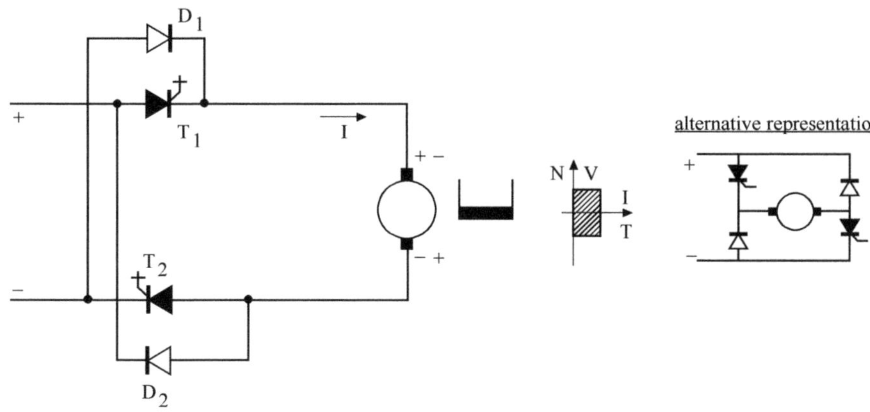

Fig. 8.9 Chopping in quadrants 1 and 4

and the other switch turned on and off periodically, operation as in Fig. 8.2 is obtained (e.g.: T_2 turned off continuously; if T_1 is turned on, the currents built up in the circuit T_1- D_2; when T_1 is turned off, current is supplied to the DC source). Note that the polarity of the machine is now reversed.

8.5 Resonant Circuits for DC-DC Converters

Resonant circuits are frequently used to assist switching. In the past, resonant circuits were required to turn off the thyristors (in *killer circuits*), but with the advent of switches such as Mosfets, IGBTs or GTOs, this became history. For low-power applications, however, resonant circuits are frequently used to realise high-frequency circuits with reduced switching losses (as these high frequencies are advantageous to reduce the dimensions and weights of filter inductances and transformers). Thousands of resonant circuits have been reported in the literature, see e.g. [5]. Here we discuss two basic circuits for DC-DC converters.

8.5.1 Series-Loaded Half Bridge

In the circuit in Fig. 8.10, the load (represented by the resistance R) is connected in series with the inductance L and the capacitor C between the mid-point of the half bridge and the artificial midpoint of the source (realised by two large capacitors C_{DC1} and C_{DC2}). The two switches are controlled complementarily, each with a duty ratio of 0.5 and an angular frequency ω around the natural angular frequency, $\omega_c = 1/\sqrt{LC}$ (Fig. 8.11).

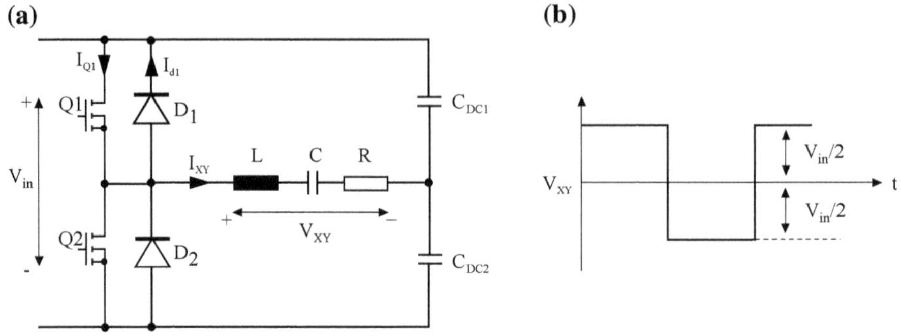

Fig. 8.10 Series-loaded resonant circuit

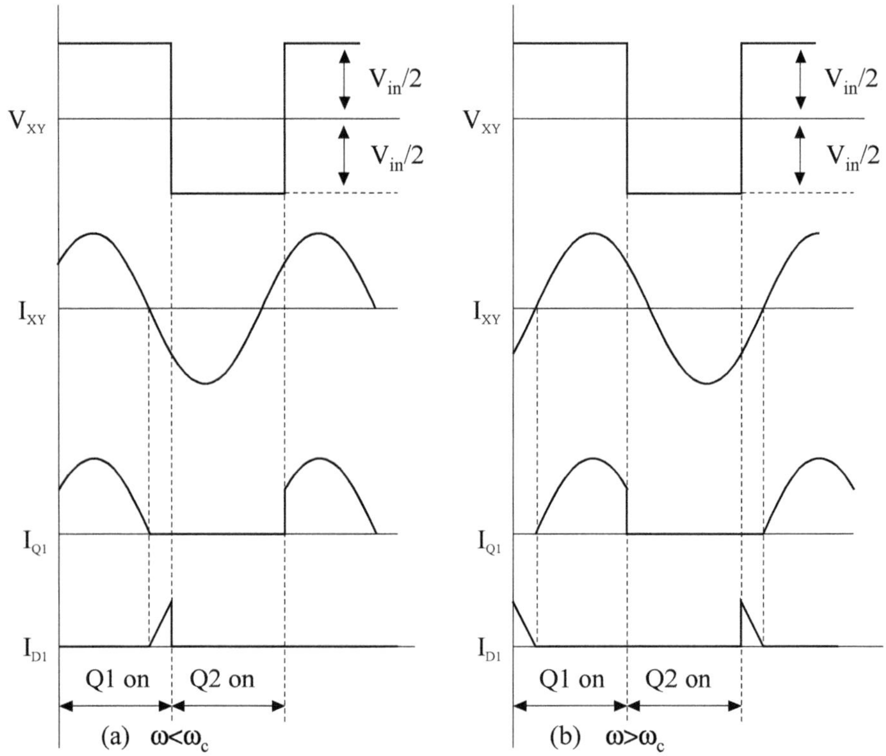

Fig. 8.11 Voltage and current for series-loaded resonant circuit

As a result, the voltage V_{xy} becomes a block wave with amplitude $V_{in}/2$ and angular frequency ω.

The output current I_{xy} results from the impedance of the series impedance Z_{LCR}:

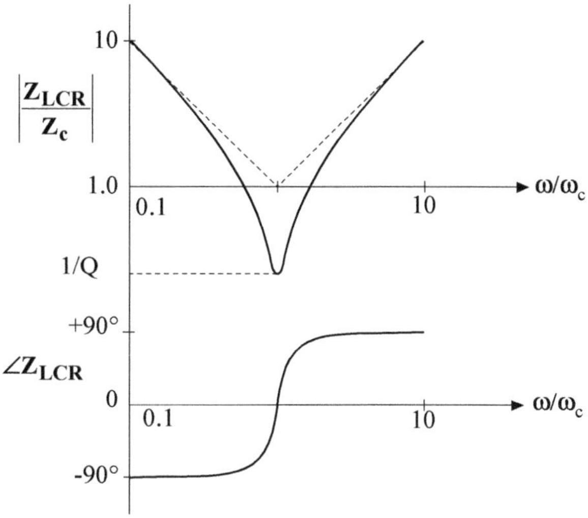

Fig. 8.12 Resonant impedance

$$Z_{LCR} = R + j\omega L + \frac{1}{j\omega C} = Z_c \left[\frac{1}{Q} + j\left(\frac{\omega}{\omega_c} - \frac{\omega_c}{\omega} \right) \right]$$

with the characteristic impedance $Z_c = \sqrt{L/C}$, the natural frequency $\omega_c = 1/\sqrt{LC}$ and the quality factor $Q = \omega_c L/R$.

Figure 8.12 illustrates the amplitude and phase characteristics of the impedance Z_{LCR}. The impedance behaves as inductive for frequencies higher than ω_c, but capacitive for frequencies lower than ω_c. For $\omega = \omega_c$, the impedance is purely resistive and attains its minimum value. Note that the amplitude variation is quite large around this characteristic frequency.

In addition to a fundamental current with frequency ω, the resulting current I_{xy} will contain the same higher harmonics (orders 3, 5, 7,...) as the voltage block wave. However, as the impedance becomes quite large for frequencies higher than ω_c, these higher harmonics in the current can be disregarded and the output current can be considered as purely sinusoidal.

The output load AC voltage, $R \cdot I_{xy}$, can be adjusted by varying the switching frequency around the resonance frequency ω_c. The output AC voltage can subsequently be adapted using a transformer (thus at $\omega \approx \omega_c$) and rectified using a diode rectifier to obtain a DC voltage. Since the AC voltage can be adjusted by a slight variation of the switching frequency, so can the rectified voltage. The frequencies used are mostly in the MHz range, permitting smaller inductive components and transformers. The power of such converters usually ranges from a few watts to hundreds of watts.

In most cases, the circuit is utilised in either the range $\omega \leq \omega_c$ (where the impedance is capacitive) or the range $\omega \geq \omega_c$ (where the impedance is inductive).

In the former range, the current switches sign before the voltage switches sign. At the instant the switch (e.g. T_1) is turned off, the current will therefore already been transferred to the *anti-parallel diode of this same switch* (e.g. D_1). This is

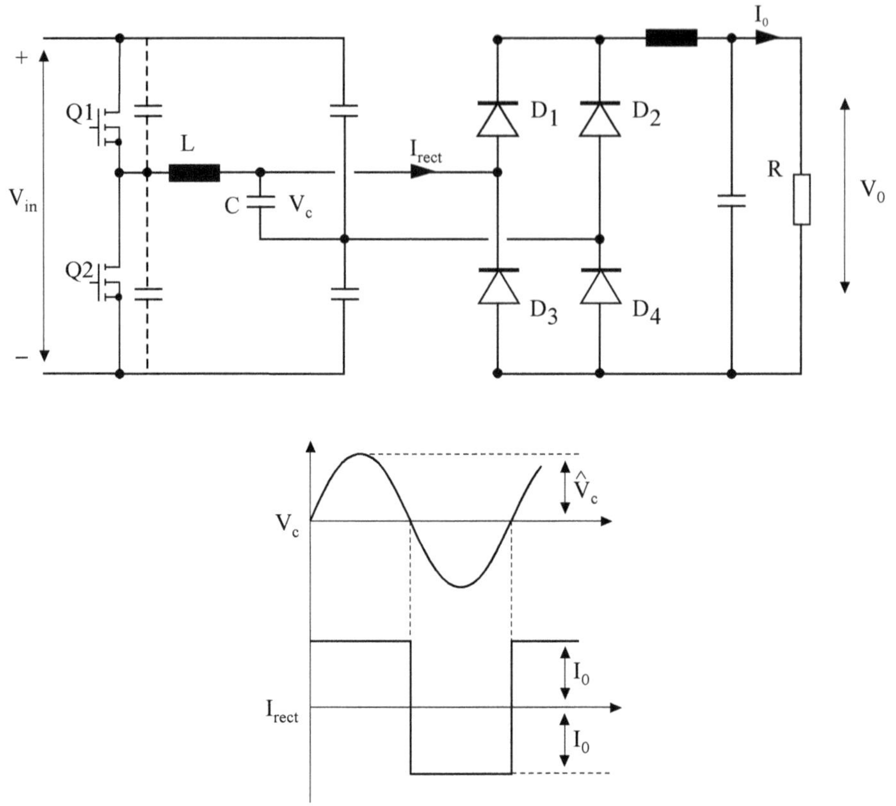

Fig. 8.13 Parallel-loaded resonant circuit

called zero-current switching, or zero-current turn off. Turn-off of this switch will not result in any losses but turn-on of the complementary switch will result in a remarkable transient of the diode (i.e. "recovery transient"). Snubbers are required and the switching loss is not negligible.

In the latter range, the current will have to switch to the anti-parallel diode (e.g. D_2) *of the complementary switch* at the instant the first switch (e.g. T_1) turns off. The turn-off of the first switch will result in losses due to the shift of the current to the anti-parallel diode of the complementary switch. In contrast, switching on the complementary switch will not incur any losses as the current will automatically commutate from the diode to this switch if the current reverses sign. This is called zero-voltage switching, or zero-voltage turn-on.

8.5.2 Parallel-Loaded Resonant Converter

Another frequently used topology is the parallel-loaded resonant converter (see Fig. 8.13). This figure also depicts the diode rectifier with which a DC-DC converter is obtained. The operating principle is dual to the one in the previous section. The output current is now rectangular, while the output voltage V_c (and the input voltage of the rectifier) is - or should be - approximately sinusoidal. A transformer may be inserted between the resonant bridge and the rectifier to adapt the voltage and to obtain galvanic insulation.

There are numerous other resonant converter circuits, some combining the advantages of series- and parallel-loaded converters (i.e. hybrid or series-parallel-loaded converters).

Chapter 9
AC Chopper

Abstract The aim of the AC chopper (also called *AC line control* or *phase control*) is somewhat analogous to the DC chopper, i.e. to obtain a variable voltage. Here, the output is an AC voltage with the same fundamental frequency as the mains supply. An important difference compared to the DC chopper is that also the current is now bi-directional. This chapter discusses the basic circuits and analyses in more detail a pure inductive load, both single-phase and three-phase.

9.1 Basic Principle

Figure 9.1A shows the basic circuit of a single-phase AC chopper. The source is a sinusoidal voltage with constant amplitude and frequency. In series with the load (here an R-L load), a triac or a pair of anti-parallel thyristors is connected.

First, we consider the case of a purely resistive load. If the triac gets a gate signal at the start of each positive (t_1) and negative (t_2) half period of the voltage (or a continuous gate signal), then the load gets the full voltage and the resulting current is sinusoidal (and in phase with the voltage). If, instead, the gate signal is sent only later in the half period (t_2, t_2',...), then we get a current as in (B, b) in Fig. 9.1. The load voltage is shown in (B, c). Of course, neither voltage nor current are sinusoidal but the frequency of the fundamental is still the mains frequency, albeit with a reduced fundamental amplitude.

If the load is not purely resistive, but resistive-inductive, for example, then for load current and voltage the curves of (B, d) and (B, e) are obtained, respectively. Load voltage and current are not sinusoidal but the frequency of the fundamental is still the grid frequency, of course with reduced amplitudes of these fundamentals. The higher harmonics are more or less harmless as the inductive load filters out these higher current harmonics.

In principle, the control of chopping can be based on the hold-off angle γ, on the delay angle α or on the extinction angle β (angles α and β are measured with respect to the zero-crossings of the voltage). However, in many cases a control based on β generates dynamic problems.

J. A. Melkebeek, *Electrical Machines and Drives*, Power Systems,
https://doi.org/10.1007/978-3-319-72730-1_9

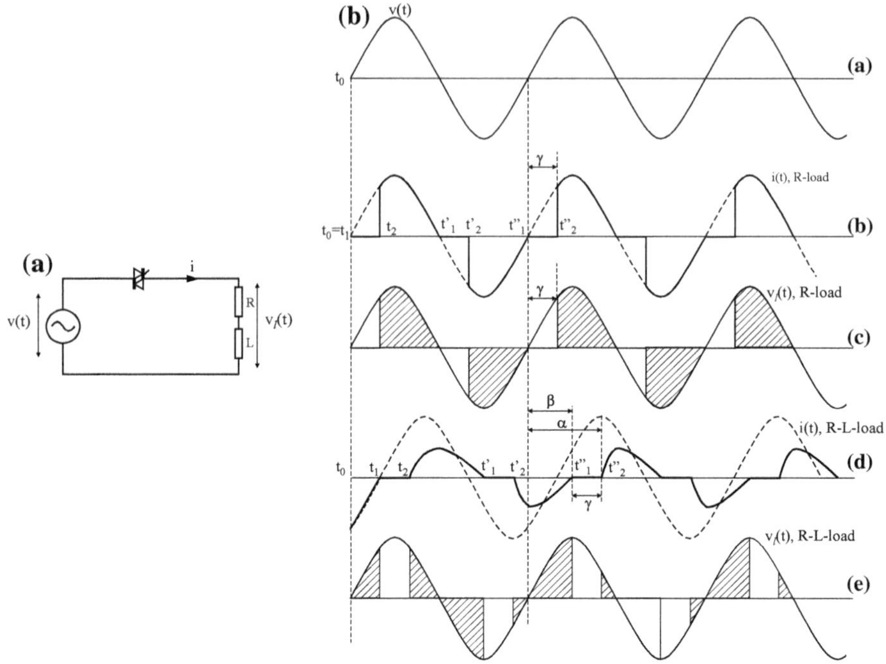

Fig. 9.1 Single-phase AC chopper

9.2 Phase Control of a Single-Phase Inductance

For a purely inductive load (Fig. 9.2), phase control can easily be modelled analytically, at least as to the fundamental behaviour.

From the waveforms in Fig. 9.2B - for steady-state - it is easy to derive (assuming lossless triac or thyristors) that $\alpha + \beta = \pi$. Using $\alpha - \beta = \gamma$, it follows that $\alpha = (\pi + \gamma)/2$ and $\beta = (\pi - \gamma)/2$.

Fourier analysis of the voltage over the switch (the *control voltage*) yields for the amplitude:

$$\hat{V}_{t1} = \frac{2}{\pi} \int_{-\gamma/2}^{\gamma/2} (\hat{V} \cos \omega t) \cdot \cos \omega t \cdot d(\omega t) = \hat{V} \frac{\gamma + \sin \gamma}{\pi} \tag{9.1}$$

The fundamental of the load voltage is therefore (in effective value):

$$V_l = V - V_{t1} = V \left[1 - \frac{\gamma + \sin \gamma}{\pi} \right] \tag{9.2}$$

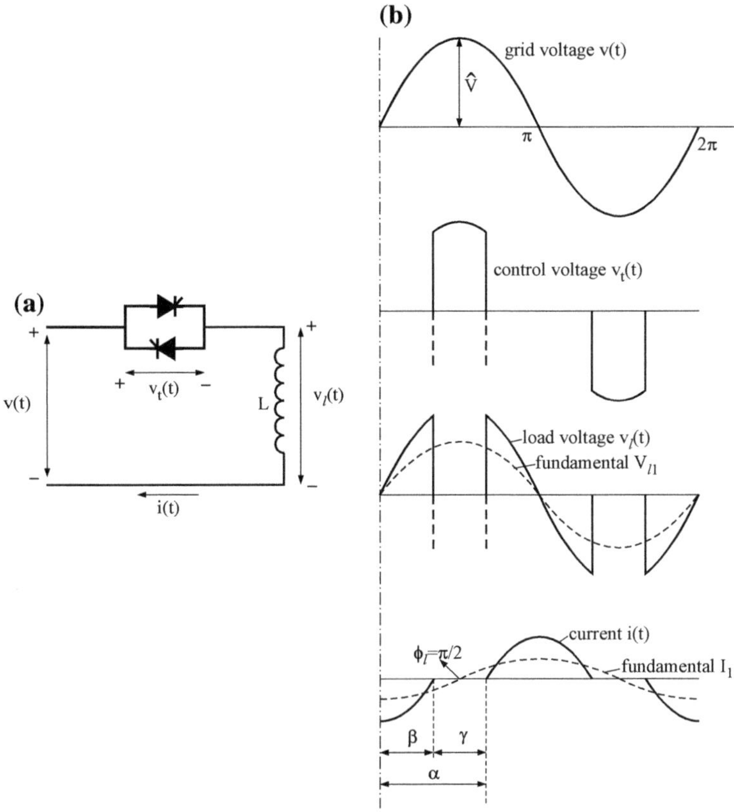

Fig. 9.2 Phase control of a single-phase inductance

The fundamentals of load voltage and control voltage are in phase with the grid voltage, while the fundamental of the load current is $\pi/2$ lagging. Choosing the real axis along the voltage, the phasor representation of the current fundamental is

$$\underline{I}_1 = \frac{V}{jX_l}\left[1 - \frac{\gamma + \sin\gamma}{\pi}\right] = \frac{V}{jX_l}\left[\frac{\pi - \gamma - \sin\gamma}{\pi}\right] \tag{9.3}$$

As observed from the grid, this is equivalent to an additional inductance in series with the load inductance

$$X_t = X_l\left[\frac{\gamma + \sin\gamma}{\pi - \gamma - \sin\gamma}\right] \tag{9.4}$$

Figure 9.3 shows X_t/X_l as a function of the hold-off angle.

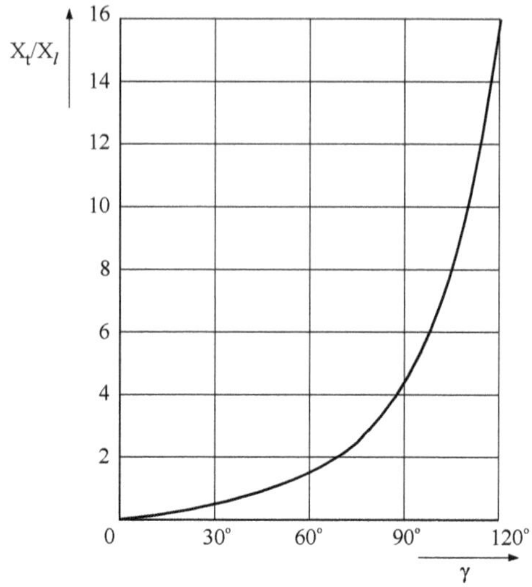

Fig. 9.3 X_t/X_l as a function of γ for a single-phase inductive load

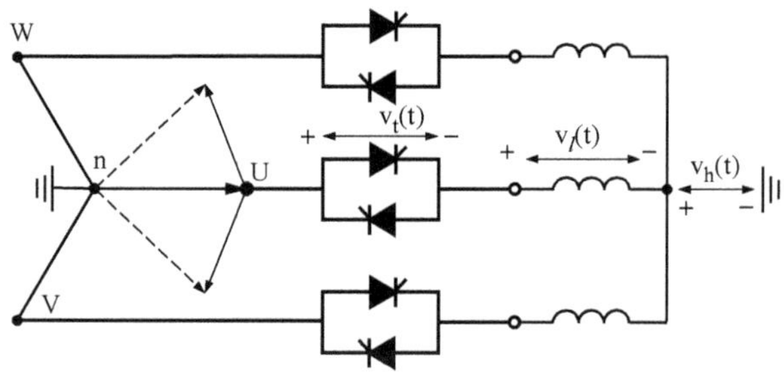

Fig. 9.4 Phase control of a three-phase inductance

Note that, in principle, a similar analysis also holds for a capacitive load.[1]

9.3 Phase Control of a Three-Phase Inductance

Figure 9.4 shows the circuit for phase control of a symmetrical three-phase inductance. The grid voltage is assumed to be sinusoidal and three-phase symmetric. Naturally, the gate signals for the three switches are also assumed to be three-phase symmetrical (i.e. shifted by $2\pi/3$ radians with respect to each other).

As the neutrals are not connected, zero-sequence load currents cannot flow. Therefore, the load voltages will not contain zero-sequence components. From the three-phase symmetry of voltages and of the load and switching sequence, it follows that the voltage difference $v_h(t)$ between the neutral points of load and grid will not contain fundamental components.

[1] Verify whether phase control of a capacitor comes down to an equivalent larger or smaller capacitor.

As the grid voltage is sinusoidal, also the sum of the voltages $v_l(t) + v_t(t) + v_h(t)$ is sinusoidal. For the phase U, we may therefore write:

$$v_{lu}(t) + v_{tu}(t) + v_h(t) = \hat{V} \cdot \sin \omega t \tag{9.5}$$

Adding the three similar equations for the three phases yields

$$v_h(t) = -\frac{1}{3}[v_{tu}(t) + v_{tv}(t) + v_{tw}(t)] \tag{9.6}$$

To calculate the control and load voltages, a distinction must be made between operation with $\gamma < \pi/3$ (at each instant 2 or 3 switches conducting) and $\pi/3 \leq \gamma \leq 2\pi/3$ (either 2 or no switches conducting).

Operation with $\gamma < \pi/3$

For $\gamma < \pi/3$, either 3 or 2 switches are conducting.

When $\beta < \omega t < \alpha$ (i.e. within the hold-off interval for phase U), only the switches for phase V and W are conducting. The load voltage for phase U is thus zero, $v_{lu}(t) = 0$ and therefore

$$v_{tu}(t) + v_h(t) = \hat{V} \sin \omega t$$

With the switches for phases V and W conducting, the voltage potential of the star point O of the load is the average of the voltage potentials of phases V and W:

$$v_h(t) = -\frac{1}{2}\hat{V} \sin \omega t \tag{9.7}$$

The control voltage for phase U is thus

$$v_{tu}(t) = \frac{3}{2}\hat{V} \sin \omega t \tag{9.8}$$

Note that Eqs. 9.7 and 9.8 are in agreement with the general Eq. 9.6.

When $0 < \omega t < \beta$ or $\alpha < \omega t < \pi$, the phase U switch is conducting and therefore

$$v_{tu}(t) = 0 \tag{9.9}$$

and

$$v_{lu}(t) + v_h(t) = \hat{V} \sin \omega t \tag{9.10}$$

If, during this interval, the switches in the other phases conduct, then

$$v_h(t) = 0 \tag{9.11}$$

$$v_{lu}(t) = \hat{V} \sin \omega t \tag{9.12}$$

If only one of the other switches is conducting, then (analogous to Eq. 9.7)

$$v_h(t) = -\frac{1}{2}\hat{V}\sin(\omega t \pm \frac{2\pi}{3}) \tag{9.13}$$

and thus

$$v_{lu}(t) = \hat{V}\sin\omega t + \frac{1}{2}\hat{V}\sin(\omega t \pm \frac{2\pi}{3}) \tag{9.14}$$

In the left part in Fig. 9.5, (c) shows the load voltage for phase U over a period and (d) illustrates the load current.

Control and load fundamental voltage can be derived from the previous results using Fourier analysis. If the switches are lossless, then the fundamental of the current is lagging $\pi/2$ with respect to the voltage and thus $\alpha + \beta = \pi$. Using $\alpha - \beta = \gamma$ it follows that $\alpha = (\pi + \gamma)/2$ and $\beta = (\pi - \gamma)/2$.

The fundamental of the control voltage is therefore

$$\hat{V}_{t1} = \frac{2}{\pi}\int_{-\gamma/2}^{\gamma/2} (\frac{3}{2}\hat{V}\cos\omega t)\cdot\cos\omega t\cdot d(\omega t) = \frac{3}{2}\hat{V}\frac{\gamma + \sin\gamma}{\pi} \tag{9.15}$$

The fundamental of the load voltage is therefore (in effective value):

$$V_{l1} = V - V_{t1} = V\left[1 - \frac{3}{2}\frac{\gamma + \sin\gamma}{\pi}\right] \tag{9.16}$$

The fundamentals of load voltage and control voltage are in phase with the grid voltage, while the fundamental of the load current is $\pi/2$ lagging. If we choose the real axis along the voltage, the phasor representation of the current fundamental is

$$\underline{I}_1 = \frac{V}{jX_l}\left[1 - \frac{3}{2}\frac{\gamma + \sin\gamma}{\pi}\right] = \frac{V}{jX_l}\left[\frac{\pi - \frac{3}{2}\gamma - \frac{3}{2}\sin\gamma}{\pi}\right] \tag{9.17}$$

As observed from the grid, this is equivalent to an additional inductance in series with the load inductance

$$X_t = X_l\left[\frac{\gamma + \sin\gamma}{\frac{2}{3}\pi - \gamma - \sin\gamma}\right] \tag{9.18}$$

Operation with $\gamma > \pi/3$

For $\gamma > \pi/3$, the hold-off intervals overlap, in which case no switches conduct. For each phase there are then two current pulses, each with a duration of $2\pi/3 - \gamma$ (see (a) in the right part of Fig. 9.5). The load voltage is then half the line voltage (see (c) in the same figure).

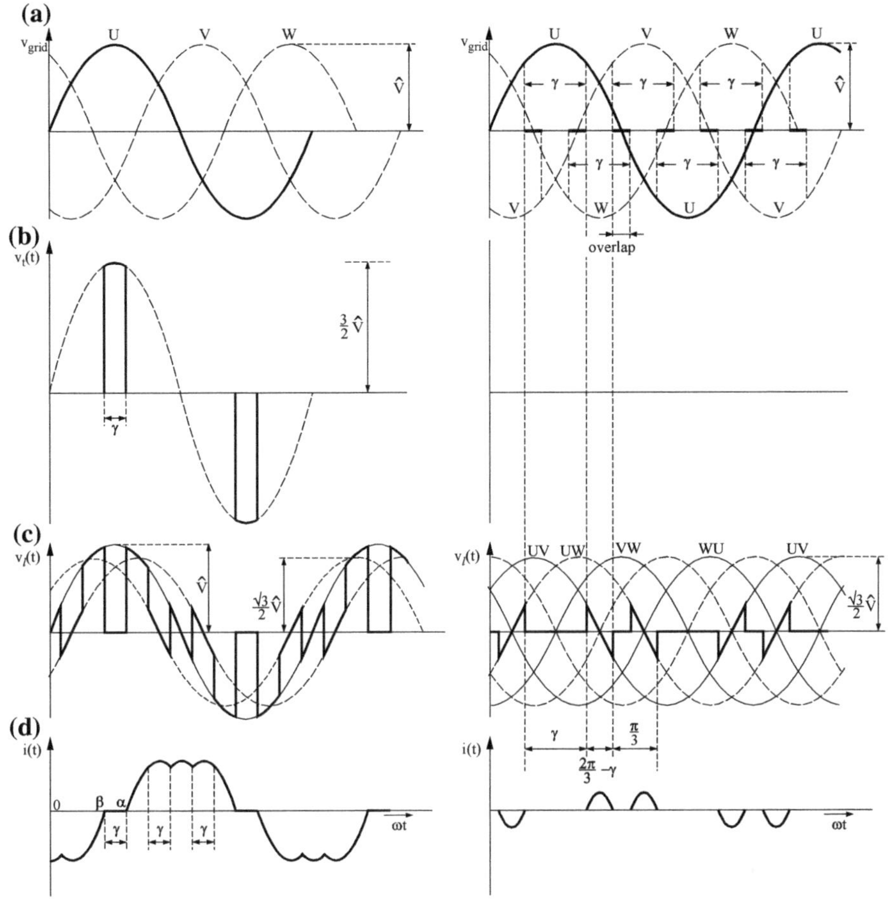

Fig. 9.5 Waveforms for phase control of a three-phase inductance load

The fundamental load voltage is:

$$\hat{V}_{t1} = \frac{2}{\pi} \int_{\gamma/2-\pi/6}^{\pi/2-\gamma/2} \left[\frac{\sqrt{3}}{2} \hat{V} \sin(\omega t - \frac{\pi}{6}) \right] \cdot \sin \omega t \cdot d(\omega t)$$

$$+ \frac{2}{\pi} \int_{\gamma/2+\pi/2}^{7\pi/6-\gamma/2} \left[\frac{\sqrt{3}}{2} \hat{V} \sin(\omega t + \frac{\pi}{6}) \right] \cdot \sin \omega t \cdot d(\omega t) \tag{9.19}$$

or, after rearrangement of the integrals:

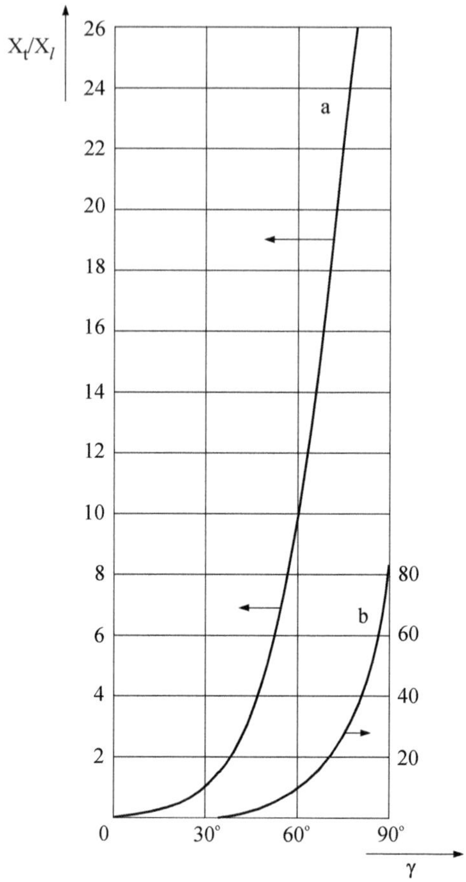

Fig. 9.6 X_t/X_l as a function of γ for a three-phase inductance load

$$\hat{V}_{t1} = \frac{4}{\pi}\sqrt{3}\int_0^{\pi/3-\gamma/2}\left[\frac{\sqrt{3}}{2}\hat{V}\sin\omega t\right]\cdot\sin\omega t\cdot d(\omega t) = \frac{3}{2}\hat{V}\frac{(\frac{2\pi}{3}-\gamma)-\sin(\frac{2\pi}{3}-\gamma)}{\pi} \tag{9.20}$$

Therefore the equivalent additional reactance is

$$X_t = X_l\left[\frac{\gamma+\sin(\frac{2\pi}{3}-\gamma)}{(\frac{2\pi}{3}-\gamma)-\sin(\frac{2\pi}{3}-\gamma)}\right] \tag{9.21}$$

The evolution of X_t/X_l as a function of γ is illustrated in Fig. 9.6.

9.4 Phase Control of a General Load

An analytic modelling of phase control for a general load is not straightforward. In many cases, however, the load can be approximated by a reactance in series with an inductance, as is the case in an induction machine (see Sect. 13.4.3).

Chapter 10
Cycloconverter

Abstract Variable speed operation of rotating field machines requires a variable frequency supply. Nowadays, a variable frequency supply is usually realised by means of an inverter (see Chap. 11), which accomplishes this starting from a DC source (i.e. a DC battery or the rectified grid voltage). The cycloconverter offers another solution, however limited to rather low output frequencies. Nowadays, the cycloconverter is (or was) mainly used for high output power (and low output frequencies). In this chapter we review the operating principle, provide some examples of practical three-phase circuits and discuss the main control methods. Important aspects are the output voltage harmonic content and the input current harmonic content and reactive power requirements.

10.1 Introduction

In contrast to an inverter, a cycloconverter starts directly from the AC grid voltage and creates a variable frequency output voltage by combining suitable segments of the input grid voltage. As will be shown below, it does this through the principle of the controlled rectifier (with thyristors as switches). A consequence is that the attainable output frequencies are limited to a fraction of the input frequency. In addition, much reactive power is required (as is also the case with the controlled rectifier).

For many years, another kind of direct converter has been widely researched, the matrix converter. In short, this converter can be envisaged as a matrix of nine bidirectional switches (for three-phase input and three-phase output), where every switch may connect each input line to each output line. The switches can be turned on and off (e.g. Mosfets or IGBTs) and the matrix converter is not subjected to the same restrictions as the cycloconverter. Despite these advantages, industrial applications of the matrix converter remain rare.

J. A. Melkebeek, *Electrical Machines and Drives*, Power Systems,
https://doi.org/10.1007/978-3-319-72730-1_10

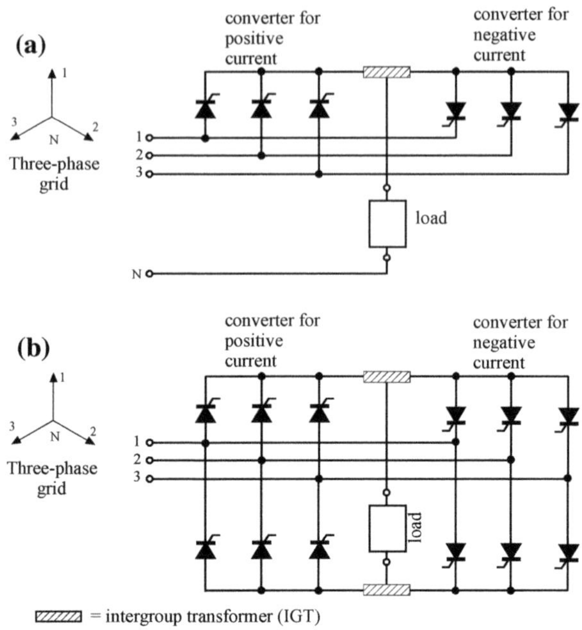

Fig. 10.1 Basic cycloconverter circuits

10.2 Operating Principle

As explained in Chap. 7, a controlled rectifier can convert the AC voltage into a variable positive or negative output DC voltage, $V_{d\alpha} = V_{do} \cdot \cos\alpha$. A reversal of the current direction can also be realised by means of an anti-parallel or cross-connection. In this way, a four-quadrant operation in the output $V - I$ plane is obtained.

Consider the rectifier circuits in Fig. 10.1. Suppose that the delay angle is controlled as $\alpha = \omega_o t + k\pi$ (with $\omega_o \ll \omega_i$ where ω_i is the input or grid angular frequency). The *average* output voltage then becomes $V_{d\alpha} = V_{do} \cdot \cos\omega_o t$. It can be proven that this is also generally the fundamental harmonic of the output voltage, with a few rare exceptions.

If the delay angle is controlled as $\cos\alpha = r \cdot \cos\omega_o t$ (with $0 < r \leq 1$), then the amplitude of the output can be varied as well.

In nearly all practical cycloconverters, the output is also three-phase (or multi-phase) to avoid DC and zero-sequence components in the input current. Figure 10.2 illustrates the input and output voltages for a three-phase input and a three-phase output; the delay angles for the three output phases are of course shifted by $2\pi/3$ with respect to each other. It is clear from the figure that the output voltages are composed of appropriate segments of the three input voltages. The output voltage, however, contains many higher harmonics in addition to the fundamental. This limits the ratio of the output to the input frequency ω_o/ω_i (e.g. 30 to 60%). The input current will contain many harmonics as well.

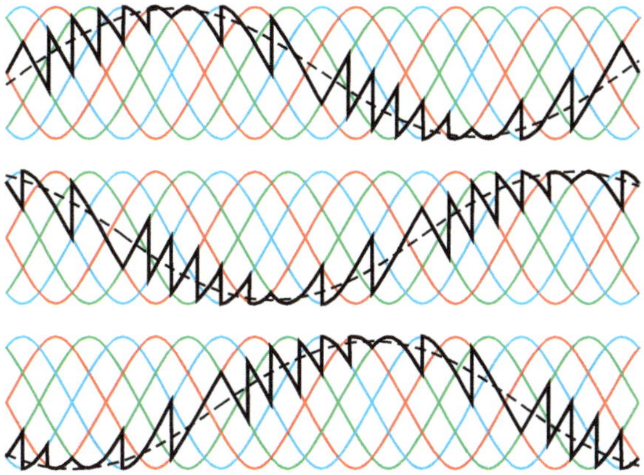

Fig. 10.2 Input and output voltages for three-phase to three-phase bridge cycloconverter

Note that the *average delay angle* over one half period of one of the output phase voltages is much larger than 0. Obviously, such a cycloconverter will need much reactive power from the grid (as will be expounded in detail later on).

10.3 Examples of Some Practical Cycloconverter Circuits

Cycloconverters may be based on either wye or bridge circuits. In general, wye circuits (e.g. Fig. 10.3) will result in much larger over-dimensioning of (the secondary of) the transformer, as can be seen in the tables below the illustrations. On the other hand, a three-phase bridge rectifier will require either an isolated load or three separate secondaries to avoid short circuits between the input phases (see Figs. 10.4 and 10.5). The disadvantage of a non-isolated three-phase load is the large over-dimensioning of the secondaries (see Figs. 10.4, 10.5 and 10.6).

It is intuitively clear that the higher the pulse number of the basic rectifier circuit, the better the output waveform and the better the input current waveform. A six-pulse wye rectifier (Fig. 7.25) will, however, require interphase transformers (IPT) in order to have each switch conducting during $2\pi/3$ (not $\pi/3$). Twelve-pulse circuits are quite common for high-power applications (see Figs. 10.7 and 10.8).

In all circuits, the intergroup transformers (reactors) may be omitted if the short-circuit currents of the anti-parallel rectifier circuits are mitigated (as explained in Chap. 7 and further on in Sect. 10.5.2).

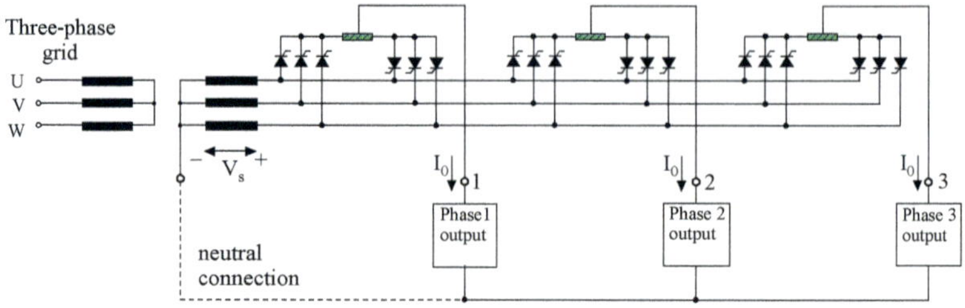

sinusoidal modulation		output power factor $\cos\varphi_o=1$			
		$\cos\phi$ input	apparent power input	apparent power primary trafo	apparent power secondary trafo
single-phase output	r = 1.0	0.843	$1.91V_{wmax}I_0$	$1.91V_{wmax}I_0$	$2.42V_{wmax}I_0$
	r = 0.1	0.078	$1.91V_{wmax}I_0$	$1.91V_{wmax}I_0$	$2.42V_{wmax}I_0$
three-phase output	r = 1.0	0.843	$1.32(3V_{wmax}I_0)$	$1.32(3V_{wmax}I_0)$	$1.32(3V_{wmax}I_0)$
	r = 0.1	0.078	$1.32(3V_{wmax}I_0)$	$1.32(3V_{wmax}I_0)$	$1.32(3V_{wmax}I_0)$

V_{wmax} = maximum rms output voltage
I_0 = rms output current
V_s = rms wye voltage secondary trafo
(hatched) = intergroup transformer or IGT (not always required)

Fig. 10.3 Three-phase wye cycloconverter

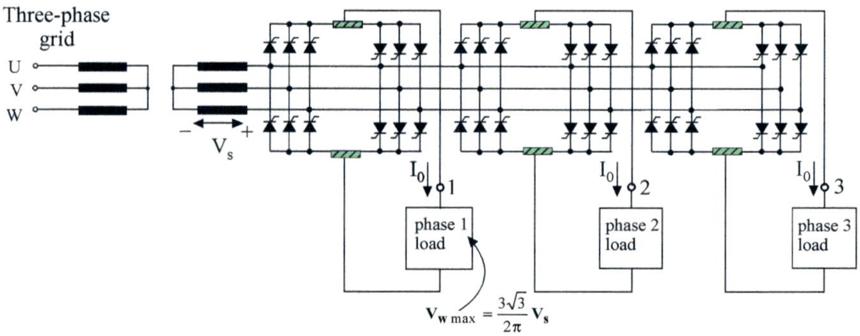

sinusoidal modulation		Output power factor $\cos\varphi_o=1$			
		$\cos\phi$ input	apparent power input	apparent power primary trafo	apparent power secondary trafo
single-phase output	r = 1.0	0.843	$1.48V_{wmax}I_0$	$1.48V_{wmax}I_0$	$1.48V_{wmax}I_0$
	r = 0.1	0.078	$1.48V_{wmax}I_0$	$1.48V_{wmax}I_0$	$1.48V_{wmax}I_0$
three-phase output	r = 1.0	0.843	$1.21(3V_{wmax}I_0)$	$1.21(3V_{wmax}I_0)$	$1.21(3V_{wmax}I_0)$
	r = 0.1	0.078	$1.32(r3V_{wmax}I_0)$	$1.32(3V_{wmax}I_0)$	$1.32(3V_{wmax}I_0)$

V_{wmax} = maximum rms output wye voltage
I_0 = rms output current
V_s = rms wye voltage secondary trafo
(hatched) = IGT (not always required)

Fig. 10.4 Three-phase delta cycloconverter with isolated load

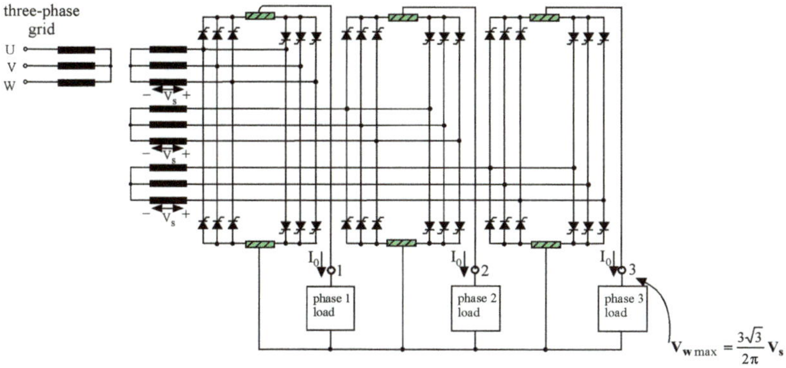

sinusoidal modulation		output power factor $\cos\varphi_o = 1$			
		$\cos\phi$ input	apparent power input	apparent power primary trafo	apparent power secondary trafo
single-phase output	r = 1.0	0.843	$1.48V_{wmax}I_0$	$1.48V_{wmax}I_0$	$1.48V_{wmax}I_0$
	r = 0.1	0.078	$1.48V_{wmax}I_0$	$1.48V_{wmax}I_0$	$1.48V_{wmax}I_0$
three-phase output	r = 1.0	0.843	$1.21(3V_{wmax}I_0)$	$1.21(3V_{wmax}I_0)$	$1.48(3V_{wmax}I_0)$
	r = 0.1	0,078	$1.32(3V_{wmax}I_0)$	$1.32(3V_{wmax}I_0)$	$1.48(3V_{wmax}I_0)$

V_{wmax} = maximum rms output wye voltage
I_0 = rms output current
V_s = rms wye voltage secondary transformer
= IGT (not always required)

Fig. 10.5 Three-phase delta cycloconverter with non-isolated load

10.4 Control Methods

As explained above, the principle of a cycloconverter is that appropriate segments of the (sinusoidal) input voltage are switched to the output so as to obtain a fair approximation of a sinusoidal output voltage, with the desired frequency and amplitude. The choice of these segments is determined by the variable delay angle of the converters. However, there are many different algorithms to obtain the proper delay angles. Each of these methods has its own advantages and disadvantages (in terms of e.g. harmonics, reactive power) as will be discussed below.

10.4.1 Sinusoidal Modulation (Open Loop)

Undoubtedly, the sinusoidal modulation is the most natural method to obtain the delay angles. As explained above, for the positive converter the switching angles are

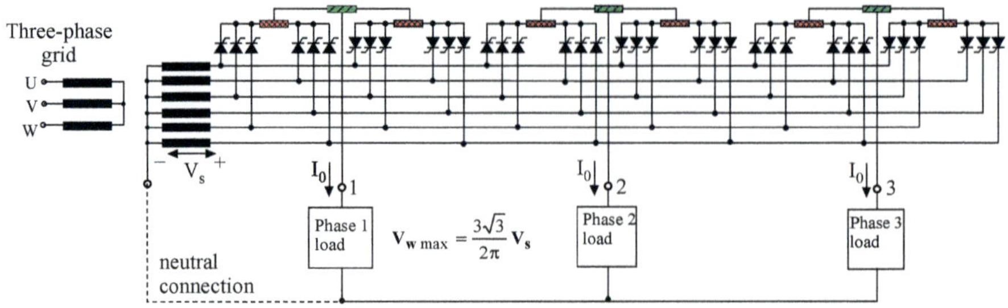

sinusoidal modulation		Output power factor $\cos\varphi_o=1$			
		$\cos\phi$ input	apparent power input	apparent power trafo, primary	apparent power trafo, secondary
single-phase output	r = 1.0	0.843	$1.48V_{wmax}I_0$	$1.48V_{wmax}I_0$	$2.09V_{wmax}I_0$
	r = 0.1	0.078	$1.48V_{wmax}I_0$	$1.48V_{wmax}I_0$	$2.09V_{wmax}I_0$
three-phase output	r = 1.0	0.843	$1.21(3V_{wmax}I_0)$	$1.21(3V_{wmax}I_0)$	$1.32(3V_{wmax}I_0)$
	r = 0.1	0.078	$1.32(3V_{wmax}I_0)$	$1.32(3V_{wmax}I_0)$	$1.32(3V_{wmax}I_0)$

V_{wmax} = maximum rms output voltage
I_0 = rms output current
V_s = rms wye-voltage secondary trafo
(cross-hatched symbol) = Interphase transformer (IPT)
(hatched symbol) = IGT (not always required)

Fig. 10.6 Six-phase wye cycloconverter

obtained by comparing the $\cos\alpha$ curves (called ignition waves) with the sinusoidal reference waves $r \cdot \cos\omega_o t$. For the negative converter, the $\cos\alpha'$ waves must be compared with the reference wave $-r \cdot \cos\omega_o t$. This process can be somewhat simplified by comparing $\cos\alpha$ and $-\cos\alpha'$ with the same reference wave $r \cdot \cos\omega_o t$, as is illustrated in Fig. 10.9. The switching instants are therefore given by the intersections of the waves $\cos\alpha = \cos(\omega_i t - j\frac{2\pi}{p})$ and the reference waves $r \cdot \cos(\omega_o t - k\frac{2\pi}{n} - \xi)$, with $j = (0 \ldots p-1)$ the input phase number, $k = (0 \ldots n-1)$ the output phase number and ξ the desired phase angle of the output reference phase (i.e. the phase $k = 0$).

To demonstrate that the sinusoidal modulation is the most natural modulation method, consider Fig. 10.10. In this figure, both the ignition waves and the reference waves are referred to the same value as the intersection point of two subsequent sine curves of the input, i.e. $\hat{V} \cdot \sqrt{3}/2$. A closer look at the ignition waves shows that these ignition waves are equal to the average of two subsequent input sine waves. As a result, the crossing of the reference wave and the ignition wave, and thus the switching to the next thyristor, occurs in such a way that the distance between the *sinusoidal* reference wave and the input sine waves always remains minimal. Moreover, it can be proven that the fundamental of the resulting output voltage is indeed equal[1] to the (sinusoidal) reference wave.

[1] Exception: a difference of a few percent for some discrete values of the input to output frequency.

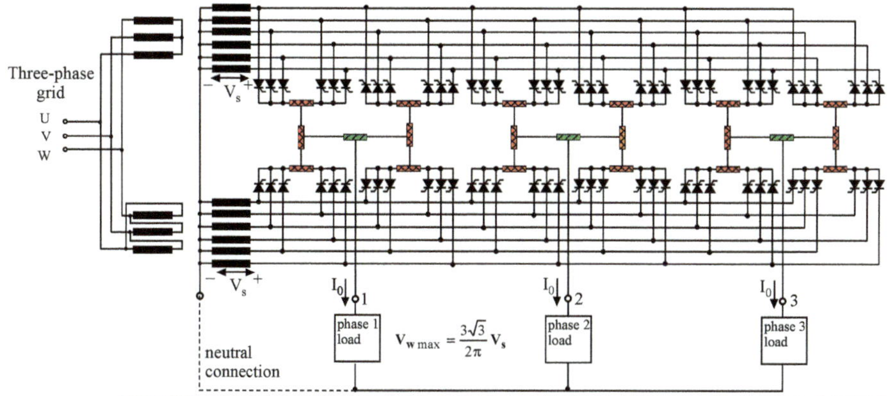

sinusoidal modulation		output power factor $\cos\varphi_o = 1$			
		$\cos\phi$ output	apparent power input	apparent power primary trafo	apparent power secondary trafo
single-phase output	r = 1.0	0.843	$1.48V_{wmax}I_0$	$1.48V_{wmax}I_0$	$2.09V_{wmax}I_0$
	r = 0.1	0.078	$1.48V_{wmax}I_0$	$1.48V_{wmax}I_0$	$2.09V_{wmax}I_0$
three-phase output	r = 1.0	0.843	$1.19(3V_{wmax}I_0)$	$1.21(3V_{wmax}I_0)$	$1.32(3V_{wmax}I_0)$
	r = 0.1	0.078	$1.29(3V_{wmax}I_0)$	$1.32(3V_{wmax}I_0)$	$1.32(3V_{wmax}I_0)$

V_{wmax} = maximum rms output wye voltage
I_0 = rms output current
V_s = rms wye voltage secondary transformer
= IPT
= IGT (not always required)

Fig. 10.7 Twelve-phase wye cycloconverter

Although sinusoidal modulation seems the most obvious modulation principle, there are some disadvantages. A first disadvantage is the bad input power factor as the *average* delay angle in half a period of the input is much larger than 0 (even more so as the reduction factor is smaller). A second disadvantage is that the output voltage contains a lot of harmonics, including sub-harmonics.[2]

10.4.2 Trapezoidal Modulation (Open Loop)

To address the poor power factor of sinusoidal modulation, trapezoidal modulation may offer some opportunities. With a trapezoidal reference wave, the delay angles remain small during a much longer interval. It is intuitively clear that the *average* power factor will be improved.

---

[2]Although these are, strictly speaking, not real harmonics as their frequencies are not pure multiples of the input frequency but combinations of input and output frequency.

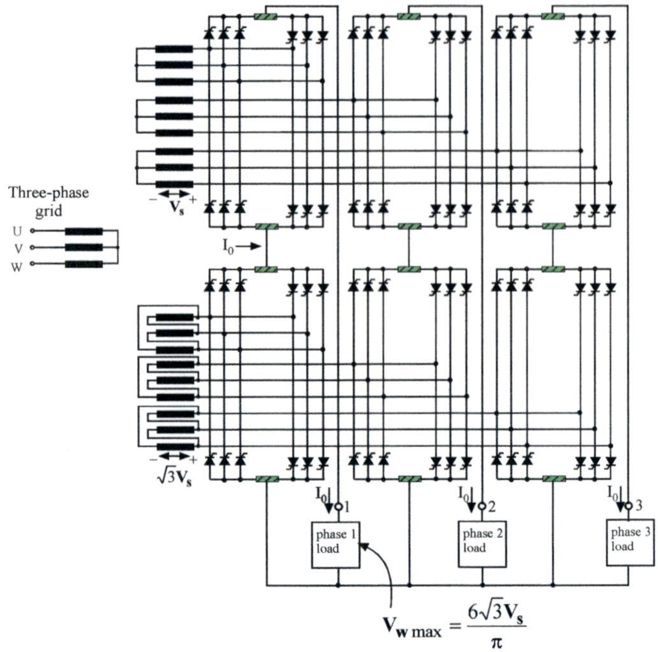

sinusoidal modulation		output power factor $\cos\varphi_o = 1$			
		$\cos\phi$ input	apparent power input	apparent power primary trafo	apparent power secondary trafo
single-phase output	r = 1.0	0.843	$1.43V_{wmax}I_0$	$1.43V_{wmax}I_0$	$1.48V_{wmax}I_0$
	r = 0.1	0.078	$1.43V_{wmax}I_0$	$1.43V_{wmax}I_0$	$1.48V_{wmax}I_0$
three-phase output	r = 1.0	0.843	$1.19(3V_{wmax}I_0)$	$1.19(3V_{wmax}I_0)$	$1.48(3V_{wmax}I_0)$
	r = 0.1	0.078	$1.29(3V_{wmax}I_0)$	$1.29(3V_{wmax}I_0)$	$1.48(3V_{wmax}I_0)$

V_{wmax} = maximum rms output voltage
I_0 = rms output current
V_s = rms wye voltage secondar transformer
(hatched box) = IGT (not always required)

Fig. 10.8 Twelve-phase delta cycloconverter

However, the output voltage will contain much larger harmonics of lower order (e.g. 3, 5, 7,...).[3] Figure 10.11 compares the output waveforms for voltage and current of sinusoidal and trapezoidal modulation. It is clear that the output current is much more distorted for the trapezoidal modulation.

Note that this figure also is interesting as it shows which converter operates at each instant, as well as the rectifying or regenerating operation.

[3]For three-phase systems triple harmonics do not harm; then a trapezoidal waveform with limited fifth harmonic will be chosen.

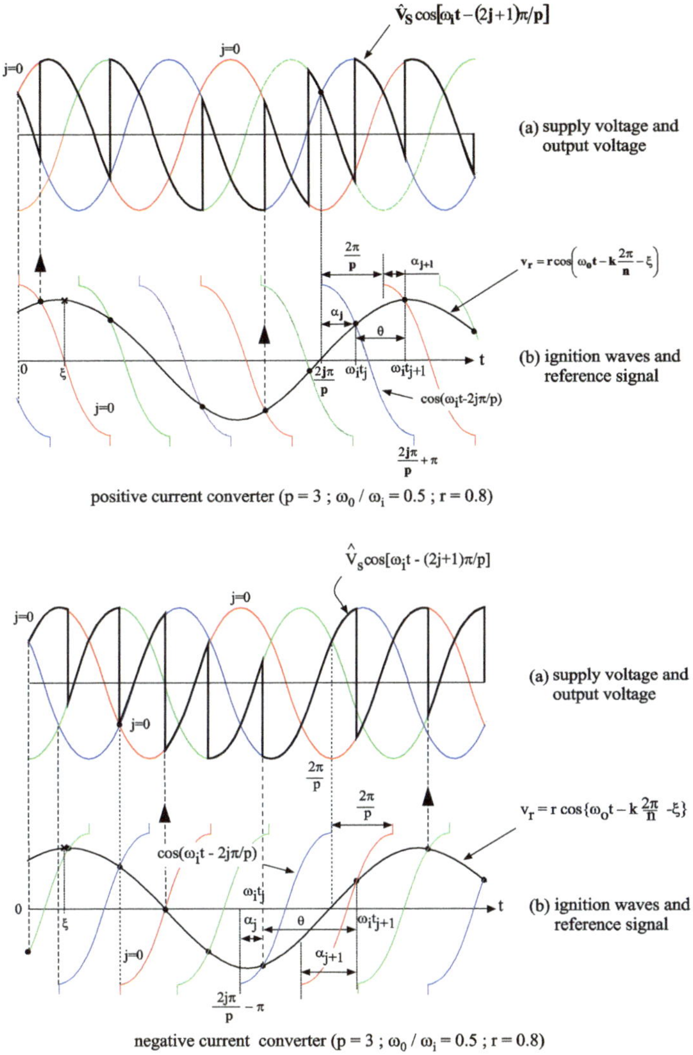

Fig. 10.9 Sinusoidal modulation

10.4.3 Closed-Loop Control

To address the harmonics in the output, closed loop control with suppression of disturbing harmonics is sometimes used. Some harmonics are, however, inherent to the basic cycloconverter operating principle, and trying to eliminate these may lead to instability.

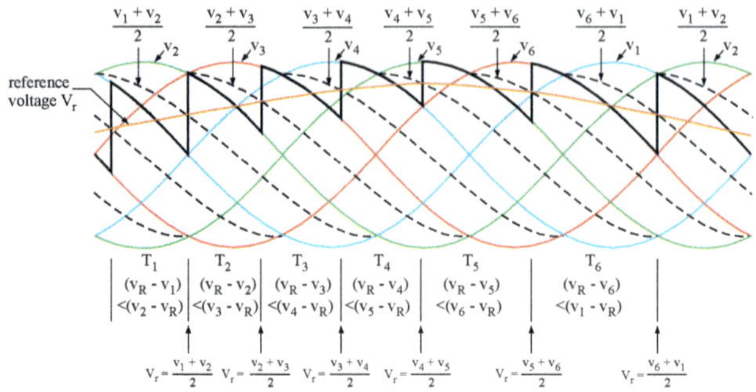

Fig. 10.10 Sinusoidal modulation as most natural

Fig. 10.11 Trapezoidal modulation versus sinusoidal modulation

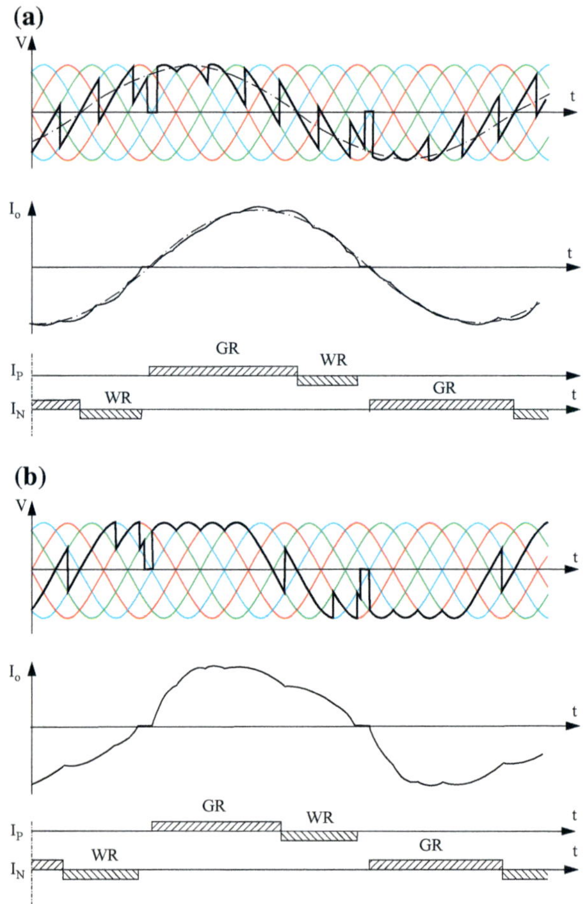

10.5 Cycloconverter Circuits with or Without Circulating Current

10.5.1 Cycloconverters with Free Circulating Current

Although the average output voltages of the anti-parallel converters are equal, the instantaneous voltages differ and unidirectional circulating currents will flow (see also Sect. 7.3 in Chap. 7). Like the rectifier with free circulating current, these currents are to be limited by inserting inductances between the anti-parallel converters.

In an anti-parallel rectifier, the circulating currents are intermittent (pulsating) for small delay angles.

However, in a cycloconverter the sinusoidal output current results in a second type of circulating current, superposed on the pulsating circulating current components. This phenomenon, called the *self-induced circulating current*, can be explained as follows. Figure 10.12 schematically represents the two anti-parallel switches of an input phase. The two switches are connected by the intergroup transformer (IGT) to limit the short-circuit currents. Suppose that for $t \leq 0$ (only) the positive converter switch is conducting. The load current can be assumed to be purely sinusoidal (due to the filtering of the load). At $t = 0$, the load current reaches a maximum and the mmf of the transformer at that instant is $w \cdot i_p = w \cdot \hat{I}_o$. From $t = 0$ on, the load current starts decreasing. From $t = 0^+$ on, di/dt becomes negative and induces a voltage in the secondary of the IGT, so that the switch of the negative converter becomes positively polarised and starts conducting. From then on, both switches conduct and both sides of the IGT are connected to the same voltage (disregarding the instantaneous voltage differences of the anti-parallel rectifiers). Thus, the voltage difference over the IGT is zero and therefore the mmf of the IGT is constant: $w \cdot i_p + w \cdot i_n = w \cdot \hat{I}_o$. Moreover, as $i_p - i_n = \hat{I}_o \cdot \sin \omega_o t$, we obtain $i_p = \hat{I}_o \cdot (1 + \sin \omega_o t)/2$ and $i_n = \hat{I}_o \cdot (1 - \sin \omega_o t)/2$. This implies that from that instant on, both switches are continuously conducting and there is a continuous circulating current. Except for one discrete instant in the period of the load current, there is never more than an instantaneous tendency for the switches to block.

Figure 10.13 illustrates the instantaneous voltages of the positive and negative converters, their average and difference, and the resulting ripple current superposed on the continuous circulating current.

The most important advantage of the free circulating current is that the output load voltage is much smoother[4] than the separate voltages (see (c) in Fig. 10.13). A second advantage is that a detection of the current direction is not required, as the appropriate converter will take over the current automatically. The most important disadvantage is that the thermal loading of the switches is much higher than in case the circulating current is suppressed.

---

[4]Because it is the average of the voltages of the positive and negative converters.

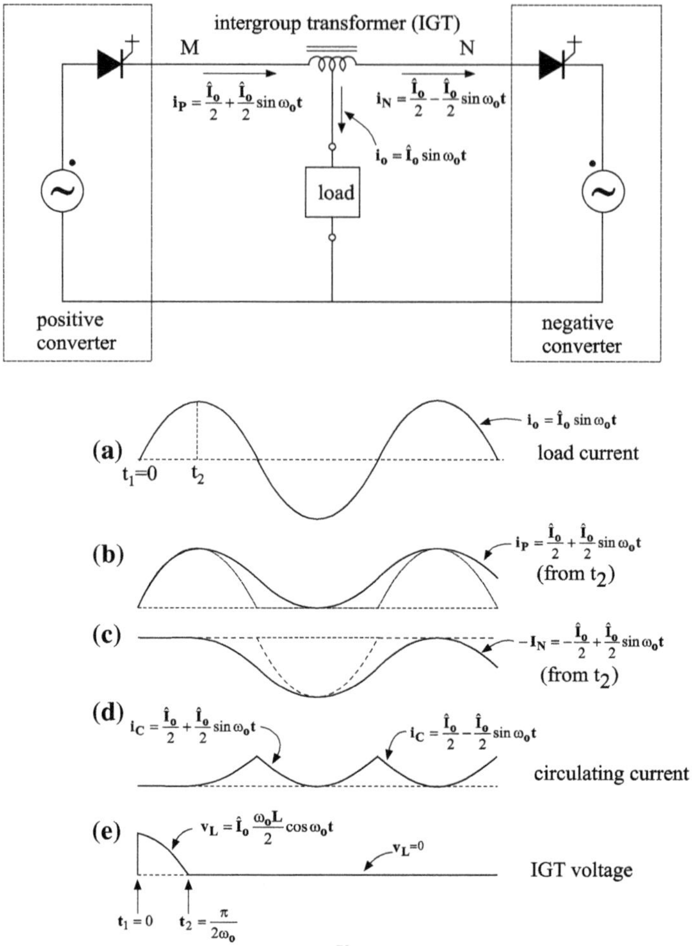

Fig. 10.12 Simplified circuit for circulating current

10.5.2 Cycloconverters Without Circulating Current

To avoid the additional thermal loading of the switches and the corresponding Joule losses, often one of the converters is blocked, depending on the current direction. However, in addition to the added complexity of sensors and logical circuits for detecting the current direction, this operation without circulating current brings along some other disadvantages as well. Because the output voltage is not smoothed as the average of positive and negative converter, its harmonic content is much higher than in the case of a free circulating current. The output load voltage and thus its harmonic content in particular now also depend on the power factor of the load.

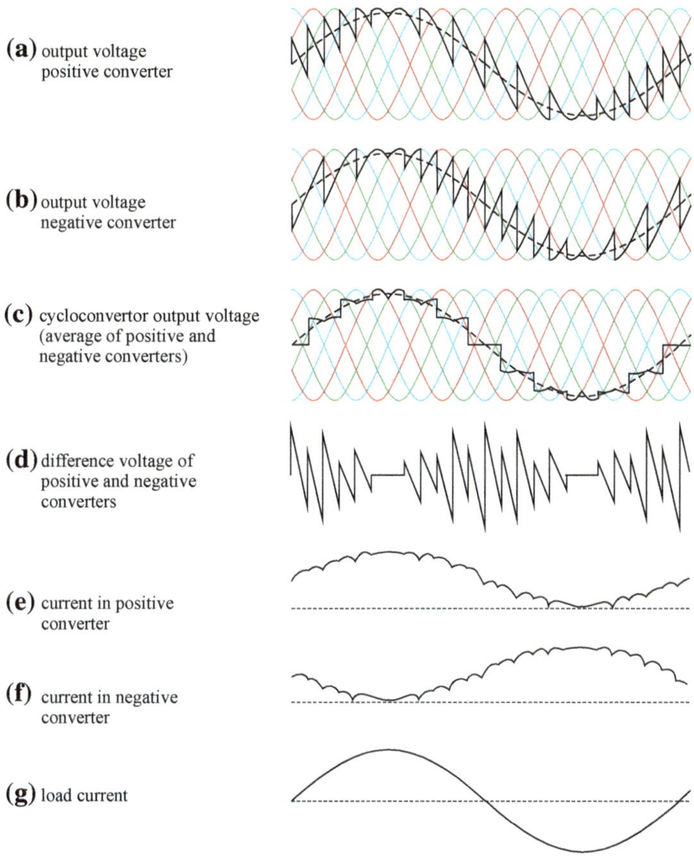

Fig. 10.13 Voltages and currents with circulating current

The delay (and/or uncertainty as to the average current direction) when the converters switch may also lead to discontinuous current and thus additional distortion of the output voltage, especially for low resistive and capacitive loads as illustrated in Figs. 10.14 and 10.15.

To mitigate some of these problems, modified circulating-current limiting algorithms can be applied:

1. time-limited circulating current: it is often difficult to determine the exact instant of reversal of the load current as the current is not smooth (due to the harmonics). Therefore, both anti-parallel converters receive a gate signal during an interval around the *estimated* reversal instant. The start of this interval can be estimated in several ways: the interval is assumed to start when a first zero-crossing of the current is detected or when the average current descends below a threshold value, or it can also be deduced from the average currents in the two other phases, for example.

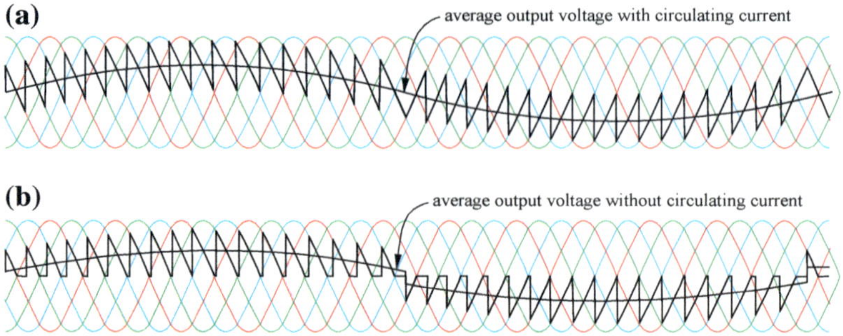

Fig. 10.14 Distortion due to discontinuous current, R-load

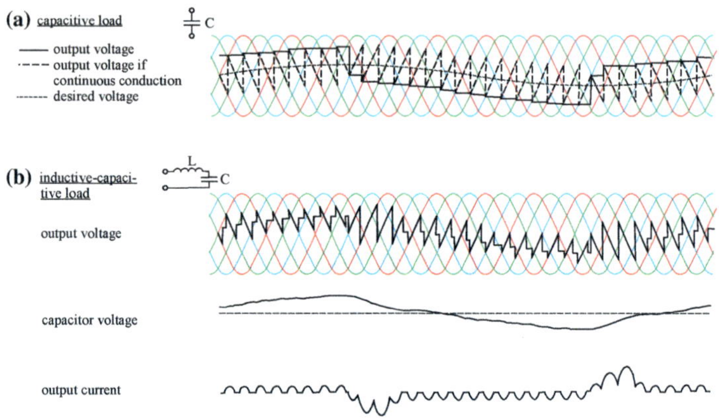

Fig. 10.15 Distortion due to discontinuous current, C-load

2. amplitude-limited circulating current: both converters continuously receive a gate signal, but only the converter whose current direction corresponds to the actual average load current direction receives the normal gate signal; the other converter receives a modified current-limiting gate signal so that the circulating current is limited. When the average current direction changes, these gate signals are reversed from one converter to the other.
3. closed-loop control of the voltage: this forces the converters to follow the desired value.

10.6 Output Voltage Harmonic Content

The output voltage of a cycloconverter is not at all sinusoidal. Its harmonic content mainly depends on the pulse number of the converters, the ratio of output to input frequency, the reduction factor r of the output voltage, the control principle, free circulating current or suppressed circulating current and the power factor of the load in case of operation without circulating current.

The first four factors determine the output of the separate positive and negative converters. The last two determine which of these harmonics will appear in the output voltage.

First, we discuss the harmonic content of the separate converters. We will limit ourselves to sinusoidal modulation. In addition to the desired fundamental,

$$s \cdot \frac{\sin \pi/m}{\pi/m} \cdot \hat{V}_i \cdot r \cdot \sin \omega_o t \tag{10.1}$$

the output voltages v_p and v_n of the positive and negative converters contain the following families of harmonic frequencies (for three-phase systems):

$$f_{h1} = |3(2k - 1) \cdot f_i \pm 2l \cdot f_o| \tag{10.2}$$

$$f_{h2} = |6k \cdot f_i \pm (2l + 1) \cdot f_o| \tag{10.3}$$

with $1 \leq k \leq \infty$ and $0 \leq l \leq \infty$. The pulse number may restrict the whole numbers k and l. For a pulse number of six, only the harmonics f_{h2} exist; for a pulse number of twelve, only those in f_{h2} with k even.

Which harmonics occur in the output voltage depends on whether a circulating current is present and, if no circulating current is present, on the load power factor. In case of a free circulating current, the output voltage is the average of the positive and negative converter voltages: $v_o = (v_p + v_n)/2$. As has been illustrated in Fig. 10.13, the output voltage is much smoother. Indeed, many harmonics are cancelled out. Of the harmonics in Eqs. 10.2 and 10.3, only a limited number remain in the output, as now the inequalities $2l \leq 3(2k - 1 + 1$ and $2l + 1 \leq 6k + 1$ apply to the families f_{h1} and f_{h2}, respectively.

For a cycloconverter without circulating current, the harmonics are not cancelled out. Moreover, the output voltage depends on the power factor of the load: $v_o = v_p \cdot u_p + v_n \cdot u_n$ where $u_p = 1$ when the positive converter conducts, i.e. for $\varphi_o < \omega_o t < \varphi_o + \pi$ and zero elsewhere; $u_n = 1$ when $\varphi_o + \pi < \omega_o t < \varphi_o + 2\pi$ and zero elsewhere. The harmonics in the output thus depend on the power factor of the load.

The effect of the circulating current on the harmonics is illustrated graphically in Figs. 10.16 and 10.17. The y-axis shows harmonic frequencies while the x-axis represents the fundamental output frequency. The ordinates of the lines starting at the y-axis represent the harmonic frequencies as a function of the fundamental output frequency (x-axis). Without circulating current each vertical line starting from a point on the x-axis (the fundamental frequency) intersects a two-fold infinite number of harmonic frequency lines. With circulating current, only a finite number of lines start at each point on the y-axis, and thus the number of harmonic frequencies is much reduced. Of course, the graph does not tell us anything about the amplitude of

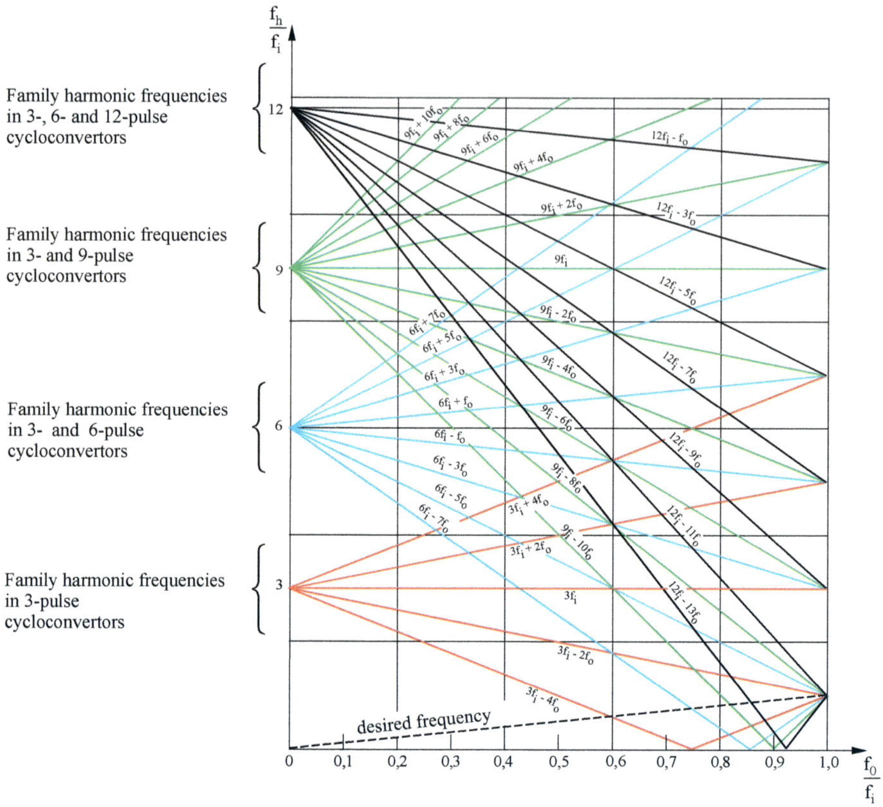

Fig. 10.16 Voltage harmonics - free circulating current

these harmonics. It is clear, however, that without circulating current, the potential of an important harmonic with a frequency below the fundamental frequency (cf. the dashed-line) is much greater.

To avoid important sub-harmonics (i.e. with an amplitude larger than 2.5% of the fundamental), the ratio of output to input frequencies will usually be limited to 33% for three-pulse converters, 50% for six-pulse converters, and 75% for twelve-pulse converters (for $r = 1$ and with free circulating current).

10.7 Input Current Power Factor and Harmonic Content

A cycloconverter does not contain energy storage elements, and thus at each instant input and output active power are equal. If the grid voltage can be considered as sinusoidal, then only the fundamental input current is responsible for active power. In addition to the fundamental, the output voltage also contains many harmonics.

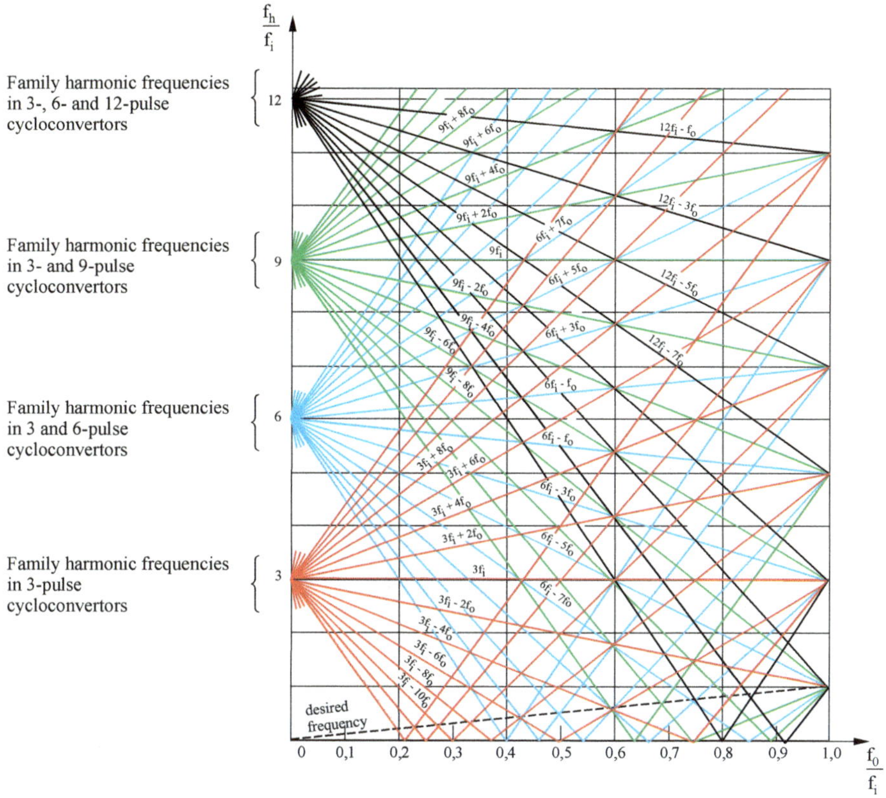

Fig. 10.17 Voltage harmonics - no circulating current

If the load is sufficiently inductive, the output current may be considered as almost sinusoidal (i.e. due to the filtering of the load). We may then write

$$p_o(t) \approx P_o = \frac{1}{2} n \hat{V}_o \hat{I}_o \cos \varphi_o \tag{10.4}$$

with n as the number of output phases (only cycloconverters with multiphase output are considered).

Therefore, the input active power, equal to P_o, is also constant. This permits us to derive the active component $\hat{I}_{iw}$ of the fundamental input current $\hat{I}_i$

$$\hat{I}_{iw} = n \cdot r \cdot \frac{s}{\pi} \cdot \sin \frac{\pi}{m} \cdot \hat{I}_o \cos \varphi_o \tag{10.5}$$

Next to this active (fundamental) component, the input current also contains a reactive component and higher harmonics.

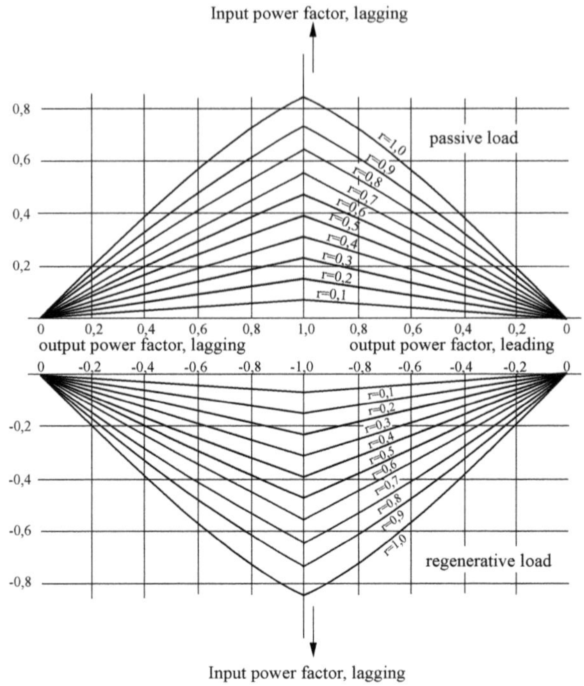

Fig. 10.18 Input power factor for a cycloconverter

The higher harmonics do not result in average active power as the supply (grid) is assumed to be purely sinusoidal. However, they result in pulsating power and therefore additional losses in the supply lines. The orders of these harmonics correspond to the harmonics in the output voltage, Eqs. 10.2 and 10.3. Important orders are the 'characteristic harmonics', $f_i \pm 6kf_o$ for a three-phase output ($f_i \pm 2kf_o$ for a single-phase output).

More important is the input reactive component and the input reactive power. As the operating principle of a cycloconverter is similar to that of a controlled rectifier, a cycloconverter requires much reactive power: indeed, the *average* delay angle over half a period of the input is much larger than 0 radians.

First, consider a cycloconverter with a purely resistive load. The output voltage is $\hat{V}_o$, output current is $\hat{I}_{o,R}$ and output active power is $P_o = \frac{1}{2}n \cdot \hat{V}_o\hat{I}_{o,R}$. As the output reactive power $Q_o = 0$, output apparent power $S_o = P_o$. The input active power is of course equal to the output active power, $P_i = P_o$, but because of the operating principle of a cycloconverter there is a non-negligible input reactive power $Q_{i,R}$: $S_{i,R} = \frac{1}{2}m \cdot \hat{V}_i\hat{I}_{i,R} > P_i = \frac{1}{2}m \cdot \hat{V}_i\hat{I}_{i,R} \cdot \cos\varphi_i$.

Next, we connect an inductance in parallel with the output resistance. As a result, the cycloconverter will have to provide the reactive power for this inductive loading, $Q_{o,L} > 0$. While the output active power remains the same, the output current and apparent power have of course increased compared to the case with a purely resistive load, $S_{o,L} = \frac{1}{2}n \cdot \hat{V}_o\hat{I}_{o,L} > P_o = \frac{1}{2}n \cdot \hat{V}_o\hat{I}_{o,R}$. At the input as well, the active power is still the same but the current amplitude has increased as the currents in a cycloconverter are directly transferred, $\hat{I}_{i,L} > \hat{I}_{i,R}$. The input reactive power has therefore increased, $Q_{i,L} > Q_{i,R}$.

If, instead of an inductance, we connect a capacitor in parallel with the output resistance, we arrive at the same conclusions. At the output the cycloconverter has to absorb the reactive power provided by the capacitor load. The output current amplitude and apparent power increase compared to the resistive case, $S_{o,C} = \frac{1}{2}n \cdot \hat{V}_o\hat{I}_{o,C} > P_o = \frac{1}{2}n \cdot \hat{V}_o\hat{I}_{o,R}$. As output and input currents are directly related by the cycloconverter, also the input current amplitude increases compared to the resistive case, i.e. the reactive current and reactive power drained from the grid have increased. It is important to note that in this case, reactive power is drawn from both sides.

Indeed, there is no such thing as conservation of reactive power. Reactive power is only a mathematical concept[5] to quantify the excess current required when the load is not purely resistive.

Figure 10.18 gives the input reactive power as a function of the output reactive power. Either inductive or capacitive power at the output results in the same input reactive power requirements. This is also independent of the direction of active power. Note also that the greater reactive power demand is, the smaller the output voltage will be, i.e. the smaller the reduction factor r (which can be explained by the larger *average* delay angle with smaller r).

[5]Only for linear time-invariant circuits is reactive power conserved.

Chapter 11
Inverter

Abstract Rectifiers convert AC into DC. With a controlled rectifier, energy can also be transferred from the DC-side to the AC-side. However, the AC source (i.e. the grid) always remains necessary as it is responsible for the commutation of the switches (using the emf of the source or, put differently, using the reactive power of the source, to switch off the thyristors). Controlled rectifiers can therefore not be used to convert DC into AC with variable frequency. Nevertheless, the only energy-efficient way to obtain variable speed operation of rotating field machines (i.e. induction and synchronous machines) is by feeding them from a variable frequency source. Inverters are able to convert DC into AC with variable frequency and, in most cases, also variable amplitude. Contrary to controlled rectifiers, inverters require switches that can be turned on **and off** at will at any instant. Nowadays, the switches used in inverters are mainly Mosfets (for lower power), IGBTs or IGCTs (for very high power).

11.1 Single-Phase Inverter

A first basic scheme for a single-phase inverter[1] is illustrated in (a) in Fig. 11.1 (for the time being, disregard the optional large inductance L_d). When the upper switch is "on" (conducting), the load gets the voltage +E; with the lower switch "on", the load sees the voltage −E, as is shown in the upper curve (A, a) in Fig. 11.2. By varying the "on" times of upper and lower switches (always equal to each other, to achieve symmetry), the frequency of the output can be varied at will. Note that, without the optional inductance L_d, the upper and lower switches may never be "on" at the same time.

However, this configuration requires a DC source with midpoint available. To avoid the DC source with midpoint, the configuration (b) in Fig. 11.1 can be used. In fact, the AC-side transformer now realises a load with a midpoint. Both configurations permit stepwise waveforms with variable frequency as (A, a) in Fig. 11.2 (for

[1] In most figures the symbol used for these switches is that of a GTO, but this may also represent any other switch.

J. A. Melkebeek, *Electrical Machines and Drives*, Power Systems,
https://doi.org/10.1007/978-3-319-72730-1_11

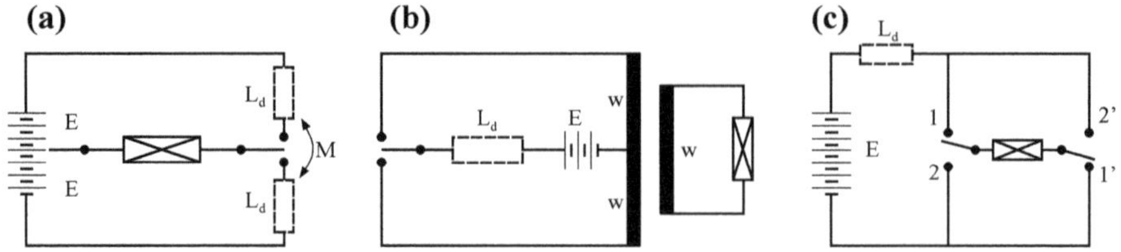

Fig. 11.1 Single-phase inverter circuits

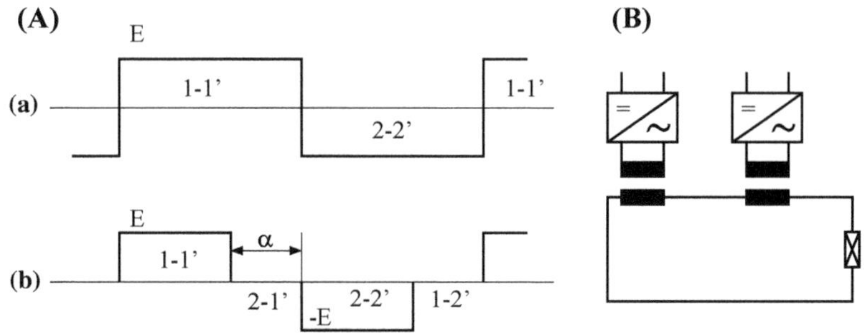

Fig. 11.2 **A** Voltage output for single phase inverter; **B** Combination of inverters

most (machine) applications only the fundamental is useful of course). However, the amplitude cannot be varied.

The configuration (c) in Fig. 11.1 requires four switches but does not require a source with midpoint. Moreover, connecting both left and right sides of the load to the same AC terminal (i.e. 1 and 2' on or 1' and 2 on) gives the load a zero voltage (see the lower waveform (A, b) in Fig. 11.2). In this way, the fundamental amplitude can be varied, together with the frequency. Another possibility is a series connection (using transformers) of two phase-delayed inverters (see (B) in Fig. 11.2).

However, a major disadvantage of the waveform (A, b) in Fig. 11.2 is that the higher harmonics are not always reduced in the same proportion, and thus one may end up with a relatively higher harmonic content (higher harmonics only lead to extra losses). Pulse width modulation (PWM) is a better option to obtain a variable fundamental amplitude, as is demonstrated in Sect. 11.3.

If a large inductance L_d is present, the inverter is called a current source inverter or CSI (in contrast to a voltage source inverter or VSI, i.e. without the large inductance). Although this is not a "real" current source, as the amplitude of the current still depends on the AC load, the inverter now switches a current with virtually constant amplitude to the AC side.

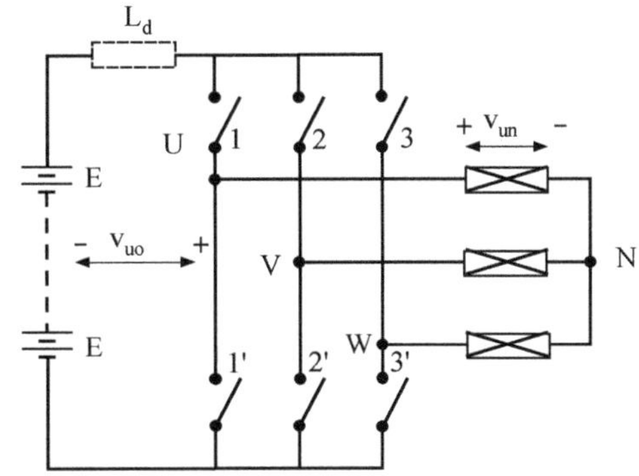

Fig. 11.3 Three-phase inverter scheme

Table 11.1 Switching sequence for 120° conduction

Interval	I	II	III	IV	V	VI
+side	1	1	2	2	3	3
−side	2'	3'	3'	1'	1'	2'

11.2 Three-Phase Six-Step Inverters

To obtain a three-phase inverter, three single-phase inverters could be combined. However, the three-phase inverter in Fig. 11.3 is a better option as it requires a smaller number of switches. Again, without the inductance L_d it is called a VSI and with a large inductance L_d it is referred to as a CSI.

There are several possible switching sequences to obtain a three-phase symmetrical output. However, two basic switching sequences are widely used because of the favourable output voltage or current, i.e. 120° and 180° switching.

11.2.1 The 120° *Switching Sequence*

In the 120° switching sequence, each switch remains closed (conducting) during 120°. Taking into account three-phase symmetry, we obtain the switching sequence of Table 11.1.

During each interval of 60°, two AC output lines are connected to the DC source while the third remains open. In case there is no inductance L_d, the connected output lines have the voltage (potential difference) of $+E$ or $-E$ with respect to the midpoint of the DC source, while the voltage of the third is zero (*if the load is symmetrical and passive*). For the voltages of the output lines, this yields the upper three curves in Fig. 11.4. These are also the phase voltages (v_{un}, v_{vn}, v_{wn}) of the load if this load is symmetrical and wye connected. The lower curve shows the line voltage v_{uw} (equal to the phase voltage for a delta-connected load).

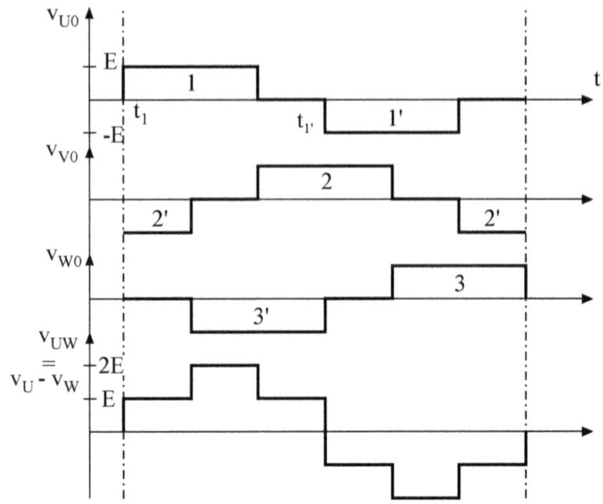

Fig. 11.4 Output voltages for 120° switching

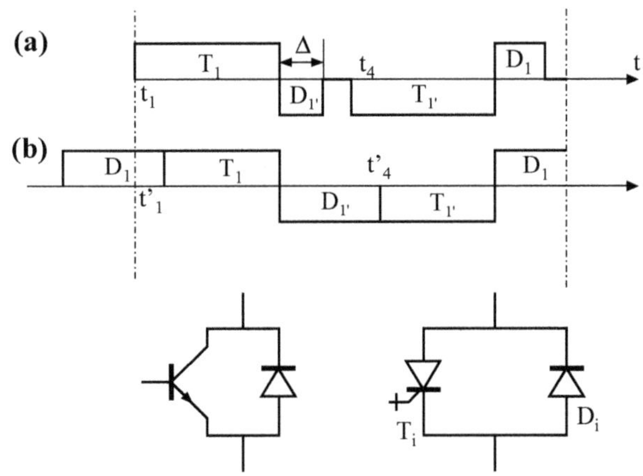

Fig. 11.5 Conduction length change with inductive load

However, if the load is not symmetrical, the voltage of the not-connected line is not zero and takes on a value dependent on the degree of asymmetry of the load. This implies that the 120° switching sequence is vulnerable to asymmetry of the load if used with a DC voltage source. Another disadvantage of the 120° sequence is that an inductive load will change the conduction length. With an inductive (or capacitive) load, it becomes necessary to provide anti-parallel diodes (Fig. 11.5) to allow the anti-parallel diode of the opposite switch to take over the current when a switch is turned off (e.g. the anti-parallel diode of switch 1' for turning off switch 1). However, this changes the conduction length, as illustrated in (a) in Fig. 11.5. In extreme cases the conduction length becomes 180° (see (b) in Fig. 11.5).

When the DC source is a current source (e.g. a DC source with a large inductance L_d), these problems do not occur as it is the current itself that is switched off (and

Table 11.2 180° switching table

Interval	VI	I	II	III	IV	V
+side	1 + 3	1	1 + 2	2	2 + 3	3
−side	2'	2' + 3'	3'	1' + 3'	1'	1' + 2'

the load voltage may lead or lag the current without any change to the conduction length). Further, anti-parallel diodes are not required (and even unwanted) in this case. Note that the commutation takes place between switches at the same DC side: for example, switch 2 takes over the current from switch 1, similar to a controlled rectifier.

The 120° sequence is therefore only used for current source inverters. *A clear advantage of this inverter scheme is that, at each instant, the current in each of the output lines is unambiguously related (i.e. +I, −I or 0) to the input current.*[2]

11.2.2 The **180°** *Switching Sequence*

In the 180° switching sequence, each switch remains closed (conducting) during 180°. Taking into account three-phase symmetry, we obtain the switching sequence of Table 11.2. Note that the commutation now occurs between upper and lower switches of an output line (e.g. from interval I to interval II from switch 3 to switch 3').

During each interval of 60°, all three AC output lines are connected to the DC source, i.e. two lines on one side and the third on the other side. If there is no inductance L_d, the connected output lines have the voltage (potential difference) of $+E$ or $-E$ with respect to the midpoint of the DC source (see the upper three curves (v_{uo}, v_{vo}, v_{wo}) in Fig. 11.6). *With a DC voltage source the voltages of all output lines are therefore uniquely determined at each instant.* For a delta-connected load, we obtain phase voltages like curve d for V_{uv} in the figure.

For a wye-connected load, the phase voltages of the load (v_{un}, v_{vn}, v_{wn}) differ somewhat from the voltages v_{uo}, v_{vo}, v_{wo} because of the potential difference of the star point with respect to the midpoint of the DC source. Indeed, for a symmetrical load and without neutral current (the neutral is almost never connected), the voltage between the neutral of the load and the midpoint of the source is given by curve e in Fig. 11.6. This voltage only contains the zero-sequence components (i.e. the third harmonics) of the voltages of curves a, b, c. This zero-sequence voltage is in fact equal to 1/3rd of the sum of all three voltages of curves a, b and c. Subtracting this zero-sequence voltage from the voltage curve v_{u0} yields the phase voltage v_{un} of curve f. Like the voltage for a delta-connected load (curve d), the voltage v_{un} of

[2]In reality, pure DC current sources do not exist and the current value will depend on the load; however, the shape is rectangular if the inductance is large enough.

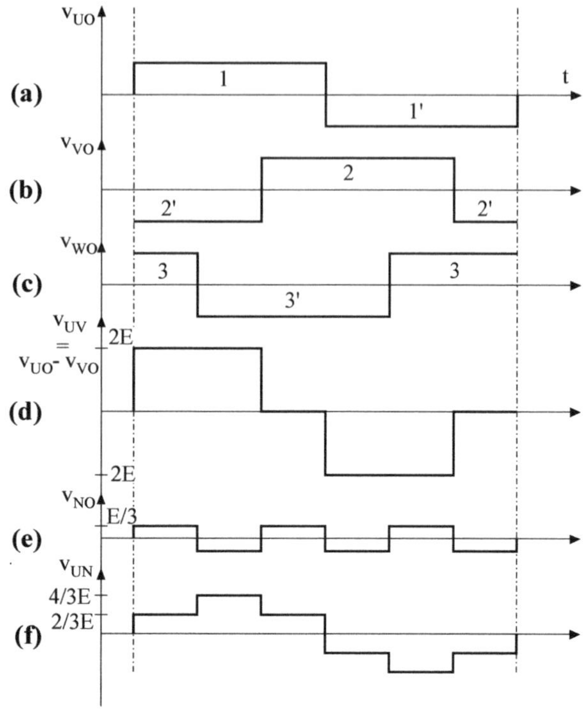

Fig. 11.6 Output voltages for the 180° switching sequence

curve f is rid of the third harmonics. However, the fundamental harmonic of curve f is the same as the fundamental of curve a, i.e. $4E/\pi$.

If a large inductance L_d is present, the input behaves like a DC current source. However, the division of this DC current between the two output lines that are connected together to one of the poles of the DC source is not known as it depends on the load. Therefore (and also for reasons of commutation), the 120° switching sequence is preferably used for current source inverters, not the 180° switching sequence.

11.2.3 *The Six-Step Voltage Source Inverter (VSI)*

11.2.3.1 Basic Configuration

The basic configuration of a voltage source inverter or VSI is the one in Fig. 11.3 (but without the inductance L_d), using the 180° switching sequence and a DC voltage source. Nowadays the switches[3] are mostly IGBTs (or sometimes Mosfets for very small power ratings and IGCTs or GTOs for very high power ratings), always with anti-parallel diodes.

[3]In most figures, these switches are depicted as GTOs.

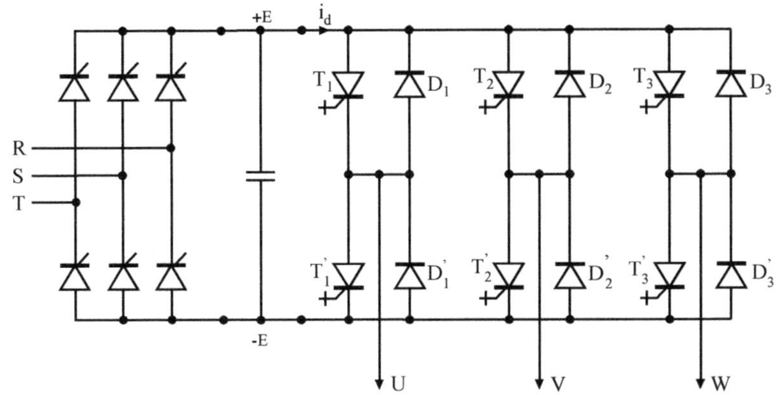

Fig. 11.7 VSI with rectifier as DC source

The DC voltage source is either a battery or, for mains supply, a (controlled) rectifier with a capacitor at the DC side. In the latter case, the combination of rectifier and inverter (see Fig. 11.7) is also sometimes called VSI.

With the 180° switching sequence, we obtain output voltages as in Fig. 11.8. These voltages are independent of the load, whether capacitive or inductive. Their fundamental amplitude is given by $\hat{V}_1 = 4E/\pi = 2V_d/\pi$, as mentioned before (where $2E = V_d$). However, now the nature of the load determines the shape of the current in the DC link.

Curve (d) in Fig. 11.8 illustrates the DC link current for an inductive load. In interval I, only switch T_1 is conducting at the plus side of the source, and the current i_d in the DC link therefore equals the then (lagging) AC current in phase U. In the other intervals as well, the current i_d follows the current in the corresponding phase and thus the shape of the current i_d is a repetition of this sine-wave segment. Note that in each interval, first the anti-parallel diode is conducting and the switch takes over when the current reverses sign (on condition that the gate signal for the switch is still present). Commutation to the next interval requires the switch to be turned off explicitly so that the diode of the opposite switch takes over (e.g. D_2 from $T_2^{'}$ from interval I to II). If the switch were not turned off explicitly, the DC source would short circuit.

Curve (e) in Fig. 11.8 illustrates the DC link current for a capacitive load. In interval I, only switch T_1 is conducting at the plus side of the source and thus the current i_d in the DC link equals the then (leading) AC current in phase U. Similarly, in the other intervals, the current i_d follows the current in the corresponding phases and again we obtain the shape of the current i_d as a repetition of this sine-wave segment. Note that now, in each interval, first the switch is conducting and the anti-parallel diode takes over when the current reverses sign. For the commutation to the next interval, switching on the opposite switch should, in principle, be sufficient to cause the current to commutate to this side (e.g. T_2 from $D_2^{'}$ from interval I to II). However,

interval →	VI	I	II	III	IV	V	
+ side	1+3	1	1+2	2	2+3	3	
- side	2'	2'+3'	3'	1'+3'	1'	1'+2'	

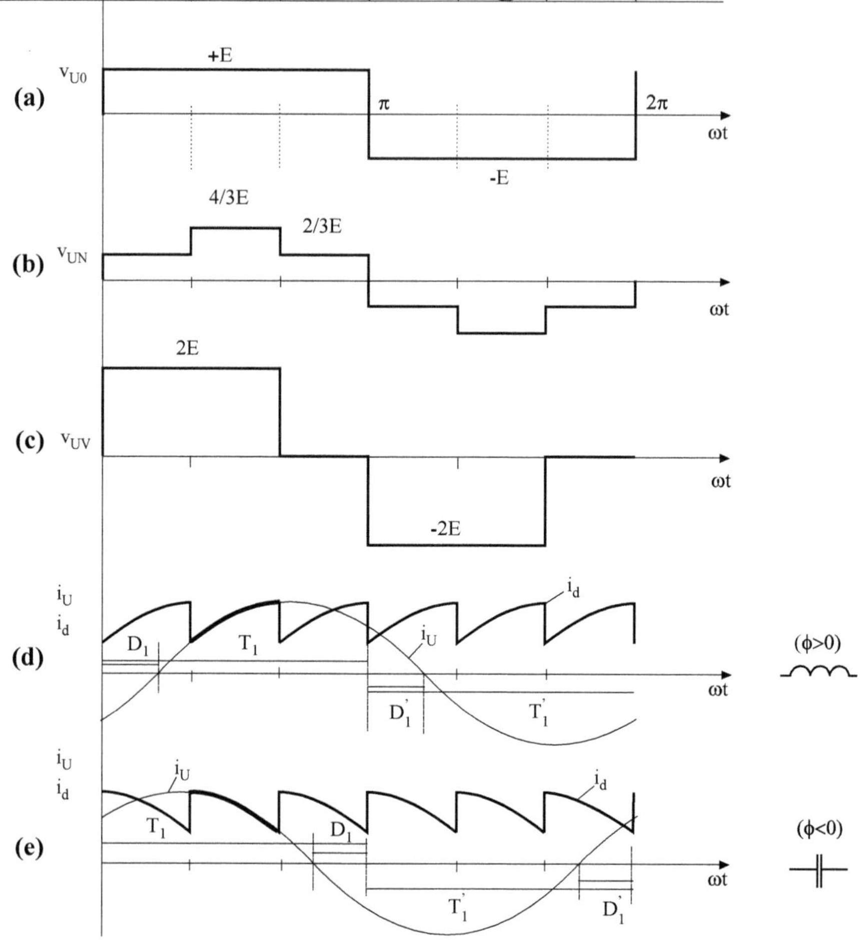

Fig. 11.8 VSI voltages and DC link current wave-shapes

an explicit turn-off signal will always be given to the previous switch (T_2 in this example) before turning on the next one ($T_2^{'}$), because otherwise a short circuit of the DC source would ensue if the load was not (or not sufficiently) leading.

Table 11.3 Instantaneous voltages and corresponding space vectors

Interval	VI	I	II	III	IV	V
Switches	S1,S2',S3	S1,S2',S3'	S1,S2,S3'	S1'.S2,S3'	S1',S2,S3	S1',S2',S3
v_{an}	$\frac{2}{3}E$	$\frac{4}{3}E$	$\frac{2}{3}E$	$-\frac{2}{3}E$	$-\frac{4}{3}E$	$-\frac{2}{3}E$
v_{bn}	$-\frac{4}{3}E$	$-\frac{2}{3}E$	$\frac{2}{3}E$	$\frac{4}{3}E$	$\frac{2}{3}E$	$-\frac{2}{3}E$
v_{cn}	$\frac{2}{3}E$	$-\frac{2}{3}E$	$-\frac{4}{3}E$	$-\frac{2}{3}E$	$\frac{2}{3}E$	$\frac{4}{3}E$
$\underline{v}$	$-\frac{4}{3}E \cdot a$	$\frac{4}{3}E$	$-\frac{4}{3}E \cdot a^2$	$\frac{4}{3}E \cdot a$	$-\frac{4}{3}E$	$\frac{4}{3}E \cdot a^2$
	$\underline{V}_6$	$\underline{V}_1$	$\underline{V}_2$	$\underline{V}_3$	$\underline{V}_4$	$\underline{V}_5$

11.2.3.2 Energy Reversal

In Fig. 11.8, the current i_d was in each case depicted with a mean value which is positive. However, for a sufficiently lagging (d) or leading (e) load, the instantaneous value may be negative and even the mean value of this DC link current could be negative. A negative average value of the DC current implies that the energy flow goes from the AC side to the DC side. This also means that for the VSI inverter, energy reversal is quite feasible.

As the DC current reverses, the DC source must be able to absorb this current. This is not possible with a single rectifier (even a controlled one): energy reversal would cause the capacitor to blow up. In small drives, a series connection of a dissipating resistor and a switch may be connected in parallel with the DC link capacitor. For large drives, usually an anti-parallel rectifier (for negative DC current) is utilised to avoid large energy dissipation. When the DC source is a battery, no problems arise as the battery will absorb the reversed current to charge the battery. It is mainly for that reason that VSIs are preferred in electric cars.

11.2.3.3 Space Vector Representation

In each output fundamental period, the instantaneous voltages of a VSI take on a different value during each of the six modes of 60° or $\pi/3$ (see Table 11.3). The instantaneous (line-to-neutral) three-phase voltages corresponding to these six modes can also be represented by a space vector (with $a = \exp(j2\pi/3)$), defined as follows:

$$\underline{v} = \frac{2}{3}(v_{an} + a \cdot v_{bn} + a^2 \cdot v_{cn}) \tag{11.1}$$

Applying Eq. 11.1 to the instantaneous voltages leads to the vectors in Table 11.3.

Representing these vectors in the complex plane yields Fig. 11.9 (where in dashed lines also the a, b, c axes of an imaginary rotating field machine are shown). In each interval, the space vector remains constant during $\pi/3$ radians. At the commutation instants, the vector jumps over $\pi/3$ radians to the next vector (here, counter-clockwise). Note that the length of each vector is equal to $\frac{4}{3}E = \frac{2}{3}V_{dc}$ (of course also

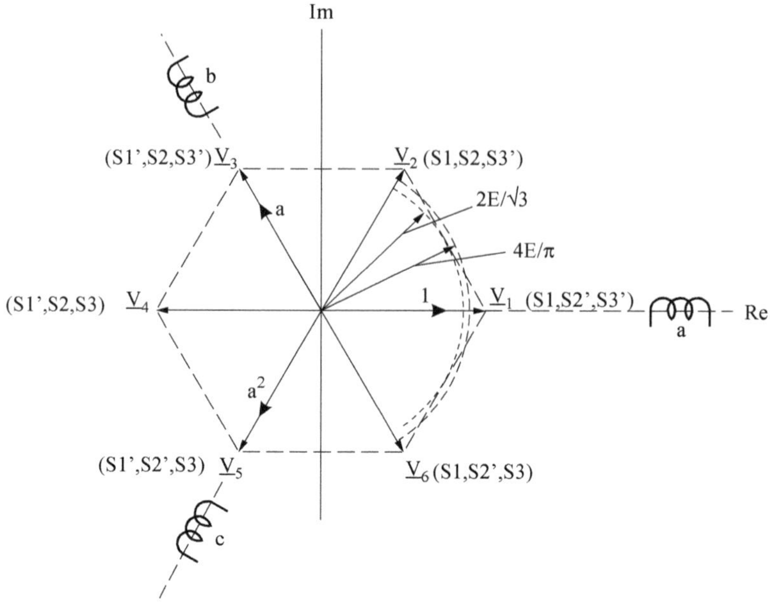

Fig. 11.9 Space vector representation for a six-step VSI

the consequence of the well-chosen factor 2/3 in Eq. 11.1). The fundamental of this six-step voltage is $\frac{4}{\pi}E$ and is thus somewhat smaller. This fundamental corresponds to a circle with this same radius, i.e. this fundamental vector has a constant amplitude and a constant rotational speed equal to the angular frequency of the sine. The radius of this circle is larger than the radius of the circle inscribed into the hexagonal formed by the six VSI-vectors. We will come back to this later on.

11.2.4 The Six-Step Current Source Inverter (CSI)

11.2.4.1 Basic Configuration

The basic configuration of a CSI is again the one in Fig. 11.10, but now the 120° switching sequence (and switches without anti-parallel diodes) is used.

The DC source can be a DC battery, or for mains supply a (controlled) rectifier; in both cases the inductance L_d at the DC side must be large, approximating a current source.

With the 120° switching sequence, we get output line currents as in Fig. 11.11. For a given DC current I_d the fundamental amplitude of the AC line currents is given by $\hat{I}_{u1} = 2\sqrt{3}I_d/\pi$ (see curve (a)). When the load is delta-connected, the phase currents are as in curve (b). Their fundamental amplitude is $\hat{I}_{u\triangle 1} = 2I_d/\pi$.

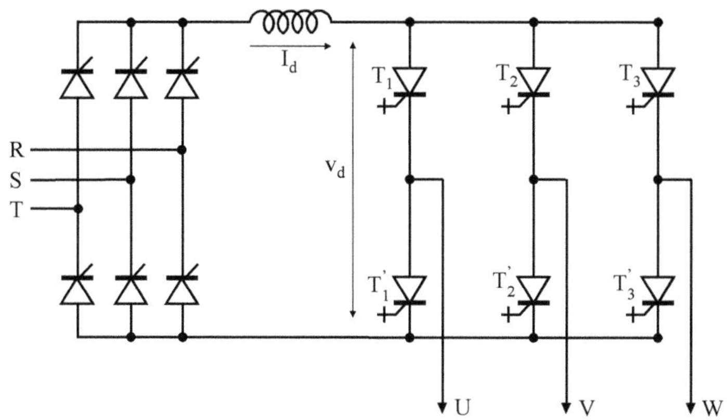

Fig. 11.10 CSI with rectifier

The shape of the input DC link voltage v_d is determined by the load. On the one hand, for an inductive load, we obtain a curve as (d) in Fig. 11.11. Its shape can be derived by considering the fundamental voltage wave-shapes in (c). The fundamental voltage v_u is leading the line current i_u and the line voltage v_{uv} leads v_u by 30°. As in interval I, the switches 1 and 2' are conducting, and the DC link input voltage is equal to that segment of the voltage v_{uv} in this interval.

For a capacitive load, on the other hand, we get a v_d waveform as in (f) which can be derived from the lagging voltage v_u depicted in (e).

As to commutation, the difference between inductive and capacitive load is crucial for a CSI. In a CSI, commutation takes place between switches at the same side of the DC source, for example between switch 1 and switch 2 at the end of the second interval, at 150° or $\omega t = 5\pi/6$. In the capacitive case, the voltage v_{uv} is positive and switching on switch 2 will automatically result in a decreasing current in phase U, while the current in phase V increases and takes over the DC current. Thus, when the load provides reactive energy, explicitly switching off the preceding switch is not necessary. Therefore, GTOs or IGCTs or IGBTs are not required and thyristors can be used. Note that this kind of commutation is completely identical to the one in a controlled rectifier, where the grid provides the reactive energy for the commutation. This kind of commutation in an inverter is commonly called load commutation.

When the load is inductive, the voltage v_{uv} is negative at the end of the second interval, and switch 1 should be switched off explicitly while switching on switch 2. Thus this operation mode requires switches like GTOs or IGCTs.

The CSI with load commutation is mostly used with over-excited synchronous machines as load (similar to classical grids that are supplied by synchronous machines). It can also be used with induction machines as load, but then a capacitor bank in parallel with the induction machine is required (even though this may give rise to unwanted ringing). Nowadays, however, a VSI with a current control loop will be preferred for basically inductive loads like induction machines, if a current supply is required.

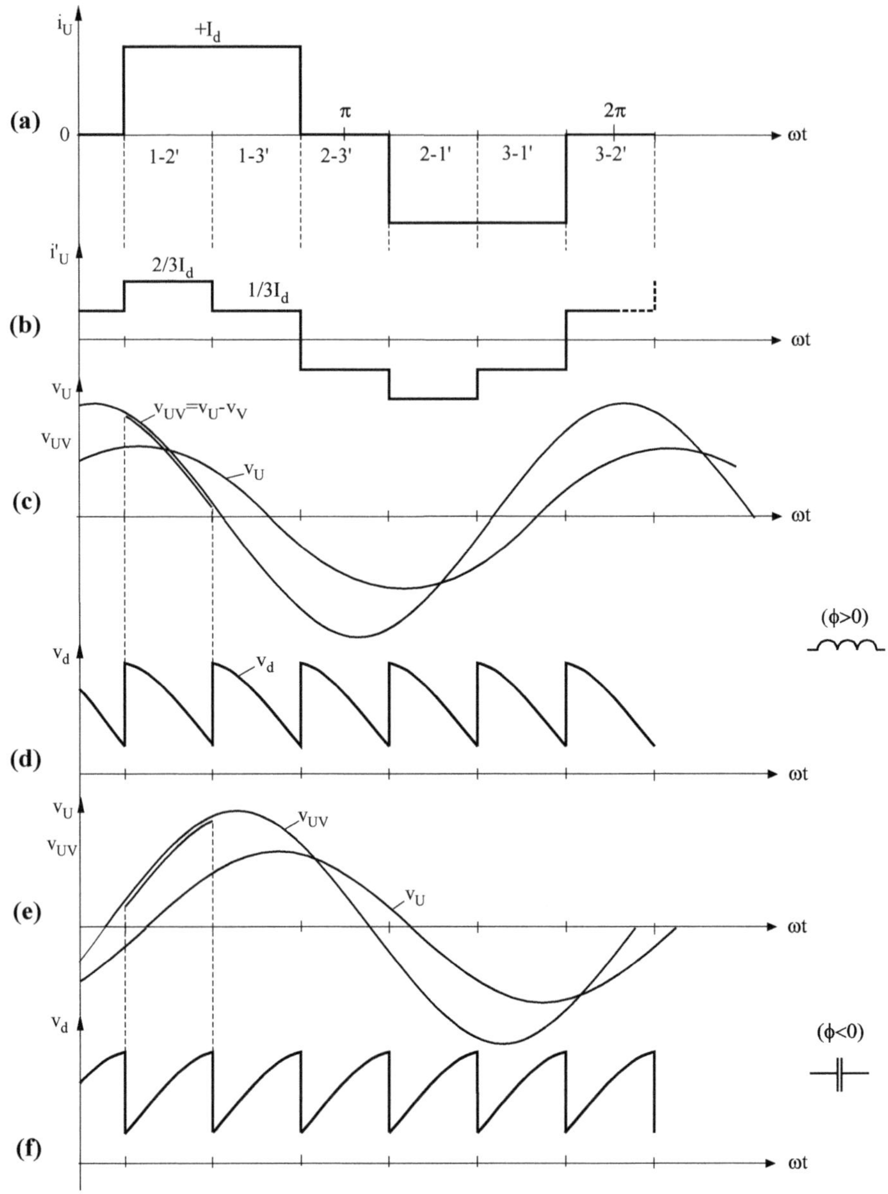

Fig. 11.11 CSI currents and DC link voltage wave-shapes

As a final remark, observe that, in contrast with a VSI, an insufficient capacitive load in a CSI will not result in a short circuit of the DC source because the large inductance L_d will limit the current.

11.2.4.2 Energy Reversal

Like for VSIs, energy reversal is not at all a problem for the inverter part of a CSI. The mean value of v_d may as well be negative, meaning that the energy may flow from the AC side to the DC side as well. Again, the limitation will be the DC source. A DC battery will not accept a reversed polarity of the voltage v_d and a polarity reversal at the DC side will be necessary. However, with a *controlled* rectifier as DC source, energy flow reversal is not a problem at all. It suffices to use a delay angle α greater than 90° to match the negative voltage v_d. Therefore, CSIs with a controlled rectifier are often applied for large industrial drives where energy recuperation is important or essential.

11.3 PWM Inverters

11.3.1 Principle: Single-Phase PWM Inverters

For variable frequency operation of AC (rotating field) electrical drives, it is desirable to vary the voltage amplitude with (and in proportion to) the frequency so as to keep the flux constant, if possible at its maximum (i.e. rated) value. Indeed, only in this way can maximum torque (or power) be produced, given the maximum current which is limited by the section of the machine winding conductors.

The method to achieve a variable voltage amplitude, as illustrated in (A, b) in Fig. 11.2, suffers from the fact that the higher harmonics do not always decrease in the same proportion as the fundamental. Also, the frequencies of the 3rd, 5th, 7th ... harmonics are too low to be adequately filtered out by the (inductive) load.

Pulse Width Modulation (PWM) offers an effective solution to these problems. The idea is to replace the single positive (or negative) pulse - and the zero interval for the case of (A, b) in Fig. 11.2 - in each half period with an adequately chosen series of smaller pulses and zeros. In this way the right and adjustable fundamental amplitude is obtained, as well as a shift of the harmonics to much higher orders.

Dependent on the hardware configuration, two output waveforms can be distinguished, the bipolar one ((a) in Figs. 11.12 and 11.13), and the unipolar one ((b) in Figs. 11.12 and 11.13). In the bipolar type, the output switches between +E and −E, which therefore results in rather severe power pulsations. In the unipolar type, the output switches between +E and 0 (or −E and 0 for the negative half cycle), thus reducing power pulsations in comparison with the bipolar type. However, the required hardware is somewhat more complex (and hence more expensive).

The classical (but not the only) way to obtain the pulse widths is to use the intersections of a sinusoidal (or, in a simplified version, a rectangular) reference wave of the desired fundamental amplitude with a high frequency carrier wave (see Fig. 11.14).

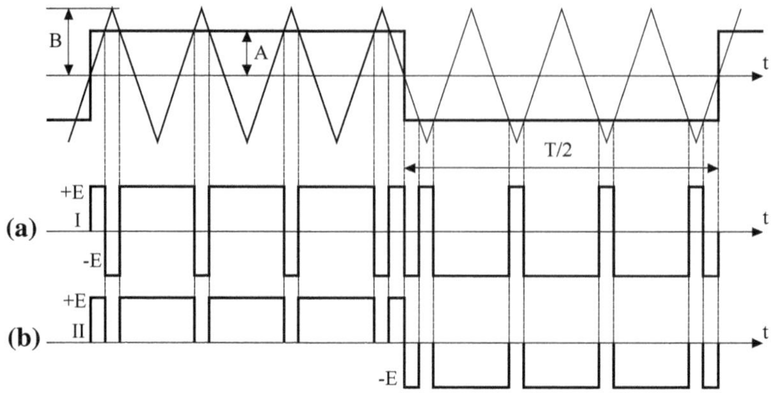

Fig. 11.12 Rectangular reference

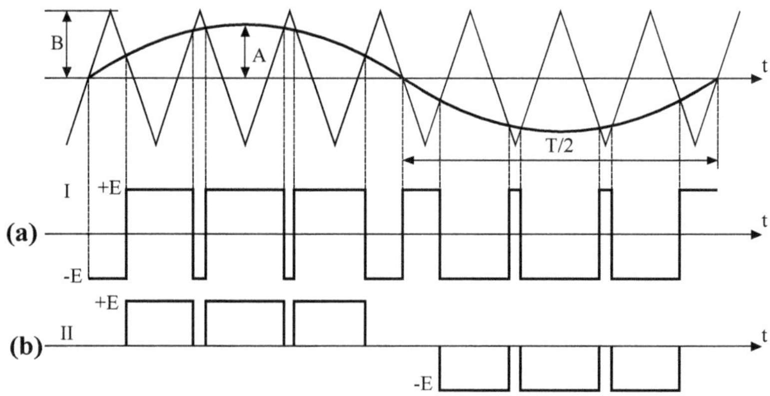

Fig. 11.13 Sine reference

It can be shown (using Bessel's functions, see Ref. [12]) that for a sinusoidal reference wave, the output fundamental amplitude is exactly proportional to the amplitude of the reference wave (at least as long as the amplitude modulation index m_a, i.e. the ratio between the amplitude of the reference wave and the amplitude of the carrier, is less than or equal to one). The *harmonics*[4] in the output depend on the frequencies of the desired fundamental and of the carrier wave, and to a minor degree also on the shape of the carrier wave (i.e. triangular or sawtooth). For a sawtooth carrier wave, more harmonics are present in the output, as will be elaborated further in Sect. 11.3.3. When the frequency modulation index m_f (i.e. the ratio between the frequency of the carrier f_c and the frequency f of the desired output) is sufficiently high (e.g. >10), one may assume that pure harmonics of the desired output fundamental are absent and that only carrier harmonics and its side-bands (of

[4]As explained further on, these are not pure harmonics.

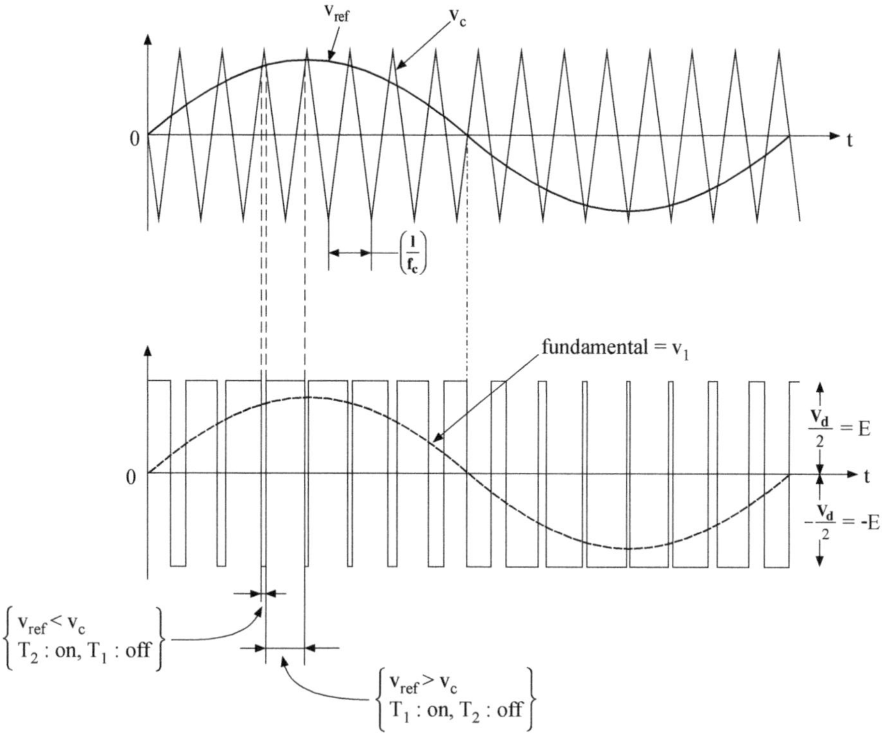

Fig. 11.14 PWM principle

the desired output frequency and its multiples) will be present. In the starting days of power electronics development, the switching frequencies of the available switches (thyristors with commutation circuits or bipolar transistors) were rather low so that the frequency modulation index was limited. At that time, an often heard and applied requirement was that m_f should be a whole number (i.e. synchronous PWM), and, depending on bipolar or unipolar hardware, odd or even. Nowadays asynchronous PWM is mostly used, with rather high m_f so that pure lower harmonics of the required fundamental are absent (see Sect. 11.3.3 for further details).

11.3.2 Three-Phase PWM Inverters

The principle for obtaining PWM in three-phase inverters is basically the same as for single-phase inverters. The reference waves for the three phases are displaced by 120° and thus the output fundamental will consist of symmetrical three-phase sinusoids. Similar conclusions as for the single-phase inverter hold true for the three-phase inverter:

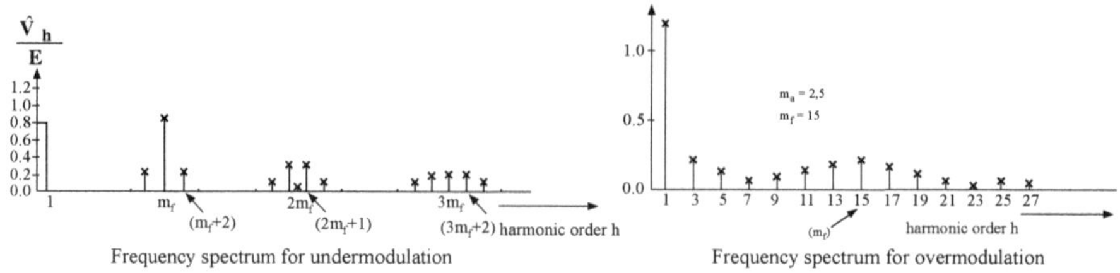

Fig. 11.15 PWM spectra

- the output fundamental amplitude is proportional to the amplitude of the reference wave if the amplitude modulation index m_a is less than 1
- if the frequency modulation index mi is large enough, pure harmonics of the fundamental will be absent.

Zero output voltage is obtained either when all upper switches are connected to the upper or positive DC rail, or when all lower switches are connected to the lower or negative DC rail.

As we have seen before, the maximum instantaneous output wye voltage amplitude for a six-step inverter is $4E/3$, with a somewhat smaller corresponding fundamental amplitude $4E/\pi \approx 1.274E$. When we limit the amplitude modulation index m_a of the PWM to one (so that no output pulses are lost, to avoid compromising the quality of the output voltage), the maximum fundamental wye output voltage is only E. To obtain higher fundamental output voltages, m_a has to be increased beyond one, but then pulses get lost and the lower harmonics of the fundamental reappear (see Fig. 11.15). In addition, the fundamental output voltage is no longer proportional to the amplitude modulation index. In Sect. 11.3.3.3, a method will be discussed to increase the fundamental output voltage somewhat further beyond E, without sacrificing the quality of the output voltage.

Similarly to the single-phase PWM, one used to be limited to low switching frequencies and thus the frequency modulation index was limited. As such, synchronous PWM was nearly always used and for the frequency modulation index, odd numbers as a multiple of three were used. The nowadays much higher switching frequencies allow us to use asynchronous PWM without any pernicious consequences for the quality of the output.

11.3.3 PWM Modulation Principles

This section will describe the modulation basics for single-phase and three-phase PWM inverters in detail. We start from the basic half-bridge converter and then apply the results for the commonly used full-bridge single-phase and three-phase inverters. There are in fact four methods to obtain the best switching instants for the pulses and zeros:

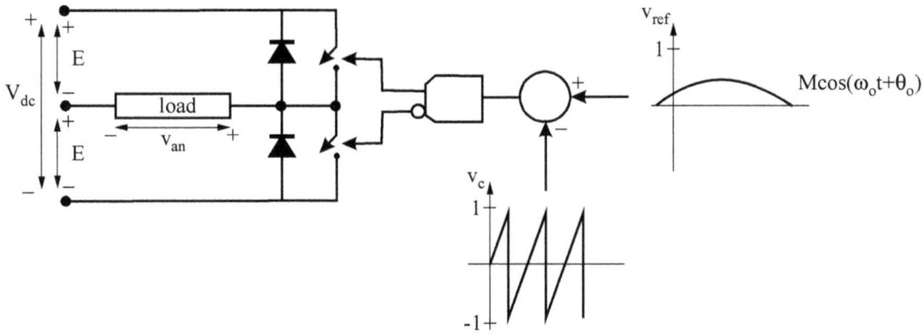

Fig. 11.16 Single-phase half-bridge inverter

1. Optimal PWM: the switching instants are calculated (real-time or tabulated) so as to obtain an 'optimal' output waveform.
2. Switching on the intersections of a sinusoidal reference wave (with the required frequency and amplitude) and a high-frequency carrier wave (naturally sampled PWM).
3. Switching on the intersections of a sampled sinusoidal reference wave (with the required frequency and amplitude) and a high-frequency carrier wave (regular sampled PWM).
4. Switching so that over a period of the carrier wave, the time integral of the output waveform equals that of the reference wave (direct PWM).[5]

In what follows, we will concentrate mainly on methods 2 and 3 (i.e. naturally sampled PWM and regular sampled PWM).

11.3.3.1 Single-Phase Half-Bridge Inverter

Naturally Sampled PWM

Consider the single-phase half-bridge inverter in Fig. 11.16.

The basic principle is to obtain the switching instances by comparing the carrier wave with a pure sinusoidal reference wave of the required amplitude (both in per-unit, for example).

For a (trailing edge) sawtooth carrier, the output voltage is illustrated in Fig. 11.17. With this half-bridge, the output voltage switches between +E and −E, i.e. a bipolar output. The frequency spectrum of this output voltage is shown in Fig. 11.18. The output consists of:

- the required output fundamental, $E \cdot M \cdot \cos \omega_o t$
- the carrier wave harmonic components, $\cos m\omega_c t$ (for $m = 1 \ldots \infty$)

[5]This can be considered as a special case of 'optimal PWM'.

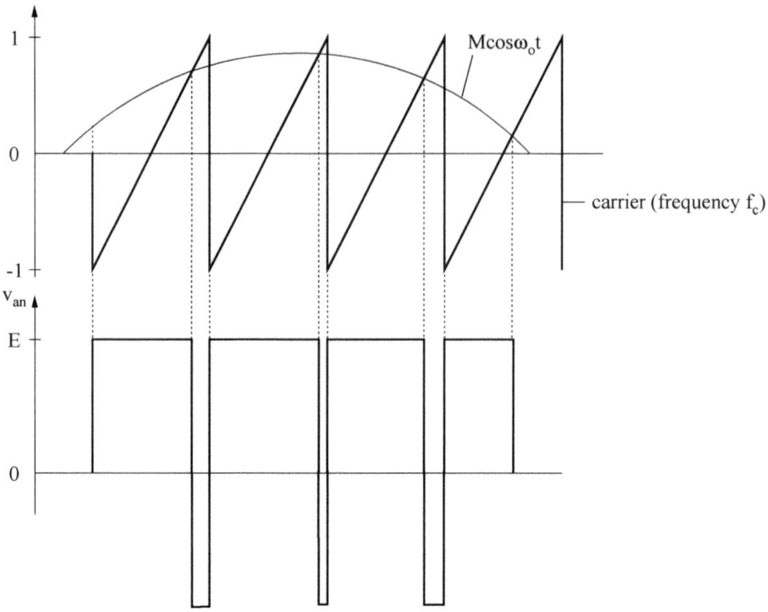

Fig. 11.17 Single-phase half-bridge naturally sampled PWM with sawtooth carrier

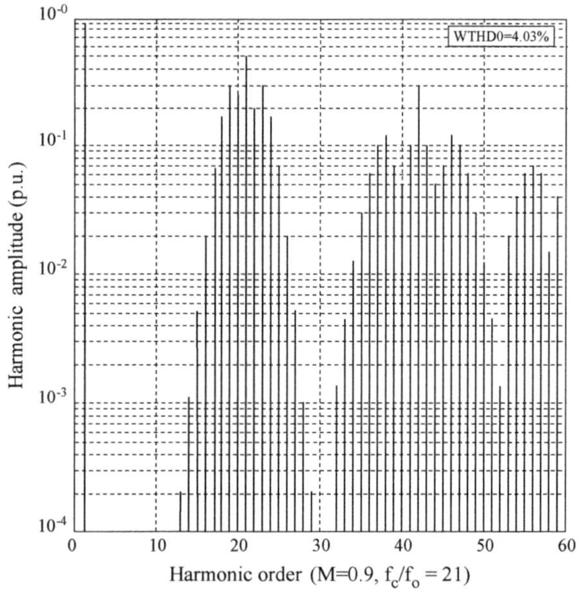

Fig. 11.18 Spectrum of a single-phase half-bridge inverter with naturally sampled sawtooth PWM

- the side-bands of the carrier harmonics, $\cos(m\omega_c t - n\omega_o t)$ (for $m = 1 \dots \infty$ and $n = -\infty \dots \infty; n \neq 0$).

The amplitudes of the carrier wave harmonic components and its side-bands can be expressed as a function of the Bessel functions of order 0 and n, $J_o(m\pi M)$ and $J_n(m\pi M)$. Note that for the general case with $\omega_c/\omega_o \neq whole\, number$, pure harmonics of the reference wave will be absent.

Instead of a sawtooth carrier, a triangular carrier wave is frequently used (sine-triangle PWM or double edge naturally sampled PWM). Fourier analysis of the resulting output shows similar components as above for the sawtooth carrier, with as additional restrictions:

- even harmonics of the carrier are absent (i.e. only those with $m = odd$ appear)
- side-bands of the carrier with $m + n = even$ are absent (i.e. only those with $m + n = odd$ appear).

Nevertheless, the total harmonic content for triangle modulation is not that much lower than for sawtooth modulation. Indeed, the most important harmonic component is still the one with carrier frequency. For polyphase inverters, in contrast, the triangular modulation results in a smaller harmonic content (see Sect. 11.3.3.3).

Regular Sampled PWM

The naturally sampled PWM does not lend itself to a digital implementation, because the intersection of the reference sinusoid and the carrier is defined by a transcendent equation and is complex to calculate. This can be solved by replacing the sine reference with a sampled sine ('sample and hold'), sampled at a sufficiently high frequency. Ideally, the sampling should be synchronised with the carrier so that the level changes of the sampled waveform occur at the positive and/or negative peaks of the carrier (and not while comparing the sampled reference with the carrier). For a trailing edge sawtooth carrier, for example, the sampling should occur at the end of the carrier period, when the carrier wave falls. For a triangular carrier, the sampling can be *symmetrical*, where the sampled reference is taken at the positive or negative peaks of the carrier, or *asymmetrical*, where the reference is re-sampled at both the negative and positive peaks of the carrier.

Figure 11.19 illustrates these three alternatives. We note that there is a delay of half a carrier period for the sawtooth or symmetrically regular sampled PWM and of one quarter of this carrier period for an asymmetrically regular sampled PWM. The delay can be compensated by leading the reference wave by half or one quarter of the carrier period, respectively.

The harmonic analysis for a sawtooth regular sampled PWM yields:

- an approximation for the required output fundamental, $E \cdot M \cdot \cos \omega_o t$
- harmonics of the required output fundamental, $\cos n\omega_o t$
- the carrier wave harmonic components, $\cos m\omega_c t$ (for $m = 1 \dots \infty$)
- the side-bands of the carrier harmonics, $\cos(m\omega_c t - n\omega_o t)$ (for $m = 1 \dots \infty$ and $n = -\infty \dots \infty; n \neq 0$).

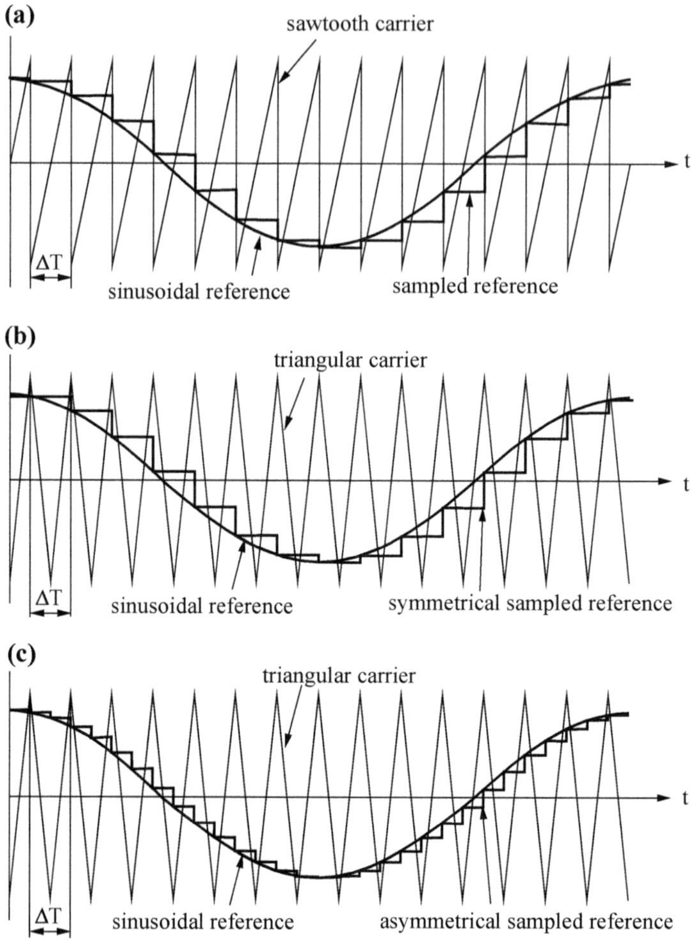

Fig. 11.19 Regular sampled PWM modulation principles

Compared to the naturally sampled sinusoidal PWM, we can also find harmonics of the fundamental. However, their amplitude is limited if the frequency modulation index is sufficiently high. Another difference is that now the side-bands of the carrier frequency harmonics are not completely symmetrical (i.e. the amplitudes of the side-bands $m\omega_c t - n\omega_o t$ and $m\omega_c t + n\omega_o t$ are not equal). In addition, the fundamental output amplitude is not exactly equal to $E \cdot M$ but the difference is minimal if f_c is large:

$$\frac{2E}{\pi} \cdot \frac{J_1(\pi M(\omega_o/\omega_c))}{(\omega_o/\omega_c)} \approx E \cdot M \tag{11.2}$$

Already for $f_c = 10$ and $M = 1$, the difference is less than 1%. Figure 11.20 shows the harmonic spectrum for $f_c = 21$ and $M = 0.9$.

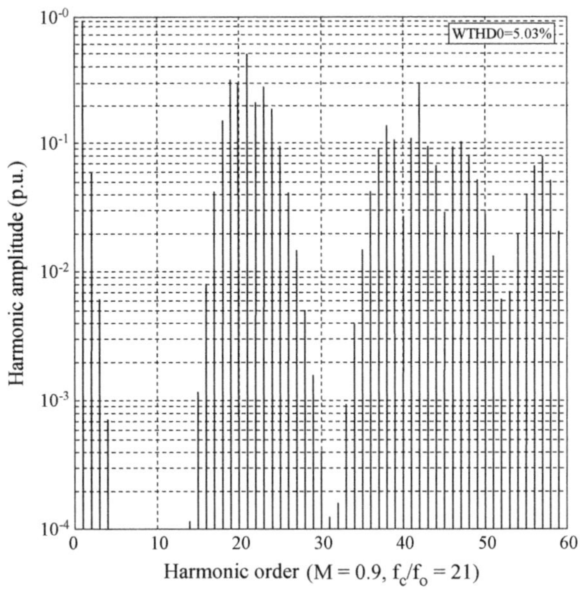

Fig. 11.20 Spectrum of half-bridge PWM inverter with regular sampled sawtooth carrier

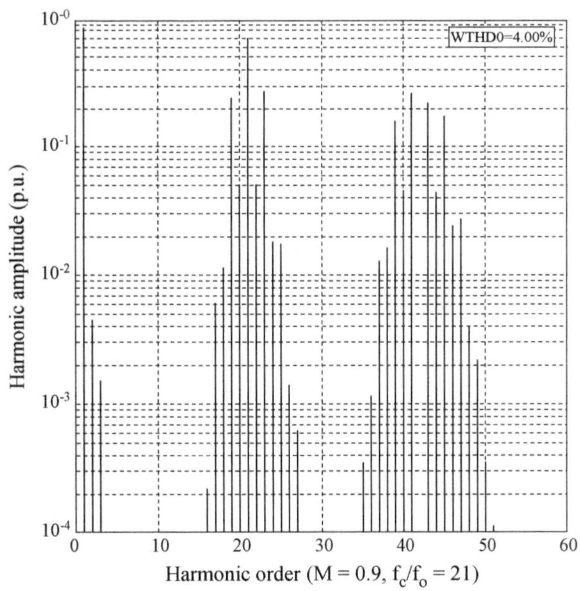

Fig. 11.21 Spectrum for half-bridge inverter with symmetrical regular sampled PWM

Figure 11.21 shows the spectrum for a symmetrical sampled PWM for $f_c = 21$ and $M = 0.9$. Compared to the previous case of a sawtooth regular sampled PWM, the approximation for the fundamental is still slightly improved, harmonics of the fundamental are significantly decreased in amplitude, the even carrier frequency

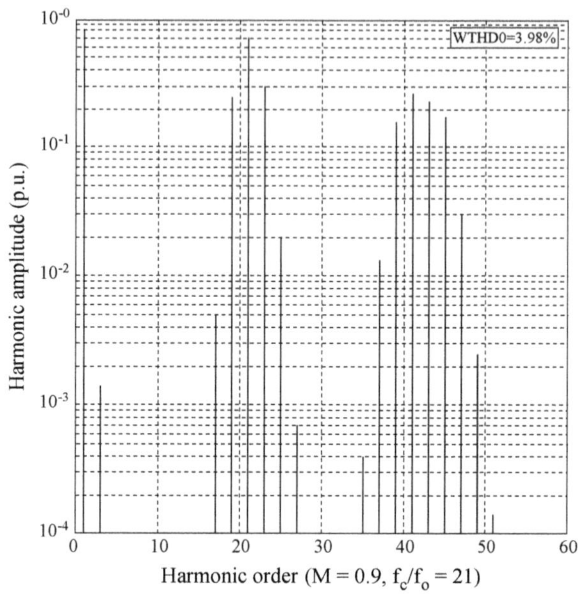

Fig. 11.22 Spectrum for half-bridge inverter with asymmetrical regular sampled PWM

harmonics are absent, and the amplitudes of the carrier side-bands with $m + n =$ *even* have significantly lower values.

For an asymmetrical regular sampled PWM (Fig. 11.22) the result of a harmonic analysis is even better:

- even harmonics of the fundamental are absent
- all side-bands with $m + n = even$ are absent.

In fact, it is intuitively clear that this asymmetrical regular sampled PWM comes down to a sampling with twice the carrier frequency.

11.3.3.2 Single-Phase Full-Bridge Inverter

The half-bridge inverter is not very practical for a single-phase inverter, as it requires a DC source with accessible centre-tap point. Therefore, single-phase inverters nearly always make use of a full bridge.

The single-phase full-bridge inverter consists of two half-bridge inverters, as shown in Fig. 11.23. With this configuration the actual output may be switched between $+E$, 0 and $-E$ (where $V_{dc} = 2E$). The switches of the two legs are controlled in a complementary pattern. The actual output voltage waveform (and its quality) will, however, depend on the actual control scheme as shown below.

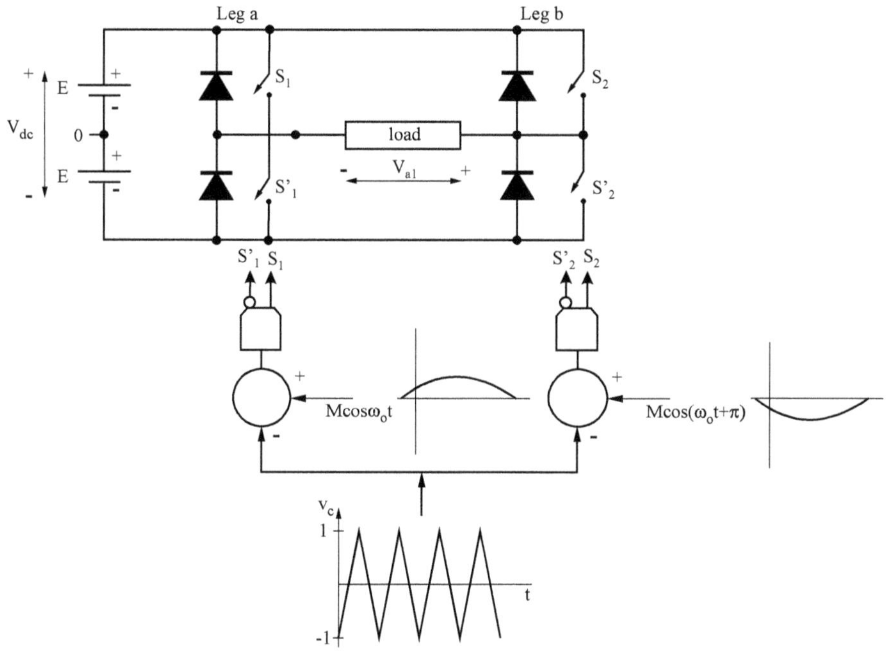

Fig. 11.23 Single-phase full-bridge PWM inverter

Modulation with Three Levels

In the example in Fig. 11.23, both legs make use of a common carrier (here a triangular carrier), but of 180° opposed reference waves. The result of this modulation scheme is shown in Fig. 11.24.

Legs a and b are switched between the + and − rails of the DC source, not simultaneously as these switching instants depend on the comparison between the (common) carrier and the different (complementary) sinusoidal reference waves. While legs a and b are switched between +E and −E, the resulting (load) output voltage waveform contains pulses 2E and −2E with intervals of zero voltage, i.e. a unipolar waveform, without the need for a DC source with centre-tap. The advantage of such a unipolar output waveform is that the harmonic content is much lower. Figure 11.25 illustrates this for a double edge naturally sampled PWM: (a) for leg a, V_a; (b) for both legs, V_{ab}.

The harmonic content for each leg follows from the analysis in the previous paragraph, i.e. the fundamental $E \cdot M \cdot \cos \omega_o t$ or $E \cdot M \cdot \cos(\omega_o t - \pi)$ respectively, the odd harmonics of the carrier and the side-bands of the carrier with $m + n = odd$. In the output voltage V_{ab}, we get as fundamental $2E \cdot M \cdot \cos \omega_o t$ and all odd harmonics (m odd) of the carrier disappear, as well as all their side-bands (n even). Thus only side-bands with m even and n odd remain. As illustrated in the expanded switching detail for a carrier period in Fig. 11.24, (cf. the lower sketch), the switching

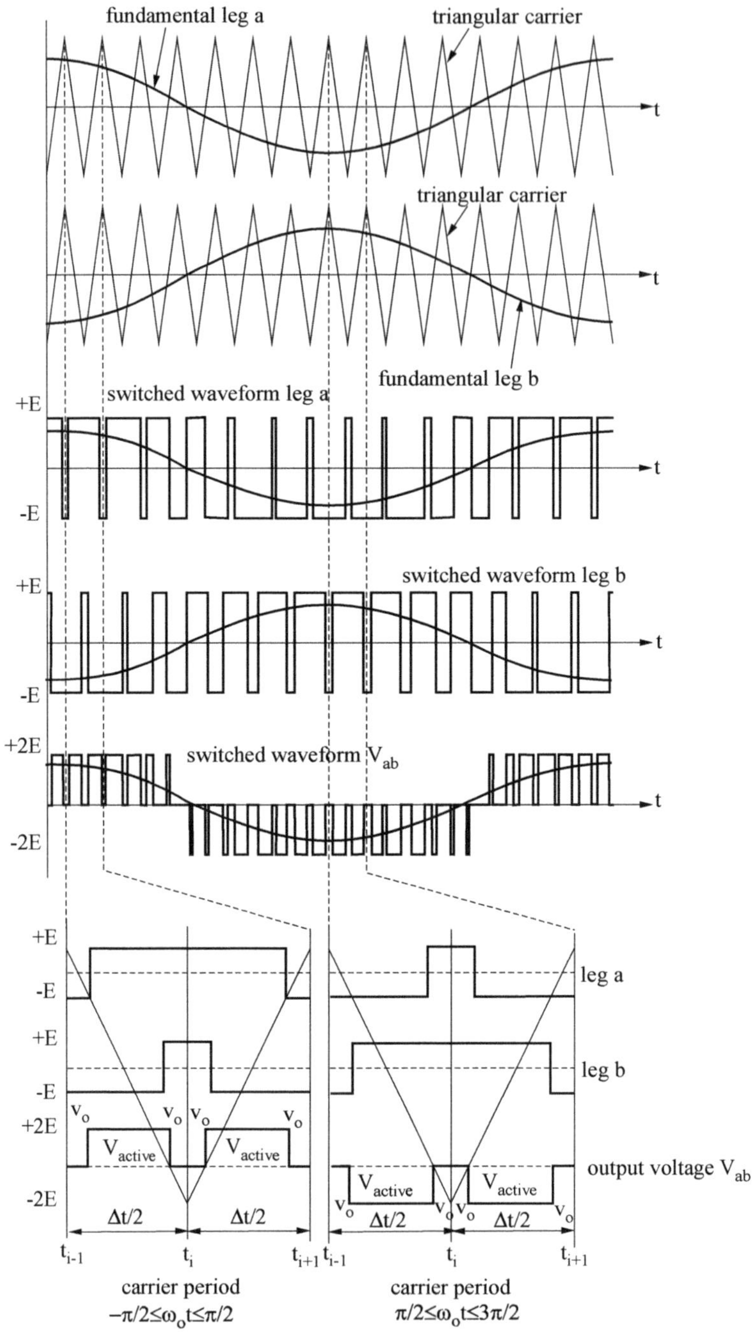

Fig. 11.24 Three-level modulation for a single-phase full-bridge naturally sampled PWM inverter with triangular carrier

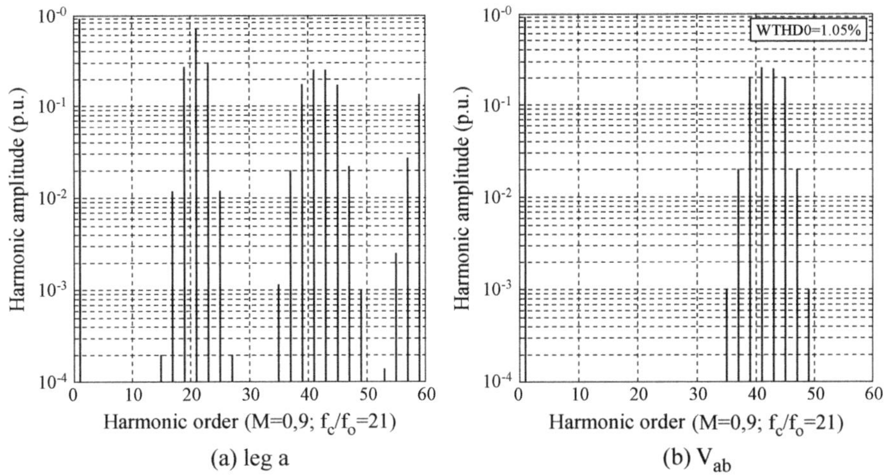

Fig. 11.25 Spectra for three-level single-phase full-bridge naturally sampled PWM inverter with triangular carrier

frequency for the output voltage is effectively doubled and in each half period the non-zero pulse is located approximately in the centre of this half period.

The same principle can also be used for regular sampled PWM (symmetrical or asymmetrical), sawtooth carrier regular sampled PWM, and other PWMs. See Ref. [12] for further details.

Remarks:

1. There are also other modulation methods to obtain a three-level output. The previous one belongs to the category of 'continuous PWM'. In discontinuous PWM methods, only one leg is PWM controlled while the other is connected to the + or − DC rail during half a period (see Ref. [12]).
2. All PWM methods result in (exactly or approximately) the same fundamental (corresponding with the reference). However, the higher harmonics may differ considerably, dependent on the control scheme. All modulation methods (with the same f_c) result in the same **width** of the pulse in a period (or half a period) of the carrier: this corresponds to the same value of the time integral of this pulse as the time integral of the reference wave in the same time span. However, they differ as to the position of the pulse within this period or (half the period). For the case studied above, the pulse is placed approximately in the centre of each half period (and exactly in the centre for a sampled PWM). Modulation methods which place the pulses closer to the centre have a superior harmonic performance than those which do not centre the pulses. Moving the pulse to the beginning or end of the period results in a worse harmonic performance.
   For example, if we move the pulse in the first half carrier interval to the end and the one in the second half carrier interval to the beginning (Fig. 11.26), we get one

Fig. 11.26 Pulse placement

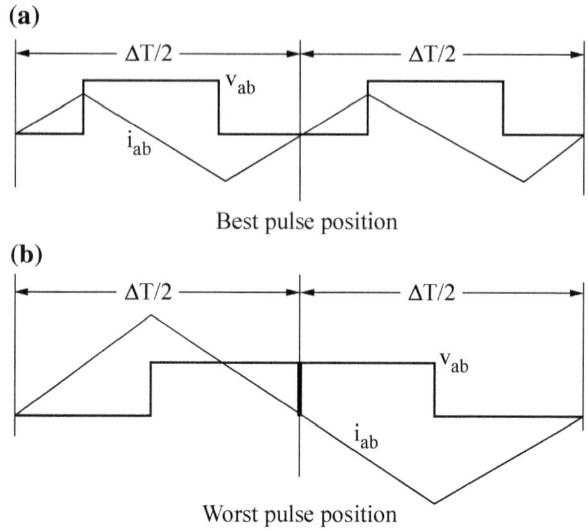

large pulse with half the original harmonic frequency and a higher amplitude. The cost of the 'optimal' placement is that both legs must be independently modulated, which is usually a more expensive implementation than other simplified methods.

3. For an inverter with two switching legs as in the case studied above, the elimination of a large part of the harmonics essentially depends on an accurate control of the two legs. If, for example, the carriers for both legs are not exactly the same or if there is a larger delay in the control of one leg, the elimination will not be complete.

Modulation with Two Levels

An inverter with two levels can be obtained for a full-bridge scheme with two switched legs, if for the switching function of one leg, the opposite wave is chosen as for the other leg (see Fig. 11.27).

The output voltage now switches between +2E and −2E in each switching period, without zero states (see Fig. 11.27). From the detail in one carrier period, we note that the pulses in each half period are shifted to the beginning or the end. The result is one large pulse in a carrier period and the effective switching frequency is f_c, instead of $2f_c$. Although the control is less complicated, the quality of the output voltage is inferior compared to the three-level inverter as the harmonic content is much higher.

11.3.3.3 Three-Phase PWM Voltage Source Inverter

The topology of a three-phase PWM voltage source inverter is the same as that for a six-step voltage source inverter (see Fig. 11.28).

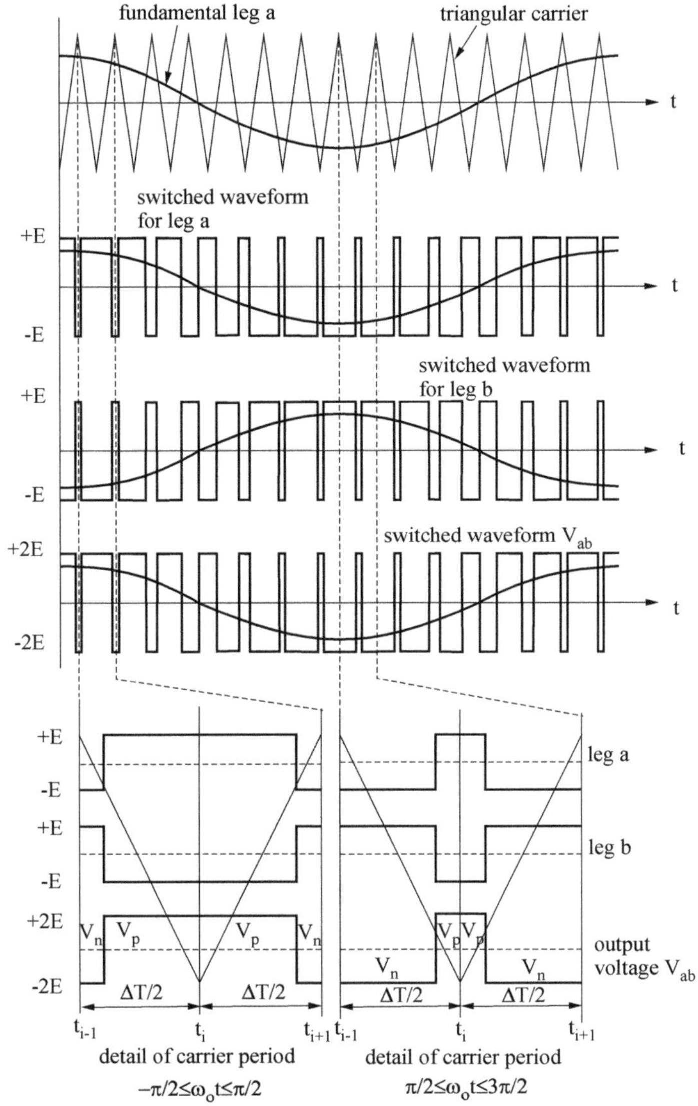

Fig. 11.27 Two-level modulation for single-phase PWM (naturally sampled with triangular carrier)

The modulation principles are essentially the same as for the single-phase inverter, but the three-phase output connection makes the output fundamental and harmonic voltages somewhat distinct. Moreover, the widespread application of three-phase PWM inverters justifies a separate discussion. In a subsequent section, the "space vector modulation" (SVM) will be discussed. It is shown that the output of SVM is equivalent to that of a classical PWM.

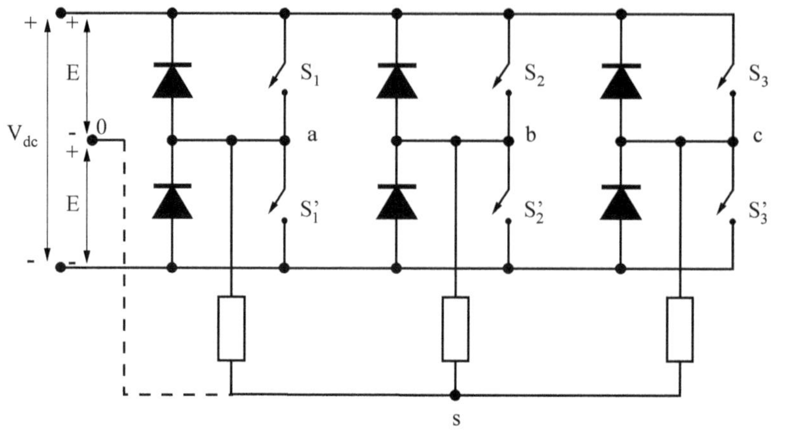

Fig. 11.28 Three-phase full-bridge VSI

Three-Phase Modulation with Sinusoidal Reference

Figure 11.29 shows the modulation for a three-phase inverter using a sinusoidal reference and a triangular carrier. As for the single-phase three-level inverter, a common carrier is used for the three phases. The reference sine waves are now of course 120° shifted with respect to each other. As is clear from Fig. 11.29, the resulting output line-to-line voltages are unipolar, i.e. switching of these voltages occurs between +2E and 0 or −2E and 0.

The fundamental and higher harmonics of the line-to-line voltages may be derived from the harmonic analysis of the corresponding half-bridge inverter. For a naturally sampled three-phase PWM, we therefore obtain:

- the required output fundamental, $\sqrt{3}E \cdot M \cdot \cos\omega_o t$
- side-bands of the carrier harmonics, $\cos(m\omega_c t - n\omega_o t)$ (for $m = 1 \ldots \infty$ and $n = -\infty \ldots + \infty$; $n \neq 0$; and $n \neq 3k$ and only those with $n + m = odd$).

Thus, compared to the single-phase full-bridge PWM inverter, for this naturally sampled PWM the fundamental line-to-line voltages have a reduced amplitude ($\sqrt{3}E$ instead of $2E$ if $M \leq 1$). Not all side-bands around the odd multiples of the carrier frequency are absent, but only the third side-bands and multiples are absent (including those with $n = 0$, i.e. the pure carrier harmonics). The side-bands with $m + n = even$ are absent as these do not appear for the half-bridge inverter. Thus, the most significant harmonics are those with frequencies $\omega_c \pm 2\omega_o, \omega_c \pm 4\omega_o, 2\omega_c \pm \omega_o, 2\omega_c \pm 5\omega_o, \ldots$ (not $2\omega_c \pm 3\omega_o$ as these are cancelled out in the line-to-line voltages). It must be remarked, however, that the THD ('total harmonic distortion'),

$$THD = \sqrt{\frac{V_{rms}^2 - V_{1rms}^2}{V_{1rms}^2}}$$

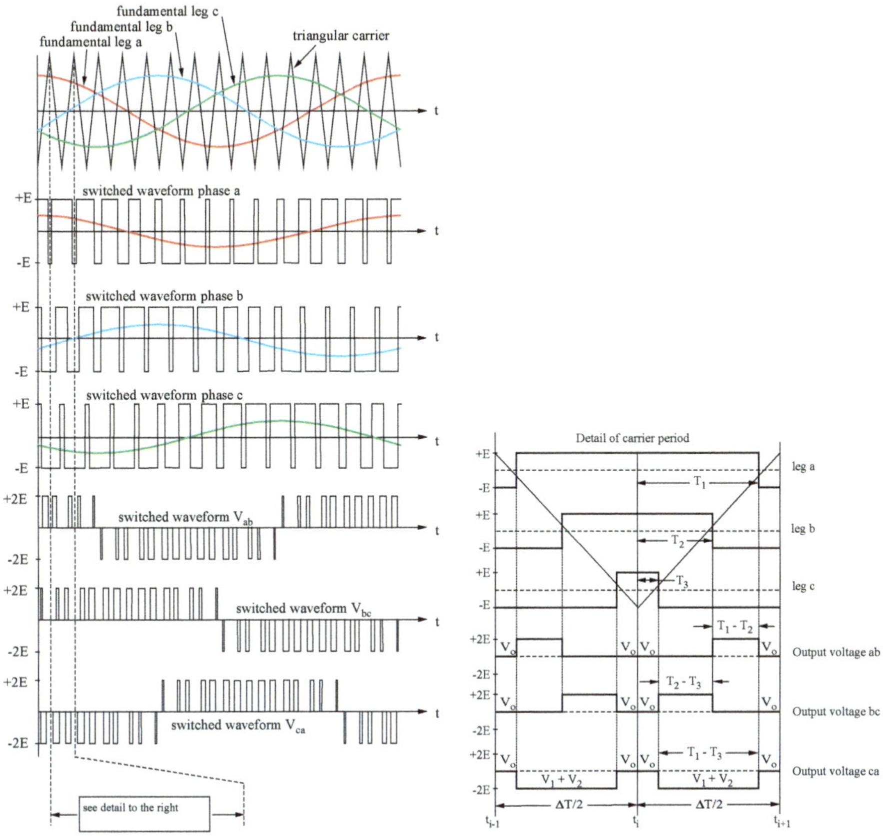

Fig. 11.29 Naturally sampled three-phase PWM inverter with triangular carrier

of a three-phase inverter with three levels is somewhat worse than that of the single-phase inverter with three levels, as the side-bands around the odd carrier harmonics do not disappear in the output.

Three-Phase Modulation with Regular Sampled Sinusoidal Reference

A three-phase VSI can of course also be modulated using a sampled PWM. The carrier can be a sawtooth or a triangular carrier, where for the latter a symmetrical or asymmetrical sampling may be used.

Like for the single-phase inverter, the best results are obtained with an asymmetrical regular sampled PWM (thus a triangular carrier). The harmonic analysis shows that the side-bands with $m + n = even$ are absent, due to the asymmetrical sampling. The side-bands $m + 2n$, however, do not disappear for the line-to-line voltages (as was the case for the single-phase full-bridge inverter), but the side-bands $m + 3n$ as well as the triple harmonics of the fundamental are now cancelled out in the line-to-line voltages. The fundamental output line-to-line voltage is not exactly equal to

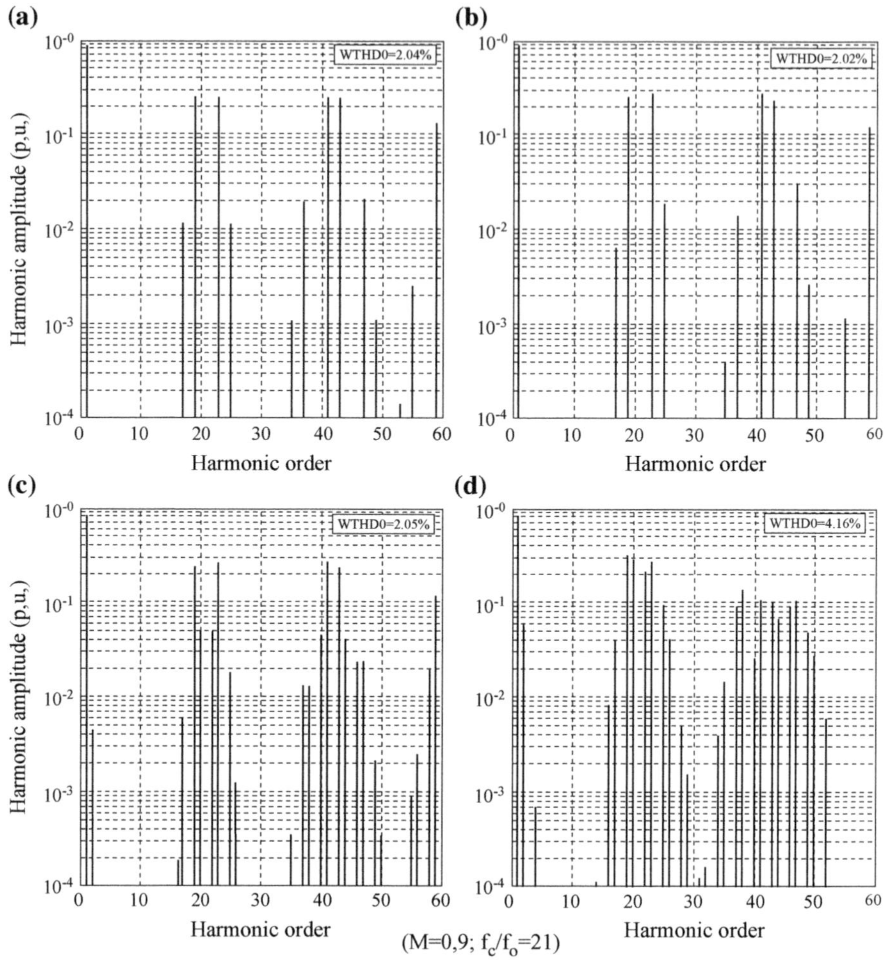

Fig. 11.30 Spectra for three-phase VSI

$\sqrt{3}E \cdot M$, but for sufficiently large values of the frequency modulation index, the difference is small (less than 1% for $m_f > 10$). If the amplitude modulation index is restricted to values $M = m_a \leq 1$, the maximum fundamental output voltage is thus $\sqrt{3}E$.

For a symmetrical regular sampled PWM, the harmonic content is much higher. The output contains for example a second harmonic of the fundamental, as well as more side-bands of the carrier harmonics. For a sawtooth sampled PWM, the second harmonic is rather high (6% of the fundamental) and this modulation principle is not at all advised for three-phase inverters (unless m_f is very large).

Figure 11.30 compares the spectra of three phase inverters for (a) a naturally sampled PWM with triangular carrier, (b) an asymmetrical regular sampled PWM,

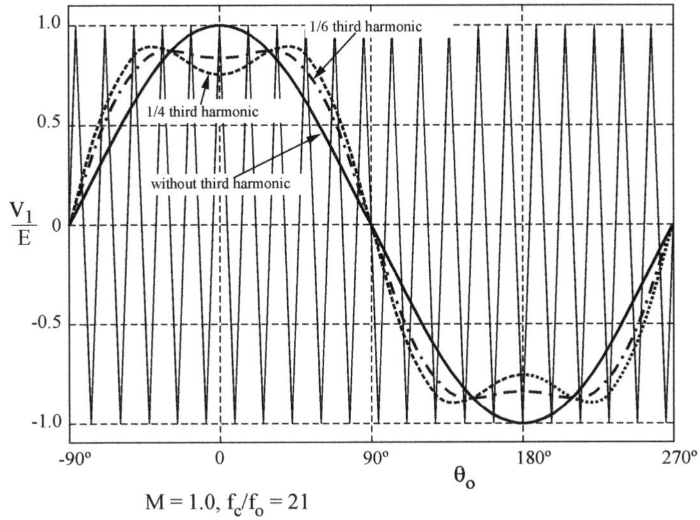

Fig. 11.31 Injection of a third harmonic

(c) a symmetrical regular sampled PWM, and (d) a regular sampled PWM with sawtooth carrier. It is clear that (a) and (b) yield by far the best results, while (c) and in particular (d) exhibit much worse harmonic behaviour.

Third Harmonic Injection

For single-phase full-bridge inverters, the maximum fundamental output voltage is $2E$ if the amplitude modulation index is restricted to $M \leq 1$. Letting $M > 1$ makes it possible to attain the fundamental voltage $\frac{4}{\pi}2E$ for a block inverter but to the detriment of the harmonic content.

For three-phase inverters, the maximum fundamental line-to-line output voltage is $\sqrt{3}E$ (or E for the line-to-neutral voltage) if the amplitude modulation index is restricted to $M \leq 1$. Letting $M > 1$ permits attaining the fundamental line-to-line voltage $\frac{4}{\pi}\sqrt{3}E$ (or $\frac{4}{\pi}E$ for the line-to-neutral voltage) of a six-step inverter, but to the detriment of the harmonic content. The cause is that an amplitude modulation index larger than 1 results in pulses falling away and being replaced by wider blocks. Lower order harmonics then reappear. For three-phase inverters, a way to solve this is to add a third harmonic to the sine reference wave in such a way that the maximum value of the combined reference wave does not exceed unity (while the sine reference may have an amplitude larger than unity, see Fig. 11.31).

In this way, the maximum fundamental line-to-line voltage may attain $2E$, or $(2/\sqrt{3})E$ for the line-to-neutral voltage. This is exactly equal to the radius of the circle inscribed in the hexagonal of the six non-zero vectors of a six-step inverter (see Fig. 11.9). Because the sine reference and the fundamental output voltage now have a higher amplitude for about the same harmonic content, the THD turns out to be lower than for $M \leq 1$. Of course, the maximum fundamental output voltage

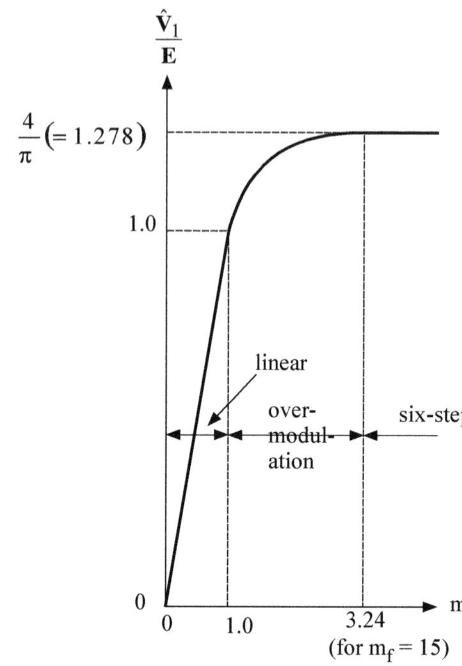

Fig. 11.32 Amplitude of fundamental as a function of the amplitude modulation index

$(2/\sqrt{3})E \approx 1.156E$ is still lower than the fundamental for a six-step inverter, i.e. $\frac{4}{\pi}E \approx 1.273E$. The third harmonics in the output (which stem from the third harmonic in the reference wave) do not have any detrimental effect as they are cancelled out for a delta-connected load, or do not result in third-harmonic current in a wye connected load (as the neutral of a wye inverter load will rarely be connected).

The optimal[6] amplitude of the third harmonic is equal to $1/6^{th}$ of the fundamental reference amplitude, i.e. $\gamma = M_3/M = 1/6$. For $\gamma = 1/6$ the maximum value of M for which no pulses disappear (i.e. the maximum value of the combined reference wave just attains 1) is equal to $M = 2/\sqrt{3} = 1.156$.

Thus, by injection of a third harmonic in the reference wave, the fundamental output voltage can attain the value corresponding to the inscribed circle in the hexagonal of the six-step VSI (see Fig. 11.9), without reducing the quality of the output voltage.

[6]This can be calculated as follows: with a third harmonic added to the sine reference, the combined reference is $v_{ref} = E \cdot M[\cos\omega_o t + \gamma\cos 3\omega_o t]$. Deriving this expression yields that the maximum of this voltage occurs for $\cos\omega_o t_o = \sqrt{\frac{3\gamma-1}{12\gamma}}$. Substituting this in the reference then yields for the maximum value $v_{ref,max} = E \cdot M[\frac{1}{3}(1-3\gamma)\sqrt{1-\frac{1}{3\gamma}}]$. This maximum value has a minimum for $\gamma = -\frac{1}{6}$ and the minimum is $(\sqrt{3}/2) \cdot E \cdot M$. Thus without losing pulses, M may increase to $2/\sqrt{3}$.

If a further increase of the fundamental is required, M can be increased, but then pulses will be lost and lower harmonics will reappear. In addition, the output voltage is no longer proportional to M, as is illustrated in Fig. 11.32 for a PWM without third harmonic.

11.4 Space Vector Modulation

From the discussion in the previous section, it became clear that nearly all natural and regular sampled PWM schemes result in the same fundamental output voltage. However, they differ as to the higher harmonic content in the output. We also observed that this voltage quality is narrowly related to the position of the PWM pulses within the switching interval (carrier period or half the carrier period).

Instead of determining the width and the placement of the pulses by comparing a reference wave with a carrier, the pulse widths for each switching interval could also be determined in such a way that the time integral of the pulse equals the time integral of the desired output sine (in that interval), i.e. "direct PWM". The position of the pulse within the interval could then be used to reduce the harmonic content.

Space Vector Modulation (SVM) relies on the eight possible combinations for the six switches of a VSI: six non-zero or active voltage vectors and two zero voltage vectors. The active vectors correspond to the six modes of a six-step VSI (see also Fig. 11.9 and Table 11.3). The two zero voltage vectors are obtained if either all upper switches are closed on the positive DC rail, or if all lower switches are closed on the negative DC rail (in both cases, the three output voltages are zero).

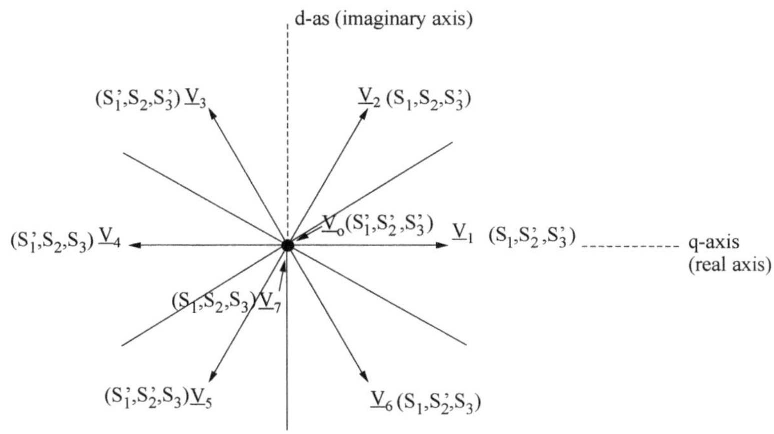

Fig. 11.33 The eight vectors of a VSI

Fig. 11.34 Space vector modulation principle

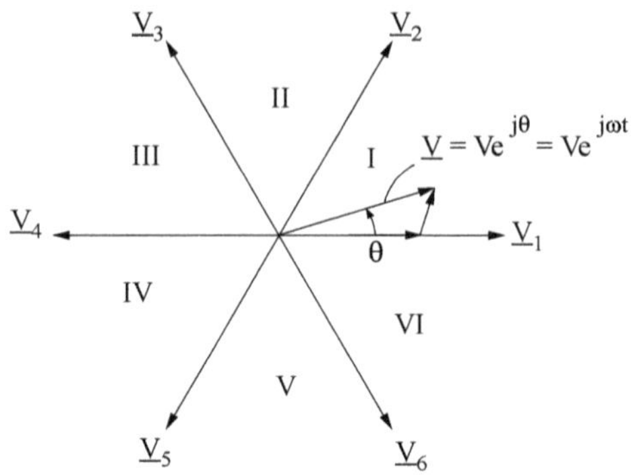

These eight 'stationary' vectors are represented in Fig. 11.33. The aim is to supply a three-phase load with a symmetrical sinusoidal three-phase voltage, i.e. corresponding with a rotating vector $V \cdot \exp(j\omega t) = V \cdot \exp(j\theta)$.

We will try to approximate this ideal rotating vector as follows. The fundamental period $T = 2\pi/\omega$ is divided into a sufficiently large number (usually denoted $2n$) of intervals (also called sub-periods) $\Delta T/2 = T/2n$. In each of these intervals, we will approximate the 'average' voltage vector in the interval (e.g. interval m) with a vector with amplitude V and instantaneous phase angle $\theta = \omega \cdot m \cdot \Delta T/2 = \omega \cdot m \cdot T/2n$, and this for all m, $0 \le m \le 2n$.

Consider for example the vector $\underline{V}$ in the first sector in Fig. 11.34. To approximate $\underline{V}$ in the mth interval centred around it, we make use of the active vectors $\underline{V}_1$ and $\underline{V}_2$. By switching on $\underline{V}_1$ and $\underline{V}_2$ during fractions $\alpha_1 = T_1/(\Delta T/2)$ and $\alpha_2 = T_2/(\Delta T/2)$, respectively, of the switching interval $\Delta T/2$, we may write

$$\underline{V} = V \cdot \exp(j\theta) = \alpha_1 \cdot \underline{V}_1 + \alpha_2 \cdot \underline{V}_2 \tag{11.3}$$

or

$$(\Delta T/2)\underline{V} = T_1 \cdot \underline{V}_1 + T_2 \cdot \underline{V}_2 \tag{11.4}$$

or

$$(\Delta T/2)V \cdot \exp(j\theta) = T_1 \cdot V_m \cdot \exp(j0) + T_2 \cdot V_m \cdot \exp(j\pi/3) \tag{11.5}$$

with $V_m = (4/3)E$ the amplitude of the voltage vectors $\underline{V}_1 \ldots \underline{V}_6$. The on-times T_1 and T_2 are thus given by:

$$T_1 = \alpha_1 \cdot \Delta T/2 = \frac{V \cdot \sin(\pi/3 - \theta)}{V_m \cdot \sin \pi/3} \cdot \Delta T/2 \tag{11.6}$$

$$T_2 = \alpha_2 \cdot \Delta T/2 = \frac{V \cdot \sin \theta}{V_m \cdot \sin \pi/3} \cdot \Delta T/2 \tag{11.7}$$

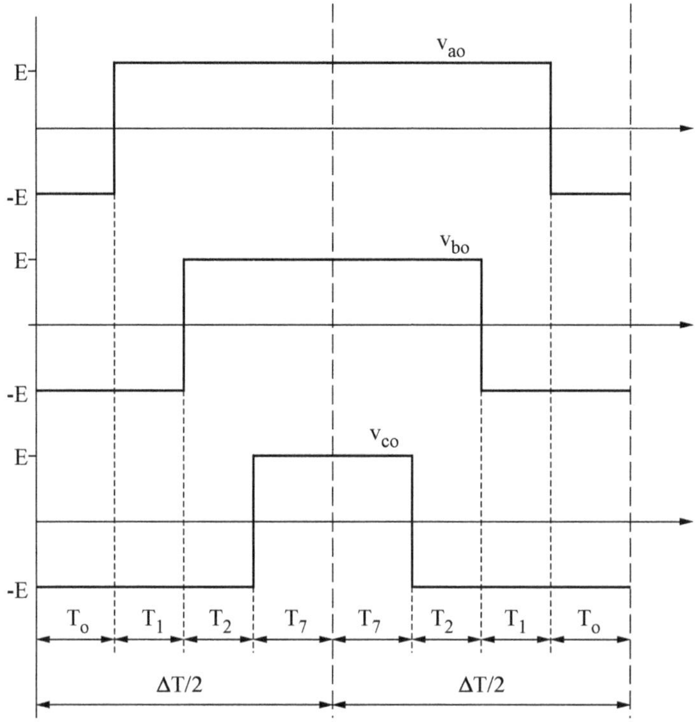

Fig. 11.35 SVM pulse position

As $0 \leq (T_1, T_2) \leq \Delta T/2$, the maximum amplitude V is indeed V_m and in the first sector this can only occur for $\theta = 0$ or $\theta = \pi/3$.

Further, as $0 \leq T_1 + T_2 \leq \Delta T/2$, the maximum value for $T_1 + T_2$ in the first sector occurs for $\theta = \pi/6$ and the corresponding maximum value for V is given by $V_{max} = V_m \sin \pi/3 = (2/\sqrt{3})E$. Thus for SVM the maximum line-to-neutral voltage is $(2/\sqrt{3})E$ (or $2E$ for the maximum line-to-line voltage). This corresponds to the radius of the circle inscribed inside the hexagonal of the voltages of a VSI, and this is also exactly the maximum voltage for a natural or regular sampled PWM with optimal third harmonic injection.

In general, however, $T_1 + T_2 < \Delta T/2$ and the remaining time in the interval $\Delta T/2$ must be filled by zero vectors, $\underline{V}_o$ or $\underline{V}_7$. The choice between $\underline{V}_o$ or $\underline{V}_7$ is such that the switching is reduced as much as possible. For example, if switching to $\underline{V}_o$ requires changing one switch and switching to $\underline{V}_7$ requires changing two switches, then $\underline{V}_o$ will be selected. The active pulses (or vectors) will also be centred in each switching interval and the remaining time will be divided between $\underline{V}_o$ and $\underline{V}_7$. For an interval in the sector $0 \leq \theta \leq \pi/3$, the resulting sequence will be $\underline{V}_o \rightarrow \underline{V}_1 \rightarrow \underline{V}_2 \rightarrow \underline{V}_7 \rightarrow \underline{V}_7 \rightarrow \underline{V}_2 \rightarrow \underline{V}_1 \rightarrow \underline{V}_o$ (see Fig. 11.35).

This sequence is exactly the same as for the asymmetrical sampled PWM (triangular carrier with sampling on both rising and descending sides), except that, in the classical PWM, the times for $\underline{V}_o$ and $\underline{V}_7$ are not necessarily equally divided (as these are determined by the algorithm without any additional degree of freedom).

Remarks:

- the reversal of the sequence, i.e. ($\underline{V}_o$, $\underline{V}_7$, $\underline{V}_7$,$\underline{V}_o$) instead of ($\underline{V}_o$, $\underline{V}_7$, $\underline{V}_o$,$\underline{V}_7$), is not necessary, but without it the SVM becomes equivalent to a regular sampled sawtooth PWM (with a somewhat worse harmonic content).
- if the on-times T_1 and T_2 are not calculated every $\Delta T/2$ but only every ΔT, SVM becomes equivalent to a symmetrical regular sampled PWM.
- it can be proven that SVM inherently adds third harmonics, as mentioned above.

Part III
Electrical Drives and Special Electric Machines

Chapter 12
DC Commutator Motor Drives

Abstract In the past, the DC machine was the only motor that could provide easy speed control. Nowadays, since the advent of power electronics, rotating field machines offer cheaper and more powerful possibilities for speed control. Nevertheless, in older industrial installations and environments, DC motor drives may still be used. As a result, and also because of their excellent characteristics, it remains instructive to shortly review the possibilities of DC machines for speed control, including braking in particular.

12.1 Basic Characteristics of DC Motors

In Chap. 2, we already discussed the basic characteristics of separately, shunt and series-excited DC motors in detail. Basically, there are two types of torque-speed characteristics:

- the (almost) linear torque-speed characteristic of separate and shunt motors
- the (almost) hyperbolic torque-speed characteristic of series-excited motors

By combining series and parallel excitation windings (i.e. compound), we may obtain intermediate characteristics. Adding series or parallel resistances allows us to somewhat modify these characteristics (albeit not without introducing extra losses).

All characteristics can be derived from the following equations (in absolute values):

$$\begin{aligned} V_a &= E_a + (R_a + R_s + R_v)I_a = E_a + R_t I_a \\ E_a &= K \cdot \Omega_a \cdot \Phi_m \\ T_{em} &= K \cdot I_a \cdot \Phi_m \\ \Phi_m &= \Phi_m(F_m) \end{aligned} \tag{12.1}$$

where: R_s is the resistance of the series excitation winding (if present); R_v (if present) is a resistance between the supply and the machine; $K = \frac{z}{2\pi a} N_p$ and $F_m = w_m I_m = \sum w_x I_x$ (armature reaction can be accounted for by deducting a term $K_a I_a$ from the mmf).

J. A. Melkebeek, *Electrical Machines and Drives*, Power Systems,
https://doi.org/10.1007/978-3-319-72730-1_12

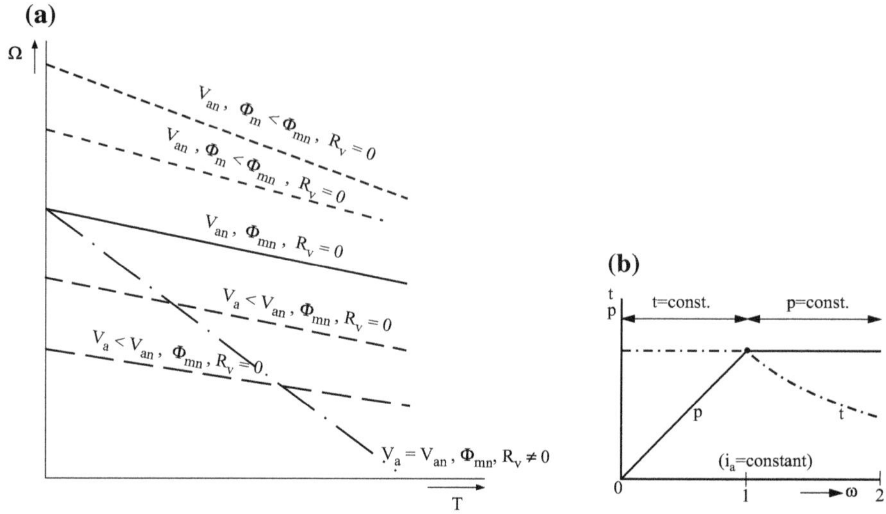

Fig. 12.1 DC motor with separate excitation: **a** torque-speed characteristics; **b** power and torque diagram

In per unit, these equations are

$$\begin{aligned} v_a &= e_a + (r_a + r_s + r_v)i_a = e_a + r_t i_a \\ e_a &= \omega_a \cdot \varphi_m \\ t_{em} &= i_a \cdot \varphi_m \\ \varphi_m &= \varphi_m(f_m). \end{aligned} \tag{12.2}$$

12.2 Torque-Speed Characteristics of Separately Excited or Shunt-Excited DC Motors

12.2.1 Basic Characteristics

Basically, the torque-speed characteristic of a separately excited (or shunt-excited) DC motor is a straight line with a slope determined by the resistance of the armature circuit (R_a or $R_a + R_v$), as illustrated[1] in (a) in Fig. 12.1.

For speeds below the rated speed (i.e. for voltages below rated voltage), speed control is achieved by varying the armature supply voltage. This results in parallel shifted characteristics. For speeds higher than rated, speed control can be achieved by reducing the flux (flux weakening). Flux weakening (for higher speeds) yields torque-

[1]The figure only shows the first quadrant, although the characteristics continue in the second quadrant for generator operation (see Sect. 12.2.2).

speed lines with a larger slope. However, note that the more the flux is weakened, the more the torque is limited (for the same maximum, i.e. rated, armature current). This range is called the constant power range (see (b) in Fig. 12.1). In Sect. 12.2.2, some implementations of such a voltage control and flux weakening are discussed in more detail.

Another method for speed control is to add resistances R_v in series with the armature, but this method is not lossless at all. For that reason, series resistances are nowadays only used for starting purposes, to reduce the starting current. Series resistances are no longer considered as an acceptable method for speed control.

Braking of shunt or separately excited DC motors is possible in quadrant II (with energy recuperation, thus as a generator) or in quadrant IV. In quadrant IV, however, braking results in high energy dissipation as not only mechanical energy but also additional electrical energy is converted into heat.

12.2.2 Ward-Leonard Drive

The basic (rotating) Ward-Leonard drive consists of a separately excited DC motor fed by a separately (or compound) DC generator, as illustrated in (a) in Fig. 12.2. The DC generator is driven by an internal combustion engine, or by an induction motor. To control speeds less than or equal to rated speed, the voltage supplied to the DC motor is varied by controlling the excitation of the generator (the curves a in (b) in Fig. 12.2). To obtain speeds higher than rated, the excitation of the DC motor can be decreased (field weakening) (see curves b in the figure).

There are no real limitations as to braking (plugging) or regenerative braking. For regenerative braking, the DC motor will work as generator while the generator

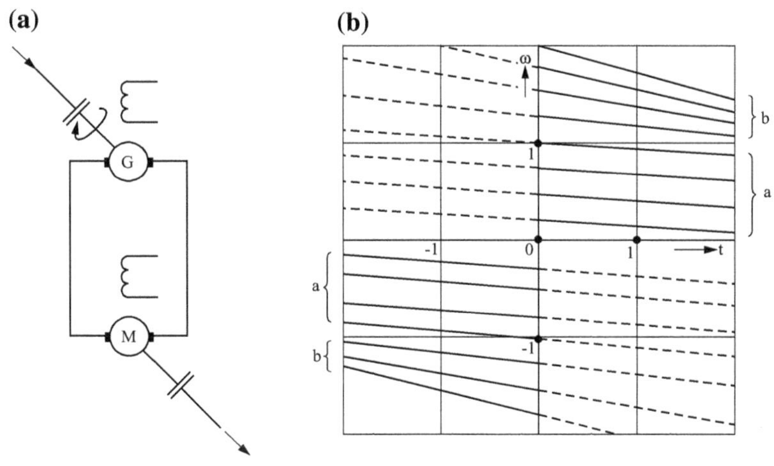

Fig. 12.2 Ward-Leonard (rotating)

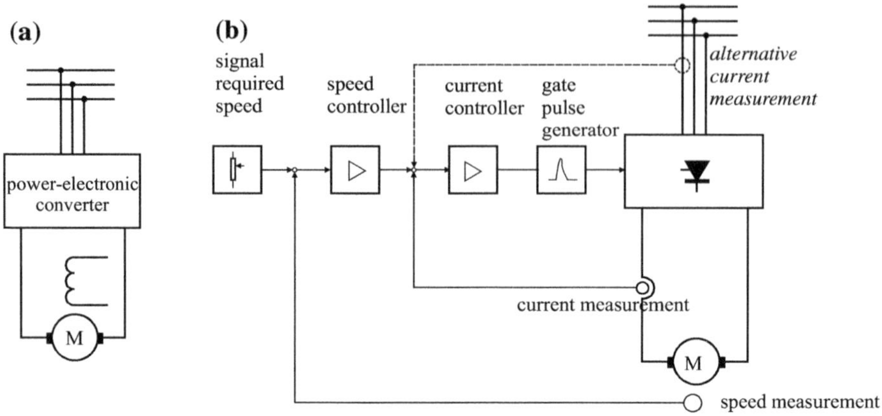

Fig. 12.3 Static Ward-Leonard **(a)** and speed control circuit **(b)**

will work as motor. However, for real energy recuperation, also the machine driving the generator must be able to convert the mechanical energy provided by the DC generator, which is now working as a motor, into useful energy. Naturally, this does not pose any problems for, for example, an induction motor driving the DC generator.

Motoring (or generating) in the other rotation direction can be achieved either by reversing the excitation of the generator, or by switching the connections between generator and motor.

Advantages of the rotating Ward-Leonard include the very accurate speed control without any significant losses, the four-quadrant operation with possible energy recuperation, and the presence of the large inertia of the driving engine and generator to absorb even large load pulsations.

Disadvantages are the high acquisition price, the large space required, the cost of the building foundations to support the weight of the three machines, the losses associated with the threefold energy conversion, and the slow response time (due to the large time constant of the generator excitation).

In the static Ward-Leonard (Fig. 12.3), the generator and driving machine (or engine) are replaced by a power-electronic converter, usually a controlled rectifier fed from the grid. In fact, such a static converter also provides the required variable supply voltage for the DC motor. As regards to braking and energy recuperation, there are important differences, however. Indeed, in a simple controlled rectifier, the current cannot reverse. For example, consider an operating point in the first quadrant (motoring). When the load torque decreases, the speed will increase to the no-load speed for that supply voltage. When the load torque reverses (i.e. it becomes a driving torque), the operating point cannot shift into the second quadrant (on the prolonged characteristic), because this would correspond with a negative armature current. In order to attain operating points in the second quadrant, we need an antiparallel rectifier to absorb the negative current (and, as a result, recuperate the energy). With a controlled rectifier in antiparallel connection, all four quadrants are attainable, just as for the rotating Ward-Leonard.

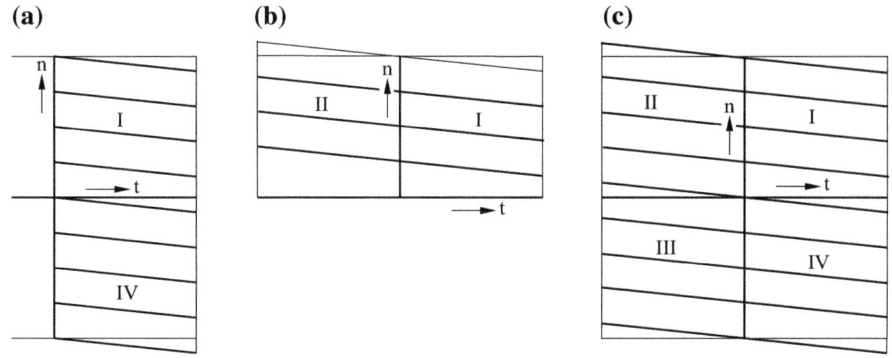

Fig. 12.4 Two- or four-quadrant operation

If, instead of the controlled rectifier, a diode rectifier with subsequent chopper is used, only the first quadrant can be attained (or the third, by switching leads). Of course, for braking, a braking resistance can always be used, albeit without energy recuperation.

Depending on the actual implementation, operation is either limited to one or two quadrants, or possible in all four quadrants (see Fig. 12.4).

Compared to the rotating Ward-Leonard, the static Ward-Leonard offers the following advantages:

- it is much cheaper
- it does not require much space or building investments
- it is easier to integrate into automatic control environments
- it allows for much faster control.

Disadvantages include:

- it is sensitive to over-currents, and thus not as robust as the rotating version
- power-electronic converters require fast built-in protections
- a controlled rectifier needs much reactive power
- grid harmonics of a rectifier
- the harmonics in the DC voltage may cause commutation problems for the DC motor and current harmonics may result in torque harmonics (i.e. vibrations, noise).

12.3 Characteristics of Series-Excited DC Motors

12.3.1 Speed Control

The hyperbolic torque-speed characteristic of a series-excited DC motor can also be altered for lower speeds by reducing the supply voltage below the rated voltage.

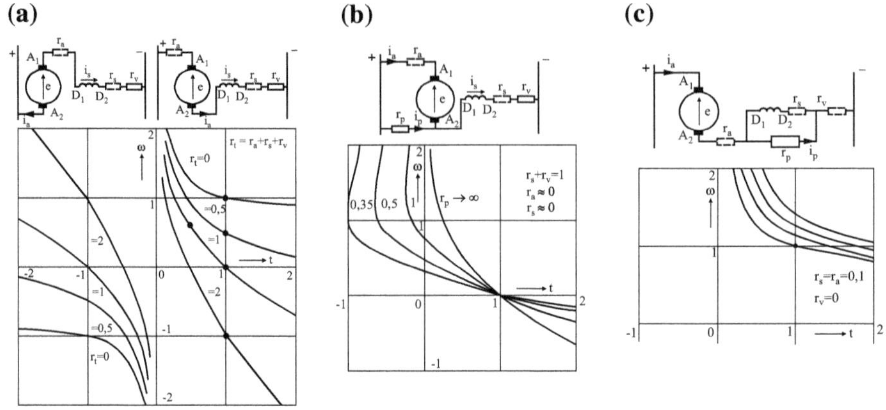

Fig. 12.5 Torque-speed characteristics of series DC motor

However, this results in much steeper curves, as is illustrated in (a) in Fig.[2] 12.5. Curves that are too steep, however, lead to control difficulties (i.e. a small torque variation results in too high a speed variation). The high losses in the series resistance constitute another disadvantage of this control method. Moreover, the speed runaway for low torques is not avoided.

The circuit (b) in Fig. 12.5 provides a solution to most of these problems. If a resistance r_p is connected in parallel with the armature, part of the excitation current is provided directly, without passing the armature. The lower the parallel resistance, the flatter the characteristics are. Note that the characteristics also enter the second quadrant (albeit without net energy recuperation).

Field weakening (for higher speeds) is achieved by a parallel resistance to the series excitation winding. It is easily shown[3] that the resulting characteristics shift parallel to higher torque values, as is illustrated in (c) in Fig. 12.5.

In the past, electrical traction was an important application of series DC motors. Nowadays, if and when series DC motors are still used in traction, power electronic control methods are applied (see below in Sect. 12.3.3).

12.3.2 Braking

Braking is needed either for shortening the run-out (i.e. reversing the torque for the same rotation direction), or for controlling the speed when the speed direction reverses for the same torque direction (for example, for lowering the load of cranes).

[2]The characteristics to the left are obtained by reversing the connection between armature and series excitation; reversing the polarity of the supply does not affect the characteristics.

[3]To prove this, consider operating points with the same flux level and therefore also the same excitation current.

12.3.2.1 Plugging

Speed reversal, from motoring in quadrant I to braking in quadrant IV, occurs when the load torque becomes too high, or, for a constant load torque, when the motor torque decreases (for example, if the armature voltage is reduced).

For braking in the same speed direction, the connection between armature and excitation winding must be reversed, i.e. connecting for motoring in the other rotation direction. The operating point then moves from quadrant I to quadrant II.

Please note that in both cases no energy recuperation[4] occurs. On the contrary, both (additional) electrical and mechanical power are converted into heat. This kind of braking is sometimes denoted as plugging.

12.3.2.2 Resistive Braking

For resistive braking, the series-excited DC motor is disconnected from the supply and connected to a load resistance. With sufficient remnant magnetism and the right rotation direction, self-excitation may occur. In self-excitation, the emf from the remnant flux must be such that the resulting current in the field winding reinforces the flux. For example, when the motor has previously operated in quadrant I (positive speed with the connections $D2 - D1 - A2 - A1$ and e.g. $A1$ connected to the plus of the supply), the remnant emf direction will be from $A2$ to $A1$ ($A1$ positive). Connecting the terminals to a resistance will result in a current which will reinforce the field if the rotation direction is now negative. To obtain braking in the positive speed direction, the connections between field and armature must first be reversed, i.e. $D2 - D1 - A1 - A2$. The characteristics are as those illustrated in (a) in Fig. 12.6. By varying the braking resistance, we can obtain different curves.

The main disadvantage of this braking method is that braking to zero speed is slow or non-existent. A second disadvantage is that the presence of a sufficient remnant flux is often uncertain.

12.3.2.3 Potentiometric Braking

To overcome the disadvantages of pure resistive braking, the circuit depicted in (b) in Fig. 12.6 can be used. In this circuit, the supply is connected via a resistance r_v. Therefore, the braking is no longer exclusively dependent on the remnant magnetism. The characteristics (here for different values of r_l) do not pass through the origin, resulting in braking even for small torques and speeds. Other characteristics, for example, for different values of r_l', are also possible. Although the braking characteristics continue in the third and fourth quadrant, energy recuperation is very limited or absent.

[4]Moreover, it can be shown that a series-excited machine is not able to recuperate electrical energy into a voltage supply.

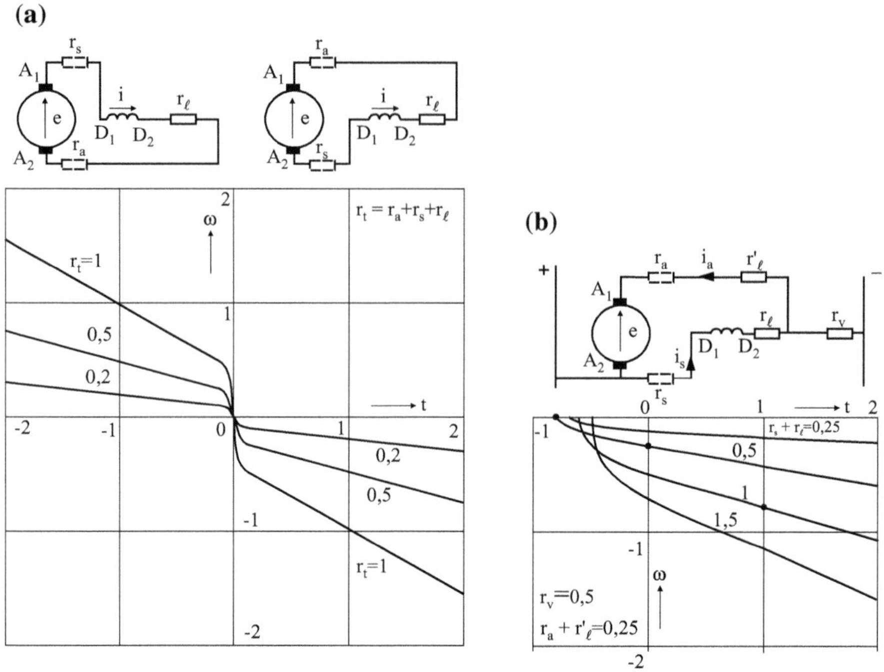

Fig. 12.6 Resistive and potentiometric braking of a series motor

12.3.2.4 Recuperative Braking

In most, if not all, of the above braking configurations, energy recuperation is not possible. For a DC machine with a pure series excitation, energy recuperation is not possible at all, as can easily be demonstrated.[5] To achieve energy recuperation, the only possibility is to remove the (exclusive) series excitation. In (b) in Fig. 12.6, the exclusive series character of the excitation is removed by the resistance r_v. A more drastic solution is illustrated in (a) in Fig. 12.7: the excitation winding is fed independently (for example, by a chopper). In this way, the motor becomes separately excited and regenerative breaking is possible.

[5]A transition from motoring in the first (or third) quadrant to generating in the second (or fourth) quadrant is impossible: for zero load torque the speed goes to infinity, and armature current and excitation current become zero. Even if there were an (artificial) 'operating point' with recuperation into a voltage supply, such an operating point would not be stable. Again, this can be proven quite easily.

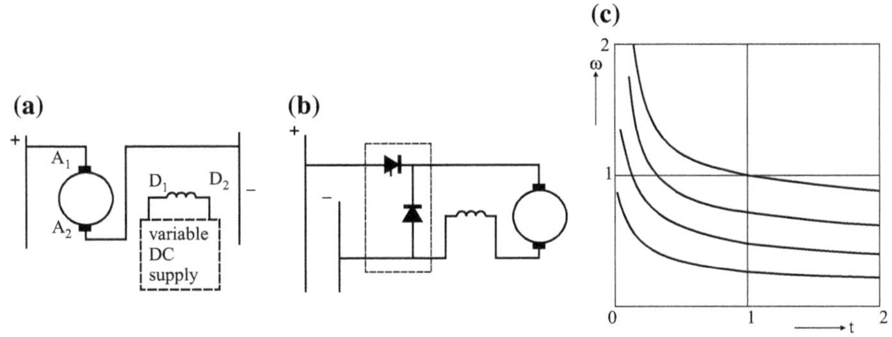

Fig. 12.7 **a** Independent supply of the series excitation winding; **b** Chopper supply; **c** Characteristics for decreasing voltage

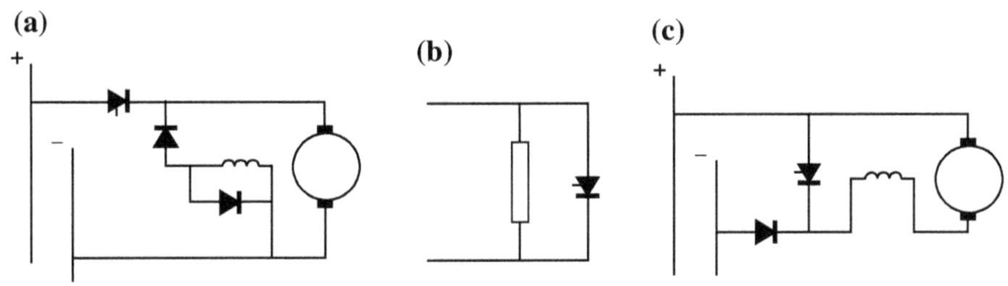

Fig. 12.8 Power-electronic supply of series DC motors

12.3.3 Power-Electronic Supply of Series-Excited DC Motors

A power-electronic supply of a series motor permits a nearly lossless variation of the motor voltage. Either a controlled rectifier or a (buck) DC chopper (as illustrated in (b) in Fig. 12.7) can be used. In the case of a DC chopper, the DC source voltage v_n is modified into a variable motor voltage $v_m = \alpha \cdot v_n$. Disregarding the resistive voltage drop over the armature and series excitation resistances, the emf of the machine will also vary proportionally to α. The resulting torque-speed characteristics are therefore shifted parallel to lower speed values ($\alpha \leq 1$), as can be seen in (c) in Fig. 12.7. Speed values higher than rated are obtained by field weakening. Figure 12.8 (a) shows a circuit for automatic field weakening, yielding a gradual field weakening when α increases.[6]

Please, note that a simple chopper does not permit negative current. Therefore, operation is limited to motoring in the first quadrant (or the third quadrant, if the connection between armature and excitation winding is converted). As such, regenerative braking is not possible either.

[6]Indeed, the larger α, the smaller the time available to increase the excitation current, and vice versa.

Advantages of chopper-supplied (and controlled-rectifier supplied) series motors include the nearly lossless speed variation and therefore also less heat dissipation, the fact that starting current and associated losses can be reduced, the fine and accurate control[7] possibilities, and the great aptitude for automation.

For braking, resistive braking can obviously be used. If the resistance is chopped, as in (b) in Fig. 12.8, better braking control can be achieved. Regenerative braking is possible by means of an anti-parallel chopper, as is shown in (c) in Fig. 12.8.

[7]This is important for the control of the adhesion between the rail and wheel in trains.

Chapter 13
Constant Frequency Voltage Supply of Rotating Field Machines

Abstract For rotating field machines supplied by a constant frequency supply, there are almost no acceptable means of speed control (and none whatsoever for synchronous machines). This section will mainly focus on starting, accelerating and braking, and speed control will be discussed in later chapter. Pure synchronous machines fed from a constant frequency supply cannot start and accelerate, as should be clear from Chap. 5 in Part 1. However, when starting and accelerating is required, the damper winding of a synchronous machine (which is usually required for stability) can be adapted to give a starting and accelerating torque, operating as an asynchronous machine. When approaching synchronous speed, the DC supply of the field winding can be switched on and the machine will synchronise when the load is not too large. The asynchronous starting, accelerating and braking discussed in the first section below is therefore also important for synchronous machines which are not too large. In the next section, speed control and cascade connections for slip-ring induction motors are reviewed. Other sections discuss the behaviour of rotating field machines at voltage variations and power-electronic voltage control for rotating field machines.

13.1 Start-Up, Accelerating and Braking of Squirrel-Cage Induction Machines

13.1.1 Accelerating Time and Power Loss

Accelerating time and power loss (heat) can be calculated starting from the equation of motion:

$$J\frac{d\Omega}{dt} = T_m - T_l \tag{13.1}$$

where J is the combined inertia of machine and load, T_m is the motor torque and T_l is the load torque.

The motor torque depends on the speed (slip) and even if the load torque is constant, integration of this equation is not straightforward.

J. A. Melkebeek, *Electrical Machines and Drives*, Power Systems,
https://doi.org/10.1007/978-3-319-72730-1_13

For a zero load torque, with only the machine inertia and *if the rated machine torque T_{nom} is available throughout the acceleration*, the resulting acceleration time τ_{nom} from standstill to synchronous speed (called the *rated* or *nominal acceleration time*) is easily calculated as follows:

$$\tau_{nom} = \frac{J_m \Omega_{sy}}{T_{nom}} \tag{13.2}$$

Obviously, this is a theoretical value which only serves as a reference (and is related to machine size).

The actual accelerating time with load inertia and under load (and with an accelerating torque $T - T_l$ which is in general dependent on the speed) can be calculated by integrating equation 13.1

$$\tau_v = (J_m + J_l) \cdot \int_{\Omega_{min}}^{\Omega_{max}} \frac{d\Omega}{T_m - T_l} \tag{13.3}$$

or in pu:

$$\tau_v = \tau_{nom} \cdot (1 + j_l) \cdot \int_{s_{min}}^{s_{max}} \frac{ds}{t - t_l} \tag{13.4}$$

with $j_l = J_l/J_m$ the relative or pu load inertia, $t = T_m/T_{nom}$ the pu torque and $s = (\Omega_{sy} - \Omega)/\Omega_{sy}$.

This can usually be approximated by a finite summation:

$$\tau_v = \tau_{nom} \cdot (1 + j_l) \cdot \sum \frac{\Delta s}{\Delta t} \tag{13.5}$$

The rated acceleration time τ_{nom} depends on the power rating of the machine[1] and varies between 0.1 s and some seconds: $\tau_{nom} \approx (1.6 \ldots 2)\tau_p$.

The relative load inertia is usually less than 1 (e.g. for pumps mostly less than 0.1), but can be higher than 10 as well.

The power dissipation during start-up and acceleration is relevant here. It can be a limiting factor: if the duration of the start-up is rather short, this might lead to an excessive temperature increase of the machine. During acceleration, the power loss in the windings can be ten times larger than the iron losses (whereas both have the same order of magnitude in rated operation). Especially the losses in the cage are important because of the skin effect, the unequal distribution of heat over the rotor bar height and the less efficient cooling of the rotor at lower speeds.

The (infinitesimal) heat dissipation in the rotor (cf. slip loss) during an infinitesimal time $\delta\tau$ is given by

---

[1]The inertia varies with the fifth power of the pole pitch, while the rated torque varies with approximately the fourth power of the pole pitch.

$$\delta W_{th} = T_m \Omega_{sy} s \cdot \delta\tau \tag{13.6}$$

From Eq. 13.1, the time interval $\delta\tau$ is expressed as a function of the speed or slip variation, which leads to

$$W_{th} = \int \delta W_{th} = \int_{s_{min}}^{s_{max}} T_m \Omega_{sy} s \cdot \frac{J_m + J_l}{T_m - T_l} \cdot \Omega_{sy} \cdot \delta s = (J_m + J_l) \cdot \Omega_{sy}^2 \cdot \int_{s_{min}}^{s_{max}} \frac{T_m}{T_m - T_l} s \cdot \delta s \tag{13.7}$$

For a start-up without load ($J_l = 0$ and $T_l = 0$), the resulting (rotor) losses are called the *rated* or *nominal starting losses.*

$$W_{nom} = \frac{1}{2} J_m \cdot \Omega_{sy}^2 \tag{13.8}$$

Apparently, for a start-up without load, these losses are equal to the kinetic energy. The energy required from the mains for this start up is thus twice the kinetic energy. It is important to note that, whereas in rated conditions only a small fraction of the supply power is converted into losses, for start-up 50% of the supplied power is dissipated.

For a start-up with load torque (and load inertia), Eq. 13.7 must be integrated (where T_m is a function of the slip). Equation 13.7 can be rewritten in pu as follows

$$W_{th} = 2W_{nom}(1 + j_l) \cdot \int \frac{s \cdot \delta s}{1 - t_l/t_m} \tag{13.9}$$

For a zero load torque, the dissipated energy is again equal to the kinetic energy: $W_{th} = \frac{1}{2}(J_m + J_l) \cdot \Omega_{sy}^2 = W_{nom}(1 + j_l)$. For a non-zero load torque, Eq. 13.7 can sometimes be approximated by averaged values

$$W_{th} = W_{nom}(1 + j_l) \cdot \sum \frac{\Delta(s^2)}{(1 - t_l/t_m)_{av}} \tag{13.10}$$

Note that W_{nom} is independent of the rotor resistance of the machine; in contrast, W_{th} does depend on the rotor resistance used for start-up (e.g. for slip-ring machines).

There exists a straightforward relationship[2] between W_{nom}, τ_{nom} and P_{nom}: $W_{nom} = \frac{1}{2} P_{nom} \tau_{nom}$.

In line with this, for a start-up with load inertia but without load torque, we get $W_{th} = \frac{1}{2}(1 + j_l) \cdot P_{nom}\tau_{nom} = \frac{1}{2} P_{nom} \tau_v$.

The product $P_{nom}\tau_{nom}$ varies with machine size (thus the pole pitch) proportionally to τ_p^5 (indeed, for $\tau_{nom} \sim \tau_p$, see above, and for $P_{nom} \sim \tau_p^4$, see Chap. 4 in Part 1).

This indicates a possible thermal problem for start-up, as the acceptable heat dissipation during start-up is proportional to the mass of the machine (rotor), $W_{ac} \sim \tau_p^3$. For large power ratings, W_{th} could become larger than W_{ac}, even without load

[2]Prove this with the definition of $P_{nom} = \Omega_{sy} T_{nom}$ and the expression for τ_{nom}.

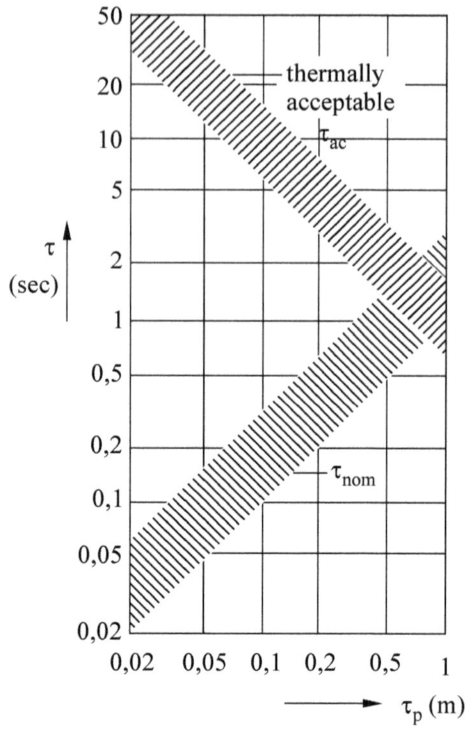

Fig. 13.1 Nominal and acceptable starting time

inertia ($W_{th} = W_{nom}$). This is illustrated in Fig. 13.1, in which both $\tau_{nom} \sim \tau_p$ and $\tau_{ac} = W_{ac}/P_{nom} \sim \tau_p^{-1}$ are plotted.

Where both curves intersect, only the machine inertia can be accelerated; for larger pole pitches the machine is not allowed to accelerate, even without load inertia (according to the figure, this occurs for large power ratings, with a pole pitch larger than approximately 1 m).

Remark: Multi-speed machines have a special stator winding that can be switched between two numbers of pole pairs (usual ratios are 1:2 or 1:3). With multi-speed machines, the starting power loss is reduced by first accelerating to the synchronous speed for the higher pole number and then further accelerating to the synchronous speed for the lower pole number.[3]

[3]Prove that, if α is the ratio between the lowest and highest pole number, the power loss is reduced to a fraction $\alpha^2 + (1 - \alpha)^2$.

13.1.2 *Traditional Starting Methods for Cage Induction Machines*

Direct starting of squirrel-cage induction machines on the mains supply may result in two important problems: a limited starting torque for high power ratings (due to low rotor resistances), and a starting current which is too high.

There is no simple standard solution for the relatively low starting torques of high-power machines on a fixed frequency supply. As will be shown in Chap. 15, variable frequency supply offers an excellent solution (also for starting purposes).

For the high starting currents in the traditional fixed frequency supply situation, there are several possible solutions. The ultimate aim of these is to reduce the starting current as much as possible, while limiting the reduction of the starting torque. These standard starting methods include wye-delta, auto-transformer, series-parallel, stator resistance, and inductance starting. The first three methods realise a mains current reduction which is equal to the torque reduction, while for the last two the torque reduction is larger than the mains current reduction (see Table 13.1).

The results in this table can be derived as follows. The torque is proportional to the square of the *winding* current (for a given slip value). For the last two methods, stator inductance and stator resistance, winding current (subscript w) and supply current (subscript s) are the same. Therefore, the following holds:

$$\frac{t_{red}}{t_{dir}} = \left(\frac{i_{w,red}}{i_{w,dir}}\right)^2 = \left(\frac{i_{s,red}}{i_{s,dir}}\right)^2 \tag{13.11}$$

Table 13.1 Standard starting methods

IEV 411	Starting method	Winding current %	Mains current %	Torque %	Interruption	Starting time critical
−22–18	Direct	–	–	–	No	No
−22–26	Series-parallel	1/2	1/4	1/4	Yes	No
−22–19	Wye-delta	$1/\sqrt{3}$	1/3	1/3	Yes	No
−22–20	Auto-transformer	α	α^2	α^2		
−22–21	With 2 switches	α	α^2	α^2	Yes	No
−22–22	With 3 switches	α	α^2	α^2	No	No
−22–23	Stator inductance	α	α	α^2	No	Yes
−22–25	Stator resistance	α	α	α^2	No	Yes

For the series-parallel, wye-delta and auto-transformer methods, the supply current reduction is equal to the square of the winding current reduction, and thus equal to the torque reduction:

$$\frac{i_{s,red}}{i_{s,dir}} = \left(\frac{i_{w,red}}{i_{w,dir}}\right)^2 = \frac{t_{red}}{t_{dir}} \tag{13.12}$$

Consider a wye-delta start-up, for example. In the direct or normal (full winding voltage) situation, the machine winding is delta-connected and each winding gets the full line voltage V_l. The winding current is, however, only $1/\sqrt{3}$ times the supply current. For initial start up, the machine winding is wye-connected and, as such, each winding only gets the voltage $V_l/\sqrt{3}$. The winding current (equal to the winding current in a wye connection) is therefore also reduced by the same factor $1/\sqrt{3}$. The torque is reduced by the square of it, i.e. $1/3$. Yet, as the supply current in the full voltage (delta) connection is $\sqrt{3}$ times the winding current, the result of Eq. 13.12 is obtained. Naturally, this solution requires the rated machine winding voltage to be equal to the line voltage.

For the auto-transformer start-up, the primary or supply current is $1/\alpha$ times the secondary or machine current (if α is the transformer ratio), whereas in the direct connection, supply and machine (winding) current are equal. If we consider the fact that in each case the torque is proportional to the machine (winding) current, we can also obtain the result of Eq. 13.12. The downside of this solution is that it requires a rather expensive auto-transformer.

For the series-parallel connection, the reasoning is similar. The disadvantage of this solution is that two machines are needed on the same shaft that can be connected electrically in series or parallel on the mains.

Remarks:

1. Starting, braking, reconnecting a machine and switching an electric circuit on and off in general results in transient phenomena. For a single-phase AC electrical circuit, the transient can be avoided by switching on exactly at the instant the expected steady-state current goes through zero. However, this is not possible for a three-phase circuit of course. Therefore, direct starting of an induction machine always results in high transient torques (see Fig. 13.2).

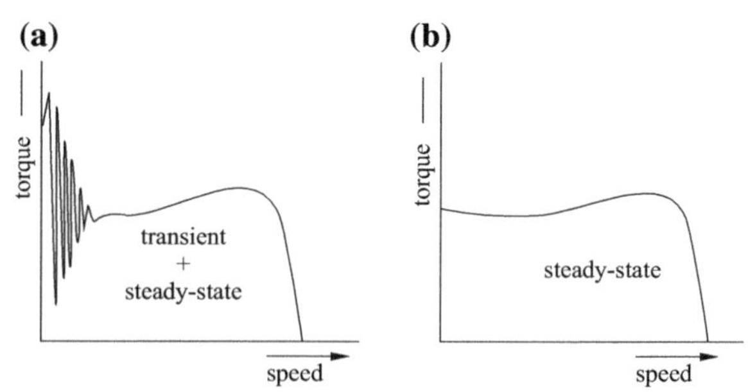

Fig. 13.2 Transient torque

With high load inertias, these pulsating transient torques may result in the shaft breaking, the rotor bars within the lamination breaking loose, ore the end windings coming loose, for example.
2. For the wye-delta start-up, there are two possibilities to switch between wye and delta (see Fig. 13.3).

Connection (1) is the preferred one, as this results in the lowest transient current at the switching instant. The cause for the difference between connections (1) and (2) lies in the voltage that is induced in the open-circuited machine winding at switch-off, due to the remnant flux in the rotor. The remnant flux in the rotor decreases with a relatively large time constant (up to seconds) and induces in the stator winding a remnant voltage with the lower frequency $(1 - s)\omega$ (corresponding to the rotor speed). As the time for switching from wye to delta is much shorter than the mains period, it is advantageous to switch to new winding voltages which are lagging (1) instead of leading (2) the old ones.

13.1.3 Braking of Induction Machines

13.1.3.1 Run-Out

If the machine is switched off and allowed to slow down without external load torque, the time elapsed until standstill may be rather large. For example, when the friction loss torque is only 10% of the rated torque (which is extremely large: usually friction and ventilation torque is less than $10^{-4}T_{nom}$), the run-out time will be 10 times the nominal starting time τ_{nom}. In many cases this is not acceptable, and braking will be required.

13.1.3.2 Electromagnetic Braking

Electromagnetic braking (IEC terminology) uses a mechanical brake to quickly slow down the machine. As such, the name is somewhat misleading, although the brake is indeed controlled by an electromagnet.

The brake can be a disk brake or drum brake, or something in between. The use of a drum brake is demonstrated in Fig. 13.4. The brake pads are pushed against the drum (or disk, in case of a disk brake) by a spring. This is in fact a "normally-on" device: in the absence of electrical power, it will brake. When power is applied to the machine, an electromagnet supplied by the same supply lines as the machine will retract the brake pads away from the drum or disk and no braking will occur.

This type of brake is used in many small applications, such as home appliances like tumble dryers.

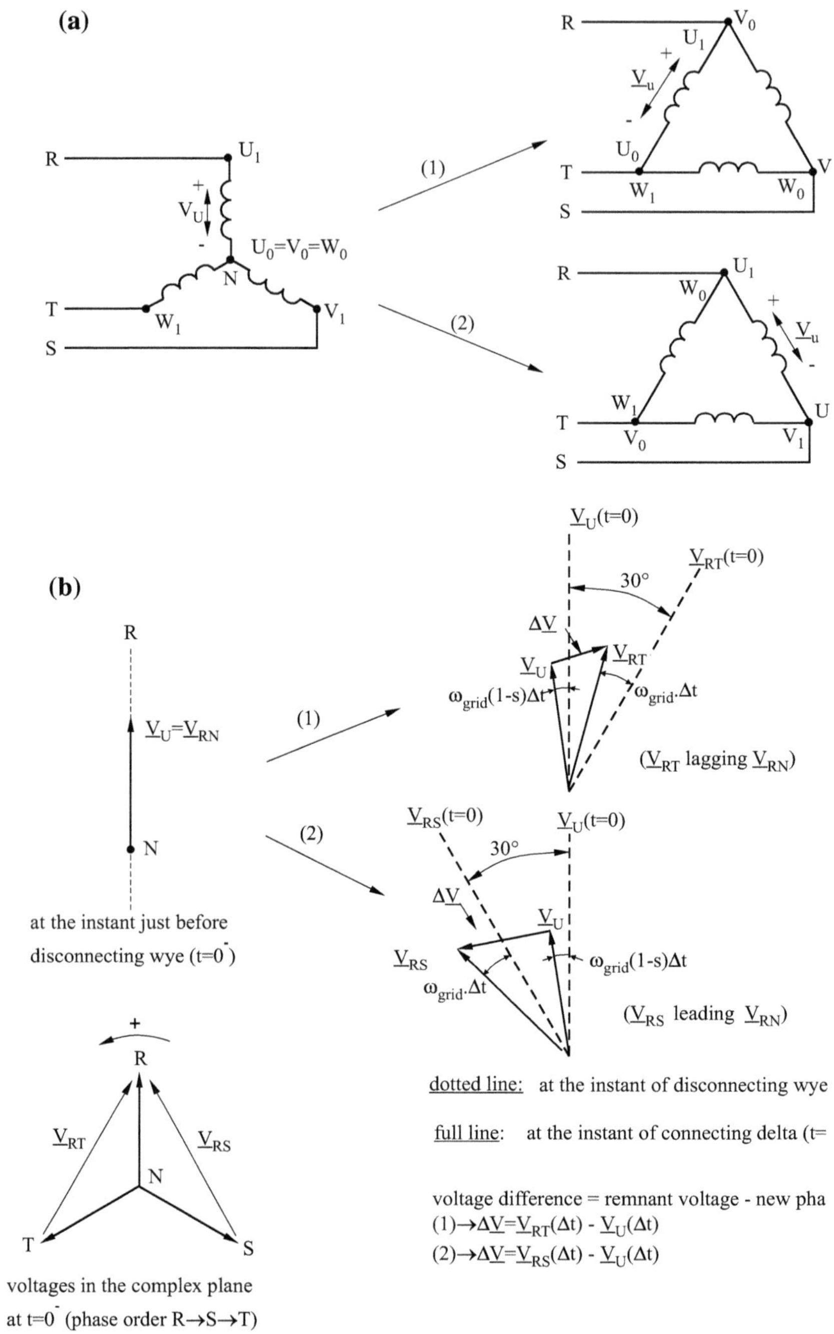

Fig. 13.3 Wye-delta switching

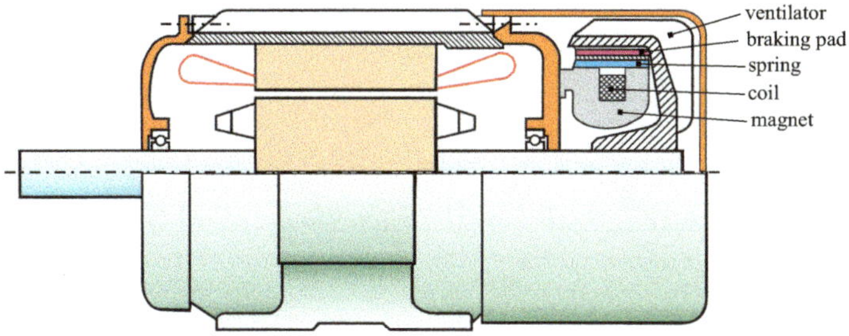

Fig. 13.4 Electromagnetic braking

13.1.3.3 Electric Braking

As to electric braking, some possibilities may be distinguished: reversal braking, braking to a lower speed for a multi-speed machine, DC-excited braking (i.e. a Foucault brake), and capacitive braking. Each will be discussed below.

(a) Reversal braking:

Fast braking can be obtained by switching the phase order, i.e. reversal of the synchronous speed. Braking is quite fast, but there is no energy recuperation and near standstill the machine must be switched off (if not, the machine will accelerate in the reverse direction, see Fig. 13.5).

(b) Braking for multi-speed machines:

When the machine is rotating at (or near) the highest synchronous speed (lower pole number), switching to the higher pole number results in regenerative braking to the lower synchronous speed (Fig. 13.6). From then on, other braking methods must be used for braking to standstill.

(c) DC-excited braking:

For DC-excited braking, the stator winding is switched off from the mains and connected to a DC supply (e.g. for a wye-connected winding, connecting phase U to the plus-side of the DC-source and V (or W) or the parallel connection of V and W to the minus side (see (a) and (b) in Fig. 13.7, respectively).

Fig. 13.5 Reversal braking

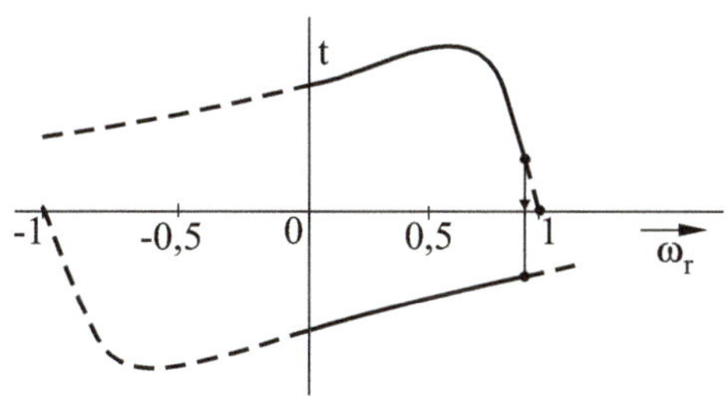

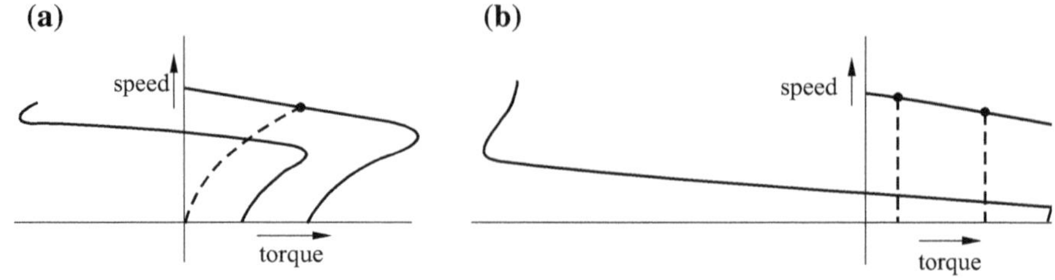

Fig. 13.6 Multi-speed braking

Fig. 13.7 DC connections

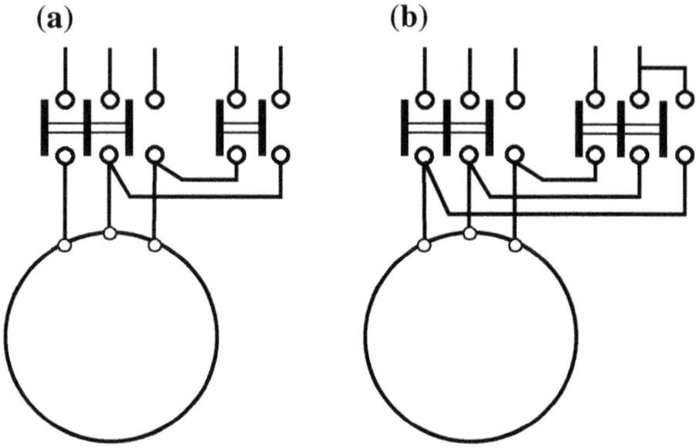

The DC current in the stator winding creates a standstill electromagnetic field in the air gap. Therefore, when the rotor is rotating, an emf is induced in the rotor windings, resulting in a rotor current and associated power loss. The kinetic energy is thus converted into heat.

The braking torque can be calculated as follows (see also Fig. 13.8). According to the simplified equivalent circuit in L, the induction machine torque can be written as

$$T = I^2 \cdot (R/s) \approx I_1^2 \cdot (R/s) \tag{13.13}$$

with I_1, I and R the stator current, the rotor current and rotor resistance, respectively. For normal AC voltage supply, the current varies with the slip (see (a) in Fig. 13.8). If the machine were supplied by an AC current source instead, the current amplitude would be constant and the corresponding torque as a function of the slip could be calculated point by point from the curves in (a), by means of Eq. 13.13 (see the result in (b) in Fig. 13.8).

For a DC current supply, the synchronous speed is zero. Therefore, the curve of (b) must be shifted to the left (see curve (c) in the figure).

The equivalence between DC and AC current amplitudes can be derived from Fig. 13.9: to the left, (a), we see the DC situation and to the right, (c), we see the instantaneous condition in AC at the instant the current in phase U is zero (time axis as in (b)). Thus, if the DC current I_{DC} is equal to $(\sqrt{3}/2)\hat{I}$, then its mmf is the same

Fig. 13.8 DC braking torque

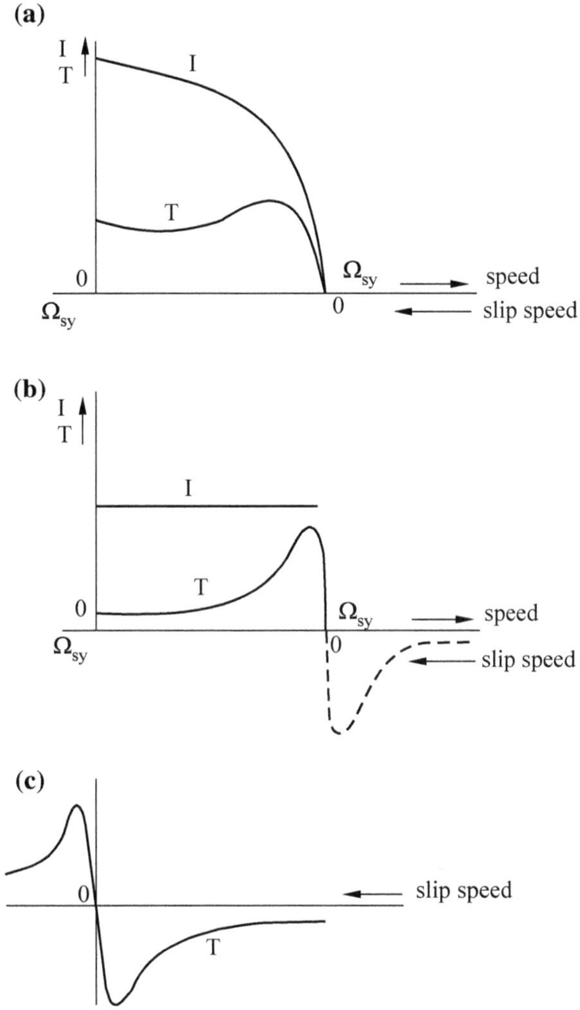

as the mmf $(3/2)\hat{I}$ of a three-phase current $\hat{I}$, which means that also the torques are equivalent. In other words, a DC current I_{DC} has the same effect as an AC current kI_n (with I_n the effective value) if $I_{DC} = (\sqrt{3/2}) \cdot kI_n$.

(d) Capacitive braking

For capacitive braking, capacitors and resistances are connected (in series or parallel) to the stator terminals, as is illustrated in Fig. 13.10. When the rotor rotates and a remnant magnetic field is present in the rotor, a small emf is induced in the stator windings. With an appropriate match of speed, capacitor and resistance values, the system may enter a self-excitation state in which an AC current converts the mechanical energy of the rotating rotor into heat in the resistances. As such, the frequency of this AC current will have to be lower than the speed (cf. negative slip for genera-

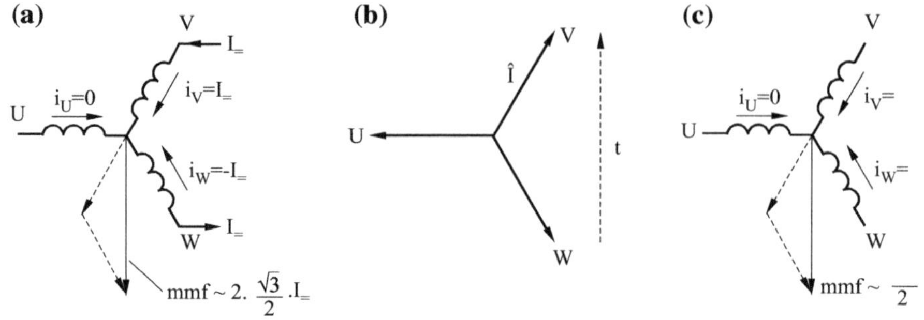

Fig. 13.9 Equivalence of DC and AC for the mmf

Fig. 13.10 Capacitive breaking-series connection

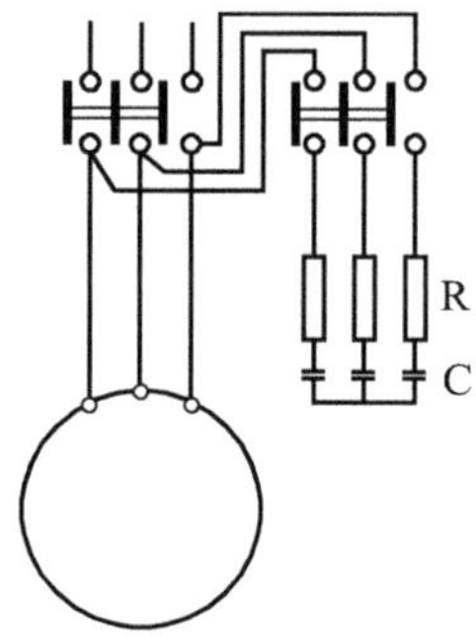

tor operation). The match between the parameters (i.e. speed, machine inductances, capacitors, resistances) will therefore also depend on the machine saturation level, since this determines the inductance values.

Although the self-excitation is inherently non-linear, the equivalent circuit offers some intuitive insight into the phenomenon. Consider the equivalent circuit in Fig. 13.11, where the stator terminals are connected to an external resistance and capacitor in series. If a non-zero current $\underline{I}$ exists in the circuit, then the following relation for the impedances must be fulfilled:

$$\underline{Z} = (R + 1/j\omega C) + \underline{Z}(\omega, s) = 0 \tag{13.14}$$

(from Kirchhoff's law $\sum \underline{V} = \underline{Z} \cdot \underline{I} = 0$), where $\underline{Z}(\omega, s)$ is the impedance of the machine which depends on the frequency and the slip. Denoting by ω_r the angular speed of the machine (in electrical radians), we can conclude that the relation between angular frequency, speed and slip is: $\omega_r = (1 - s)\omega$.

The complex equation 13.14 is equivalent to two real equations for two unknowns (i.e. slip and angular frequency). It should already be clear that, in order for a solution to exist, the slip should be negative, i.e. $\omega_r > \omega$. Instead of trying to find the general solutions for this (non-linear) equation, we will consider two extreme situations: very small slip, $|s| \approx 0$ and very large slip, $|s| \ggg 0$.

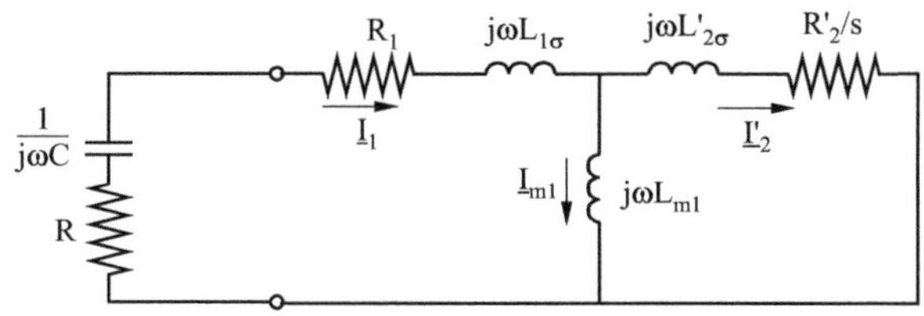

Fig. 13.11 Capacitive braking: equivalent circuit

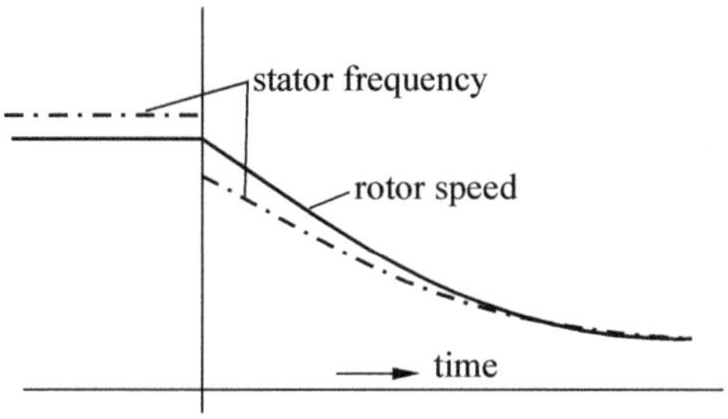

Fig. 13.12 Capacitive braking: speed and frequencies

For large slip, on the one hand, the machine impedance can be approximated to $\underline{Z}(\omega, s) \approx R_2'/s + j\omega L_{1\sigma}$ and therefore $\omega \approx 1/\sqrt{L_{1\sigma}C}$ and $s \approx -R_2'/(R + R_1)$.

For very small slip, on the other hand, the machine impedance is approximately $\underline{Z}(\omega, s) \approx L_{m1}^2 \cdot s/R_2' + j\omega L_{\sigma 1} + j\omega L_{m1}$ and therefore $\omega \approx 1/\sqrt{L_1 C}$, while $s \approx -(R + R_1)R_2'/L_{m1}^2$.

In each of these cases, the speed must correspond to $\omega_r = (1 - s)\omega$. We may therefore conclude that when capacitive braking is used to slow down a machine, the braking starts (at the highest speed) with a large slip and a frequency approximately corresponding to the resonance frequency for the leakage inductance (this frequency must and will of course be lower than the electrical speed ω_r). When the machine slows down, both the slip and the frequency (as well as the rotation speed ω_r) decrease until the frequency approaches the resonance frequency corresponding with the main inductance. As soon as that happens, braking ceases (Fig. 13.12).

Note that, according to this approximated and linearised modelling, the amplitude of voltages and currents is not determined. In reality, the magnetic circuit is non-linear and the magnetic circuit saturation actually[4] lays down the amplitudes.

[4]It can be shown that this self-excitation is in fact a limit cycle of this non-linear system; when a linearised model is used, there is a zero in the origin which corresponds to a non-existing degree of freedom for the amplitudes.

13.2 Slip-Ring Induction Machines: Start-Up, Speed Control and Energy Recuperation

As follows from the equivalent circuit in Fig. 13.13a, or its simplified version (b), the torque and current of an induction machine is determined by the ratio of slip to secondary resistance (s/r).

From the circle diagram in (a) in Fig. 13.14, the curves for current and torque in (b) as a function of s/r can be derived. By increasing the secondary resistance, we can shift the advantageous range of higher torque with lower current values (i.e. $0 < |s/r| < 2$, thus around the rated value of $|s/r| = 1$) to higher slip values.

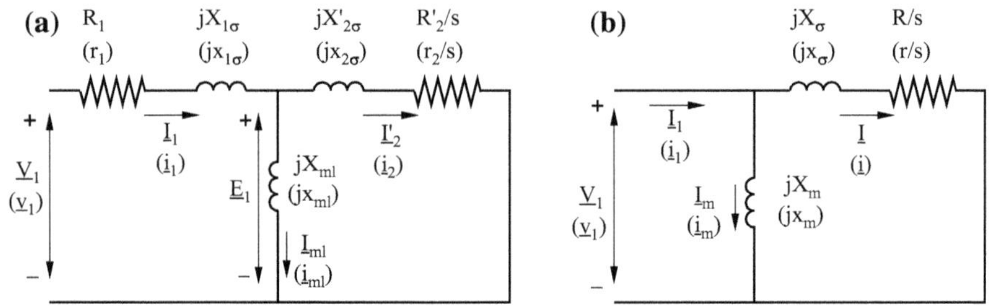

$$X_m = X_{1\sigma} + X_{ml} = (1+\sigma_1)X_{ml} = X_1$$

$$X_\sigma = X_{1\sigma}(1+\sigma_1)+X'_{2\sigma}(1+\sigma_1)^2 = \frac{\sigma}{1-\sigma}X_1$$

$$R = (1+\sigma_1)^2 R'_2$$

$$\underline{I} = \underline{I}'_2/(1+\sigma_1)$$

Fig. 13.13 Equivalent circuits of induction machines

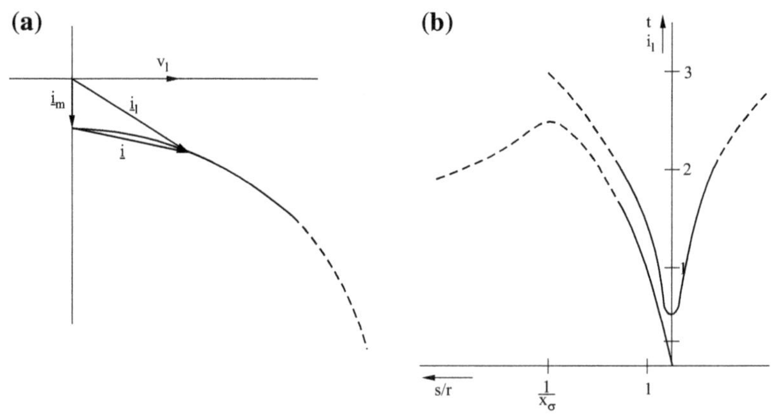

Fig. 13.14 Torque and current as a function of s/r

For slip-ring induction machines, the secondary resistance consists of the internal rotor resistance and an external resistance connected to the slip rings. By adjusting the external resistance R_{ex} (r_{ex}), we obtain that for all slip values between 1 and s_n ($\approx r_r$), the value of $|s/r|$ remains limited to $2 \cdots 5$ and the machine operates in its optimal range. This can be applied for starting and accelerating, and in theory also for speed control, but as there are more efficient methods to achieve this, it is hardly ever used for that purpose nowadays.

13.2.1 Start-Up of Slip-Ring Induction Machines

The use of secondary resistances for slip-ring induction machines eases start-up considerably, because higher torques at lower currents are obtained for lower speeds. A standard method to start slip-ring induction machines uses a battery of secondary resistances so as to obtain smooth starting between maximum and minimum torque values. As will be shown below, for starting between constant maximum and minimum machine torque values, the subsequent secondary resistance values (as well as the slip values at the minimum and maximum torque values) form a mathematical series (see Fig. 13.15).

Indeed, the torque is only determined by the value of $|s/r|$ and therefore we may write at t_{max} and t_{min}

$$\frac{s_1}{r_o} \equiv \frac{s_1}{r_r} = \frac{s_2}{r_1} = \frac{s_3}{r_2} = \ldots = \frac{s_n}{r_{n-1}} = \frac{s_{n+1}}{r_n} \tag{13.15}$$

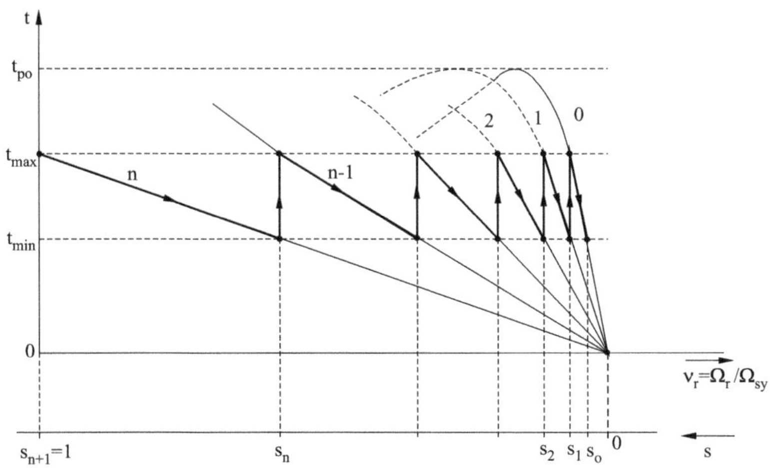

Fig. 13.15 Starting of slip-ring induction machines

$$\frac{s_o}{r_o} \equiv \frac{s_o}{r_r} = \frac{s_1}{r_1} = \frac{s_2}{r_2} = \ldots = \frac{s_{n-1}}{r_{n-1}} = \frac{s_n}{r_n} \tag{13.16}$$

respectively, where r_n, r_{n-1}, ..., r_1, $r_o = r_r$ are the consecutive secondary resistances in each step.

Dividing Eqs. 13.16 and 13.15 term by term leads to

$$\frac{s_1}{s_o} = \frac{s_2}{s_1} = \frac{s_3}{s_2} = \ldots = \frac{s_{n+1}}{s_n} \tag{13.17}$$

and

$$\frac{r_1}{r_r} = \frac{r_2}{r_1} = \frac{r_3}{r_2} = \ldots = \frac{r_n}{r_{n-1}} \tag{13.18}$$

Both slip and resistance values therefore form a mathematical series

$$\frac{s_{n+1}}{s_o} = \left(\frac{s_1}{s_o}\right)^{n+1} \tag{13.19}$$

$$\frac{r_n}{r_r} = \left(\frac{r_1}{r_r}\right)^{n} \tag{13.20}$$

with

$$\frac{s_1}{s_o} = \frac{s_{j+1}}{s_j} = \frac{r_1}{r_r} = \frac{r_{j+1}}{r_j} \tag{13.21}$$

If the machine starts from standstill, $s_{n+1} = 1$, the number of steps is given by

$$n = \frac{\log s_1}{\log s_o - \log s_1} \tag{13.22}$$

Clearly, the smaller the torque steps, the larger the required number of resistances.

Figure 13.16a illustrates a traditional configuration of resistances and switches. The required number of resistances (and switches) can be reduced by:

- asymmetrical resistances (see (b) in Fig. 13.16), creating a *Kusa*-circuit. This causes distortions of the torque-slip characteristic due to inverse rotating fields (Fig. 13.17) and is no longer used nowadays.
- a rectifier in de rotor circuit; instead of a three-phase battery of resistances, only one set of resistances is required, at the DC side (see (c) in Fig. 13.16).
- the step-wise variation of the secondary resistance can be replaced by a continuously variable liquid resistance, as is illustrated in (a) in Fig. 13.19. Besides the gradual variation of the resistance, another notable advantage is the high thermal capacity. Disadvantages are the vulnerability and required maintenance.
- a power-electronic version to replace the battery of resistances is obtained by placing a triac (or a pair of anti-parallel thyristors) in parallel or series with the external resistance (see Fig. 13.18).

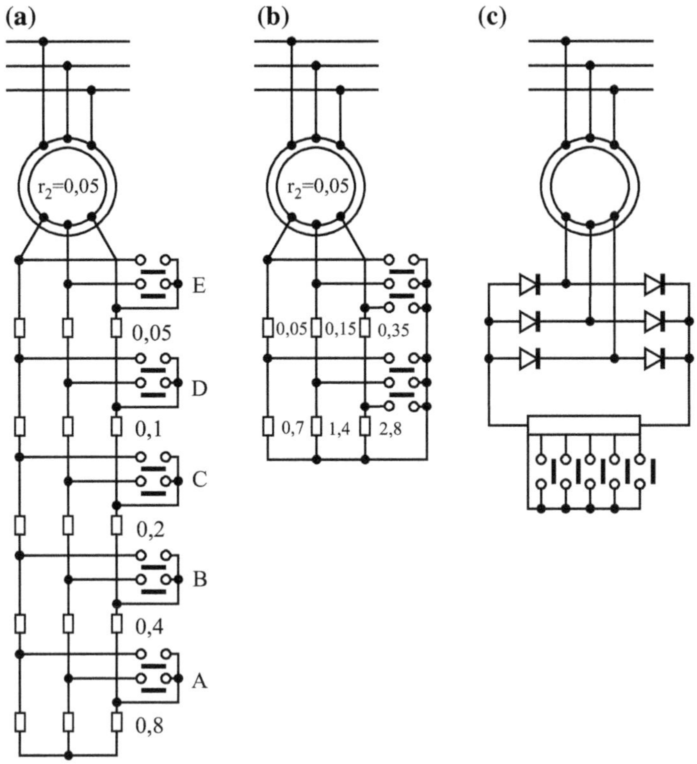

Fig. 13.16 Resistance battery for starting

Fig. 13.17 Torque deformation with asymmetrical resistances

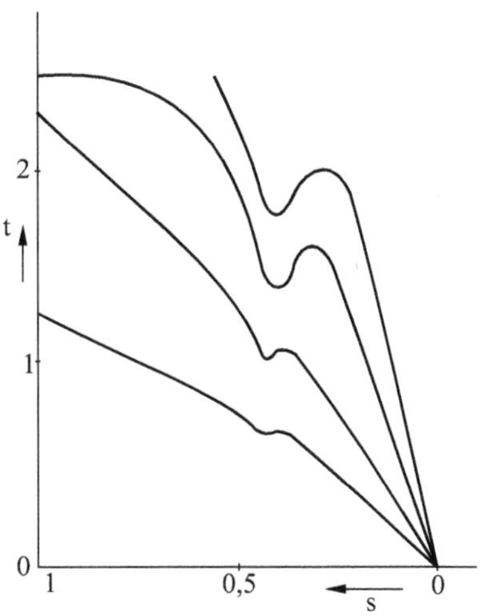

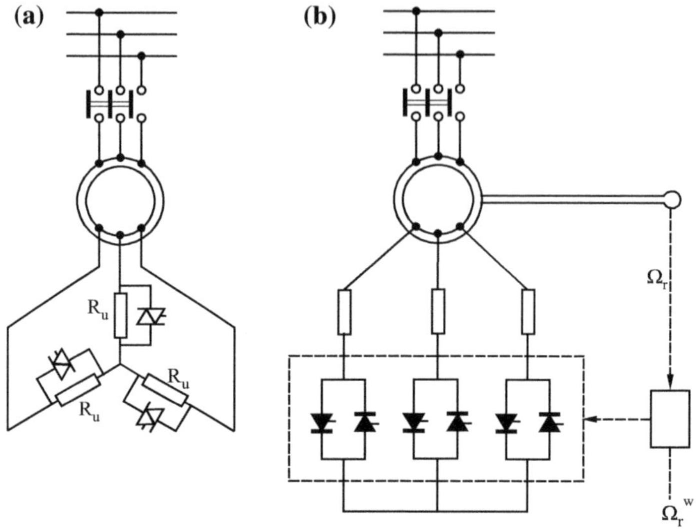

Fig. 13.18 Power-electronic resistance control

The parallel connection allows the external resistance to vary between 0 and R_{ex}. The series connection (somewhat more common) allows for an effective resistance variation between R_{ex} and ∞. These power-electronic versions also facilitate automatic speed control.

Remark: When the resistances are only used for starting, a construction was sometimes used in which, in steady state, the brush contact is replaced by short-circuit bars inserted in holes in the slip rings (see (b) in Fig. 13.19).

13.2.2 Speed Control of Slip-Ring Induction Machines Using Secondary Resistances

Secondary resistances can also be used for speed control, although this is nowadays mostly replaced by a variable frequency supply. A typical application was in the manipulation of loads, for example in harbour installations (e.g. cranes).

Figure 13.20 shows the speed torque characteristics for several values of the secondary resistances. The characteristics (1a) are intended to hoist up the loads. The characteristics (1b) can be used to hoist up small loads at low speeds, or, in the fourth quadrant, to lower them. Disadvantages are that the characteristics (1b) are rather steep, making it rather difficult to achieve a precise speed control (and lowering is limited to loads larger than approximately 67%). It is also possible to bring down loads by connecting the machine for the reverse direction (characteristics II). For this high-speed descending (in generating mode), speed control is almost entirely

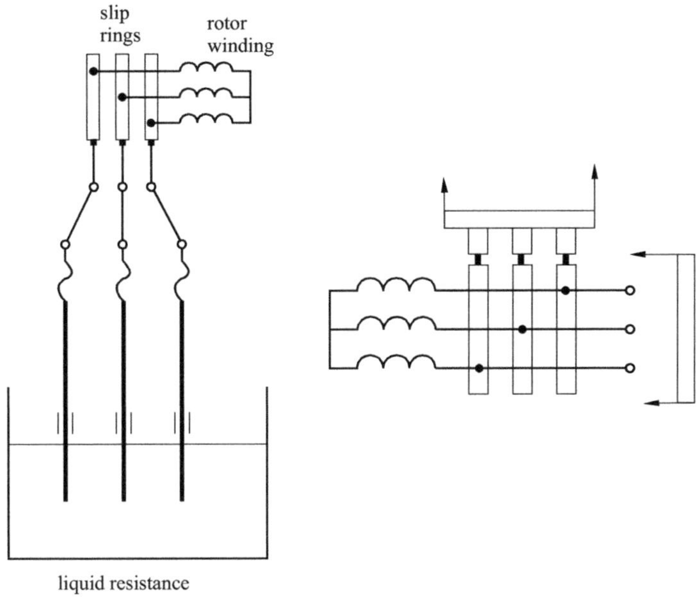

Fig. 13.19 Liquid resistance (**a**) and short-circuit bars (**b**)

Fig. 13.20 Secondary resistances for hoisting

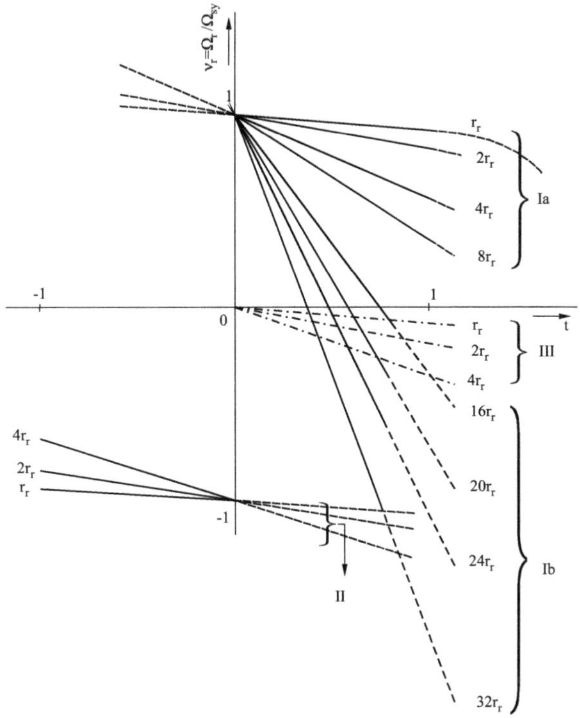

absent, however. Loads can be brought down at low speeds by means of DC braking (characteristics III), in which case speed control is available by varying the secondary resistance.

Remark: In most hoisting applications, the thermal management of the machine at these low speeds will not pose problems as periods of no-load generally alternate with periods with load. Moreover, most of the heat is produced outside the machine, in the external secondary resistances.

13.2.3 Speed Control of Slip-Ring Induction Machines by Means of Cascade Connections

13.2.3.1 Principle

A significant disadvantage of speed control by means of secondary resistances is the extremely low energy efficiency (approximately $1 - s$), as all slip losses are dissipated. Using a cascade connection, however, the slip energy can be transformed into either useful mechanical energy, or electrical energy supplied back into the grid.

The basic energy conversion equation in an induction machine can be applied here:

$$P_{em1} = (1 - s)P_{em1} + sP_{em1} \tag{13.23}$$

Part $(1 - s)P_{em1} = T \cdot (1 - s)\Omega_{sy} = T \cdot \Omega_r$ is the mechanical output power of the machine, while part sP_{em} is the secondary electrical power which is, for a shorted rotor winding or rotor connected to external resistances, dissipated as joule loss. This is also clear from the electrical equation for the secondary circuit:

$$\underline{E}_2 = s\underline{E}_1^{'} = R_2\underline{I}_2 + jsX_{2\sigma}\underline{I}_2 \tag{13.24}$$

yielding (for a three-phase rotor)

$$P_{em2} = sP_{em1} = Re(3\underline{E}_2\underline{I}_2^*) = 3R_2\underline{I}_2^2 \tag{13.25}$$

Suppose now that we connect an AC voltage source with the secondary frequency $f_2 = sf_1$ and voltage $\underline{V}_2 = \gamma\underline{E}_1^{'}$ to the secondary. Then

$$\underline{E}_2 = s\underline{E}_1^{'} = R_2\underline{I}_2 + jsX_{2\sigma}\underline{I}_2 + \underline{V}_2 = R_2\underline{I}_2 + jX_{2\sigma}\underline{I}_2 + \gamma\underline{E}_1^{'} \tag{13.26}$$

from which

$$P_{em2} = sP_{em1} = 3Re(\underline{E}_2\underline{I}_2^*) = 3R_2\underline{I}_2^2 + 3Re(\underline{V}_2\underline{I}_2^*) = 3R_2\underline{I}_2^2 + 3Re(\gamma\underline{E}_1^{'}\underline{I}_2^*) \tag{13.27}$$

Clearly, $P_2 = 3Re(\underline{V}_2\underline{I}_2^*) = 3Re(\gamma\underline{E}_1^{'}\underline{I}_2^*)$ is the power transferred to (or from) the secondary source.

As $3Re(\underline{V}_2\underline{I}_2^*) = 3Re(\gamma\underline{E}_1^{'}\underline{I}_2^*) = 3\gamma P_{em1}$, it follows from Eq. 13.27 that

$$(s-\gamma)P_{em1} = 3R_2\underline{I}_2^2 \geq 0 \tag{13.28}$$

Therefore, (with $P_{em1} = T\Omega_{sy}$) we find that $T > 0 \iff s > \gamma$ and $T < 0 \iff s < \gamma$. For $s = \gamma$ we have to revert to Eq. 13.26, which yields $\underline{I}_2 = 0$ and thus $T = 0$ or no-load.

We may therefore conclude that the external secondary source results in a shift of the no-load point from $s = 0$ to $s = \gamma$ with positive torque for $s > \gamma$ and negative torque for $s < \gamma$. Note that γ can be positive or negative, which means that we may shift the no-load point to either sub-synchronous or over-synchronous speeds - at least, if the secondary source allows this (see below).

For $1 > \gamma > 0$, the no-load speed is shifted into the sub-synchronous speed range. If $1 > s > \gamma$, we get motoring, with $P_{em1} = T\Omega_{sy} > 0$ and mechanical power $P_m = T \cdot (1-s)\Omega_{sy} > 0$ and positive torque. As to the secondary (slip) power P_{em2}, the part $P_2 = \gamma P_{em1} > 0$ is transferred to the secondary source while only the small remaining part $P_{j2} = (s-\gamma)P_{em1}$ is dissipated.

If $s < \gamma$, we are in generating mode, with $P_{em1} = T\Omega_{sy} < 0$ and negative torque and mechanical power. The secondary source power $P_2 = \gamma P_{em1}$ is now negative, which implies that the secondary source is delivering power. The secondary slip power $P_{em2} = sP_{em1}$ is negative when $s > 0$, implying that the secondary source power is used for the slip losses and the remaining part of it is transferred to the primary source. For $s < 0$, a portion of the input mechanical power is used for the slip losses (in addition to the secondary source power $P_2 = \gamma P_{em1}$).

When operating with $s > 1 > \gamma$, we get reverse braking with negative mechanical power; this dissipates a large amount of P_{j2} but still recuperates some P_2.

For $\gamma < 0$, the no-load speed is shifted to the over-synchronous speed (commonly called *super-synchronous operation*). When $1 > s > \gamma$, we again obtain motoring with positive primary electromagnetic power P_{em1} and positive mechanical power (and, of course, positive torque). However, $P_2 = \gamma P_{em1} < 0$, which implies that the secondary source now also delivers power to the drive, in addition to the primary source. When $s < \gamma$, we get generating with negative primary electromagnetic power P_{em1} and negative mechanical power (and, naturally, negative torque). However, $P_2 = \gamma P_{em1} > 0$, implying that power is also transferred *to* the secondary source, in addition to the primary source.[5] When $\gamma < 0$, operation with $s > 1$ (i.e. reverse braking) is also possible, at least in theory - it is never used in practice.

Similarly, operation with $\gamma > 1$ is also possible in theory, but does not have any practical significance.

[5]Note that the URS is used for the primary, while the GRS is used for the secondary.

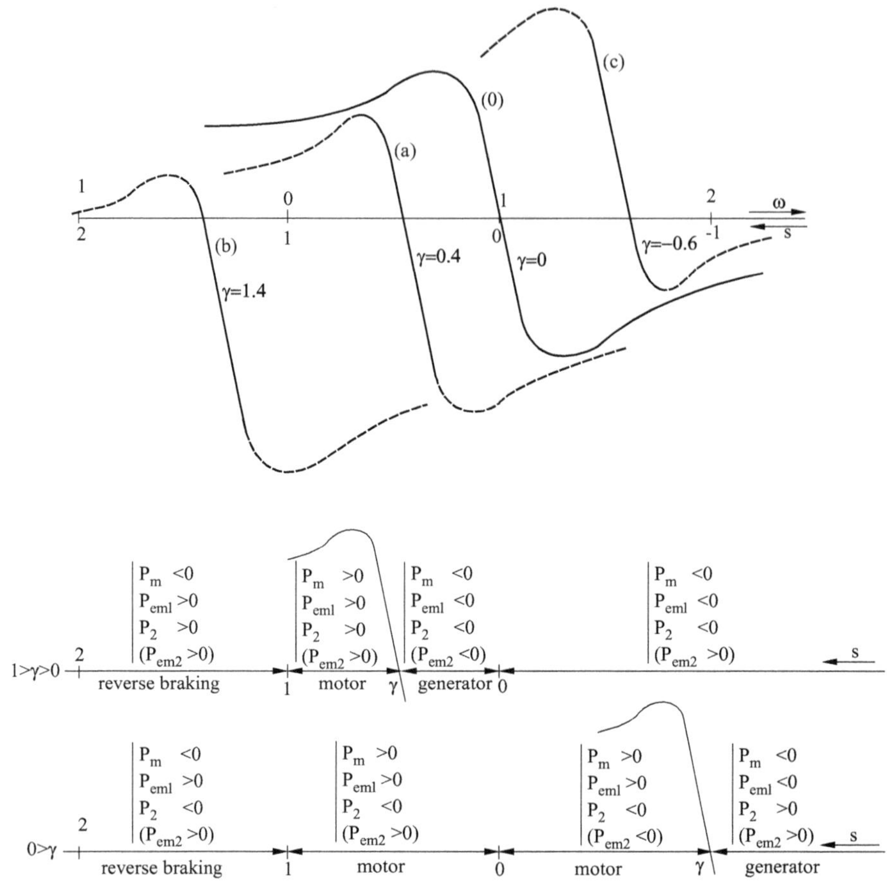

Fig. 13.21 Torque-speed characteristics for slip energy recuperation

Figure 13.21 illustrates the torque-speed characteristics and the signs of the different power quantities for these modes,[6] while Fig. 13.22 shows the Sankey diagrams for the discussed operating modes.[7]

The main advantages of a cascade connection include:

- the large range of speed control, both sub-synchronous and super-synchronous.
- the fact that slip energy is not completely dissipated as heat, and that the efficiency is quite high as a result.
- the fact that mechanical output power can be larger than the primary input power, which is useful for super-synchronous motoring.

[6]As is illustrated, the pull-out torque values are not constant, however.

[7]Prove the signs of the power flows and explain.

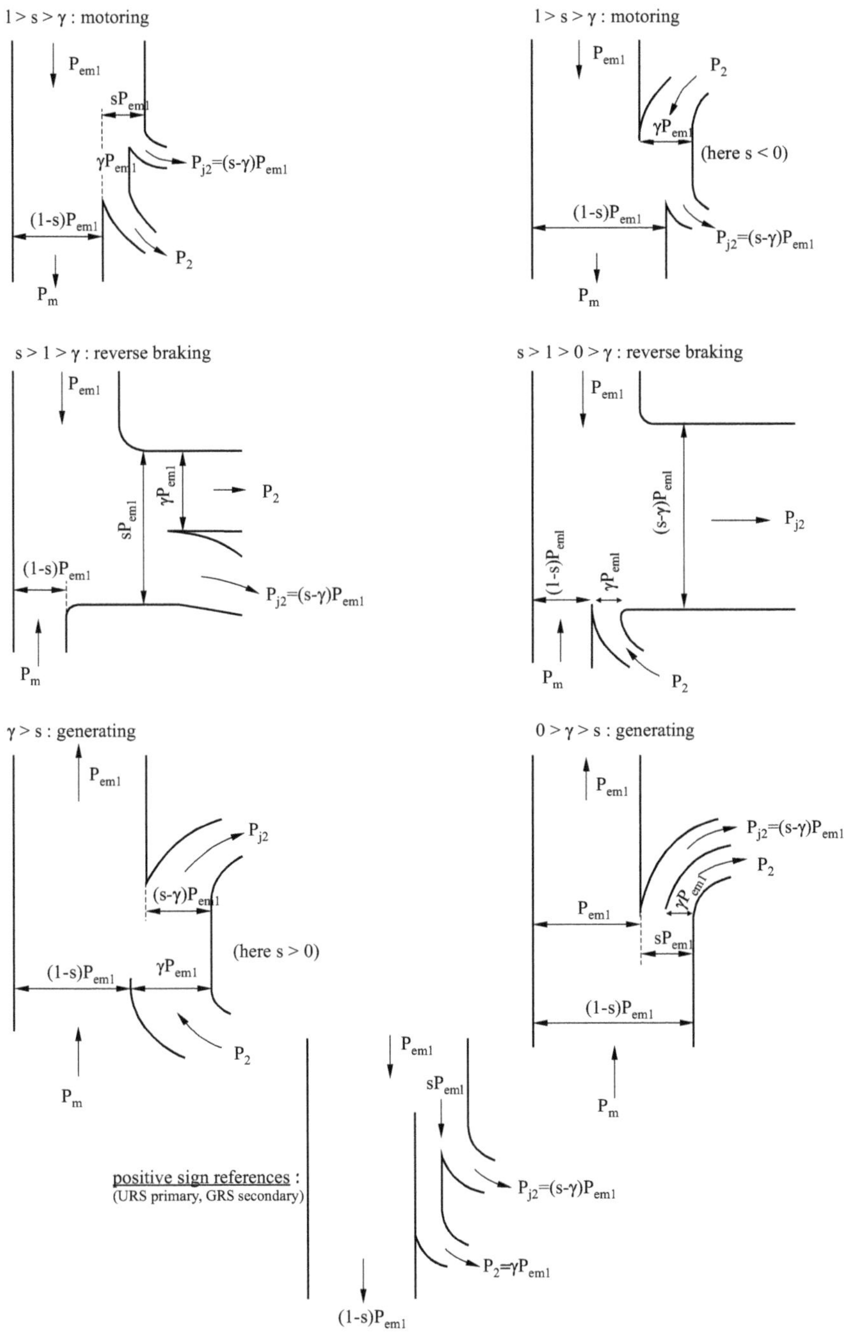

Fig. 13.22 Sankey diagrams for cascade connection

13.2.3.2 Practical Implementations

The traditional implementations use a rectifier in the secondary and a DC machine as source. In this way the (variable) secondary frequency is decoupled from the source. The induction machine sees the AC equivalent of the DC source, while the DC source sees the rectified DC equivalent of the secondary voltage of the induction machine.

In the Krämer cascade (Fig. 13.23), the DC machine is mechanically coupled to the induction machine. Speed control is achieved by adjusting the excitation of the DC machine, which alters the counter-emf (see (b) in Fig. 13.23). The power P_2 is transformed into additional mechanical power on the same shaft as the induction machine and the load. However, the diode rectifier only permits operation with positive P_2 (e.g. sub-synchronous motoring).

The disadvantage of the Krämer cascade is that, at low speeds, the secondary power converted by the DC machine requires high currents, as the emf is low at low speed. The DC machine must therefore be over-dimensioned, depending on the lowest anticipated speed. As a consequence, slip values s_o larger than 0.5 are not feasible.

Another standard scheme is the Scherbius cascade (see (a) in Fig. 13.24). The DC machine is mechanically connected to another induction machine, working as an over-synchronous generator. The secondary power is thus converted into electrical power. Although the DC machine is now operated at high speeds and lower current, this scheme requires three electrical machines. Moreover, this secondary power is not directly converted into mechanical energy for the load. On the other hand, low-speed operation ($s_o > 0.5$) is possible. Speed control is also achieved by varying the excitation of the DC machine (in the diagram in Fig. 13.24, the E_g lines are now horizontal lines).

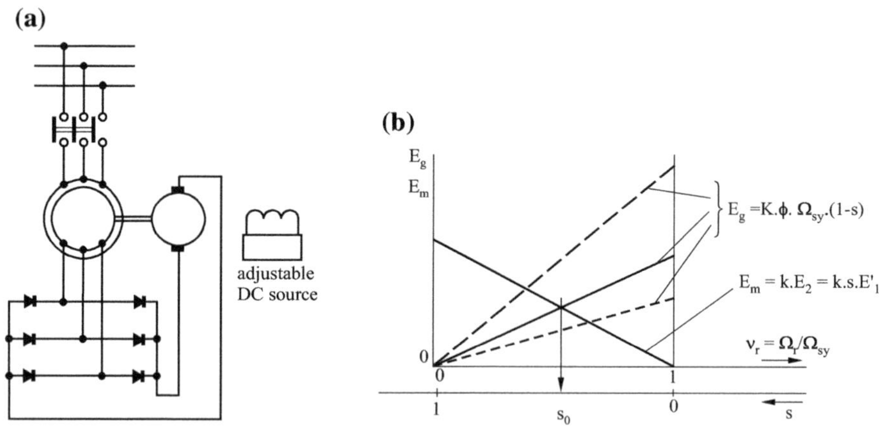

Fig. 13.23 Krämer cascade

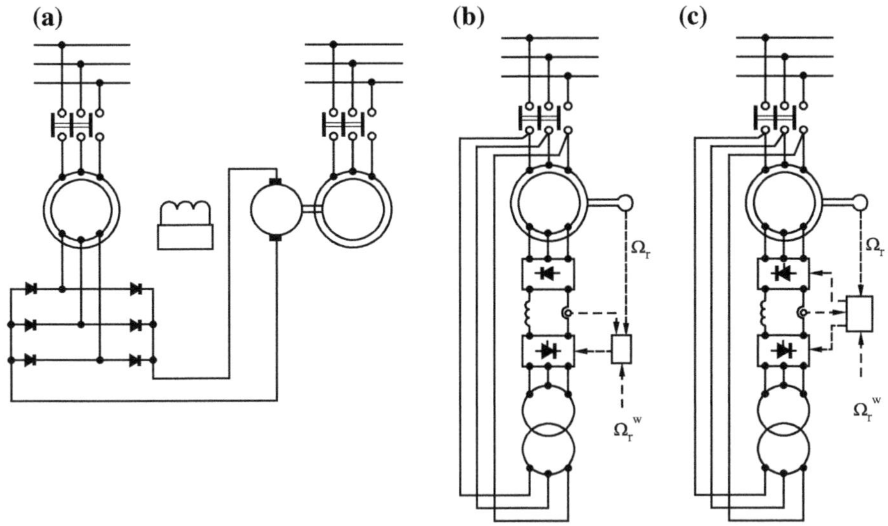

Fig. 13.24 Scherbius cascade and PE versions

Nowadays, power electronic schemes are used. As illustrated in (b) in Fig. 13.24, the counter-emf is provided by a second controlled rectifier that is connected to the grid, usually via a transformer, to adapt the voltage levels of machine and grid. For $P_2 > 0$, the machine-side converter operates as a rectifier (diode rectifier or controlled rectifier with $\alpha < \pi/2$), and the grid-side rectifier operates in inverter mode ($\alpha > \pi/2$). With two controlled rectifiers (see (c) in Fig. 13.24), it is also possible to obtain both sub- or super-synchronous operation, and both as motor and generator, but the delay angles of both rectifiers must be coordinated. It is also possible to replace the intermediate DC current circuit with a DC voltage circuit (and the CSI with a PWM-VSI).

Cascade circuits were the earliest means to obtain energy-efficient speed control for induction machines. Nowadays, they have mostly been replaced by variable frequency operation of cage induction motors (using inverters at the primary side). However, there are still special applications where the robust behaviour (i.e. the overload capacity) and the broad speed range of cascade drives are valuable, for example for dredging machines.

In recent years, a new application of cascade drives has emerged. In wind turbines, the primary frequency of the generator is fixed (grid frequency) while the (optimal) speed of the sails depends on the wind speed.[8]

[8]Note that also a wind turbine with a synchronous machine as generator requires an inverter to adapt the generator output to the grid frequency.

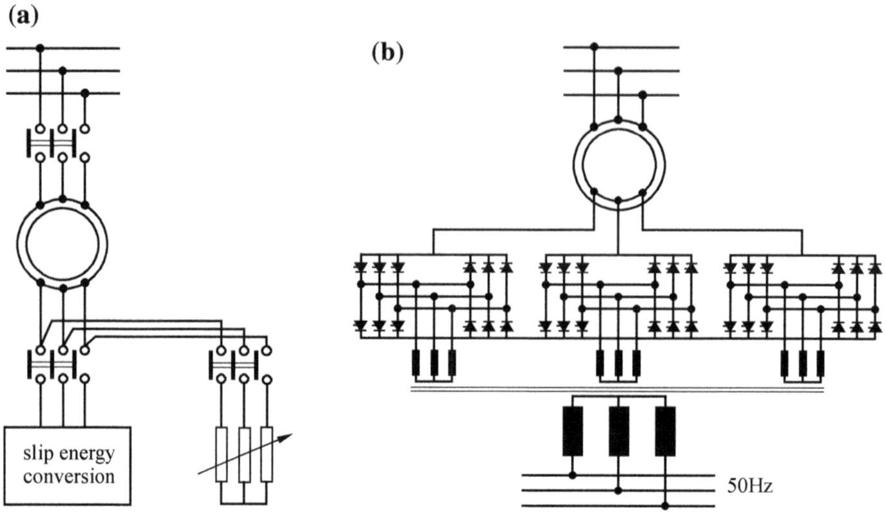

Fig. 13.25 Starting of a cascade using secondary resistors and cycloconverter version of a cascade

Remarks:

- In many cascade drives, operation at very large slip values is not permitted.[9] Therefore, usually a variable resistor in the secondary is provided for starting purposes (see (a) in Fig. 13.25).
- In the theoretical derivation, we assumed that the phase angle of the secondary voltage $\underline{V}_2$ was equal to that of the secondary emf $\underline{E}_2$; it is also possible to vary the phase angle in order to control the reactive output of the machine, which is useful for wind turbine applications.
- In the power electronic versions, the rectifier-inverter can be replaced by a cycloconverter (see (b) in Fig. 13.25), which allows four-quadrant operation. However, unless the cycloconverter control is made dependent on (i.e. follows) the slip, this kind of drive operates as a synchronous machine with the speed determined by the difference of primary and secondary frequencies.

13.3 Behaviour of Rotating Field Machines at Voltage Variations

13.3.1 Introduction

Voltage variations can be accidental (due to the grid) or deliberate (to obtain certain effects). Deliberate voltage variations are usually three-phase symmetrical. However,

[9] Why?

accidental voltage variations can be (and often are) non-symmetrical. Another difference is that deliberate voltage variations are commonly long-term, while accidental variations can also be short-term.

13.3.2 *Induction Machines at Voltage Variations*

13.3.2.1 Symmetrical Voltage Deviations

Symmetrical voltage deviations are in particular important for

- starting (i.e. the starting current and torque)
- full-load behaviour, in particular the full-load current and thermal loading of the stator winding.

To analyse the consequences of voltage variations, we will make use of the equivalent circuits in Fig. 13.26: circuit (a) or, if possible, the simplified circuits (b) or (c).

The *starting current* is mainly determined by the stator and rotor resistances and the leakage:

$$\underline{i}_1 \approx \underline{i} \approx \frac{v}{jx_\sigma + r_1 + r/s} \tag{13.29}$$

(with $r \approx r_2$). For $s = 1$, this may be simplified to

$$\underline{i}_1 \approx \underline{i} \approx \frac{v}{jx_\sigma} \tag{13.30}$$

The starting current would be proportional to the voltage (and the torque proportional to the square of the voltage), if the leakage inductance were constant. However, the leakage inductance is often saturation-dependent (especially the rotor leakage for

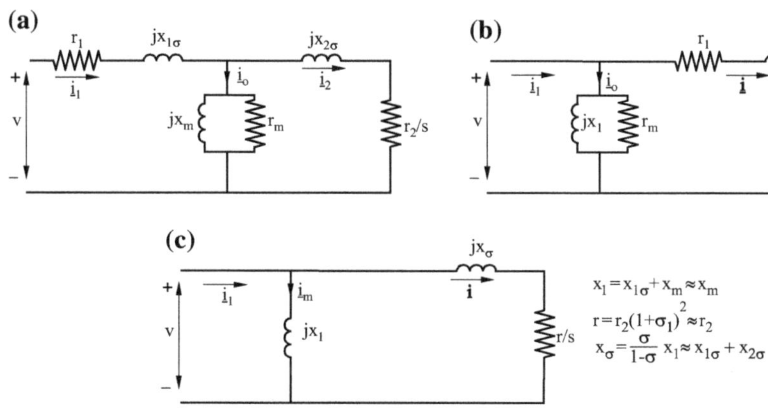

Fig. 13.26 Equivalent circuits for induction machines

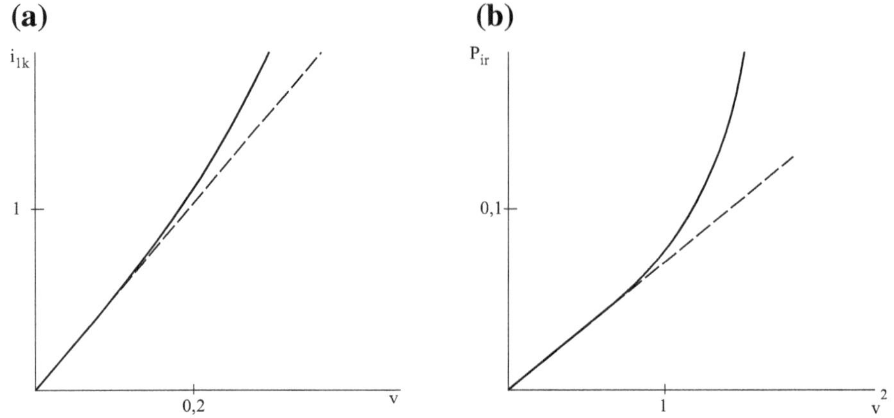

Fig. 13.27 **a** leakage saturation; **b** iron loss

cage rotor machines) and the current then varies more than proportional to the voltage (see (a) in Fig. 13.27). As a consequence, a low voltage may cause starting problems, particularly when the machine is loaded.

As far as the *full-load behaviour* is concerned, consider the case of a constant load torque. In this case, we have $v \cdot i_w = t = t_l = c^t$. For a constant load torque, the active current therefore varies in inverse proportion to the voltage. On the other hand, the reactive current increases with the voltage and, because of the main field saturation, even more than proportionally. In addition, the load torque may also vary with the varying speed (slip) when the machine voltage changes.[10] How the *total current* and the *joule loss* in the windings change will depend on the load torque, the saturation of the magnetic circuit and the $\cos\varphi$ of the machine. In general, the current and joule loss may increase with increasing voltage when the power factor of the machine is low (e.g. less than 0.7), but they may decrease with increasing voltage when the power factor is higher (e.g. higher than 0.8).

In addition to the joule loss in the windings, the iron loss must not be disregarded. Iron loss varies for higher voltage more than quadratically with the voltage (see (b) in Fig. 13.27). Although the loss component of the magnetising component is almost negligible compared to the load component in the stator current, the iron loss itself cannot be ignored.

13.3.2.2 Asymmetrical Voltage Deviations

Asymmetrical voltage deviations can be expressed as an additional inverse component, besides the normal direct component (the zero-sequence component is seldom

[10]Note that the (pull-out) torque of an induction machine varies with the square of the voltage; a voltage dip to 70% may lead to a maximum torque below the rated torque.

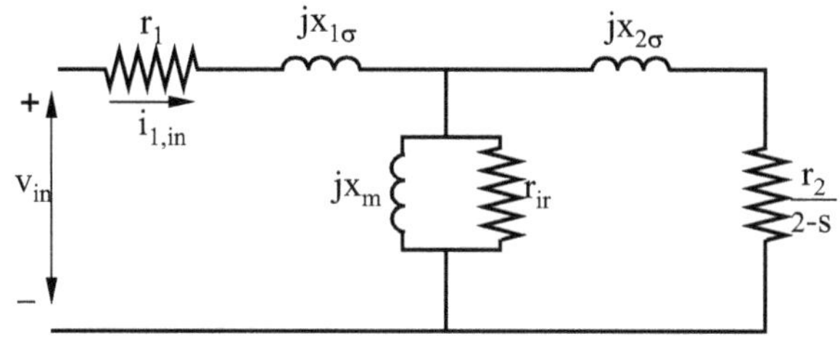

Fig. 13.28 Equivalent scheme for inverse voltage

present, as the neutral of a wye-connected winding is almost never connected). For the inverse or negative sequence component, the equivalent circuit in Fig. 13.28 is to be used.

The slip s' for the negative sequence component is very large at the normal small slip values s for the positive sequence: $s' = 2 - s$. The resulting low machine impedance leads to very high inverse currents, even for small inverse voltage components (e.g. 3% inverse voltage may lead to 20% inverse current). Although the skin effect in the rotor results in somewhat higher equivalent rotor resistances for the inverse component and therefore limits the current, the larger rotor resistance leads to significant additional joule losses that may be detrimental for the machine. Indeed, the additional rotor heat may lead to uneven expansion and fracture of the rotor bars. In addition, most of the rotor heat will be evacuated through the air gap to the stator, thus increasing the temperature of the stator winding. Another important aspect is the unequal winding currents.

As a consequence, the wide-spread use of induction machines has resulted in stringent standards for the inverse component that is allowed (e.g. less than 1–2%).

13.3.2.3 Supply Voltage Adaptation to the Load

When the load torque of an induction machine remains low during a considerable interval of the load cycle, it may be economically advantageous to reduce the voltage during these intervals.

If saturation of the main field is disregarded, i.e. for a constant x_m, a variation of the voltage with a factor α results in a variation of the currents with the same factor if the slip remains constant. The torque will then vary with a factor α^2.

Conversely, if the voltage is adapted in proportion to the square root of the torque ($\upsilon \sim \sqrt{t}$), the current will vary as $\sqrt{t}$, the slip will be constant and also the power factor will be constant - at least if resistances and inductances remain constant. The iron loss will therefore vary approximately as υ^2, in other words as the torque t. This is also true for the winding losses, which means that the efficiency of the machine will remain the same (i.e. higher) as in full-load (without voltage reduction, efficiency would be much lower).

It should be noted that keeping the slip constant and equal to the rated slip does not necessarily result in the optimal part-load efficiency. In fact, in theory we should take into account many factors, such as the actual ratio between iron losses and joule losses, the fact that for higher voltages the iron losses vary more than proportionally to the square of the voltage, or the variation of the (rotor) resistance with temperature. The lower losses could in fact also allow less cooling and consequently lower ventilation losses, although the ventilation is almost never decreased. In addition, if power electronics (i.e. AC choppers, see Sect. 13.4) are used to vary the voltage, the additional higher harmonic losses may (partly) annihilate the intended efficiency optimisation.

However, the reduction of the reactive current resulting from voltage adaptation may be more important than efficiency optimisation.

13.3.2.4 Voltage Reduction for Speed Control

Voltage reduction can also be used for some speed control of cage rotor machines, albeit with some limitations. Because the increased slip results in large secondary (slip) losses, this method should only be used for quadratic load torques (see Fig. 13.29).

Moreover, to limit the current, the rotor resistance should be higher than $0.1 \cdots 0.2$ pu. Therefore, this method is only suitable for low power ratings (some kW) and/or intermittent use.

13.3.3 Synchronous Machines at Voltage Variations

13.3.3.1 Symmetrical Voltage Variations

The most important consequences of voltage variations relate to:

- starting current and torque (asynchronous starting)

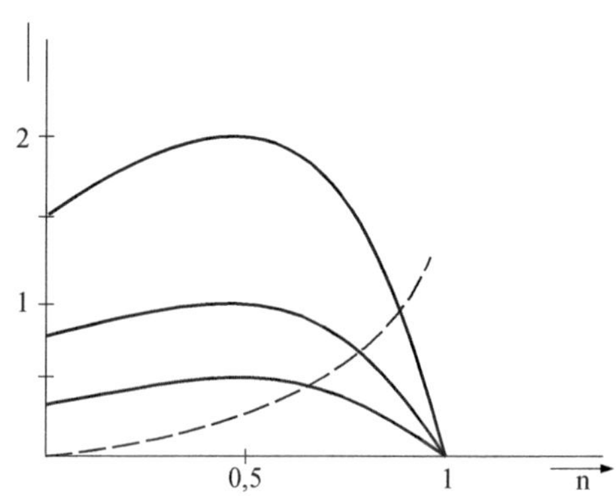

Fig. 13.29 Speed control for quadratic load

- synchronising
- full-load behaviour, in particular current and thermal loading.

Asynchronous starting (if any) is similar to the induction machine starting, as was demonstrated in the previous section. Saturation results in a more than proportional variation of the current with the voltage and therefore a more than quadratic variation of the torque.

Synchronising is rarely fundamentally affected; only in critical situations can synchronising become uncertain with voltage reductions, as it is mostly performed without load.

The full-load heating of the stator winding will mostly be affected by the stator joule losses and somewhat less by the iron losses. The active component changes inversely proportionally to the voltage. The reactive current will change in the same way as the active current does when the machine is over-excited[11] (i.e. increasing with decreasing voltage and vice versa). When the machine is under-excited, the variation of the reactive current will be in the sense of the variation of the voltage, just as for the induction machine. Therefore, only for an over-excited machine is it possible to predict the variation of the current with voltage variations. The current of an over-excited machine will decrease with increasing voltage and thus the winding temperature will decrease (and vice versa for a voltage decrease).

13.3.3.2 Asymmetrical Voltage Deviations

Expressing an asymmetrical voltage in terms of direct and inverse components, we observe that the inverse component will rotate at twice the synchronous speed with respect to the rotor. It will therefore result in induced rotor currents with twice the supply frequency. As the rotor impedance is very small for these frequencies (the *inverse reactance* of a synchronous machine is the same order of magnitude as the leakage and very small for machines with a damper cage or massive rotor), the rotor current will have a very high amplitude. This means that the rotor losses for the inverse system can be extremely high, especially for laminated rotors with damper windings or massive rotors. Just as for induction machines, the inverse component of the voltage should be limited to 1–2%.

13.3.3.3 Supply Voltage Adaptation to the Load

Let us now consider a simultaneous variation of the supply voltage and the excitation with a factor α. When the main field saturation can be disregarded (and the magnetising reactance x_m or the synchronous reactance x in the circuits in Fig. 13.30a and b, respectively, can be considered constant), then, for a constant load angle β, the current will vary with the same factor α.

[11] Prove this.

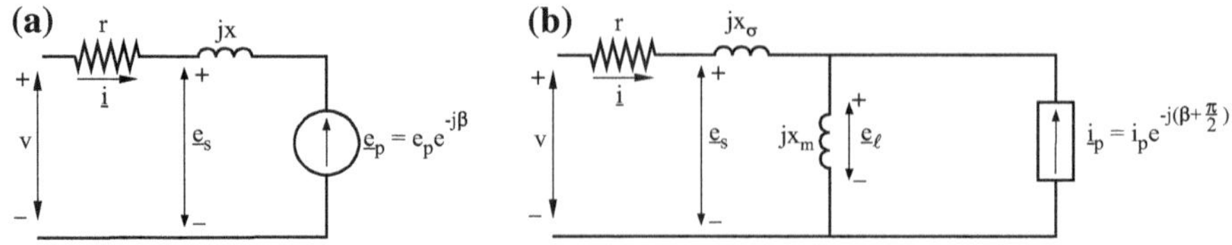

Fig. 13.30 Equivalent circuits for synchronous machines

Fig. 13.31 Torque adaptation

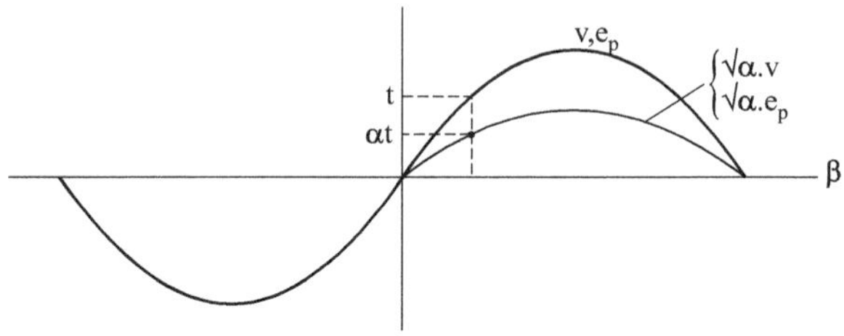

The torque will then change with a factor α^2. Therefore, a variation of both supply voltage and excitation proportional to the square root of the load torque will lead to a constant load angle as well as a constant power factor $\cos\varphi$ (see also Fig. 13.31).

In reality, the main field saturation results in a somewhat smaller power factor and a smaller primary current with decreasing torque.

The decrease of winding loss and iron loss (proportional to $e_l^2 \approx v^2$) is advantageous for the efficiency of the machine when it is partly loaded. This efficiency could, in theory, be further increased if the cooling and ventilation losses could be reduced. However, this requires a separately driven ventilator, which is only used for very high power ratings. In many cases, however, such an additional control is also considered an excessive and undesirable burden.

Note that for smaller power ratings also the adaptation of the excitation might constitute an unwanted complication, or may be impossible because of permanent-magnet excitation.

13.4 Power Electronic Starting and Voltage Adjustment of Rotating Field Machines to the Load

13.4.1 Introduction

Chapter 9 in Part 2 discussed the characteristics of AC choppers. These are frequently used as starting aid (to limit the starting current) or for voltage adjustment to the load (to increase part-load efficiency).

For (asynchronous) starting, AC choppers cannot only be used to reduce the starting current, but also to limit the transient phenomena, as will be explained in the next section.

13.4.2 Power Electronic Starting of Induction Machines

AC choppers can lower the initial starting current by temporarily reducing the voltage. These are known as basic *soft starters*. Like series resistors or inductances for starting, their disadvantage is the lower starting torque as the torque is proportional to the square of the current. Another disadvantage is that the voltages and currents are not purely sinusoidal due to the chopping.

However, most of these hardware configurations for phase control can also reduce the transient currents, which is not possible with traditional starting methods. As is well known, a single-phase R-L-load can be switched on without a transient by switching it on at the zero-crossing of the (expected) steady-state current. Of course, this is not possible for three-phase loads, but the transients can be mitigated by first switching on two phases at the zero-crossing instant of the corresponding steady-state single-phase current. The third phase is then switched on at the *optimum* instant (some 90°) later on. Note, however, that this will not completely avoid the transients, as by switching-on the third phase there will always be a transient from single phase to three-phase supply. It will simply mitigate the transients. Combining this two-step switching-on with phase control may further limit the accelerating torque jumps, as it lets the currents increase slowly.

Figure 13.32 shows some hardware configurations for phase control that can be used for phase control and *soft starting*. In the past, some of the triacs (or

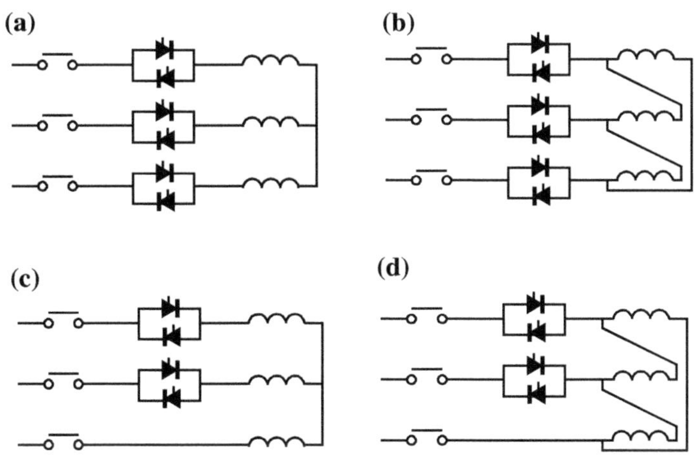

Fig. 13.32 Phase control schemes

anti-parallel thyristors) were sometimes replaced by diodes to reduce costs, but such circuits are far from optimal (asymmetric), or are limiting the switching possibilities (and semiconductors are quite cheap nowadays).

13.4.3 Power Electronic Voltage Adjustment to the Load

AC choppers are frequently used to adjust the voltage in partial-load operation. These circuits are sometimes called *Nola-converters* (Frank Nola invented the principle in the late seventies and was granted a patent in 1984). As has been explained above, the part-load efficiency can be increased because the iron and winding losses can be reduced compared to full-voltage operation. However, it is important to keep in mind that the distorted voltages and currents may introduce additional harmonic losses that could annihilate the fundamental loss reduction.

Another application of AC choppers can be speed (slip) control, but as mentioned before, this is not a preferred method for speed control unless the load torque is quadratic and the rotor resistance of the machine is sufficiently high.

In Chap. 9 of Part 2, we derived a fundamental model for AC chopping of a pure reactive load (single-phase and three-phase).

It is obvious that an induction machine is not a pure reactive load. Nevertheless, an approximate model can be used to describe AC chopping of an induction machine. Consider the simplified equivalent circuit in (a) in Fig. 13.33. The stator resistance is neglected and the model consists of the total leakage (as seen from the stator) in series with the equivalent rotor impedance. As the electrical time constants of the rotor of a 50 Hz (or 60 Hz) induction machine are quite high (80 ms . . . 1 s) compared to the time intervals of the AC chopping (less than 1/6th the mains period or 3.33 ms for 50 Hz), the rotor may be represented by a fundamental sinusoidal rotor emf e_r. For an induction machine fed by an AC chopper, this leads to the schematic representation in (b) in Fig. 13.33.

Using this simplifying assumption, we obtain a similar analysis as for a pure reactance, but now with the fictitious source $v' = v - e_r$ and as reactance the total leakage reactance x_σ. In this way, the induction machine fed by an AC chopper can be represented by an additional reactance x_t proportional to $x_b = x_\sigma$ (cf. the equations in Chap. 9). This leads to the approximate circuit in (c) in Fig. 13.33.

For the angles α' and β', which are defined with respect to the zero-crossing of the voltage v', the following relations are valid:

$$\alpha' = \frac{1}{2}(\pi + \gamma)$$

$$\beta' = \frac{1}{2}(\pi - \gamma)$$

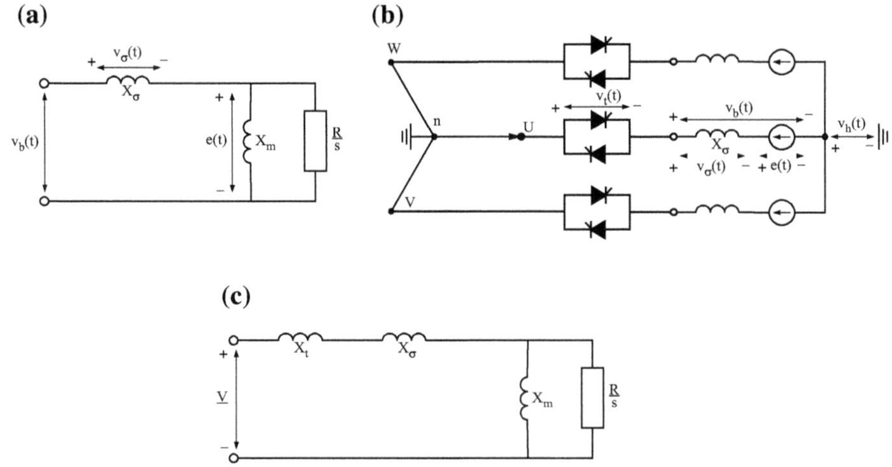

Fig. 13.33 Fundamental scheme for phase control of a three-phase induction machine

The phase displacement with respect to v' is indeed $\pi/2$. For the phase displacement with respect to the real supply voltage v, we then have

$$\varphi = \beta + \frac{\gamma}{2} = \alpha - \frac{\gamma}{2} = \frac{\alpha + \beta}{2}$$

and therefore

$$\alpha = \alpha' + \varphi - \frac{\pi}{2} = \varphi + \frac{\gamma}{2}$$

$$\beta = \beta' + \varphi - \frac{\pi}{2} = \varphi - \frac{\gamma}{2}.$$

Chapter 14
Ideal Current Supply of Rotating Field Machines

Abstract An *ideal* current supply (i.e. a current independent of the load) is not feasible. In fact, most sources come close to more or less ideal voltage sources. In the (distant) past, a DC current supply was sometimes used for the supply of a group of machines, as will be illustrated in the next section. In the subsequent two sections, a purely theoretical discussion will be given of the ideal current supply of induction and synchronous machines, respectively. Primarily, the aim of this discussion is to point out the practical issues that *would* arise if an ideal current source existed. In reality, modern controlled drives make use of a controlled current source inverter or a voltage source inverter with a current control loop, as will be discussed in the following chapters.

14.1 Current Supply of DC Commutator Machines

14.1.1 Individual Current Supply

Individual current supply of a separately excited DC machine comes down to pure torque control:

$$T = K \cdot \Phi \cdot I \tag{14.1}$$

For a given flux, the current determines the torque. With a prescribed machine torque, the speed follows from the load torque characteristic (i.e. the equality of load and machine torque).

In contrast, the usually voltage-supplied separately excited DC machines are in fact speed-controlled drives: with a negligible armature resistance, the speed is completely determined by flux and voltage.

In the past, a current or torque control was realised by a (modified) Ward-Leonard (see (a) in Fig. 14.1). The DC generator in this case is a compound (or a three-winding) generator, in which the series excitation is laid out so that a constant current output is obtained. Nowadays, this could also be achieved by a controlled rectifier (or chopper) with current control loop.

J. A. Melkebeek, *Electrical Machines and Drives*, Power Systems,
https://doi.org/10.1007/978-3-319-72730-1_14

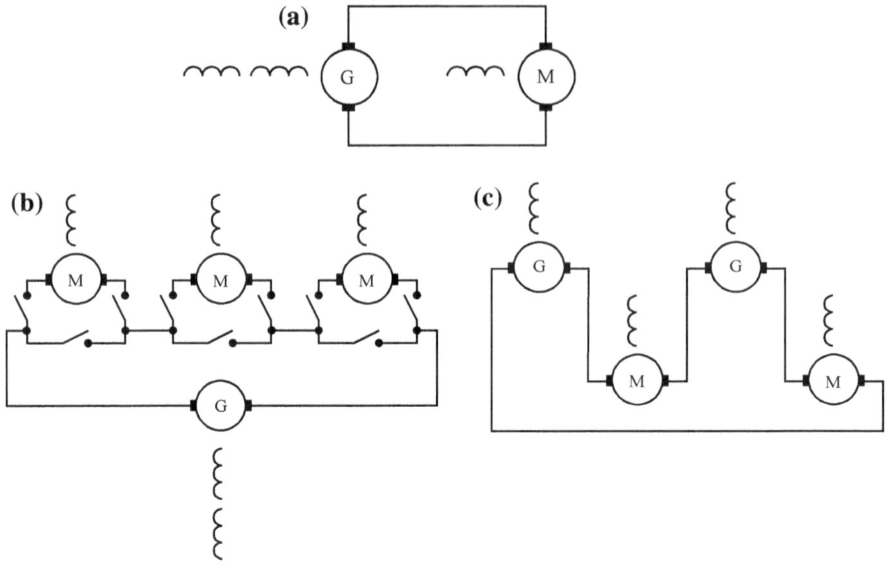

Fig. 14.1 Current supply of DC commutator machines

14.1.2 Group Current Supply

One hundred years ago, group current supply was quite common. In this supply configuration, DC motors and generators are connected in series so that each motor is fed by the same current (see (b) and (c) in Fig. 14.1). Individual (speed or torque) control of the motors is then realised by flux control. An important advantage is that the commutation of all motors is quite effectively controlled (as mainly the current is responsible for a good commutation and especially over-currents must be avoided). Another advantage compared to (group) voltage supply is that the speed of the motors is not directly affected by a variation of the supply voltage. Of course, this group current supply is not very useful if the rated power (current) of the motors differs considerably.

14.2 Ideal Current Supply of Induction Machines

14.2.1 Current, Voltage and Torque Relations

From the equivalent circuit (a) in Fig. 14.2 (assuming an impressed primary current), we derive the following relations:

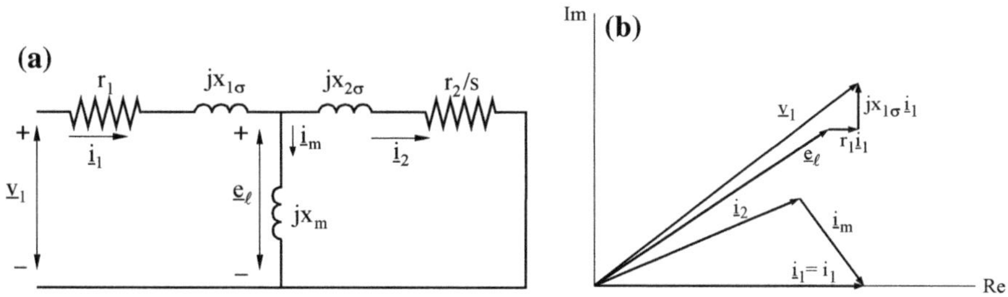

Fig. 14.2 Induction machine current supply: equivalent circuit and phasor diagram

$$\underline{i}_2 = \underline{i}_1 \cdot \frac{jx_m}{(r_2/s) + j(x_m + x_{2\sigma})} \tag{14.2}$$

$$\underline{i}_m = \underline{i}_1 \cdot \frac{jx_{2\sigma} + r_2/s}{(r_2/s) + j(x_m + x_{2\sigma})} \tag{14.3}$$

$$\underline{e}_l = \underline{i}_1 \cdot \frac{1}{[jx_m]^{-1} + [r_2/s + jx_{2\sigma}]^{-1}} \tag{14.4}$$

$$\underline{v}_1 = \underline{e}_l + \underline{i}_1 \cdot (r_1 + jx_{1\sigma}) \tag{14.5}$$

$$t = i_2^2 \cdot (r_2/s) = i_1^2 \cdot \frac{x_m^2 \cdot (r_2/s)}{(r_2/s)^2 + (x_m + x_{2\sigma})^2} \tag{14.6}$$

For current supply, the real axis is usually chosen along the primary current (see the phasor diagram (b) in Fig. 14.2).

14.2.2 Behaviour of the Induction Machine Neglecting Main Field Saturation

Suppose that the (main field) saturation is constant, i.e. $x_m = x_{mo} = constant$ (for example, at the rated condition $x_{mo} = x_{mn}$). Then, from Eq. 14.6 it follows that:

- for small slip, $|s/r_2| \ll (x_{mo}+x_{2\sigma})^{-1}$, i.e. the torque varies approximately linearly with the slip:

$$t \approx i_1^2 \cdot x_{mo}^2 \cdot (s/r_2) \tag{14.7}$$

- for large slip, $|s/r_2| \gg (x_{mo}+x_{2\sigma})^{-1}$, i.e. the torque varies approximately hyperbolically with the slip:

$$t \approx i_1^2 \cdot \frac{x_{mo}^2}{(x_{mo} + x_{2\sigma})^2} \cdot (r_2/s) \tag{14.8}$$

- in between, i.e. for $s/r_2 = \pm(x_{mo} + x_{2\sigma})^{-1}$, the torque attains the pull-out or maximum value:

$$t = \pm i_1^2 \cdot \frac{x_{mo}^2}{2(x_{mo} + x_{2\sigma})} \tag{14.9}$$

Note that the pull-out slip is much lower than for a constant voltage supply. The torque-slip characteristic is therefore much steeper as well.

This has two important consequences. First, to obtain a pull-out torque of at least 200% compared to rated torque (100%), a primary current of 117.5% or more is required (for the assumed $x_m = 3$ pu and $x_{\sigma 2} = 0.1$ pu). Secondly, the emfs are very high (and consequently also the required supply voltages, as real DC supplies are voltage sources): if there were no saturation (and thus x_m were constant), Eq. 14.4 would lead to an emf of 2.5 pu at the pull-out slip of $s/r_2 = 0.32$ (and an emf of 3 pu at slip $s = 0$).

Obviously, this is impossible and, in reality, saturation will lead to a lower (limited) x_m and hence a limited emf. Then, to obtain an overload capacity of 2, the required primary current will be much higher (e.g. $i_1 = 1.7$ for $x_m = 1.5$ leading to an emf of 1.8 for a slip value of $s/r_2 = 0.63$).

14.2.3 Behaviour of the Induction Machine Including Main Field Saturation

Saturation of the magnetising field results in:

- saturation harmonics in magnetising current and/or emf
- a magnetising current which increases much more than linearly with the emf

In what follows, the saturation harmonics will not be considered.

The saturation may, for example, be mathematically described by the following law between magnetising current and air-gap emf:

$$i_m = \frac{1}{x_{mo}} \cdot e_l + (\frac{1}{x_{mn}} - \frac{1}{x_{mo}})e_l^z \tag{14.10}$$

where x_{mo} and x_{mn} represent the magnetising reactances in unsaturated condition and rated condition ($e_l = 1$), respectively. This may also be written as

$$\frac{1}{x_m} = \frac{1}{x_{mo}} + (\frac{1}{x_{mn}} - \frac{1}{x_{mo}})e_l^{z-1} \tag{14.11}$$

with x_m denoting a variable saturation-dependent magnetising reactance. The linear term represents (mainly) the air-gap reluctance, the non-linear term the reluctance of the iron. Normal values for the exponent are $z = 6 \ldots 7$.

Fig. 14.3 Saturation characteristic $e_l(i_m)$ and $x_m(e_l)$

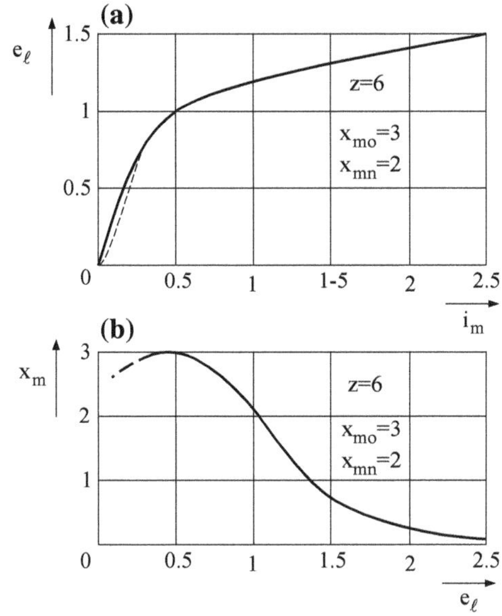

Figure 14.3 illustrates the emf and magnetising reactance as a function of the magnetising current; the dotted lines correspond to the Rayleigh interval, which is not taken into account in Eq. 14.10.

The characteristics taking into account saturation can be calculated[1] from Eq. 14.2 through 14.6 by inserting the expression 14.4 for x_m:

$$\underline{e}_l = \underline{i}_1 \cdot \frac{1}{[r_2/s + jx_{2\sigma}]^{-1} - j\left[\frac{1}{x_{mo}} + (\frac{1}{x_{mn}} - \frac{1}{x_{mo}})e_l^{z-1}\right]} \tag{14.12}$$

For $s = 0$, on the one hand, the emf attains a maximum provided by the equation

$$i_1 = \frac{1}{x_{mo}} \cdot e_{lmax} + (\frac{1}{x_{mn}} - \frac{1}{x_{mo}})e_{lmax}^z \tag{14.13}$$

For $s \to \infty$, on the other hand, the emf attains a minimum provided by

$$e_{lmin} = i_1 \cdot \left[x_{2\sigma}^{-1} + x_{mo}^{-1}\right]^{-1} \approx i_1 \cdot x_{2\sigma} \tag{14.14}$$

For small slip, the torque characteristic approaches the tangent to the torque characteristic for a fictitious constant voltage supply with an emf e_{lmax}:

[1] To find the solutions of this non-linear equation, it is easiest to start from a proposed emf value and calculate the corresponding slip.

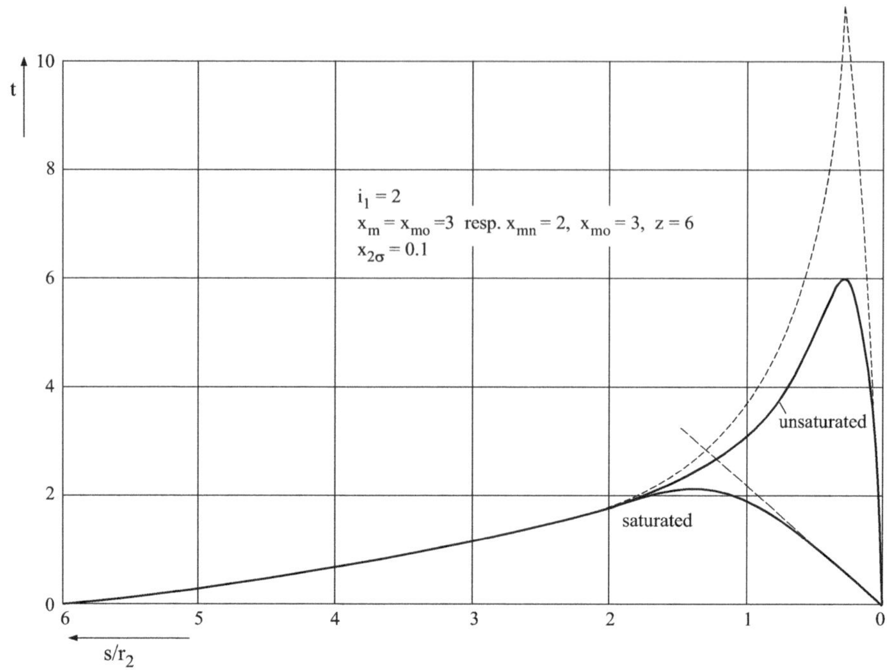

Fig. 14.4 Torque characteristic with and without saturation

$$t = e_{lmax}^2 \cdot (s/r_2) \tag{14.15}$$

while for high slip, the torque is hyperbolic with the slip according to Eq. 14.8 ($x_m = x_{mo}$ for the unsaturated condition).

In between, the torque attains a maximum that occurs approximately at the intersection of both asymptotes, i.e. approximately at

$$\frac{s}{r_2} = \frac{i_1}{e_{lmax}} \cdot \frac{x_{mo}}{x_{mo} + x_{2\sigma}} \approx \frac{i_1}{e_{lmax}} \tag{14.16}$$

Figure 14.4 illustrates the effect of saturation on the torque-slip characteristic for $i_1 = 2$ and $x_{2\sigma} = 0.1$. Note that the torque in the saturated condition ($x_{mo} = 3; x_{mn} = 2; z = 6$) is much lower compared to an unsaturated machine with $x_{mo} = x_{mn} = 3$. The pull-out torque is only slightly larger than 2, even for a primary current of twice the rated current. Figure 14.5 compares the torque characteristics for some current values ($i_1 = 1; 1.5; 2; 2.5$) for the parameter values $x_{2\sigma} = 0.2$; $x_{mo} = 3; x_{mn} = 2; z = 6$. For the sake of comparison, the figure also shows the torque characteristic (dashed line) for a constant voltage supply with $e_l = 1$. At the crossing points of the constant-current characteristics with the constant-voltage characteristic for $e_l = 1$, the corresponding air-gap emf is unity. Most of these crossing points are,

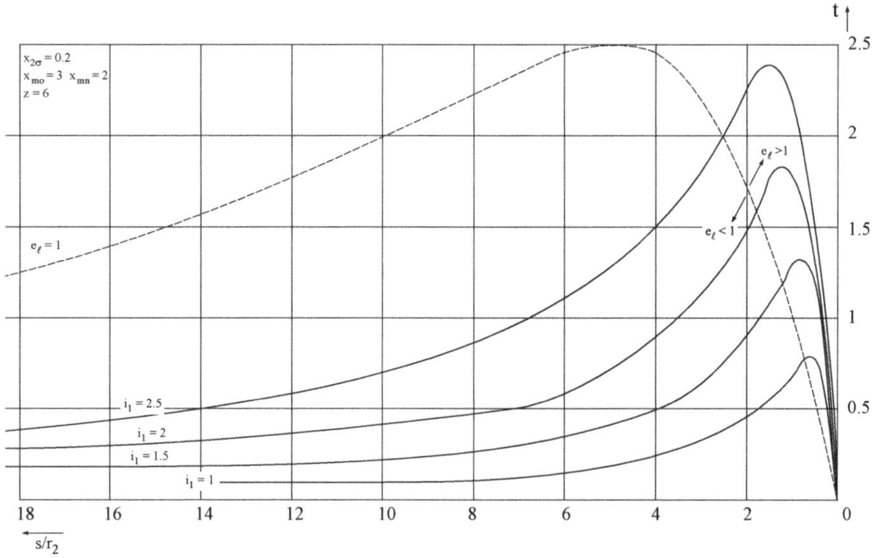

Fig. 14.5 Torque characteristics for different currents

however, with decreasing slip, on the upward part of the characteristic, which is the statically unstable part for a constant load torque. Figure 14.6 shows the torque and emf characteristics for unsaturated and saturated conditions.

From the above, it should be clear that main field saturation results in:

- a much lower pull-out torque than in the unsaturated case (and for rated current, much smaller than that for a constant voltage supply[2])
- a significant increase of the pull-out slip
- still rather high emf values[3] compared to a constant voltage supply.

Because of this (and the statically unstable behaviour for higher slip values where the emf is nevertheless acceptable), *constant current* drives are always used with (feedback) current and speed control loops, i.e. controlled current drives. In nearly all cases, variable frequency control is used as well. The use of both controlled (variable) frequency and controlled current allows for precise torque and speed control, as will be explained in Chap. 17.

---

[2]Explain why this is the case.

[3]These lead to high iron losses.

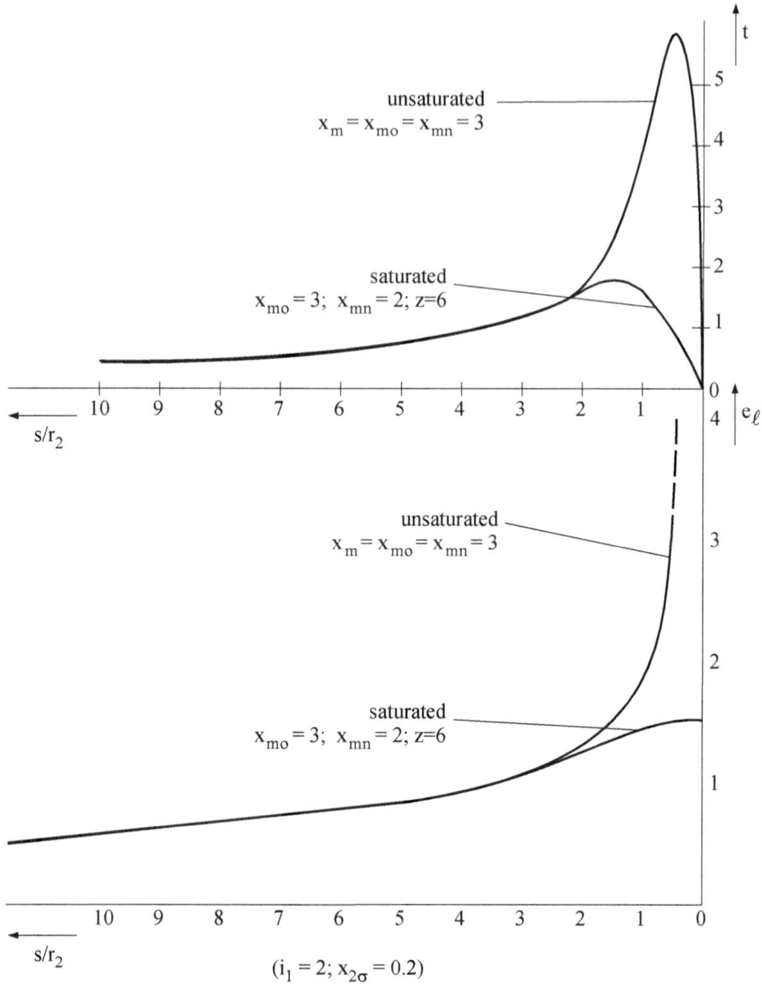

Fig. 14.6 Torque and emf curves

14.3 Ideal Current Supply of Synchronous Machines

14.3.1 Current, Voltage and Torque Relations

As we know, the usual equivalent scheme for voltage supply of a synchronous machine is that in (a) in Fig. 14.7, with a vector diagram as in (b). Here, the real axis is chosen along the supply voltage.

For current supply, however, the alternate equivalent circuit in Fig. 14.8 is much more appropriate. The superposition of mmfs (currents) rather than voltages is more

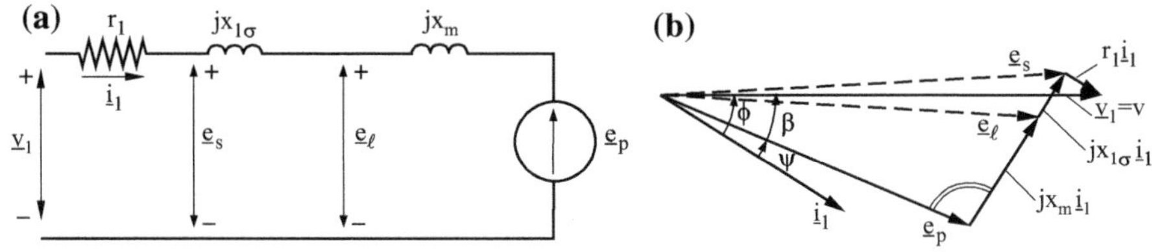

Fig. 14.7 SM equivalent circuit for voltage supply

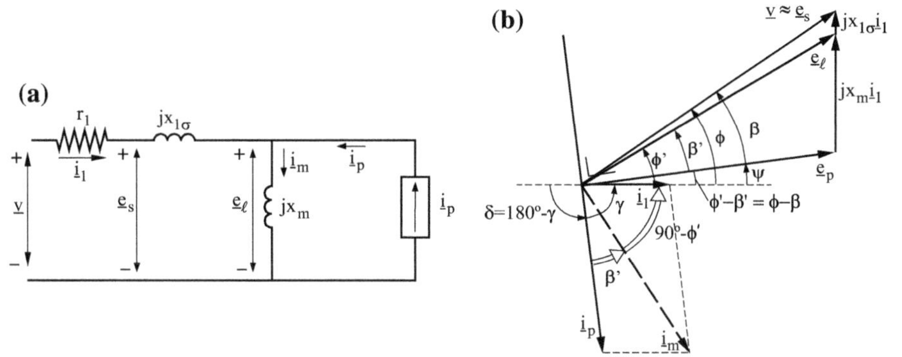

Fig. 14.8 SM equivalent circuit for current supply

apt for modelling saturation. In fact, adding emfs as in Fig. 14.7 is only meaningful if the saturation condition is almost constant, as it approximately is for a constant voltage supply.

From the circuit in Fig. 14.8, the following relations are easily derived:

$$\underline{v} = \underline{e}_s + r_1 \cdot \underline{i}_1 \tag{14.17}$$

$$\underline{e}_s = \underline{e}_l + jx_{1\sigma} \cdot \underline{i}_1 \tag{14.18}$$

$$\underline{e}_l = jx_m \cdot \underline{i}_m \tag{14.19}$$

$$\underline{i}_m = \underline{i}_1 + \underline{i}_p \tag{14.20}$$

Apparently, adding magnetising currents (or mmfs) as in Eq. 14.20 is the most correct method if the magnetic circuit is non-linear. However, note that i_m determines the saturation condition and therefore also the value of the magnetising inductance x_m. It is of course true that the last two equations are equivalent to $\underline{e}_l = \underline{e}_p + jx_m\underline{i}_1 = jx_m\underline{i}_p + jx_m\underline{i}_1$ (see Fig. 14.7), but then emfs are being added again.

Choosing the real axis along the primary current in Fig. 14.8, we can derive the following relations:

$$\underline{i}_p \equiv i_p \cdot \exp(-j\gamma) \tag{14.21}$$

$$\underline{i}_m \equiv i_m \cdot \exp[-j(\pi/2) + j\varphi'] = i_1 + \underline{i}_p \tag{14.22}$$

$$\underline{e}_p \equiv e_p \cdot \exp(j\psi) = jx_m\underline{i}_p \tag{14.23}$$

$$\underline{e}_l \equiv e_l \cdot \exp(j\varphi') = jx_m\underline{i}_m \tag{14.24}$$

$$\underline{e}_s = \underline{e}_l + jx_{1\sigma}\underline{i}_1 \tag{14.25}$$

$$\underline{v} \equiv v \cdot \exp(j\varphi) = \underline{e}_s + r_1\underline{i}_1 \tag{14.26}$$

Between the angle γ (considered positive when $\underline{i}_p$ lags $\underline{i}_1$ over less than π) and the internal displacement angle ψ, there is the relation $\gamma = \pi/2 - \psi$.

While the phase angle φ and the load angle β are defined with respect to the voltage $\underline{v}$, the phase angle φ' and the load angle β' are defined with respect to the air-gap emf $\underline{e}_l$. Here the URS is used, which is why β (β') is considered positive when $\underline{v}$ (e_l) leads $\underline{e}_p$, while φ (φ') is considered positive when $\underline{i_1}$ lags v ($\underline{e}_l$). This means that

$$\psi = \varphi - \beta = \varphi' - \beta' \tag{14.27}$$

To simplify the analysis, we will use the air-gap emf as reference.

From the above equations, it follows that

$$\gamma = \pi/2 - \varphi' + \beta' = \pi/2 - \varphi + \beta \tag{14.28}$$

and

$$i_m \sin\varphi' - ji_m \cos\varphi' = (i_1 + i_p \cos\gamma) - ji_p \sin\gamma \tag{14.29}$$

Equating the real and imaginary parts and dividing the resulting two equations side-by-side results in

$$\tan\varphi' = \frac{(i_1/i_p + cos\gamma)}{\sin\gamma} \tag{14.30}$$

From Eqs. 14.21 to 14.30, the following conclusions can be derived (see also Fig. 14.9 for a graphical derivation and Fig. 14.10 for a summary of the conclusions):

- for $i_1/i_p = 1$, Eqs. 14.21 and 14.22 yield $\varphi' = \pi/2 - \gamma/2$. Therefore, the relation between β' and γ is a straight line: $\beta' = \gamma/2$ (see Fig. 14.10). When γ varies from 0 to π, β' increases from 0 to $\pi/2$, φ' decreases from $\pi/2$ to 0 and the internal displacement angle ψ varies between $\pi/2 \cdots 0 \cdots - \pi/2$.
- for $i_1/i_p > 1$ (under-excitation), Eq. 14.21 yields $\varphi' > \pi/2 - \gamma/2$. The relation between β' and γ is a curve, as is illustrated in Fig. 14.10 for $i_1/i_p = 4/3$. Here, we always get $\beta' > \gamma/2$. When γ varies from 0 to π, β' varies between 0 and π, φ' varies between $\pi/2 \cdots \varphi'_{min} \cdots \pi/2$ (φ'_{min} corresponds with $\cos\varphi'_{min} = i_p/i_1$).
- for $i_1/i_p < 1$ (over-excitation), Eq. 14.21 yields $\varphi' < \pi/2 - \gamma/2$ while $\beta' < \gamma/2$. When γ varies from 0 to π, β' varies between $0 \cdots \beta'_{max} \cdots 0$ and φ' varies between

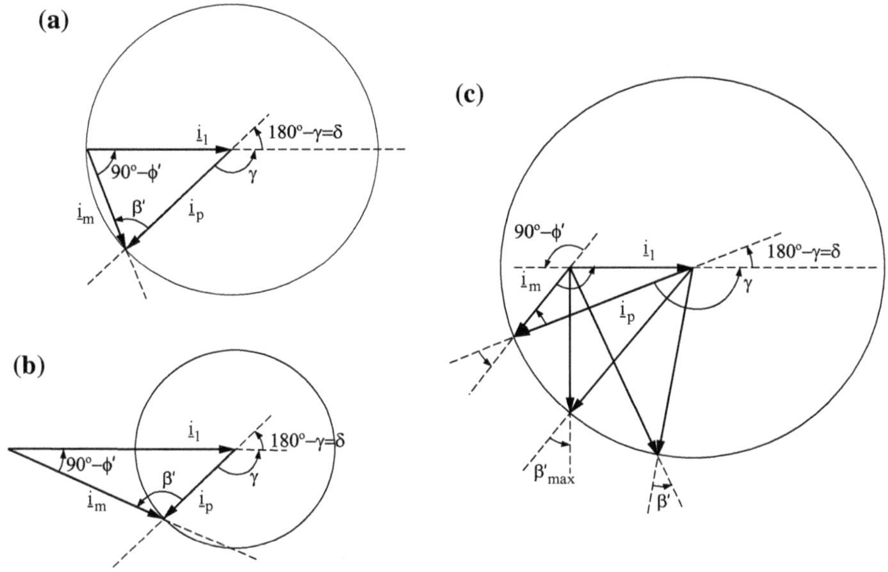

Fig. 14.9 Magnetising current as a function of γ

Fig. 14.10 Load angle as a function of γ

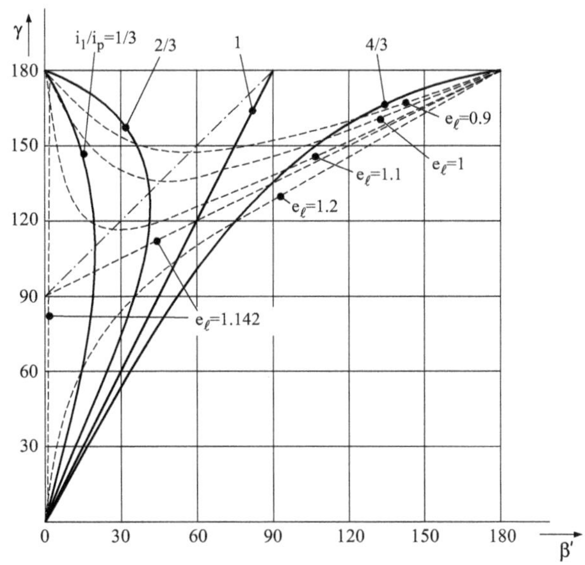

$\pi/2\cdots 0\cdots -\pi/2$. The maximum load angle β'_{max} (with $\sin\beta'_{max} = i_1/i_p$) occurs when $\varphi' = 0$, i.e. when the current triangle is right-angled.[4] Then we also have

[4]Derive this analytically from the equations above, starting from $\partial\beta'/\partial\gamma = 0$.

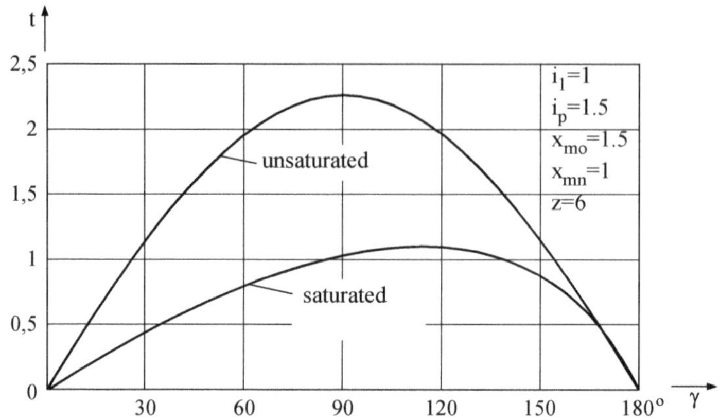

Fig. 14.11 SM: torque characteristic saturated and unsaturated

$\beta'_{max} = \gamma - \pi/2$ and thus all maximums for varying i_1/i_p are on a straight line (see the dash-dot line in Fig. 14.10).

It is particularly important to note that the highest value of the magnetising current i_m will occur for small values of γ, which are therefore likely to result in significant main field saturation. This may have some repercussions for the torque production.

Indeed, the torque or the rotating field power of a synchronous machine can be written[5] as a function of the currents i_1, i_p and the angle γ:

$$t \cdot n = x_m \cdot i_1 \cdot i_p \cdot \sin\gamma = x_m \cdot i_p \cdot i_q$$

with i_q the q-axis component of the primary current. Saturation will affect the magnetising reactance x_m and thus directly the torque.

14.3.2 Behaviour of the Synchronous Machine Neglecting Main Field Saturation

If we disregard main field saturation, x_m is constant, for example equal to the unsaturated value x_{mo}. The torque then varies sinusoidally with the angle γ, similarly to the sinusoidal variation with the load angle β for constant voltage supply (see the upper curve in Fig. 14.11).

Note, however, that the magnetising current i_m varies substantially with the angle γ ($i_m = |i_1 + \underline{i}_p| = |i_1 + i_p \cdot \exp(-j\gamma)|$). Consequently, also the air-gap field and emf e_l will vary with γ. As discussed in the previous section, the increase of the emf might be particularly relevant for small values of γ.

---

[5]Derive this from the power equation $t \cdot n = e_l \cdot i_1 \cdot \cos\varphi' = e_l \cdot i'_p \cdot \sin\beta'$.

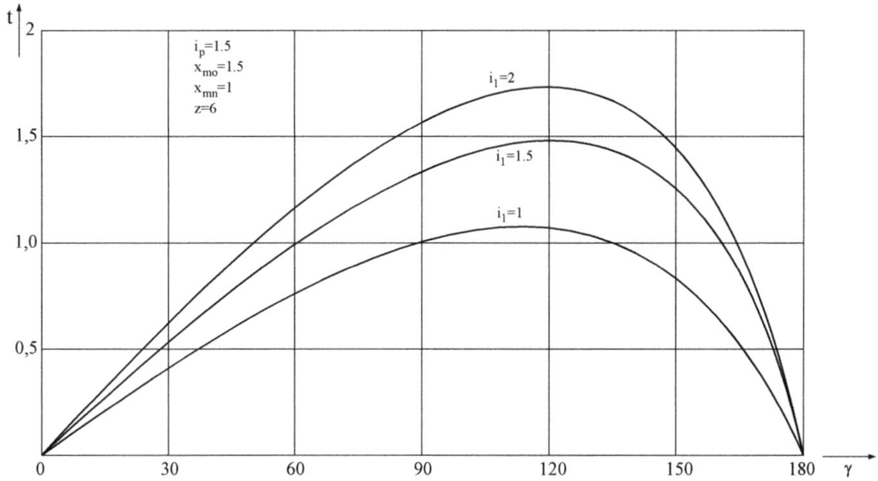

Fig. 14.12 Torque as a function of γ for different currents (saturated)

14.3.3 Behaviour of the Synchronous Machine Including Main Field Saturation

Main field saturation may be modelled as for the induction machine, i.e. by means of Eq. 14.10 or 14.11. The only difference is that now the reactances are much smaller, for example, for the unsaturated value of the magnetising reactance $x_{mo} = 1.5$ and for the saturated value $x_{mn} = 1 \ldots 1.2$.

As mentioned before, saturation will be especially important for small angles γ (see the torque characteristic (b) in Fig. 14.11). For small angles, the torque may be substantially reduced compared to the unsaturated case. The maximum torque is also reduced and the pull-out angle is now larger than $\pi/2$.

To calculate this saturated torque characteristic, we may start from a given angle γ and calculate i_m, which allows us to calculate e_l from the non-linear[6] equation 14.10. Equation 14.11 then gives the value of x_m.

Figure 14.12 shows the torque characteristics as a function of γ for $i_p = 1.5$ and for $i_m = 1.5;\ 1;\ 2$. For current supply, the angle γ plays the analogue role as the load angle β for constant voltage supply. Saturation decreases the torque but apparently increases the statically stable region for a constant load torque *if the drive were operated in an open loop (which is rarely the case for current supply)*. However, as will be explained in later chapters, operation at large angles of γ is in fact field weakening. This may be useful for high-speed operation, especially for permanent magnet excitation, which can be considered as a constant excitation current i_p.

[6]To avoid the solution of the non-linear equation, a value of e_l may be taken as a starting point, somewhere between its minimum and maximum value, between $\gamma = \pi$ and $\gamma = 0$, respectively.

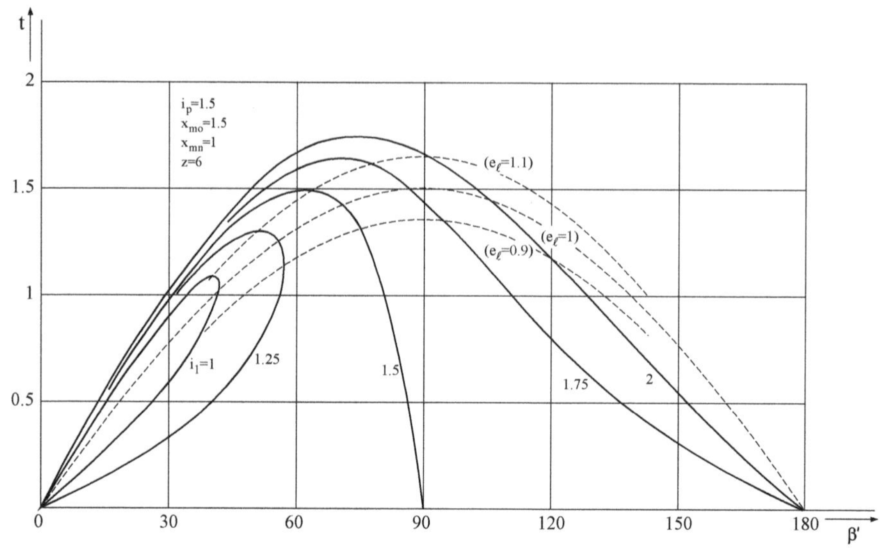

Fig. 14.13 Torque as a function of the load angle (saturated)

It is interesting to draw these torque characteristics for constant current supply as a function of the load angle β' (see Fig. 14.13). For $i_1/i_p < 1$, the torque is not an unambiguous function of β' (the maximum value of β' has been discussed above). What is also interesting is the comparison with the torque characteristics for constant voltage e_l (cf. the dashed lines). If the constant-current characteristics are above the constant-voltage characteristic for $e_l = 1$, high saturation levels may be expected.

Chapter 15
Variable Frequency Voltage Supply of Rotating Field Machines

Abstract Traditional rotating field machines have no (for synchronous motors) or poor and limited (for induction motors) possibilities for variable speed operation when power is supplied from a fixed frequency source. In this chapter, we will discuss the open-loop variable frequency voltage supply of rotating field machines. These kinds of drives are commonly denominated as V/f drives, because both the voltage and the frequency are controlled together in open loop, so as to control both the speed and the flux. The aim of the open-loop variable frequency operation of rotating field machines is to obtain a cheaper and more reliable variable speed drive than the expensive and high-maintenance DC drive. Drives that are even more demanding in terms of speed, position or torque control will be discussed in later chapters.

15.1 Introduction

In Part 1, we discussed induction and synchronous machines supplied by the (fixed frequency) grid. However, in principle there are no stringent limitations for the frequency of the supply (apart from possible mechanical limitations for the speed, for example). As the sections below will demonstrate, the models and equivalent circuits derived in Part 1 can also be used for variable frequency supply, if reactance notations jX are replaced by inductance notations $j\omega L$. For the pu notation, $j\nu l$ instead of jx will be used. In this case, the reference value for the angular frequency is normally the rated angular frequency ω_n of the machine. Note that choosing a reference frequency also implies a corresponding reference value ω_n/N_p for speed, and a reference value for time in dynamical studies; in most cases, a pu time $\tau = \omega_n t$ is also used. The reference value for inductances follows from $X_{ref} = \omega_{ref} L_{ref} = \omega_n L_{ref}$.

Varying the frequency implies that the supply voltage level will also have to vary, however. Indeed, the power output of a machine is determined by flux level and current. Maximum power is obtained for the maximum flux level (limited by saturation to the rated flux) and the maximum current (rated current). The air-gap flux is related to the air-gap emf and the frequency: $E_l = \omega\psi_l$

J. A. Melkebeek, *Electrical Machines and Drives*, Power Systems,
https://doi.org/10.1007/978-3-319-72730-1_15

If we disregard the resistive and leakage voltage drop in the armature, the air-gap emf is approximately equal to the voltage. For a constant flux equal to the rated flux, the voltage should therefore vary proportionally to the frequency: $V \approx E_l = \omega\psi_{ln}$. Alternatively, as rated flux corresponds to rated voltage at the rated frequency, $V/V_n = \omega/\omega_n$.

Please note that certain limits apply:

- for frequencies higher than the rated frequency, the voltage cannot increase beyond the rated voltage, because of insulation limitations of the winding and/or limitations to the available DC voltage. For these high frequencies, the voltage will be limited to the rated voltage, and the flux will decrease inversely proportionally to the frequency (i.e. field weakening)
- for very low frequencies, the resistive voltage drop in the armature can no longer be disregarded (compared to the reactive voltages), and an additional voltage bias that is more or less equal to the product of resistance and (real part of the) rated current has to be added.

15.2 Variable Frequency Supply of Induction Machines

To study the behaviour of induction machines with variable frequency supply, we will use the equivalent circuits in Fig. 15.1 (a and c for absolute values, b and d for pu values). Compared to the equivalent circuits for grid supply discussed in Part 1, the reactances (X, x) have been replaced by $\omega L = \nu\omega_n L$ and $\nu l = (\omega_n/\omega)l$. The slip s as always refers to the *actual* frequency or *actual* synchronous speed, $s = (\Omega - \Omega_r)/\Omega = (\omega - \omega_r)/\omega$, where $\Omega = \omega/N_p$ is the synchronous speed corresponding to the *actual* supply frequency ω. We may also write $s = (1 - \nu_r/\nu)$, where $\nu_r = \omega_r/\omega_n = (1 - s)\nu$ is the speed in pu.

Note that the resistive voltage drop becomes relatively important compared to the reactive voltage drops when the frequency is very low. Therefore, for low-frequency operation, the stator resistive voltage drop cannot always be ignored.

For frequencies that are not too low, we may use the simplified L-circuits (b) and (d) in Fig. 15.1 and we obtain the following equations for currents, power and torque (in absolute values and pu, respectively):

$$\underline{I}_1 = \underline{I}_m + \underline{I} = \frac{\underline{V}_1}{j\omega L_m} + \frac{\underline{V}_1}{R/s + j\omega L_\sigma} \tag{15.1}$$

$$\underline{i}_1 = \underline{i}_m + \underline{i} = \frac{\underline{v}_1}{j\nu l_m} + \frac{\underline{v}_1}{r/s + j\nu l_\sigma} \tag{15.2}$$

$$P_{em1} = 3V_1^2 \cdot \frac{R/s}{(R/s)^2 + (\omega L_\sigma)^2} \tag{15.3}$$

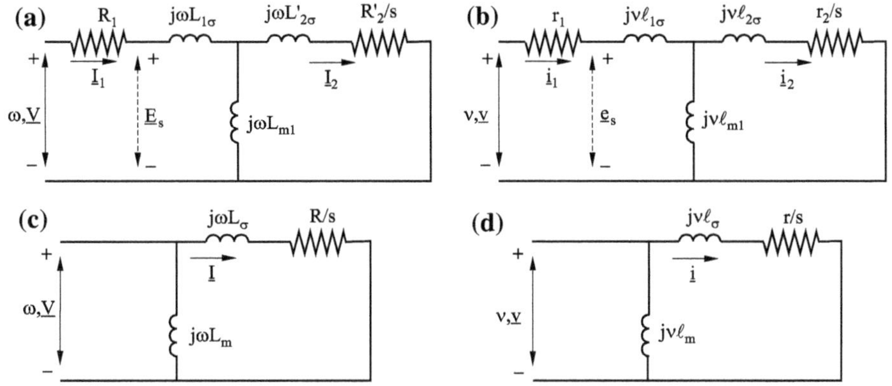

$R = R'_2 \cdot (1+\sigma_1)^2$

$X_{1\sigma} = \omega_n L_{1\sigma}$

$L_\sigma = \frac{\sigma}{1-\sigma} L_1 = \frac{\sigma}{1-\sigma} (L_{1\sigma} + L_{m1})$

$X'_{2\sigma} = \omega_n L'_{2\sigma}$

$L_m = L_1 = L_{1\sigma} + L_{m1}$

$X_\sigma = \omega_n L_\sigma$

$\underline{I} = \underline{I}'_2/(1+\sigma_1)$

$X_m = \omega_n L_m$

Fig. 15.1 Induction machine equivalent circuits for variable frequency

$$p_{em1} = v_1^2 \cdot \frac{r/s}{(r/s)^2 + (\nu l_\sigma)^2} \tag{15.4}$$

$$P_m = (1-s) P_{em1} \tag{15.5}$$

$$p_m = (1-s) p_{em1} \tag{15.6}$$

$$T = \frac{3}{\Omega} V_1^2 \cdot \frac{R/s}{(R/s)^2 + (\omega L_\sigma)^2} \tag{15.7}$$

$$t = \frac{1}{\nu} v_1^2 \cdot \frac{r/s}{(r/s)^2 + (\nu l_\sigma)^2} \tag{15.8}$$

Maximum torque (pull-out torque) is always obtained for the *same (pull-out) slip frequency* $s\nu\omega_n = s\omega = R/L_\sigma$ or $s\nu = r/l_\sigma$ (in other words, the pull-out *slip* diminishes with increasing frequency). The pull-out torque is given by

$$T_{po} = \frac{3}{\Omega} V_1^2 \cdot \frac{1}{2\omega L_\sigma} = \frac{3}{2} N_p \cdot \frac{V_1^2}{\omega^2 L_\sigma} \tag{15.9}$$

$$t_{po} = \frac{v_1^2}{2\nu^2 l_\sigma} \tag{15.10}$$

Rated active current ($I_w = I_n$ or $i_{1w} = i_w = 1$) is obtained for

$$i_w = v_1 \cdot \frac{r/s}{(r/s)^2 + (\nu l_\sigma)^2} = 1 \tag{15.11}$$

or

$$\frac{r}{s} = \frac{v_1}{2}\left(1 \pm \sqrt{1 - \frac{4\nu^2 l_\sigma^2}{v_1^2}}\right) \approx \frac{v_1}{2}\left(1 \pm \left[1 - \frac{2\nu^2 l_\sigma^2}{v_1^2}\right]\right) \tag{15.12}$$

Only the largest solution corresponds to a useful small slip value, which means that

$$\frac{r}{s} \approx v_1\left(1 - \frac{\nu^2 l_\sigma^2}{v_1^2}\right) \tag{15.13}$$

From now on, we will distinguish between operation at frequencies below or above the rated frequency.

For *frequencies below (or equal to) the rated frequency,* the voltage is varied proportionally to the frequency to obtain a constant flux (see Sect. 15.1). The pull-out torque is therefore constant, independent of the frequency (see Eq. 15.9 or 15.10). For rated flux, $v_1 = \nu$ and consequently

$$t_{po} = \frac{1}{2l_\sigma} \tag{15.14}$$

Maximum power therefore increases linearly with the frequency.

For the slip values corresponding to the rated active current (Eq. 15.13), we obtain a *constant slip frequency* $\nu s \approx r \approx s_n$ in this frequency range. For the corresponding electromagnetic power, we obtain $p_{em1} = v_1 \cdot i_w = \nu \cdot 1 = \nu$. The corresponding torque is thus the rated torque ($t = p_{em1}/\nu = 1$). In this *constant torque range* (or constant flux range), the ratio of maximum torque to rated torque is also constant.

For *frequencies above the rated frequency,* the voltage must be limited to the rated value (see Sect. 15.1). The flux will therefore decrease in inverse proportion to the frequency. From Eqs. 15.9 to 15.10, it follows that the maximum torque in this range will decrease inversely proportionally to the square of the frequency. This means that in this range, the maximum power decreases inversely proportionally to the frequency.

As to the slip values corresponding to Eq. 15.13, we now obtain constant slip values $s \approx r \approx s_n$ (at least if ν is not too large) and, as a result, increasing slip frequency values with increasing frequency. With the constant voltage $v_1 = 1$, we obtain $p_{em1} = 1$ for the corresponding electromagnetic power; the corresponding torque therefore decreases in inverse proportion to the frequency. This frequency range is called *the constant power range* or the *field weakening region.*

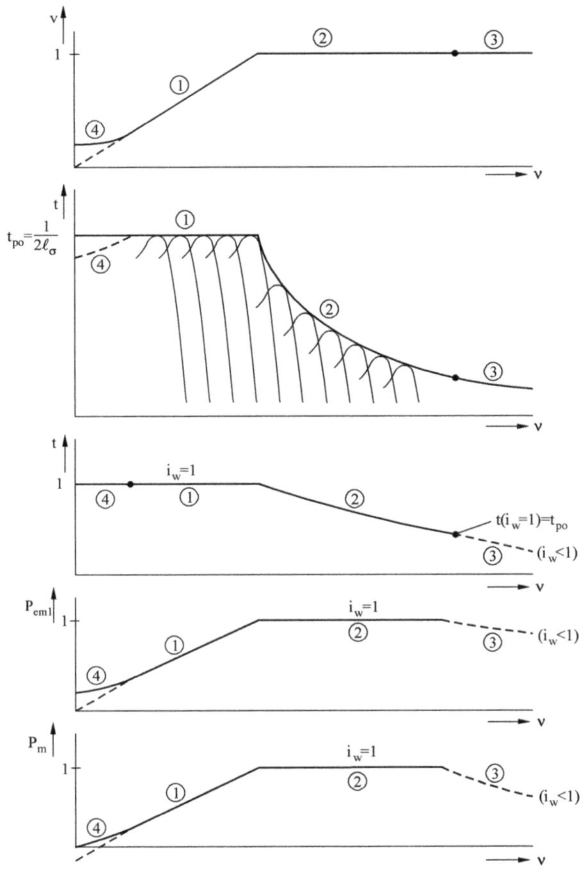

Fig. 15.2 Torque and power as a function of frequency

However, as the maximum (pull-out) torque decreases faster than this torque for rated active current, for very high frequencies (sometimes called the *over-speed range*), the rated torque can no longer be maintained.[1]

The above analysis is summarised in Fig. 15.2. Range 1 is the constant torque region; range 2 is the constant power or field weakening region. In range 3 (over-speed range), power decreases inversely proportionally to the frequency. Range 4 is the very low frequency region where a voltage bias is required to still obtain a constant torque.

Figure 15.3 illustrates some torque-speed characteristics in the normal ranges (1 and 2). Usually, the control will attempt to keep the slip sufficiently low, in the linear part of the torque characteristics shown in bold (this requires slip control or current control loop).

Negative speeds (with negative torques) imply a reversal of the phase sequence, which is easily accomplished with inverters.

[1]Calculate the corresponding maximum frequency.

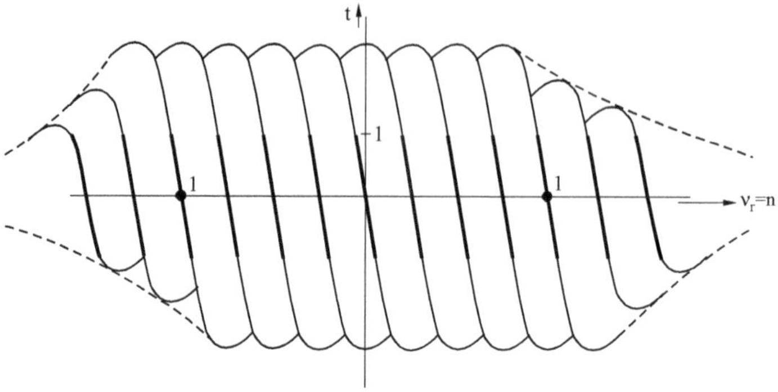

Fig. 15.3 Torque-speed characteristics in a large speed range

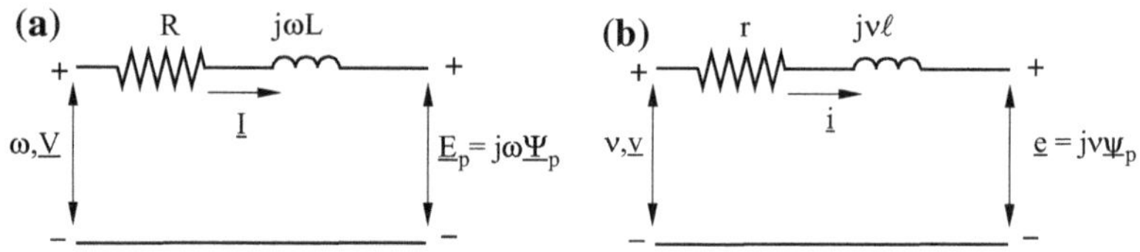

Fig. 15.4 Equivalent circuits for variable speed operation

Negative torques for positive speeds or positive torques for negative speed imply generator operation. This is not a problem for the inverter, but the DC supply might prohibit the power reversal (see Chap. 7 in Part 2).

15.3 Variable Frequency Supply of Synchronous Machines

In the equivalent circuits in Fig. 15.4, the reactances (X, x) have again been replaced by $\omega L = \nu\omega_n L$ and $\nu l = (\omega_n/\omega)l$. Note that the rotor emf depends on the frequency. For constant rotor excitation, the emf is proportional to the frequency.

For frequencies that are not too low, we may disregard the resistive voltage drop. We then obtain for the pu values:

$$\underline{v} = \underline{e}_p + j\nu l\underline{i} \tag{15.15}$$

Choosing the voltage along the real axis, $\underline{v} = v$, the emf (in the users reference system or URS) is $\underline{e}_p = e_p \cdot \exp(-j\beta)$. We then obtain (here in pu only)

$$\underline{i} = (\underline{v} - \underline{e}_p)/j\nu l = [v - \nu\psi_p \exp(-j\beta)]/j\nu l \tag{15.16}$$

$$\underline{i}_w = (\psi_p \sin\beta)/l \tag{15.17}$$

$$\underline{i}_b = -j(v - \nu\psi_p \cos\beta)/\nu l \tag{15.18}$$

$$p_{em} = t \cdot \nu = \frac{v \cdot e_p}{\nu l} \sin\beta = \frac{v \cdot \nu\psi_p}{\nu l} \sin\beta \tag{15.19}$$

$$t = \frac{v \cdot e_p}{\nu^2 l} \sin\beta = \frac{v \cdot \nu\psi_p}{\nu^2 l} \sin\beta \tag{15.20}$$

Pull-out torque is obtained for $\beta = \pi/2$ and is equal to

$$t_{po} = \frac{v \cdot \psi_p}{\nu l} \tag{15.21}$$

For *frequencies below (or equal to) the rated frequency,* the voltage is varied in proportion to the frequency, $v = \nu$ and we find a constant torque (for constant β), a constant pull-out torque and electromagnetic power proportional to the frequency:

$$t = p_{em}/\nu = \frac{\psi_p}{l} \sin\beta \tag{15.22}$$

$$t_{po} = \frac{\psi_p}{l} \tag{15.23}$$

For example, for $\psi_p = 2, l = 1.25$, there is a pull-out torque of 1.6; $\beta = \pi/6$ results in $|i| = 1, \underline{i}_w = 0.8, \underline{i}_b = j0.6, t = 0.8$ and $p_{em} = 0.8\nu$. Note in particular that in this *constant torque range,* the reactive current is constant (for constant β), independent of the frequency.

For *frequencies above the rated frequency,* the voltage is kept constant at the rated value, $v = 1$.

If the excitation can be controlled (which is the case for DC excitation), the excitation current will be varied inversely proportionally to the frequency, $\psi_p = \psi_{pn}/\nu$, resulting in a constant rotor emf e_p. In that case, the maximum electromagnetic power will decrease inversely proportionally to the frequency and the pull-out torque decreases as $1/\nu^2$. The rated active current can be maintained by increasing the load angle β, until the frequency for which $\beta = \pi/2$ is attained. In this range, the electromagnetic power corresponding with this rated active current will remain constant, but the torque will decrease with the frequency as $1/\nu$ (*constant power range*). Yet larger frequencies will result in a torque decreasing with the frequency as $1/\nu^2$.

If the machine has permanent-magnet excitation, the rotor emf will increase with the frequency (while the terminal voltage now remains constant at the rated value). The pull-out torque will therefore only decrease as $1/\nu$ and the maximum power remains constant. In principle, the active current component can be maintained at its rated value for the same load angle. However, the reactive current component will increase with increasing frequency (the current will become more and more leading). As a result, the total current will increase and the active current component will be more and more limited. For a given frequency (and beyond), no active current will be possible and no torque will be produced.[2]

To date, open-loop operation of V/f-supplied synchronous machines is rarely applied because of possible dynamical problems: when the frequency is adjusted too quickly, the machine may stall; an induction motor, on the other hand, will just increase the slip. In recent literature, however, several solutions have been proposed.

Note that in most applications, variable frequency operation of synchronous machines is based on current supply and current control (see Chaps. 16 and 17).

[2]Calculate this maximum frequency.

Chapter 16
Modelling of Inverter Supplied Rotating Field Machines

Abstract In this chapter (which is mainly based on Novotny, Equivalent circuit steady state analysis of inverter driven electric machines, [30]), we will study a method for the fundamental harmonic modelling of inverter supplied rotating field machines. First, we derive fundamental harmonic equivalent circuits for the inverter. These fundamental harmonic circuits are then combined with the steady-state equivalent circuits of rotating field machines to derive the basic characteristics of these inverter-supplied machines. For the modelling of the inverters, the resistive voltage drop over the switches will not be taken into account. Commutation will also be simplified, as most transients will be disregarded.

16.1 Fundamental Harmonic Models of VSI and CSI

16.1.1 Review of the Basic Inverter Schemes

The inverter types that are commonly used for the supply of rotating field machines are the (unmodulated) voltage source inverter or VSI, the modulated voltage source inverter or PWM-VSI, and the (unmodulated) current source inverter or CSI.

The notations used in this chapter are the following:

- subscript e for equivalent quantities
- subscript i for input quantities of the inverter
- subscript l for line quantities (line voltage, line current)
- subscript f for phase quantities
- subscript y for wye quantities (real wye connection or equivalent wye).

16.1.1.1 The Unmodulated Voltage Source Inverter

Figure 16.1(1a) shows the principal scheme of the VSI. The switches contain an antiparallel diode as the load can be reactive. The capacitor in the DC link serves to absorb temporarily negative DC link currents. Figure 16.1(1b) illustrates the six

J. A. Melkebeek, *Electrical Machines and Drives*, Power Systems,
https://doi.org/10.1007/978-3-319-72730-1_16

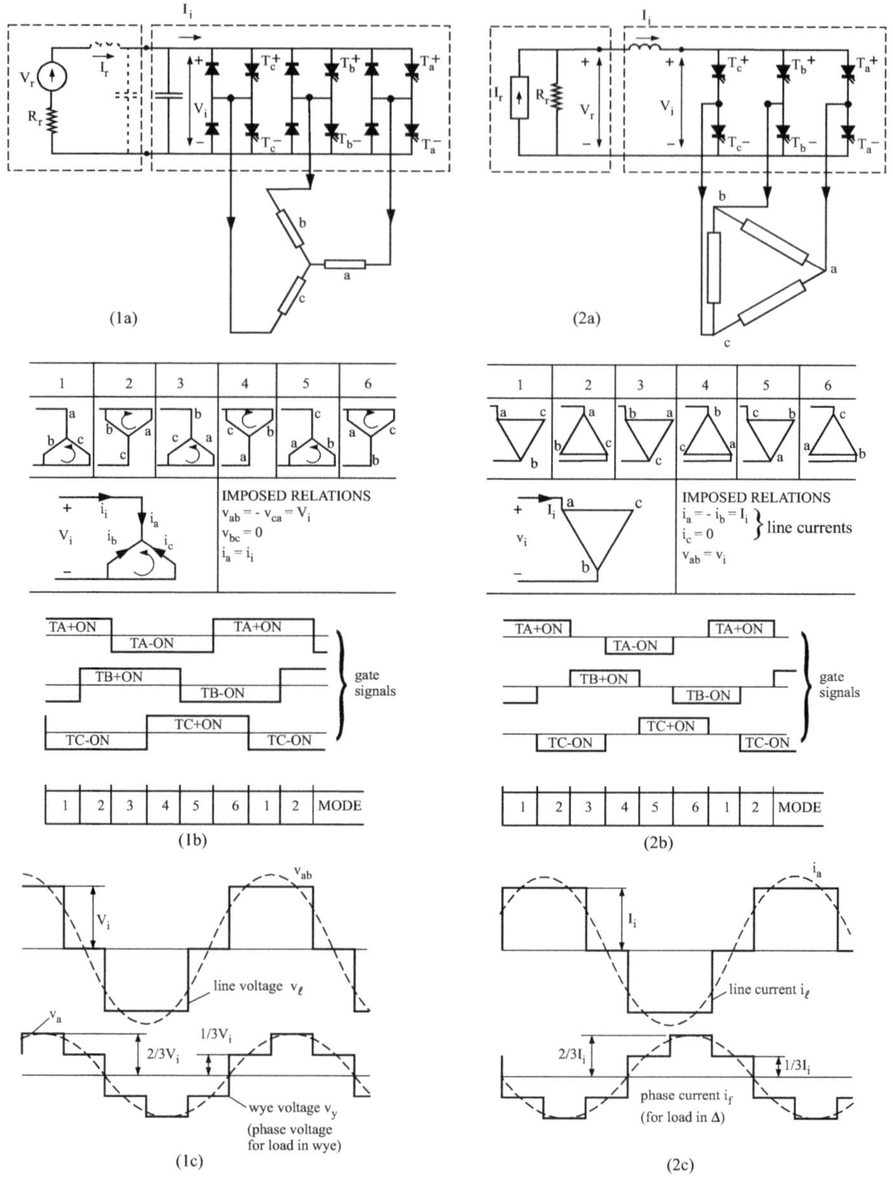

Fig. 16.1 VSI and CSI circuits, modes and waveforms

conducting modes and the corresponding gating signals, while Fig. 16.1(1c) shows the instantaneous line voltages and the wye voltages, as well as their fundamentals (cf. dashed lines). For a wye-connected load, the wye voltages are the phase voltages, while for a delta connected load the phase voltages equal the line voltages.

As a result of the 180° conduction and the antiparallel diodes, each of the output phases is at every instant connected to either the plus or the minus side of the DC source. Commutation takes place between the upper and lower switches of the same output phase (the required dead time between the switching-off of one switch and the switching-on of the opposing switch is disregarded here). In other words, there is a one-to-one relationship between the DC voltage and the AC phase voltages. In contrast, the AC side currents are not completely determined by the circuit itself, as at each instant two output phases are connected to one of the DC side terminals. In fact, as figure (1b) shows, at each interval there is a circulating current in the two phases connected to the same DC side terminal; obviously, this circulating current is determined by the load, but not by the inverter circuit.

16.1.1.2 The Unmodulated Current Source Inverter

The principal scheme of the CSI is shown in (2a) in Fig. 16.1. The inductance in the DC link serves to absorb temporarily negative DC link voltages. Figure (2b) illustrates the six conducting modes and the corresponding gating signals. Figure (2c) shows the instantaneous line currents and the phase currents for a load in delta, as well as their fundamentals (cf. dashed lines). For a wye-connected load, the line currents are also the phase currents.

As a result of the 120° conduction, only two of the output phases are connected to the DC source - one at the plus side and one at the minus side - and the third output phase is not connected. Commutation takes place between the switches connected to a DC terminal. This means that there is a one-to-one relationship between the DC current and the AC phase currents. In contrast, the AC side voltages are not completely determined by the circuit itself, as at each instant one output phase is not connected. As figure (2b) shows, the voltage in the not-connected phase is not a priori determined by the inverter circuit but instead is determined together with the load.

16.1.1.3 The Modulated Voltage Source Inverter

The PWM-VSI is a variant of the VSI, where in each mode sub-intervals are inserted with the three AC terminals shorted on one of the DC terminals. As Fig. 16.2 illustrates, at each instant there also is a one-to-one relationship between the output voltages and the DC voltages. Again, the distribution of the currents between the output phases is only partially determined by the inverter. The load will determine the missing relations.

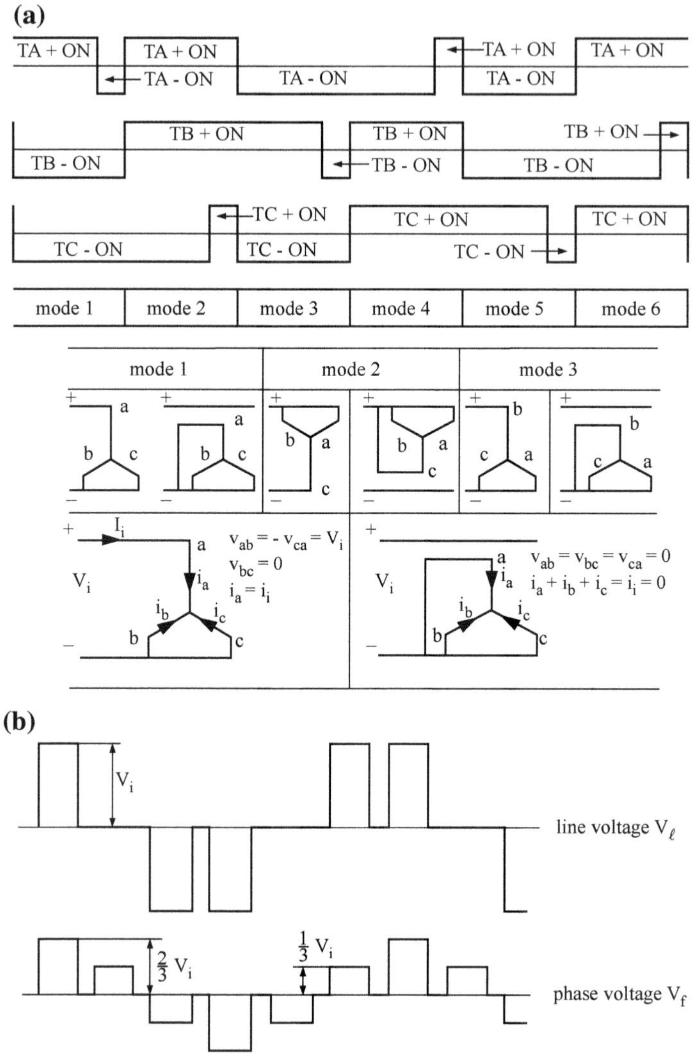

Fig. 16.2 PWM-VSI modes and waveforms

16.1.2 Idealised Output Waveforms

Only steady state will be considered. The input quantities of the inverter, v_i and i_i, will be replaced by their constant (or average) DC values V_i and I_i, and the commutation phenomenon will be simplified as much as possible.

With these assumptions, the output waveforms of the voltages of the VSI and the currents of the CSI are determined only by the inverter-imposed relations.

For the VSI, the Fourier expansion of the line and wye voltages yields:

$$v_l = \frac{2\sqrt{3}}{\pi} V_i \left[\sin \vartheta_l - \frac{1}{5} \sin 5\vartheta_l - \frac{1}{7} \sin 7\vartheta_l + \frac{1}{11} \sin 11\vartheta_l + \ldots \right] \tag{16.1}$$

$$v_y = \frac{2}{\pi} V_i \left[\cos \vartheta_y + \frac{1}{5} \cos 5\vartheta_y - \frac{1}{7} \cos 7\vartheta_y - \frac{1}{11} \cos 11\vartheta_y + \ldots \right] \tag{16.2}$$

respectively, where $\vartheta_x = \omega t + \vartheta_{xo}$ and with ω determined by the inverter and ϑ_{xo} determined by the chosen time reference instant (x can be either l or y).

For the CSI, we obtain for the line current and for the phase current (if load in delta):

$$i_l = \frac{2\sqrt{3}}{\pi} I_i \left[\sin \vartheta_l - \frac{1}{5} \sin 5\vartheta_l - \frac{1}{7} \sin 7\vartheta_l + \frac{1}{11} \sin 11\vartheta_l + \ldots \right] \tag{16.3}$$

$$i_f = \frac{2}{\pi} I_i \left[\cos \vartheta_f + \frac{1}{5} \cos 5\vartheta_f - \frac{1}{7} \cos 7\vartheta_f - \frac{1}{11} \cos 11\vartheta_f + \ldots \right] \tag{16.4}$$

For the PWM-VSI, the waveform depends on the modulation principle. Besides the fundamental harmonic and multiples, there are also carrier harmonics and side-bands. In general, we may write

$$v_l = \frac{2\sqrt{3}}{\pi} V_i \left[a_1 \sin \vartheta_l - \frac{a_5}{5} \sin 5\vartheta_l - \frac{a_7}{7} \sin 7\vartheta_l + \frac{a_{11}}{11} \sin 11\vartheta_l + \ldots \right] \tag{16.5}$$

$$v_y = \frac{2}{\pi} V_i \left[a_1 \cos \vartheta_y + \frac{a_5}{5} \cos 5\vartheta_y - \frac{a_7}{7} \cos 7\vartheta_y - \frac{a_{11}}{11} \cos 11\vartheta_y + \ldots \right] \tag{16.6}$$

where the coefficients depend on the modulation principle. In the above expressions, it is clear that a_1 will be smaller than unity.

It is often sufficient to consider only the fundamental harmonic, for example for studying the average electromagnetic behaviour without paying attention to harmonic torques or losses. Of course, harmonics may also contribute to the average torque but their contribution is rather small.

The fundamental harmonics for the three inverter types are in effective value:

- for the VSI:

$$V_l = \frac{\sqrt{6}}{\pi} V_i \tag{16.7}$$

$$V_y = \frac{\sqrt{2}}{\pi} V_i \tag{16.8}$$

- for the CSI (the mentioned phase quantity is valid only for a load in delta):

$$I_l = \frac{\sqrt{6}}{\pi} I_i \tag{16.9}$$

$$I_f = \frac{\sqrt{2}}{\pi} I_i \tag{16.10}$$

- for the PWM-VSI:

$$V_l = \frac{\sqrt{6}}{\pi} a_1 V_i \tag{16.11}$$

$$V_y = \frac{\sqrt{2}}{\pi} a_1 V_i \tag{16.12}$$

16.1.3 Secondary Quantities

Voltages for the VSI and currents for the CSI are primary quantities, determined by the inverter only. The currents for a VSI and the voltages for a CSI, which we will call the secondary quantities, are (mainly) determined by the load. The exact transfer relations between DC and AC values may only be determined through a detailed analytical study. However, for a fundamental harmonic approach, the power balance between the DC and AC sides provides sufficient information.

For the VSI (and PWM-VSI), we know the exact relation between input (DC) and output (AC) voltages. To derive the relation between input and output currents, we suppose that the (internal) inverter losses can be disregarded and that the total DC input power is transformed into fundamental harmonic AC output power:

$$V_i I_i = 3 V_y I_l \cos\varphi \tag{16.13}$$

where φ is the angle by which the fundamental harmonic output load current is lagging the output voltage. Substituting the expression for V_y from Eq. 16.8 results in

$$I_i = \frac{3\sqrt{2}}{\pi} I_l \cos\varphi \tag{16.14}$$

For a wye-connected load, $I_f = I_l$. For a load in delta, $I_f = I_l/\sqrt{3}$ and thus $I_i = \frac{3\sqrt{6}}{\pi} I_f \cos\varphi$.

At the DC-side, we see only the active component of the AC load current. The reactive component of the AC load current must therefore be provided by the inverter, as the circulating current between the two phases that are shorted by the inverter in each of the six modes.

By analogy, for the CSI we find:

$$V_i = \frac{3\sqrt{6}}{\pi} V_y \cos\varphi \tag{16.15}$$

where $V_y \equiv V_f$ for a load in wye, and $V_i = \frac{3\sqrt{2}}{\pi} V_f \cos\varphi$ for a load in delta.

For the PWM-VSI, the only difference with respect to the six-step VSI is the reduction factor a_1, thus

$$I_i = \frac{3\sqrt{2}}{\pi} a_1 I_l \cos\varphi \tag{16.16}$$

with $I_f \equiv I_l$ for a load in wye, and

$$I_i = \frac{3\sqrt{6}}{\pi} a_1 I_f \cos\varphi \tag{16.17}$$

for a load in delta.

16.1.4 Fundamental Harmonic Equivalent Circuits

Based on the above relations, we may derive fundamental harmonic equivalent circuits for the VSI and CSI. We will refer all equivalent circuits to the AC phase quantities (V_f and I_f).

For a VSI, we may write:

$$V_f = \frac{\sqrt{2}}{\pi} V_i \tag{16.18}$$

$$I_f \cos\varphi = \frac{\pi}{3\sqrt{2}} I_i \tag{16.19}$$

Equation 16.18 expresses the phase voltage in an equivalent DC voltage, while Eq. 16.19 expresses the active current component in the equivalent DC current. Figure 16.3a shows a simple equivalent circuit representing Eqs. 16.18 and 16.19. To the right, we find the output terminals of the inverter, connected to the load. As we have chosen the real axis along the output voltage, the load current is $\underline{I}_f = I_f \exp(-j\varphi)$. In parallel with the load, the inverter has a variable reactive element (i.e. a capacitor for a reactive load) that will continuously compensate the reactive power requirements of the load. This can be regarded as being continuously

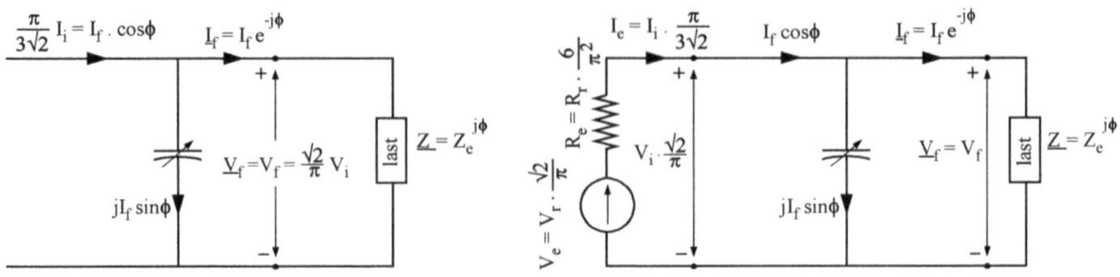

(a) VSI equivalent circuit (wye load)

(b) VSI and source equivalent circuit (wye load)

Fig. 16.3 Equivalent circuit for a VSI

in parallel resonance with the load.[1] To the left of this element, we find the input current of the inverter, referred to the AC side (by the factor $\pi/3\sqrt{2}$). The voltage at the input and the output of the inverter is the same, $V_f = (\sqrt{2}/\pi)V_i$.

We can complete this inverter scheme by adding to the left the Thévenin equivalent circuit for the DC source, comprising an ideal voltage source in series with the internal resistance of the DC source (see (b) in Fig. 16.3). The (rectifier) DC voltage V_r has to be referred to the AC side as $V_e = V_r \cdot (\sqrt{2}/\pi)$; the internal resistance also has to be referred to the AC side as $R_e = R_r \cdot (6/\pi^2)$. We can easily verify that this (single-phase) equivalent circuit correctly represents the three-phase power balance:

$$V_e \cdot I_e = R_e \cdot I_e^2 + V_f \cdot I_f \cos\varphi \tag{16.20}$$

or, indeed,

$$\frac{1}{3}V_r \cdot I_i = \frac{1}{3}R_r \cdot I_i^2 + V_f \cdot I_f \cos\varphi \tag{16.21}$$

The equivalent scheme for a CSI can be derived in a similar way (see (a) in Fig. 16.4). To obtain a complete duality with the VSI, here a load in delta has been considered together with a Norton equivalent for the DC source. The real axis is now chosen along the inverter current. Instead of a parallel reactive element as for the VSI, now a series reactive element should be inserted. The currents to the left and right of this element are the same, i.e. the inverter output phase current I_f, equal to the input DC current referred to the AC side, $I_i \cdot (\sqrt{2}/\pi)$.

The voltage to the left of this element is the input DC voltage referred to the AC side. It is equal to the real part of the AC voltage to the right of this reactive element. The reactive part of the AC voltage is absorbed by the reactive element, which can be a capacitor or an inductance. This reactive voltage can be seen as the voltage of the not-connected phase in each mode of the inverter; for a rotating field machine as load, this voltage is the emf induced by the rotating field. The Thévenin equivalent circuit for the DC source is demonstrated in (b) of Fig. 16.4.

[1]This reactive current may be interpreted as corresponding to the short-circuit currents in the two phases that are shorted by the inverter, in each of the six modes.

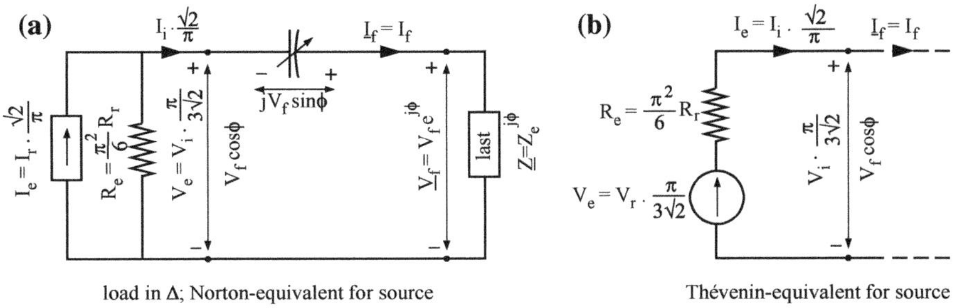

Fig. 16.4 Equivalent circuit for a CSI (load in delta)

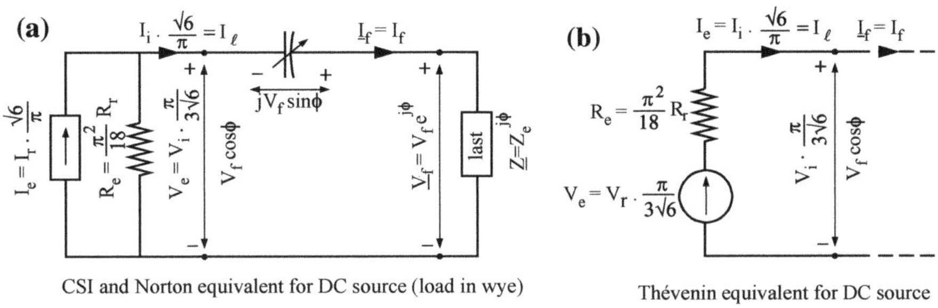

Fig. 16.5 Equivalent circuit for a CSI (load in wye)

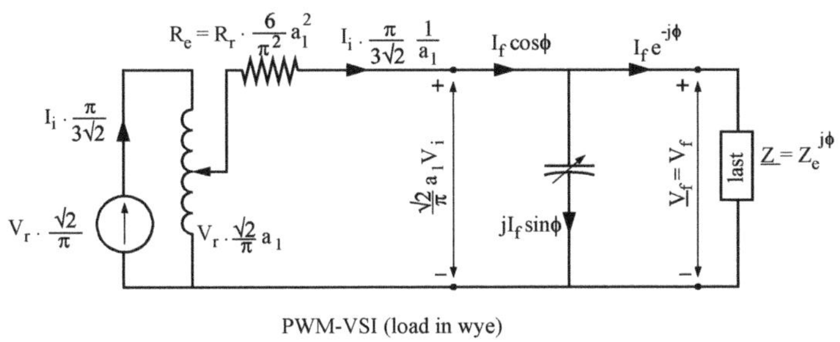

Fig. 16.6 Equivalent circuit for a PWM-VSI (load in wye)

The CSI equivalent circuit for a load in wye is shown in Fig. 16.5 ((a) for a Norton and (b) for a Thévenin equivalent for the DC source).

Figure 16.6 presents the equivalent circuit for a PWM-VSI. The difference compared to the VSI lies only in the reduction factor a_1 for the fundamental voltage (symbolically represented by an autotransformer). This reduction factor also needs to be accounted for in the equivalent values for the DC source current (and its internal resistance).

16.1.5 Discussion of the Equivalent Circuits

The equivalent circuits of the VSI and CSI may, in theory, be used to study the steady-state behaviour when any symmetrical three-phase load is supplied. In the next sections, we will apply these circuits when induction and synchronous machines are supplied.

First, however, we should pay attention to the assumptions used to derive these equivalent circuits, more particularly the fact that we disregarded the commutation phenomenon.

For a CSI, commutation takes place between switches connected to the same DC terminal (e.g. from T_a^+ to T_b^+, see (2b) in Fig. 16.1). When a CSI supplies a capacitive load (such as an over-excited synchronous machine), its operation is similar to that of a controlled rectifier with a control angle $\alpha > 90°$. The AC side supplies the reactive power required for the commutation: during commutation, two phases are shorted by the inverter and the emfs of the AC side have the right polarity to transfer the current to the next switch. There is only a small delay due to the inductance in the shorted phases, which is actually the (small) leakage inductance when a synchronous machine is supplied. After this short commutation interval, the previous phase is open-circuited and the emf of the load (a synchronous machine, for example) will induce a voltage in this open-circuited phase. It may therefore be assumed that the reactive power is transferred from the load to the inverter first during the short commutation interval and subsequently as the voltage induced in the open-circuited phase. It is mainly the latter which is represented by the variable reactive element in the inverter equivalent circuit.

For a VSI supplying an inductive load, the situation is somewhat dualistic with the previous case. Commutation takes place between switches connected to the same output phase. For example, for the commutation from T_a^+ to T_a^- (from mode 2 to mode 3 in (1b) in Fig. 16.1), switch T_a^+ receives the gate signal to switch off. The inductive current will then switch to the antiparallel diode D_a^- of the opposing switch. At approximately the same instant, T_a^-will receive a gate signal to conduct. Because the voltage in the phase a is reversed, the current in phase a will first decrease to zero and then reverse (through switch T_a^-). The reactive energy is now transferred, on the one hand, during the short commutation form T_a^+ to D_a^-/T_a^-, and on the other hand during the remainder of mode 3, where two output phases (a and c in this case) are shorted to the minus terminal of the DC source. The first part of the reactive energy is quite small and corresponds to the reactive energy in the leakage field. Normally, the second part is much larger and corresponds to the reactive energy transferred from phase to phase by the main rotating field.

For a CSI with capacitive load and a VSI with inductive load, the fundamental harmonic circuits provide fair approximations. Indeed, the usual load for a VSI (and PWM-VSI) is a reactive load like an induction machine. A VSI relies on forced commutation and uses switches that can be turned on and off. In contrast, the natural load for a CSI is a capacitive load, for example an over-excited synchronous machine, as a traditional CSI relies on natural commutation (for inverter operation, this is also referred to as *load commutation*). Indeed, for a CSI in its basic configuration, switches that can only be turned on will suffice, as is the case in a controlled rectifier. In fact, that is what it really is: the over-excited synchronous machine acts like the grid which supplies reactive energy for the commutation.

In contrast, for a CSI with inductive load, the current will not automatically transfer to the next switch, because the sign of the voltage ($V_b - V_a$ in this case) has the wrong sign. This requires switches that can be turned off as well, or thyristors with a killer circuit. At the switching instant from $T_a^+$to T_b^+, a very high di/dt can be observed, and therefore also very high peak voltages over the load will appear.

For a VSI with capacitive load, the current will have switched from T_a^+ to D_a^+ before the end of mode 2. At the end of mode 2, switch T_a^- receives a gate signal and the current will transfer automatically to T_a^- using the reactive energy of the load.[2] However, this will result in a large dv/dt over the load and therefore also high peak currents.

16.2 Inverter Supply of Induction Machines (Open Loop)

16.2.1 Induction Motor Supplied by a VSI or PWM-VSI

When an induction motor is supplied with a VSI or PWM-VSI, the harmonics in the voltage waveform will result in harmonic currents, besides the fundamental current. As the slip of the machine for these higher harmonics is close to one (for actual machine speeds between 0 and the synchronous speed for the fundamental), the impedance of the machine for these harmonics is more or less equal to the short-circuit impedance, i.e. approximately the leakage inductance: $Z_{k1} = R_{k1} + j\omega_i L_{k1} \approx jX_\sigma \cdot (\omega_i/\omega_1)$. The low leakage inductance results in relatively high harmonic currents, although limited by the high order of the harmonics. Figure 16.7 illustrates the current wave-shapes in no-load and full-load for a normal leakage of 0.2 pu and for an artificially low leakage of 0.06.

In the analysis below, we will concentrate on the fundamental harmonics. Combining the equivalent circuit of the VSI and the equivalent circuit of the induction machine creates the circuit in Fig. 16.8. Here, ν denotes the pu fundamental frequency (with respect to the rated frequency of the machine). Reactances X_{xy} are also referred to the rated frequency of the machine.

The phase voltage of the machine (chosen along the real axis) is equal to the rectifier voltage referred to the AC side:

$$\underline{V}_f = V_f = \frac{\sqrt{2}}{\pi} V_i = V_e - R_e I_e = \frac{\sqrt{2}}{\pi} V_r - \frac{6}{\pi^2} \cdot R_r \cdot \frac{\pi}{3\sqrt{2}} I_i \tag{16.22}$$

[2]For a VSI switches are always used that can be turned off because a wrong commutation would result in a short circuit of the DC voltage source.

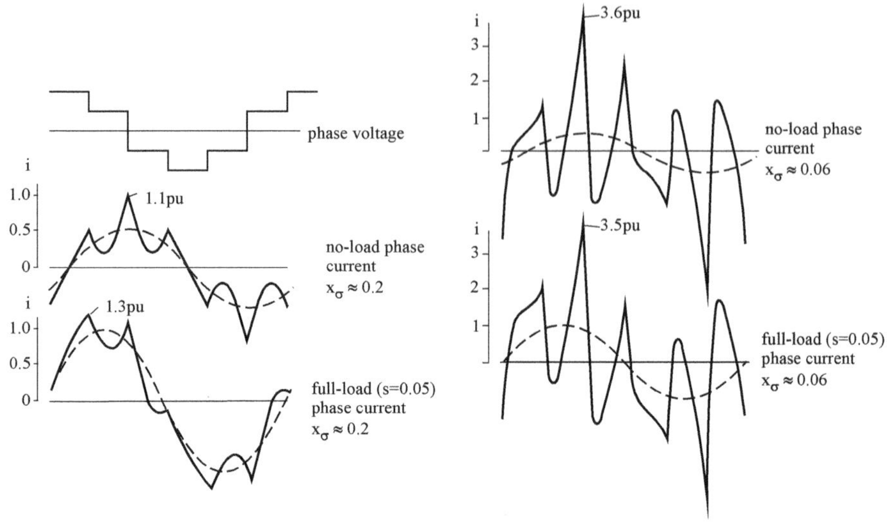

Fig. 16.7 Current waveforms for induction machine VSI supply

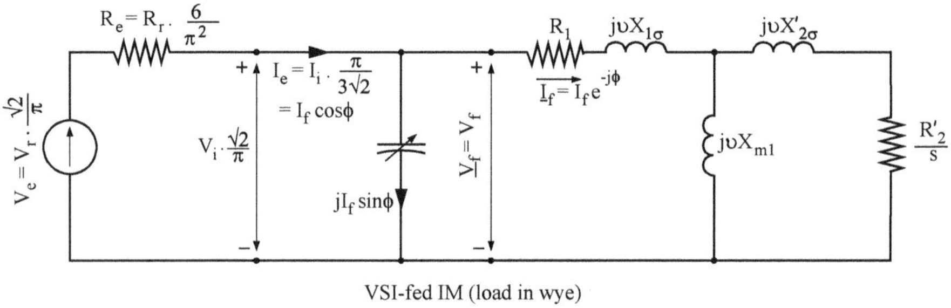

Fig. 16.8 Equivalent circuit for VSI fed induction machine (wye connection)

The input current of the inverter, referred to the AC side, is equal to the active component of the motor current

$$I_e = \frac{\pi}{3\sqrt{2}} I_i = I_f \cos\varphi \tag{16.23}$$

From the equivalent circuit we learn that, with regard to the fundamental, the operation of an induction motor fed by a VSI (or PWM-VSI) is not very different from the operation from the grid. The only difference lies in the DC source resistance which is not completely equivalent to a higher stator resistance. Indeed, the voltage drop over the resistance R_e is proportional to the active component of the stator current, while the voltage drop over the stator resistance is proportional to the total stator current. A higher stator resistance leads to a reduction of the torque in motoring (see Part1). However, the torque reduction due to the internal resistance of the rectifier is

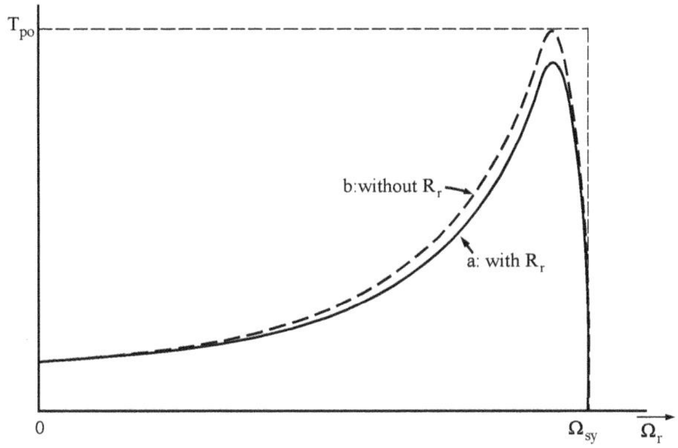

Fig. 16.9 Effect of the stator resistance on VSI-fed induction machines

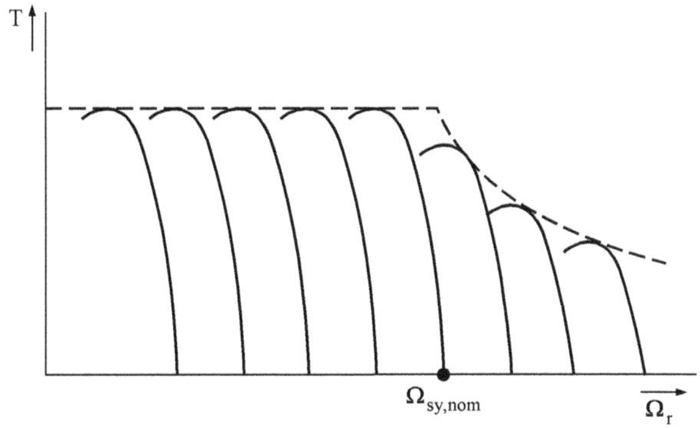

Fig. 16.10 Torque-speed characteristics for VSI-fed induction motor

only somewhat similar[3] for those slip values where the power factor is high. In other words, this is not the case for very low or very high slip values (see Fig. 16.9).

As we know, the machine voltage is varied proportionally to the frequency for frequencies below the rated frequency, and it is kept constant for higher frequencies. For a VSI, this variable voltage is realised by varying the rectifier voltage V_r, while for a PWM-VSI it is realised by the PWM. The resulting torque-speed curves are as illustrated in Fig. 16.10. For a PWM-VSI, the fundamental behaviour is the same, but the current harmonics are of a higher order and are therefore much smaller in amplitude.

16.2.2 Induction Motor Fed by a CSI

With an ideal CSI, the induction motor gets block current waves of 120° (see Fig. 16.11). Such a three-phase current corresponds to a space vector of constant

[3]For PWM-VSI, the effect is even lower due to the additional reduction factor a_1^2 for the resistance.

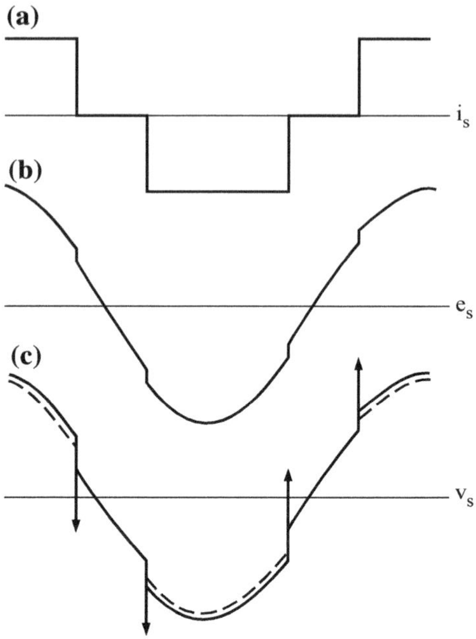

Fig. 16.11 Voltage waveforms for a CSI-fed induction machine

amplitude whose phase angle shifts each sixth of the period by 60° (cf. the hexagonal for the output phasors of an inverter).

In each sixth of the period, the induction machine sees a constant DC current in two phases (and zero current in the third), i.e. a constant current layer in the stator windings. The rotor, rotating with an electrical speed of $(1-s)\omega$, will then see exponential air-gap emf terms with an exponent comprising $j(1-s)\omega t$, thus sinusoidal segments with a frequency of $(1-s)\omega$. At the switching instants, a sudden phase shift of $s\pi/3$ is added so that the average frequency[4] of the emf is ω. To calculate the stator voltage, the resistive and inductive voltage drops over the stator resistance and stator leakage must be added. In theory, the current jumps and the leakage inductance results in impulse voltages in the stator voltage, but in practice these voltage peaks are limited because of the non-ideal current source.

We will only consider the fundamentals of the current and voltages. For these, we will use the equivalent circuit of a CSI combined with the equivalent circuit of the induction machine.

However, a distinction should be made between a (more or less ideal) DC current source and a (more or less ideal) voltage DC source.

For a *current supplied CSI,* we use the combined equivalent circuit in Fig. 16.12, where the parallel source resistance R_e is very large (or infinite). It is easy to see that for $R_e = \infty$, the induction motor fundamental behaviour is exactly the same as if it were fed by an ideal sinusoidal current source (see Chap. 14). In this case,

[4]Indeed: $6 \cdot (1-s)\pi/3 + 6 \cdot s\pi/3 = 2\pi$.

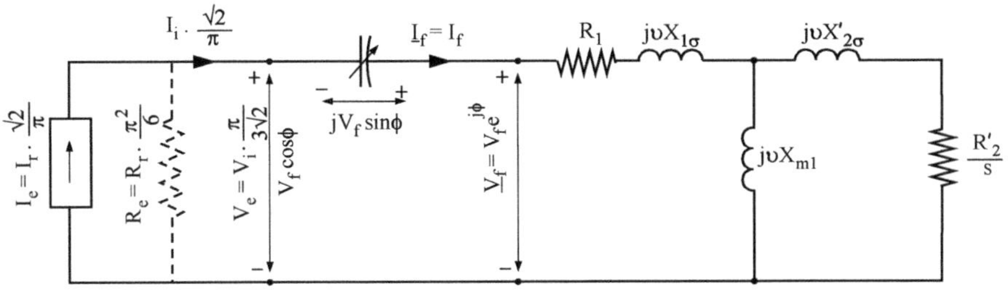

Fig. 16.12 CSI-fed induction machine (delta connected) with DC current source

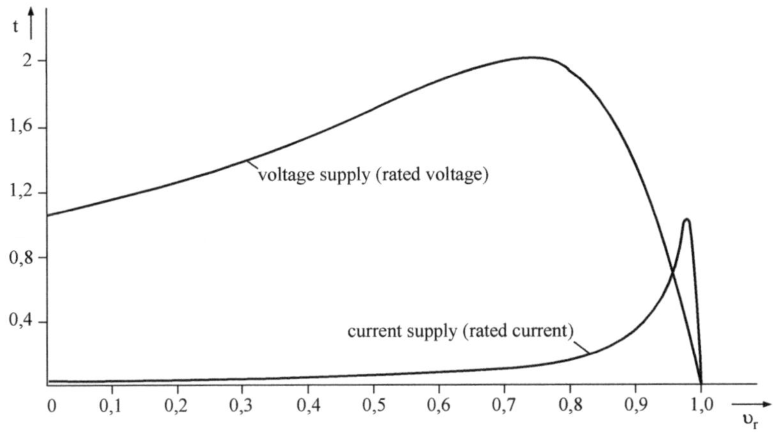

Fig. 16.13 Torque-speed characteristic for CSI-fed induction motor with DC current source

$I_f = I_i \cdot \sqrt{2}/\pi = I_e \equiv I_r \cdot \sqrt{2}/\pi$. The only difference compared to an ideal sinusoidal supply is that the reactive energy is now provided by the inverter, represented by the variable reactive element in the circuit (which is, for an induction machine, actually a variable capacitor). The machine phase voltage results from the machine equations, but only the real part of it is visible at the inverter input (i.e. the DC terminals).

The resulting torque-speed characteristic (see Fig. 16.13) is very steep with a rather limited pull-out torque at rated current. The slip values in the statically stable region (slip smaller than the pull-out slip) are very small and the emf values are quite high, resulting in saturation levels that are too high. Therefore, the CSI-fed induction motor is normally used at slip values larger than the pull-out slip, requiring a closed loop control for stability (see Chap. 17). The DC current can also be adjusted according to needs and can be temporarily above rated current, if necessary.

Clearly, this is exactly the dual case of the ideal VSI-supplied induction motor.

For a *voltage supplied CSI,* the scheme in Fig. 16.14 applies, as a result of combining the CSI equivalent circuit with a Thévenin equivalent circuit for the DC source and the induction motor equivalent circuit. The difference compared to the current

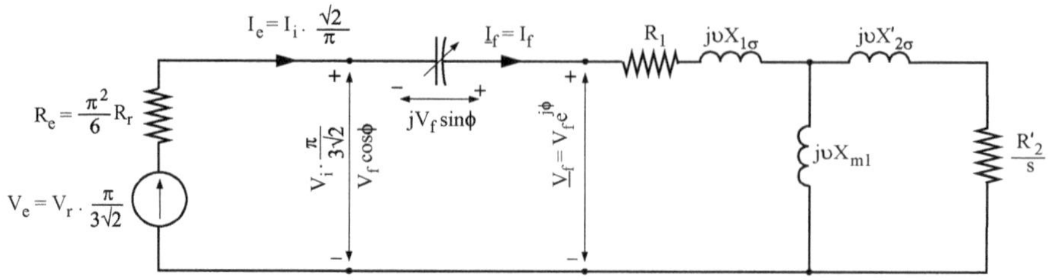

Fig. 16.14 CSI-fed induction machine (delta connected) with DC voltage source

supplied CSI is that here the voltage is controlled independently (and switched to the AC side by the inverter).

The equivalent source (rectifier) resistance can be added to the stator resistance. Motor current and equivalent DC current are equal, but only the real (i.e. active) component of the AC voltage is reflected to the DC side. The variable reactive element (in this case, the capacitor) compensates the reactive component of the motor voltage. As a result, the current is completely determined by the DC source voltage and the total equivalent resistance:

$$I_f = \frac{V_e}{R_e + R(s)} = \frac{V_r \cdot \frac{\pi}{3\sqrt{2}}}{\frac{\pi^2}{6} R_r + R(s)} \tag{16.24}$$

where R(s) is the real part of the motor impedance:

$$R(s) = R_1 + \frac{(R_2'/s) \cdot \nu^2 X_{m1}^2}{(R_2'/s)^2 + (\nu X_2')^2} \tag{16.25}$$

$(X_2' = X_{2\sigma}' + X_{m1})$.

For motor current and torque, we then find:

$$I_f = \frac{\pi}{3\sqrt{2}} \cdot \frac{V_r}{R} \cdot \frac{1 + \left(\frac{s\nu X_2'}{R_2'}\right)^2}{1 + \left(\frac{s\nu X_2'}{R_2'}\right)^2 + \frac{s\nu^2 X_{m1}^2}{R R_2'}} \tag{16.26}$$

$$T = \frac{N_p}{\nu\omega_n} \cdot \frac{\pi^2}{6} \cdot \left(\frac{V_r}{R}\right)^2 \cdot \frac{\frac{s\nu^2 X_{m1}^2}{R_2'} \cdot \left(1 + \left(\frac{s\nu X_2'}{R_2'}\right)^2\right)^2}{\left(1 + \left(\frac{s\nu X_2'}{R_2'}\right)^2 + \frac{s\nu^2 X_{m1}^2}{R R_2'}\right)^2} \tag{16.27}$$

in which $R = R_1 + \frac{\pi^2}{6} R_r$ and $\nu\omega_n / N_p = \Omega_{sy}$.

The curve for the current as a function of the pu rotor speed $\nu_r = (1 - s)\nu$ is illustrated in Fig. 16.15. The current becomes very large for slip zero, even infinite for

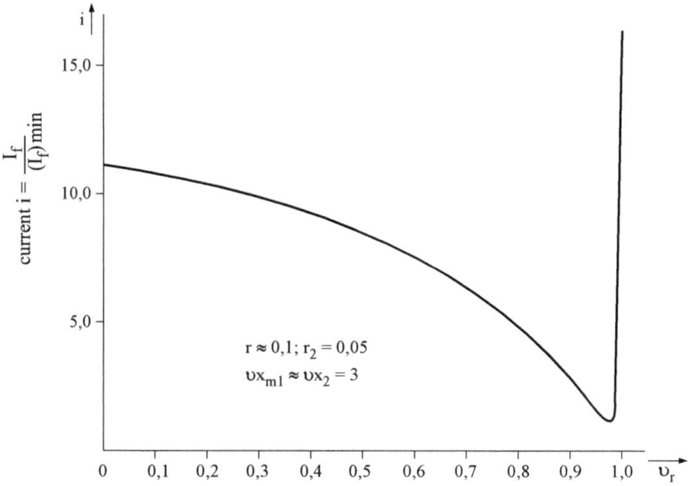

Fig. 16.15 CSI-fed induction machine: current as a function of speed

a small negative slip value. Indeed, the current is determined by the total resistance in the circuit, which is $R_e + R_1$ for zero slip.

For increasing slip (decreasing speed), the current reaches a minimum value for $s = R_2'/\nu X_2'$, after which it increases again with increasing slip. The minimum value of the current is attained when the real part of the motor impedance is at a maximum. This is exactly the same slip value as when the torque of a current-fed induction machine is at a maximum, thus $s_{min} = R_2'/\nu X_2'$. For both cases, this maximum real part of the motor impedance is given by:

$$\max[R(s)] = R_1 + \frac{\nu^2 X_{m1}^2}{2\nu X_2'} \tag{16.28}$$

For current supply, this corresponds to the maximal air-gap power (or torque), whereas for a voltage supply it corresponds with the minimum current.

The torque characteristic shows a minimum and two (equal) maximums, see Fig. 16.16. The minimum torque is at the same slip value where the current is minimal and is given by

$$T_{min} = N_p \cdot \frac{\pi^2}{3} \cdot V_r^2 \cdot \frac{L_2'}{\nu^2 \omega_n^2 L_{m1}^2} \cdot \left(1 - 4\frac{L_2' R}{\nu \omega_n L_{m1}}\right) \tag{16.29}$$

The maximum torque is given by

$$T_{max} = \frac{N_p}{\nu \omega_n} \cdot \frac{\pi^2}{24} \cdot \frac{V_r^2}{R} \tag{16.30}$$

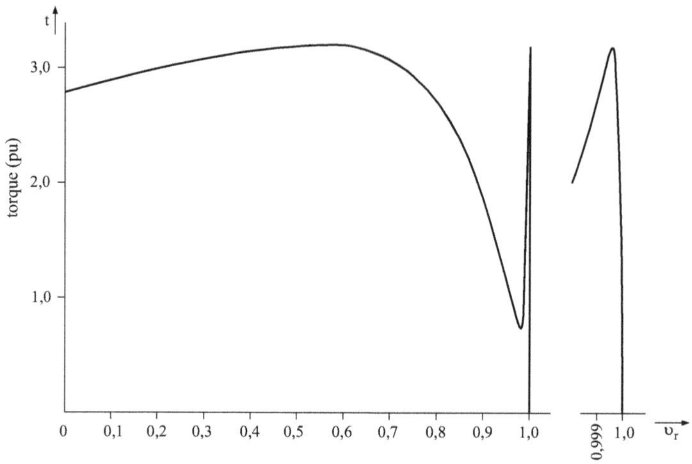

Fig. 16.16 Torque-speed characteristic for CSI-fed induction motor with DC voltage source

for the slip values

$$s_{max,1} = \frac{R_2}{\nu X_2'} \cdot \frac{\nu^2 X_{m1}^2}{2R\nu X_2'} \cdot \left(1 + \sqrt{1 - 4\left(\frac{R\nu X_2'}{\nu^2 X_{m1}^2}\right)^2}\right) \approx \frac{R_2'}{R} \cdot \frac{L_{m1}^2}{L_2'^2} \tag{16.31}$$

$$s_{max,2} = \frac{R_2}{\nu X_2'} \cdot \frac{\nu^2 X_{m1}^2}{2R\nu X_2'} \cdot \left(1 - \sqrt{1 - 4\left(\frac{R\nu X_2'}{\nu^2 X_{m1}^2}\right)^2}\right) \approx \frac{R_2' R}{\nu^2 \omega_n^2 L_{m1}^2} \tag{16.32}$$

The above discussion illustrates that the reactive current compensation by the inverter dominates the behaviour of the system: current (and torque) are determined by the real part of the impedances only.

We should note, however, that an open-loop operation of an induction motor with a voltage-supplied CSI is not possible. It can be shown that, even in the statically stable slip range, there are complex eigenvalues in the right half plane. Nevertheless, such a drive can be used in a closed loop where frequency and DC voltage are adapted depending on the slip (or slip frequency) and the required speed (see Chap. 17). The closed-loop control will also adjust the required flux level for the machine, dependent on the frequency.

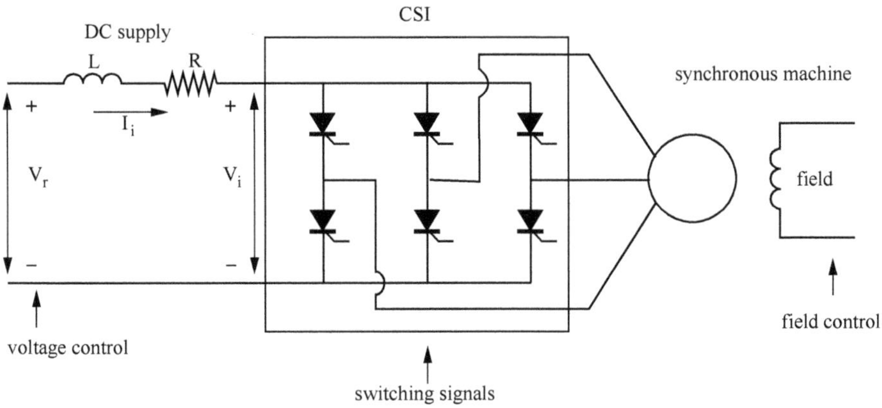

Fig. 16.17 CSI-fed synchronous machine controls

16.3 Inverter Supply of Synchronous Machines

16.3.1 Introduction

Nowadays, most induction machine drives use a voltage source inverter (VSI or PWM-VSI), frequently with a current control loop to imitate a current supply. For very large power ratings, current source inverters are sometimes used, although less and less frequently.

For synchronous machine drives, a CSI supply is much more common, although a VSI with current control loop is also often applied for smaller power ratings. An important reason to use a CSI for synchronous machine drives is that the synchronous machine can deliver the reactive power for the inverter, while at the same time excellent dynamic behaviour comparable with that of a DC machine can be obtained.

Figure 16.17 illustrates a general scheme including DC source control, inverter switching control and field regulation (which is absent in case of permanent-magnet excitation). The obtained characteristics mainly depend on the control used for the switching, but also on the DC control and the excitation control. As will be discussed below, there are some differences between smooth-rotor synchronous machines and salient-pole machines as well.

16.3.2 CSI-Fed Synchronous Machine with Smooth Rotor

Figure 16.18 shows the equivalent scheme of a smooth-rotor synchronous machine, combined with the equivalent scheme for a CSI. For the source, a Thévenin equivalent has been used. $\underline{V}$ and $\underline{I}$ are phase quantities of the machine. The variable reactive

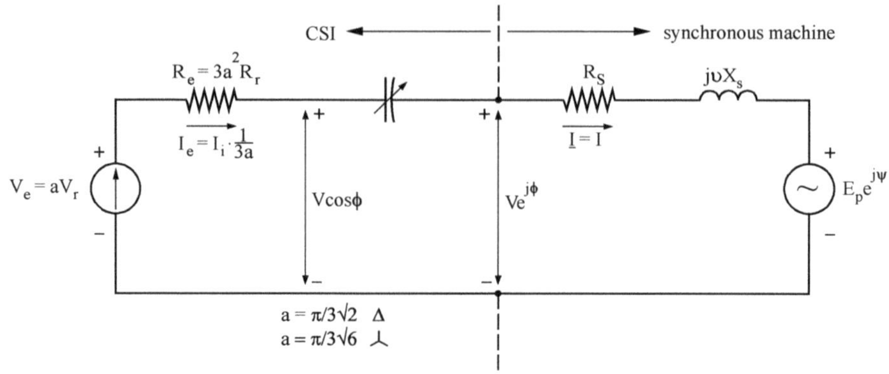

Fig. 16.18 CSI-fed synchronous machine equivalent circuit (DC V-source)

element (in this case probably an inductance: i.e. load commutation for an over-excited synchronous machine) compensates the reactive part of the motor impedance. The output power of the source (rectifier) is equal to three times the single-phase power corresponding with the scheme

$$P_r = V_r \cdot I_r = 3V_e \cdot I_e = 3V_e \cdot I \tag{16.33}$$

At the input of the inverter, we have:

$$P_{in} = P_r - R_r I_r^2 = P_r - 3R_e I^2 = 3V \cdot I \cdot \cos\varphi \tag{16.34}$$

From this power input to the inverter, the power converted by synchronous machine into mechanical power is

$$P_{in} - R_s I^2 = 3V \cdot I \cdot \cos\varphi - R_s I^2 = 3E_p \cdot I \cdot \cos\psi \tag{16.35}$$

The rotor emf depends on the excitation and the speed (frequency):

$$E_p = \omega \cdot \Psi_p = \omega \cdot L_m \cdot I_p \tag{16.36}$$

Depending on the control circuit (i.e. inverter switching signals, excitation), different drive characteristics may be obtained, as is illustrated in the examples below.

16.3.2.1 Constant ψ Control

For a constant internal angle ψ and a constant rectifier voltage V_r, we get the phasor diagrams in Fig. 16.19.

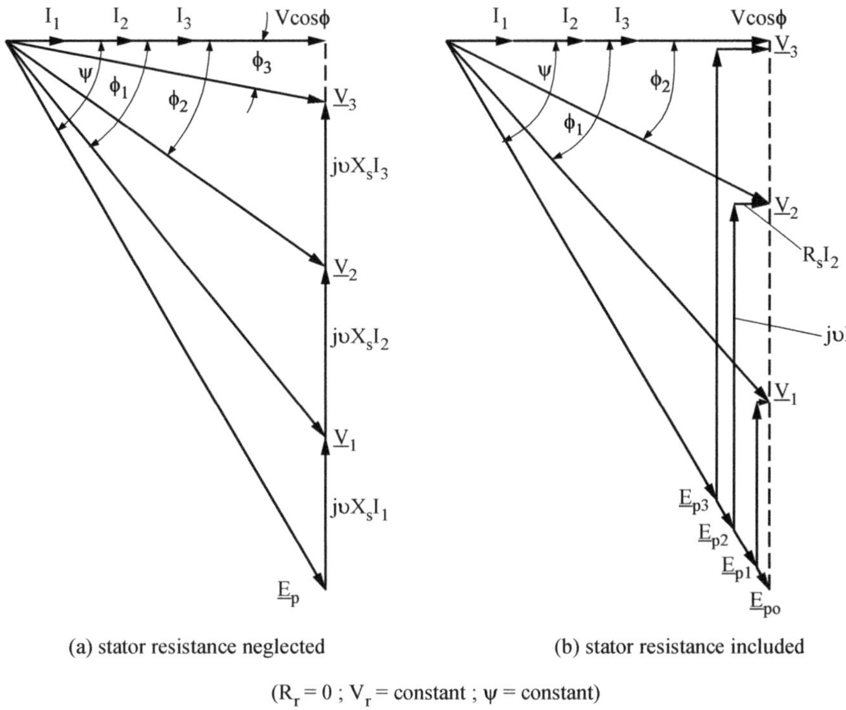

Fig. 16.19 Phasor diagrams for constant ψ control

As $V\cos\varphi$ is constant because of the constant DC voltage, all voltage phasors end on the same vertical line. In the phasor diagram to the left, the stator resistance is assumed to be zero or negligible in any case. Because the machine voltage drop is now purely reactive, the rotor emf $\underline{E}_p$ turns out to be a constant phasor. For a constant excitation, the speed (or frequency) is therefore constant, independent of the current or load torque. This means that the torque-speed characteristic is a horizontal line, similar to a separately excited DC machine with negligible armature resistance. Note that the machine voltage decreases with increasing current.

From the phasor diagram to the right in Fig. 16.19, we can derive that a non-zero stator resistance results in a decrease of E_p with increasing current. The torque-speed characteristic for constant excitation is a straight line with decreasing speed for increasing load (as for a separately excited DC machine).

As the reactive component of the load is continuously compensated by the variable reactive element of the inverter, the current is only determined by the real components of the voltages in the circuit. Consequently, the equivalent circuit may be simplified to that in Fig. 16.20 and all reactive components may be omitted.

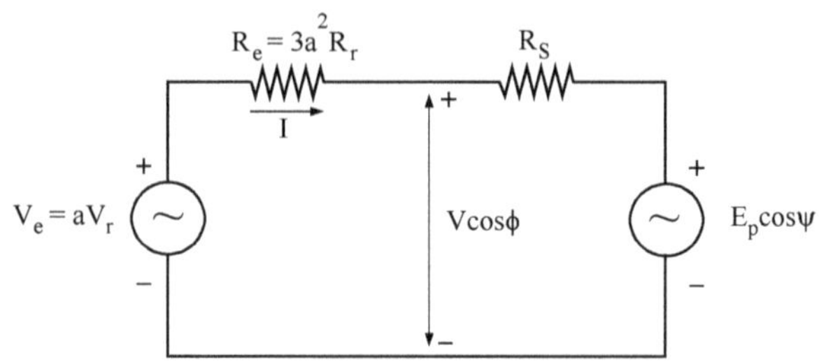

Fig. 16.20 Reduced equivalent circuit for a CSI-fed synchronous machine (smooth rotor)

The voltage equation for this scheme reads

$$V_e = (R_e + R_s) \cdot I + E_p \cos\psi = (R_e + R_s) \cdot I + \cos\psi \cdot \Psi_p \cdot \omega \qquad (16.37)$$

similar to the armature voltage equation of a DC machine. For the three-phase power, we get

$$P_{in} = 3V_e \cdot I = 3(R_e + R_s) \cdot I^2 + 3E_p \cdot I \cdot \cos\psi = 3(R_e + R_s) \cdot I^2 + 3\cos\psi \cdot \Psi_p \cdot \omega \cdot I \qquad (16.38)$$

From the total input power, the last term on the right hand side of this equation is the electro-mechanical power:

$$P_{em} = T \cdot \Omega = 3\cos\psi \cdot \Psi_p \cdot I \cdot \omega \qquad (16.39)$$

which yields for the torque

$$T = 3N_p \cdot \Psi_p \cdot I \cdot \cos\psi \qquad (16.40)$$

The practical implementation of the constant ψ control requires a position sensing of the rotor. Indeed, the inverter switching signals determine the phase angle of the current (i.e. ψ) with respect to the rotor flux or the rotor position. In this kind of drive, the inverter actually follows the instantaneous speed of the machine and load.[5]

The complete control circuit of the drive will also contain a current control (for controlling the torque) and an excitation control (for controlling the flux dependent on the speed). The excitation may also be varied to optimise some other aspects of the behaviour of the drive. In addition, one may depart from a fixed internal angle ψ (e.g. zero or even a slightly positive angle ψ, to obtain field orientation, see Chap. 17).

Figure 16.21 shows some characteristics of a (smooth rotor) synchronous machine drive. As discussed before, the torque-speed characteristic is a straight line with decreasing speed for increasing load torque, with a slope dependent on the stator resistance.

[5]In contrast, in a traditional (grid) voltage fed drive, the machine and load speed follow the frequency of the supply.

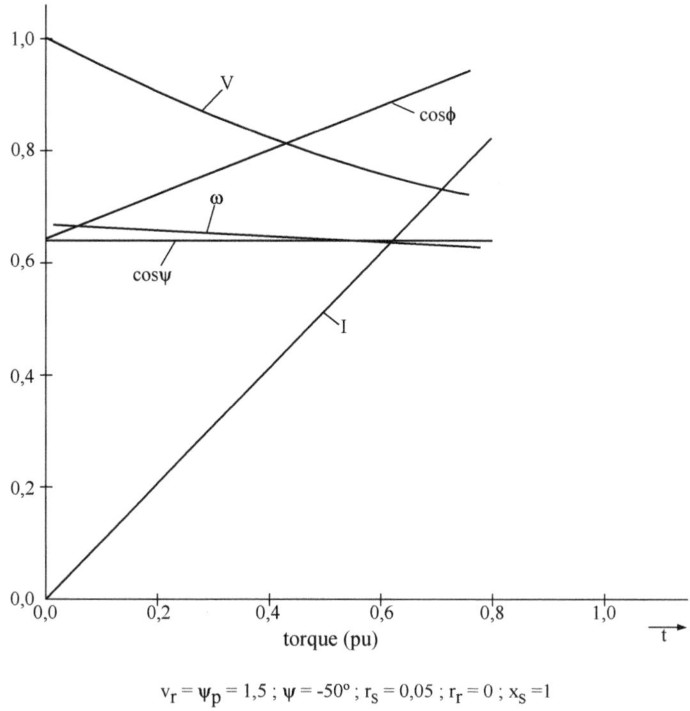

Fig. 16.21 Characteristics of a CSI-fed synchronous motor with smooth rotor for constant ψ

In general, also external control circuits will be added, for example for speed or position control. These will determine the desired value of the torque (and thus the current). In later chapters, this will be discussed in more detail.

16.3.2.2 Constant φ Control

A significant problem with the constant ψ control is that it requires a rotor position sensor, unless sensorless methods are used. An alternative solution is a constant φ control, thus avoiding a mechanical position sensor and using an electronic sensor instead.

Figure 16.22 shows phasor diagrams for a synchronous machine with a smooth rotor. The diagram shows that increasing the stator current from I_1 to I_2 results in an increase of ψ and E_p and a decrease of $E_p \cos\psi$. For a constant excitation, the increase of E_p (with an increasing current or torque), also causes the speed to rise (see the characteristics in Fig. 16.23). This statically unstable characteristic (e.g. for a constant load torque) does not create any instability as such a drive is always operated in closed loop control. Nevertheless, the excitation may also be programmed to increase with increasing load so as to avoid a rising speed.

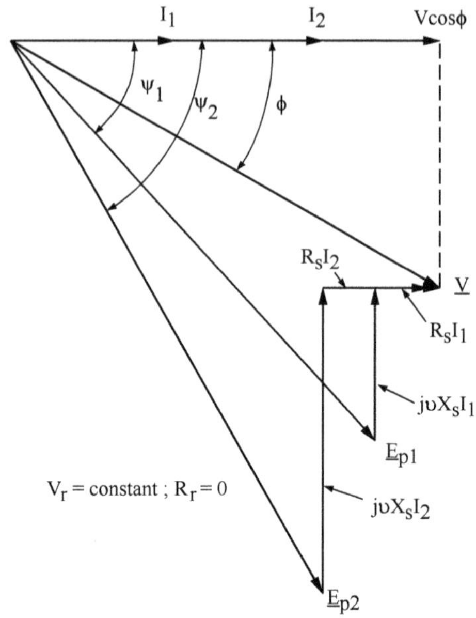

Fig. 16.22 Phasor diagram of CSI-fed synchronous motor for constant φ

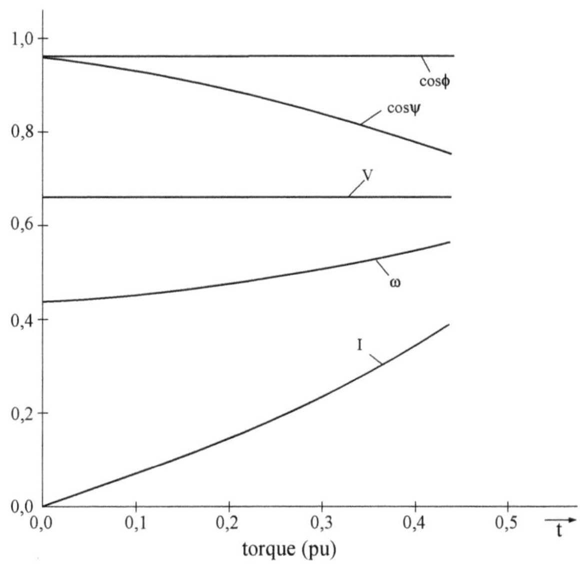

Fig. 16.23 Characteristics of CSI-fed synchronous motor for constant φ

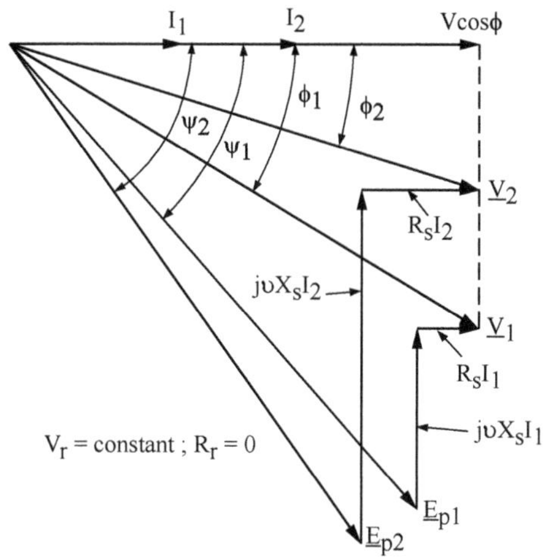

Fig. 16.24 Open-loop control of CSI-fed synchronous motor

There are also other advantages of this constant φ control. Indeed, the constant ψ control method inherently disregards the main field saturation: in the case of main field saturation, a stator current variation will affect the rotor flux and, as such, it is more appropriate to express the torque as a function of the air-gap flux and the current (i.e. it seems more logical to align the stator current with respect to the resulting or air-gap flux). This will also be further discussed in Chap. 17.

16.3.2.3 Constant-Speed Control

With constant-speed control, the synchronous machine operates as in a traditional grid operation. The inverter switching signals determine the frequency and the synchronous machine follows the variable frequency supply synchronously. The phasor diagrams derived above can still be used. For a constant frequency (and excitation), the rotor emf E_p has a constant amplitude. Load changes will then result in a variation of the internal angle ψ (see Fig. 16.24).

16.3.3 CSI-Fed Salient-Pole Synchronous Machines

A similar analysis for a synchronous machine with saliency in the rotor is much less straightforward. Indeed, there is no equivalent circuit for the *phasor* quantities of a synchronous machine when the reactances in d- and q-axes are distinct.[6]

[6]When the stator resistance is zero or negligible, two scalar equivalent schemes exist but still no equivalent circuit for the complex phasor quantities exists.

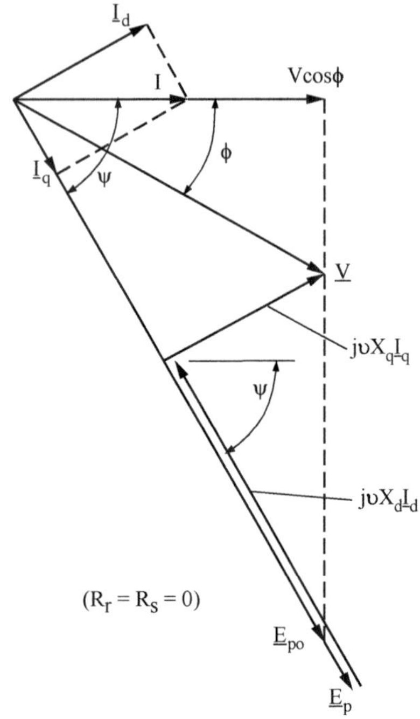

Fig. 16.25 Phasor diagram for CSI-fed salient pole synchronous motor

Therefore, we have to start from the phasor diagram, as in Fig. 16.25.

From the fundamental equivalent circuit of the CSI, we know that the machine current $\underline{I} = I$ is equal to the equivalent source current I_e and that the real part of the load (machine) voltage $V = Re(\underline{V}) = V\cos\varphi$ equals the equivalent input voltage V_i. As the rotor emf $\underline{E}_p$ differs from the machine terminal voltage by $R_s \cdot \underline{I} + j\nu X_q \cdot \underline{I}_q + j\nu X_d \cdot \underline{I}_d \approx j\nu X_q \cdot \underline{I}_q + j\nu X_d \cdot \underline{I}_d$, we obtain, for a constant internal angle ψ, the vector diagram in Fig. 16.25 (the stator and source resistances are disregarded here). We note that, when the machine is loaded, the amplitude of the rotor emf varies (and increases if $X_d > X_q$), even for a constant input voltage. For a constant excitation current, the speed will therefore increase with increasing load (see Fig. 16.26).

This observation indicates that the saliency may be modelled by an equivalent resistance, depending on the saliency. Indeed, from the phasor relations for the machine

$$\underline{V} = \underline{E}_p + R_s \cdot \underline{I} + j\nu X_q \cdot \underline{I}_q + j\nu X_d \cdot \underline{I}_d \approx \underline{E}_p + j\nu X_q \cdot \underline{I}_q + j\nu X_d \cdot \underline{I}_d \tag{16.41}$$

After projection on the real axis, we find

$$\begin{aligned} V\cos\varphi &= E_p\cos\psi + R_s \cdot I - \nu X_q \cdot I\sin\psi\cos\psi + \nu X_d \cdot I\sin\psi\cos\psi \\ &= E_p\cos\psi + R_s \cdot I + \frac{1}{2}\nu(X_d - X_q) \cdot I\sin 2\psi \end{aligned} \tag{16.42}$$

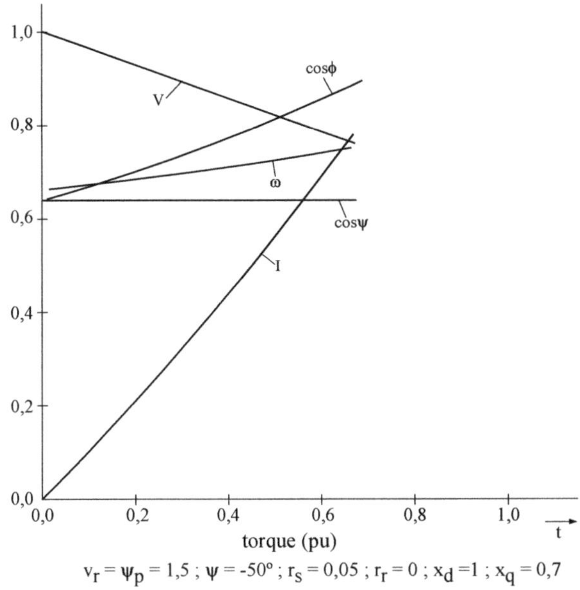

Fig. 16.26 Characteristics for CSI-fed salient-pole synchronous motor

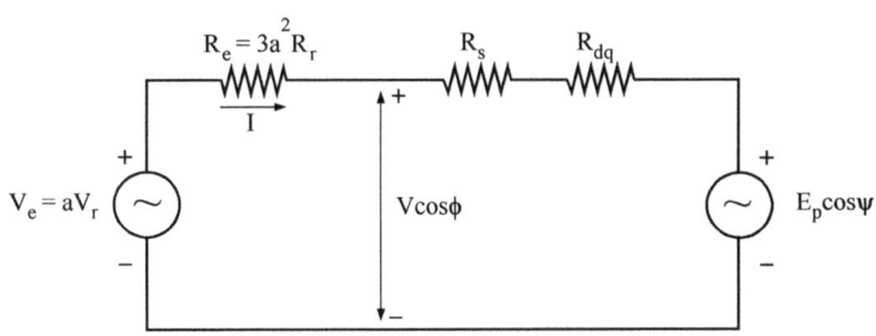

Fig. 16.27 Reduced equivalent circuit for salient-pole CSI-fed synchronous motor

or, for a negligible stator resistance,

$$V \cos \varphi \approx E_p \cos \psi + \frac{1}{2}\nu(X_d - X_q) \cdot I \sin 2\psi \tag{16.43}$$

This means that the saliency can be considered as an additional resistance

$$R_{dq} = \frac{1}{2}\nu(X_d - X_q) \cdot \sin 2\psi$$

Here, the internal angle is negative, so the resistance is also negative.

This leads to the *reduced* equivalent scheme in Fig. 16.27, which takes into account both the stator resistance and the source resistance. The power *dissipated* in the equivalent resistance R_{dq} in reality corresponds with the reluctance torque and power for a salient-pole synchronous machine:

$$T = \frac{3}{\Omega} \cdot \left[E_p I \cos\psi + \frac{1}{2}\nu(X_d - X_q) \cdot I^2 \sin 2\psi \right].$$

16.4 Effect of the Commutation Delay

In both the grid-side rectifier and the machine-side CSI, the commutation overlap results in deviations from the idealised case or model. As we know, the commutation overlap leads to a voltage drop and a shift of the fundamental of the AC current (which in reality has a trapezoidal shape instead of a rectangular one due to the commutation overlap).

For the grid-side rectifier, the only significant effect is a small DC voltage reduction. For a three-phase bridge rectifier, this voltage drop can be described by the drop over a fictitious resistance[7] R_{kr}

$$\Delta V_{kr} = R_{kr} \cdot I_r = \frac{3}{\pi}\omega_n L_{kr} \cdot I_r \tag{16.44}$$

with L_{kr} (mainly) the short-circuit inductance of the transformer at the grid side and ω_n the grid frequency.[8]

For the machine side CSI, both the voltage drop and the commutation delay resulting from the overlap may have important repercussions. Indeed, a CSI is in fact a rectifier operating at a control angle $\alpha > 90°$(in other words, with energy flow from the DC side to the AC side).

The voltage drop can again be described by a resistance determined by the short-circuit inductance of the synchronous machine. For this short-circuit inductance, we should use the subtransient inductance $L_k^{"}$ of the synchronous machine (per phase in wye or equivalent wye).[9]

$$\Delta V_{ki} = R_{ki} \cdot I_i = \frac{3}{\pi}\omega_i L_k^{"} \cdot I_i \tag{16.45}$$

or, for the equivalent voltage values in the CSI-machine equivalent circuit (for a stator winding in wye),

$$\Delta V_f = \frac{\pi}{6}\omega_i L_k^{"} \cdot I_f = \frac{\pi}{6}\nu X_k^{"} \cdot I_f = R_{ke} \cdot I_f \tag{16.46}$$

---

[7] See Chap. 7: the voltage drop is equal to $smfL_k I_d$ with $s = 2$ for a bridge.

[8] Prove that the corresponding resistance in the equivalent circuit is given by $R_{kre} = \frac{\pi}{2}\omega_n L_{kr}$ and $R_{kre} = \frac{\pi}{6}\omega_n L_{kr}$ for a load in delta or wye, respectively.

[9] This subtransient inductance is the very small equivalent inductance for a sudden short circuit of a synchronous machine (see Part 4).

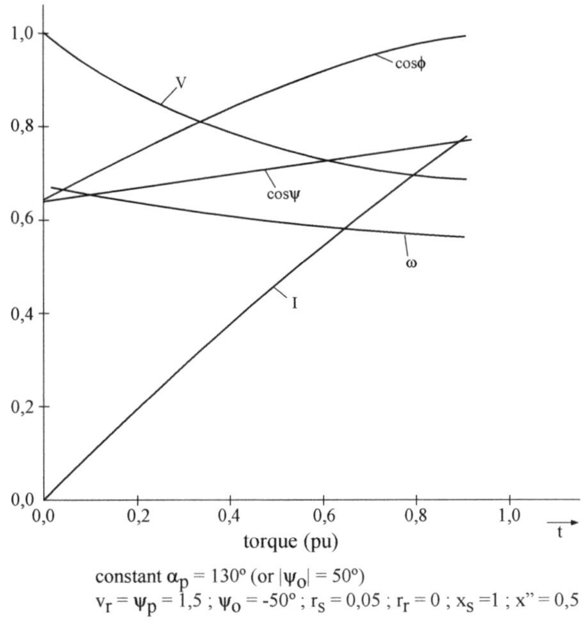

Fig. 16.28 Effect of commutation voltage drop on the speed

This causes the machine AC voltage to slightly decrease, as the machine side converter works with a control angle larger than 90°. This voltage drop may be taken into account by an additional series resistance[10] in the equivalent circuit.

The two additional resistances (from the rectifier and CSI overlap) result in a rather significant speed drop with increasing load, as is demonstrated by the characteristics in Fig. 16.28.

The commutation delay in the CSI may also affect the control considerably. Suppose a switching command α (with respect to the voltage for φ control) or α_p (with respect to the rotor emf E_p for ψ control). The commutation overlap μ causes an additional lagging of the current with respect to the voltage or the emf. For overlap angles that are not too large, this lagging angle may be approximated[11] by $\Delta\varphi = \mu/2$. In the diagram in Fig. 16.29, α and α_p represent the switching commands for constant φ (with respect to the voltage vector $\underline{V}$) and constant ψ control (with respect to the rotor emf vector $\underline{E}_p$), respectively. The current ($-\underline{I}$ when considered as a rectifier) will then lag by an additional angle $\Delta\varphi = \mu/2$.

The overlap angle μ (or the lagging angle $\Delta\varphi = \mu/2$) may be calculated as follows. The voltage drop, Eq. 16.45, can also be written as a function of the subtransient emf E_y'' of the synchronous machine (which is the emf behind the subtransient reactance in the subtransient model of the synchronous machine, see part 4):

[10] However, this resistance is dependent on the output frequency.

[11] Cf. the trapezoidal current wave form.

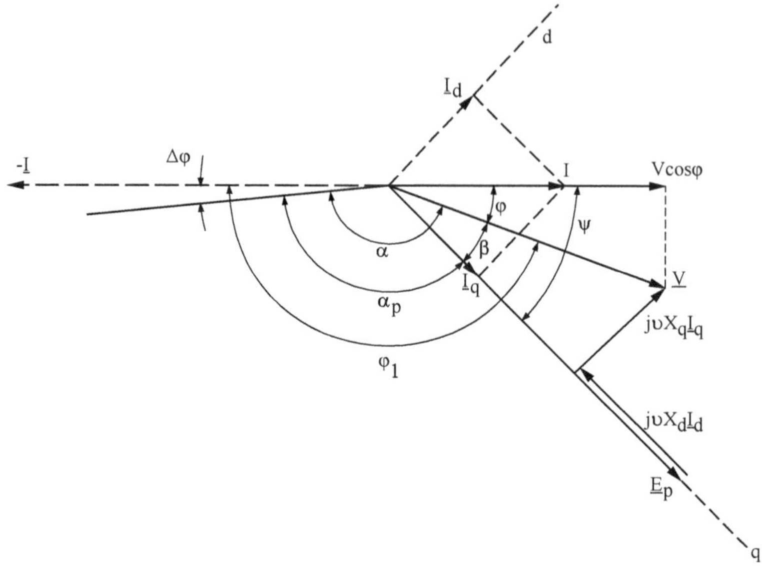

Fig. 16.29 Commutation delay

$$\Delta V_{ki} = \frac{3}{\pi}\omega_i L_k^{"} \cdot I_i = \frac{3}{\pi}\sqrt{6} \cdot E_y^{"} \cdot \frac{1}{2}[\cos\alpha - \cos(\alpha + \mu)] \tag{16.47}$$

($\frac{3}{\pi}\sqrt{6} \cdot E_y^{"}$ is the rectifier DC voltage V_{do} for a bridge with an AC voltage amplitude $\hat{V}_s = \hat{E}_y^{"} = \sqrt{2} \cdot E_y^{"}$ in wye or equivalent wye).

The subtransient emf can be calculated using, on the one hand, the equality of the (fundamental) active component of the synchronous machine voltage in steady state and, on the other hand, the AC output voltage of the rectifier expressed by means of the subtransient emf $E_y^{"}$ (per phase) with the commutation voltage drop:

$$E_p \cos\psi + I \cdot \nu(X_d - X_q)\cos\psi \sin\psi = -R_{ke} \cdot I - E_y^{"} \cos(\alpha + \mu) \tag{16.48}$$

To avoid commutation failure, the angle $\alpha + \mu$ must be below π. For low output frequencies $\omega_i = \nu\omega_n$ this is not possible because it causes a relatively large resistive voltage drop in the commutation circuit.

Chapter 17
Basics of Controlled Electrical Drives

Abstract As is clear from the previous chapters, DC commutator machines are ideal machines to obtain controlled torque and speed. To obtain similar properties, rotating field machines require more complicated supplies and control circuits. In this chapter we derive the basic control methods for rotating field machines, with as starting point the DC commutator machine characteristic properties.

17.1 Introduction: DC Machine Analogy

Both fixed frequency (mains) supply and variable frequency supply (V/f control) of rotating field machines are open-loop controls. The supply is to deliver a suitable voltage and frequency, and the load torque together with the machine torque and the inertia for transient states will determine the speed and other operating parameters. This is the analogue to the voltage supply of separately excited DC machines. For the separately excited DC machine, the DC voltage determines the speed for a given flux level:

$$V_a = R_a I_a + E = R_a I_a + K\Phi_m \Omega_r \approx K\Phi_m \Omega_r \tag{17.1}$$

In reality, the resistive voltage drop will result in a slight decrease of the speed with increasing load. As the machine torque is given by $T = K\Phi_m I_a$, the load torque will determine the required armature current and thus also the resistive voltage drop and speed drop. If required, the speed drop can of course be compensated by feedback control and by adapting the supply voltage.

A (somewhat) dualistic method to obtain a controlled DC drive is to supply the machine with a given current. Now, the machine torque is determined by this current together with the given flux:

$$T = K\Phi_m I_a \tag{17.2}$$

The characteristic of the load torque as a function of the speed determines the speed. Feedback control is used to adjust the current so as to obtain the desired speed (see

J. A. Melkebeek, *Electrical Machines and Drives*, Power Systems,
https://doi.org/10.1007/978-3-319-72730-1_17

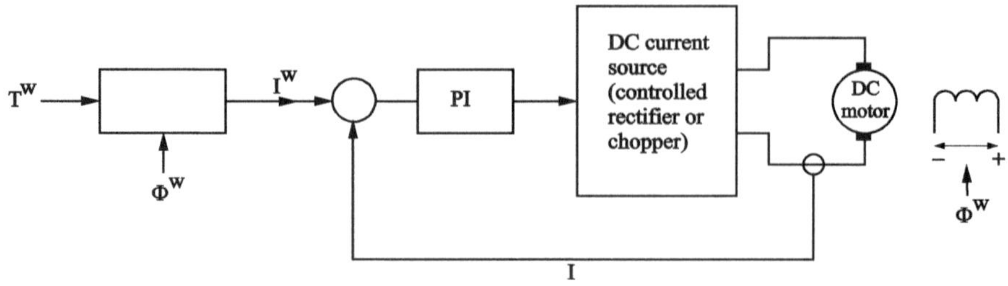

Fig. 17.1 DC machine drive with torque control

Fig. 17.1). With this kind of drives, it is possible to obtain fast dynamics and very high precision, in the order of 1:1000 (e.g. for a speed of 1500 rpm, a precision of 1 rpm).

Although DC machine drives are quite straightforward as far as control implementation is concerned, DC machines have many disadvantages, such as high maintenance costs for the commutator and brushes, a high sensitivity to vibrations (due to the commutation), a high acquisition cost, the limitation of the product of power and speed to 1000 MW · rpm (e.g. 750 kW at 1500 rpm).

Because of these disadvantages, the DC machine is nowadays mostly replaced by rotating field machines. However, the same control principles may be used, albeit with a somewhat higher complexity.

17.2 V/f Control of Rotating Field Machines

17.2.1 Introduction

As discussed in Chap. 15, both induction and synchronous machines are quite apt to operate with a variable frequency supply. V/f control of rotating field machines is in some way analogous to DC machine control with voltage supply, although for rotating field machines the main input is the frequency and not the amplitude of the voltage. As discussed above, in order to maintain the desired flux level, the voltage has to be adapted proportionally to the frequency (for frequencies lower than the rated one) or needs to be limited (i.e. the flux-weakening region for higher frequencies) because of insulation limitations and/or available voltage.

17.2.2 V/f Control of Induction Machines

V/f control of induction motors is very often used in lower cost drives and/or cases where the required speed accuracy is not that high, often in open-loop control. By means of feedback speed control, the speed accuracy can be considerably improved to compensate the slip.

An additional advantage is that the variable frequency can also be applied for smooth starting, eliminating the need for traditional or power electronic soft starting methods.

For drives with higher demands concerning accuracy and dynamics, vector control or similar control methods have to be used.

17.2.3 V/f Control of Synchronous Machines

In principle, V/f speed control is equally applicable to synchronous motors as induction motors. The only problem is that a synchronous machine has to remain synchronous all the time (there is no slip), meaning that the frequency should be varied sufficiently slowly, so as to not exceed the pull-out torque in accelerating or decelerating (if there is no cage in the rotor). In the past, V/f control was not frequently used for that reason. Recent scientific papers have shown, however, that appropriate control methods may make V/f control of synchronous motors quite feasible and practical.

Still, vector control or similar methods are much more common for synchronous machine drives.

17.3 Vector Control of Rotating Field Machines

17.3.1 Principle

The torque of an ideal DC commutator machine (with a brush axis electrically orthogonal to the flux axis and no armature reaction) is proportional to the product of flux and armature current (see Eq. 14.1). This is because the symmetry axis of the armature current layer is co-incident with the field flux axis, and the torque is at a maximum for a given armature current and flux (Fig. 17.2).

At the same time, if there is no armature reaction,[1] a variation of the armature current will not affect the flux and transients in the armature current will not influence the torque-producing flux. As a consequence, there will be no additional transients.

In rotating field machines, the torque results from the interaction of a rotating sinusoidal field distribution $b(x, t)$ and a rotating sinusoidal current layer $a(x, t)$, as was explained in Chap. 3 of Part 1. As the speeds of both field distribution and

[1] For a brush axis in the neutral position, this requires a compensation winding or the absence of saturation-induced armature reaction; however, a brush axis which is not in the neutral position may result in armature reaction for non-rated operating conditions.

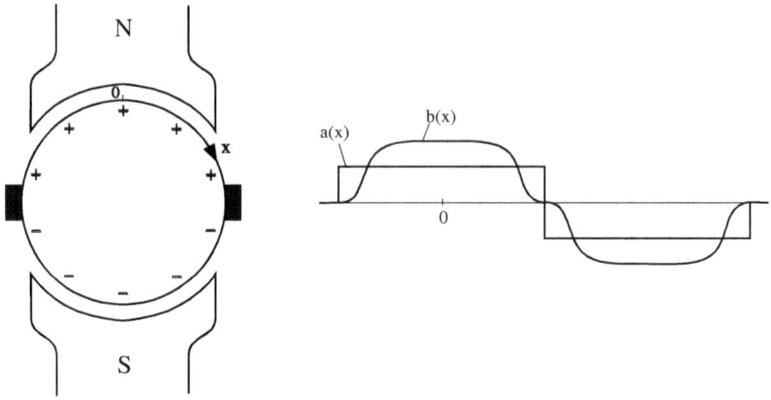

Fig. 17.2 Field and current layer orientation in a DC machine

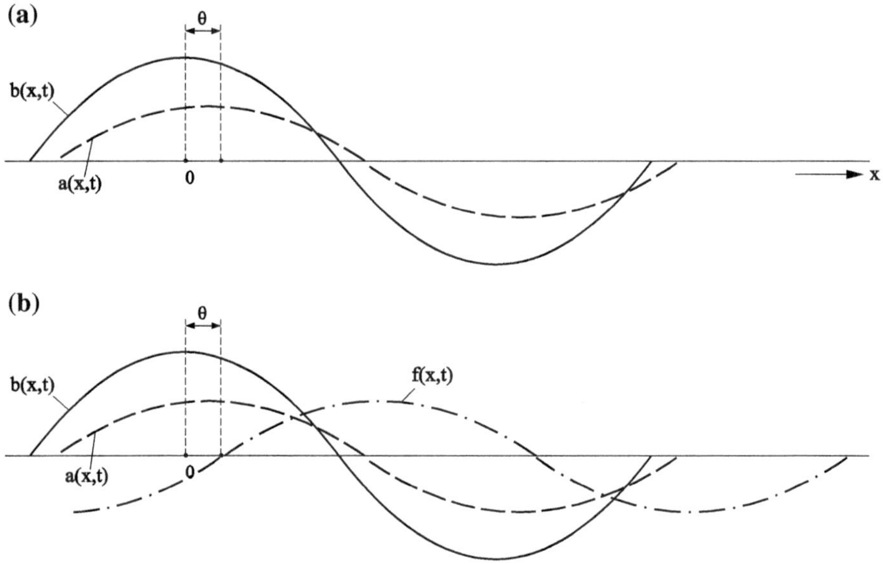

Fig. 17.3 Field and current distributions in a rotating field machine

current layer are the same, this torque is constant in time, but its magnitude depends on the angle ϑ between the symmetry axes of field and current layer (Fig. 17.3):

$$T \sim \hat{A} \cdot \hat{B} \cdot \cos \vartheta \tag{17.3}$$

For given current layer and field amplitudes, the torque is at a maximum when the symmetry axes of field and current layer are co-incident, i.e. $\vartheta = 0$. When we draw the mmf distribution $f(x, t)$ corresponding with the current layer $a(x, t)$, it

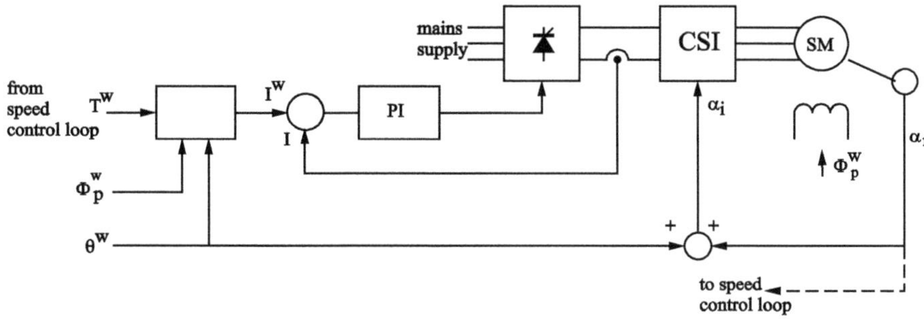

Fig. 17.4 CSI synchronous machine drive

becomes clear that the mmf distribution due to the torque-producing current layer will not affect[2] the field distribution if the symmetry axes of field and current layer are co-incident ($\vartheta = 0$).

We can therefore conclude that similar behaviour to that of the DC commutator machine can be obtained with a rotating field machine, on these conditions:

1. the field can be controlled independently
2. the torque-producing current can be controlled independently
3. the angle ϑ can be controlled continuously at $\vartheta = 0$

This operating condition, called field orientation, creates an ideal dynamic behaviour just like the DC commutator machine. In some cases, other limitations preclude an angle $\vartheta = 0$. In that case, the angle ϑ can be controlled at another fixed value. This is commonly called vector control, in which case transients of the torque producing current will obviously affect the flux and result in higher order transients.

17.3.2 Vector Control and Field Orientation of Synchronous Machines

Figure 17.4 illustrates a basic control scheme for a synchronous machine drive with a CSI. There are two control loops: the control of the current amplitude ($I = I^w$) and the control of the instantaneous angle of the current ($\alpha_i = \alpha_r + \vartheta^w$, with α_r the instantaneous rotor angle measured by an encoder and ϑ^w the desired displacement between field axis and current layer axis). In addition, except for permanent magnet excitation, there is a flux control circuit (not shown), with the desired rotor excitation *(rotor flux level Φ_p^w)* depending on the speed (rated rotor flux for speeds below the rated speed and field weakening for higher speeds).

---

[2]In the absence of saturation.

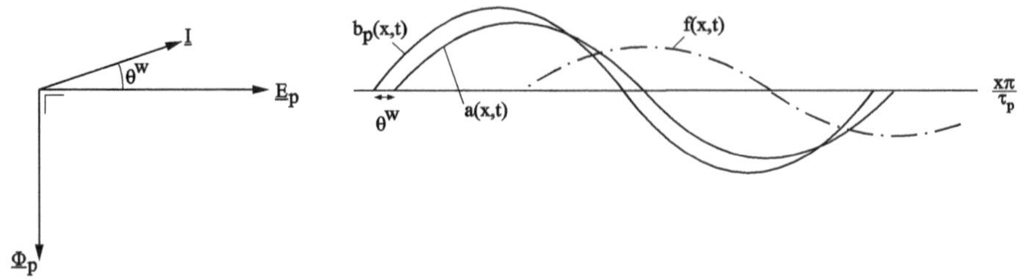

Fig. 17.5 $b_p(x, t)$ and $a(x, t)$ for leading current

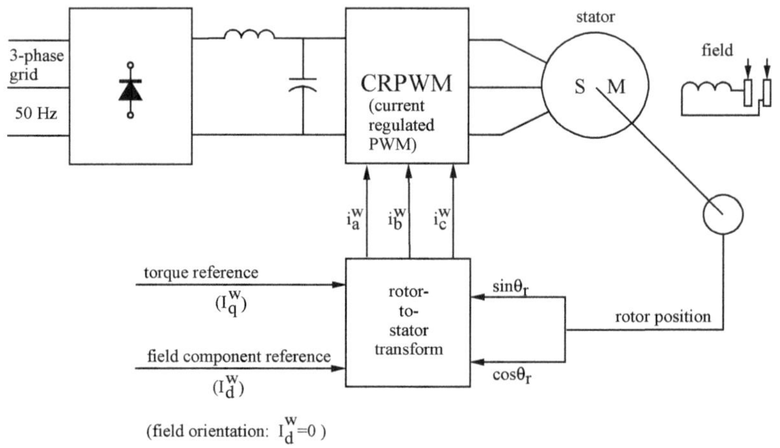

Fig. 17.6 VSI synchronous machine drive

The desired current magnitude depends on the desired torque: for example, the latter is determined by an external speed control circuit (not shown).

The desired displacement between field and current layer is zero for field orientation. However, for a CSI-supply with load commutation, the current should lead the voltage and therefore the current layer should lead the field (Fig. 17.5). In that case, there is vector control and no ideal field orientation.

It is obvious that for a supply by a CSI with forced commutation, field orientation (with $\vartheta = 0$) is the preferred control.

Moreover, for a VSI supply, there is no requirement for a leading current. The control principle illustrated in Fig. 17.6 uses a VSI with a current control loop,

Here, the separate control loops for magnitude and angle of the current are replaced by a (real-time) control with the desired current amplitude and current angle as input and the desired switching angles of the PWM-VSI as output (i.e. a CRPWM, cf. Fig. 17.7).

In fact, the control in Figs. 17.4 and 17.6, in which the current vector position is controlled with respect to the *rotor flux*, corresponds to the ψ-control explained in

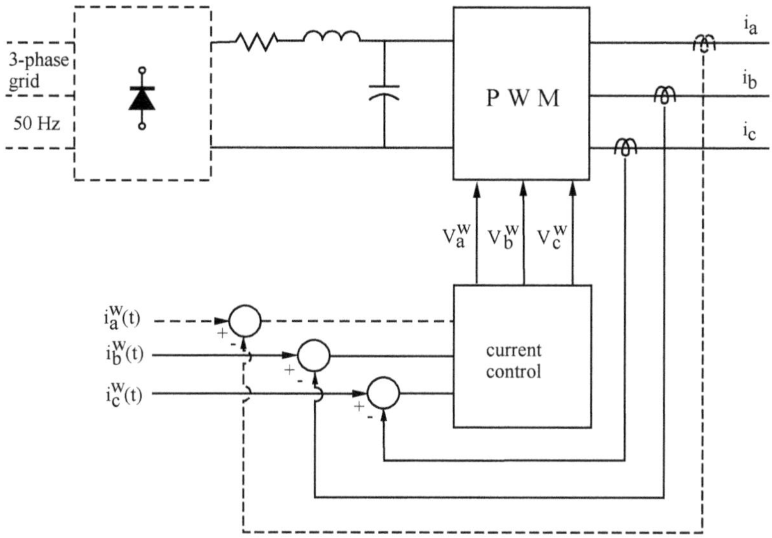

Fig. 17.7 Current-regulated PWM (CRPWM)

Chap. 16 - and the expression for the torque as a function of I_p and I (see Part1). This would be correct if there were no saturation, but in all practical machines saturation will cause a variation of the flux with variations of the armature current.[3] Another practical disadvantage is that this method requires a position sensor of the rotor. Such sensors reduce the reliability of the drive and therefore some vector control schemes control the position of the stator current vector with respect to the stator voltage (see also the φ-control of Chap. 16). This alternative solution requires an electronic voltage sensor for the voltage, which is normally less prone to faults. The required phase angle $\theta^w = \varphi$ is now added to the angle α_v of the voltage to obtain the required angle α_i of the current. In terms of torque control, this method is quite comparable because, for a negligible stator resistance (and iron losses), the torque can be written as $T \cdot \Omega_{sy} = P_{em} = 3 \cdot Re(E_r \cdot I^*) \approx 3 \cdot V \cdot I \cdot \cos\varphi$. As far as field orientation is concerned (and specifically the avoidance of higher order transients), it might also be argued that an orientation of the current with respect to the resultant emf E_r corresponds better with physical reality in the case of main field saturation. However, although the stator resistance might be negligible in most cases, the stator leakage usually is not.

[3] In fact, also note that fluxes cannot be added mathematically if there is any saturation.

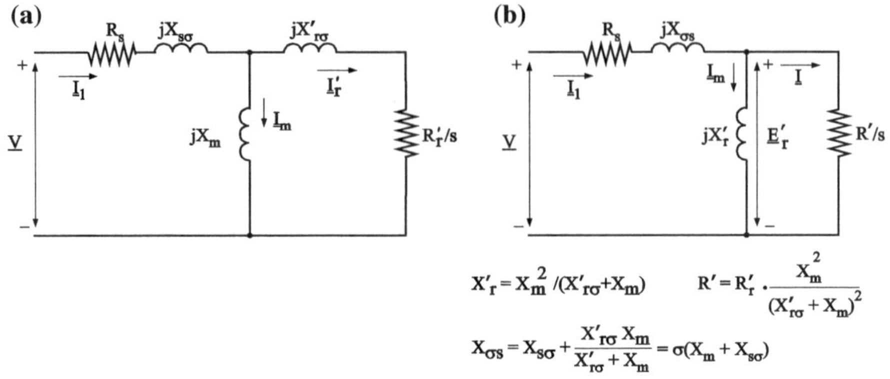

Fig. 17.8 Equivalent circuits for an induction machine

17.3.3 Vector Control and Field Orientation of Induction Machines

The principles of field orientation or vector control also apply to induction machines, i.e. field orientation requires a controllable flux, a controllable current magnitude and a controllable displacement angle between flux and current. Nevertheless, it is much more complicated to actually put this into practice, as the rotor position is not representative of the flux position. In addition, there is no such a thing as an excitation winding. Instead, there is only the stator current which determines the torque producing current and, together with the slip, also the air-gap flux.

To explain the basic principle of field orientation for the induction machine, consider the equivalent circuits in Fig. 17.8. From the general "T" circuit in (a), the "L" circuit in (b) is easily derived. In this circuit, $X_{\sigma s}$ is the total leakage referred to the stator.[4] The rotor resistance is now directly in parallel with the rotor (*magnetising*) reactance X_r.

Apparently, the stator current is split up into two orthogonal components $\underline{I}_m$ and $\underline{I}$, with $(R'/s)\cdot\underline{I} = jX_r'\cdot\underline{I}_m = \underline{E}_r'$. As the torque can be written as a function of both orthogonal components, $T = \frac{3}{\Omega_{sy}}I^2\cdot(R'/s) = \frac{3}{\Omega_{sy}}Re(\underline{E}_r'\cdot\underline{I}^*) = 3N_p\cdot L_r'\cdot I_m\cdot I$, the condition of orthogonality for field orientation seems to be met.[5]

The other conditions for field orientation stipulate that these two components of the stator current must be controlled independently. Note, however, that $(R'/s)\cdot\underline{I} = j\omega L_r'\cdot\underline{I}_m$, or the two components must always satisfy the slip equation (here in static form)

$$R'\cdot\underline{I} = (\omega s)\cdot jL_r'\cdot\underline{I}_m \quad (17.4)$$

[4] In the "L" circuit of Part 1, we referred the total leakage to the rotor.

[5] Is it now safe to conclude that there is always field orientation in an induction machine?

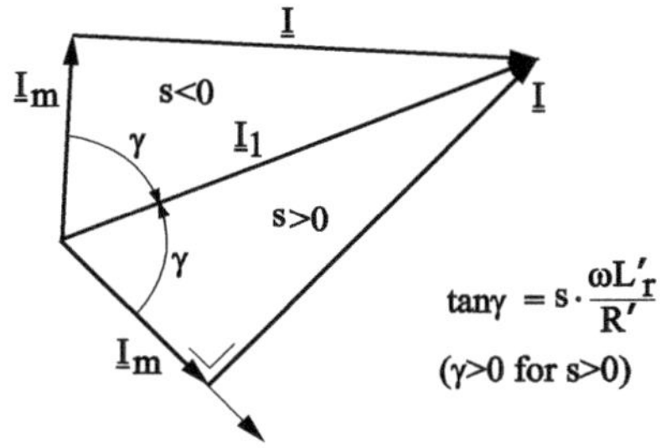

Fig. 17.9 Slip equation: sign of γ

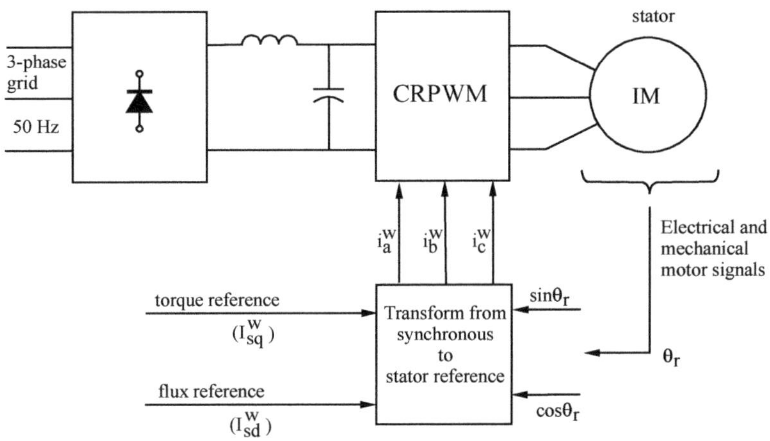

Fig. 17.10 Direct field orientation for induction machine

The two components are therefore narrowly related to the slip frequency, while their vectorial sum is equal to the stator current (see also Fig. 17.9).

If we succeed in controlling both components of the stator current, then both torque and flux are controlled. However, in contrast to a synchronous machine, the flux of an induction machine cannot be measured directly, as has already been mentioned (e.g. the flux rotates at slip frequency with respect to the rotor and thus the rotor position is not representative of the flux position).

To realise field orientation for an induction machine, there are two methods: direct field orientation and indirect field orientation.

In the **direct method,** position and magnitude of the flux are measured. From the measured value of the flux and its desired value (which may depend on the speed, cf. field weakening), the required (correction for the) magnetising current $(\Delta)\underline{I}_m$ is derived. The desired value of the torque then determines the desired value of the torque-producing component $\underline{I}$. The required stator current $\underline{I}_1 = \underline{I}_m + \underline{I}$ is now known. A schematic for a possible implementation with a VSI-PWM and CRPWM is shown in Fig. 17.10 (the d-axis corresponds to the flux axis).

For the flux measurement, there are two possibilities:

- direct measurement by means of Hall sensors or flux coils in the stator
- measurement of stator voltage and current with correction for the stator resistance and leakage voltage drop (and integration to obtain the flux)

The direct measurement with Hall sensors or flux coils is rather expensive because it requires modifications to the stator of the machine. Furthermore, it yields the air-gap flux and not the rotor flux, which means that another correction for the rotor leakage[6] is required.

The advantage of the measurement of voltage and current is that these sensors are mostly already available for the power-electronic converter. The disadvantage of this method is that the stator resistance (required for the calculation of the flux) is temperature-dependent. Finally, the leakage correction is not very straightforward, because it is saturation-dependent.

The **indirect method** to obtain field orientation for an induction machine relies on the slip equation. Indeed, it can be shown that the slip equation is not only a necessary condition for field orientation,[7] but also a sufficient one (at least in dynamical form, i.e. augmented by a term corresponding to a possible flux variation, cf. Chap. 29 in Part 4).

Most implementations use a combination of feedback and feedforward as follows:

- the required flux determines $|\underline{I}_m|$
- the required torque then yields $|\underline{I}|$
- from $|\underline{I}_m|$ and $|\underline{I}|$ and the slip equation, the slip frequency required for field orientation can be calculated
- the rotor speed is measured (or estimated for sensorless versions)
- measured rotor speed and required slip frequency lay down the required stator frequency for field orientation; this frequency and phase is used in feedforward for the inverter (both I_m and $\underline{I}$ are valid in a reference frame fixed to the rotor flux)

The most important disadvantage of the indirect method is that the slip equation relies upon the rotor resistance, which is dependent on temperature and skin effect.

In the discussion above, we concentrated on field orientation based on the rotor flux. There are also field orientation methods that are based on the air-gap flux or the stator flux. The advantage of these methods is that a correction to the rotor flux (using the saturation-dependent leakage) is not required. The disadvantage is that neither stator flux nor air-gap flux field orientation offer the same automatic orthogonality between the flux- and torque-producing components. In Part 4, this is explored in further detail.

---

[6]However, in most cases, the rotor leakage of a cage motor is heavily saturation-dependent.

[7]This is shown above for steady state.

17.4 Other Torque Control Methods for Rotating Field Machines

In contrast to vector control and field orientation, where current control is used to control the torque (and flux), Direct Torque Control (DTC) and related methods like Direct Self Control (DSC) directly steer the output voltage vector of a PWM-VSI in response to deviations of the flux and torque from their desired values.

The principle is as follows. As is well known, a PWM-VSI switches one of the six non-zero voltage vectors $\underline{u}_1 \ldots \underline{u}_6$ or one of the zero voltage vectors $\underline{u}_0, \underline{u}_7$ (see Fig. 17.11) to the output during a given interval, for example the PWM period of a traditional PWM. We recall that in space-vector PWM (SVM), each sector of $\pi/3$ is subdivided into a large number of subintervals. In each subinterval, the desired output vector is formed by switching between the two adjacent non-zero vectors and one of the zero vectors, each for an appropriate time.

DTC is somewhat similar to SVM, but in DTC the proper voltage vectors are switched dependent on the deviation of actual and desired flux vectors and/or torque.

The (stator) ***flux control*** is based on the terminal voltage equation:

$$\underline{V}_s = R_s \underline{I}_s + \frac{d\underline{\Psi}_s}{dt}$$

Or, if we disregard the stator voltage drop

$$\frac{d\underline{\Psi}_s}{dt} \approx \underline{V}_s$$

When the stator voltage is equal to one of the non-zero vectors $\underline{u}_1 \ldots \underline{u}_6$, the flux will thus evolve in the direction of this voltage vector (see Fig. 17.12). For the flux

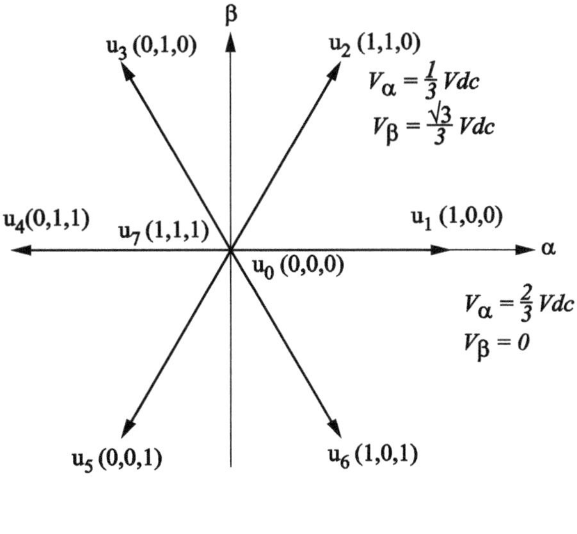

Fig. 17.11 The six non-zero vectors and the two zero vectors of a VSI

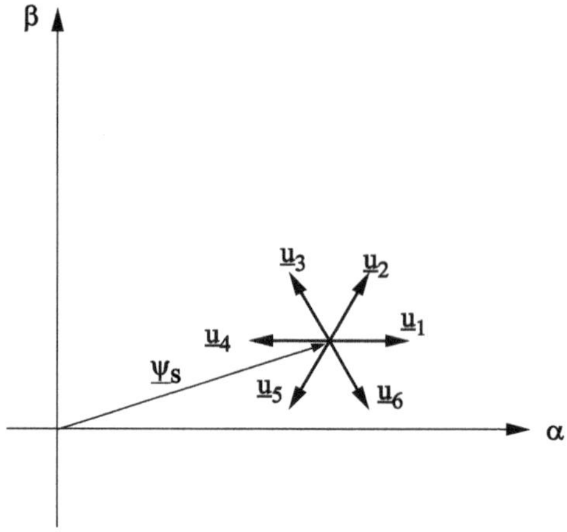

Fig. 17.12 Direct flux control in sector 1

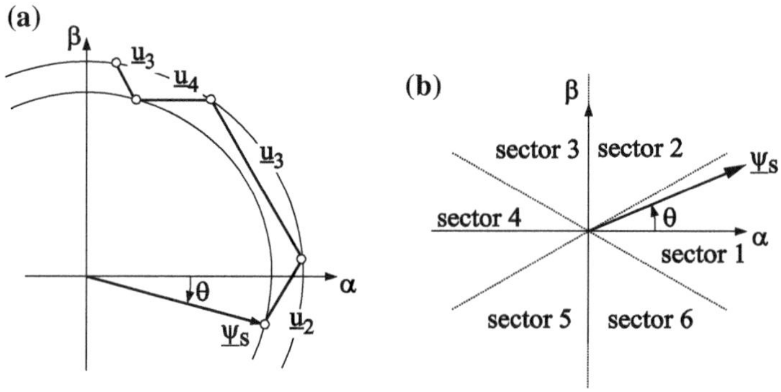

Fig. 17.13 Direct flux control: all sectors and hysteresis band

vector $\underline{\Psi}_s$ in the figure, a voltage vector $\underline{u}_3$ will cause a counter-clockwise rotation, without a noticeable variation in amplitude. A voltage vector $\underline{u}_2$ will cause both a counter-clockwise rotation and an increase of the amplitude. When a zero voltage vector is switched, the flux will remain more or less constant (in fact, it will slightly decrease because of the resistance).

In the usual implementation of DTC, the complex plane is divided into six sectors of $\pi/3$, each centred around one of the non-zero vectors (see Fig. 17.13). We will now turn to the case where the flux vector is in the first sector ($-\pi/6 \le \vartheta \le \pi/6$) and the required rotation direction is positive (i.e. counter-clockwise), as is the case in (b) in Fig. 17.13. Which vector is switched depends on the deviation of the angle and amplitude of the actual flux vector with respect to the desired one (if hysteresis controllers are used). If the actual flux vector is behind (lags) the desired one, either $\underline{u}_2$ or $\underline{u}_3$ can be switched to advance the flux. When the actual amplitude is lower than the desired one, $\underline{u}_2$ should be switched; when the actual amplitude is higher,

Table 17.1 Switching vectors for flux control

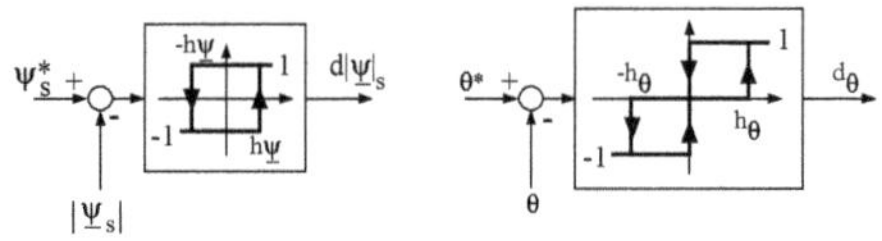

d_θ	$d\vert\Psi\vert_s$	Sector 1	Sector 2	Sector 3	Sector 4	Sector 5	Sector 6
1	1	$u_2(1,1,0)$	$u_3(0,1,0)$	$u_4(0,1,1)$	$u_5(0,0,1)$	$u_6(1,0,1)$	$u_1(1,0,0)$
	-1	$u_3(0,1,0)$	$u_4(0,1,1)$	$u_5(0,0,1)$	$u_6(1,0,1)$	$u_1(1,0,0)$	$u_2(1,1,0)$
0	1	$u_7(1,1,1)$	$u_0(0,0,0)$	$u_7(1,1,1)$	$u_0(0,0,0)$	$u_7(1,1,1)$	$u_0(0,0,0)$
	-1	$u_0(0,0,0)$	$u_7(1,1,1)$	$u_0(0,0,0)$	$u_7(1,1,1)$	$u_0(0,0,0)$	$u_7(1,1,1)$
-1	1	$u_6(1,0,1)$	$u_1(1,0,0)$	$u_2(1,1,0)$	$u_3(0,1,0)$	$u_4(0,1,1)$	$u_5(0,0,1)$
	-1	$u_5(0,0,1)$	$u_6(1,0,1)$	$u_1(1,0,0)$	$u_2(1,1,0)$	$u_3(0,1,0)$	$u_4(0,1,1)$

$\underline{u}_3$ is switched. At some point, the actual flux vector may start to lead the desired one. Then, normally one of the zero vectors ($\underline{u}_0$ or $\underline{u}_7$) will be switched, causing the actual flux vector to stand still (and slightly decrease). In principle, either one of the zero voltage vectors can be chosen, but the one resulting in the lowest number of switchings will be preferred. To avoid too high a switching frequency, a wide hysteresis band can be utilised (see (a) in Fig. 17.13).

Note that for an actual flux vector leading the desired one, $\underline{u}_5$ or $\underline{u}_6$ could have been used as well. However, this would increase the switching frequency too much and is thus best avoided.

Repeating this for the other sectors and also adding the sequence for the negative rotation direction results in the switching Table 17.1. As mentioned above, in steady state, only the first and second rows are used for the positive (counter-clockwise) rotation direction, while only the third and second rows are used for the negative (clockwise) rotation. The hysteresis boundaries for amplitude and angle can be modified to obtain the required accuracy or limit the switching frequency.

A similar principle can be used for ***direct torque control***. As we will see, this also controls the flux level at the same time.

To explain the principle, we express the dynamic equations for the induction machine in an instantaneous reference system fixed to the rotor flux[8]:

$$0 = R_r I_{rq} + \frac{d\Psi_{rq}}{dt} - (\omega - \omega_r)\Psi_{rd} \tag{17.5}$$

$$0 = R_r I_{rd} + \frac{d\Psi_{rd}}{dt} + (\omega - \omega_r)\Psi_{rq} \tag{17.6}$$

with

$$\Psi_{rq} = L_r I_{rq} + L_m I_{sq} \tag{17.7}$$

[8]In the following equations in this section, all currents and fluxes are amplitude values (without the hat notation); see also Part 4.

$$\Psi_{rd} = L_r I_{rd} + L_m I_{sd} \tag{17.8}$$

For example, suppose the rotor flux is fixed to the d-axis ($\underline{\Psi}_r = \Psi_{rq} + j\Psi_{rd} \equiv j\Psi_{rd}$). For the rotor flux, on the one hand, we may then derive:

$$\frac{d\Psi_{rd}}{dt} = -\frac{R_r}{L_r}\Psi_{rd} + R_r\frac{L_m}{L_r}I_{sd} \tag{17.9}$$

$$\frac{d\Psi_{rq}}{dt} = 0; \quad \Psi_{rq} = L_r I_{rq} + L_m I_{sq} \equiv 0 \tag{17.10}$$

On the other hand, the stator flux can be written as

$$\Psi_{sd} = \sigma L_s I_{sd} + \frac{L_m}{L_r}\Psi_{rd} \tag{17.11}$$

$$\Psi_{sq} = \sigma L_s I_{sq} \tag{17.12}$$

while the torque can be written as

$$T = \frac{3}{2}N_p \cdot Im(\underline{\Psi}_s^* \cdot \underline{I}_s) = \frac{3}{2}N_p \cdot Im(\underline{\Psi}_r \cdot \underline{I}_r^*) \tag{17.13}$$

or,

$$T = -\frac{3}{2}N_p \cdot \frac{L_m}{L_r} \cdot \Psi_{rd} \cdot I_{sq} = -\frac{3}{2}N_p \cdot \frac{L_m}{\sigma L_s L_r} \cdot \Psi_{rd} \cdot \Psi_{sq} = \frac{3}{2}N_p \cdot \frac{1-\sigma}{\sigma L_m} \cdot |\underline{\Psi}_r| \cdot |\underline{\Psi}_s| \cdot \sin\gamma \tag{17.14}$$

γ is the angle between the fluxes $\underline{\Psi}_s$ and $\underline{\Psi}_r$, see Fig. 17.14. This torque equation shows that the torque can be controlled by controlling the stator and rotor flux *amplitudes* and the *angle between both vectors*. From Eqs. 17.9 and 17.10, we observe that the rotor flux has a very large time constant L_r/R_r (of the order of seconds for medium power machines). The stator flux, in contrast, does not only comprise the rotor flux (corrected with the winding factor-like ratio L_m/L_r), but also the leakage flux (which may be varied with a small time constant). The stator flux, and thus its amplitude as well as the angle γ, can be changed very quickly by the supply.

The direct torque control is based on these principles: the *amplitude* of the flux is controlled in the same way as for the direct flux control described above, while the torque is controlled by the lead angle γ of the stator flux with respect to the rotor flux. For the positive rotation direction, in sector 1 the voltage vectors $\underline{u}_2$ and $\underline{u}_3$ are used to increase the torque. When the torque is higher than the desired value, one of the zero vectors is switched (switching the vectors $\underline{u}_5$ or $\underline{u}_6$ to reduce the angle would increase the switching frequency too much). The switching sequence for all sectors and both rotating directions is shown in Table 17.2.

The hysteresis band can be decreased to improve the desired accuracy or increased to limit the switching frequency.

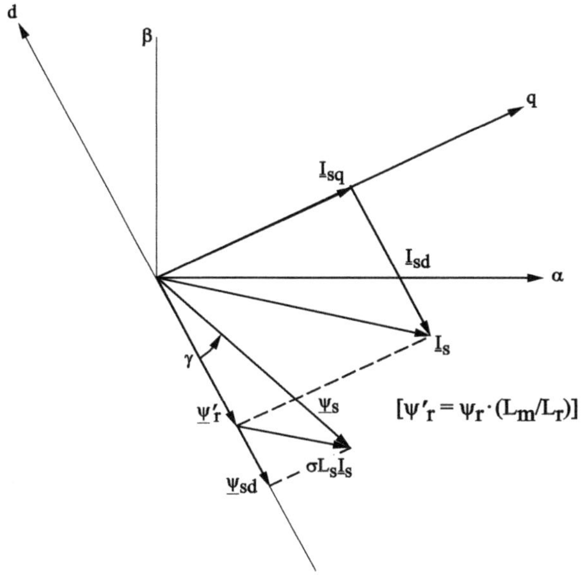

Fig. 17.14 Flux vectors for torque control

Table 17.2 Switching table for DTC

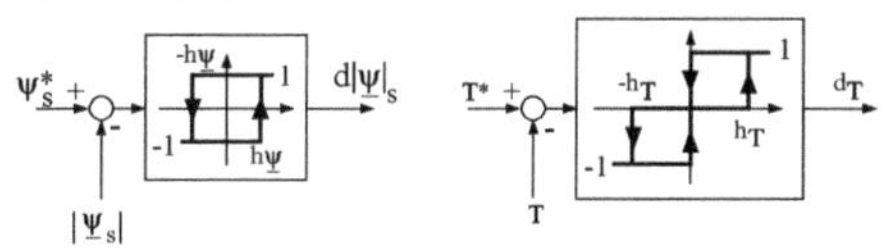

d_T	$d\|\psi\|_s$	Sector 1	Sector 2	Sector 3	Sector 4	Sector 5	Sector 6
1	1	$u_2(1,1,0)$	$u_3(0,1,0)$	$u_4(0,1,1)$	$u_5(0,0,1)$	$u_6(1,0,1)$	$u_1(1,0,0)$
	-1	$u_3(0,1,0)$	$u_4(0,1,1)$	$u_5(0,0,1)$	$u_6(1,0,1)$	$u_1(1,0,0)$	$u_2(1,1,0)$
0	1	$u_7(1,1,1)$	$u_0(0,0,0)$	$u_7(1,1,1)$	$u_0(0,0,0)$	$u_7(1,1,1)$	$u_0(0,0,0)$
	-1	$u_0(0,0,0)$	$u_7(1,1,1)$	$u_0(0,0,0)$	$u_7(1,1,1)$	$u_0(0,0,0)$	$u_7(1,1,1)$
-1	1	$u_6(1,0,1)$	$u_1(1,0,0)$	$u_2(1,1,0)$	$u_3(0,1,0)$	$u_4(0,1,1)$	$u_5(0,0,1)$
	-1	$u_5(0,0,1)$	$u_6(1,0,1)$	$u_1(1,0,0)$	$u_2(1,1,0)$	$u_3(0,1,0)$	$u_4(0,1,1)$

Figure 17.15 shows an actual recording of the torque as a function of time for a sudden start-up from standstill to the no-load speed at 40 Hz. The direct torque control was implemented digitally using a DSP; due to the peculiarities of this digital control, there are some instances where the torque briefly leaves the hysteresis band.

Although our derivation was in a synchronous reference frame, DTC does not require any axis transformation. Instead, the whole control is executed in the standstill stator coordinates.

The stator flux (estimation) follows from the stator voltage by integrating

$$\underline{\Psi}_s = \int (\underline{V}_s - R\underline{I}_s)dt \approx \int \underline{V}_s dt \tag{17.15}$$

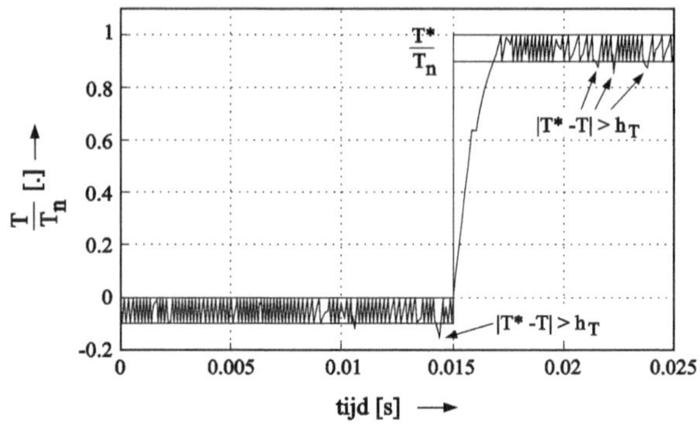

Fig. 17.15 Start-up with DTC from standstill to no-load speed

If the effect of the stator resistance cannot be disregarded (in particular at low speed), this requires the measurement of both voltage and current. A pure integration may also lead to stability problems such as drift. Therefore, the integration is sometimes replaced by a low-pass filter, although this limits the lowest usable frequency for DTC.

Once the stator flux is known, the actual torque can be calculated from the stator voltage and current in stator coordinates ($\alpha - \beta$ components):

$$T = \frac{3}{2} N_p \cdot [\Psi_{s\alpha} I_{s\beta} - \Psi_{s\beta} I_{s\alpha}] \tag{17.16}$$

When we compare DTC with field orientation, it should be clear that in an ideal case (i.e. without saturation), field orientation will avoid most transients, except those resulting from a flux change. Obviously, this is not the case with DTC. Whereas field orientation controls the torque producing current (and also the flux), DTC directly steers the voltage supplied by the inverter to control the torque (and the flux). Recent scientific papers have shown (experimentally) that both methods usually produce almost equivalent results in terms of acceleration.

The main advantages of DTC are the fast dynamics, the simple control, and the fact that neither reference axis transformations nor a PWM modulator are required. Disadvantages include the potentially high switching frequencies, the real-time calculations that require fast processors, the high torque ripple, some occasional problems at low speeds, and the variable switching frequency.

Chapter 18
Small Electric Machines and Their Power Electronic Control

Abstract Electric machines cover an extremely wide range of power ratings, from 1 mW (10^{-3} W) or less, to 1 GW (10^9 W), which is a ratio of $1 : 10^{12}$. The power range of 1 kW may be considered as the boundary between small and large machines.

In addition to size and power rating, small and large machines differ in many other respects:

- manufacturing: mass production to series production and singular production
- MTBF: 1000 h (or less) to (much) more than 10,000 h (e.g. 100,000 h for power plant generators)
- construction: integrated units (into the application) to autonomous units
- disturbance effects compared to the main effect: large versus very small
- design variations: numerous versus minor.

The description of small machines is especially complicated due to the large disturbance effects, such as relatively large resistances, large losses and vibration torques, and the many construction variants. Further, the steady-state operation common for large machines is now replaced by servomotor operation with high dynamics (often leading to a special construction).

18.1 Small DC Commutator Machines

18.1.1 Introduction

Small DC commutator machines (i.e. those < 1 kW) differ from their big brothers by a non-negligible armature resistance,[1] a small number of poles (usually only two), a small number of armature coils and commutator segments, a relatively large air gap, a lack of compensation windings, often also a lack of commutation windings, and series excitation or (in many cases) permanent-magnet excitation instead of field windings. Apart from these machine-side differences, small DC machines are

---

[1]This is the result of scaling laws.

J. A. Melkebeek, *Electrical Machines and Drives*, Power Systems,
https://doi.org/10.1007/978-3-319-72730-1_18

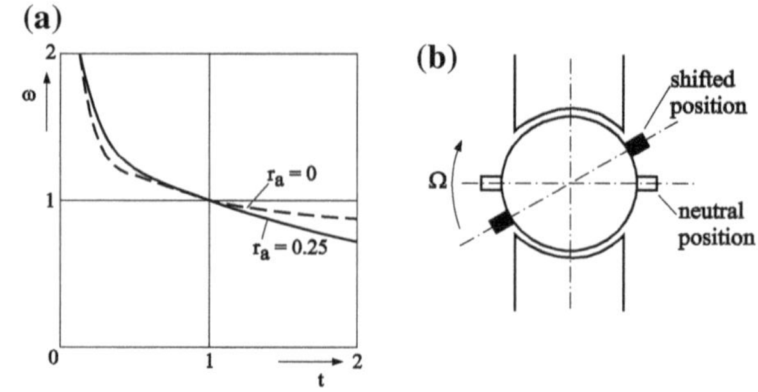

Fig. 18.1 Small series-excited DC machine: torque characteristic and brush axis shift

often supplied by low-power choppers or rectifiers, also bringing along some specific peculiarities.

18.1.2 Series-Excited DC Machine

Due to the relatively larger resistance of the armature winding and the series excitation winding, the torque-speed characteristic differs somewhat from those of larger machines (see (a) in Fig. 18.1).

Because auxiliary poles are absent, the brush axis is sometimes rotated to improve the commutation (see (b) in Fig. 18.1). As the brush axis should be rotated contrary to the rotation direction (for motoring) in order to aid the commutation, this limits its use to a single rotation direction.

The inductance of the series excitation increases the total inductance, which is advantageous for limiting the starting current. The large inductance is also an advantage for power electronic supply of the machine (rectifier or chopper), as this reduces the current ripple. However, for power electronic supply the stator also needs to be laminated (in which case the machine resembles an AC commutator machine).

18.1.3 Permanent-Magnet Excited DC Machine

There are many variants as to the construction of permanent-magnet DC machines. All these variants have as an advantage that the reluctance of the main magnetic circuit is enlarged due to the permanent magnet. This reduces the armature reaction (the effect of the armature mmf on the main flux due to saturation).

The simplest construction type is one where the magnets are placed directly in the air gap, without any pole shoes, as is demonstrated in (a) in Fig. 18.2. There are, however, some disadvantages of this construction:

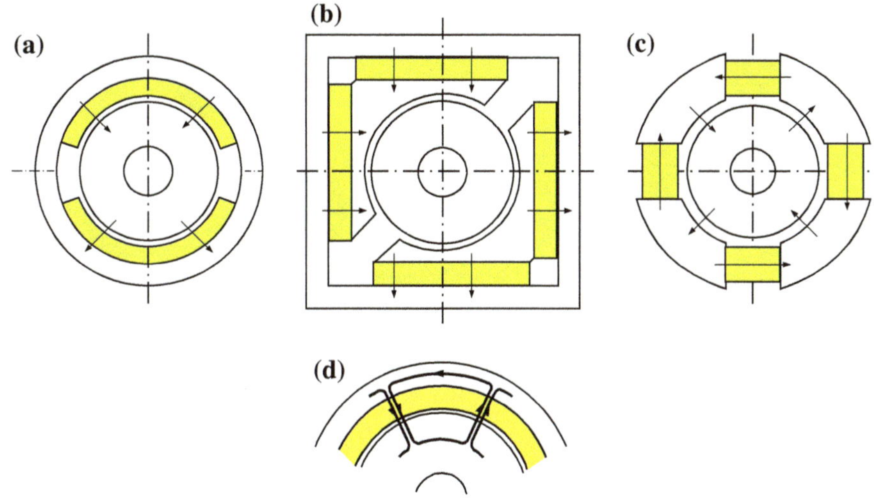

Fig. 18.2 Small permanent magnet DC machines

- the required curvature of the magnets
- the lack of a flux-concentration effect
- although the armature reaction is low, the armature mmf may cause demagnetisation of the permanent magnets (e.g. if the machine stalls or overloads), as is shown by the flux lines in (d) in Fig. 18.2.

A more complicated construction type uses iron pole shoes in the air gap (see (b) and (c) in Fig. 18.2). Advantages of this construction variant are the following:

- flat magnets can be used
- the magnetic field in the air gap can be larger than in the magnets due to flux-concentration
- the magnets are shielded from the armature mmf.

On the other hand, the good magnetic conductivity of the pole shoes will increase somewhat the armature reaction.

Sometimes the permanent magnets are placed in between the ferromagnetic pole shoes, as in (c) in Fig. 18.2.

The small armature reaction results in an almost ideal linear torque-speed characteristic for voltage supply,

$$\Omega = \frac{V_a}{k \cdot \Phi} - \frac{R_a}{(k \cdot \Phi)^2} \cdot T$$

Compensation windings can also be placed in the pole shoes (if present) to further reduce the armature reaction.

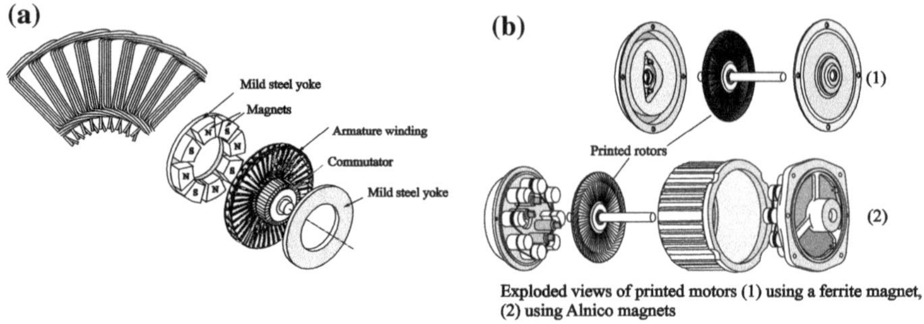

Fig. 18.3 Small axial field DC machines

Such small DC machines are very common in cars, for example as window, roof or mirror motors or as windscreen wiper motors.

When fast dynamics are required, which implies a low inertia, either a machine with a large ratio of armature length to diameter can be used or the (rotating) armature winding can be detached from the rotor yoke (remaining stationary). The latter design frequently utilises an axial field with radial windings (see Fig. 18.3). The armature winding can be a printed winding as well. The disadvantage of such a design is that the overload capacity of the winding is very small and there is almost no thermal inertia.

18.1.4 Power Electronic Supply of (Small) DC Machines

The small series-excited DC commutator machine is often supplied by a controlled rectifier (see (A) in Fig. 18.4). The large machine inductance is an advantage as it reduces the current ripple for the relatively low switching frequency. This configuration can also be used for (resistive) braking, i.e. with a negative machine voltage. The large machine inductance may result in a continuous current even in case of zero average output voltage of the rectifier ($\alpha = \pi/2$), as is illustrated in (A, c) in Fig. 18.4. The emf of the machine is assumed to be negative here (i.e. resistive braking in the reverse rotation direction).

Small permanent-magnet excited DC machines are preferably supplied by a chopper (see (B) in Fig. 18.4). To limit the current ripple, a sufficiently high switching frequency can be used.

Remark: The universal motor, a series-excited AC commutator machine (see Chap. 19) is often supplied by an AC chopper (e.g. in vacuum cleaners or cheaper washing machines) (see (a) in Fig. 18.5). As for its cousin, the series-excited DC commutator machine, the large series inductance is beneficial to limit the higher current harmonics. From the equivalent circuit, (b) in Fig. 18.5, we can note that

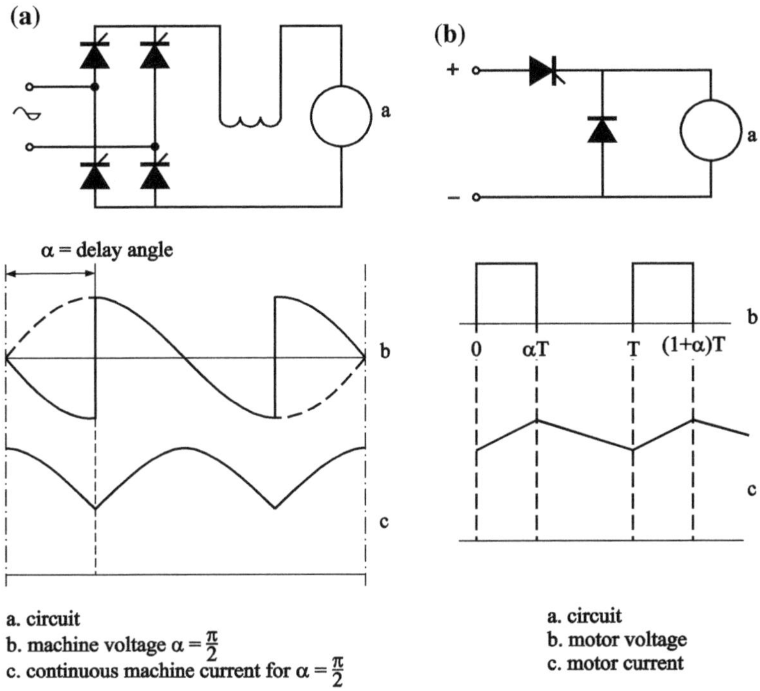

Fig. 18.4 DC machine: PE supply

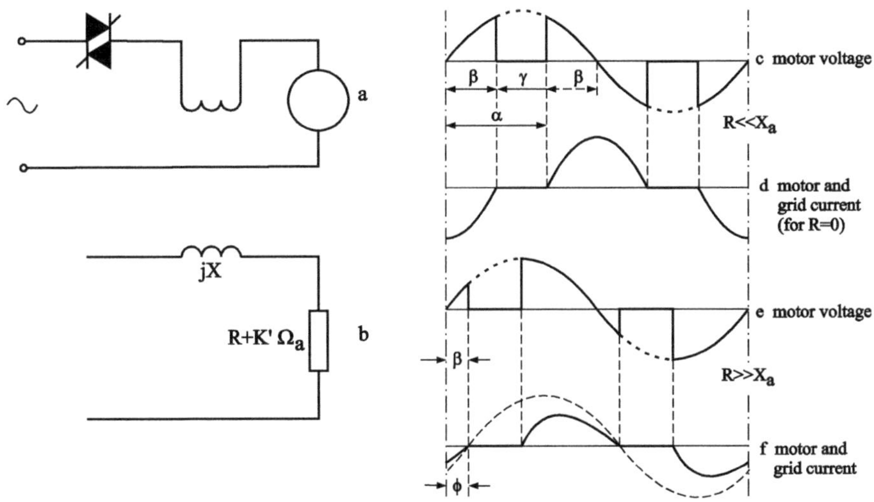

Fig. 18.5 Universal motor: chopper supply

- for low speeds the machine impedance is almost purely reactive, leading to voltage and current curves as in (c) and (d) in Fig. 18.5
- for high speeds the machine impedance comprises an important resistive component and the voltage and current curves are as in (e) and (f) in the figure (ϕ is the displacement angle without chopping; the power factor angle increases in case of chopping).

18.2 Small Induction Machines

Small induction machines can be three-phase but are in many cases single-phase or two-phase. Common peculiarities of small induction machines are

- a larger effect of stator and rotor resistances: because of the scaling laws, the pu resistances are much larger
- a large pu magnetising current, mainly because of the relatively larger air gap
- larger effects of mmf harmonics on current, torque and losses.

18.2.1 Three- and Two-Phase Induction Machines

Small three-phase machines usually also have a distributed winding, but because they have few slots, the mmf harmonics are more important than for greater power ratings. These mmf harmonics result in rather significant harmonic torque components. The low magnetising inductance and high resistance values lead to rather low efficiencies (e.g. less than 40% for power ratings below 0.5 kW).

Two-phase induction motors suffer from similar or even more problems regarding mmf harmonics as the stator winding is often a concentrated winding. As two-phase supplies are not very common, the second phase is commonly an auxiliary winding supplied by means of a capacitor (for the capacitor motor, see Chap. 4 in Part 1). In Sect. 18.2.3, a cheap two-phase servomotor (with power-electronic supply) is described.

18.2.2 Single-Phase Induction Motors

As was discussed in Chap. 4 of Part 1, pure single-phase induction motors have no starting torque and are therefore mostly useless. With an auxiliary winding and a starting or continuous capacitor, they actually become two-phase machines, albeit supplied by a single-phase supply. Adding phase control makes these motors even more versatile, as is also explained in Sect. 18.2.3. Another variant of the single-phase fed induction motor is the shaded pole motor, which was also described in Chap. 4.

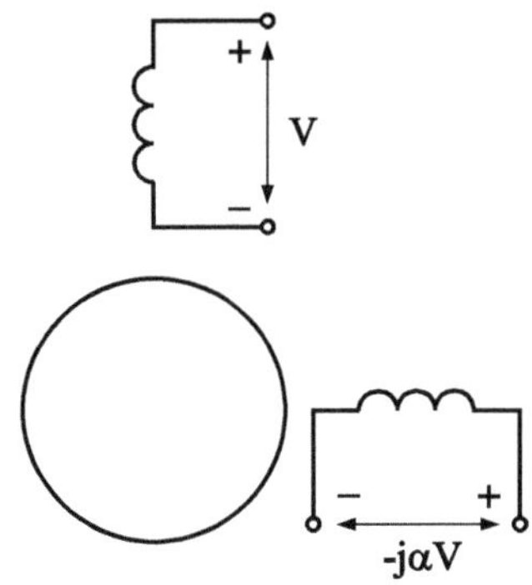

Fig. 18.6 Two-phase servo induction machine

18.2.3 Power-Electronic Supply of Small Induction Motors

Similar to their big brothers, small induction motors may be supplied by inverters, which transform them into rather versatile variable speed drives. Still, less expensive solutions are often called for.

One of these solutions is a two-phase motor in which the auxiliary phase is supplied by the grid in series with an AC chopper. The two-phase supply consists of one phase supplied directly by the grid (voltage V), while the secondary phase is supplied by the grid in series with an AC chopper, resulting in a voltage that is $\pi/2$ leading and with a variable amplitude (voltage $-j\alpha V$, see Fig. 18.6).

This supply can be split up into direct and inverse components as follows[2]:

$$V_d = \frac{1}{2}[V + j(-j\alpha V)] = \frac{1}{2}V(1+\alpha) \tag{18.1}$$

$$V_i = \frac{1}{2}[V - j(-j\alpha V)] = \frac{1}{2}V(1-\alpha) \tag{18.2}$$

For $\alpha = +1$ and $\alpha = -1$, a direct and inverse voltage is obtained, respectively. Supposing that the torque-slip characteristic can be simplified to a straight line (as in the small slip range $|s| < |s_{po}|$), we may find the total torque as the superposition of a direct and an inverse torque component as follows:

$$T_d = \left[\frac{V_d}{V_n}\right]^2 \cdot 2\frac{T_{po}}{s_{po}} \cdot \left[1 - \frac{\Omega}{\Omega_{sy}}\right] \tag{18.3}$$

$$T_i = -\left[\frac{V_i}{V_n}\right]^2 \cdot 2\frac{T_{po}}{s_{po}} \cdot \left[1 + \frac{\Omega}{\Omega_{sy}}\right] \tag{18.4}$$

---

[2]The two-phase symmetrical components transformation is similar to the three-phase one, but instead of $a = \exp(j2\pi/3)$ we now have $a = \exp(j\pi/2) = j$).

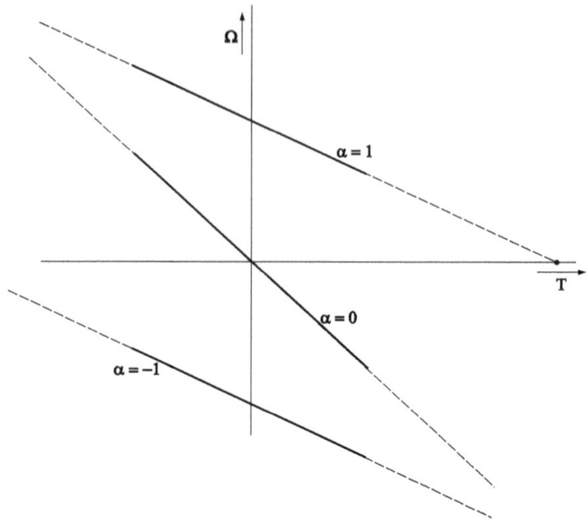

Fig. 18.7 Torque for the 2-phase servo

$$T = T_d + T_i = 2\frac{T_{po}}{s_{po}} \cdot \left[\frac{V}{V_n}\right]^2 \cdot \left[\alpha - \frac{1}{2}(1+\alpha^2)\frac{\Omega}{\Omega_{sy}}\right] \tag{18.5}$$

or

$$\frac{\Omega}{\Omega_{sy}} = \frac{2}{(1+\alpha^2)} \cdot \left[\alpha - \frac{T}{2T_{po}/s_{po}}\right] \tag{18.6}$$

According to Eq. 18.6, every speed between $+\Omega_{sy}$ and $-\Omega_{sy}$ should be attainable by controlling α between $+1$ and -1 (see also Fig. 18.7). This is of course only valid if our approximation, the linearisation of the torque characteristic for small slip, were valid in the whole speed range. However, this is not the case unless the pull-out slip is very large (i.e. larger than 1); if it is smaller than 1, the speed range in which these equations apply is limited to $|\frac{\Omega}{\Omega_{sy}}| > 1 - |s_{po}|$.

The maximum available torque is also rather limited at low speeds. To obtain fast dynamics, the inertia must therefore be limited. A solution to this predicament is the Ferraris motor, in which the rotor is a conducting cup (aluminium) and the secondary iron yoke is at standstill (see Fig. 18.8).

Another application of power-electronics for small induction motors is the AC chopper supply of single-phase fed motors.

A first example is the pure single-phase motor as in (a) in Fig. 18.9. If the effects of the inverse field are disregarded, similar equations as for the three-phase chopper supply (see Chap. 9 in Part 2) can be used. This is mainly useful to adapt the supply voltage to the load or for limited speed control with a quadratic load torque (as long as the motor temperature remains limited). In scheme (b) in Fig. 18.9, the AC chopper is used for a capacitor-start motor, which may reduce the starting current. However,

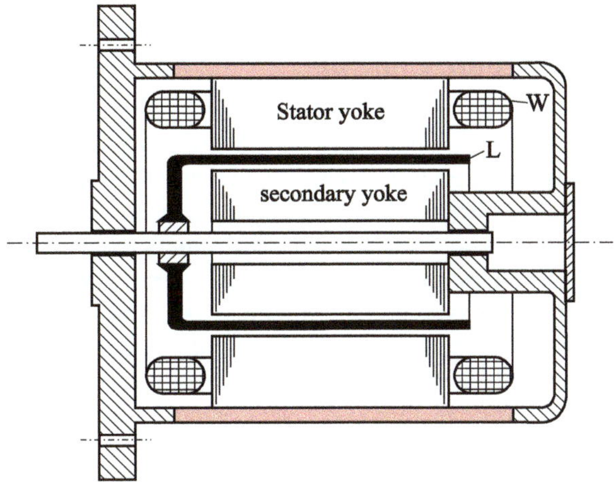

Fig. 18.8 Ferraris motor

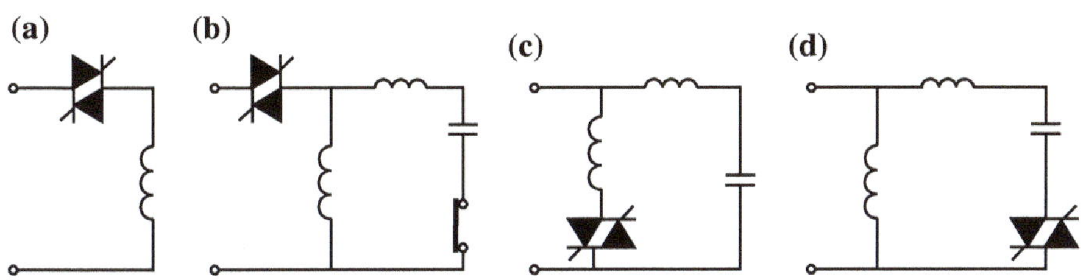

Fig. 18.9 AC chopper applications for single-phase fed induction motors

the starting torque is reduced by the square of the current reduction. Adaption to the load torque and speed control are only possible above the speed at which the centrifugal switch opens. Normally, AC chopper control will not be used for motors with a continuous capacitor (i.e. a capacitor motor), as the already small starting torque would be reduced even further. However, it could be acceptable if the AC chopper could be inserted in series with the main winding only (see (c) in Fig. 18.9). Yet, for this, separate connections to the main and auxiliary windings need to be accessible.

Another useful application of the AC chopper is shown in (d) in Fig. 18.9. Here, the chopper is inserted in series with the auxiliary winding only. This allows us to vary the effective capacitive impedance, for example from starting to steady state. Additional advantages are that only one capacitor is required, without a (centrifugal or current) switch for the starting capacitor, and that rotation in the reverse direction is possible.

18.3 Small Synchronous Machines and Their Power-Electronic Control

Small synchronous machines usually lack an excitation winding but have permanent magnet excitation instead. Because of the large variety of small synchronous motors, their construction will be dealt with separately in Chap. 20.

Here, we will concentrate on the power-electronic supply of these permanent-magnet synchronous motors and particularly the inverter supply, as AC choppers are not frequently used for synchronous motors. In case of inverter supply, a rotor cage is not required (as for asynchronous starting) and is usually not desired (unless the control cannot provide sufficient stability, in which case a cage is still required for damping purposes). Instead, the power electronic supply and control will mostly require a rotor position sensor, unless modern sensorless methods are used, which estimate the rotor position from electrical signals.

Depending on the permanent-magnet arrangement, i.e. surface permanent magnets or interior permanent magnets (see Chap. 20), different kinds of power-electronic supply and control are preferred.

The arrangement of the permanent magnets at the surface often results in a rather rectangular distribution of the air-gap induction over the circumference. Interior permanent magnets are most likely to cause an almost sinusoidal distribution of the air-gap induction over the circumference.

A sinusoidal distribution of the air-gap induction will create an almost sinusoidal emf in time (see (a) in Fig. 18.10), in which a sinusoidal current will lead to an optimal, ripple-free constant three-phase torque or air-gap power (see Chap. 3 in Part 1).

A rectangular air-gap induction distribution will create a trapezoidal[3] emf in time. The optimal current wave-shape (in time) to obtain a constant ripple-free torque or air-gap power is then a block wave. For three-phase symmetry, the constant part of the emf and current waves should be $2\pi/3$, as has been assumed in (b) in Fig. 18.10.

Let us now compare the output power (or torque) for both arrangements, starting from equal maximum values of the air-gap induction (i.e. the same material and same saturation level) and thus the same maximum emf ($\sqrt{2}$ in the figure - in per unit -). For the sinusoidal case, we assume a current of 1 per unit in effective value, or with an amplitude of $\sqrt{2}$. The current amplitude for the rectangular case, resulting in the same joule losses in the primary winding as for the sinusoidal case (assuming the same winding resistances), is then[4] $\sqrt{3/2}$.

For the sinusoidal case, this leads to a constant three-phase air-gap power of 3, while for the trapezoidal case the three-phase air-gap power amounts to $2\sqrt{3}$.

[3]It is trapezoidal if there is a distributed winding with $q > 1$; for a concentrated winding, the emf also has a block shape.

[4]Prove this.

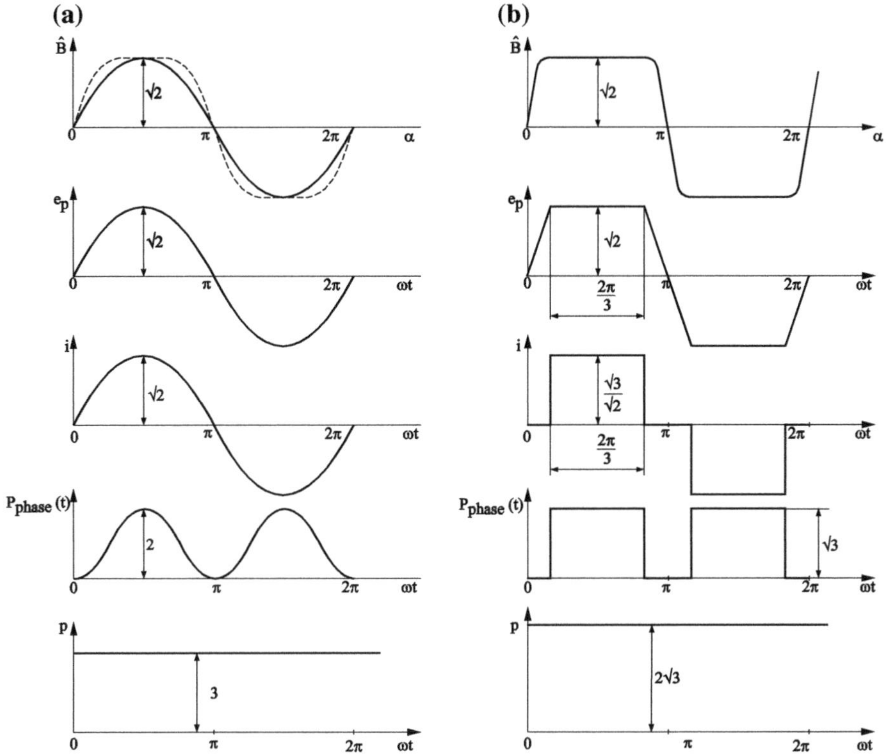

Fig. 18.10 Brushless AC and DC emf, current and power

In other words, the trapezoidal motor theoretically leads to some 15% higher output power than the sinusoidal motor for the same maximum induction and joule losses in the stator winding. In reality, this only holds for speeds that are not too high, as for the trapezoidal case the higher harmonics in the emf and current lead to harmonic losses (in the winding and also in the permanent magnets, especially for rare-earth permanent magnets).

For both (high dynamic) drive types, rotor position information is required to synchronise the current with the emf. For optimal torque production (and minimal transients, i.e. field orientation), the maximum of the current layer and the maximum of the emf should be in phase. This is valid for both the trapezoidal and the sinusoidal motors. Rotor position information and inverter together act now as an electronic commutator, just like the mechanical commutator in a DC machine. This is also why the term *Brushless DC motor* is used for the trapezoidal pm-motor. This terminology sometimes also refers to the sinusoidal pm-motor, but given the sinusoidal current and field distributions, a more correct term is the *Brushless AC motor*.

However, for the sinusoidal pm-motor a much higher resolution for the position is required than for the trapezoidal motor. The sinusoidal pm-motor requires a resolution of 5.6°(in an electrical degrees). For the trapezoidal motor, a resolution of about 60° is usually sufficient.

Between sinusoidal and trapezoidal pm-motors, there are many other differences too:

- the stator winding of a sinusoidal pm-motor is mostly a distributed winding ($q > 1$), while for the stator winding of a trapezoidal motor a concentrated winding ($q = 1$) is often preferred.
- The rotor construction of a sinusoidal pm-motor with iron pole shoes (and internal permanent magnets) leads to a q-axis reactance x_q larger than the d-axis reactance x_d; the large reluctance of the surface permanent magnets of a trapezoidal pm-motor and their pole width of about π electrical radians leads to more or less equal reactances x_q and x_d. As long as current and field axes are orthogonal, $i_d = 0$, the reluctance torque component $t = (x_d - x_q)i_d i_q$ does not contribute to the total torque. For the sinusoidal pm-motor, a positive d-axis component of the current will result in a positive contribution of the reluctance torque, which is thus very useful in the high-speed and thus field-weakening speed range (the only way to obtain field weakening for a pm-motor is to apply a positive d-axis current component[5]).
- The torque ripple of the sinusoidal pm-motor is very small, but for a trapezoidal motor the torque ripple can be significant.

In addition to the motor differences, the converters differ considerably as well:

- for trapezoidal pm-motors, the preferred current blocks favour the use of a CSI (current blocks of 120°) while the preferred sinusoidal current distribution for a sinusoidal pm-motor requires a VSI-PWM. Nowadays, however, for smaller trapezoidal pm-machines also a voltage source type inverter is used quite frequently, but with a current control loop that mimics a block current wave.
- The trapezoidal motor with CSI requires only one current sensor (in the DC-link) while the sinusoidal motor needs two sensors, i.e. in two phases.
- As mentioned above, the rotor position resolution requirements are much lower for the trapezoidal motor.

As a result, the practical application of the two types of pm-motors is somewhat different:

- for high-end servo applications that require a small torque ripple, the sinusoidal pm motor is mainly used; also the more accurate position rotor information can be an advantage for the application itself
- for applications with a high speed range, only sinusoidal pm-motors qualify. Indeed, high speeds imply high emf values, so the available DC voltage might not be sufficient to provide the required current. For the trapezoidal type, the inverter starts to function as a voltage source inverter (interestingly, with 120° conduction). Some recovery of the torque can be obtained by leading the current ($i_d > 0$, thus counteracting the p.m. flux) but this only causes a minor increase of

[5]In the URS, the PM excitation is supposed to be along the **negative** d-axis, which corresponds to a rotor emf along the positive q-axis.

the speed range and actually worsens the torque ripple. Somewhat better results can be obtained by switching to 180° conduction, in combination with a leading current.

Far better results can be obtained with a sinusoidal pm-motor and its VSI-PWM converter with current control. Here, too, the current control starts to saturate at higher speeds (and the VSI-PWM goes from PWM to a six-step operation). With a leading current, $i_d > 0$, i.e. field weakening, some recovery of the torque production can be obtained by the reluctance torque component, especially when x_q is sufficiently larger than x_d. If the motor is well-designed ($x_q \gg x_d$), sinusoidal pm-motors can yield a much higher speed range than trapezoidal pm-motors.

Chapter 19
Single-Phase AC Commutator machines

Abstract Single-phase AC commutator machines are nowadays mostly used in (small) household machines, such as coffee grinders, mixers, or vacuum cleaners. Larger AC commutator machines also used to be applied in traction, for example in trains, but are now replaced by inverter-fed rotating field machines. In this chapter we review the basic properties of these single-phase AC commutator machines.

19.1 Introduction

Small single-phase AC commutator machines have much in common with small DC commutator machines, for example:

- they usually have no more than two poles
- they both have a rather large air gap
- there are generally no auxiliary poles

However, for larger power ratings, which is the case in traction, auxiliary poles as well as compensation windings used to be applied.

Compared to DC commutator machines, the construction of both small and large AC commutator machines exhibits some differences: due to the alternating flux, the stator core is completely laminated, and the alternating flux induces voltages in the (short-circuited) armature coils in commutation, so that the number of windings of the armature coils has to be limited (and, as a result, there has to be a large number of armature coils).

19.2 Motional EMF, Transformer EMF and Torque

19.2.1 Motional EMF

The flux configuration (see Fig. 19.1) differs from the DC commutator machine as the longitudinal field is not constant but a time varying (sinusoidal) field $B(t)$. A rotation

J. A. Melkebeek, *Electrical Machines and Drives*, Power Systems,
https://doi.org/10.1007/978-3-319-72730-1_19

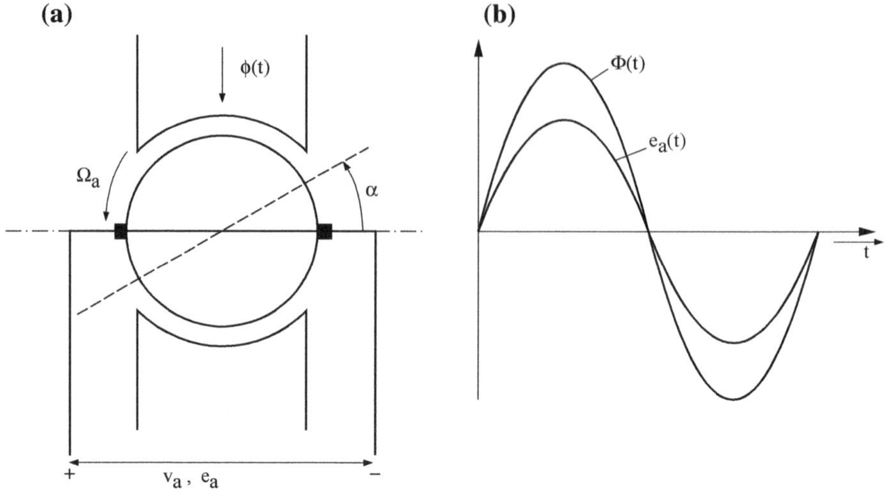

Fig. 19.1 AC commutator machine

of the armature results in an emf of motion induced in each armature conductor under a pole shoe which is a (sinusoidal) function of time:

$$e_{co}(t) = B(t) \cdot l_a \cdot v \tag{19.1}$$

with $v = 2\pi \cdot r \cdot N_a = 2N_p\tau_p \cdot N_a = (N_p\tau_p/\pi) \cdot \Omega_a$ the linear speed of a point on the armature surface and l_a the armature length.

For the total induced emf in a parallel branch, we obtain:

$$e_a(t) = \frac{z}{2\pi\,\mathrm{a}} \cdot N_p \cdot \Omega_a.\Phi(t) \tag{19.2}$$

with z being the total number of conductors, $2a$ the number of parallel branches and $\Phi(t)$ the flux over a pole pitch (i.e. the integral of $B(t)$ between two successive brushes). Note that this flux decreases if the brushes are shifted away from the geometrically neutral position.

Except for the time dependence of the emf (and flux), this is essentially the same expression as for the DC commutator machine. For an excitation current varying sinusoidally with time, the flux $\Phi(t)$ is also alternating, varying sinusoidally with time if saturation effects are negligible. The emf is thus also an alternating voltage with the same frequency as the flux or excitation current and is in phase or shifted 180° with respect to the flux, depending on the rotation direction. The effective value of the emf depends on the amplitude of the flux according to

$$E_a = \frac{z}{2\pi\,\mathrm{a}} \cdot N_p \cdot \Omega_a \cdot (\hat{\Phi}/\sqrt{2}) \tag{19.3}$$

For the relation between the (time) phasors of emf and flux, the following is valid

$$\underline{E}_a = \pm\frac{z}{2\pi \mathrm{a}} \cdot N_p \cdot \Omega_a \cdot (\underline{\hat{\Phi}}/\sqrt{2}) = \pm\frac{1}{\sqrt{2}} K\Omega_a \cdot \underline{\hat{\Phi}} \tag{19.4}$$

The (±) sign depends on the direction of rotation. The rotation direction in which the emf and flux are in phase will be referred to as the positive rotation direction.

19.2.2 Transformer EMF

Because the longitudinal flux is alternating, an emf will be induced in each of the armature coils by transformer action. Whether or not a transformer voltage can be measured at the brushes depends on the position of the brushes.

In general, this transformer voltage can be calculated according to

$$e_{a,t} = \frac{d\Psi(t)}{dt} \tag{19.5}$$

with $\Psi(t)$ the flux coupled with the complete armature winding.

This coupled flux depends on the angle between the brush axis and the neutral position. It is zero for $\alpha = 0$ and is maximum for $\alpha = 90°$. However, the calculation of this coupled flux is not straightforward. Only for the coils with the conductors in the neutral position is the coupled physical flux equal to the pole flux. For the other coils, the coupled flux depends on the shape of the induction curve along the armature circumference. This shape is mostly not sinusoidal (as in rotating field machines), nor is it constant under the pole shoes.

However, with the brushes in the geometrically neutral position, this voltage at the brushes is no cause for concern, as the voltages of the (series) connected coils will be cancelled out by symmetry.

On the other hand, the transformer-induced voltage is also present in the coils which are commutating, and it will have a negative effect on the commutation (see below).

19.2.3 Torque

Between two consecutive brushes, the current layer is constant in space (see (a) in Fig. 19.2) as each conductor carries the same current $i_a/2a$ at a given instant

$$A = \frac{i_a \cdot z}{2a \cdot 2N_p\tau_p} \tag{19.6}$$

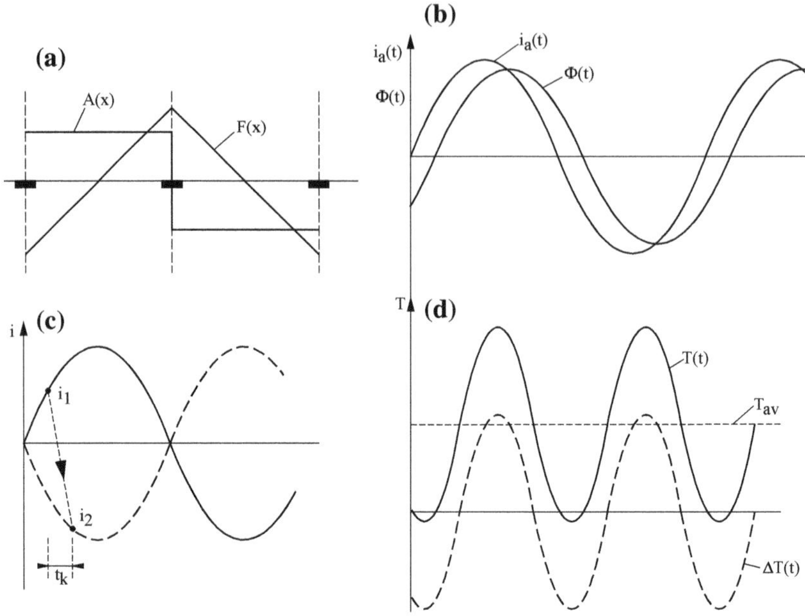

Fig. 19.2 Armature current, layer and mmf; commutation and torque

Naturally, current and current layer amplitudes vary in time (see (b) in Fig. 19.2). For the instantaneous value of the torque, we find the same expression as for the DC machine:

$$T(t) = \frac{z}{2\pi a} N_p \cdot i_a \cdot \Phi(t) \tag{19.7}$$

This can also be derived from the general expression for the torque effect of an (air gap) induction field distribution $B(x, t)$ on a current distribution $a(x, t)$ (see Chap. 3)

As flux and current are sinusoidal time functions here

$$i_a(t) = \hat{I}_a \sin \omega t \tag{19.8}$$

$$\phi(t) = \hat{\Phi} \sin(\omega t + \varphi) \tag{19.9}$$

the torque is not constant in time but contains a pulsating term with twice the supply frequency (see (d) in Fig. 19.2)

$$T(t) = \frac{z}{2\pi a} N_p \cdot \hat{I}_a \cdot \hat{\Phi} \cdot \frac{1}{2}\Big[\cos\varphi - \cos(2\omega t + \varphi)\Big] = \frac{1}{\sqrt{2}} K \cdot I_a \cdot \hat{\Phi}[\cos\varphi - \cos(2\omega t + \varphi)] \tag{19.10}$$

As a consequence, the average torque is

$$T(t) = \frac{z}{2\pi a} N_p \cdot I_a \cdot \hat{\Phi} \cdot \frac{1}{\sqrt{2}} \cos\varphi = \frac{1}{\sqrt{2}} K \cdot I_a \cdot \hat{\Phi} \cdot \cos\varphi \tag{19.11}$$

From the preceding, it is clear that the time shift between flux and current is equal to the time shift between the emf of motion and the current. From Eqs. 19.4 and 19.11, we may write for the electromagnetic power (i.e. the input power minus the Joule losses):

$$E \cdot I \cdot \cos\varphi = T \cdot \Omega \tag{19.12}$$

In other words, the mechanical power is indeed equal to the electromagnetic power.

Note that the average torque (or power) is but $1/\sqrt{2}$ or 70% of the torque (power) of a DC machine for the same maximal flux (i.e. flux amplitude, determining the saturation level) and the same armature current (i.e. effective value, responsible for the losses).

The pulsating term in the torque has twice the supply frequency, so the speed variation will remain quite small, even for a very small inertia. However, it might give rise to noise and vibrations.

19.2.4 Commutation

Similar to DC commutator machines, the commutation is affected by

- the resistances of the coils, their connection with the commutator and the contact resistance with the brushes
- the leakage inductances of the short-circuited coils (giving rise to a *reactance voltage*)
- the voltage induced by the quadrature field (i.e. the voltage induced by the quadrature field of the other armature coils)
- the auxiliary poles (if present)

For the quadrature field and the auxiliary poles, the only difference compared to DC machines is that the corresponding voltages are AC (with supply frequency) instead of DC.

The reactance voltage can be calculated as follows. If we denote the leakage inductance of the short-circuited coils by L_σ, the reactance voltage in the short-circuited coil is given by

$$e_{kr} = L_\sigma \frac{di}{dt} \tag{19.13}$$

If during the commutation time t_k the current changes from i_1 to i_2 (the latter with an opposite sign, see (c) in Fig. 19.2), then the average reactance voltage is

$$e_{kr} = L_\sigma \frac{i_2 - i_1}{t_k} \tag{19.14}$$

(if the commutation is approximately linear). With the average brush current during t_k given by $i_b \equiv \frac{i_a}{a} \approx i_2 - i_1$and with $t_k = b_b/(b_k.k.N_a)$ - where b_b= brush width; b_k= commutator segment width; N_a= speed in rev/sec - we obtain

$$e_{kr} \approx -\left[L_\sigma \frac{k.b_k}{a.b_b}\right] \cdot N_a.i_a(t) = c_r \cdot N_a \cdot i_a(t) \tag{19.15}$$

It is clear that the reactance voltage is proportional to the speed and the (*average*) armature current during the commutation.

In addition, the quadrature field (i.e. the field created by the other armature coils) as well induces in the short-circuited coils a voltage proportional to the speed and the armature current. Both voltages oppose (i.e. slow down) the commutation but can be compensated by auxiliary poles. The field of these auxiliary poles induces a motional voltage in the commutating coils given by

$$e_{kh} = \frac{z}{k} B_h(t) \cdot l_a \cdot v_a = [2\frac{z}{k} \cdot N_p \cdot \tau_p \cdot l_a] \cdot N_a \cdot B_h(t) \tag{19.16}$$

with z = total number of armature conductors; z/k = number of conductors in each commutating coil. As the auxiliary poles are connected in series with the armature, the reactance and quadrature voltages can be compensated completely.

For an AC commutator machine, the transformer voltage induced in the commutating coils also needs to be taken into account. For a brush axis in the geometric neutral position (i.e. orthogonal to the excitation flux axis), the axis of the commutating coils is coincident with the axis of the excitation winding. As a consequence, the alternating excitation field induces a voltage in these short-circuited coils given by:

$$e_{kt} = \frac{d\Psi_k}{dt} \tag{19.17}$$

in which Ψ_k is the flux coupled with the short-circuited coils. This voltage is not proportional with the speed, however; instead, it is proportional with the supply frequency, which means that it cannot be compensated by auxiliary poles. In order to reduce the detrimental effect of this emf on the commutation, the design of the machine will reduce the number of turns per coil as much as possible, i.e. the number of segments of the commutator should be as large as possible. The total induced voltage in the commutating coils (at rated conditions) should be limited to $1.5 \cdots 2$V or $2.0 \cdots 2.5$V depending on the material of the brushes.

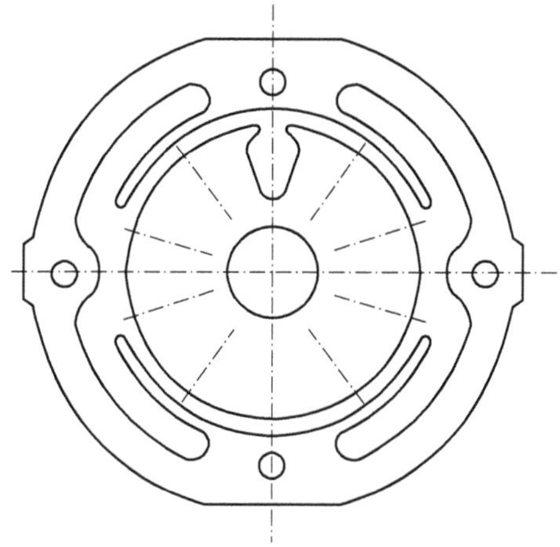

Fig. 19.3 Universal motor

19.3 The Single-Phase AC Commutator Motor (Universal Motor)

19.3.1 Introduction

In the past, the single-phase AC commutator motor with series excitation was the motor of choice for traction purposes (particularly trains). The power electronics revolution brought about a profound change. For railways where the supply was AC, the universal motor was first replaced by the rectifier-fed DC commutator machine. Later on, with the (r)evolution of power electronics, it was replaced by inverter fed induction and synchronous machines, almost everywhere in traction.

As has already been mentioned, the construction of the (series-excited) AC commutator motor is similar to the DC commutator machine, but the magnetic core is completely laminated (the stator included). For traction applications, the excitation of the universal motor is a series excitation; auxiliary poles and commutation poles are used for higher power ratings only.

The single-phase AC commutator motor with series excitation is still widely used, however, but only for lower power ratings, as in household machines. It is then referred to as a universal motor (Fig. 19.3). Actually, this name itself indicates that it can be used for both AC and DC supply.

19.3.2 Operating Characteristics

For emf and torque, the general Eqs. 19.4 and 19.11 derived in the preceding section remain valid for the universal motor. Here, however, the flux $\underline{\hat{\Phi}}$ is in phase with and proportional to - if saturation is neglected - the current:

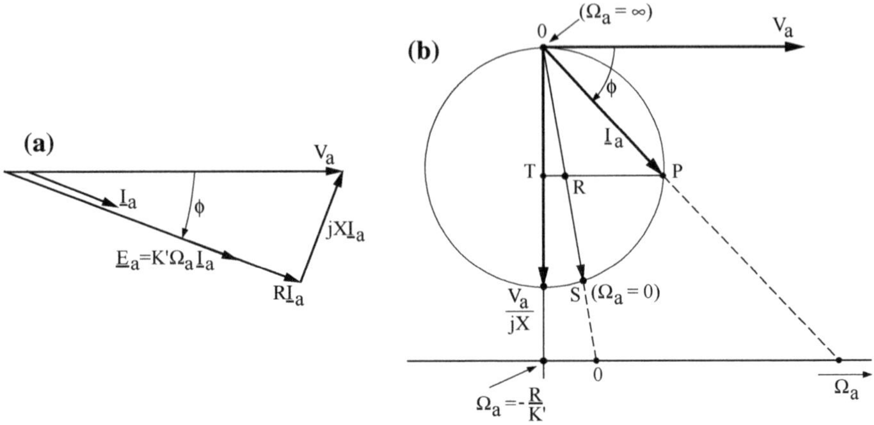

Fig. 19.4 Phasor diagram and current locus

$$\underline{\hat{\Phi}} = k' \cdot \underline{I}_a \tag{19.18}$$

Therefore:

$$\underline{E}_a = \frac{1}{\sqrt{2}}.K.k'.\Omega_a.\underline{I}_a = K'.\Omega_a.\underline{I}_a \tag{19.19}$$

$$T(t) = \frac{1}{\sqrt{2}}.K.k'.I_a^2[1 - \cos 2\omega t] = K'.I_a^2[1 - \cos 2\omega t] \tag{19.20}$$

For the relation between supply voltage, emf and current, we may write:

$$\underline{V}_a = (R + jX)\underline{I}_a + \underline{E}_a = (R + K'\Omega_a)\underline{I}_a + jX\underline{I}_a \tag{19.21}$$

The corresponding phasor diagram is drawn in (a) in Fig. 19.4.

For a constant voltage, the endpoint of the phasor follows a circle for variable speed, as is illustrated in (b) in Fig. 19.4. The diameter of the circle V_a/jX, corresponds to a slightly negative speed $\Omega_a = -R/K'$. For increasing speed, the current amplitude decreases and the $\cos\varphi$ increases.

As K' is proportional to the product of the number of turns of the excitation winding and the armature winding, we may write

$$\tan\varphi = \frac{X}{R + K'\Omega_a} \approx \frac{X}{K'\Omega_a} \sim \frac{aw_a^2 + bw_b^2}{w_a.w_b} \sim \frac{w_b}{w_a} \tag{19.22}$$

In order to obtain an acceptable $\cos\varphi$ *(for a given rated speed) the leakage reactance X of the armature will have to be limited as much as possible. The ratio of the number of turns of the excitation winding to that of the armature will also be kept as low as possible. Note that a lower supply frequency will help improve the* $\cos\varphi$ *(cf. the old 16 2/3 Hz traction supply in many countries).*

Similar to the circle diagram of an induction machine, a linear speed scale can be obtained by drawing a line parallel to the line connecting the origin and the operating point for infinite speed. Here, this is a line parallel to the tangent in the origin or thus the real axis. The circle diagram also allows us to derive graphically the power conversion: TP = supply power, TR = joule losses in the armature, and RP = mechanical power. The line OS connects the origin with the standstill operating point.

From Eq. 19.21, we obtain for the current:

$$\underline{I}_a = \frac{Va}{(R + K'\Omega_a) + jX} \tag{19.23}$$

or

$$I_a^2 = \frac{V_a^2}{(R + K'\Omega_a)^2 + X^2} \tag{19.24}$$

As a consequence, the average torque is given by:

$$T = K'.\frac{V_a^2}{(R + K'\Omega_a)^2 + X^2} \tag{19.25}$$

The maximum torque is obtained for $\Omega_a = -R/K'$ and is given by

$$T_{max} = K'.\frac{V_a^2}{X^2} \tag{19.26}$$

We may also rewrite the torque equation as

$$\frac{T}{Tmax} = \frac{1}{1 + \left(\frac{R}{X} + \frac{K'}{X}\Omega_a\right)^2}$$

or,

$$\Omega_a = \frac{X}{K'}\sqrt{\frac{1 - T/T_{max}}{T/T_{max}}} - \frac{R}{K'} \approx \frac{X}{K'}\sqrt{\frac{1 - T/T_{max}}{T/T_{max}}} \tag{19.27}$$

For most universal motors, the value of the constant X/K' *turns out to be between* $(0.35...0.75)\omega/N_p$*, with* ω *the supply frequency. Therefore,* ω/N_p *is sometimes called the synchronous speed, although* ***the notion of synchronous speed has no physical meaning for an AC commutator machine****. Indeed, there is no connection between speed and supply frequency for an AC commutator machine.*

Curve (a) in Fig. 19.5 illustrates the torque as a function of speed (for $X/K' = 0,5\omega/N_p$ and *R/X*=0,2). Speed or torque control is achieved by varying the voltage

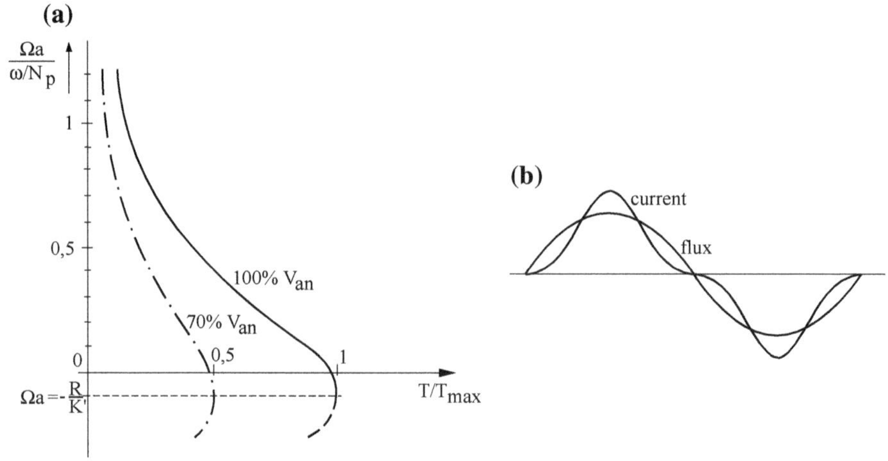

Fig. 19.5 Torque characteristic; third current harmonic

amplitude. Note that the torque is a quadratic function of the voltage. For 70% voltage, the maximum torque is therefore only 50% (cf. the dash-dotted line in (a) in Fig. 19.5, where T_{max} still represents the maximum torque for 100% voltage).

Braking is obtained by self-excitation with a resistance (as for the DC series commutator machine) or with a series connection of a resistance and a capacitor. With a resistance, self-excitation is achieved with DC current. With a capacitor in series with the resistance, self-excitation with AC current may occur. However, this depends on the value of the capacitor and the saturation of the magnetic core; *if* it works, the frequency corresponds to $1/\sqrt{L.C}$, which may be derived by the equilibrium of reactive power.

Nevertheless, braking with energy recuperation is not possible, for the same reasons as for a series-excited DC commutator machine. If braking with recuperation is desired, the excitation should be changed to an independent excitation.

19.3.3 Remarks

1. As mentioned above, the reactance X will be limited as much as possible in order to reduce the required reactive power. However, a low reactive voltage drop implies that the emf and therefore also the flux will be almost sinusoidal, also if the magnetic core is saturated. Thus, with a saturated core (which is almost always the case), the current will not be sinusoidal and will contain third harmonics (see (b) in Fig. 19.5). On the other hand, the low reactance will result in reduced differences between the torque-speed characteristics for AC or DC supply.
2. Wear and life of the brushes require special attention in the design, construction and use of AC commutator machines. The two most important items affecting wear are mechanical factors (e.g. vibrations, eccentricity of the commutator) and

Fig. 19.6 Small grinding motor

sparking. Sparking is a rough adaptation of the current in the short-circuited coil when the contact of the brush with the commutator is broken. As explained before, due to the transformer emf, a perfect commutation is not quite feasible in AC commutator machines. This is particularly true for small machines where auxiliary poles are unusual - and thus the reactance voltage and the quadrature field voltage are not compensated. Sparking can be reduced by utilising a large number of commutator coils, a brush width somewhat larger than the width of the commutator segments, and an odd number of coils per pole, so that the commutation of both brushes is shifted in time. In fact, all these measures reduce the reactance of the commutating coil.

3. It is important to stress the absence of a relation between frequency and speed again. For 50 Hz, supply speeds higher than 3000 rpm are quite feasible (e.g. 10000 rpm for grinding machines, see Fig. 19.6).

19.4 Special Single-Phase Commutator Machines

In the past, other types of single-phase commutator machines were also relatively common. Two of these are the repulsion motor and the Déri motor, whose moveable brush axis makes variable speeds possible. Of course, with the advent of power electronics, these machines became obsolete.

19.4.1 The Repulsion Motor

The stator of a repulsion motor has an excitation winding supplied directly from the AC grid. For somewhat higher power ratings, this winding was sometimes a distributed winding, other times a concentrated winding as in a normal commutator machine. The rotor (armature) contains a commutator winding with a moveable brush axis and short-circuited brushes (see (a) in Fig. 19.7).

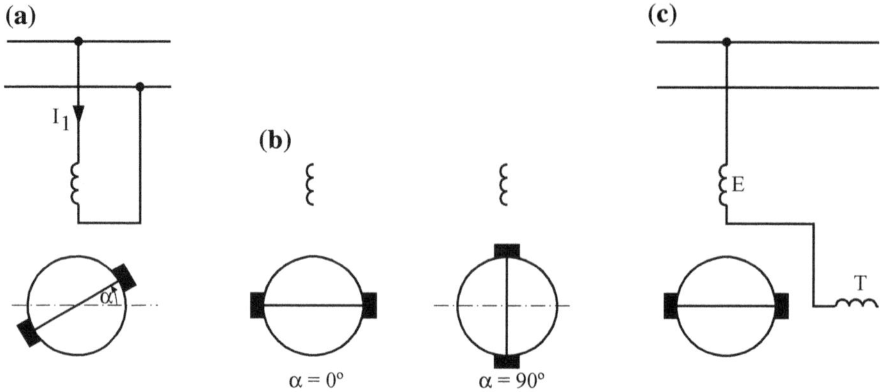

Fig. 19.7 Repulsion motor

The armature is not in contact with the supply and power can only be transferred to it by transformer action. The two extreme positions of the brush axis are shown in (b) in Fig. 19.7. For $\alpha = 0$ (called no-load), there is no coupling between armature and stator, but a large short-circuit current will be induced in the coils short-circuited by the brushes. For $\alpha = 90°$ (called short circuit), the armature is maximally coupled with the stator by transformer action. For an arbitrary brush axis situation α, we may consider splitting up the stator field (which is assumed to be sinusoidally distributed) into an excitation part E proportional to $\cos\alpha$ and a transformer part T proportional to $\sin\alpha$ (see (c) in Fig. 19.7). The transformer part induces a current I_a in the armature. The corresponding current distribution and the excitation part will result in a torque. As a first approximation, we may suppose that the current distribution I_a completely compensates the transformer field T. This means that, for the current distribution, we may write $I_a \sim I_1 \sin\alpha$. Thus, with the excitation field E proportional to $I_1 \cos\alpha$, we obtain for the torque

$$T \sim I_1 \sin\alpha \cdot I_1 \cos\alpha \sim I_1^2 \sin 2\alpha \tag{19.28}$$

The electromagnetic energy conversion (and the torque) can also be explained as follows. In Fig. 19.8, all windings are supposed to have right turns. A stator current I_1 gives rise to a flux $\Phi_1$with the indicated direction. A counter-clockwise or clockwise rotation of the brush axis results in an armature current with the direction shown in (a) and (b) in Fig. 19.8, respectively. From the direction of the flux and the current, we may derive that the torque direction is clockwise in (a) and counter-clockwise in (b). In other words, the direction of rotation for motoring is opposite to the direction of the twist of the brush axis.

In a similar way, we may also easily derive how the torque varies with current and twist angle of the brush axis. From the equivalence of electrical and mechanical power, we have

$$T.\Omega = E_d.I_a.\cos\psi \tag{19.29}$$

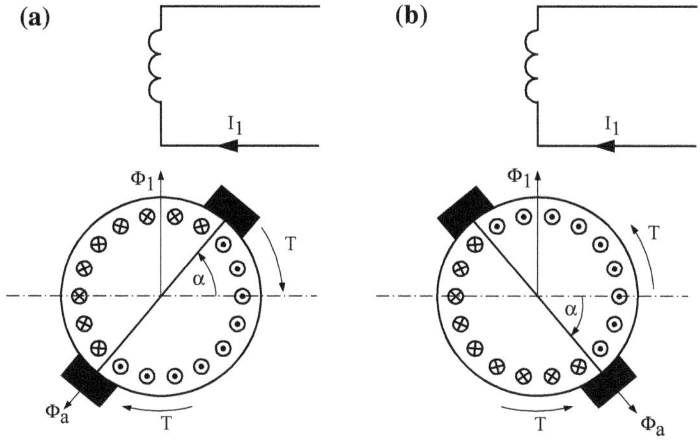

Fig. 19.8 Repulsion motor (symbolic representation)

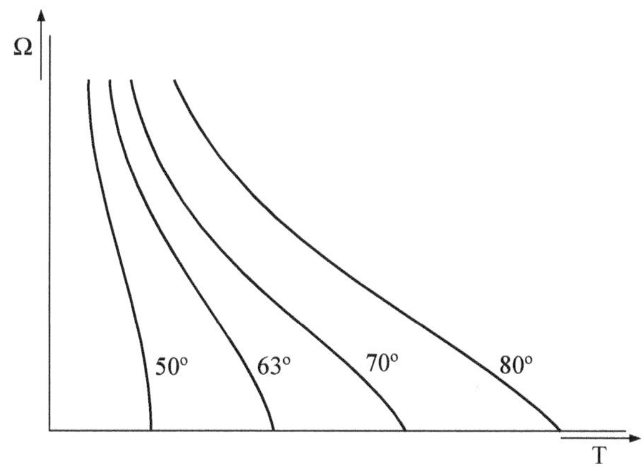

Fig. 19.9 Repulsion motor torque characteristic

where E_d is the motional emf and ψ is the phase lag between E_d and I_a. The emf $\underline{E}_d$ is proportional to the field component orthogonal to the brush axis, i.e. $\underline{I}_1 \cos\alpha$. We may suppose that the armature current distribution completely compensates the field component along the brush axis, and therefore we may write $I_a \sim I_1 \sin\alpha$. This also implies that $\cos\psi \approx 1$, so the same expression as before arises (Eq. 19.28).

Note, however, that the stator current I_1 depends on the armature current I_a and therefore also on the speed. Indeed, the stator is partially coupled with the armature by transformer action and the armature current also depends on the motional emf which is proportional to speed. The torque-speed characteristics are quite similar to those of the commutator motor with series excitation (see Fig. 19.9). For lower speeds, however, the torque approaches a (lower) limit value faster.

In normal operation, the α- value is around 75°. Maximum torque occurs between 80 and 85°.

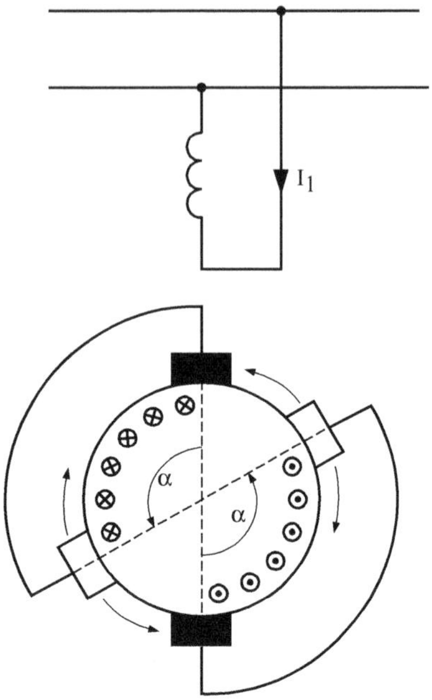

Fig. 19.10 Déri motor

The biggest advantage of the repulsion motor is (was) its straightforward speed control. Because of this, the design of the repulsion motor could concentrate more on obtaining good commutation behaviour, without any concerns about the supply voltage. On the other hand, a disadvantage is that auxiliary poles are not possible because of the moveable brush axis. Furthermore, it is worth mentioning that, with respect to the rotor, the field is an elliptical rotating field that becomes circular when the speed is approximately ω/N_p.

19.4.2 The Déri Motor

The Déri motor is in fact a repulsion motor with four brushes for each pair of poles. Each pair of brushes is (electrically) independent of the other. Of each pair, one brush is fixed along the excitation winding axis. The other two brushes are located along a rotatable axis (see Fig. 19.10).

Compared to the repulsion motor, the advantages of the Déri motor are the following:

- under no-load ($\alpha = 0°$) the armature current is zero, also for the coils short-circuited by the brushes. As a consequence, the motor may remain supplied with full voltage without any danger

- the larger control range (180° compared to 90°) permits a more accurate speed control

However, similar to the repulsion motor, the Déri motor is only interesting from a historical perspective and is no longer relevant for current usage.

Chapter 20
Small Synchronous Motors

Abstract For large synchronous machines, the excitation is usually provided by a DC-fed field winding in the rotor. For small synchronous machines, however, this is undesirable, on the one hand because of the complications for supplying DC to the rotor (i.e. slip rings, rectifier), on the other hand because the smaller the dimensions, the less efficient a field winding is in producing a magnetic flux.

For small machines, synchronous operation can (preferably) be achieved in other ways:

- by permanent magnet excitation (in the rotor)
- by reluctance variation in the rotor (i.e. saliency)
- by using a ring of hysteretic material in the rotor.

20.1 Synchronous Machines with Excitation by Permanent Magnets

20.1.1 Permanent Magnet Material

Compared to DC excitation of synchronous machines, the advantages of permanent magnet (PM) excitation are the following:

- speeds can be comparable to those of cage induction machines (of the same dimensions)
- joule losses in the rotor are eliminated, which is particularly advantageous for lower power ratings, where efficiency is generally low
- for lower power ratings, permanent magnet excitation results in a lower volume of the required material, i.e. a more compact machine.[1]

[1]From scaling laws, it can be shown [21] that the ratio of the volumes of material required for permanent magnet excitation to that for DC excitation varies proportionally to the pole pitch; thus for smaller power ratings, permanent magnet excitation allows for a smaller machine.

J. A. Melkebeek, *Electrical Machines and Drives*, Power Systems,
https://doi.org/10.1007/978-3-319-72730-1_20

There are numerous different rotor concepts, depending on the shape and location of the magnets, the presence of a damper cage, and many other factors. The rotor construction is narrowly related to the type of permanent magnet material used. The most important permanent magnet materials are

- cast materials (e.g. Alnico, Ticonal),
- ceramic materials (e.g. Barium Ferrite),
- rare earth materials (e.g. $SmCo_5$), and
- amorphous materials (e.g. NdFeB).

The magnetic properties are characterised by the demagnetising curve, i.e. the segment of the relation B-H in the second quadrant ((a) in Fig. 20.1). The main properties of a hard magnetic material are the following:

- the intersection points of the demagnetising characteristic with the y-axis and the x-axis, i.e. the remanent induction B_r (for $H = 0$) and the coercive force H_c (for $B = 0$)
- the temperature coefficients of B_r and H_c
- the shape and specifically the slopes of the minor loops that determine the repetitive demagnetising and magnetising; these minor loops can be approximated by the *recoil lines*, i.e. straight lines with a slope almost equal to the slope of the demagnetising curve in B_r.

Consider now the basic magnetic circuit (b) in Fig. 20.1. For a *non-stabilised* magnet, the no-load point P_o is the intersection of the demagnetising curve of B_r and the *air-gap line*

$$H_{pm} \cdot l_{pm} + H_\delta \cdot l_\delta = 0$$

or

$$H_{pm}.l_{pm} + \frac{\Sigma_{pm} \cdot B_{pm}}{\Sigma_\delta \cdot \mu_o} l_\delta = 0 \tag{20.1}$$

where Σ_{pm} and Σ_δ are the cross-sections of the permanent magnet material and the air gap, respectively; l_{pm} and l_δ are the magnetic path lengths in the magnetic material and the air gap, respectively. A normal range for the slope $(l_{pm}.\Sigma_\delta)/(l_\delta.\Sigma_{pm})$ of the air-gap line is approximately 10–100.

The operating point can be (irreversibly) changed by opposing ampere-turns (e.g. in the stator of a machine). The operating point will then (temporarily) move to Q as the air-gap line will shift to the left by the opposing ampere-turns. When the opposing current disappears, the air-gap line shifts to the right again but the new operating point becomes P (on the minor loop through Q, although this can be approximated by the recoil line). The induction is then reduced from B_o to B. In order to minimise the

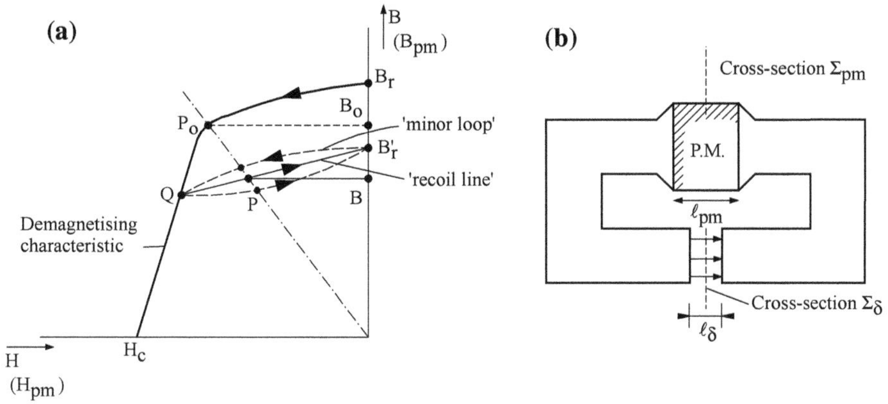

Fig. 20.1 Demagnetising characteristic (**a**) and basic magnetic circuit (**b**)

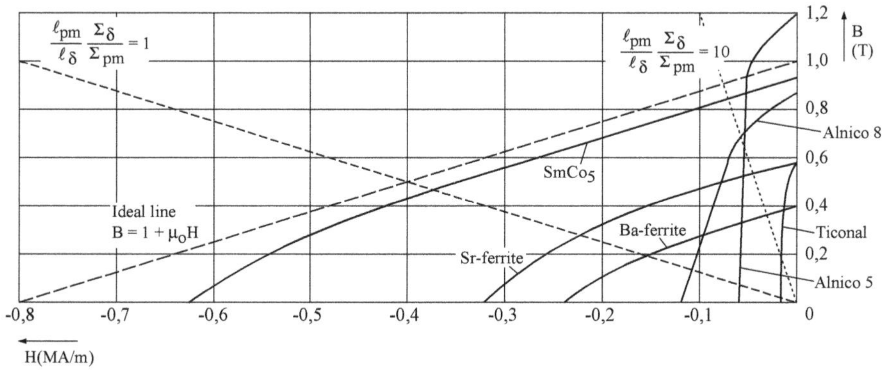

Fig. 20.2 Demagnetising characteristics for some PM materials

reduction of B, the demagnetising characteristic should approach a straight line with slope μ_o.[2,3]

In this respect, it is quite instructive to compare the various permanent magnet materials (see Fig. 20.2):

- cast materials (Alnico, Ticonal) yield a high induction but are rather susceptible to demagnetisation; the temperature coefficients are quite low (-2.10^{-4} for B_r and $(-7 \ldots +3).10^{-4}$ for H_c)
- ceramic materials (Sr- and Ba-ferrites) are less susceptible to demagnetisation but offer much lower induction values. Moreover, their temperature coefficients are rather high (-2.10^{-3} for B_r and $(+2 \ldots +5).10^{-3}$ for H_c)

[2] In practice, a machine will always have an already stabilised operating point at delivery.

[3] Note that also disassembling and removing the rotor from the stator without magnetically short-circuiting the rotor will result in an irreversible reduction of the induction.

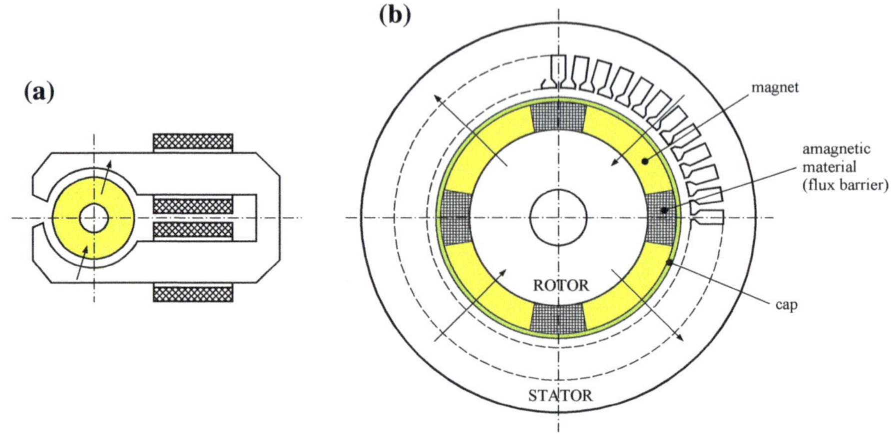

Fig. 20.3 Surface PM motor

- rare earth ($SmCo_5$) and amorphous (*NdFeB*) permanent magnet materials show demagnetising characteristics that are relatively close to the ideal demagnetising line with $Br \gtrapprox 1$ and $dB/dH \approx \mu o$, but they are quite expensive. Their temperature coefficients are quite low (-5.10^{-4} for B_r and -3.10^{-4} for H_c).

20.1.2 Rotor Configurations

In the *single-phase motor* in (a) in Fig. 20.3, the permanent magnet is directly fixed on the shaft. The laminated stator core and the pole shoes are asymmetrical, so that a rotating field component can be obtained (cf. the shaded pole/jump pole induction motor). At standstill, this results in a slant angle of the rotor with respect to the vertical position, facilitating the run-up. At the same time, this asymmetry may allow a more straightforward assembly of the coils.

The *rotating field motor* in (b) in Fig. 20.3 has a laminated stator core with a three-phase distributed winding (in slots). The rotor has a laminated core as well, around which permanent magnet strips are fixed, separated from each other by a-magnetic spacings. The assembly is held together by a support cap, which may provide an asynchronous starting torque if the cap is conducting.

In this surface PM configuration (and also somewhat in the one of (a)), there is no appreciable difference between the reluctances (and thus the main field inductances) in the direct (along the N-S axis of the magnet) and the quadrature (orthogonal to the magnet axis) axes.

Figure 20.4 shows rotor constructions with a flat PM, which is the preferred shape for PMs as the material is rather hard and cannot easily be machined. A cage is provided in order to obtain an asynchronous starting torque. The cage also yields

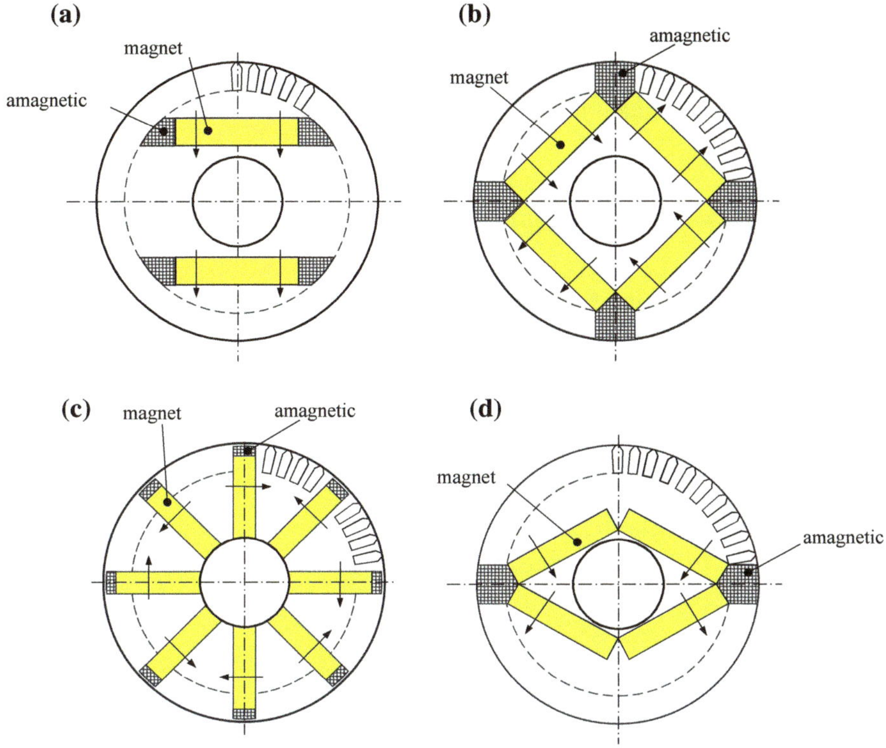

Fig. 20.4 Interior PM machines

damping for transients. Although the zone with teeth and slots (indicated by the dashed line in the figures) protects the PM against demagnetisation, it reduces the space available for the permanent magnets and thus the air-gap flux. Note that the shaft also reduces the space for the magnets. In addition, the mechanical strength of these anisotropic rotors requires special attention, as protection against centrifugal forces is required. Particularly with ferrite as a permanent magnet material, the space available can be problematic as the maximal induction of ferrite is smaller than the normally desired induction in the air gap. The air-gap flux can be improved by using PM material with higher induction or by a special shaping of the magnetic circuit in the rotor. In the rotor configuration (c) in Fig. 20.4, the magnets are placed radially, resulting in an air-gap induction higher than the PM induction, especially for a high pole number. However, this requires an a-magnetic shaft, to keep the magnets from short-circuiting. A variant on this is the rotor configuration (d) where the magnets are positioned somewhat obliquely with respect to the radial direction.

In the configurations in Fig. 20.4, the main inductance in the d-axis is mostly smaller than the main inductance in the q-axis, as the permeability of the PM material is more or less the same as that for vacuum. This results in an opposite sign of the reluctance torque compared to traditional synchronous machines. However, this can be used advantageously for current supply. An additional advantage of the configurations in Fig. 20.4 is that the iron pole shoe shields the permanent magnet material against demagnetisation by the stator currents.

20.1.3 Electromagnetic Behaviour and Torque of PM Motors

The electromagnetic behaviour and torque of permanent magnet motors is rather comparable with that of DC-excited synchronous motors, as will be shown below. The main difference lies in the (potential) sign reversal of the reluctance torque.

Figure 20.5 represents schematically the air-gap fields for armature current distributions A_{ad} in the d-axis and A_{aq} in the q-axis.

To begin with, consider the case of no-load (zero armature current, I = 0). For a flux line along the d-axis, we have

$$\oint H \cdot dl \equiv 2H_\delta \cdot l_\delta + 2H_{pm} \cdot l_{pm} = 0$$

or

$$F_\delta \equiv H_\delta . l_\delta = -H_{pm} \cdot l_{pm} \tag{20.2}$$

The interpretation of Eq. 20.2 is that the permanent magnet mmf drop $H_{pm} \cdot l_{pm}$ provides the magnetic field in the air gap (as $-H_{pm}.l_{pm} > 0$).

With

$$\Sigma_\delta \cdot B_\delta = \Sigma_{pm} \cdot B_{pm} \tag{20.3}$$

and

$$B_\delta = \mu_o H_\delta \tag{20.4}$$

we then obtain

$$B_{pm} = -\mu_o \cdot \frac{\Sigma_\delta . l_{pm}}{\Sigma_{pm} . l_\delta} . H_{pm} = \frac{1}{\Sigma_{pm}} . \mu_o \cdot \frac{\Sigma_\delta .}{. l_\delta} . (-H_{pm} l_{pm}) \tag{20.5}$$

Next, consider an armature current I, or the corresponding current sheet A_a. For the same integration path along the d-axis (see (b) in Fig. 20.5), we get

$$2H_\delta \cdot l_\delta + 2H_{pm} \cdot l_{pm} = \int_0^{\tau_p} A_a(x).dx = \int_0^{\tau_p} A_{ad}(x).dx \tag{20.6}$$

or

$$F_\delta \equiv H_\delta \cdot l_\delta = -H_{pm} \cdot l_{pm} + F_{ad} \tag{20.7}$$

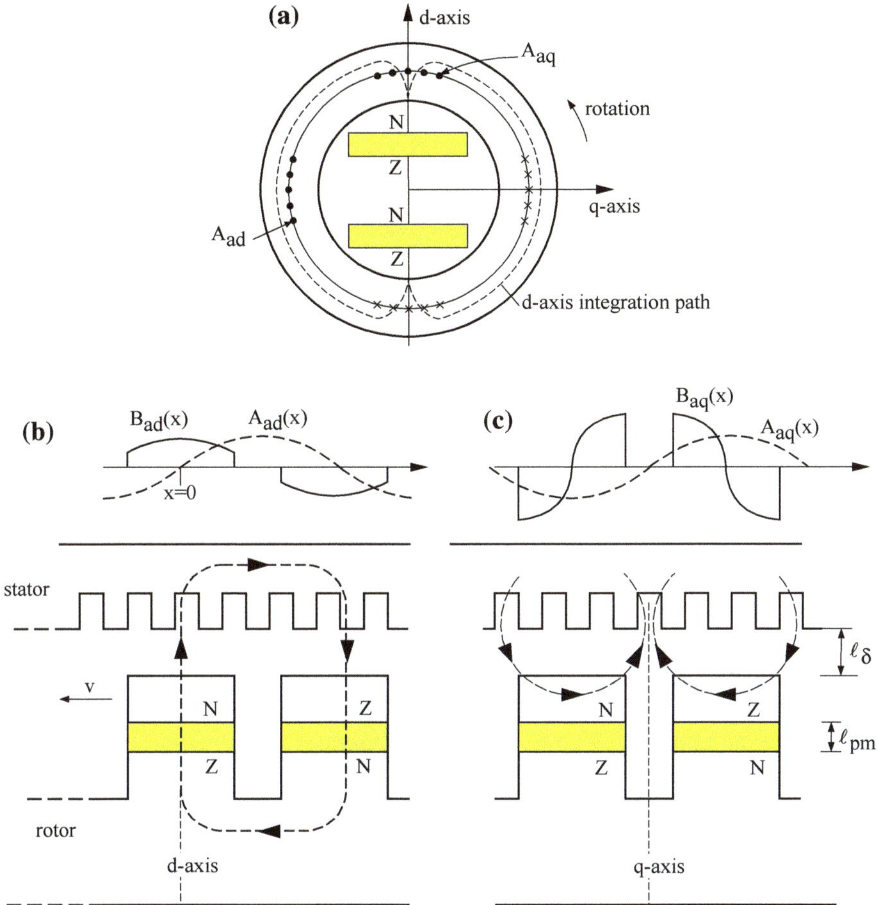

Fig. 20.5 PM excitation and armature current

with F_{ad} the armature (air-gap) mmf along the d-axis

$$F_{ad} = \frac{1}{2} \int_{0}^{\tau_p} A_{ad}(x) \cdot dx \tag{20.8}$$

Two sources are now responsible for the mmf along the d-axis: the PM and the d-axis component of the armature current.

Equations 20.7, 20.3 and 20.4 result in

$$B_{pm} = \frac{1}{\Sigma_{pm}} \cdot \mu_o \cdot \frac{\Sigma_{\delta} \cdot}{l_{\delta}} \cdot (F_{ad} - H_{pm} l_{pm}) \tag{20.9}$$

If the minor loop (or recoil line) for the operating point (H_{pm}, B_{pm}) can be linearised as (see Fig. 20.1)

$$B_{pm} = \mu_m(H_r + H_{pm}) = B'_r + \mu_m H_{pm} \tag{20.10}$$

(where $B'_r = \mu_m H_r$ defines H_r), then

$$B_{pm} = \frac{1}{\Sigma_{pm}} \cdot (H_r l_{pm} + F_{ad}) \cdot \Lambda_d \tag{20.11}$$

or also

$$\Phi_d = (H_r l_{pm} + F_{ad}) \cdot \Lambda_d \tag{20.12}$$

with Λ_d the d-axis permeance

$$\Lambda_d^{-1} = \Lambda_{pm}^{-1} + \Lambda_\delta^{-1} \tag{20.13}$$

and

$$\Lambda_{pm} = \mu_m \cdot \frac{\Sigma_{pm}}{l_{pm}} \quad and \quad \Lambda_\delta = \mu_o \cdot \frac{\Sigma_\delta}{l_\delta} \tag{20.14}$$

the permeances of magnet and air gap, respectively.[4]

Equation 20.12 again demonstrates that the d-axis flux results from the equivalent magnet mmf $H_r l_{pm}$ and the armature mmf F_{ad}. The magnet mmf $H_r l_{pm}$ can be treated like the mmf F_p of a DC-excited synchronous machine. All equations for the DC-excited synchronous machine remain valid. With the mmf $F_p = H_r l_{pm}$, the emf E_p corresponds; for the d-axis armature mmf F_{ad}, the armature emf E_{aq} is the corresponding value When adding the mmfs F_p and F_{ad} to obtain the total fundamental d-axis emf, we must take into account the difference of the fundamental harmonic magnetic conductivity along the d-axis between an approximately square mmf F_p and an approximately sinusoidal mmf F_{ad}. This difference can be accounted for by including a reduction factor k_d for the armature mmf F_{ad}

$$\underline{\hat{F}}_d = \underline{\hat{F}}_p + k_d . \underline{\hat{F}}_{ad} \tag{20.15}$$

The vector (or phasor) sign (underscore) could in fact be omitted as all these mmfs are in the d-axis; the hat sign has been used to indicate that for fundamental harmonic mmfs (as well as for induction values or fluxes), we always use amplitude values.

If saturation can be disregarded, we may also add the corresponding emfs

$$\underline{E}_q = \underline{E}_p + \underline{E}_{aq} = \underline{E}_p + jX_{ad}\underline{I}_d \tag{20.16}$$

---

[4]When the reluctance of the iron core is not negligible, this reluctance must be added in the denominator of Eq. 20.13.

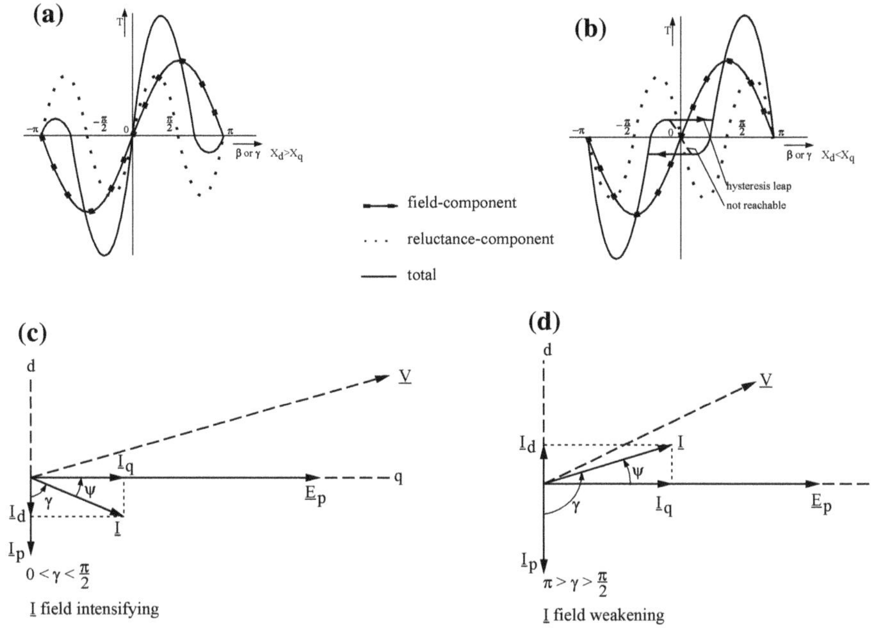

Fig. 20.6 Excitation torque and reluctance torque

where $jX_{ad}\underline{I}_d = j\omega(w\xi_1) \cdot k_d \cdot \underline{\hat{F}}_{ad} \cdot \Lambda_d \cdot (2\sqrt{2}/\pi)$ and $\underline{E}_p = j\omega(w\xi_1) \cdot \underline{\hat{F}}_p \cdot \Lambda_d \cdot (2\sqrt{2}/\pi)$.

In the q-axis, the permeance of the magnet does not come into play - at least not for designs with an assembly pole shoe (see Fig. 20.5). The q-axis component of the armature mmf, F_{aq}, gives rise to an emf along the d-axis

$$\underline{E}_d = jX_{aq}\underline{I}_q \tag{20.17}$$

The reactance X_{aq} corresponds to the permeance of the flux lines shown in (c) in Fig. 20.5. For synchronous machines with DC excitation, the reactance X_{aq} is always smaller than X_{ad}. For permanent magnet excitation, the q-axis reactance is usually larger than the d-axis reactance because the permeability of the permanent magnet μ_m is only of the same order of magnitude as the permeability of vacuum μ_o.

For the torque as well, we have the same equations as for traditional synchronous machines,

$$T = \frac{3}{\Omega} \cdot \left[\frac{E_p V \sin\beta}{X_d} + \frac{1}{2} V^2 \left(\frac{1}{X_q} - \frac{1}{X_d} \right) \sin 2\beta \right] \tag{20.18}$$

or

$$T = \frac{3}{\Omega} \cdot \left[E_p I \sin\gamma + \frac{1}{2} I^2 \left(X_d - X_q \right) \sin 2\gamma \right] \tag{20.19}$$

for voltage and current supply, respectively. In these equations, $X_d = X_\sigma + X_{ad}$; $X_q = X_\sigma + X_{aq}$; $\beta = \arg(\underline{V}, \underline{E}_p)$; $\gamma = \arg(\underline{I}, \underline{I}_p)$.

Note that the reluctance torque changes sign with respect to the *normal* (for a DC-excited machine) situation when $X_d < X_q$. For $X_d > X_q$, the reluctance torque has the same sign as the excitation torque for small $|\beta|$ or $|\gamma|$ (i.e. $< \pi/2$) and therefore increases the useful torque for these small load angles (see (a) in Fig. 20.6). In contrast, for $X_d < X_q$ the sign of the reluctance torque is opposite to that of the excitation torque for small load angles ($|\beta|$ *or* $|\gamma| < \pi/2$). As a consequence, for these small load angles the useful torque decreases and may even become negative for small load angles (see (b) in Fig. 20.6). The dashed section in this figure, where $dT/d\beta$ or $dT/d\gamma$ is negative, is unstable and cannot be used in open loop. This is called a hysteresis jump (see the horizontal lines with an arrow in (b) in Fig. 20.6).

However, this (at first sight) negative characteristic can be turned into an advantage depending on the supply and operating point. To this end, we rewrite Eq. 20.19 in terms of currents as follows:

$$T = \frac{3}{\Omega} \cdot \left[X_{ad}.I_p.I_q - \left(X_d - X_q\right) I_d I_q\right] \tag{20.20}$$

with $I_q = I \cos\psi = I \sin\gamma$ and $I_d = -I \sin\psi = -I \cos\gamma$.

A positive torque component due to the PM excitation requires $I_q > 0$ or $0 \leq \gamma \leq 180°$. If $X_d > X_q$, then the reluctance torque acts in the same direction as the PM torque if $I_d < 0$. Note that in that case the stator current also intensifies the PM excitation (see also (c) in Fig. 20.6).

In contrast, if $X_d < X_q$ (as is mostly the case for PM excitation), the reluctance torque acts in the same direction as the PM torque if $I_d > 0$. In this case, the armature current is weakening the PM excitation (see (d) in Fig. 20.6). In other words, this is advantageous to still obtain a sufficiently high torque for high speeds (in the field weakening area). It is also important to keep in mind that for current supply (or current control) in the range of $I_d > 0$, thus $\gamma > 90°$, saturation is reduced and the higher magnetising inductance also causes a higher (i.e. less reduced) torque.

20.1.4 Axial Flux Permanent Magnet Motors

The permanent magnet machines we considered above are radial flux machines, meaning that the main flux path is radial, just as in most traditional synchronous machines with a DC rotor excitation winding. Most radial flux machines have an inner rotor, although there exist also machines with outer rotor (called *outer-runners*). The advantage of outer-runners is that, for the same tangential force, the torque is larger as it is produced on a larger diameter.

Axial flux machines, in contrast, have their flux path in the axial direction. The stator and rotor are disks, as is illustrated in Fig. 20.7. This means that the air gap is

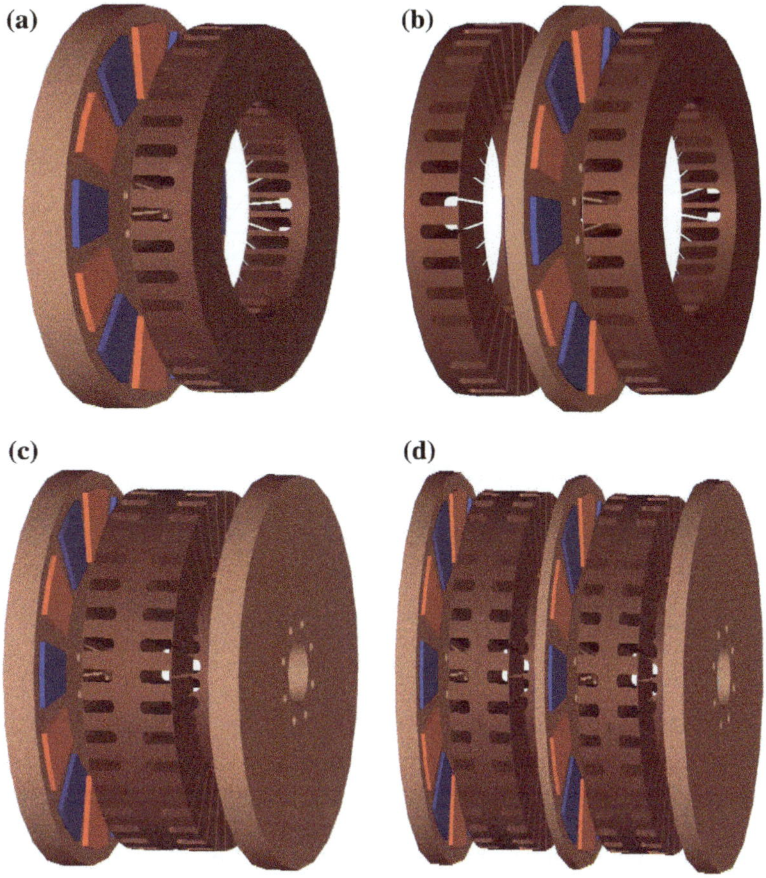

Fig. 20.7 Axial flux machines; **a** single stator and rotor; **b** double stator and single rotor; **c** single stator and double rotor; **d** multi-stage with two stator modules and three rotor modules

flat, not a cylinder as in radial flux machines. It is obvious that the electromagnetic energy conversion principle is the same as that for radial flux machines.

These axial flux machines may offer larger torques, similar to the radial outer-runners. Their relatively high torque and the small axial length render these machines quite suitable as wheel motors for electric vehicles. However, a disadvantage of axial flux machines are the sometimes large axial forces, which may bend the stator and rotor and may necessitate special bearings.

There is quite a wide range of different construction types. In addition to the single stator - single rotor machine (see (a) in Fig. 20.7), there are also double stator - single rotor machines (as illustrated in (b) in Fig. 20.7), single stator - double rotor machines (see (c) in Fig. 20.7), and multistage versions as illustrated in (d) in Fig. 20.7.

The single stator - single rotor version has as a disadvantage that an iron yoke is required on both stator and rotor. Therefore, the inertia of the rotor can be rather

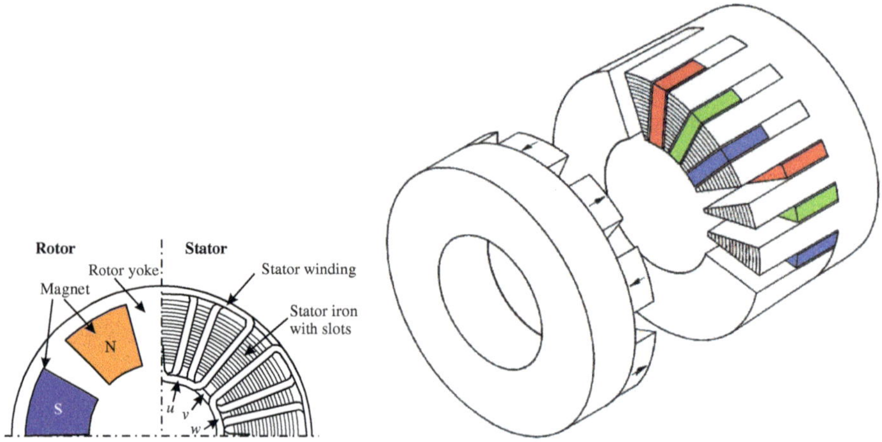

Fig. 20.8 Single stator - single rotor axial flux PM machine

large, as are the axial forces. Figure 20.8 illustrates a possible configuration with a three-phase stator winding in slots (here a concentrated winding, i.e. $q = 1$) and rotor permanent magnets glued onto an iron backplate. The stator iron is a helicoid of a thin iron sheet, possibly an amorphous metal, with slots punched in an appropriate way (e.g. with a constant tooth width, or a constant slot width).

Symmetrical types, for example the double stator - single rotor depicted in (b) in Fig. 20.7, create symmetrical axial forces on the rotor, ideally completely cancelling out each other. In addition, this machine does not require a rotor iron, resulting in relatively low rotor inertia (cf. its application as wheel motors and servo drives).

The single stator - double rotor depicted in (c) in Fig. 20.7 requires iron rotor yokes, increasing the inertia. Yet, as the permanent magnets are in an outer position, this facilitates their cooling. Indeed, rare earth and NdFeB magnets are prone to demagnetisation when subjected to temperatures above 120° to 180° (the exact critical temperature depends on the type of magnets). The stator yoke serves as flux path for both sides. As illustrated in (a) in Fig. 20.9, the flux paths of both sides remain separate, which means that this machine is actually a cascade or multistage connection of two machines, resulting in a larger output for the same diameter.

The configuration depicted in (d) in Fig. 20.7 is a further extension of the idea of using multistage axial motors to obtain higher output power. It should be noted, however, that increasing the diameter yields a more efficient machine than the multistage arrangement.

Starting from the configuration in (c) in Fig. 20.7, we might consider removing the stator back iron and rearranging the stator winding and rotor. This would create a configuration as depicted in (b) in Fig. 20.9. The stator winding is then a concentrated winding around segmented teeth, and a north pole opposes a south pole on the other end of each tooth. Obviously, the flux paths are also different from those in (a) in Fig. 20.9. A few of the advantages of this configuration are its high efficiency (there

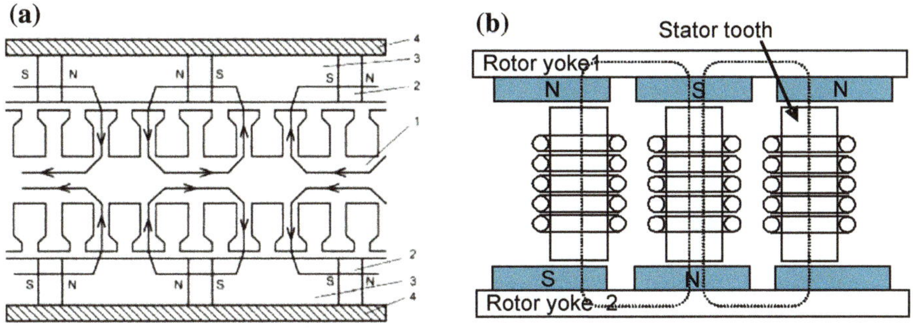

Fig. 20.9 **a** Single stator - double rotor; **b** YASA machine

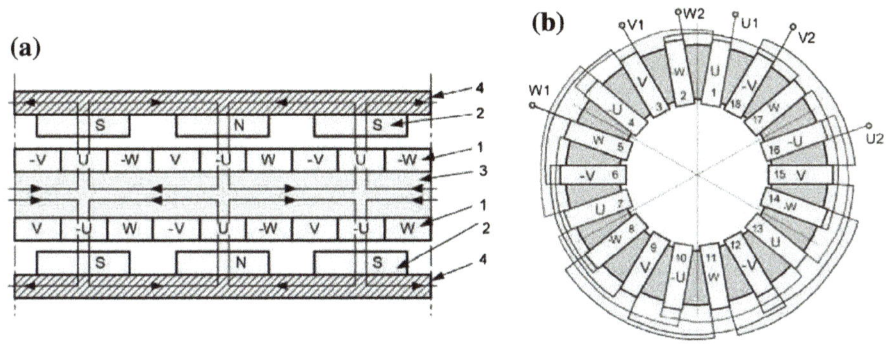

Fig. 20.10 Torus axial flux machine

is less iron loss because of the reduced stator iron), its straightforward construction as the teeth with windings can be constructed before assembly, and its assembly using composite material in the space between the windings. This machine type is sometimes referred to as a YASA machine (Yokeless And Segmented Armature).

Another simplified construction is illustrated in Fig. 20.10. Here, stator slots are completely absent. Instead, the three-phase winding is wound around a toroidal yoke (i.e. a torus). Although the rotors are similar to those in the YASA machine, here they are positioned so that north poles face north poles and south poles face south poles. Needless to say, the flux paths are also different. Despite its simple construction, some disadvantages of the air-gap winding are the forces on the windings, the rather large air gap and therefore also a large magnetic reluctance, and the excess winding losses due to transversal flux.

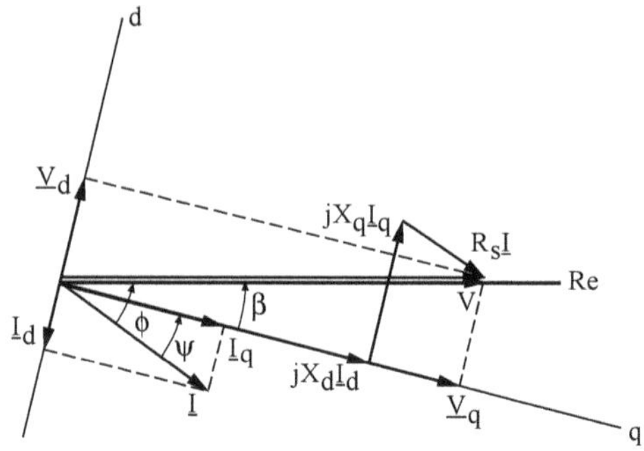

Fig. 20.11 Phasor diagram

20.2 Reluctance Motors

20.2.1 Introduction

Reluctance motors are synchronous motors that only utilise the reluctance torque for electromagnetic energy conversion. They usually also have a rotor cage to provide asynchronous starting and create better stability.

For the same frame size, the power that can be converted by the reluctance torque is lower than that converted by the asynchronous torque or PM torque; it can be proven[5] that for the same maximum induction in the air gap, the converted power is only one third of that for comparable induction motors or synchronous motors with DC or PM excitation. On the other hand, because rotor losses (in the cage of induction motors, i.e. about 1/3 of the total losses) are absent, higher losses can be allowed in the stator and thus the maximum induction can be increased by some 20% ($\approx\sqrt{3/2}$). In this way, the power converted can be increased up to two-thirds or even three quarters of that of comparable induction motors or synchronous motors.

20.2.2 Current and Torque: Effect of the Stator Resistance

Basically, the torque expression is that of the synchronous machine, which means that Eqs. 20.18, 20.19 or 20.20 apply, but without the excitation torque component ($E_p = I_p = 0$). However, reluctance motors mostly have small power ratings, in which case the stator resistance is not negligible.

For the current and torque we obtain the following expressions, taking the stator resistance into account (see Fig. 20.11):

[5] Cf. Reluktanzmotoren kleiner Leistung, Hans-Joachim Gutt; Etz-Archiv, Bd.10, 1988, H.11, pp. 345–354.

$$\underline{I} = \frac{V}{R_s^2 + X_d X_q} \cdot \left\{ R_s - j\frac{X_d + X_q}{2} + \frac{1}{2} j(X_d - X_q) \cdot e^{-j2\beta} \right\} \tag{20.21}$$

or

$$\underline{I} \approx V \cdot \left\{ \frac{R_s}{X_d X_q} - \frac{1}{2} j \left[\frac{1}{X_q} + \frac{1}{X_d} \right] + \frac{1}{2} j (\frac{1}{X_q} - \frac{1}{X_d}) \cdot e^{-j2\beta} \right\} \tag{20.22}$$

and

$$T \cdot \Omega_{sy} = \frac{3}{2} V^2 . \frac{1}{\left[1 + \frac{R_s^2}{X_d X_q}\right]^2} \cdot \left[\frac{1}{X_q} - \frac{1}{X_d} \right] \cdot \left\{ \left[1 - \frac{R_s^2}{X_d X_q} \right] \sin 2\beta + R_s \left[\frac{1}{X_q} + \frac{1}{X_d} \right] \cos 2\beta - R_s \left[\frac{1}{X_q} - \frac{1}{X_d} \right] \right\} \tag{20.23}$$

or, if we define $\tan \varrho_d = R_s / X_d$ and $\tan \varrho_q = R_s / X_q$,

$$T.\Omega sy = \frac{3}{2} V^2 \cdot \left[\frac{1}{X_q} - \frac{1}{X_d} \right] \cdot \frac{\cos \rho_d \cos \rho_q}{\cos^2(\varrho_q - \rho_d)} . \left\{ \sin(2\beta + \rho_d + \rho_q) - \sin(\varrho_q - \rho_d) \right\} \tag{20.24}$$

or, as $\cos \rho_d \approx \cos \rho_q \approx 1$,

$$T.\Omega sy \approx \frac{3}{2} V^2 \cdot \left[\frac{1}{X_q} - \frac{1}{X_d} \right] \cdot \left\{ \sin(2\beta + \rho_d + \rho_q) - \sin(\varrho_q - \rho_d) \right\} \tag{20.25}$$

In the complex plane, the current vector describes a circle as shown in Fig. 20.12 ($Z = R_s^2 + X_d X_q \approx X_d X_q$). The most important effect of the stator resistance is a horizontal shift $VR_s / X_d X_q$ of the circle, i.e. an additional active current component. The reactive current remains almost unchanged; it is at a minimum for $\beta = 0$ (i.e. V/X_d) and at a maximum for $\beta = \pi/2$ (i.e. V/X_q). When the current vector is orthogonal to the radius of the circle, we obtain the optimal power factor, i.e. for $\cos \varphi = cos 2\beta \approx (1 - X_q/X_d)/(1 + X_q/X_d)$ (disregarding the terms in R_s).

On the torque, the effect of the stator resistance is much more pronounced, as is clear from the vertical and horizontal displacement of the torque curve in Fig. 20.13. Just as for the induction machine, the effect of the stator resistance is a reduction of the torque for motoring and an increase for generating. There is also a horizontal shift, resulting in a non-zero torque for $\beta = 0$. Note that the maximum torque is proportional to $(1/X_q - 1/X_d)$.

20.2.3 Design and Construction

The reactances X_d and X_q are the most important design parameters. Naturally, these reactances increase in absolute value with the dimensions (or power rating)

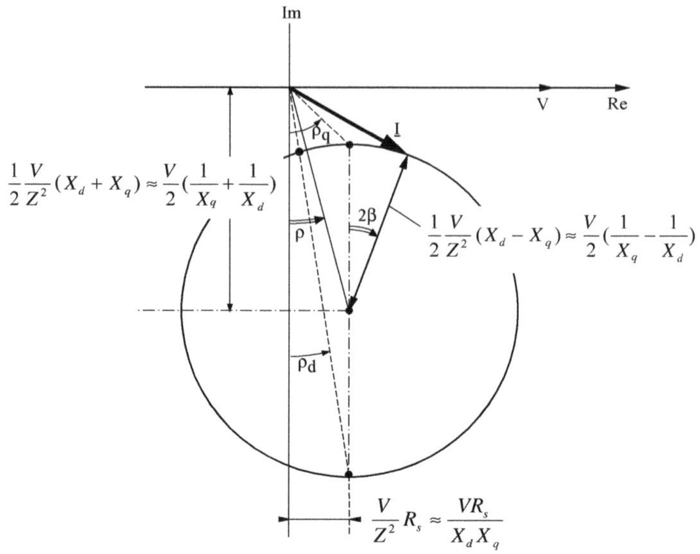

Fig. 20.12 Current locus

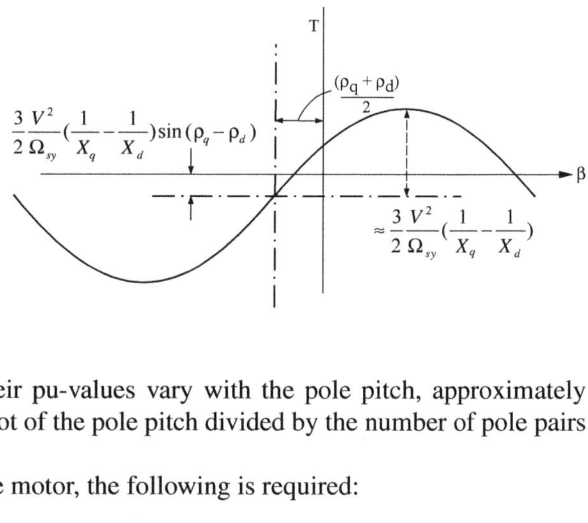

Fig. 20.13 RM torque characteristic

of the machine; but also their pu-values vary with the pole pitch, approximately proportional to the square root of the pole pitch divided by the number of pole pairs $(\sqrt{\tau_p/N_p})$.[6]

For an efficient reluctance motor, the following is required:

- a sufficiently high value for x_d as this determines the (minimum) reactive current, i.e. at no-load
- a high value for x_d/x_q, in order to improve the $\cos\varphi$
- a large difference between x_d and x_q, i.e. a high value for $(1/x_q - 1/x_d)$, to obtain a large (maximum) torque.

[6]How do the absolute values of the reactances vary with the pole pitch? Remember the variation of the rated impedance Z_n with the pole pitch for induction machines.

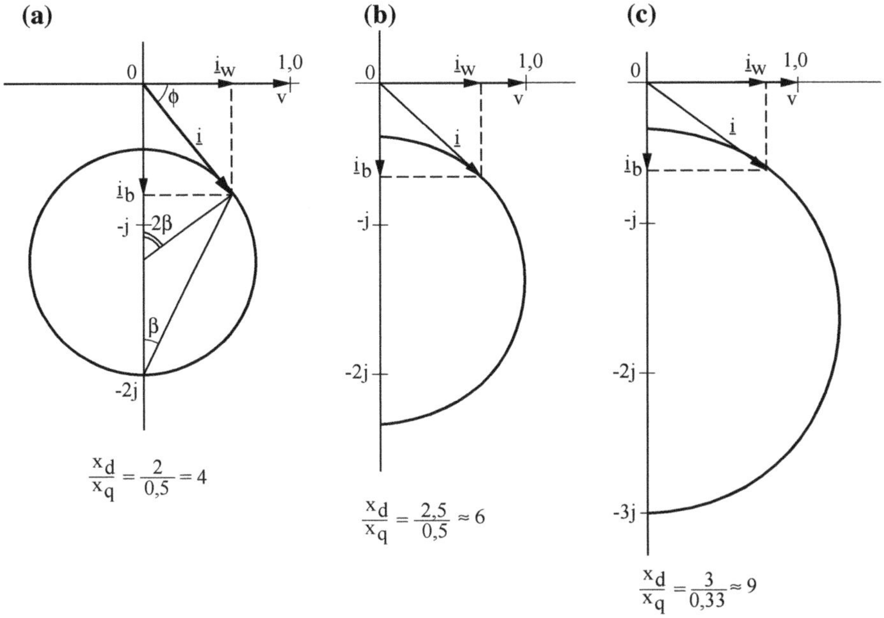

Fig. 20.14 Circle diagrams for different x_d and x_q

Figure 20.14 illustrates the effect of x_d and x_q. Clearly, the larger the ratio x_d/x_q, the higher the pull-out torque and the better the $\cos\varphi$.

Some designs are depicted in Fig. 20.15. The most straightforward (and oldest) design is the crenellated rotor depicted in (A). Rotor sheets of an induction machine can be used as a basis, after which the gaps in the q-axes are cut to obtain saliency. This has as an additional advantage that a starting cage is present if the induction motor rotor sheets already contain the slots for the cage (and if these slots are filled, see (A-a)). Moreover, the gap can be filled with electrically conducting material so that the cage is again (more or less) *complete*. However, the ratio of x_d/x_q, that can be obtained in this way is limited to approximately 2. Indeed, if x_q is reduced by an increase in the width or depth of the gap, x_d will also decrease. In most cases, a pole width of around $0.5\tau_p$ is selected.

The barrier rotor designs in (B) and (C) allow x_d/x_q to increase to more than 2 without compromising the two other requirements. Figure (B-a) shows a two-pole version and (B-b) a four-pole version. In figures (C) (sometimes called a segmented rotor), only one barrier is used; a cage can also be provided as illustrated in (C-b). A clear disadvantage of this construction principle (which was presented a long time ago[7]) is that it has a rather complicated assembly.

[7]J.K. Kostko, Polyphase reaction synchronous motors, JAIEE, 1923, pp. 1162–1168.

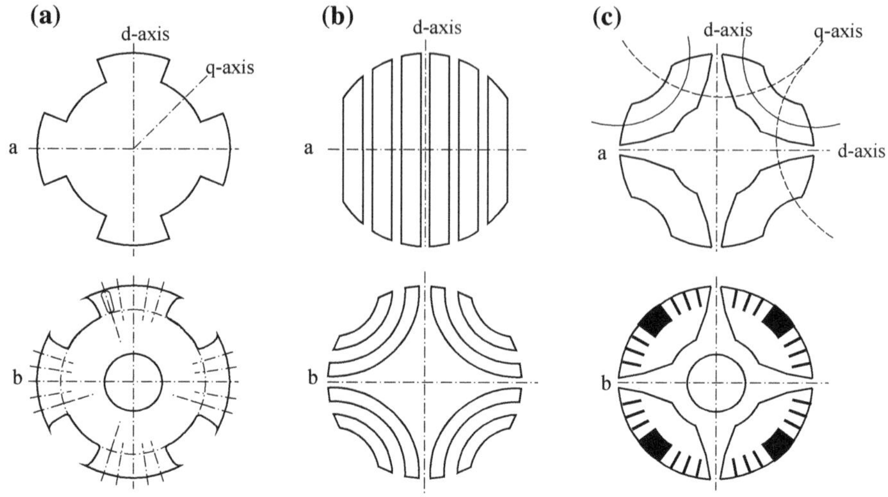

Fig. 20.15 RM rotor designs

The larger the number of barriers, the greater the ratio x_d/x_q that can be obtained. For one barrier, the theoretical maximum for x_d/x_q is approximately 5, as can be derived as follows.

Consider a sinusoidal armature current distribution resulting in a q-axis mmf (or magnetic potential) $M_{aq}(x)$ in Fig. 20.16; (a) shows the *flat* representation of one half of the q-axis, so half of the complete q-axis circuit depicted in (c); (b) and (d) show the mmf distributions.

The segments which are magnetically isolated with respect to each other (and to the stator) take on a (non-zero) magnetic potential M_x that can be derived from the equivalence of the flux crossing the air gap and the barrier:

$$\mu_o \frac{\tau_p l}{\delta}\left[\frac{2}{\pi}\hat{M}_{aq} - M_x\right] = 2M_x \cdot (\Lambda_{bar}/2) \cdot l \tag{20.26}$$

with $(2/\pi)\hat{M}_{aq}$ the average stator mmf over one pole pitch, l the armature length and Λ_{bar} the permeance of *half a barrier* per meter of armature length (in fact, the permeance for *one* pole pitch per meter). Therefore, we obtain for the magnetic potential M_x

$$M_x = \frac{2}{\pi}\hat{M}_{aq} \cdot \left[1 + \frac{\delta}{\mu_o \tau_p} \cdot \Lambda_{bar}\right]^{-1} \tag{20.27}$$

The *fundamental* value of the mmf difference between $M_{aq}(x)$ and M_x determines the *fundamental* of the q-axis (air-gap) mmf and thus the fundamental q-axis flux (see also (d) in Fig. 20.16),

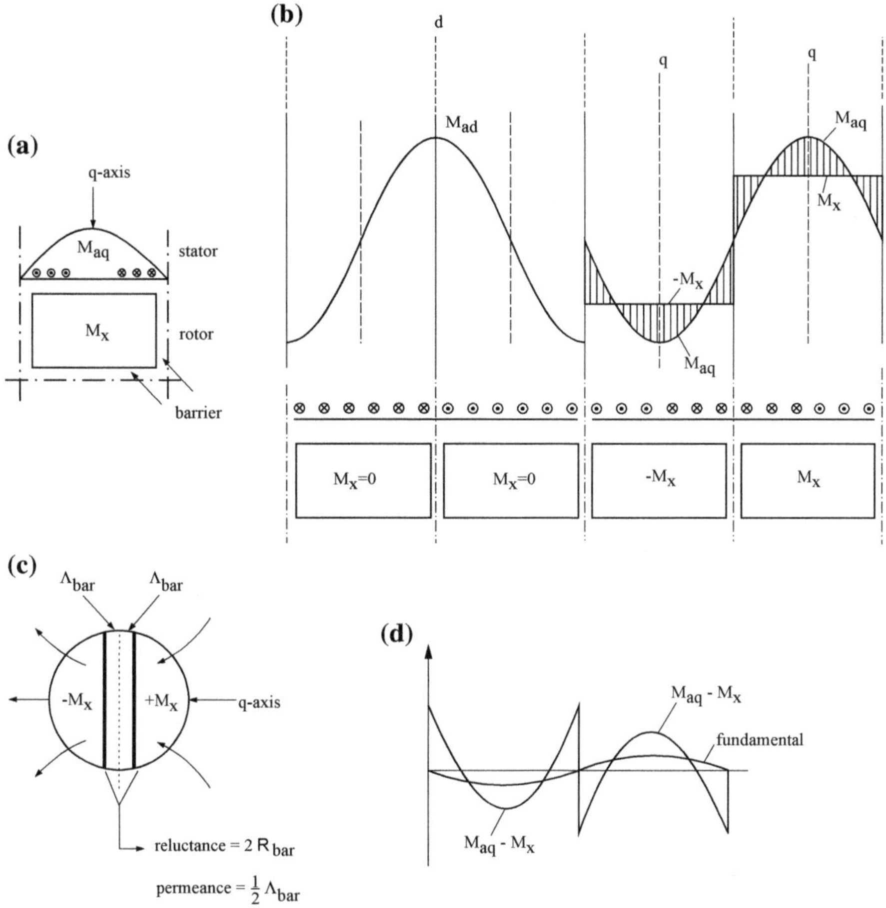

Fig. 20.16 Mmfs for a barrier rotor

$$\hat{F}_{q1} = \hat{M}_{aq} - \frac{4}{\pi} M_x = \hat{M}_{aq} \cdot \left\{ 1 - \frac{8}{\pi^2} \left[1 + \frac{\delta}{\mu_o \tau_p} \cdot \Lambda_{bar} \right]^{-1} \right\} \tag{20.28}$$

It is clear that the lower Λ_{bar}, the lower the value of $\hat{F}_{q1}$; a theoretical minimum value would occur for $\Lambda_{bar} = 0$

$$\hat{F}_{q1,min} = \hat{M}_{aq} \left\{ 1 - \frac{8}{\pi^2} \right\} \tag{20.29}$$

For a (sinusoidal) d-axis mmf $M_{ad}(x)$, on the other hand, the segments remain magnetically neutral, i.e. $M_x = 0$, and thus

$$\hat{F}_{d1} = \hat{M}_{ad} \tag{20.30}$$

The reluctance ratio x_d/x_q is equal to the ratio of the fundamental air-gap mmfs $\hat{F}_{d1}$ and $\hat{F}_{dq1}$ for identical armature mmf values $\hat{M}_{ad} = \hat{M}_{aq}$:

$$x_d/x_q = \left\{1 - \frac{8}{\pi^2}\left[1 + \frac{\delta}{\mu_o \tau_p} \cdot \Lambda_{bar}\right]^{-1}\right\}^{-1} \tag{20.31}$$

with as theoretical maximum

$$\left(x_d/x_q\right)_{max} = \left\{1 - \frac{8}{\pi^2}\right\}^{-1} \approx 5.28$$

This value corresponds to the maximum value of the magnetic potential M_x for a given q-axis armature mmf (i.e. the average of a sine).

For a higher number of barriers, much higher values for the reluctance ratio are possible. Indeed, a higher number of barriers results in a rotor-mmf consisting of several steps that fit better with the sinusoidal M_{aq}. The difference between M_{aq} and the fundamental of the stepwise rotor mmf M_x then becomes smaller.

20.3 Hysteresis Motors

20.3.1 Construction

The energy conversion principle of hysteresis motors is related to both induction and synchronous machines. In the rotor, the hysteresis motor uses “hysteretic” material, i.e. a semi-hard magnetic material with a sufficiently wide hysteresis loop. In this hysteretic material, two types of magnetisation occur: alternating magnetisation and rotational magnetisation. In the former case, the amplitude and sign of the induction vary but the direction remains constant, while in the latter case the direction of the induction varies but the amplitude remains constant.

There are basically two configurations, as is shown in Fig. 20.17. To favour alternating magnetisation in both of them, the thickness of the hysteretic ring is kept small, i.e. about $0.2\ \tau_p$. Between the ring and the shaft, either an a-magnetic material is used so that the flux is mainly tangential (a), or a ferromagnetic material is chosen so that the flux is practically radial (b).

Fig. 20.17 Hysteresis motor basic configurations

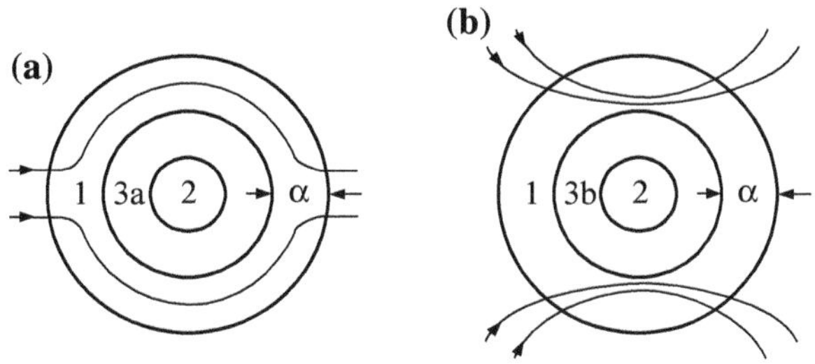

1 = hysteretic material
2 = shaft
3a = amagnetic material
3b = ferromagnetic material

20.3.2 Principle

In the analysis of the working principle of induction machines, it is usually assumed that only the rotor winding (or cage) is responsible for the conversion of the primary rotating field power, i.e. mechanical power and some slip losses. In reality, some of the primary power is also converted to the rotor through the rotor iron, i.e. by eddy currents in the iron and by hysteresis losses.

At a given slip s, we have

$$T \cdot \Omega_{sy} \equiv P_{em1} = (1-s)P_{em1} + sP_{em1} = P_{mech} + sP_{em1} \tag{20.32}$$

and also

$$sP_{em1} = P_{j2} + P_{m2} = P_{j2} + P_{f2} + P_{h2} \tag{20.33}$$

with P_{em1} the primary rotating field power ($P_{em1} = P_1 - P_{j1} - P_{m1}$), P_{j2} the joule losses in the rotor windings, P_{f2} the Foucault losses in the rotor iron and P_{h2} the hysteresis losses in the rotor iron. Of course, P_{j2}, P_{f2} *and* P_{h2} are dependent on the slip and the air-gap flux.

For P_{j2}, we know from Chap. 4 that

$$P_{j2} = 3\frac{E_1^2}{(R_2'/s)^2 + (X_{2\sigma}')^2}R_2' \approx 3s^2 \cdot \frac{E_1^2}{R_2'} \tag{20.34}$$

The approximation to the right is valid for sufficiently small slip (much smaller than the pull-out slip). This means that for small slip E_1 is about constant (i.e. not quite reduced by the rotor currents) so that the corresponding torque is approximately proportional to the slip:

$$T_j = \frac{3}{\Omega_{sy}}(E_1^2/R_2') \cdot s \approx \frac{3}{\Omega_{sy}}(V_1^2/R_2') \cdot s \tag{20.35}$$

Eddy current (Foucault) loss is proportional to the square of the induction and the square of the frequency. In the rotor iron, the frequency is the slip frequency $s \cdot f$. As a result,

$$P_{f2} = s^2 \cdot P_f(E_1 , f) \tag{20.36}$$

with $P_f(E_1 , f)$ the eddy current loss in the rotor iron for the supply frequency f and for the same air-gap induction (or the same stator emf E_1). The corresponding torque component is then also proportional to the slip (if the slip is not too large so that $E_1 \approx V_1$, i.e. the air-gap induction is not affected by the rotor currents):

$$T_f = \frac{1}{\Omega_{sy}} P_f(E_1 , f) \cdot s \approx \frac{1}{\Omega_{sy}} P_f(V_1 , f) \cdot s \tag{20.37}$$

It is not surprising that this Foucault torque is similar to the normal induction machine torque: both torque components result from currents (in rotor iron or cage winding, respectively) and the corresponding mechanical power is compensated by the primary power, together with the corresponding rotor joule loss.

Hysteresis loss, on the other hand, is proportional to the frequency and the area of the hysteresis loop. The area of the hysteresis loop is a function of the induction amplitude and is usually assumed to be proportional to $\hat{B}^{\beta}$ (with $\beta \approx 1.6 \ldots 2$). For the rotor hysteresis loss, we may thus write

$$P_{h2} = |s| \cdot P_h(E_1 , f) \tag{20.38}$$

with $P_h(E_1 , f)$ the rotor hysteresis loss for s = 1 and the same air-gap induction (or stator emf).[8]

The hysteresis torque corresponding to this loss component can be written as

$$T_h = \frac{1}{\Omega_{sy}} P_h(E_1 , f) \cdot sign(s) \tag{20.39}$$

For not too large a slip ($|s| < s_k$ so that $E_1 \approx V_1$), this hysteresis torque is constant, independent of the slip magnitude. The hysteresis torque, however, changes sign with the sign of the slip: motoring for $s > 0$ and generating for $s < 0$. Obviously, the hysteresis loss and the corresponding mechanical power are also compensated by the electrical mains power, just like the other torque and rotor loss components. Figure 20.18 shows the hysteresis torque and the Foucault torque as a function of the slip. For induction machines, these torque components are usually negligible in comparison with the normal induction motor torque. Nevertheless, in an ideal no-load experiment with sufficiently accurate equipment, the hysteresis jump around s = 0 can be measured.

For a hysteresis motor, i.e. without a rotor cage, the Foucault torque and in particular the hysteresis torque are the only torque components. As there is no rotor

[8] Why is the loss proportional to |s| and not simply s?

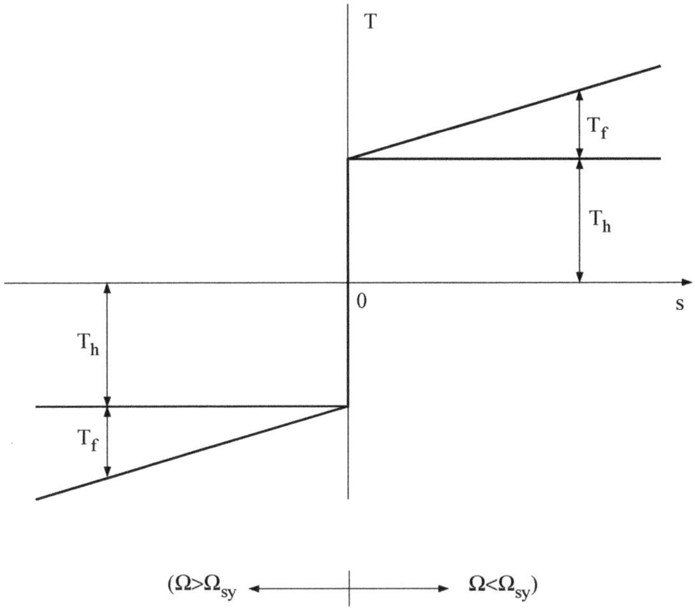

Fig. 20.18 Hysteresis torque characteristic

winding or cage to counteract the field, the derived behaviour of these torque components remains valid for a broad range of the slip, i.e. $0 < |s| < 1$.

An equivalent circuit for the hysteresis motor is shown in Fig. 20.19. The equivalent resistance $R_h = sign(s) \cdot |R_h|$ corresponds with the hysteresis torque. The magnetising current for the rotor iron can be represented by a reactance X_h in parallel with the normal magnetising reactance X_m - or by a lower value of X_m. The possible eddy current component can be represented by the resistance R_f/s.

From this equivalent circuit, the following power equations can be derived:

$$P_{em1} = 3 \cdot P_h + 3 \cdot \frac{E_1^2}{R_f/s} = \Omega_{sy}.(T_h + T_f) \tag{20.40}$$

$$P_{h2} = 3 \cdot s \cdot \frac{E_1^2}{R_h} = s \cdot \Omega_{sy} \cdot T_h \tag{20.41}$$

$$P_{f2} = 3 \cdot s^2 . \frac{E_1^2}{R_f} = s \cdot \Omega_{sy} \cdot T_f \tag{20.42}$$

The physical background for the Foucault torque is evident: it is the torque due to the air-gap field and the eddy currents in the rotor iron.

The preceding discussion, however, does not provide an adequate physical explanation for the hysteresis torque, nor does it explain what happens for synchronism. Physically, hysteresis loss is due to microscopic currents in the grain boundaries.

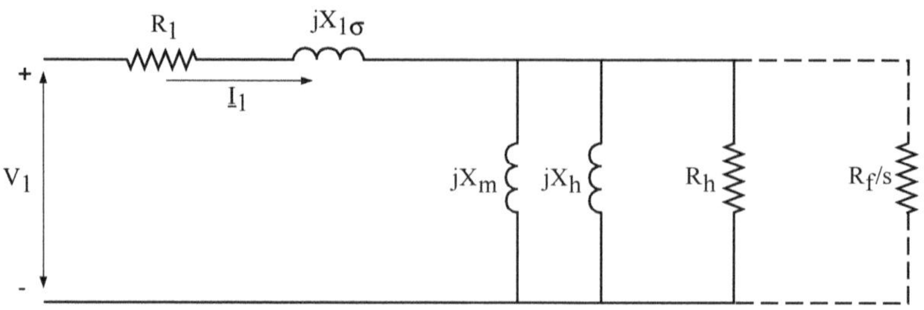

Fig. 20.19 Equivalent circuit

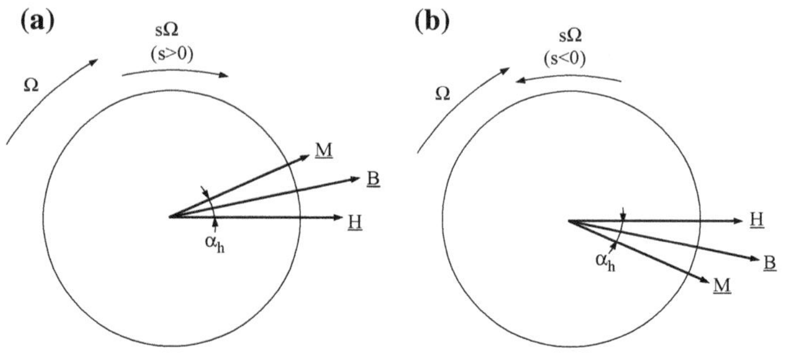

Fig. 20.20 Hysteresis torque action

Thus, the corresponding torque can be understood by the torque exerted by the imposed field on these microscopic currents.

The hysteresis torque, its sign and its behaviour at synchronism can also be explained as follows. When the rotor speed differs from the synchronous speed Ω of the rotating induction field, the induction vector $\underrightarrow{B}$ imposed by the stator slips with respect to the rotor iron with a speed $s\Omega$ (see Fig. 20.20). Seen from the rotor, the iron in the rotor describes a hysteresis loop.

In the time domain, the time phasor $\underline{H}$ *leads* the time phasor $\underline{B}$, as can easily be derived graphically (see e.g. Chap. 1). I n the space domain, the corresponding magnetic field vector $\underrightarrow{H}$ then leads $\underrightarrow{B}$ (with respect to the rotor iron). As $\underrightarrow{B} = \mu_o(\underrightarrow{H} + \underrightarrow{M})$, it follows that $\underrightarrow{M}$ is lagging the field vector $\underrightarrow{H}$ (in space). As the torque at a magnetic moment is given by $\underrightarrow{T} = \underrightarrow{M} \times \mu_o \underrightarrow{H}$, it follows that the torque is clockwise (driving) for $s > 0$ and counter-clockwise (generating) for $s < 0$ (see Fig. 20.20). The angle α_h between $\underrightarrow{M}$ and $\underrightarrow{H}$ is constant and depends only on the width of the hysteresis loop (Fig. 20.21), which means that the hysteresis torque is constant for a given induction as long as there is a speed difference between field and iron.

When the slip decreases to zero (synchronism), the magnetic situation that existed just before synchronism becomes instantaneously frozen with respect to the iron. Then the (macroscopic) magnetic moment $\underrightarrow{M}$ becomes fixed to the iron: the rotor iron becomes permanently magnetised. If the instantaneous angle between the rotor

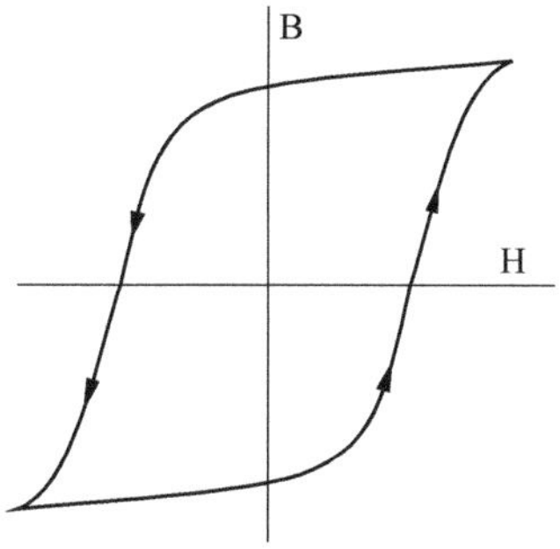

Fig. 20.21 Hysteresis loop

and the imposed induction $\underset{\rightarrow}{B}$ varies slightly, the angle between $\underset{\rightarrow}{B}$ and $\underset{\rightarrow}{M}$ will change from α_h to an angle α (with $-\alpha_h < \alpha < \alpha_h$) and the torque becomes $T = B \cdot M \cdot \sin\alpha$, just as the torque of a synchronous machine.

20.3.3 Properties

The almost speed-independent torque and the resulting high torque at standstill and starting makes the hysteresis motor quite apt for loads with high inertia. However, the hysteresis motor is rather expensive compared to an induction motor, due to the expensive hysteretic material and the rather low power density (i.e. power or torque per unit machine volume).

Indeed, for the hysteresis losses per cycle and per unit volume, we may write

$$W_h = \eta \cdot 4 \cdot \hat{B}\hat{H} \approx \hat{B}.\hat{H} \approx 10^4\,\mathrm{J/m^3} \tag{20.43}$$

For a supply frequency f and slip s, we obtain for the tangential force per square meter of air-gap area

$$s.v_{sy}.F_h^{\nabla} = s.f.d.W_h \tag{20.44}$$

with the synchronous speed $v_{sy} = 2f\tau_p$. If we approximate the thickness d of the hysteresis ring by $d \approx 0.2\tau_p$, we get for the force per square meter

$$F_h^{\nabla} \approx 10^3\,\mathrm{N/m^2} \tag{20.45}$$

This is, however, much less than the corresponding value for induction machines, i.e. $F_h^{\nabla} \approx 3.5\,10^3 \ldots 7\,10^4\,\mathrm{N/m^2}$ (see the Chap. xxx on induction machines).

20.3.4 *Final Remarks*

1. Because of the strongly non-linear B-H characteristic, a wye connection of the stator windings is preferred to avoid third harmonics in the line currents.
2. To obtain a higher starting torque, an increased voltage is sometimes applied for starting, as it increases the area of the hysteresis loop.
3. The stator slotting may result in a somewhat reduced torque capability and increased losses due to secondary hysteresis loops and eddy currents by mmf harmonics.

20.4 Small Motors for Special Applications

There are many other variants of small synchronous motors, used for example in watches or as auxiliary drives for control or measurement apparatus. Some of these examples will be discussed below.

20.4.1 *Impulse-Field Motor (Not Self Starting)*

The impulse-field motor (see Fig. 20.22) is a synchronous motor which is not self starting and is (was) frequently used in clockworks.

The rotor has an even number (30–40) of teeth and is enclosed by the two pole shoes of the stator. The width of each pole shoe should be equal to a broken number of tooth pitches; the pole shoes may have teeth as well (with the same tooth pitch) but this is not essential (although it does increase the torque). Between the two pole shoes, there is a single-phase excitation winding.

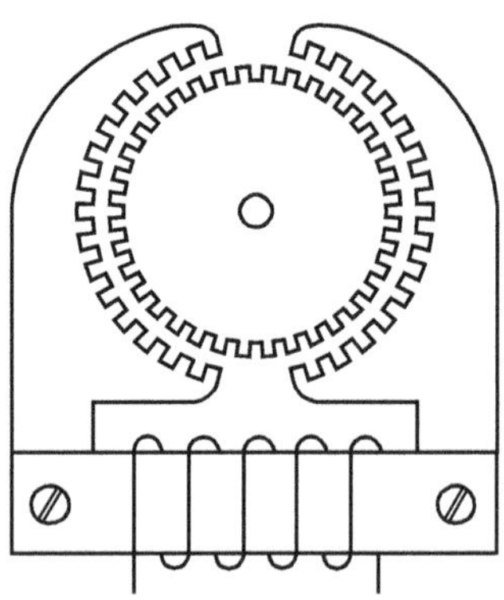

Fig. 20.22 Impulse field motor

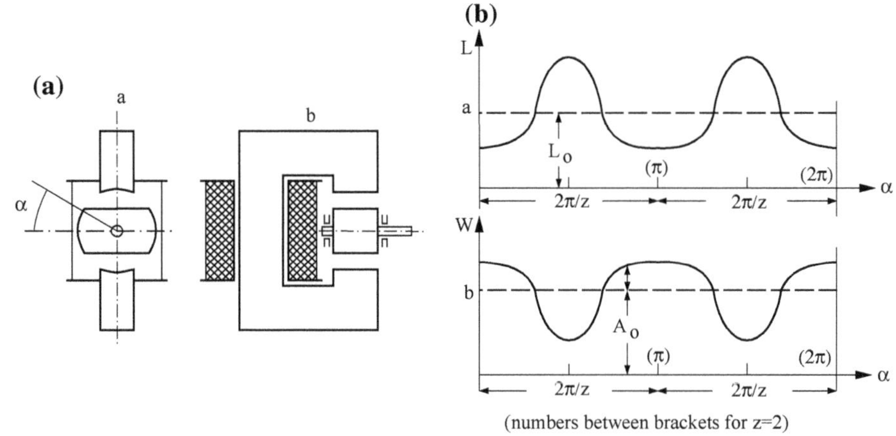

Fig. 20.23 Inductance variation

With z denoting the number of rotor teeth and α the rotor position, it should be intuitively clear that the inductance will vary with the rotor position α and with a period $2\pi/z$, i.e. $L = L(z\alpha)$ (for example, see Fig. 20.23 for z = 2).

Expanding $L(z\alpha)$ in its Fourier components

$$L(z\alpha) = L_o + \sum_{\nu=1}^{\infty} L_\nu \cdot \sin(\nu.z\alpha)$$

then it is clear (e.g. from virtual work) that a non-zero torque can only exist if $\nu.z.\alpha = 2\omega t$, i.e.

$$\Omega_\nu = \frac{d\alpha}{dt} = \frac{2\omega}{\nu z}$$

The largest possible speed corresponds to the fundamental period of the inductance, i.e. $\nu = 1$:

$$\Omega = \frac{2\omega}{z}$$

For example, if $z = 30$ and $f = 50\,\text{Hz}$ then for the speed we get $N = 200\,\text{rpm}$.

Depending on the harmonics in $L(z\alpha)$, lower speeds are possible in theory. In practice, we need to make sure that the fundamental harmonic is dominant, which means that only the highest speed is stable. Note that this motor is not self starting and that either direction of rotation is possible. One will, however, limit the rotational direction to one direction, by mechanical means.

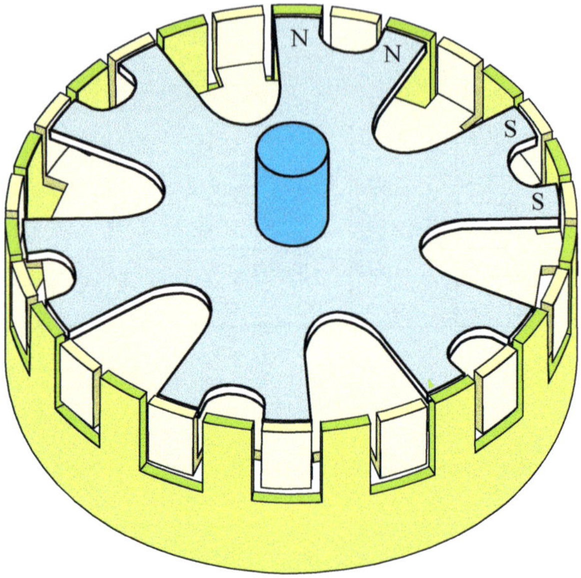

Fig. 20.24 Self-starting impulse-field motor

20.4.2 *Self-starting Impulse-Field Motor*

Figure 20.24 depicts a self-starting impulse-field motor. The stator contains 30 poles of alternating polarity excited by a single-phase ring winding. The rotor has six arms with permanent magnets; on each arm there are alternating two north poles or two south poles. Between two arms, two stator poles are skipped while the two poles of each arm skip one stator pole. When the stator is excited, the polarity of the stator poles changes with the mains frequency. If the rotor is originally at standstill, it will start swinging. The oscillations will increase until they exceed a pole pitch and then the rotor will synchronise with either the clockwise or counter-clockwise rotating field. The rotation will be mechanically limited to one direction.

20.4.3 *Other Single-Phase Synchronous Motors*

In addition to the impulse-field motor, other single-phase synchronous motors include single-phase hysteresis motors, single-phase reluctance motors or single-phase permanent magnet motors. Using partial screening of the stator poles, as in the shaded-pole motor, we obtain an (elliptic) rotating field and the motor becomes self starting (see Fig. 20.25).

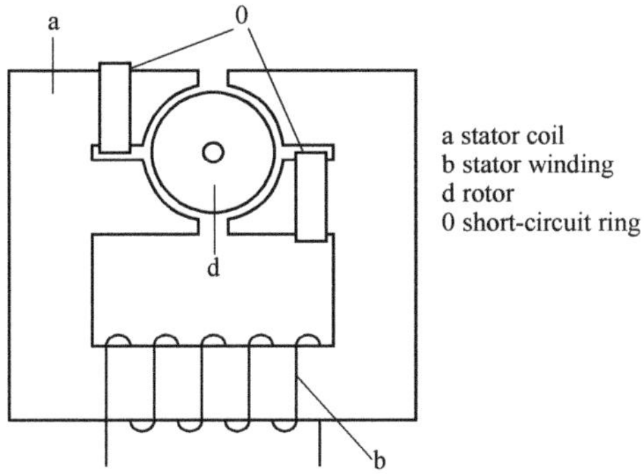

Fig. 20.25 Single-phase hysteresis motor

20.5 Electrostatic Motors

All motors treated so far use electromagnetic fields for energy conversion. Indeed, for dimensions larger than around 1 mm, electromagnetic fields are much more advantageous than electrostatic fields in terms of power (or energy or force or torque) pu volume. However, for very small dimensions, i.e. less than about 1 mm, electromagnetic fields become rather inefficient in terms of force per unit volume. In addition, the largest part of the electrical energy is dissipated as heat, so cooling problems arise for these small dimensions.

That electrostatic fields can be more efficient than electromagnetic fields for smaller dimensions. This can be explained qualitatively as follows.

For electromagnetic fields, we consider the force on a current-carrying conductor. Taking into account the practical limits for current density ($J \leq 4\text{A/mm}^2$) and induction ($B \leq 1.5\,\text{T}$) as well as the expression for the synchronous speed $v = 2\tau_p f$, we obtain for the electromagnetic tangential force and the electromagnetic power

$$F_{em} = I \cdot l \cdot B = J \cdot \Sigma \cdot l \cdot B \sim J \cdot B \cdot D^3 \sim D^3$$

$$P_{em} \sim F_{em} \cdot v \sim D^4$$

For electrostatic fields, we may write for the electrostatic tangential force (considering $E < 3.10^6\,\text{V/m}$)

$$F_{es} = q \cdot E = C \cdot V \cdot E = \varepsilon(\Sigma/l) \cdot V \cdot E \sim \epsilon \cdot \Sigma \cdot E^2 \sim D^2$$

$$P_{es} \sim F_{es} \cdot v \sim D^3$$

Therefore

$$P_{em}/P_{es} \sim D^4/D^3 \sim D$$

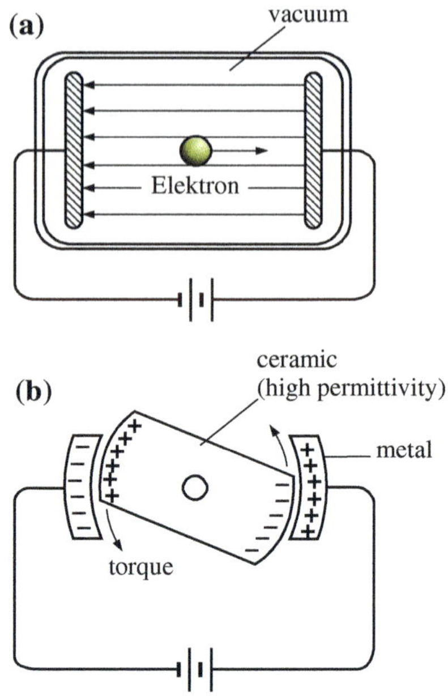

Fig. 20.26 Principle of electrostatic motor

For larger dimensions, electromagnetic energy conversion is more efficient in terms of power per unit of volume. For small dimensions, however, electrostatic energy conversion becomes advantageous. As the proportionality factor in the power above is approximately 10^4*, electrostatic power conversion becomes advantageous for* $D < 10^{-4}\,m$.

20.5.1 Electrostatic Stepping Motor

The energy conversion in an electrostatic stepping motor is the force on a charge in an electric field: $F = q \cdot E$. Figure 20.26 shows a simplified configuration.

The rotor is built from ceramic material with high permittivity and can pivot between two metal electrodes. A potential difference between the two electrodes results in charges induced in the ceramic rotor. The resulting electric field in the air gaps will attempt to align the rotor.

Figure 20.27 illustrates two real implementations on chip level[9] of electrostatic stepping motors. A typical rotor diameter is approximately 60 μm with a thickness of around 20 μm.

[9]The dimensions shown are micrometers.

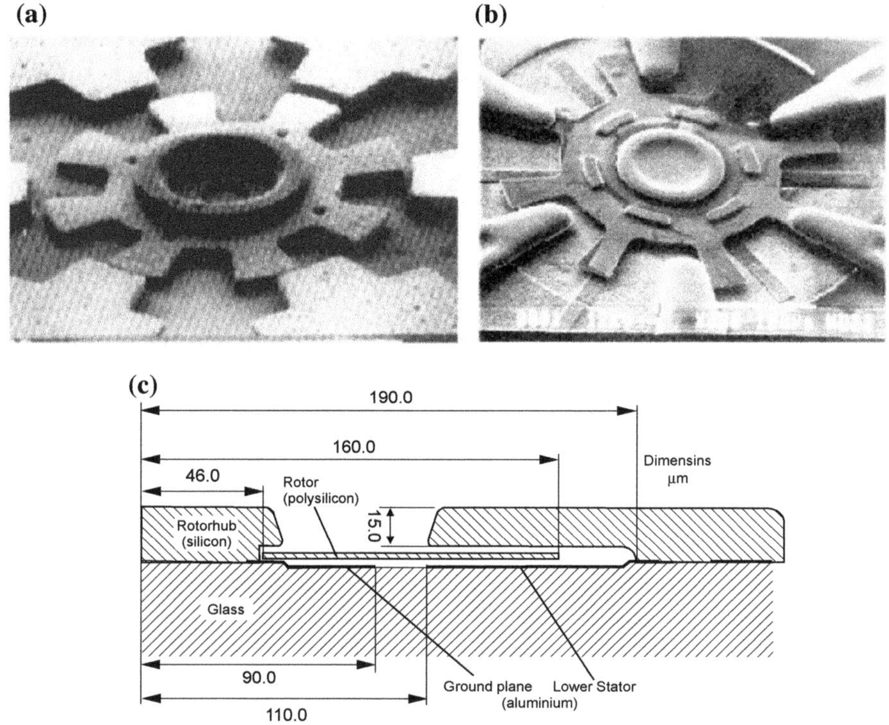

Fig. 20.27 Electrostatic stepping motor

20.5.2 *Piezo-Electric Actuators*

The operation principle of piezo-electric actuators or motors is based on the inverse piezo-electric effect. When a force is exerted on a crystalline material, an electric potential difference between the end surfaces is generated. This is the (direct) piezo-electric effect. On the other hand, when an electric potential difference is applied between the faces of a piezo-electric material, the crystals will attempt to align with the electric field and an elongation of the material will follow (see (b) in Fig. 20.28). In this way, relatively large forces can be obtained, albeit with rather small displacements.

Piezo-electric actuators are used in ink-jet printers.

20.5.3 *Ultrasonic Actuators and Motors*

Ultrasonic actuators and motors also utilise the (inverse) piezo-electric effect, but with excitation at ultrasonic frequencies. Rotational motion can be obtained by

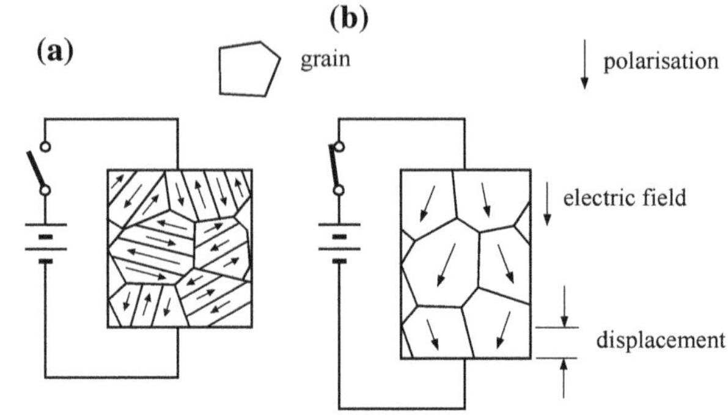

Fig. 20.28 Piezo-electric effect

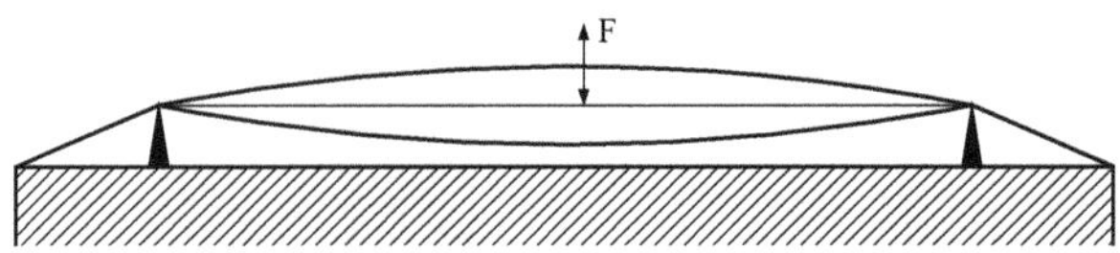

Fig. 20.29 Standing wave

making use of travelling waves in elastic material and the resulting elliptical trajectory of the points on the surface.

Travelling fields in Elastic Material

Consider an elastic beam or string (see Fig. 20.29).

When a sinusoidal force is exerted in a point of the beam or string, the beam or string will exhibit a sinusoidal steady-state vibration with the excitation frequency. As both ends of the string (or beam) are restricted, this motion is a standing wave. We also know that a standing wave can be considered as the superposition of two travelling waves, with the same amplitude but in opposite directions. What we need for rotational motion is (1) to suppress the reflections at both ends, and (2) to impose two excitations with the same frequency displaced in space and time.

This can be obtained by arranging the beam in the form of a ring and by exciting the ring in two places with both time and space shifts of 90°.

Elliptical Motion

When a travelling wave appears in an elastic material, each point on the surface performs an elliptical motion (see Fig. 20.30). If a stiff, inflexible body is pushed against this surface, the body will be dragged along by friction and will move with the elliptic motion on the surface.

Piezo-electric Actuators and Motors

For the mechanical excitation, piezo-electric actuators are fed by ultrasonic voltage sources. Ultrasonic frequencies are used because of two reasons: (1) frequencies above 20 kHz are imperceptible to the ear and do not cause any irritation;, and (2) as the piezo-electric displacement is rather small (e.g. 2 μm), a sufficiently high

Fig. 20.30 Elliptical motion

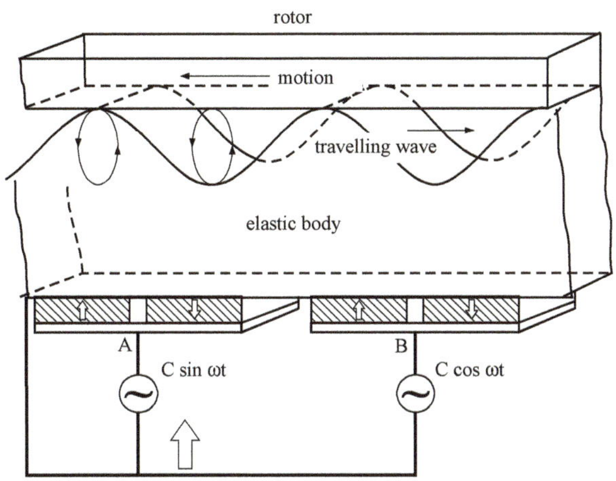

Fig. 20.31 Typical layout of an USM

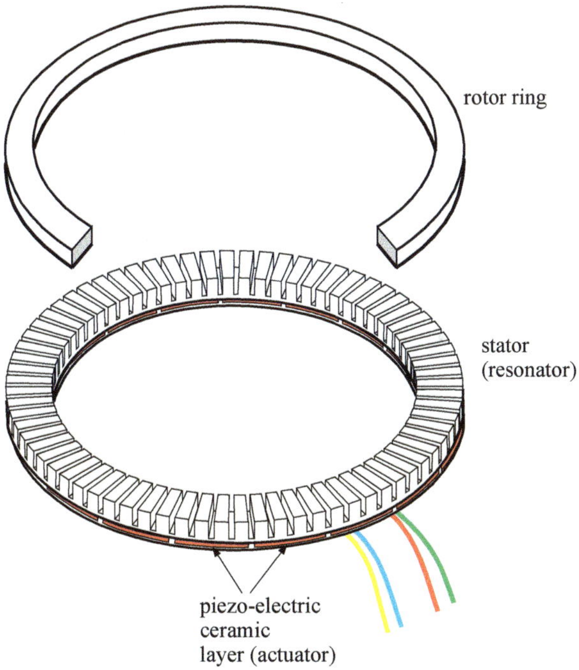

frequency is necessary to obtain practical speeds (2 μ, with a frequency of 20 kHz gives a speed of 4 cm/s).

Figure 20.31 shows the assembly of a piezo-electric ultrasonic motor (USM). It consists of

- a piezo-electric ceramic ring as actuator,
- an elastic resonator fixed on the actuator, and
- a rotor ring pressed against the resonator.

Chapter 21
Stepping Motors

Abstract In contrast with the continuous motion of usual induction or synchronous machine drives, stepping motors produce a controlled, stepwise motion without any need for position measurement and feedback. The main characteristic quantities and properties of stepping motors are described. Thereafter, the different types of stepping motors are discussed.

21.1 Introduction: Stepping Motion Versus Continuous Motion

Depending on the operation cycle, we may distinguish for electrical drives (valid for both motors and generators):

- continuous operation ('S1' according to IEC 34-1)
- short operation ('S2' according to IEC 34-1)
- intermittent operation ('S3' according to IEC 34-1)

Another possible classification makes a distinction between

- switched operation,
- controlled drives, and
- servo drives.

It is logical that servo drives and controlled drives are only relevant for motoring.

Controlled drives can be defined as drives where speed and/or torque are (continuously) controlled or adjustable. Servo drives are also controlled drives, but differ somewhat from other controlled drives in that they have a practically uninterrupted variation of position, speed, acceleration. A typical example is a drive for positioning a mechanical tool. Mostly (and historically), servo operation only concerns small drives.

Stepping motors are a special kind of servo motor. In contrast to other servo drives, they give a controlled, stepwise motion without any need for position measurement and feedback. Accurate positioning is thus realised without a costly and vulnerable position sensor and a complicated feedback of the rotor position. The electromagnetic

J. A. Melkebeek, *Electrical Machines and Drives*, Power Systems,
https://doi.org/10.1007/978-3-319-72730-1_21

construction and the excitation result in a recognisable step angle. For that reason, the stepping motor is also an ideal partner of switching power electronic circuits.

There are three kinds of stepping motors: permanent magnet, variable reluctance and hybrid stepping motors; these will all be discussed in the next paragraphs.

21.2 Characteristic Quantities and Properties

Two types of characteristics can be distinguished: static and dynamic characteristics. The explanation of the dynamic characteristics is given in Sect. 21.2.3.

21.2.1 Static Characteristics

An important static characteristic is the static T/θ (or F/x) characteristic which gives the static torque (or force) versus the rotation angle (or displacement) for a given excitation current (see Fig. 21.1).

To measure this characteristic, the equilibrium position for the given current and no-load first needs to be established. From this no-load equilibrium position on, a rotation over an angle θ away from it corresponds to a load torque value required to maintain the new position. This relation between the angle and the torque is called the static T/θ (or F/x for linear motion) characteristic, and the maximum torque is referred to as the holding torque. The difference angle between the no-load point and the point of maximum torque is not necessarily exactly one step angle but it should be equal to or larger than the step angle (see Fig. 21.1).

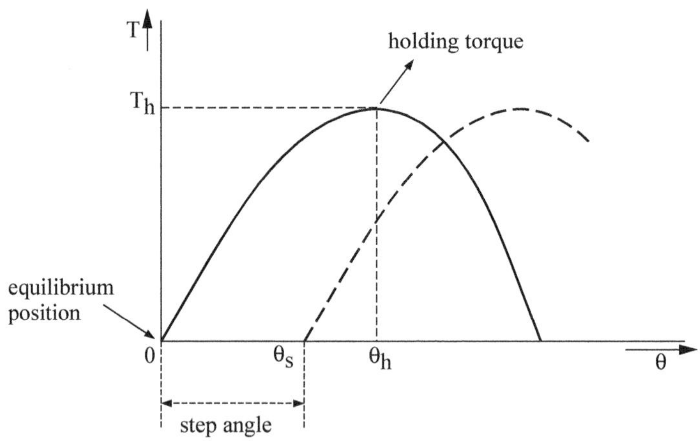

Fig. 21.1 Static torque - angle characteristic

There are also stepping motors that exhibit a non-zero torque for zero excitation current. In that case, the maximum torque for the characteristic under zero current is called the detent torque. The no-load point(s) for zero excitation current and the angle for the detent torque do not necessarily correspond to those for non-zero excitation current.

Another static characteristic gives the magnitude of the holding torque as a function of the current. This characteristic can be (approximately) linear for permanent magnet stepping motors, or quadratic for reluctance stepping motors.

21.2.2 Dynamic Characteristics

The dynamic characteristics relate to stepping motors in motion or starting. Two essential dynamic characteristics are

1. the *pull-in torque characteristic* or *starting characteristic* gives the range for the load (or friction) torque where the stepping motor will start and run *without losing any steps,* as a function of the frequency of the excitation pulse train (see Fig. 21.2).
2. The *pull-out torque characteristic* or *slewing characteristic* gives the range for the load (or friction) torque where the already running motor continues running *without losing any steps,* as a function of the frequency of the excitation pulse train (see Fig. 21.2).

Both characteristics depend heavily on the inertia of the load and the characteristics of the electrical supply circuit; both should therefore be specified. The *maximum starting frequency* is the maximum frequency of the pulse train at which the unloaded motor can start without losing steps. Similarly, the *maximum slewing frequency* is

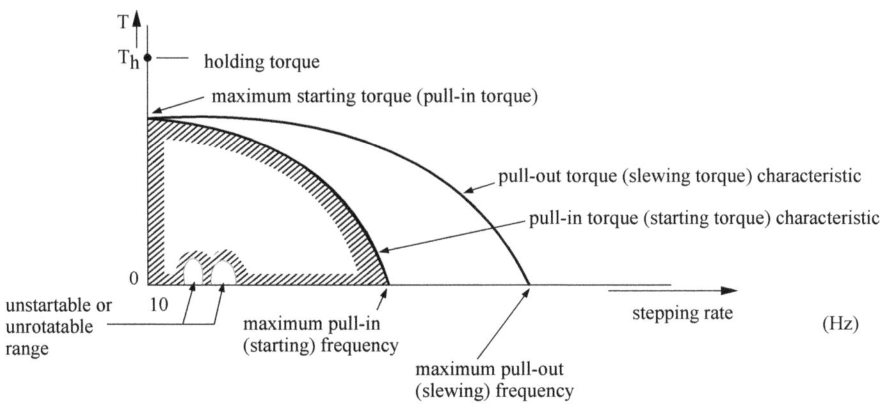

Fig. 21.2 Dynamic characteristics

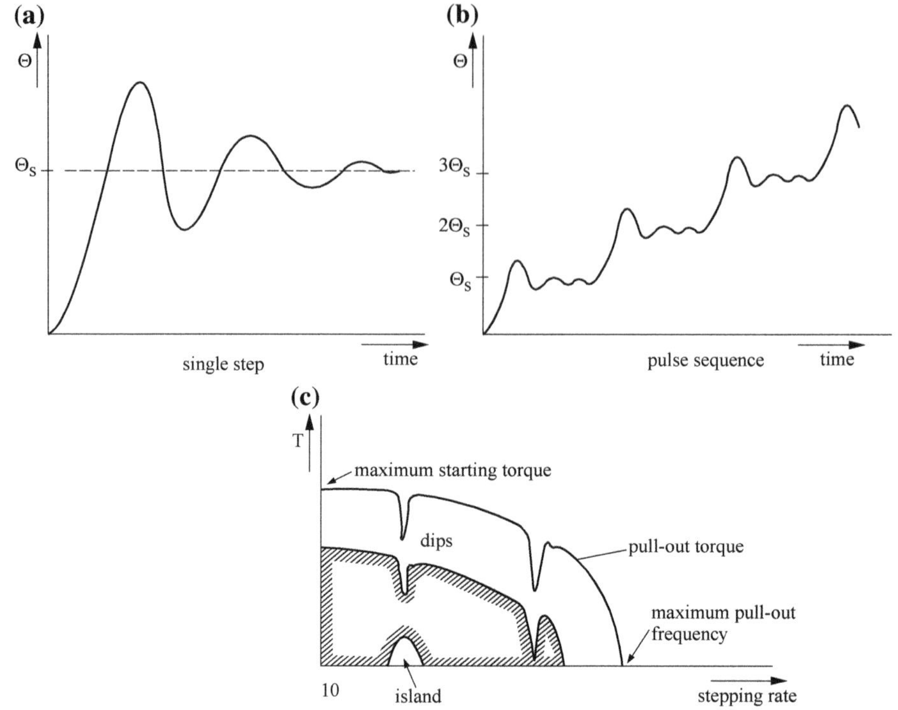

Fig. 21.3 Resonance

the maximum frequency of the pulse train at which the unloaded motor will continue running without losing any steps.

It is, however, also possible that certain areas (islands or dips) exist within the startable or runnable range, in which the motor ceases to run faultlessly. These are caused by resonances (see below).

The *maximum starting torque* or *maximum pull-in torque* is the maximum load (or friction) torque for which the motor can start and synchronise with a pulse train of sufficiently low frequency (e.g. 10 Hz).

21.2.3 Eigen Frequency, Damping, Resonance

The system consisting of converter, motor and load (and their inertias) is in fact a dynamical system of higher order (≥ 2). For a (sufficiently slow) stepwise excitation the rotor will tend to oscillate, with its dominant eigen frequency and damping, around the new equilibrium position at each step, similar to a spring-mass system (see Fig. 21.3).

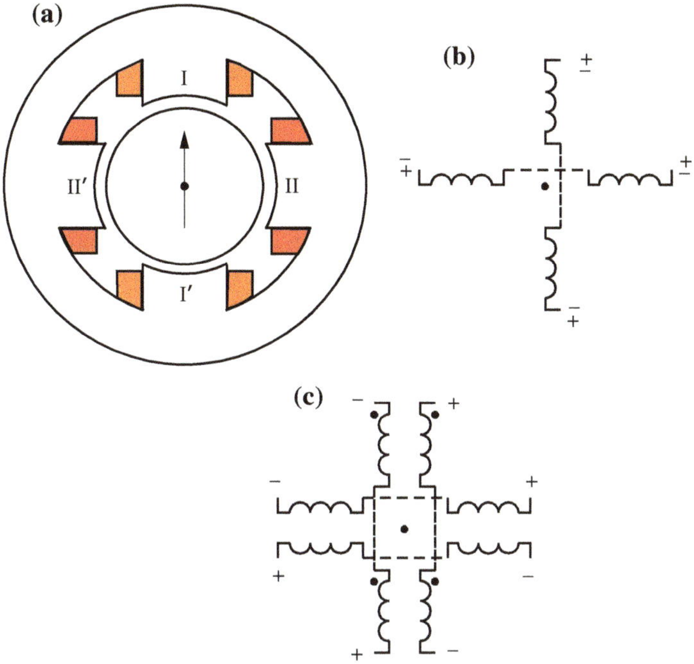

Fig. 21.4 Two-phase PM stepping motor

When, in a continuous motion, the frequency of the pulse train corresponds to (i.e. is equal to or near to) an eigen frequency of the system, there is a risk of steps being by-passed (or surpassed), in which case the rotor will not follow the pulse train accurately. This is the explanation of the islands (mostly near no-load) or dips (loaded). This risk of resonance can be suppressed by appropriate damping, either internally or externally.

21.3 The Permanent Magnet Stepping Motor

Figure 21.4 shows the cross-section of a two-phase permanent-magnet stepping motor. The stator has four poles; the rotor contains a two-pole permanent magnet.

With permanent-magnet excitation, the polarity of the stator poles must be reversible. Two designs are possible. In design (b) in Fig. 21.4, each pole pair contains one winding (the coils of poles I and I' are connected in series, just like those of II-II') but the supply should provide both polarities. When the excitation is such that pole I is a south pole (and thus I' a north pole), the rotor will align as in (a) in Fig. 21.5 (this will be referred to as a positive excitation). Switching off pole pair I-I' and a positive excitation of the pole pair II-II' will result in a rotation of 90°

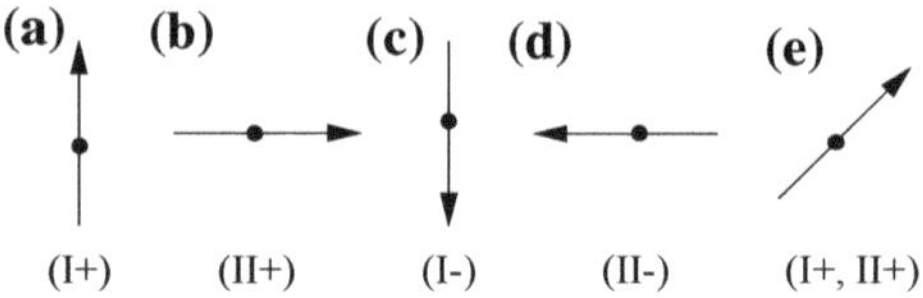

Fig. 21.5 Steps for the two-phase PM stepping motor

to the right and the alignment of the rotor as in (b). Then the negative excitation of successively I-I' and II-II' will result in further rotation to the right with steps of 90° as in (c) and (d). An intermediate position as in (e) is possible as well, if both pole pairs are excited at the same time, but the holding torque will be different than when only one pole pair is excited. A particular property of permanent magnet stepping motors is that, also without excitation, the rotor will align with one of the pole pairs. Naturally, the corresponding detent torque differs from the holding torque.

The two polarities for the supply, which requires a more complex power electronic converter, can be avoided by means of a bifilar winding. In that case, two windings are provided on each pole pair, one for the *positive* and one for the *negative* polarity. This mono-polar supply requires more copper for the motor but may result in a less complicated converter. The main disadvantage might be the inefficient use of the available space for the windings (i.e. for a given machine size, the power will be more limited).

A smaller step angle can be obtained with more phases (and pole pairs) in the stator. For example, a three-phase stator (with six poles) has a step angle of 60° and a half-step value of 30°. Doubling the number of stator phases from two to four (or from four to eight stator poles) and doubling the number of rotor poles to four results in a step angle of 45°. However, as stepping motors generally have small power ratings and sizes, the number of poles with their windings in the stator is usually a limiting factor. This constructional restriction also applies to some extent to the permanent magnets in the rotor.

21.4 The Variable-Reluctance Stepping Motor

The variable reluctance motor in Fig. 21.6 has six stator poles (with excitation windings) and four salient rotor poles (without windings or permanent magnets). The coils of opposite poles are connected in series to form three pairs of north-south poles (1-1', 2-2', 3-3').

Excitation of the pole pair 1-1' causes the rotor to rotate, resulting in a minimum reluctance position corresponding to a minimum of the magnetic energy. Subsequent excitation of 2-2' and 3-3' leads to a clockwise rotation with a step angle of 30° (see Figs. 21.7 and 21.5).

Note that the polarity of the poles (or excitation) is of no importance, which means that the supply can be mono-polar. In contrast to permanent magnet stepping motors, there is no detent torque.

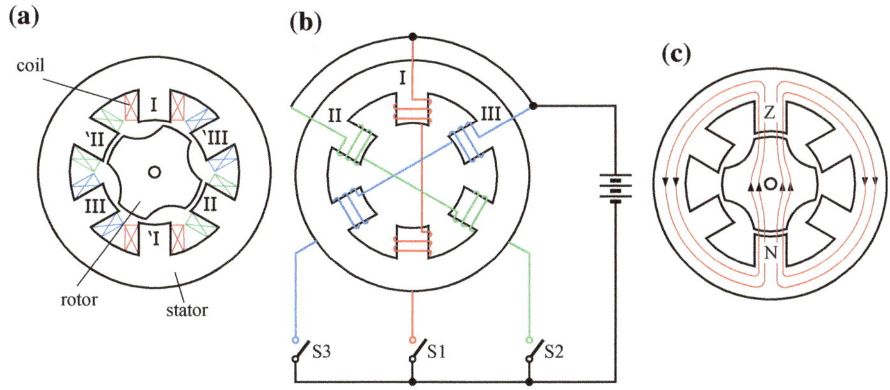

Fig. 21.6 Variable-reluctance stepping motor

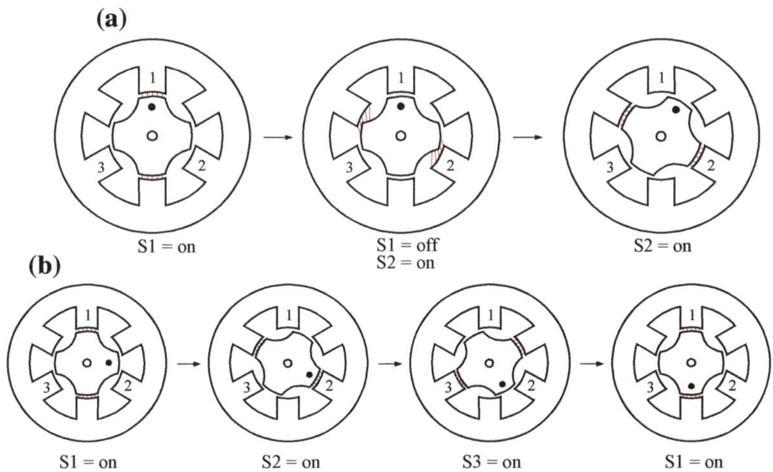

Fig. 21.7 Rotation of a 6-4 VR stepping motor

A smaller step angle can be obtained by increasing the number of stator poles and rotor poles, like in the three-phase version (a) in Fig. 21.8 with 12 stator poles and eight rotor poles. Another example is depicted in (b) in Fig. 21.8, i.e. a four-phase motor with eight stator poles and six rotor poles. Both designs offer a step angle of 15°. As is the case for the PM stepping motor, a small motor size (or power rating) may limit the number of wound stator poles.

However, the step angle can also be reduced by increasing the number of rotor poles, leaving the number of stator poles unchanged (see Fig. 21.9). In these designs, the number of stator teeth has been doubled as well in order to increase the torque and to improve the step resolution, but this does not have any effect on the step angle as long as the number of stator poles remains the same.

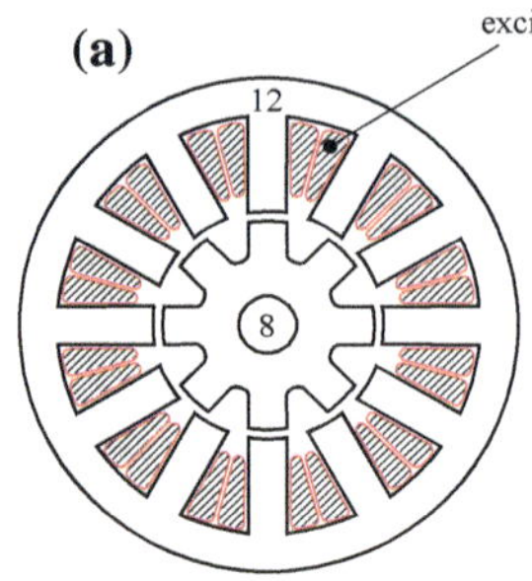

Fig. 21.8 12/8 and 8/6 stepping motors

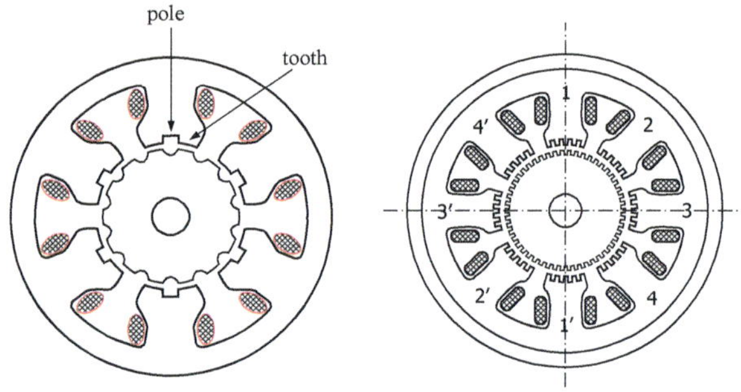

Fig. 21.9 Toothed rotor and stator

The step angle and the number of steps per revolution only depend on the rotor pole pitch and the number of stator phases, not on the number of stator teeth. Indeed, the operation principle of the motor is based on the position change of the rotor teeth with respect to the subsequently excited stator poles. Mathematically, the step angle ϵ_r can be derived as follows. Call N_r and N_s the number of rotor *teeth* and stator *teeth*, respectively. By q we represent the number of stator teeth per stator pole (which means that the number of stator poles is N_s/q). As a consequence, the rotor tooth pitch is $\lambda_r = 2\pi/N_r$ while the *average* stator tooth pitch is given by $\lambda_s = 2\pi/N_s$. The step angle ϵ_r, i.e. the displacement of the rotor at the excitation of the next stator phase, can be written as

$$\epsilon_r = \min_n[|q.\lambda_s - n.\lambda_r|] = \min_n\left[|q.\frac{2\pi}{N_s} - n.\frac{2\pi}{N_r}|\right] \tag{21.1}$$

This step angle will be a fraction of the rotor tooth pitch (and smaller than $\lambda_r/2$). Equation 21.1 can usually be simplified to

$$\epsilon_r = 2\pi/mN_r \tag{21.2}$$

Indeed, N_s/q is an even multiple of the number of phases: $N_s/q = 2k \cdot m$. A displacement of one rotor tooth pitch thus requires λ_r/ϵ_r steps and the required number of steps per revolution of the rotor is $S = N_r.\lambda_r/\epsilon_r$.

21.5 Multi-stack Stepping Motors

All stepping motors in the preceding paragraphs are single-stack types. In these, the magnetic situation is the same over the whole armature length and all phases are in a single plane.

Another type is the multi-stack or cascade type. This type of motor consists of several layers lengthways. The ferromagnetic circuit in each layer is separate from the other layers and corresponds to one phase of the stator. Here, the stator and rotor have the same tooth pitch, but there is a shift of $(1/m)$th of a tooth pitch between either the stator layers ***or*** the rotor layers. In general, the selected number of teeth is relatively high (for example, six, eight or ten).

There are two type of multi-stack motors. Figure 21.10a depicts a three-phase homopolar (variable reluctance) type in which each (phase) layer also contains two planes or layers, one for the *north* poles and one for the *south* poles. The flux passes from the stator north poles to the rotor, axially through the rotor core to the stator south poles, and then closes axially to the other side. The advantage of this configuration is that for the excitation a simple ring winding can be used (Fig. 21.11), without the need to put a winding around each pole as in a heteropolar type.

In the heteropolar type (b) in Fig. 21.10, the (stator) poles are alternating north and south poles in the same plane. This necessitates a coil around each stator pole, i.e. a more complicated winding structure with space restrictions as well.

For both types, all poles of stator and rotor of the excited phase will align. For a three-phase motor, each layer is displaced by $(1/3)$rd of a pole pitch and the excitation of the next phase will cause the rotor to rotate further by $(1/3)$rd of the pole pitch. A small step angle thus requires a large number of poles (and a small pole pitch), for example 15° for a three phase motor with eight teeth.

There are also multi-stack stepping motors with permanent magnets in the rotor. Figure 21.12 depicts a four-phase claw pole PM stepping motor. There are two layers (not four: an even number of phases always is rather special), but the stator excitation must provide both polarities as it is a PM motor. Here, the rotor PM layers are not displaced, but the stator layers are displaced by half a pole pitch (see also Fig. 21.13). The bipolar supply generates four-phase behaviour. With the claw pole configuration,

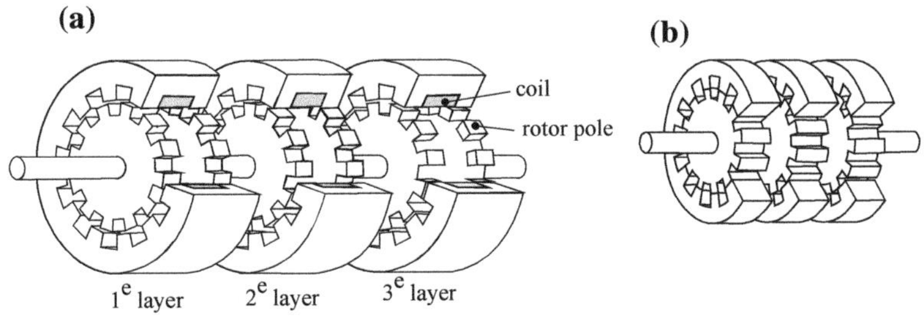

Fig. 21.10 Multi-stack VR stepping motor: **a** homopolar, **b** heteropolar

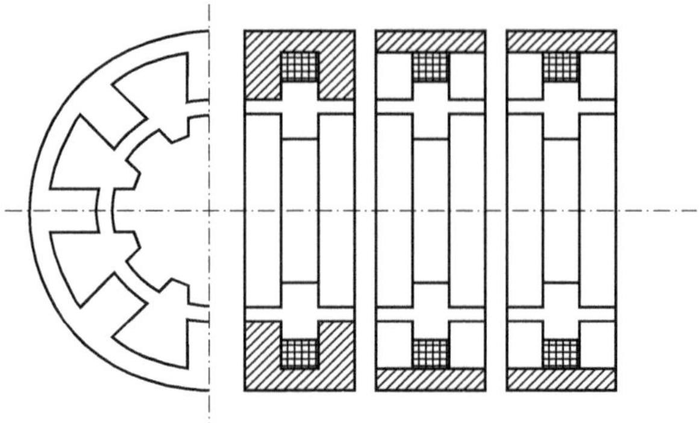

Fig. 21.11 Multi-stack: three-phase homopolar layout

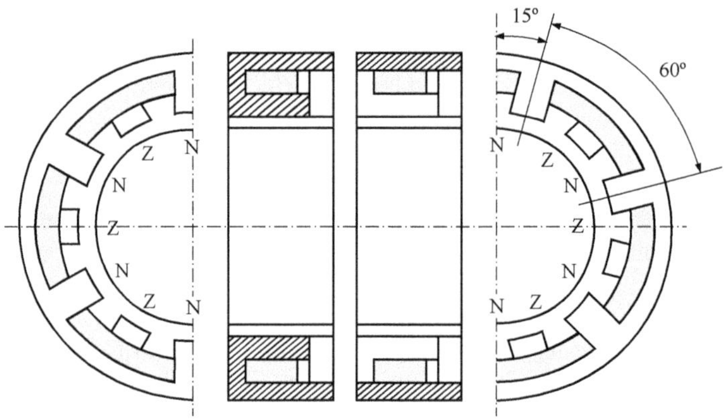

Fig. 21.12 Multi-stack: four-phase homopolar layout

a simple ring winding can be used. However, a normal winding with coils around each pole can be utilised as well and is better from an electromagnetic point of view: a claw pole results in an important 3-D magnetic field which is undesirable, for example because of iron losses.

The selected number of poles can be quite high in order to obtain a small step angle (e.g. the motor in Fig. 21.12 or 21.13 has 12 poles and thus a step angle of only 15°).

21.6 Hybrid Stepping Motors

Hybrid stepping motors combine the torque effect of reluctance and permanent magnets. Adding permanent magnets to a reluctance stepping motor aims at increasing the torque. Generally, however, the step angle will not be affected. On the other hand, a bipolar excitation or a bifilar winding will normally be required.

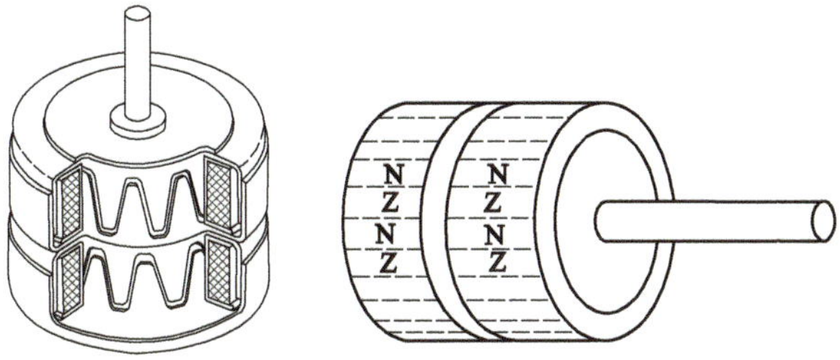

Fig. 21.13 Multi-stack four-phase claw pole stator

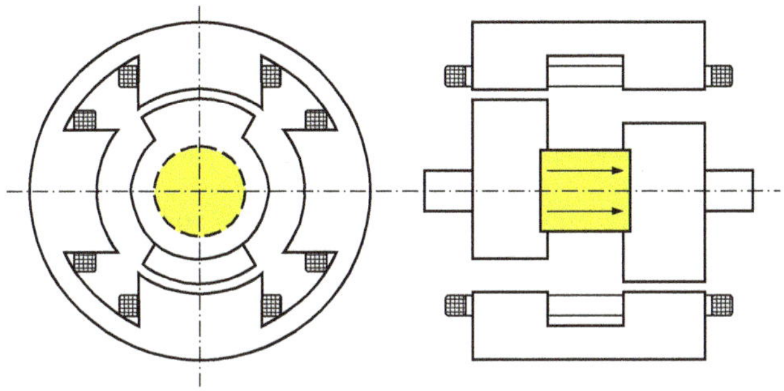

Fig. 21.14 Hybrid 2-pole stepping motor

Figures 21.14, 21.15, 21.16, 21.17, 21.18 and 21.19 illustrate the construction and operating principles of some hybrid stepping motors with permanent magnets in the rotor. Figure 21.14 is a four-pole (i.e. two-phase bipolar and thus four-phase supply) configuration. The stator is identical to the one in Fig. 21.4. The rotor contains two toothed sections (with one tooth and one slot in each section), but displaced with respect to each other by half a tooth pitch. Between both rotor sections, an axial permanent magnet is positioned. Figure 21.15 shows an eight-pole, four-phase version. The stator structure is the same as the one in (b) in Fig. 21.8, but the windings and connections between the coils are different: the two coils on each pole are assigned to distinct phases. The rotor consists of two sections with the same tooth pitch as in the stator, but displaced with respect to each other over half a tooth pitch; between these two sections, an axial permanent magnet is placed. There are also versions where the two stator sections are displaced instead of the rotor sections (see Fig. 21.17).

Field distribution and torque production are illustrated in Figs. 21.18 and 21.19. The torque results from the interaction of the two excitation types. The upper half in Fig. 21.19 corresponds to the south pole end of the permanent magnet, while the lower half corresponds to the north pole end. In the depicted situation, poles 1 and 3 are excited so as to yield a north and a south pole, respectively. The heteropolar stator field and the homopolar rotor field combine so that under the south pole of the PM (upper part of the figure) the field is intensified under the north pole (1) and

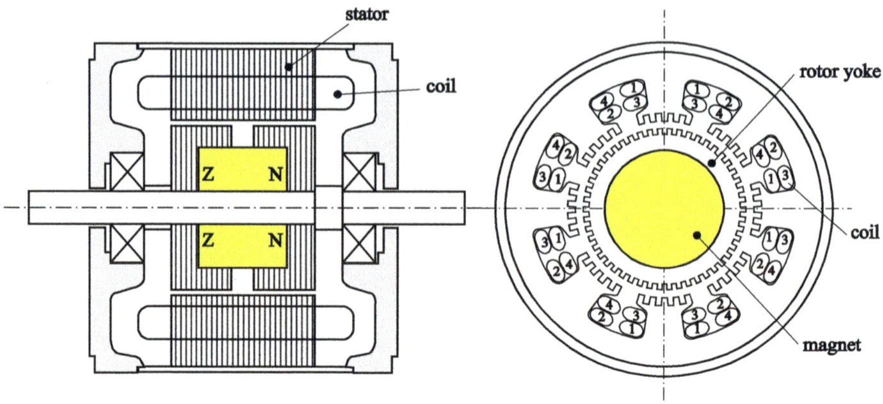

Fig. 21.15 Hybrid eight-pole stepping motor with toothed rotor and stator

Fig. 21.16 Toothed PM rotor of a hybrid stepping motor

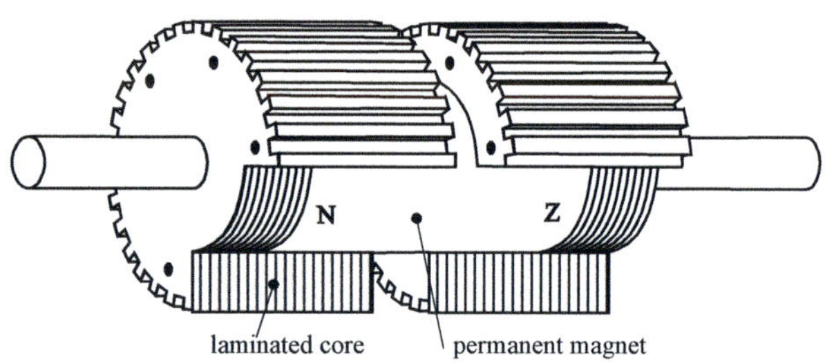

Fig. 21.17 Teeth layout of a toothed rotor

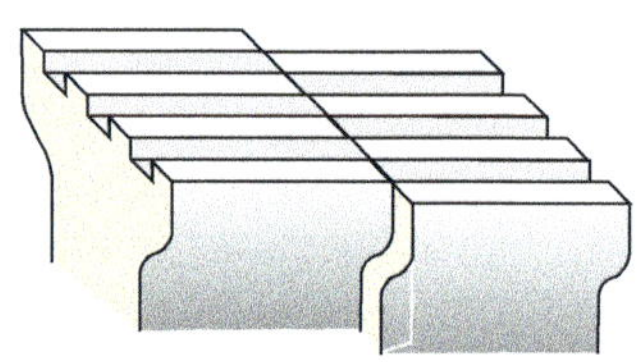

Fig. 21.18 Field lines of a hybrid stepping motor

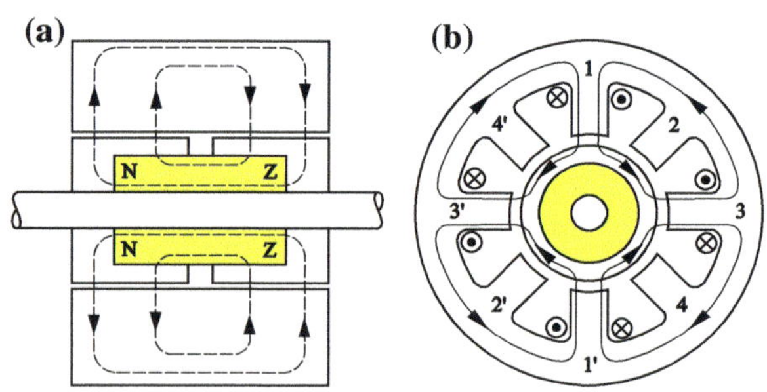

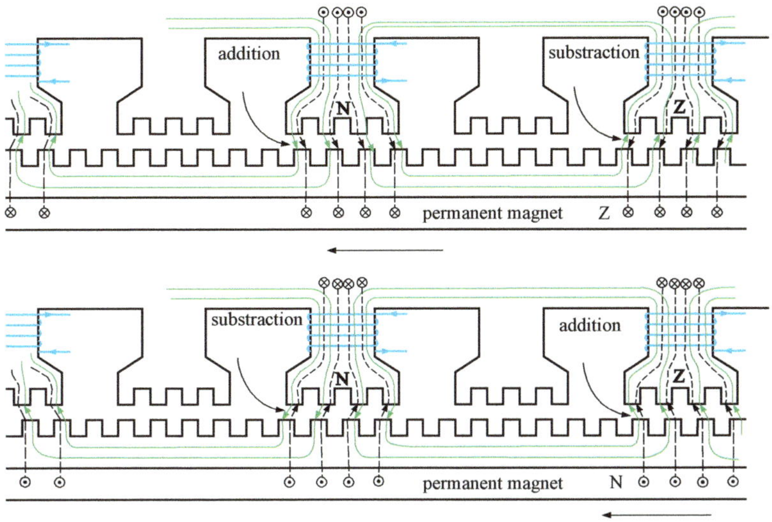

Fig. 21.19 Torque production in a hybrid stepping motor

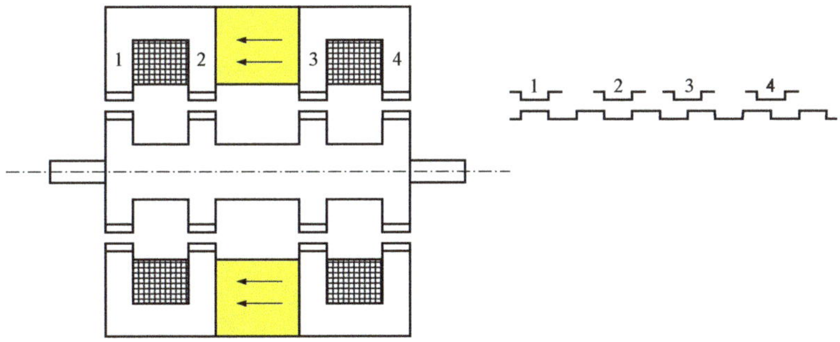

Fig. 21.20 Hybrid stepping motor with PM in the stator

weakened under the south pole (3). Under the north pole of the PM (cf. the lower part of the figure), the reverse applies. As a result, a force or torque originates that works in the same direction on both sides of the PM, i.e. to the left in the figure, as indicated by the arrows below the figures. When the teeth align, the force or torque is reduced to zero. The excitation of the next pair of stator poles will then cause further rotation.

Note that the permanent magnet is crucial for the torque or force, but that the step angle only depends on the toothed structure of stator and rotor.

Figure 21.20 shows a version with a permanent magnet in the stator. For the excitation of the stator, a ring coil is used and both sections therefore have a homopolar field. The relative toothing of stator and rotor in parts 1, 2, 3 and 4 is different, however, as is shown in the figure. Subsequent *bipolar* excitation results in a step angle of $(1/4)$th of a tooth pitch.

Chapter 22
Switched Reluctance Machines

Abstract The switched reluctance machine (SRM or also SWRM) started to receive much attention at the end of the 20th century. Its operation principle had already been known since 1838 but had not been able to find a practical use because no (fast) power electronic switches were available. Nowadays, the switched reluctance motor is applied in many industrial applications like washing machines or looms and even in more demanding applications like the starter-generator of jet turbines in airplanes. The most important advantages of the SRM are its simple and rugged construction, its inherent redundancy, and its suitability for high speeds. Its most negative aspect is the high level of noise and vibrations that it brings along.

22.1 Operation Principle

The switched reluctance motor operation principle is based on the variable reluctance effect, similar to the variable-reluctance stepping motor. Its construction is also identical to that of the variable-reluctance stepping motor (see for example Fig. 22.1 for a 6-4 type). The 6-4 SRM has six stator poles (three phases) and four rotor poles. The coils of opposite stator poles (i.e. from the same phase) are connected in series to form a coil pair for this phase. Excitation of a phase results in opposite north and south poles. The four salient poles on the rotor do not contain any windings.

Similar to the variable-reluctance stepping motor, excitation of a phase (1-1') results in a rotation of the rotor so that a minimum reluctance position is attained (see Fig. 22.1).

Subsequent excitation of the pole pairs 2-2', 3-3' results in a further clockwise rotation of the rotor, each time with a step angle of 30° (see Fig. 22.2).

As the torque is solely due to the reluctance effect, the polarity of the excitation is not important and the supply can be mono-polar. As was the case for the variable-reluctance stepping motor, the step angle can be reduced by increasing the number of stator and rotor poles, or by providing teeth on the rotor poles. Corresponding teeth on the stator poles may increase the torque, but will not change the step angle. For the SRM, such additional teeth are rather uncommon, however.

The step angle can be calculated as for the stepping motor (see paragraph 21.4).

J. A. Melkebeek, *Electrical Machines and Drives*, Power Systems,
https://doi.org/10.1007/978-3-319-72730-1_22

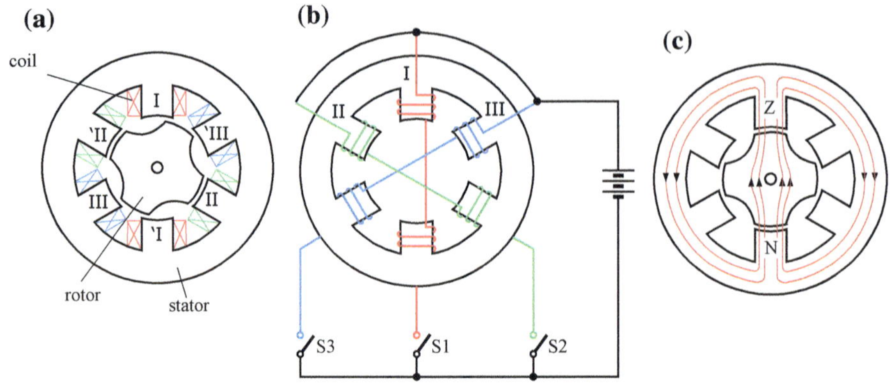

Fig. 22.1 6-4 switched reluctance motor

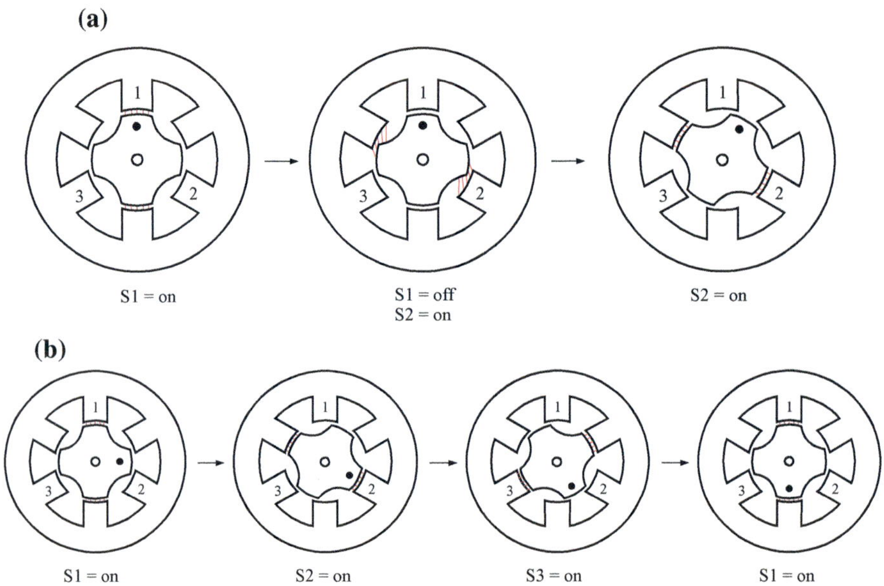

Fig. 22.2 Rotation of a 6/4 SRM

The main difference compared to the stepping motor can be found in the supply and the control. For the stepping motor, it is *assumed* that the rotor will rotate exactly one step angle at each excitation of a subsequent phase, but there is no feedback of the rotor position. Furthermore, the supply of a stepping motor provides a pulse train with voltage pulses of constant amplitude. It is assumed that the pulse train and the torque will lie within the stable region of the motor (cf. the dynamic characteristics).

In contrast, the switched reluctance motor has a rotor position measurement and feedback. Dependent on the rotor position, the appropriate phase will be supplied with an appropriate current or voltage at that instant. In most cases, a voltage supply with current control is used, i.e. it behaves as a current control. The SRM therefore has a controlled torque and can be used in a controlled drive system, such as an induction or synchronous motor drive with variable frequency supply.

22.2 Electromagnetic and Electrical Analysis

An ***aligned position*** is when a pair of rotor poles is aligned with a pair of stator poles, for example phase 1 (see (a) in Fig. 22.3). In this position, a current in phase 1 will not result in a torque as it is the minimum reluctance position for this phase. When the rotor is rotated away from this position, either to the right or to the left (see (c) and (d), respectively, in the figure), a torque will ensue that will attempt to realign the rotor poles with the stator (supposing of course that there is current in phase 1).

Fig. 22.3 Aligned and unaligned positions of a 6/4 SRM

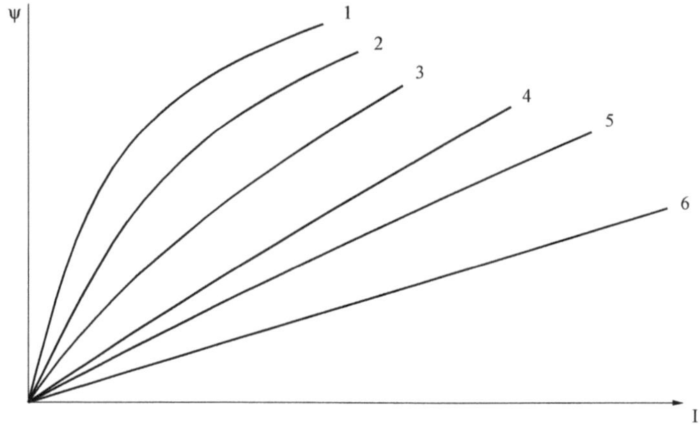

Fig. 22.4 Saturation characteristics of one phase for different rotor positions

In the ***unaligned position***, see (b) in Fig. 22.3, the reluctance of the rotor with respect to the poles 1-1' of phase 1 is at a maximum. If a current is injected in phase 1 in this rotor position, the rotor may rotate either to the left or to the right (in the *exact* unaligned position, the rotor might remain stationary but this is an unstable condition).

If the flux of stator phase 1 is drawn as a function of the current for some rotor positions, we obtain a graph as in Fig. 22.4. Curve 1 is for the completely aligned position and curve 6 for the unaligned position, while the other curves correspond to intermediate positions. In the aligned position, there is usually[1] (heavy) saturation of the magnetic circuit (poles and possibly also core). In the unaligned position, the air gap is so large that there is usually no saturation. In intermediate positions, some saturation occurs, mainly in the overlapping tips of the poles.

Instead of drawing flux versus current, we may also draw a figure showing inductance versus rotor position, with the current as parameter (see Fig. 22.5). The inductance can be the chord inductance, defined by:

$$L(\theta, i) = \psi(\theta, i)/I \tag{22.1}$$

Alternatively, the incremental or tangent-slope inductance can be used, as defined by:

$$L_t(\theta, i) = \partial\psi(\theta, i)/\partial I|_{\theta=ct} \tag{22.2}$$

The tangent-slope inductance is useful when we consider the dynamic electrical equations, especially for chopped supply (see below).

[1]As we will see later on, saturation in the aligned position improves the energy conversion of an SRM.

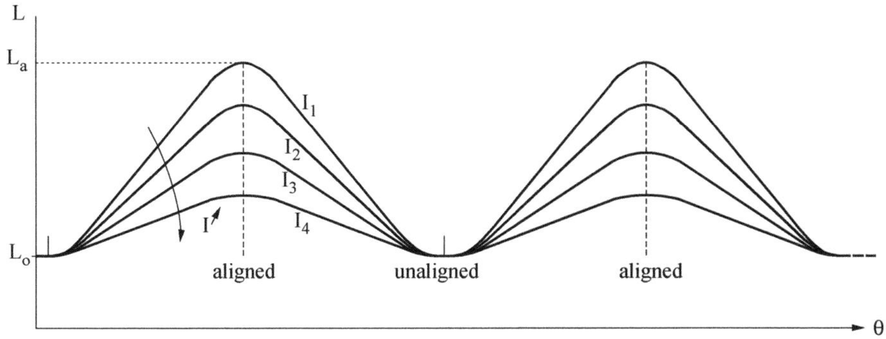

Fig. 22.5 Inductance profiles for variable rotor position and various current levels

From an energetic point of view, the energy conversion can be explained as follows. Suppose we start from the unaligned position (or around it) for phase 1. When at a given instant a voltage V is connected to phase 1, the current will increase according to

$$V = R \cdot i + \frac{d\psi}{dt} \tag{22.3}$$

or

$$V = R \cdot i + \frac{\partial\psi}{\partial i} \cdot \frac{\partial i}{\partial t} + \frac{\partial\psi}{\partial\theta} \cdot \omega \tag{22.4}$$

where $\omega = d\theta/dt$.

At the same time, there will be a tangential force and thus a torque, which means that the rotor may start rotating. Subsequently, the operating point may move from O to B along curve OAB (see (a) in Fig. 22.6).

The third term in Eq. 22.4 can be called the emf of motion. A high value of the emf of motion may result in a maximum value of the current along the curve OAB.

When the operating point moves along curve OAB, the supply delivers an amount of electrical energy W_e equal to the area {OABD}.

Consider now the state curve OEB through point B, i.e. the characteristic $\psi(i, \theta_B)$ for a rotor position corresponding to $\theta = \theta_B$. The area {OEBD} is the magnetic energy W_m stocked in the motor for the position corresponding to point B (i.e. for position angle θ_B and current i_B). The electrical energy supplied for the rotation from O to B can be written as:

$$W_e = \{OABD\} = \{OABE\} + \{OEBD\} = W_{em1} + W_m \tag{22.5}$$

This means that $W_{em1} = \{OABE\}$ is the electromechanical work during the movement from O to B.

In B, the rotor is not yet aligned with the corresponding stator pole pair. However, the electrical supply voltage is assumed to be turned off in B to ensure that the

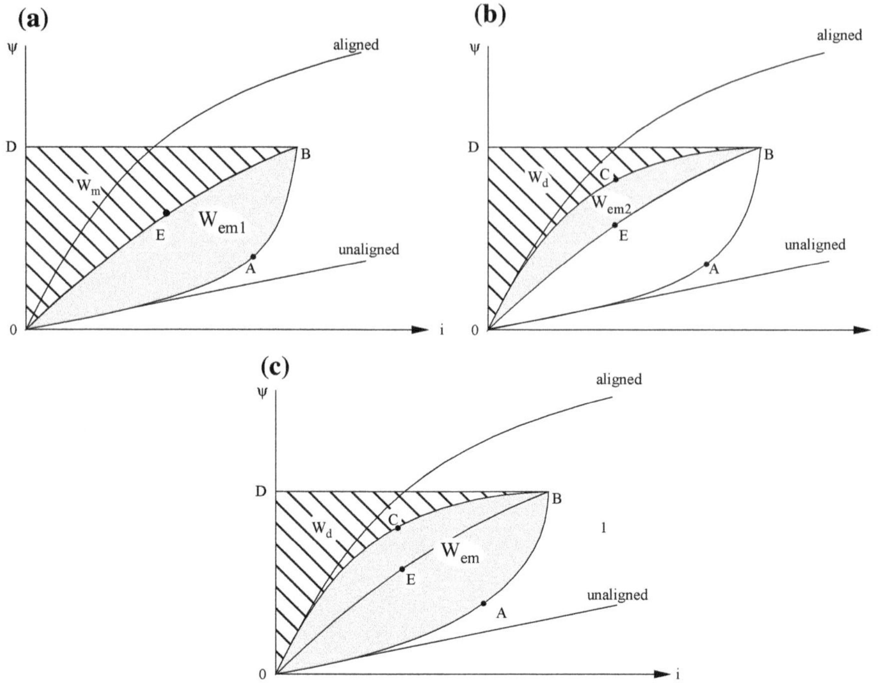

Fig. 22.6 Energy diagram for motoring

current will be reduced to zero in the aligned position (a non-zero current in the aligned position would create a negative torque when the rotor rotates away from the aligned position). If the supply voltage is turned off in B, usually anti-parallel diodes are provided so that the current (in the inductive circuit) is reduced to zero and at the same time returns to the DC source, i.e. energy recuperation. During the rotation from B to the aligned position, the state of the system follows the segment BCO. During this transition, an amount of electrical energy W_d is returned to the source equal to the area $OCBD = W_d$. In O, current and flux are zero and the magnetic energy in the system is therefore zero. During the transition from B to O, an amount of electromechanical work W_{em2} must have been delivered to the load, equal to the area {OEBC}.

The total amount of electromechanical energy delivered to the load during the cycle OABCO is therefore

$$W_{em} = W_{em1} + W_{em2} = \{OABE\} + \{OEBC\} = \{OABC\} \tag{22.6}$$

If the joule losses (in the switches and windings, i.e. $R \approx 0$ in Eq. 22.4) are small, the energy W_d is recuperated completely by the source. However, also when the losses are small, it is important that W_d is kept as small as possible. Indeed, the ratio of the

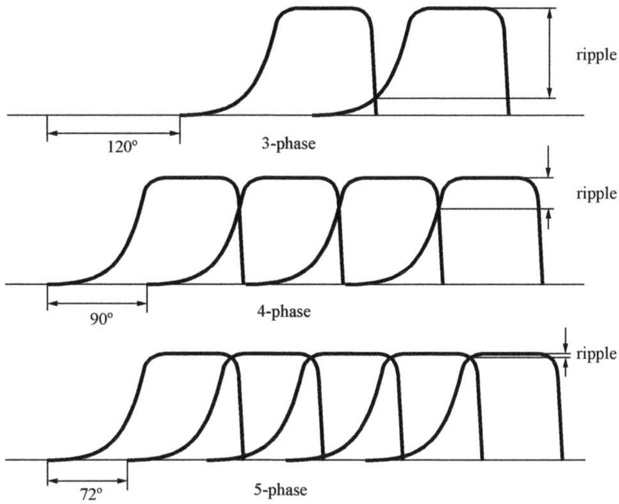

Fig. 22.7 Torque profiles for 3, 4 and 5 phase SRMs

energy usefully converted, to the energy to be disposed by the source in each cycle is given by

$$\gamma = \frac{W_{em}}{W_{em} + W_d} = \frac{W_{em}}{W_e} \tag{22.7}$$

From Fig. 22.6, it can be seen that in order to obtain a γ close to 1, the degree of saturation in the magnetic circuit should be large. The larger γ is, the larger the converted energy is for a given machine frame size and power source.

Naturally, W_{em} is the energy converted per cycle for only one stator phase. The mechanical power and torque can be derived as follows. If T_{em} is the average electromagnetic torque, the electromagnetic energy converted in one revolution of the rotor is $2\pi T_{em}$. For one revolution we need $m \cdot N_r$ cycles and as for each cycle the electromagnetic energy is W_{em}, we obtain

$$T_{em} = \frac{m.N_r}{2\pi} \cdot W_{em} \tag{22.8}$$

In contrast with induction machines, for example, the torque is not quite constant (see Fig. 22.7).

The ripple is rather pronounced when the number of phases and rotor poles is low. A higher number of stator and rotor poles offers a lower ripple, but requires a much higher supply frequency for a given speed (see also Sect. 22.5).

22.3 Converters for Switched Reluctance Machines

The converter for a switched reluctance machine should be able to:

- switch on and off the supply voltage to the appropriate phase winding at the right instant
- offer the possibility to control the current for the SRM (especially at low speeds)
- return the stocked electromagnetic energy to the supply when the phase is switched off (or, as a low cost solution, at least dissipate this energy).

Figure 22.8 shows the configuration of a standard converter for a three-phase machine.

For the current built-up in a given phase (e.g. phase 1) both upper and lower switches for this phase winding (Q1 and Q1') are switched on. For motoring, the appropriate instant is at the start of the overlapping of these stator poles and the appropriate rotor pole pair. Together with the current, the torque increases and the rotor starts to align - or aligns further - but only if the load torque is not too large.

In most cases, a controlled (average) torque is expected, implying that the (increase of the) current needs to be controlled. This is achieved by chopping the supply: one of the two switches (say Q1) is given a PWM chopping control signal, while the other (Q1') is continuously on. If both Q1 and Q1' are on, the current increases. If Q1 is off (while Q1' stays on), the current in the phase winding circulates through Q1' and D1' and will not further increase (in fact decreases a bit). Switching Q1 on again permits the current to increase (see Fig. 22.9).

This chopping is especially required for speeds that are not too high, i.e. when the back-emf is small. At high speeds, the back-emf is high and will limit the current built-up. In fact, at high speeds the current and thus also the torque might not attain

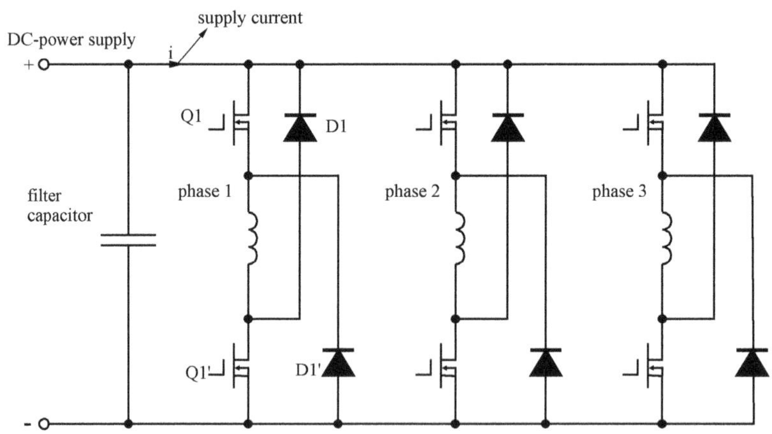

Fig. 22.8 Basic three-phase power-electronic converter with 2 switches per phase

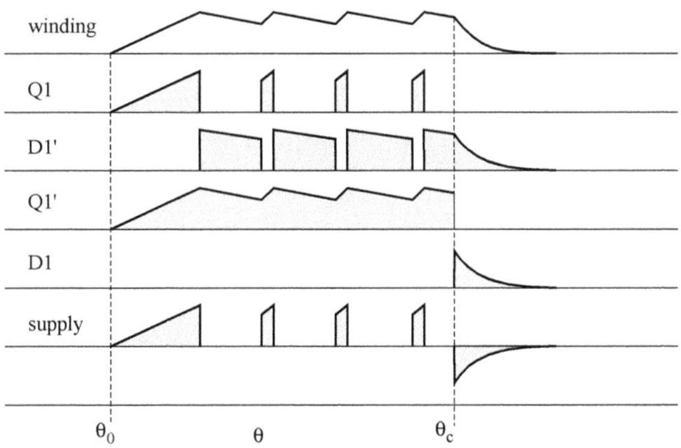

Fig. 22.9 Current profiles for the converter with 2 switches per phase

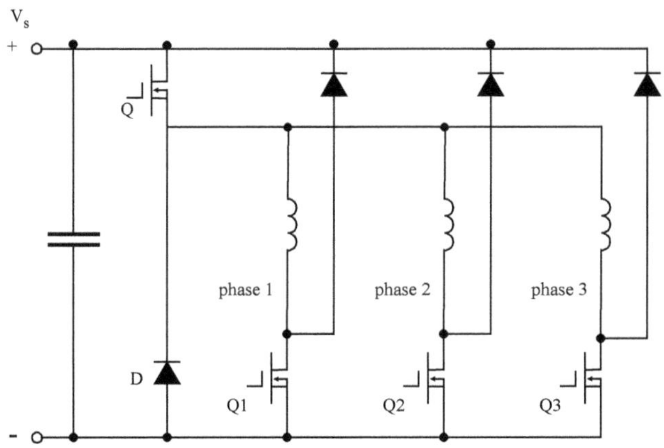

Fig. 22.10 Simplified power-electronic converter (n + 1 switches for n phases)

its rated value. This is indeed field weakening, similar to e.g. the DC machine or induction machine.

Before the rotor poles reach the completely aligned position, the current has to be switched off as fast as possible to avoid negative torques when the rotor moves beyond the aligned position. In the configuration in Fig. 22.8, this is obtained by switching off both Q1 and Q1'. The current in the winding then returns to the source through the diodes D1 and D1'. Because the source voltage is now connected in opposite direction to the winding, the current will decrease quickly and the energy W_d will be recuperated by the source.

In addition to the basic configuration shown in Fig. 22.8, many other converter configurations have been developed in recent years. An alternative configuration for a three-phase SRM is shown in Fig. 22.10. The switches Q1, Q2 and Q3 are used for chopping their particular phase current, while switch Q is common for the three

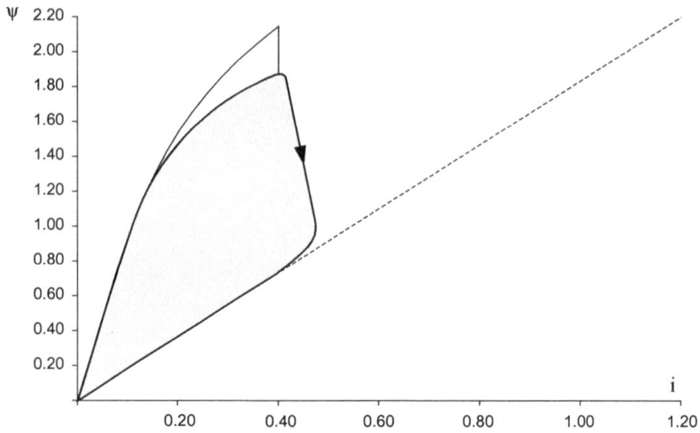

Fig. 22.11 Energy diagram for generating (moving clockwise)

phases. Q remains on for chopping but is switched off together with Qi to reduce the current of phase i to zero and return the energy W_d to the source. The disadvantage of this circuit is that a faulty switch Q results in a complete failure. In the case of Fig. 22.8, there is some redundancy as the motor may continue delivering some torque and power when one switch is defective.

With a converter as in Fig. 22.8, the SRM can also convert mechanical energy into electrical energy, i.e. generating. For that, the current already has to be built up at the instant of complete alignment. When aligned, the switches are turned off and the current from the inductive windings will flow via the diodes. When the rotor moves away from the aligned position, the torque of the SRM is negative but the positive mechanical torque from the driving machine *("load")* balances it out. From a mechanical point of view, mechanical energy is delivered (as mechanical torque is in the rotation direction) to the SRM, which converts it to electrical energy for the DC source. Indeed, the diodes now conduct, because the switches are turned off and thus the current is negative with respect to the DC source. The mechanical energy is therefore converted into electrical energy, which is transferred to the DC source.

In the $\psi - i$ diagram (Fig. 22.11), the cycle is completed in clockwise direction.

In some specific cases, the current may still increase when the rotor moves away from the aligned position. Yet, this requires the back-emf to be sufficiently large compared to the DC source voltage. Indeed, during the return of electromagnetic energy to the source we have

$$-V = R \cdot i + \frac{\partial \psi}{\partial i} \cdot \frac{\partial i}{\partial t} + \frac{\partial \psi}{\partial \theta} \cdot \omega \qquad (22.9)$$

As the DC source is now connected in reverse polarity to the windings (i.e. through the diodes), we find $-V$ on the left. On the right, the last term, the motional emf, is also negative as the flux is decreasing when the rotor moves away from the aligned position. To obtain an (at least initially) increasing current, the absolute value of this motional emf should be larger than the source voltage. This might pose a problem, as for motoring the source voltage should be much larger than the motional emf (for a given speed). A solution to this problem is to provide two voltage sources, a higher voltage for motoring and a lower one for generating, or to use a buck converter between the machine and source for generating.

22.4 Control of an SRM

For an efficient operation of an SRM, the following is required:

- an accurate control (i.e. switching on and off) of the switches
- a sufficiently accurate position measurement of the rotor, required for controlling the switches
- a control of the (average) current in order to control the torque
- a speed (and/or position) control loop of the drive system in order to generate the value of the desired torque and thus of the (average) current.

Figure 22.12 shows a basic control scheme for an SRM. What is inside the dashed lines can be considered as the actual SRM drive (i.e. SRM, converter and its control). The remaining is related to the application, i.e. the load and its control.

Only the SRM drive itself will be discussed here. An SRM requires quite an accurate position control, with 0.5° or even 0.25° as a desired accuracy level.

For motoring, to obtain an efficient torque built-up, the switches of the active phase should be turned on at the right instant. For example for normal speeds that are not too high, this will be at the start of the overlap of the rotor poles and the stator poles. The switches should also be turned off at the right instant: not too early, to avoid losing positive torque, and not too late, to avoid negative torques beyond alignment.

For a 6/4 SRM, normal control angles (on/off) are:

- 52.5°/82.5° for normal PWM mode
- 37.5°/67.5° for boost mode (high speed)
- 22.5°/67.5° for advanced mode (very high speed)

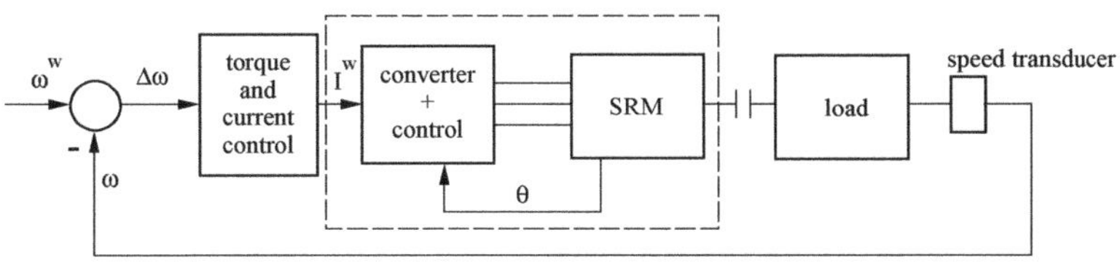

Fig. 22.12 Control scheme for an SRM

where we define 90° as complete alignment and 45° as unaligned (see (a) and (b) in Fig. 22.3).

Normal mode is for low speeds, for which PWM control can be required to control the current. Normally, turn-on and -off angles are then constant. Boost mode is for higher speeds, where the back-emf is already rather large and the switches therefore need to be switched on early to ensure enough time for the current to build up. In that case, current control is achieved by modifying the turn-on and turn-off angles. Advanced mode is similar but for very high speeds.

For generating (braking), the turn-on instant is a few degrees before alignment and the turn-off instant at or some degrees after alignment (e.g. 82.5°/112.5°). In generating mode, PWM control of the current might be necessary but the already limited back-emf will often reduce the current. The optimal turn-on and turn-off angles can be found at the machine design stage (using finite element simulation of the machine together with a model of the converter), but generally experimental verification or correction will be used as well.

For some applications, a much more sophisticated control of the drive may be desirable. By appropriately choosing the turn-on and turn-off angles and modifying the PWM control of the current, we may mitigate the torque pulsations and the noise. More theoretical and experimental research and trials are warranted to study this further.

To measure the rotor position, encoders based on the Hall effect or optical encoders may be used. However, an encoder adds to the complexity and the cost of the drive and decreases reliability. As a consequence, much research on sensorless control has been conducted in the last few years. For example, a *non-excited* phase winding may be used to estimate the rotor position: for example, a small high frequency signal is fed to this non-excited phase and from the measured current the rotor position is estimated. Other methods may use an observer to derive the position from the voltage and the current of the *active* phase.

22.5 SRM Types and Applications

In the previous paragraphs, we mainly concentrated on the 6/4 SRM. However, many other types are common:

- the three-phase 12/8 SRM, see (a) in Fig. 22.13
- the three-phase 12/10 SRM, see (b) in Fig. 22.13
- the four-phase 8/6 SRM, see (c) in Fig. 22.13
- the four-phase 8/10 SRM, see (d) in Fig. 22.13
- the two-phase 4/2 SRM, see (e) in Fig. 22.13 (note: the air gap contains a step to obtain an unambiguous rotation direction).

Important considerations to choose the number of phases and the number of poles on stator and rotor are the following:

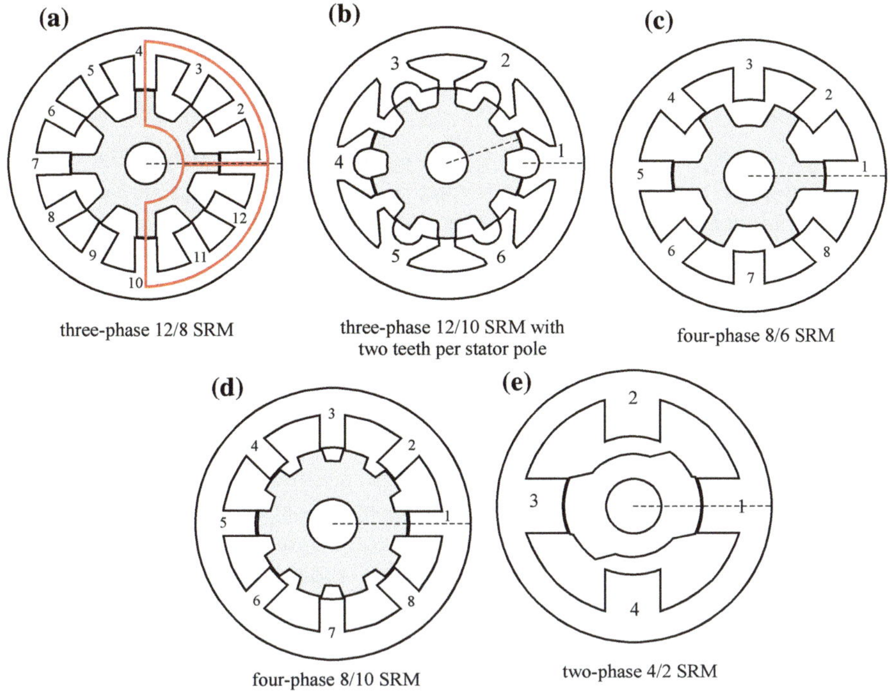

Fig. 22.13 Other SRM types

- the starting possibilities (e.g. a single-phase SRM cannot start if rotor and stator poles are aligned)
- an unambiguous rotation direction (e.g. an 4/2 SRM has no preferential rotation direction if the air gap does not contain a step as in (e) in Fig. 22.13)
- the torque pulsations
- the reliability (the larger the number of phases, the greater the reliability: an 8/6 SRM can continue working if one phase is down)
- the cost (the larger the number of phases, the higher the cost of the converter)
- the efficiency (this is related to the operating frequency required for a given speed).

A larger number of phases reduces the torque pulsations (see Fig. 22.7), improves the reliability but necessitates a higher operating frequency for a given speed (which may be imposed by the application) as the step angle ϵ_r is then reduced (see Eq. 21.2). The number of steps for one revolution is thus $m \cdot N_r$. As each step requires the excitation of a next phase, N_r excitations are required per phase. For a speed of N (1/s), the required stator frequency is $f_s = N \cdot N_r$.

For the above-mentioned SRM types, this gives:

- $f_s = 4 \cdot N$ for the 6/4
- $f_s = 8 \cdot N$ for the 12/8

- $f_s = 10 \cdot N$ for the 12/10
- $f_s = 6 \cdot N$ for the 8/6
- $f_s = 10 \cdot N$ for the 8/10
- $f_s = 2 \cdot N$ for the 4/2.

For a speed of N = 00/s (or 6000 rpm), this results in 400, 800 Hz, 1 kHz, 600 Hz, 1 kHz and 200 Hz respectively. This frequency greatly affects the efficiency of the drive. The higher the frequency, the higher the switching losses in the converter and the iron losses in the stator core. Moreover, the frequency in the rotor also increases, which in turn causes the iron losses in the rotor core to rise.

For high speeds, the 6/4 SRM is therefore a good compromise: the iron losses are relatively low, the motor can start and rotate in both directions and there is some redundancy. In addition, the low number of poles creates a large reluctance effect (or a large difference between the aligned and unaligned inductances) and therefore the energy conversion is quite efficient. Indeed, a high number of poles (and thus narrow pole widths) limits the inductance ratio. An additional advantage is that some power electronic modules for traditional three-phase machines might be used. The main disadvantages of the 6/4 SRM remain the torque pulsations and the noise.

The SRM can be used in many applications where standard variable-frequency drives (rotating field machines) are common. It is especially suited for high speeds, but its noise and vibrations must be acceptable for this specific application. Some industrial applications include the following:

- highly dynamic drives in looms as replacement for induction machine drives with field orientation (Picanol, Belgium)
- as starter-generator in jet turbines for airplanes (General Electric, USA)
- as servo drives for very high speeds (Brother Man. Co, Japan).

Part IV
Dynamics of Electrical Machines and Drives

Chapter 23
Stability and Dynamics

Abstract In this chapter we shortly review the well known principles and definitions of stability and dynamics of systems.

23.1 Introduction: Definition of Stability

Generally, a system is regarded as stable when small deviations between the actual operating conditions and the operating conditions for which it was designed do not fundamentally affect the system behaviour.

Actually, most people will intuitively feel whether a system exhibits stable behaviour or not. However, for practical applications, some more specific definitions of stability are required.

23.2 Classifications of Stability

23.2.1 Stability of an Equilibrium Point

The first type of stability involves the behaviour of a system when disturbed in an equilibrium state. A typical example is that of a ball resting on a surface and subjected to gravity (see Fig. 23.1).

In case (a) in Fig. 23.1, the ball will oscillate around its initial position with a limited amplitude (depending on the initial deviation) when it is moved away from its equilibrium position. In case (b), the ball will remain in its new position when moved away from its initial equilibrium position, as the new position is an equilibrium position too. In both cases, this behaviour can be called *stable*: the deviation from the initial position remains *bounded* and can be deliberately made small by making the initial displacement sufficiently small.

In case (c) in Fig. 23.1, the ball will definitely move away from its initial position when displaced from this position, whatever small the displacement. The equilibrium state (c) is therefore called *unstable*.

J. A. Melkebeek, *Electrical Machines and Drives*, Power Systems,
https://doi.org/10.1007/978-3-319-72730-1_23

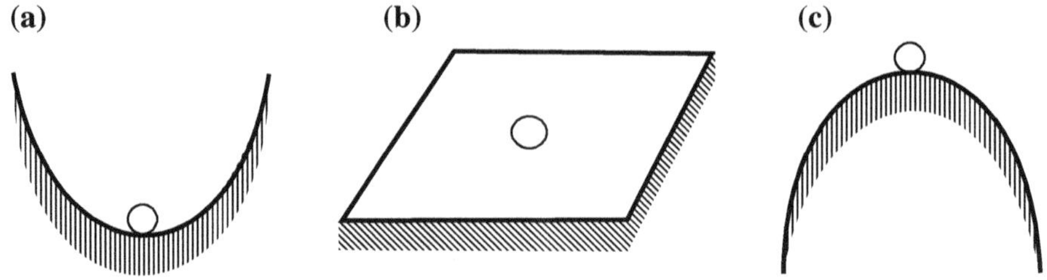

Fig. 23.1 Equilibrium states

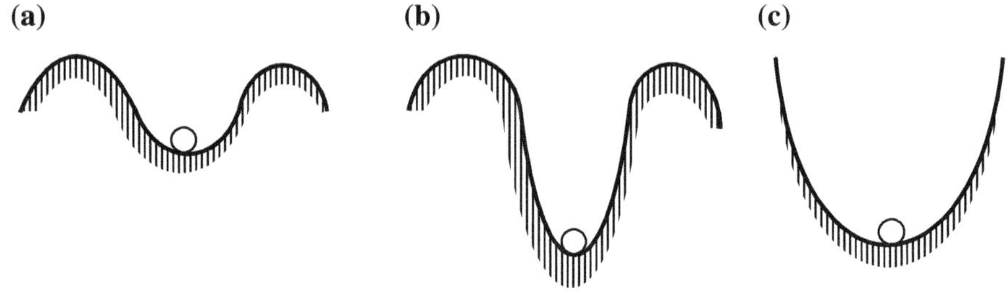

Fig. 23.2 Boundaries for stability

If friction is taken into account, in case (a) the ball will return to its original position after some damped oscillations, or even without any oscillations. This equilibrium state is therefore called *asymptotically stable*. In case (b), the ball will never return to its original position. This equilibrium state is stable but not asymptotically stable, and it is asymptotic stability that is required in practical systems.

In the previous section, we disregarded the magnitude of the deviation from the equilibrium position. Consider now the situations depicted in Fig. 23.2. In case (a), the neighbourhood of the initial position where the asymptotic stability holds is much smaller than in case (b) in Fig. 23.2. In case (c), asymptotic stability is ensured for every displacement. For practical applications, it is not sufficient that the equilibrium state is stable within a, possibly very small, neighbourhood of this equilibrium (which is called *local asymptotic stability* or *stability in the small*). The stability properties should also remain valid within a sufficient range (i.e. *the extent of [asymptotic] stability*) around the considered equilibrium state.

In an analogous way, the stability of a motion may be studied, i.e. the trajectory of the output of a (non-linear) system. This implies that it should be known how far the actual trajectory may deviate from the trajectory for which it was designed.

23.2.2 Input–Output Stability

A second class of stability concerns the behaviour of the system when subjected to a given input signal. In this case, the question is whether the outputs y of the system,

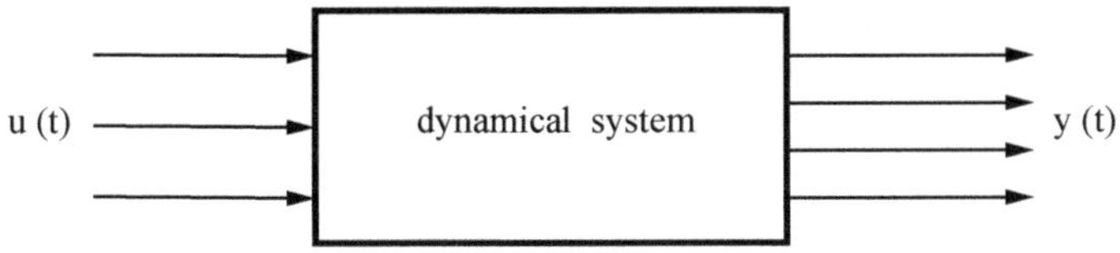

Fig. 23.3 Input–output stability

regarded as a black box, correspond to the expected behaviour of the system with the signals u as inputs (see Fig. 23.3). The expected behaviour can (for example) be described as an output between prescribed boundaries $y_{min} < y < y_{max}$ for an input between given boundaries $u_{min} < u < u_{max}$. If it is possible to transform the output variable y into a variable y' so that the boundaries become $-\infty$ and $+\infty$, respectively, and if the input variable u can be transformed into a variable u' with as boundaries $-\infty$ and $+\infty$, respectively, we obtain an equivalent dynamic system in which a bounded input needs to yield a bounded output if input–output stability is to be achieved.

23.3 Mathematical Tools to Explore the Stability of a System

For a linear system, the (asymptotic) stability can be investigated by looking at the eigenvalues of the system. If all eigenvalues are in the left half of the complex plane, the system is asymptotically stable. Eigenvalues in the right half plane indicate an unstable system. If eigenvalues on the imaginary axis are found and all other eigenvalues are in the left half plane, the system is stable but not asymptotically stable. For linear systems, the extent of stability is the whole state plane. At the same time, input–output stability is ensured.

For non-linear systems, the system may be linearised. When the linearised system is asymptotically stable, there is a neighbourhood of the equilibrium state in which the non-linear system is stable. However, this neighbourhood might be very small, and to explore the extent of stability methods like Lyapunov's method need to be applied. Note, however, that Lyapunov's method only offers sufficient conditions for stability, not necessary conditions.

If the linearised system shows eigenvalues in the right half plane, the non-linear system is unstable as to the envisioned equilibrium point. If the linearised system has eigenvalues on the imaginary axis, no conclusions can be drawn for the non-linear system.

For some classes of non-linear systems, more specific methods exist (e.g. the circle criterion). For more details, see Ref. [39].

Chapter 24
Transient Phenomena in Simple Electrical Circuits

Abstract In this chapter we discuss some transient phenomena in simple electrical circuits (resistive, inductive, capacitive), as an introduction to the later chapters on (local) stability and dynamics of electrical machines and drives.

24.1 Switching On or Off a Resistive-Inductive Circuit

Consider the resistive-inductive circuit (a) in Fig. 24.1. R and L are assumed to be constant (e.g. independent of current or frequency). The circuit is therefore described by the linear time-invariant differential equation

$$v(t) = R \cdot i(t) + L \cdot \frac{di(t)}{dt} \tag{24.1}$$

First, we will examine the case in which the initially current-less circuit is connected to a voltage source $v_s(t) = \hat{V}\cos(\omega t + \varphi)$. The solution of Eq. 24.1 with boundary conditions $i(t) = 0$ for $t \leq 0^-$ and with $v(t) = v_s(t)$ for $t \geq 0^+$ consists of two parts:

- the particular or steady-state solution

$$i(t) = \frac{\hat{V}}{\sqrt{R^2 + \omega^2 L^2}} \cdot \cos(\omega t + \varphi - \arctan \omega L/R) \tag{24.2}$$

- the transient solution

$$i(t) = I \cdot \exp(-t/\tau) \tag{24.3}$$

with $I = -\frac{\hat{V}}{\sqrt{R^2+\omega^2 L^2}} \cdot \cos(\varphi - \arctan \omega L/R)$ and $\tau = L/R$.
Note that the time constant of the system corresponds with the eigenvalue of the system Eq. 24.1, with $i(t)$ considered as state variable (and $v(t)$ as input). This time constant corresponds to the single energy storage of the system, i.e. the magnetic energy in the coil. This solution can also be found using the Laplace transform.

J. A. Melkebeek, *Electrical Machines and Drives*, Power Systems,
https://doi.org/10.1007/978-3-319-72730-1_24

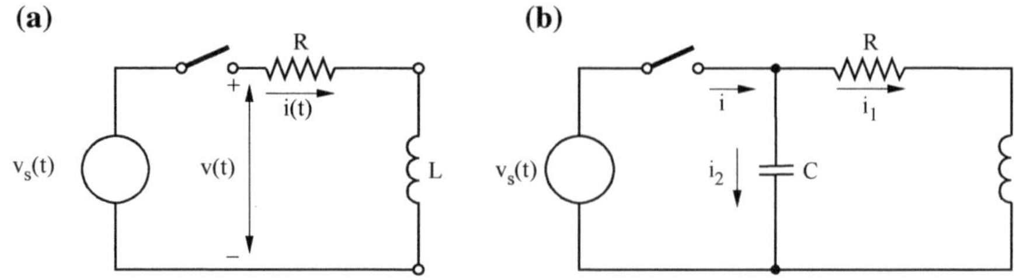

Fig. 24.1 R-L- and R-L-C-circuit

Next, we will analyse the case of an interruption of a (steady-state) DC current I_o in the circuit. The state variable is now the voltage $v(t)$, while the input is $i(t)$. For $t \leq 0^-$, $i(t) = I_o$ and $v(t) = V_o = RI_o$. At $t = 0$, the switch is opened. Suppose that the switch is ideal, i.e. $i(t) = 0$ for $t \geq 0^+$.

In order to use the (one-sided) Laplace transform, we have to transform the variables, i.e. $i'(t) = i(t) - I_o$ and $v'(t) = v(t) - V_o$. In terms of the new variables, the equations and boundary conditions become:

$$v'(t) = R \cdot i'(t) + L \cdot \frac{di'(t)}{dt} \tag{24.4}$$

$$i'(t) = -I_o \cdot u(t) \tag{24.5}$$

with $u(t)$ the unit step function: $u(t) = 0$ for $t \leq 0^-$ and $u(t) = 1$ for $t \geq 0^+$.

The Laplace transform yields

$$V'(p) = R \cdot I'(p) + pL \cdot I'(p) \tag{24.6}$$

$$I'(p) = -I_o/p \tag{24.7}$$

The solution is

$$V'(p) = -\frac{R \cdot I_o}{p} - L \cdot I_o \tag{24.8}$$

and thus in the time domain:

$$v'(t) = -R \cdot I_o \cdot u(t) - L \cdot I_o \cdot \delta(t) \tag{24.9}$$

or

$$v(t) = R \cdot I_o[1 - u(t)] - L \cdot I_o \cdot \delta(t) \tag{24.10}$$

The Dirac term in the voltage is the result of two (unrealistic) assumptions: an ideal switch and a coil with a negligible capacitance between the turns of the coil. In reality, the high voltage between the contacts of the switch will result in a spark, assuring the continuity of the current.

The capacitance between the turns of the coil can (approximately) be modelled by a lumped capacitor as in the circuit (b) in Fig. 24.1. The system equations are now

$$v(t) = R \cdot i_1(t) + L \cdot \frac{di_1(t)}{dt} \tag{24.11}$$

$$\frac{dv(t)}{dt} = \frac{1}{C} i_2(t) \tag{24.12}$$

$$i(t) = i_1(t) + i_2(t) \tag{24.13}$$

with boundary conditions $i_1 = I_o, i_2 = 0$, and $v = R \cdot I_o$ for $t \leq 0^-$ and $i_1 + i_2 = 0$ for $t \geq 0^+$.

The solution is now

$$i_1(t) = I_o \left\{ [1 - u(t)] + \frac{R/L - p_2}{p_1 - p_2} \exp(p_2 t) - \frac{R/L - p_1}{p_1 - p_2} \exp(p_1 t) \right\} \tag{24.14}$$

$$v(t) = R \cdot I_o \left\{ [1 - u(t)] + \frac{R/L - 1/RC - p_2}{p_1 - p_2} \exp(p_2 t) - \frac{R/L - 1/RC - p_1}{p_1 - p_2} \exp(p_1 t) \right\} \tag{24.15}$$

with p_1 and p_2 the eigenvalues of this second-order system[1]

$$p_{1,2} = \frac{R}{2L} \left(-1 \pm \sqrt{1 - \frac{4L}{R^2 C}} \right) \tag{24.16}$$

i.e. the zeros of the eigenvalue equation $LCp^2 + RCp + 1 = 0$.

The capacitance between the turns limits the voltage between the contacts of the switch (although sparks are still possible).

24.2 Single-Phase Transformer

If we disregard saturation and skin effects, the single-phase transformer is described by Eqs. 24.17 and 24.18:

$$v_1(t) = R_1 \cdot i_1(t) + L_1 \cdot \frac{di_1(t)}{dt} + M \cdot \frac{di_2(t)}{dt} \tag{24.17}$$

[1]Please analyse the case in which $4L < R^2C$ and $4L > R^2C$ and, in particular, the case in which $R = 0$.

$$v_2(t) = R_2 \cdot i_2(t) + L_2 \cdot \frac{di_2(t)}{dt} + M \cdot \frac{di_1(t)}{dt} \tag{24.18}$$

If the turns ratio $a = w_1/w_2$ is known, we may also rewrite these equations as

$$v_1(t) = R_1 \cdot i_1(t) + L_{1\sigma} \cdot \frac{di_1(t)}{dt} + L_{m1} \cdot \frac{d}{dt}[i_1(t) + i_2'(t)] \tag{24.19}$$

$$v_2'(t) = R_2' \cdot i_2'(t) + L_{2\sigma}' \cdot \frac{di_2'(t)}{dt} + L_{m1} \cdot \frac{d}{dt}[i_1(t) + i_2'(t)] \tag{24.20}$$

with $L_{1\sigma} = L_1 - aM$, $L_{2\sigma} = L_2 - M/a$, $L_{m1} = aM$ and with the prime indicating the secondary variables referred to the primary.

The eigenvalues of the free system, with the currents as state variables and the voltages as external inputs (assumed to be zero, for example), are the zeros of

$$\det \begin{vmatrix} L_1 p + R_1 & Mp \\ Mp & L_2 p + R_2 \end{vmatrix} = 0 \tag{24.21}$$

or

$$(L_1 L_2 - M^2)p^2 + (L_1 R_2 + L_2 R_1)p + R_1 R_2 = 0 \tag{24.22}$$

This can be written as

$$\sigma p^2 + \left(T_{m1}^{-1} + T_{m2}^{-1}\right) p + T_{m1}^{-1} T_{m2}^{-1} = 0 \tag{24.23}$$

or also as

$$p^2 + \left(T_1^{-1} + T_2^{-1}\right) p + \sigma T_1^{-1} T_2^{-1} = 0 \tag{24.24}$$

where $T_{m1} = L_1/R_1$, $T_{m2} = L_2/R_2$ are the main field or open-circuit time constants of primary and secondary, $T_1 = \sigma L_1/R_1$, $T_2 = \sigma L_2/R_2$ are the leakage field or short-circuit time constants of primary and secondary, and σ is the total leakage coefficient of the transformer.

For σ that are not too large, the eigenvalues can be approximated by

$$p_1 = -\left(T_1^{-1} + T_2^{-1}\right) \tag{24.25}$$

$$p_2 = -\left(T_{m1} + T_{m2}\right)^{-1} \tag{24.26}$$

These two eigenvalues correspond to the main and leakage fields of the transformer, i.e. the two ways for magnetic energy storage. The eigenvalue p_1 corresponds to that of an R-L-circuit with as resistance the sum of the primary and secondary resistances (referred to the same winding) and the total leakage as seen from this winding:

$$p_1 \approx -\frac{R_1 + R_2^{'}}{\sigma L_1} = -\frac{R_1^{'} + R_2}{\sigma L_2} \tag{24.27}$$

The eigenvalue p_2 corresponds to that of an R-L-circuit with as resistance the parallel connection of primary and secondary resistances (referred to the same winding) and the main field inductance seen from this same winding:

$$p_2 \approx -\frac{R_1 R_2^{'}/(R_1 + R_2^{'})}{L_{m1}} \tag{24.28}$$

To apply this, we will study the transient when a transformer, loaded with a resistor R_{2e} (denoting $R = R_{2e} + R_2$) at the secondary and fed by the grid (V_1, f_1) at the primary, is disconnected from the grid at $t = 0$. Just before the switch is opened, the primary and secondary currents are I_{10} and I_{20}, respectively. The secondary current and the primary voltage for $t \geq 0^+$ will be calculated in three ways.

Method n°1: Time Domain
At the secondary side, we have at each instant

$$Ri_2(t) + L_2 \cdot \frac{di_2(t)}{dt} + M \cdot \frac{di_1(t)}{dt} = 0 \tag{24.29}$$

For $t > 0^+$, the primary current as well as its derivative are zero. However, the flux coupled with the secondary has to remain continuous. Therefore

$$L_2 i_2(t = 0^+) = L_2 \cdot I_{20} + M \cdot I_{10} \tag{24.30}$$

or

$$i_2(t = 0^+) = I_{20} + (M/L_2) \cdot I_{10} \tag{24.31}$$

For $t > 0^+$, the secondary current has to satisfy equation 24.29 with $i_1(t) = di_1(t)/dt \equiv 0$. Thus

$$i_2(t > 0^+) = [I_{20} + (M/L_2) \cdot I_{10}] \exp(-t/\tau) \tag{24.32}$$

with $\tau = L_2/R$. For $t > 0^-$, we may write

$$i_2(t > 0^-) = I_{20}[1 - u(t)] + [I_{20} + (M/L_2) \cdot I_{10}]u(t) \cdot \exp(-t/\tau) \tag{24.33}$$

The primary voltage for $t > 0^-$ can be calculated from Eq. 24.17 with $i_2(t)$ according to Eq. 24.33 and $i_1(t) = I_{10}[1 - u(t)]$.

For $t > 0^-$, we have

$$v_1(t) = R_1 \cdot i_1(t) + \sigma L_1 \cdot \frac{di_1(t)}{dt} + M \cdot \frac{d}{dt}\left[i_2(t) + \frac{M}{L_2} i_1(t)\right] \tag{24.34}$$

or

$$v_1(t) = R_1 \cdot i_1(t) + \sigma L_1 \cdot \frac{di_1(t)}{dt} - \frac{M}{L_2} R \cdot i_2(t) \tag{24.35}$$

Using $di_1(t)/dt \equiv 0$ for $t > 0^-$ and substituting $i_2(t)$ from Eq. 24.33 yields

$$v_1(t) = R_1 \cdot I_{10}[1-u(t)] - \sigma L_1 \cdot I_{10}\delta(t) - \frac{M}{L_2} R \cdot I_{20}[1-u(t)] - \frac{M}{L_2} R \cdot [I_{20} + \frac{M}{L_2} \cdot I_{10}]u(t) \cdot \exp(-t/\tau) \tag{24.36}$$

For $t > 0^+$, we therefore have

$$v_1(t) = -\frac{M}{L_2} R \cdot [I_{20} + \frac{M}{L_2} \cdot I_{10}] \exp(-t/\tau) \tag{24.37}$$

The voltage for $t > 0^+$ corresponds to the main flux coupled with the secondary that is fading away with the secondary open-circuit time constant. When the switch opens at $t = 0$, the secondary takes over the magnetising part of the flux corresponding to the primary current (flux continuity). The Dirac voltage for $t = 0$ corresponds to the leakage flux of the primary which is not coupled to the secondary and cannot be compensated by a jump in the secondary current. This Dirac voltage will give rise to a spark in the switch.

If for $t < 0^-$ the transformer was fed by a DC voltage $v_o = R_1 I_{10}$ (with $I_{20} = 0$), we find for the primary voltage

$$v_1(t) = R_1 \cdot I_{10}[1 - u(t)] - \sigma L_1 \cdot I_{10}\delta(t) - \frac{M^2}{L_2^2} R \cdot I_{10} u(t) \cdot \exp(-t/\tau)$$

and for $t > 0^+$

$$v_1(t) = -\frac{M^2}{L_2^2} R \cdot I_{10} \exp(-t/\tau) = -(1-\sigma)\frac{L_1/R_1}{L_2/R} \cdot v_o \cdot \exp(-t/\tau) \tag{24.38}$$

In other words, the initial value is the DC voltage reduced by the factor $1 - \sigma$ and transformed with the ratio of the primary and secondary time constants.

Method n°2: Single-side Laplace Transform
To apply the single-side Laplace transform, we have to replace the currents i_1 and i_2 with fictitious currents $i_1^* = i_1 - I_{10}$ and $i_2^* = i_2 - I_{20}$ which are zero for $t \leq 0^-$. For these new variables, the secondary transformer equation becomes

$$Ri_2^*(t) + L_2 \cdot \frac{di_2^*(t)}{dt} + M \cdot \frac{di_1^*(t)}{dt} = -RI_{20} \tag{24.39}$$

and after the Laplace transform

$$RI_2^*(p) + pL_2 \cdot I_2^*(p) + pM \cdot I_1^*(p) = -R\frac{I_{20}}{p} \tag{24.40}$$

As

$$I_1^*(p) = -\frac{I_{10}}{p} \tag{24.41}$$

we find for the secondary current

$$I_2^*(p) = \frac{1}{R + pL_2}\left(-R\frac{I_{20}}{p} + MI_{10}\right) \tag{24.42}$$

Or, in the time domain:

$$i_2(t) = i_2^*(t) + I_{20} = I_{20}\,(1 - u(t)) + \left(I_{20} + \frac{M}{L_2}I_{10}\right) u(t) \cdot \exp(-t/\tau) \tag{24.43}$$

The primary voltage[2] can be calculated in a similar way.
Method n°3: A variation of Method n°2
Define the new variables $i_1^+(t)$ and $i_2^+(t)$ with $i_1^+(t) = 0$ for $t < 0^-$, $i_1^+(t) = i_1(t)$ for $t > 0^-$, $i_2^+(t) = 0$ for $t < 0^-$, $i_2^+(t) = i_2(t)$ for $t > 0^-$.

The transformer secondary equation is now

$$Ri_2^+(t) + L_2 \cdot \frac{di_2^+(t)}{dt} + M \cdot \frac{di_1^+(t)}{dt} = 0 \tag{24.44}$$

and after Laplace transform

$$RI_2^+(p) + L_2\left(pI_2^+(p) - I_{20}\right) + M\left(pI_1^+(p) - I_{10}\right) = 0 \tag{24.45}$$

With $I_1^+(p) = 0$, we obtain

$$I_2^+(p) = \frac{1}{R + pL_2}\left(L_2 I_{20} + MI_{10}\right) \tag{24.46}$$

[2] Can you also find a Dirac function in the voltage?

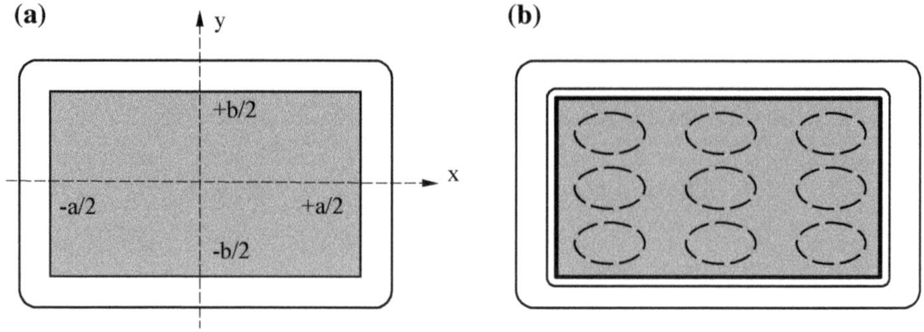

Fig. 24.2 **a** Coil with massive core **b** circulating currents

or, in the time domain for $t > 0^-$:

$$i_2^+(t) = \left(I_{20} + \frac{M}{L_2} I_{10}\right) u(t) \cdot \exp(-t/\tau). \tag{24.47}$$

24.3 Coil with Massive Iron Core

A coil with a massive iron core can be regarded as a special case of magnetically coupled coils. We consider an infinitely long coil with a core with rectangular cross-section and sides a (x-direction) and b (y-direction), as illustrated in (a) in Fig. 24.2. The coil is uniformly distributed along the core length (z-direction).

The general transient solutions have to satisfy the following equations (from Maxwell's laws):

$$E_x = \rho_{Fe} \cdot J_x = \frac{\rho_{Fe}}{\mu} \cdot \frac{\partial B_z}{\partial y} \tag{24.48}$$

$$E_y = \rho_{Fe} \cdot J_y = -\frac{\rho_{Fe}}{\mu} \cdot \frac{\partial B_z}{\partial x} \tag{24.49}$$

$$\frac{\partial E_y}{\partial x} - \frac{\partial E_x}{\partial y} = -\frac{\partial B_z}{\partial t} \tag{24.50}$$

with B the magnetic field (induction), E the electric field, J the current density, ρ_{Fe} the electric resistivity of the iron core and μ the permeability of the iron.

Eliminating the electric field components yields:

$$\frac{\partial^2 B_z}{\partial x^2} + \frac{\partial^2 B_z}{\partial y^2} = \frac{\mu}{\rho_{Fe}} \cdot \frac{\partial B_z}{\partial t} \tag{24.51}$$

As possible (transient) solutions for Eq. 24.51, we propose

$$b(x,y,t)=\sum_{\alpha,\beta} b_{\alpha\beta} \tag{24.52}$$

with

$$b_{\alpha\beta}=B_{\alpha\beta}\cos\alpha x\cdot\cos\beta y\cdot\exp(-\gamma_{\alpha\beta}t) \tag{24.53}$$

Because $x=0$ and $y=0$ are symmetry axes, only cosine functions are considered.

We will only take into account cases where the current in the coil is zero for $t>0^+$. From $t>0^+$ on, the magnetic field strength $\underrightarrow{H}$ as well as the induction $\underrightarrow{B}$ outside the core are zero. Because of the continuity of the tangential component of $\underrightarrow{H}$, the magnetic field strength and the induction at the boundaries $x=\pm a/2$ and $y=\pm b/2$ are zero from $t>0^+$ on.[3] Therefore, $\alpha=\alpha_m=m(\pi/a)$ and $\beta=\beta_n=n(\pi/b)$ with m and n odd. Substitution of the solutions of Eqs. 24.53 in 24.51 yields

$$\alpha^2+\beta^2=\frac{\mu}{\rho_{Fe}}\gamma_{\alpha\beta} \tag{24.54}$$

and thus

$$\gamma_{\alpha\beta}=\gamma_{mn}=\frac{\rho_{Fe}}{\mu}\left\{\left(m\frac{\pi}{a}\right)^2+\left(n\frac{\pi}{b}\right)^2\right\} \tag{24.55}$$

The proposed solution is

$$b(x,y,t)=b_z(x,y,t)=\sum_{m,n} B_{mn}\cos\alpha_m x\cdot\cos\beta_n y\cdot\exp(-\gamma_{mn}t) \tag{24.56}$$

and consists of the sum of mode (m,n). Each mode corresponds to transient circulating currents, as shown in (b) in Fig. 24.2 for $m=3$ and $n=3$. These currents follow from

$$J_x=\frac{\partial H_z}{\partial y} \tag{24.57}$$

$$J_y=-\frac{\partial H_z}{\partial x} \tag{24.58}$$

[3]This is only the transient solution. For other problems (e.g. switching on a sinusoidal excitation) steady-state solutions should be considered separately. Even with a non-zero current in the coil, however, the induction outside the coil is negligible because of the high iron permeability.

In addition to the boundary conditions in space, there are also boundary conditions in time.

We will investigate the case of a coil with very large diameter (so that its curvature can be disregarded) and with w_0 turns per meter. For $t < 0$ there is a DC current I_0 in the coil, giving rise to a steady-state induction B_0 in the core. The current is switched off at $t = 0$. From $t > 0^+$ on, there is no current in the coil and the magnetic field strength $\vec{H}$ as well as the induction $\vec{B}$ outside the core are zero.

The steady-state induction for $t < 0$ follows from Ampere's law:

$$\frac{B_0}{\mu}l = w_0 \cdot l \cdot I_0 \tag{24.59}$$

As a transient solution for[4] $-a/2 < x < a/2$ and $-b/2 < y < b/2$ for $t > 0^-$, we propose Eq. 24.56. The continuity at $t = 0$ then requires

$$b(x, y, t = 0) = b(x, y, 0) = \sum_{m,n} B_{mn} \cos \alpha_m x \cdot \cos \beta_n y = B_0 = \mu \cdot w_0 \cdot I_0 \tag{24.60}$$

Writing the two-dimensional block function B_0 as a product of two Fourier series (one in x and one in y) yields

$$B_{mn} = \left(\frac{4}{\pi}\right)^2 \frac{1}{m \cdot n}(-1)^{\frac{m+n-2}{2}} \cdot B_0 \tag{24.61}$$

with m and n odd. Therefore

$$b(x, y, t) = \left(\frac{4}{\pi}\right)^2 B_0 \cdot \sum_{m,n} \frac{1}{m \cdot n}(-1)^{\frac{m+n-2}{2}} \cos\left(m\pi\frac{x}{a}\right) \cdot \cos\left(n\pi\frac{y}{b}\right) \cdot \exp(-\gamma_{mn}t) \tag{24.62}$$

with γ_{mn} given by Eq. 24.55.

The current densities in the iron for $-a/2 < x < a/2$ and $-b/2 < y < b/2$ are

$$J_x = -\left(\frac{4}{\pi}\right)^2 \frac{B_0}{\mu} \cdot \frac{\pi}{b} \cdot \sum_{m,n} \frac{1}{m}(-1)^{\frac{m+n-2}{2}} \cos\left(m\pi\frac{x}{a}\right) \cdot \sin\left(n\pi\frac{y}{b}\right) \cdot \exp(-\gamma_{mn}t) \tag{24.63}$$

$$J_y = \left(\frac{4}{\pi}\right)^2 \frac{B_0}{\mu} \cdot \frac{\pi}{a} \cdot \sum_{m,n} \frac{1}{n}(-1)^{\frac{m+n-2}{2}} \sin\left(m\pi\frac{x}{a}\right) \cdot \cos\left(n\pi\frac{y}{b}\right) \cdot \exp(-\gamma_{mn}t) \tag{24.64}$$

[4] The edges are excluded as the sum is not uniformly convergent for $t = 0$.

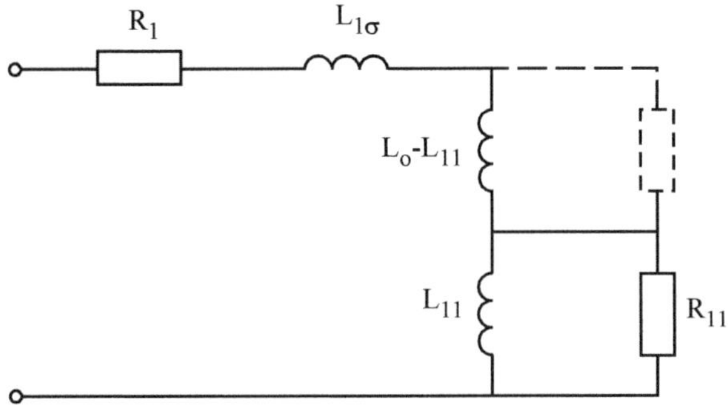

Fig. 24.3 Massive core: equivalent circuit for the fundamental mode

To calculate the flux, we will first have to integrate over $-(a-\epsilon)/2 < x < (a-\epsilon)/2$ and $-(b-\delta)/2 < y < (b-\delta)/2$ and then take the limit for $\epsilon, \delta \to 0$:

$$\Phi_{\epsilon\delta} = \left(\frac{4}{\pi}\right)^2 B_0 \cdot \int_{-\frac{(a-\epsilon)}{2}}^{\frac{(a-\epsilon)}{2}} dx \int_{-\frac{(b-\delta)}{2}}^{\frac{(b-\delta)}{2}} dy \sum_{m,n} \frac{1}{m \cdot n} (-1)^{\frac{m+n-2}{2}} \cos\left(m\pi \frac{x}{a}\right) \cdot \cos\left(n\pi \frac{y}{b}\right) \cdot \exp(-\gamma_{mn} t) \tag{24.65}$$

$$\Phi_{\epsilon\delta} = \left(\frac{4}{\pi}\right)^2 B_0 \cdot \frac{4ab}{\pi^2} \sum_{m,n} (\frac{1}{m \cdot n})^2 (-1)^{\frac{m+n-2}{2}} \sin\left(m\pi \frac{a-\epsilon}{2a}\right) \cdot \sin\left(n\pi \frac{b-\delta}{2b}\right) \cdot \exp(-\gamma_{mn} t) \tag{24.66}$$

and thus

$$\Phi = \lim_{\epsilon,\delta \to 0} \Phi_{\epsilon\delta} = \left(\frac{4}{\pi}\right)^2 \left(\frac{2}{\pi}\right)^2 \cdot ab \cdot B_0 \cdot \sum_{m,n} (\frac{1}{m \cdot n})^2 \cdot \exp(-\gamma_{mn} t) \tag{24.67}$$

The equivalent self-inductance (per meter) of the coil for mode m, n can be calculated in an analogous way as the total self-inductance

$$w_0 \Phi_0 = L_0 I_0 \tag{24.68}$$

The inductance L_{mn} results from $w_0 \Phi_{mn} = L_{mn} I_0$:

$$L_{mn} = \left(\frac{4}{\pi}\right)^2 \left(\frac{2}{\pi}\right)^2 \cdot ab \cdot \mu \cdot w_0^2 \cdot (\frac{1}{m \cdot n})^2 = \frac{64}{\pi^4 m^2 n^2} \cdot L_0 \tag{24.69}$$

with $L_0 = w_0^2 \mu ab$ the total self-inductance of the coil.

A time constant $1/\gamma_{mn} = L_{mn}/R_{mn}$ corresponds with each mode, with

$$R_{mn} = \frac{64}{\pi^2} \varrho_{Fe} w_0^2 \cdot \left(\frac{1}{n^2} \cdot \frac{b}{a} + \frac{1}{m^2} \cdot \frac{a}{b}\right) \tag{24.70}$$

The largest time constant corresponds with the mode (1, 1). This time constant is commonly called the transient time constant and the corresponding component (mode) the transient component. The other modes, with smaller time constants, are called subtransient components (corresponding with subtransient time constants). Figure 24.3 shows a possible approximate model (*equivalent circuit*); R_1 is the coil resistance and $L_{1\sigma}$ its leakage inductance (which corresponds here to field lines coupled with the coil but outside the iron core); R_{11} and L_{11} correspond with the transient component; the remaining modes are *lumped* in the dashed part.

24.4 Quasi-stationary Modelling of Rotating Machines

For rotating machines, there are not only the electrical transients but also mechanical transients (when the speed is variable).

In principle, electrical and mechanical transient phenomena have to be treated together: the variable speed affects the electrical transients (e.g. via the emf of motion) and the electrical transients affect the torque and, therefore, via the equation of motion, the speed.

In many cases, however, the mechanical time constants are one or more orders of magnitude larger than the *pure* electrical time constants (i.e. those that would occur at constant speed). If these time constants differ by an order of magnitude, it may be permitted to treat the electrical and mechanical transients separately: the electrical transients as if the speed were constant and the mechanical transients as if the electrical circuit were in steady state. For the analysis of the mechanical phenomena, we may then use the steady-state currents, voltages and torques that can be derived from the steady-state equations or equivalent circuits.

An example will be discussed in the next chapter on pulsating loads for an induction machine. As the mechanical time constant of this high-inertia drive is much larger than the electrical time constants of the machine and than the period of the pulsating torque, we may ignore the electrical transients and calculate everything as if the machine were in steady state. For small slip, the torque equation may also be approximated by its almost linear part:

$$T = \frac{3V^2}{\Omega_{sy}} \cdot \frac{R/s}{(R/s)^2 + X_\sigma^2} \approx \frac{3V^2}{\Omega_{sy}} \cdot \frac{s}{R} \tag{24.71}$$

The equation of motion can then be approximated by (using $\Omega_r = (1-s)\Omega_{sy}$):

$$-J\Omega_{sy} \cdot \frac{ds}{dt} = \frac{3V^2}{\Omega_{sy}} \cdot \frac{s}{R} - T_l \tag{24.72}$$

However, we need to keep in mind that, actually, the electrical transients may give rise to rather large transient torques, which are entirely disregarded here.

Chapter 25
Induction Machines with Pulsating Loads

Abstract For many induction machines with pulsating loads, the pulsation frequency is rather low. Therefore, a quasi-stationary approach can be used to study the behaviour of the induction machine. In this chapter it is shown that a simple mathematical averaging cannot be used for dimensioning the machine.

25.1 Introduction

The torque of the load driven by an electrical machine is not always constant. In many applications, pulsating loads are common, for example in punching and for reciprocating compressors and pumps. What is typical of these kinds of applications is the periodic character of the pulsating loads, where a rather high load torque T_b is applied during a relatively short time t_b, a fraction (e.g. 10%) of the load period t_c. The frequency of the torque pulsations is much smaller than the supply frequency, typically $1 : 100$ or $1 : 1000$. An idealised waveform is shown in (b) in Fig. 25.1. The figure also shows a simplified schematic of such a drive in (a), which typically includes a high inertia to smooth out the torque.

A main problem is the dimensioning of the drive, including the choice of the (required) inertia, the mechanical transmission and gear box and the choice of the induction machine.

Suppose the load torque (or force) is known, as is the required frequency of the torque pulsation (e.g. 30 strokes per minute). The gear box ratio and the number of pole pairs of the induction machine follow from the required frequency of the torque pulses, preferably avoiding two-pole machines (because of their inferior characteristics) and choosing the more standard four-pole or six-pole machines. These are the easiest design parameters.

Choosing the inertia and the rated power of the induction machine is much more complicated. A traditional approach starts from an estimation of the energy per torque pulse $E_n = \Omega_{av} \cdot T_b \cdot t_b$, in which the average speed Ω_{av} is known from the required frequency of the torque pulses.

The inertia has to be sufficiently high to smooth out the torque pulse over the cycle duration. However, at the same time, the allowable speed variation has to be

J. A. Melkebeek, *Electrical Machines and Drives*, Power Systems,
https://doi.org/10.1007/978-3-319-72730-1_25

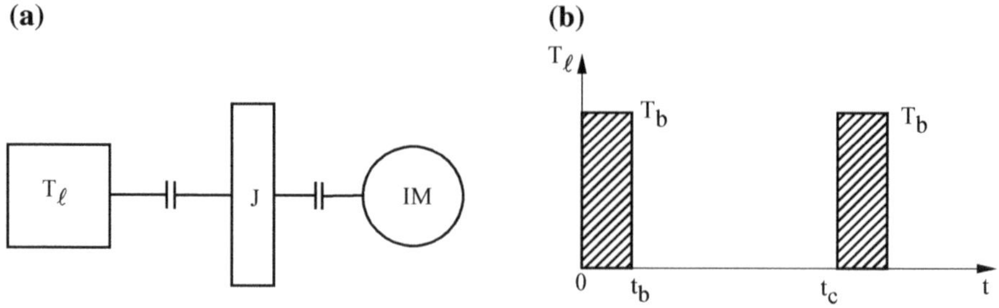

Fig. 25.1 Pulsating load torque

sufficiently large. The large speed (and this slip) variation requires the rotor resistance of the induction machine to be sufficiently high, i.e. in pu at least $r_r \approx s_n = 5 \ldots 10\%$ or even up to 30%. After choosing an acceptable speed variation $\Delta\Omega$, we may then derive the inertia from

$$E_n = \Omega_{av} \cdot T_b \cdot t_b = \frac{1}{2} J \cdot (\Omega_{max}^2 - \Omega_{min}^2) = J \cdot \Omega_{av} \cdot \Delta\Omega \tag{25.1}$$

In traditional approaches, the chosen induction machine rated power is (was) equal to the average power requirement of the load. As will be shown in the next sections, this may lead to overheating of the machine.

25.2 Quasi-stationary Analysis

For the following analysis, we suppose that the electrical transients have a much smaller time constant than the mechanical time constants and the period of the torque pulsations, allowing us to disregard them. We will also assume that the induction machine always operates in the quasi-linear range $s \ll r_r/x_\sigma$ of the torque-slip characteristic, with $T = (T_n/s_n) \cdot s$. With $\Omega = \Omega_{sy}(1 - s)$, the equation of motion can then be written as

$$J\Omega_{sy} \frac{ds}{dt} + \frac{T_n}{s_n} \cdot s = T_l \tag{25.2}$$

Introducing

$$\kappa = \frac{s_n}{T_n} \cdot \frac{\Omega_{sy}}{t_c} \cdot J = \frac{s_n \tau_i}{t_c} \tag{25.3}$$

the equation of motion can be written as

$$\frac{T_l}{T_n} = \frac{s}{s_n} + \kappa \frac{d(s/s_n)}{d(t/t_c)} \tag{25.4}$$

or, in terms of T/T_n, as

$$\frac{T_l}{T_n} = \frac{T}{T_n} + \kappa \frac{d(T/T_n)}{d(t/t_c)} \tag{25.5}$$

Note that the point (s_n/T_n) determines the slope of the torque-slip characteristic, but is not necessarily the rated operating point (s_{nom}/T_{nom}). For a slip-ring induction motor, for example, an external resistance can be used to increase the pull-out slip, i.e. to reduce the slope of the torque-slip characteristic.

Using the boundary conditions $T_l = T_b$ for $0 \le t \le t_b$ and $T_l = 0$ for $t_b < t \le t_c$, we obtain

- for $0 \le t \le t_b$

$$\frac{T}{T_n} = \frac{s}{s_n} = \frac{T_b}{T_n} - \left(\frac{T_b}{T_n} - \frac{T_{min}}{T_n}\right) \exp(-t/\kappa t_c) \tag{25.6}$$

- for $t_b < t \le t_c$

$$\frac{T}{T_n} = \frac{s}{s_n} = \frac{T_{max}}{T_n} \exp[-(t - t_b)/\kappa t_c] \tag{25.7}$$

where, denoting the relative loading time as $\varrho = t_b/t_c$,

$$T_{max} = T_b \frac{1 - \exp(-\rho/\kappa)}{1 - \exp(-1/\kappa)} \tag{25.8}$$

$$T_{min} = T_{max} \cdot \exp[-(1 - \rho)/\kappa] \tag{25.9}$$

From Eqs. 25.6 and 25.7, we may derive

$$T_{av} = \rho T_b \tag{25.10}$$

$$T_{rms} = T_b \sqrt{\rho \left[1 - \frac{\kappa}{\rho}\left(\frac{T_{max} - T_{min}}{T_b}\right)\right]} \tag{25.11}$$

Figure 25.2 illustrates the motor torque and the speed as a function of time, corresponding to Eqs. 25.6 and 25.7.

Figure 25.3 presents some relations between T_b, T_{max}, T_{rms}, T_{av} as a function of the parameters κ and ρ. The corresponding slip values are easily calculated using $s/T = s_n/T_n = s_{max}/T_{max} = s_{min}/T_{min} = s_{av}/T_{av} = s_{rms}/T_{rms}$.

From the previous analysis results, we may derive the average slip (or secondary) losses and grid (supply) power during one period of the load:

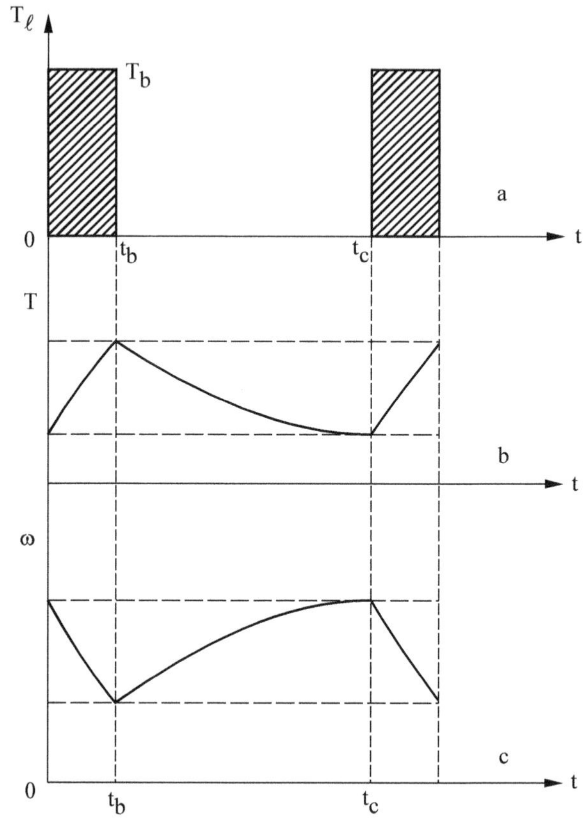

Fig. 25.2 Torque and speed as function of time

$$P_{jr} = \frac{1}{t_c}\int_0^{t_c} T \cdot s \cdot \Omega_{sy} dt = \left(\frac{T_{rms}}{T_n}\right)^2 T_n \Omega_{sy} s_n = s_{rms} T_{rms} \Omega_{sy} \tag{25.12}$$

$$P_{grid} = \frac{1}{t_c}\int_0^{t_c} T \cdot \Omega_{sy} dt = T_{av} \Omega_{sy} = \rho T_b \Omega_{sy} \tag{25.13}$$

Clearly, the slip or secondary losses are determined by the rms value of torque (and slip), while the supply power is determined by the average value of the torque. The average supply power is not influenced by the machine (disregarding the effect of the speed and the machine torque characteristic on the relative loading time ρ). However, as the secondary losses are narrowly related to the rms value of the torque, they are also narrowly related to κ and the slope of the torque-slip characteristic.

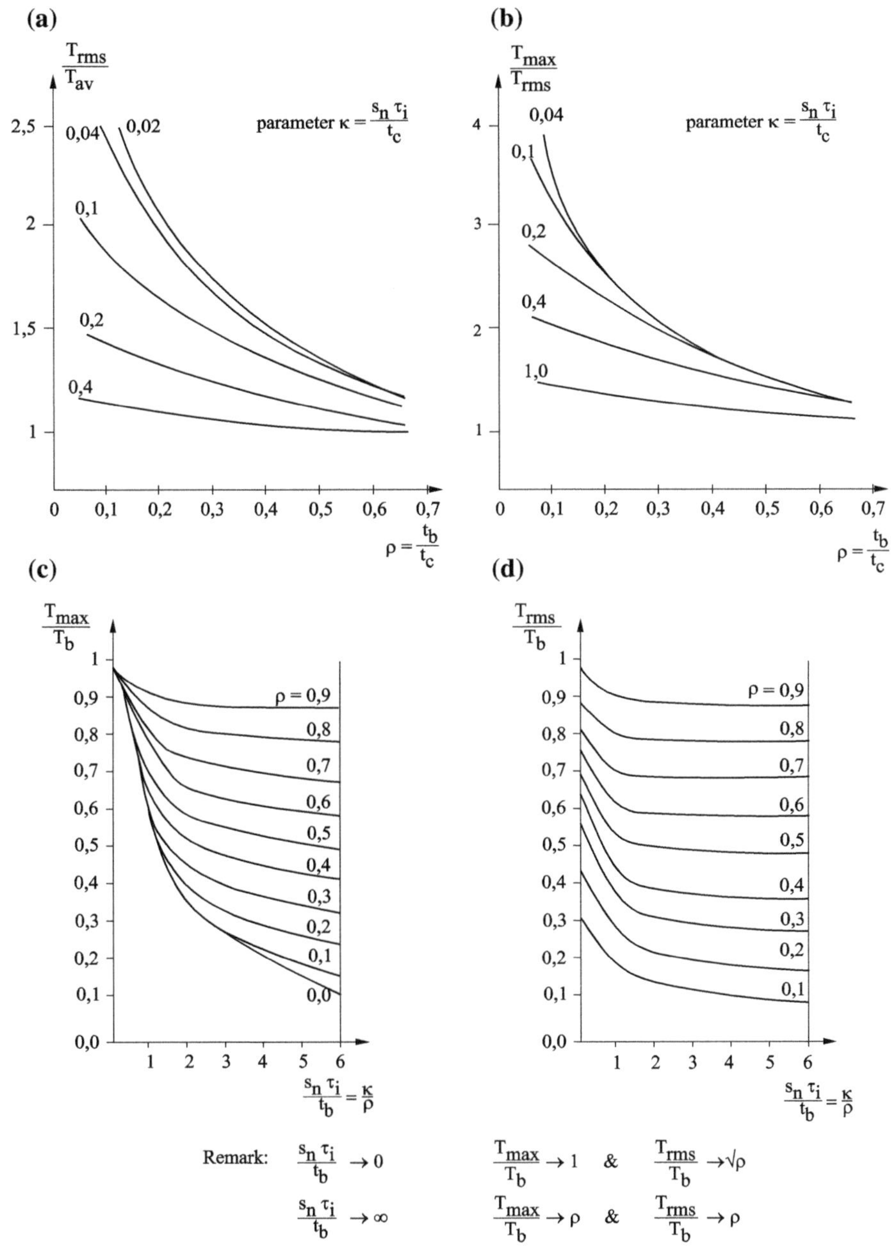

Fig. 25.3 Characteristics of maximum, rms and average torque

The rotor losses are a measure[1] for the thermal loading of the machine and will therefore determine the required power rating of the machine. Indeed, the rating of a machine is based on the maximum losses that are acceptable in steady state for the given machine size (the equilibrium between the losses and the dissipated heat determines the operating temperature of the machine). For a slip-ring machine, not all these secondary losses are dissipated as heat inside the machine. If R_{ex} is the external resistance connected at the slip rings then, with $k = (R_{ex} + R_r)/R_r$, only P_{jr}/k is dissipated inside the machine.

25.3 Drive Dimensioning

After the gear box transmission ratio and the associated number of poles of the induction machine have been chosen (see above), all that remains to be done in terms of design is dimension the inertia and choose the rated power rating of the induction machine. For this, two criteria are decisive:

- the joule loss inside the machine during the load cycle determines the power rating of the induction machine
- the maximum motor torque during the cycle should remain sufficiently below the pull-out torque of the machine, for example no more than 2/3 of the pull-out torque[2]

Mathematically, these criteria can be summarised as follows:

$$P_{jr} = s_{rms} T_{rms} \Omega_{sy} \tag{25.14}$$

$$P_{jr,int} \leq P_{jr,nom} = s_{nom} T_{nom} \Omega_{sy} \tag{25.15}$$

$$T_{max} \leq \frac{2}{3} T_{po} \approx \frac{5}{3} T_{nom} \tag{25.16}$$

$$\frac{s_n}{T_n} = \frac{s_{rms}}{T_{rms}} = \frac{s_{max}}{T_{max}} = k \cdot \frac{s_{nom}}{T_{nom}} \tag{25.17}$$

with $P_{jr,int} = P_{jr}$ for a cage induction motor, or $P_{jr,int} = P_{jr}/k$ for a slip-ring induction motor (external resistance increase).

---

[1] We disregarded the stator losses but the total copper joule losses may be approximated by doubling the rotor losses.

[2] So that the operating point always remains in the linear part of the characteristic, and to keep a safety margin as well.

As described above, for an acceptable design, the inertia should be large enough[3] to smooth out the load torque, but at the same time the speed variation should be sufficiently large. The latter in turn requires the rotor resistance to be sufficiently large. This is often a problem with standard cage induction motors, as these are designed for high efficiency, which requires small (rotor) resistances. To remedy this, sometimes a standard machine frame (i.e. standard stator and rotor laminations, slot dimensions and stator windings) is used, but the rotor cage material is replaced by an alloy with higher resistance. A slip-ring induction motor, on the other hand, makes it possible to add external resistances to flatten the torque-slip characteristic as desired. Two disadvantages of this approach are that slip-ring machines are more expensive and more difficult to maintain.

In other words, three options can be distinguished:

- a standard cage induction motor with a sufficiently high rotor resistance can be used right away. From Eqs. 25.14–25.17, we obtain (with $k = 1$): $s_{rms} T_{rms} \leq s_{nom} T_{nom}$, $T_{rms} \leq T_{nom}$, $s_{rms} \leq s_{nom}$, $T_{max} \leq (5/3) T_{nom}$. In these expressions, T_{rms}, T_{max} and s_{rms} depend on the choice of κ, and as the slope of the torque-slip characteristic is known, they depend on the choice of the inertia J. This inertia needs to be chosen so that the resulting T_{max} fulfils $T_{max} \leq (5/3) T_{nom}$. If multiple machine choices are possible, the most economical solution is the one closest to the equalities, i.e. $T_{rms} = T_{nom}$, $s_{rms} = s_{nom}$. It should be stressed, however, that such a solution is only possible when an induction motor with a sufficiently high rotor resistance is available.
- with a slip-ring induction motor, the secondary resistance can easily be adapted to obtain the optimal slope of the torque-slip characteristic. An additional advantage is that the part of the secondary loss corresponding to the external resistance is developed as heat outside the motor. From Eqs. 25.14–25.17, we now obtain: $s_{rms} T_{rms} \leq k \cdot s_{nom} T_{nom}$, $T_{rms} \leq T_{nom}$, $s_{rms} \leq k \cdot s_{nom}$, $T_{max} \leq (5/3) T_{nom}$ as $P_{jr,int} = P_{jr}/k$. If we choose $k \gg 1$, the required inertia for the same value of κ (which determines s_{rms}, T_{rms}) can be much smaller. Keep in mind, however, that too large a value of s_{rms} will reduce the relative loading time ρ as well as the average speed and the frequency of the load cycle.
- for a cage induction motor with a modified (increased) rotor resistance, all secondary slip losses are developed inside the machine. From Eqs. 25.14–25.17, we obtain $s_{rms} T_{rms} \leq s_{nom} T_{nom}$, $T_{rms} \cdot \sqrt{k} \leq T_{nom}$, $s_{rms} \leq \sqrt{k} \cdot s_{nom}$, $T_{max} \leq (5/3) T_{nom}$. As a result of the additional losses inside the machine due to the increased rotor resistance, the rated torque (or power) of the machine should now be higher than the calculated rms value by a factor $\sqrt{k}$. Indeed, the standard induction machine has been designed for the standard rotor resistance and the associated loss.

Clearly, this solution is only useful if the rms value of the torque calculated using the higher κ (with the higher rotor resistance) is lower by at least a factor $\sqrt{k}$ than

[3]Note, however, that the machine should be able to accelerate the inertia from standstill without overheating, so there is a limit.

the value that would be obtained for the standard resistance. Considering Fig. 25.3, this can only be the case if $\rho \ll 1$.

In some practical cases, the above design procedure may require some iterative calculations, as a chosen machine and inertia may lead to an unacceptable low or high average speed (i.e. the cycle period of the torque pulses).

The above calculation method is quasi-stationary, i.e. the electrical dynamics of the drive are ignored. In nearly all practical cases this will be justified, on the one hand because the frequency of the torque pulsations is much lower than the electrical time constants, and on the other hand because the electrical transients turn out to be aperiodically damped due to the large rotor resistance (see Chap. 27).

Chapter 26
Modelling and Dynamic Behaviour of DC Machines

Abstract As an introduction to the later chapters on the dynamics of induction and synchronous machines, in this chapter we study the modelling and dynamic behaviour of DC commutator machines. The traditional machine model, using a simplified modelling of the main field saturation, permits to derive the basic dynamic properties. A more accurate model for main field saturation is also presented. It is shown that the dynamic properties derived using this model are slightly different from those using the traditional model, but also somewhat more realistic.

26.1 Standard Dynamic Model of the DC Machine

26.1.1 Basic Assumptions and Equations

We will consider the most basic form of the DC commutator machine, i.e. without auxiliary poles and commutation windings (see Fig. 26.1).

The electrical model consists of two equations, one for the armature and one for the excitation, here with the URS (Users Reference System) applied:

$$V_a = R_a I_a + \frac{d}{dt}\Psi_a + K\Omega_r \Psi_{fm} \tag{26.1}$$

$$V_f = R_f I_f + \frac{d}{dt}\Psi_f \tag{26.2}$$

Subscript a stands for the armature, subscript f for an independent excitation. In the armature equation, K is the emf or torque constant. Ψ_f is the total d-axis excitation flux. The main flux linkage Ψ_{fm} is that part of the total excitation flux Ψ_f which is linked with the armature. The remaining part of the flux is the leakage flux $\Psi_{f\sigma}$, linked only with the excitation winding:

$$\Psi_f = w_f \cdot \Phi_f(I_f,\ I_a) = \Psi_{fm}(I_f,\ I_a) + \Psi_{f\sigma}(I_f) \tag{26.3}$$

$$\Psi_a = \Psi_a(I_a,\ I_f) = w_a \cdot \Phi_a(I_a,\ I_f) = \Psi_{am}(I_a,\ I_f) + \Psi_{a\sigma}(I_a) \tag{26.4}$$

J. A. Melkebeek, *Electrical Machines and Drives*, Power Systems,
https://doi.org/10.1007/978-3-319-72730-1_26

Fig. 26.1 DC machine

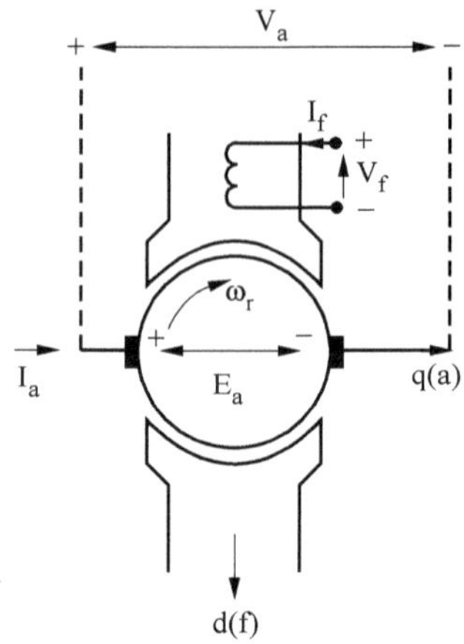

The notation Φ denotes flux per turn (whereas Ψ represents the total flux linkage). In general, both armature and excitation fluxes are non-linear functions of the currents.

In addition to the electrical equations, the dynamics also require the mechanical equation of motion:

$$J\frac{d}{dt}\Omega_r = T - T_l \tag{26.5}$$

where the torque is given by $T = K\Psi_{fm}I_a$.

The steady-state equations (subscript o) are similar to Eqs. 26.1–26.5:

$$V_{ao} = R_a I_{ao} + K\Omega_{ro}\Psi_{fmo} \tag{26.6}$$

$$V_{fo} = R_f I_{fo} \tag{26.7}$$

$$T_{lo} = T_o = K\Psi_{fmo}I_{ao} \tag{26.8}$$

The equations for small deviations around the steady state are derived by subtracting the steady-state equations from the general equations and only considering small variations (i.e. ignoring second-order variations). The equations are indeed inherently non-linear due to the emf and torque expressions, involving products of state variables, but non-linearity can generally also be attributed to the saturation of the magnetic state. Linearisation then results in:

$$\Delta V_a = R_a\Delta I_a + \frac{d}{dt}\Delta\Psi_a + K\Omega_{ro}\Delta\Psi_{fm} + K\Psi_{fmo}\Delta\Omega_r \tag{26.9}$$

$$\Delta V_f = R_f\Delta I_f + \frac{d}{dt}\Delta\Psi_f \tag{26.10}$$

$$\Delta T = K\Psi_{fmo}\Delta I_a + K I_{ao}\Delta\Psi_{fm} = J\frac{d}{dt}\Delta\Omega_r + \Delta T_l = J\frac{d}{dt}\Delta\Omega_r + K_w\Delta\Omega_r + \Delta T_l' \tag{26.11}$$

The load torque (or driving torque for a generating DC machine) has been split up into a speed-dependent part, linearised as $K_w\Delta\Omega_r$, and an external input $\Delta T_l'$.

26.1.2 Per-Unit (pu) or Relative Description

In some cases, a pu description is very practical, for example to compare the behaviour of machines with different power ratings. In a pu description, voltages, currents, torques, and speed, among other things, are referred to a reference value. To avoid introducing unwanted proportionality constants, these reference values should obey the basic physical laws such as conservation of energy or power. As a consequence, only three reference values can be chosen freely. For a machine that is meant to or designed to work mainly as a generator, the reference values are usually rated voltage V_n, rated current I_n and rated speed Ω_n. Usually, for a motor, the basic reference values are rated torque T_n, rated speed Ω_n, and rated current I_n.

In pu notation, the previous equations become the following:

$$v_a = r_a i_a + \frac{d}{d\tau}\varphi_a + \nu_r\varphi_{fm} \tag{26.12}$$

$$v_f = r_f i_f + \frac{d}{d\tau}\varphi_f \tag{26.13}$$

$$\varphi_f = \varphi_{fm}(i_f, i_a) + \varphi_{f\sigma}(i_f) \tag{26.14}$$

$$\varphi_a = \varphi_a(i_a, i_f) \tag{26.15}$$

$$\tau_n\frac{d}{d\tau}\nu_r = t - t_l \tag{26.16}$$

$$t = \varphi_{fm} i_a \tag{26.17}$$

with $\nu_r = \Omega_r/\Omega_n$ the pu speed; $\tau = \Omega_n t$ the pu time; $\tau_n = J\Omega_n^2/T_n$ the pu inertia time constant.

26.1.3 Modelling of Saturation and Armature Reaction

In rated conditions, nearly all practical electrical machines have saturated magnetic circuits. The main raison is that this yields an optimal condition as to power conversion for a given machine size and weight. Other reasons might be to obtain stable behaviour for self-generating (e.g. in shunt-excited DC machines).

In a DC machine, when the magnetic circuit is saturated, there is a coupling between the two orthogonal axes (i.e. the d-axis or main field axis and the q-axis or armature axis). A well-known consequence is the armature reaction, where an increasing armature current results in a decreasing main field linkage. As is well known, an independently excited DC motor may then become unstable for high loads. For an independently excited generator, in contrast, armature reaction results in current limiting (i.e. stabilising).

Mathematically, saturation of the magnetic circuit results in a non-linear dependence of the fluxes on the currents in the two axes, as mentioned above (Eqs. 26.3 and 26.4). For a linear magnetic circuit, the fluxes in each of the two (supposed orthogonal) axes are only dependent on the current in that axis. When the circuit is saturated, common (saturated) parts in the two flux linkages make each flux dependent on both currents. In Refs. [22, 24], a flux model for a DC machine is elaborated on, with the following expressions for the two flux linkages:

$$\Phi_{fm}(I_f, I_a) = \Phi(I_f) + \frac{1}{3!} \cdot \frac{d^2}{dI^2}\Phi(I_f) \cdot (kI_a)^2 + \cdots \tag{26.18}$$

$$\Phi_{am}(I_a, I_f) = \frac{1}{2}\frac{b_p}{\tau_p}\left\{\frac{1}{3}\frac{d}{dI}\Phi(I_f) \cdot (kI_a) + \frac{1}{5 \cdot 3!} \cdot \frac{d^3}{dI^3}\Phi(I_f) \cdot (kI_a)^3 + \cdots\right\} \tag{26.19}$$

$\Phi(I_f)$ represents the useful d-axis flux in the absence of armature current, $\Phi(I_f) = \Phi_{fm}(I_f, 0)$.

These expressions are derived supposing an air-gap mmf which is proportional to the armature current and varies linearly with the coordinate x along the circumference of the armature

$$B(\cdot) = B(I_x) = B(I_f + kI_a x) \tag{26.20}$$

with $k = (w_a/w_f)/\tau_p$. The non-linear function $B(\cdot)$ is the saturation characteristic of the iron of the armature teeth and the pole shoes (the yokes are assumed not to saturate). $\Phi_{fm}(\cdot) = l_a \cdot b_p \cdot B(\cdot)$, with l_a representing the armature length. τ_p and b_p are the pole pitch and pole shoe width, respectively. Note that $\Phi_{fm}(\cdot) = \Phi_{fm}(I_f) = \Phi_{fm}(I_f, 0)$ represents the (measured) no-load characteristic of the DC machine.

The magnetising fluxes coupled with excitation winding and armature are $\Psi_{fm} = w_f \Phi_{fm}$ and $\Psi_{am} = w_a \Phi_{am}$, respectively. It can easily be shown that $\partial\Psi_{fm}/\partial I_a = \partial\Psi_{am}/\partial I_f$: also for saturated magnetic circuits, the reciprocity property holds [24].

Equations 26.18 and 26.19 can be approximated by keeping only the first two terms in the series. As the second derivative of the no-load characteristic $\Phi_{fm}(I_f)$ is negative, we may write (with $k_f > 0$):

$$\Phi_{fm}(I_f, I_a) = \Phi_{fm}(I_f) - k_f(I_f) \cdot I_a^2 \tag{26.21}$$

Indeed, as we know, whatever the sign of the armature current, an (increasing) armature current will always result in a flux decrease (armature reaction).

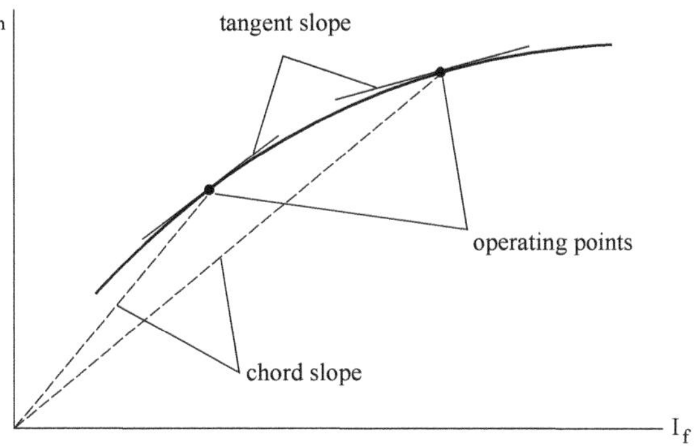

Fig. 26.2 Chord- and tangent-slope inductances

From the equations above, a rather practical mathematical flux model can be derived.

For steady state and no load, the main flux can be written using the chord-slope inductance $\psi_{fmo}(I_{fo}) = L_{fmo}(I_{fo}) \cdot I_{fo}$. As a consequence, for a loaded machine,

$$\Psi_{fmo}(I_{fo}, I_{ao}) = L_{fmo}(I_{fo}) \cdot I_{fo} - w_f k_f(I_{fo}) \cdot I_{ao}^2 = L_{fmo}(I_{fo}) \cdot I_{fo} - K_f(I_{fo}) \cdot I_{ao}^2 \tag{26.22}$$

Small variations around an equilibrium point, on the other hand, are described using incremental or tangent-slope inductances (see also Fig. 26.2)

$$\Delta\Psi_{fm} = L_{fmt}(I_{fo}) \cdot \Delta I_f - 2K_f(I_{fo}) \cdot I_{ao} \cdot \Delta I_a = L_{fmt}(I_{fo}) \cdot \Delta I_f - M_{fa}(I_{fo}, I_{ao}) \cdot \Delta I_a \tag{26.23}$$

The inductance L_{fmt} is the tangent-slope inductance of the no-load characteristic for the excitation current I_{fo}. Note that the dynamic (or incremental) mutual inductance M_{fa} can be positive or negative, dependent on the sign of the armature current: $M_{fa}(I_{fo}, I_{ao}) = 2K_f \cdot I_{ao} = 2K_f \cdot |I_{ao}| \cdot sign(I_{ao})$.

For the total excitation flux, the leakage flux needs to be added (the leakage flux can be assumed to be unsaturated).

In a similar way, the armature flux can be described. For steady state, we have

$$\Psi_{ao}(I_{ao}, I_{fo}) = L_{ao}(I_{ao}, I_{fo}) \cdot I_{ao} \tag{26.24}$$

For small variations around steady state

$$\Delta\Psi_a = L_{at}(I_{ao}, I_{fo}) \cdot \Delta I_a - M_{af}(I_{ao}, I_{fo}) \cdot \Delta I_f \tag{26.25}$$

with $L_{at} = \left(\frac{\partial\Psi_a(I_a,I_f)}{\partial I_a}\right)_{I_{ao},I_{fo}}$ and $M_{af} = -\left(\frac{\partial\Psi_a(I_a,I_f)}{\partial I_f}\right)_{I_{ao},I_{fo}}$. As mentioned above, $M_{af} = M_{fa} = -\left(\frac{\partial\Psi_{fm}(I_f,I_a)}{\partial I_a}\right)_{I_{ao},I_{fo}}$.

The flux model above can be combined with Eqs. 26.9–26.11 for the DC machine. However, due to the interaction between armature current and magnetising flux, such a dynamic analysis is rather complicated. The results of such an analysis will be presented at the end of this chapter.

In the next section, we will discuss the analysis of the dynamic behaviour by means of a simplified standard model that uses the chord-slope inductance in the operating point for both the steady-state flux and flux variations, and that disregards the armature reaction. This simplified model is thus obtained by combining Eqs. 26.9–26.11 with the following flux model:

$$\Delta\Psi_{fm} = L_{fmo} \cdot \Delta I_f \tag{26.26}$$

$$\Delta\Psi_f = L_{fo} \cdot \Delta I_f = L_{fmo} \cdot \Delta I_f + L_{f\sigma} \cdot \Delta I_f \tag{26.27}$$

$$\Delta\Psi_a = L_{ao} \cdot \Delta I_a \tag{26.28}$$

The subscript “o” for the (chord-slope) inductances will, however, be omitted in the analysis below.

26.2 Characteristic Dynamic Behaviour According to the Standard Model

As an example, we will study the dynamic behaviour of an independently excited DC machine (in a similar way, a series-excited DC machine can be analysed). If we rewrite Eqs. 26.9–26.11 in the Laplace domain and combine them with the simplified saturation model, this results in the following:

$$\Delta V_a = R_a \Delta I_a + sL_a \Delta I_a + K\Omega_{ro}\Delta\Psi_{fm} + K\Psi_{fmo}\Delta\Omega_r \tag{26.29}$$

$$\Delta V_f = R_f \Delta I_f + sL_f \Delta I_f \tag{26.30}$$

$$\Delta\Psi_{fm} = L_{fm}\Delta I_f \tag{26.31}$$

$$\Delta T = K\Psi_{fmo}\Delta I_a + KI_{ao}\Delta\Psi_{fm} = Js\Delta\Omega_r + \Delta T_l = Js\Delta\Omega_r + K_w\Delta\Omega_r + \Delta T_l' \tag{26.32}$$

For motoring, the output variable is the speed, while the input is the armature voltage and the disturbance input is the external load torque $\Delta T_l'$. For a constant excitation, the system becomes of second order. It can also be represented by the block scheme in Fig. 26.3. Clearly, the emf acts as an internal feedback.

The characteristic equation is

$$1 + \frac{K^2\Psi_{fmo}^2}{(R_a + sL_a)(K_w + sJ)} = 0 \tag{26.33}$$

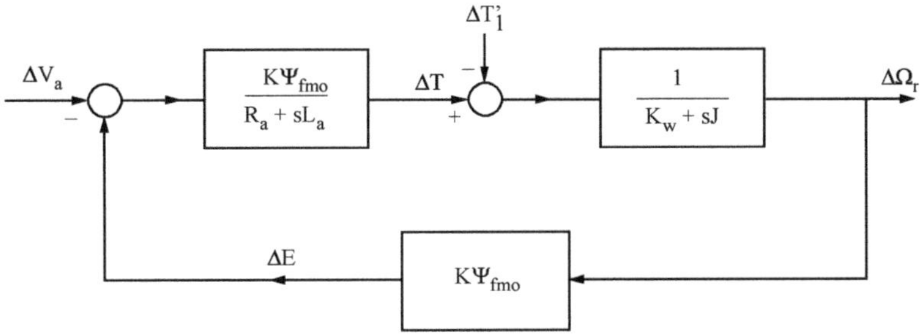

Fig. 26.3 Block scheme for motoring

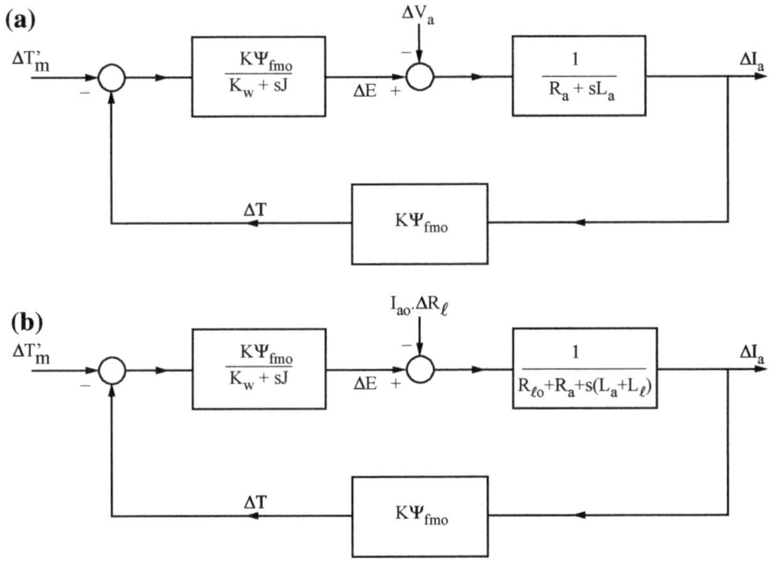

Fig. 26.4 Block schemes for generating

If the excitation is not constant, the eigenvalues remain unchanged but in addition to the external input, another input has to be added to ΔV_a:

$$\frac{\left[R_a - K\Psi_{fmo}\Omega_{ro}/I_{ao} + sL_a\right]\left(L_{fm}I_{ao}/\Psi_{fmo}\right)}{R_f + sL_f}\Delta V_f \tag{26.34}$$

For generating, the input variable is the external driving torque, and the output is the armature current, while the disturbance input is the DC mains voltage (for generating in a DC grid, see (a) in Fig. 26.4), or the load resistance (for generating in island mode, see (b) in Fig. 26.4). The eigenvalues for generating are, however, the same as in motoring mode (for island generating, the inductance and resistance of the load have to be added to the armature inductance and resistance).

The number of parameters to describe the (typical) dynamic behaviour can nevertheless be reduced to three dimensionless parameters (or even two, if the mechanical damping can be disregarded):

$$\tau_a = \varrho_a^{-1} = \frac{\Omega_n L_a}{R_a} = \frac{l_a}{r_a} \tag{26.35}$$

$$\tau_m = \varrho_m^{-1} = \frac{J \Omega_n R_a}{K^2 \Psi_{fmo}^2} = \frac{\tau_n r_a}{\varphi_{fmo}^2} \tag{26.36}$$

$$\tau_w = \varrho_w^{-1} = \frac{J \Omega_n}{K_w} = \frac{\tau_n}{k_w} \tag{26.37}$$

These armature, electromechanical and damping time constants are made dimensionless by using $1/\Omega_n$ as time reference.

With these parameters, the characteristic equation can be written as

$$1 + \frac{\varrho_a \varrho_m}{(\varrho_a + p)(\varrho_w + p)} = 0 \tag{26.38}$$

where p represents the pu Laplace operator ($p = s/\Omega_n \iff d/d\Omega_n t$). The open-loop poles of this second-order system are the negative inverse of the armature time constant $p = -\varrho_a = -\tau_a^{-1}$ and the mechanical damping time constant $p = -\varrho_w = -\tau_w^{-1} \approx 0$.

Referring the Laplace operator to the armature time constant, $\lambda = p\tau_a$, we may obtain an even more convenient representation

$$1 + \frac{\tau_a/\tau_m}{(1 + p\tau_a)(\tau_a/\tau_w + p\tau_a)} = 0 \tag{26.39}$$

or

$$1 + \frac{\tau_a/\tau_m}{(1 + \lambda)(\tau_a/\tau_w + \lambda)} = 0 \tag{26.40}$$

or, if the mechanical damping is negligible

$$1 + \frac{\tau_a/\tau_m}{(1 + \lambda)\lambda} = 0 \tag{26.41}$$

Open-loop poles are now $\lambda = -1$ and $\lambda = -\tau_a/\tau_w$. The gain is the ratio of the (electrical) armature time constant and the electromechanical time constant. Figure 26.5 shows the root locus.

For small gain values, we can find two negative real eigenvalues. For a very small gain (i.e. a large electromechanical time constant compared to the armature time

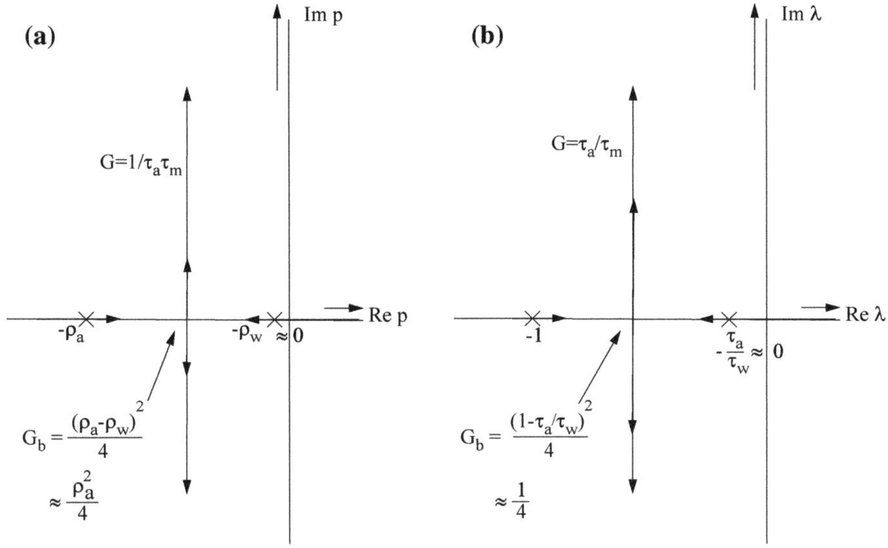

Fig. 26.5 Root locus for the independently excited DC motor

constant), the dominant pole of the closed loop is $\lambda = -\tau_a/\tau_m$ or $p = -1/\tau_m$. The dynamics are then completely determined by the inertia.[1]

For somewhat larger gain values, the two real eigenvalues approach the breakaway point $\lambda \approx -1/2$. For very large gain values ($\tau_a/\tau_m > 1/4$), the eigenvalues become complex conjugate with a real part equal to approximately (minus) half the inverse armature time constant ($\lambda = p\tau_a \approx -1/2$).

Clearly, the dynamic behaviour of the independently excited DC machine is quite stable, at least when the armature reaction is absent (see also the next section).

26.3 Characteristic Dynamic Behaviour Taking into Account Saturation and Armature Reaction

When saturation and armature reaction are taken into account, the analysis becomes much more complex. When the magnetic circuit is saturated, there are not only two inductances (i.e. chord-slope and tangent-slope) in the equations for each axis to complicate matters, but the cross-saturation (i.e. armature reaction) in particular renders the machine model rather complicated.

[1]The electromechanical time constant is proportional to the ratio of the inertia and the slope of the stationary torque-speed characteristic, as can be shown from a quasi-stationary model of the machine.

With saturation taken into consideration, the basic equations for small deviations around a steady state become (see also Sect. 26.1.3):

$$\Delta V_a = R_a \Delta I_a + L_{at} s \Delta I_a - M_{af} s \Delta I_f + K \Omega_{ro} \Delta \Psi_{fm} + K \Psi_{fmo} \Delta \Omega_r \quad (26.42)$$

$$\Delta V_f = R_f \Delta I_f + L_{f\sigma} s \Delta I_f + s \Delta \Psi_{fm} \quad (26.43)$$

$$\Delta \Psi_{fm} = L_{fmt} \Delta I_f - M_{fa} \Delta I_a \quad (26.44)$$

$$\Delta T = K \Psi_{fmo} \Delta I_a + K I_{ao} \Delta \Psi_{fm} = J s \Delta \Omega_r + \Delta T_l = J s \Delta \Omega_r + K_w \Delta \Omega_r + \Delta T_l' \quad (26.45)$$

or in pu

$$\Delta v_a = r_a \Delta i_a + l_{at} p \Delta i_a - m_{af} p \Delta i_f + \nu_{ro} \Delta \varphi_{fm} + \varphi_{fmo} \Delta \nu_r \quad (26.46)$$

$$\Delta v_f = r_f \Delta i_f + l_{f\sigma} p \Delta i_f + p \Delta \varphi_{fm} \quad (26.47)$$

$$\Delta \varphi_{fm} = l_{fmt} \Delta i_f - m_{fa} \Delta i_a \quad (26.48)$$

$$\Delta t = \varphi_{fmo} \Delta i_a + i_{ao} \Delta \varphi_{fm} = \tau_n p \Delta \nu_r + \Delta t_l = \tau_n p \Delta \nu_r + k_w \Delta \nu_r + \Delta t_l' \quad (26.49)$$

If the independent excitation can be considered as current-fed, the system remains of second order, but the open-loop poles change compared to the unsaturated model. The system equations in block-matrix form are in pu:

$$\begin{vmatrix} \Delta v_a \\ -\Delta t_l' \end{vmatrix} = \begin{vmatrix} (r_a - \nu_{ro} m_{fa}) + l_{at} p & \varphi_{fmo} \\ -(\varphi_{fmo} - m_{fa} i_{ao}) & \tau_n p + k_w \end{vmatrix} \cdot \begin{vmatrix} \Delta i_a \\ \Delta \nu_r \end{vmatrix} \quad (26.50)$$

For a voltage-fed independent excitation (as in a realistic case), the cross saturation causes additional feedbacks from the excitation equation and the system becomes of third order:

$$\begin{vmatrix} \Delta v_a \\ \Delta v_f \\ -\Delta t_l' \end{vmatrix} = \begin{vmatrix} (r_a - \nu_{ro} m_{fa}) + l_{at} p & l_{fmt} \nu_{ro} - m_{af} p & \varphi_{fmo} \\ -m_{fa} p & r_f + l_{ft} p & 0 \\ -(\varphi_{fmo} - m_{fa} i_{ao}) & -l_{fmt} i_{ao} & \tau_n p + k_w \end{vmatrix} \cdot \begin{vmatrix} \Delta i_a \\ \Delta i_f \\ \Delta \nu_r \end{vmatrix} \quad (26.51)$$

Figure 26.6 shows the block scheme for the case of current-fed excitation in motoring and generating. The characteristic equation is

$$1 + \frac{(\varphi_{fmo} - m_{fa} i_{ao}) \varphi_{fmo}}{[(r_a - \nu_{ro} m_{fa}) + p l_{at}](k_w + p \tau_n)} = 0 \quad (26.52)$$

If we introduce similar dimensionless parameters as for the simplified saturation model, i.e. $\tau_a = l_{at}/r_a$; $\tau_m = r_a \tau_n / \varphi_{fmo}^2$; $\varrho_w = \tau_w^{-1} = k_w/\tau_n$ and in addition the armature reaction parameter $\kappa = m_{fa}/l_{at}$, we may rewrite the equation as follows:

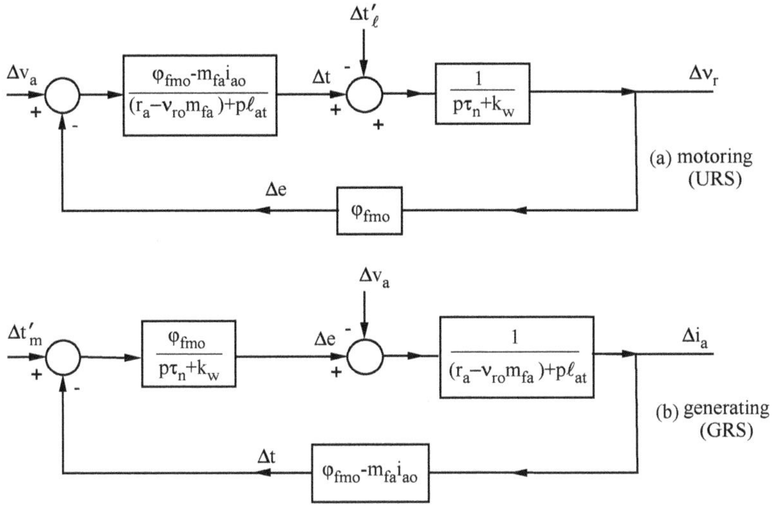

Fig. 26.6 Block schemes including saturation for current excitation

$$1 + \frac{\tau_m^{-1}\tau_a(1 - \kappa l_{at} i_{ao}/\varphi_{fmo})}{[(1 - \nu_{ro}\kappa\tau_a) + p\tau_a](\rho_w\tau_a + p\tau_a)} = 0 \tag{26.53}$$

Due to the armature reaction, the effective armature time constant (see also Eq. 26.52) changes to the incremental value

$$\tau_{ai} = \frac{l_{at}}{(r_a - \nu_{ro} m_{fa})} = \frac{\tau_a}{(1 - \nu_{ro}\kappa\tau_a)} \tag{26.54}$$

Also the slope of the torque-speed characteristic changes to

$$\frac{\Delta t}{\Delta \nu_r} = -\frac{(\varphi_{fmo} - m_{fa} i_{ao})\varphi_{fmo}}{(r_a - \nu_{ro} m_{fa})} \tag{26.55}$$

As a result, the actual electromechanical time constant also changes to the following incremental value:

$$\tau_{mi} = \tau_m \cdot \frac{(1 - \nu_{ro}\kappa\tau_a)}{1 - \kappa l_{at} i_{ao}/\varphi_{fmo}} \tag{26.56}$$

With these incremental time constants, the characteristic equation can be written as

$$1 + \frac{\tau_{ai}/\tau_{mi}}{(1 + p\tau_{ai})(\tau_{ai}/\tau_w + p\tau_{ai})} = 0 \tag{26.57}$$

The gain now becomes τ_{ai}/τ_{mi}. The relation between this incremental gain and the gain in the standard model is

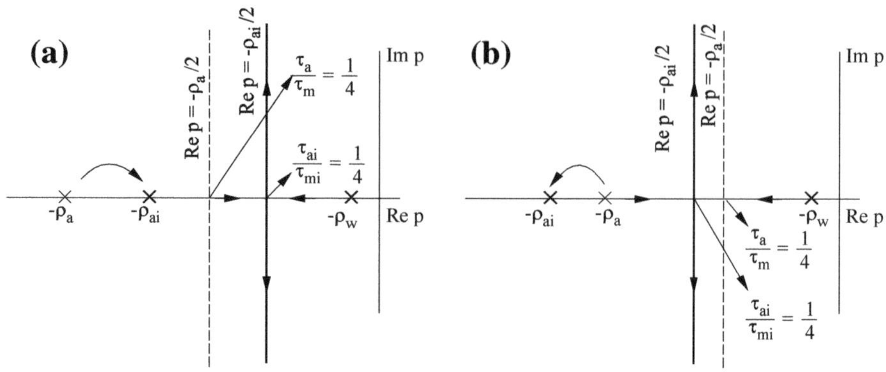

Fig. 26.7 Root loci with saturation and armature reaction

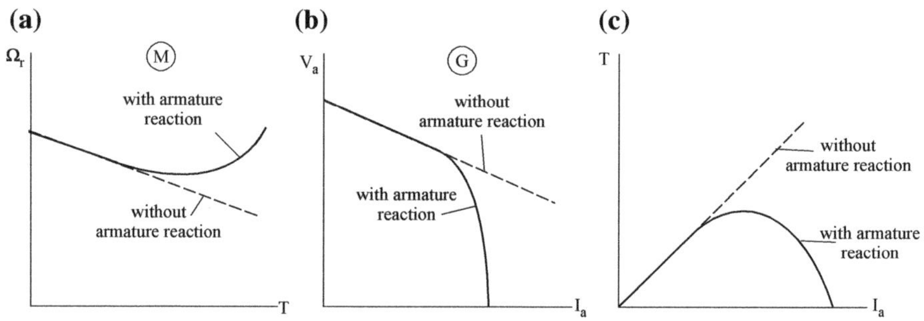

Fig. 26.8 Steady-state effect of armature reaction

$$\frac{\tau_{ai}}{\tau_{mi}} = \frac{\tau_a}{\tau_m} \cdot \frac{1 - \kappa l_{at} i_{ao} / \varphi_{fmo}}{(1 - \nu_{ro} \kappa \tau_a)^2} \tag{26.58}$$

Note that κ is positive for motoring and negative for generating. As a consequence, the root loci for motoring and generating are now as in Fig. 26.7 ((a) for motoring and (b) for generating).

In motoring, saturation results in a less stable behaviour, while for generating the damping increases. This is in line with the steady-state characteristics including armature reaction, as is illustrated in Fig. 26.8 (dashed lines are used for the unsaturated case and full-drawn lines for the saturated case). For a very large armature current (or high torque values), a motor may become unstable. As to the root loci, this corresponds with a sign change of the gain and/or an infinite value of τ_{ai}.

For a voltage-fed excitation, the block diagram is much more complex (see Fig. 26.9). Introducing three additional parameters $\tau_f = l_{ft}/r_f$, $\sigma_f = l_\sigma/l_{ft}$ and $\lambda = m_{af}/l_{ft}$, the characteristic equation can be written as follows:

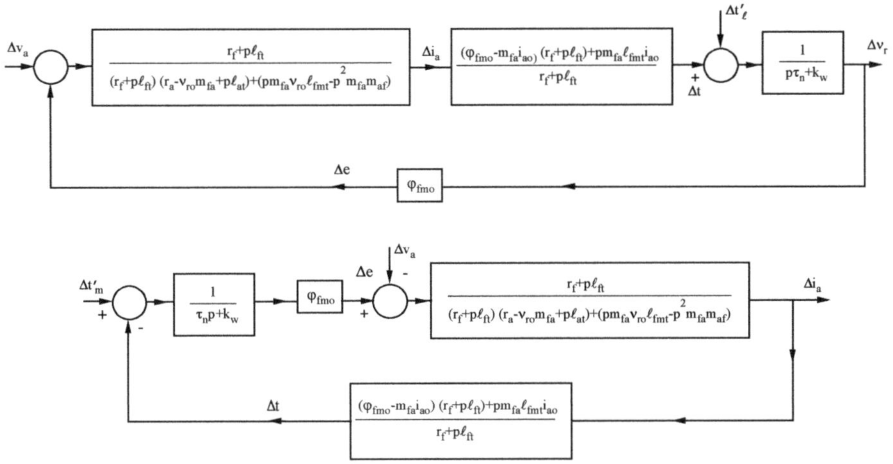

Fig. 26.9 Block schemes including saturation for voltage-fed excitation

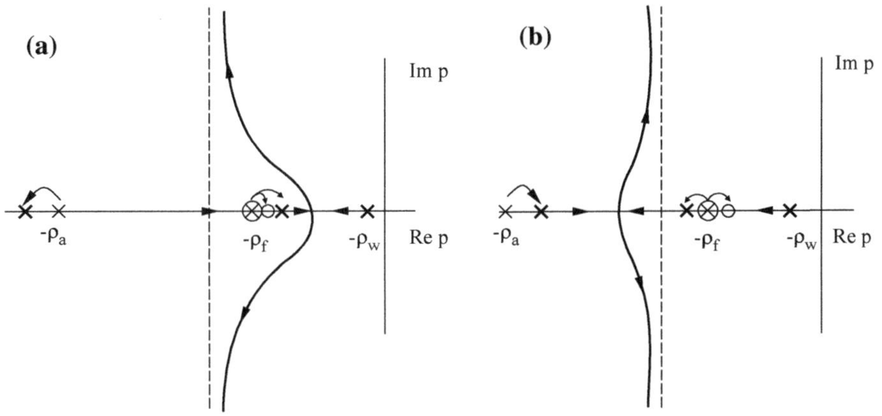

Fig. 26.10 Root loci with saturation for voltage-fed excitation

$$1 + \frac{(\tau_{ai}/\tau_{mi})\left\{1 + p\tau_f[\sigma_f + (\tau_{ai}\tau_{mi}/\tau_a\tau_m)(1-\sigma_f)]\right\}}{\left[(1+p\tau_{ai})(1+p\tau_f) + p\tau_f\tau_{ai}(1-\sigma_f)\kappa(\nu_{ro}-\lambda p)\right](\tau_{ai}/\tau_w + p\tau_{ai})} = 0 \tag{26.59}$$

There is now a finite zero around $p \approx -\tau_f^{-1}$. The open-loop poles include the mechanical damping pole and a pole around $-\tau_a^{-1}$ as well as a third open-loop pole around the same value as the finite zero, i.e. $p \approx -\tau_f^{-1}$ (without saturation and armature reaction, this pole and this zero coincide so that the system then becomes second order).

Possible root loci for motoring and generating are depicted in Fig. 26.10 ((a) for motoring and (b) for generating). As is the case for a current-fed excitation, the saturation and armature reaction stabilise the behaviour for generating, while for motoring the damping (of the dominant eigenvalue) decreases.

Chapter 27
Modelling and Dynamic Behaviour of Induction Machines

Abstract With the advent of variable frequency supply of rotating field machines in the second half of the twentieth century, some cases of hunting of induction machines fed with a low frequency supply became apparent. This was the starting point of research into the causes of these instabilities. In this chapter we analyse the stability behaviour of induction machine drives for variable frequency supply. A traditional model with constant saturation is used to analyse the dynamic behaviour. Using well-chosen dimensionless parameters, the characteristic dynamic behaviour of induction machines can be represented in a handy way. Because induction machines obey scaling laws quite narrowly, it is possible to predict the dynamic behaviour of a typical machine.

27.1 Introduction: Modelling of Rotating Field Machines Without Saliency

For the steady-state electrical modelling of rotating field machines without saliency, it is common to use complex phasor equations of voltages, currents and/or fluxes. A symmetrical three-phase induction machine with symmetrical three-phase supply, for example, is described by two complex equations, one for the stator reference phase and one for the rotor reference phase. Obviously, higher space harmonics of current layers, mmf and field are disregarded in this model.

Such a model can also be used if there is an asymmetrical supply. In that case, a set of such equations has to be used for each of the symmetrical components. Superposition of the positive, negative and zero-sequence components of voltages and currents then yields the total phasor quantities. For an induction machine, the positive and negative sequence equations are the same except for the synchronous speed and the slip (e.g. s for the positive sequence and $2-s$ for the negative sequence).

The simplifying assumptions for the steady-state model are as follows:

- a three-phase (or polyphase) symmetrical machine (both stator and rotor)
- sinusoidal distributed windings (no higher space harmonics of current layer, mmf or field)
- no skin effect

J. A. Melkebeek, *Electrical Machines and Drives*, Power Systems,
https://doi.org/10.1007/978-3-319-72730-1_27

- no saturation
- no slot-effects nor any rotor (or stator) saliency
- steady state only.

As mentioned above, there is no restriction to a symmetrical sinusoidal three-phase (or multiphase) supply. If the supply voltage contains harmonics, then for each harmonic such a model has to be used.

The model we will use for the dynamic analysis is based on the same simplifying assumptions, except for the steady state. To derive the model, we will start from the (dynamical) electrical equations for the three stator and the three rotor phases, as well as the equation of motion.

Such equations are very apt for digital simulation as the number of equations is not very relevant for this method.

For an analytical study it is, however, important to reduce the number of equations (and variables). For a wye-connected machine, the neutral is rarely connected and thus zero-sequence current cannot flow; for a delta-connected machine, the zero-sequence voltages are inherently zero. This implies that the zero-sequence equations and components can be omitted in most or all cases. Therefore, the six basic electrical equations will be split up into their symmetrical components, and after the zero-sequence equations have been omitted, the remaining electrical equations can be simplified to two complex equations or four real equations. The next section will illustrate this, in addition to showing that the equations can be made stationary (i.e. not explicitly dependent on time, whereas the basic equations are time-dependent).

Finally, it is important to stress that neither the model nor the analysis method are restricted to a symmetrical sinusoidal supply. Voltages and currents may in principle contain harmonics and may be asymmetrical.

27.2 The Standard Dynamic Model of an Induction Machine

27.2.1 Derivation of the Dynamic Model

We will derive the model for an idealised symmetrical and sinusoidal induction machine (see the assumptions in the previous section). Stator and rotor are assumed to be three phase.[1]

The differential equations between the instantaneous values of voltages and currents are

$$\mathbf{V}^i = \mathbf{R}^i \cdot \mathbf{I}^i + \frac{d}{dt}\left(\mathbf{L}^i(\theta) \cdot \mathbf{I}^i]\right) \tag{27.1}$$

[1]It can be shown that a rotor cage can always be replaced by an equivalent three-phase sinusoidally distributed rotor winding.

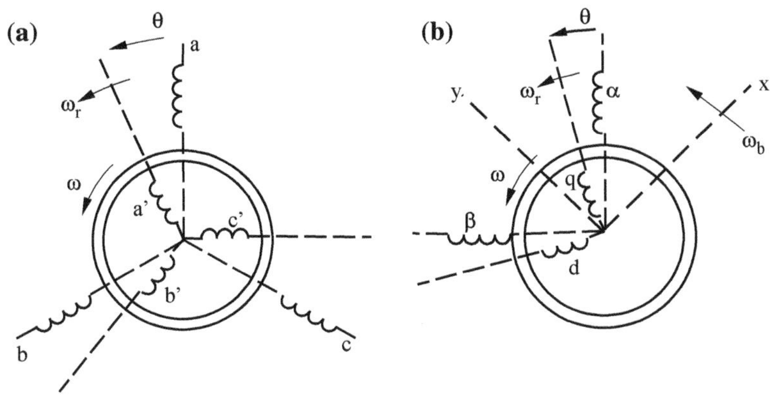

Fig. 27.1 Three-phase induction machine and two-phase model

$\mathbf{R}^i$ and $\mathbf{L}^i$ are the resistance and inductance matrices, respectively:

$$\mathbf{R}^i = diag(\mathbf{R}_s^i, \mathbf{R}_r^i) \tag{27.2}$$

$$\mathbf{L}^i(\theta) = \begin{bmatrix} \mathbf{L}_s^i & \mathbf{L}_{sr}^i(\theta) \\ \mathbf{L}_{rs}^i(\theta) & \mathbf{L}_r^i \end{bmatrix} \tag{27.3}$$

θ is the instantaneous angle between the stator and rotor reference phase (see Fig. 27.1). Voltages and currents are in principle[2] the real instantaneous functions of time.

If the machine is three-phase symmetrical with a constant air gap and sinusoidally distributed windings, then we may write for the block matrices

$$\mathbf{R}_s^i = diag(R_s, R_s, R_s) \tag{27.4}$$

$$\mathbf{R}_r^i = diag(R_r, R_r, R_r) \tag{27.5}$$

$$\mathbf{L}_\mathbf{s}^\mathbf{i} = \begin{bmatrix} L_s & M_s & M_s \\ M_s & L_s & M_s \\ M_s & M_s & L_s \end{bmatrix} \tag{27.6}$$

$$\mathbf{L}_\mathbf{r}^\mathbf{i} = \begin{bmatrix} L_r & M_r & M_r \\ M_r & L_r & M_r \\ M_r & M_r & L_r \end{bmatrix} \tag{27.7}$$

[2]However, the model can also be derived using the complex time phasors (see below).

$$\mathbf{L^i_{sr}}(\theta) = \begin{bmatrix} \cos\theta & \cos(\theta + 2\pi/3) & \cos(\theta + 4\pi/3) \\ \cos(\theta + 4\pi/3) & \cos\theta & \cos(\theta + 2\pi/3) \\ \cos(\theta + 2\pi/3) & \cos(\theta + 4\pi/3) & \cos\theta \end{bmatrix} \cdot M^i_{sr} = [\mathbf{L^i_{rs}}(\theta)]^T \tag{27.8}$$

The superscript T indicates matrix transposition. In the absence of saturation, $L_s, L_r, M_s, M_r, M_{sr}$ are constants.

The basic equations in space vector form are now derived in two steps:

1. The symmetrical components transformation $\mathbf{X}' = diag(\mathbf{T_s^{-1}}, \mathbf{T_s^{-1}}) \cdot \mathbf{X^i}$ with

$$\mathbf{T_s^{-1}} = \frac{1}{3}\begin{bmatrix} 1 & 1 & 1 \\ 1 & a & a^2 \\ 1 & a^2 & a \end{bmatrix} \tag{27.9}$$

transforms both stator and rotor three-phase systems into their *symmetrical components*.[3] Equation 27.1 becomes

$$\mathbf{V}' = \mathbf{R}' \cdot \mathbf{I}' + \frac{d}{dt}\left(\mathbf{L}'(\theta) \cdot \mathbf{I}'\right) \tag{27.10}$$

The six voltages and currents in this equation are the *zero-sequence, positive and negative sequence components* of stator and rotor, respectively.

In a wye-connected stator, the zero-sequence currents cannot flow (the neutral is rarely connected). As the zero-sequence voltages are zero in a delta-connected stator, we may omit the zero-sequence equations as there cannot be any zero-sequence power (except in asymmetrical rotor conditions). Equation 27.10 can thus be reduced to a set of four complex equations.

When real-time functions are used for voltages and currents in the transformation Eq. 27.9, the resulting positive and negative sequence voltages and currents are complex conjugate. The transformation of the original real system equation (applying Eq. 27.9 to 27.1) yields two complex conjugate equations for the positive and negative sequence components.[4] By adding the positive sequence components and the complex conjugate of the negative sequence components, we actually obtain space vectors (see also Appendix A in Part1)

$$\underline{V} = \underset{\rightarrow}{V} = V_+ + V_-^* \tag{27.11}$$

The resulting equations relating the space vectors of voltages and currents for stator and rotor are identical to those for the positive sequence components in Eq. 27.10. The outcome is a set of two complex equations

---

[3]In fact, this transformation yields the usual zero-sequence, positive and negative sequence components when the inputs are the *phasors* or *time phasors* of the voltages or currents. We apply it here to the *real-time functions*, and the resulting components are not the symmetrical components as usually defined. However, the *space vectors* we will derive afterwards will be identical (see also Appendix C).

[4]Check also $(\mathbf{T_s^{-1}RT_s})^+ \equiv \mathbf{T_s^{-1}RT_s}$ and $(\mathbf{T_s^{-1}LT_s})^+ \equiv \mathbf{T_s^{-1}LT_s}$.

$$\mathbf{V}'' = \mathbf{R}'' \cdot \mathbf{I}'' + \frac{d}{dt}\left(\mathbf{L}''(\theta) \cdot \mathbf{I}''\right) \tag{27.12}$$

with

$$\mathbf{V}'' = [\underline{V}_s''\ \underline{V}_r'']^T \tag{27.13}$$

$$\mathbf{I}'' = [\underline{I}_s''\ \underline{I}_r'']^T \tag{27.14}$$

$$\mathbf{R}'' = diag(R_s, R_r) \tag{27.15}$$

$$\mathbf{L}''(\theta) = \begin{bmatrix} L_s & M\exp(j\vartheta) \\ M\exp(-j\vartheta) & L_r \end{bmatrix} \tag{27.16}$$

and where $R_s = R_s^i$, $R_r = R_r^i$, $L_s = L_s^i - M_s^i$, $L_r = L_r^i - M_r^i$ and $M = (3/2)M_{sr}^i$.

2. The Eq. 27.12 are time-dependent, as θ is a function of time. The reason for this is that in this equation the stator reference frame is at standstill, while the rotor reference frame is rotating with rotor speed. The transformation

$$\mathbf{T_b^{-1}} = diag\left[\exp(-j\theta_b), \exp\left(-j(\theta_b - \theta)\right)\right] \tag{27.17}$$

to a common reference frame with speed $\omega_b = d\theta_b/dt$ will make the equations stationary

$$\mathbf{V} = \mathbf{R} \cdot \mathbf{I} + \frac{d}{dt}(\mathbf{L} \cdot \mathbf{I}) + j\omega_b(\mathbf{L} \cdot \mathbf{I}) - j\mathring{\theta}(\mathbf{M_r} \cdot \mathbf{I}) \tag{27.18}$$

with

$$\mathbf{V} = [\underline{V}_s\ \underline{V}_r]^T \tag{27.19}$$

$$\mathbf{I} = [\underline{I}_s\ \underline{I}_r]^T \tag{27.20}$$

$$\mathbf{R} = diag(R_s, R_r) \tag{27.21}$$

$$\mathbf{L} = \begin{bmatrix} L_s & M \\ M & L_r \end{bmatrix} \tag{27.22}$$

$$\mathbf{M} = \begin{bmatrix} 0 & 0 \\ M & L_r \end{bmatrix} \tag{27.23}$$

The last term in Eq. 27.18 represents the emf of motion.
The torque is given by (for a proof, see for example Ref. [14]):

$$T = \frac{3}{2} N_p \cdot Im\left(\mathbf{I}^+ \cdot \mathbf{M_r} \cdot \mathbf{I}\right) = \frac{3}{2} N_p \cdot Im\left(\underline{I}_s \cdot M \cdot \underline{I}_r^*\right) \tag{27.24}$$

where N_p represents the number of pole pairs and the + sign indicates the Hermitic conjugate. The complete dynamical model therefore consists of the Eq. 27.18, together with the equation of motion

$$(J/N_p)\frac{d^2\theta}{dt^2} = T - T_l$$

with the torque T given by Eq. 27.24 and with T_l the load torque.

We have transformed the original six real electrical Eq. 27.1 into a set of two complex (or four real) equations. By referring to a common reference frame, the resulting equations are stationary. However, these equations are still non-linear. Indeed, the non-linearity is inherent to electric machines: the torque is proportional to a product of currents (or flux and current) and the emf of motion is proportional to the product of speed and flux.

In many cases, we will use the real form of these equations by projecting the complex equations of stator and rotor on the real and imaginary axes of the common reference frame. In a general reference frame with speed $\omega_b = d\theta_b/dt$, the components of voltages and currents will be denoted by the subscript x and y. In a standstill (stator) reference frame ($\omega_b = 0$), the subscripts α and β are typically used. If the speed of the reference frame is chosen equal to the rotor speed, the reference frame is called a rotor reference frame. Often, a reference frame synchronous with the flux (instantaneously) is chosen, in which case the subscripts q, d are commonly used.

Remarks:

1. The same equations can be obtained by first applying the Clarke transformation to Eq. 27.1 and then the rotation to a common reference frame. This is normally the case for the synchronous machine (see Chap. 28).
2. If the Eq. 27.18 are written as two complex equations in a standstill reference frame, then for a purely sinusoidal supply and for steady state, we obtain equations which are almost identical to the common steady-state equations for the induction machine (but now in *time-phasor* notation and with amplitude values instead of effective values). In time-phasor representation, we have for a sinusoidal voltage $\underline{V}_s"(t) = \hat{V}_s \cdot \exp(j\omega t + j\varphi_s)$ and $\underline{V}_r"(t) = \hat{V}_r \cdot \exp(j\omega t + j\varphi_r)$, instead of $\underline{V}_s = V_s \cdot \exp(j\varphi_s) = (1/\sqrt{2}) \cdot \hat{V}_s \cdot \exp(j\varphi_s)$ and $\underline{V}_r = V_r \cdot \exp(j\varphi_r) = (1/\sqrt{2}) \cdot \hat{V}_s \cdot \exp(j\varphi_r)$ as in the common notation with phasors for steady state (using effective values). If we transform these time phasors into a reference frame synchronous with the supply frequency, we get the same equations as for steady state but with amplitude values. In fact, the transformation

Eq. 27.17 is now $\mathbf{T_b^{-1}} = diag[\exp(-j\omega t), \exp(-js\omega t)]$ which is equivalent to omitting the $\exp(j\omega t)$ and $\exp(js\omega t)$ in the phasor representation.

3. The resulting Eq. 27.18 are much more general than the steady-state equations, as they are not limited to steady state or sinusoidal supply.
4. If we start (Eq. 27.1) from the complex time phasors of the phase voltages and currents, instead of their real-time values, then the symmetrical components transformation yields exactly the usual zero-sequence, positive- and negative-sequence components (except for the exponential time dependences, e.g. $\exp(j\omega t)$). It is easily shown that after the transformation Eq. 27.11 the same Eq. 27.12 are obtained, as space vectors. The equivalence - or non-equivalence - of (time) phasors and space vectors has been extensively covered in the literature, but in the author's opinion, both are exactly equivalent when applied to symmetrical rotating field machines.

27.2.2 Equations for Steady State and for Small Deviations Around an Equilibrium State

As mentioned above, the induction machine equations are non-linear, both the electrical equations and the torque in the equation of motion. To study the dynamic behaviour, we will therefore revert to small deviations around a steady-state equilibrium.

For steady state (subscript o) with a sinusoidal voltage supply, the Eq. 27.18 become

$$\mathbf{V_0} = \mathbf{R} \cdot \mathbf{I_0} + j\omega_0 \left(\mathbf{L} \cdot \mathbf{I_0}\right) - j\overset{\circ}{\theta}_0 \left(\mathbf{M_r} \cdot \mathbf{I_0}\right) \tag{27.25}$$

$$T = \frac{3}{2} N_p \cdot Im\left(\mathbf{I_0^+} \cdot \mathbf{M_r} \cdot \mathbf{I_0}\right) = \frac{3}{2} N_p \cdot Im\left(\underline{I}_{s0} \cdot M \cdot \underline{I}_{r0}^*\right) = T_{l0} \tag{27.26}$$

The voltage (or current) vectors are (for a constant ω_b) of the form

$$\mathbf{V_0} = [\,\underline{V}_{so}\ \underline{V}_{ro}]^T = \left[\,V_{so} \exp\left(j\left(\omega_0 - \omega_b\right)t - j\varphi_s\right)\ V_{ro} \exp\left(j\left(\omega_0 - \omega_b\right)t - j\varphi_r\right)\right]^T \tag{27.27}$$

For $\omega_b = \omega_0$, these are equal to the usual time phasors (but here with amplitude values instead of effective values).

Subtracting Eq. 27.25 from 27.18 and linearising (omitting second-order variations), we obtain

$$\mathbf{\Delta V} = \mathbf{R} \cdot \mathbf{\Delta I} + \frac{d}{dt}\left(\mathbf{L} \cdot \mathbf{\Delta I}\right) + j\omega_{b0}\left(\mathbf{L} \cdot \mathbf{\Delta I}\right) + j\Delta\omega_b \left(\mathbf{L} \cdot \mathbf{I_0}\right) - j\overset{\circ}{\theta}_0 \left(\mathbf{M}_r \cdot \Delta\mathbf{I}\right) - j\Delta\overset{\circ}{\theta}\left(\mathbf{M}_r \cdot \mathbf{I_0}\right) \tag{27.28}$$

In a similar way, we obtain for the equation of motion (the load torque variation is written as the sum of a speed-dependent variation and an external variation):

$$(J/N_p)\frac{d^2\Delta\theta}{dt^2} = \frac{3}{2}N_p \cdot Im\left(\Delta\mathbf{I}^+ \cdot \mathbf{M_r} \cdot \mathbf{I}_0 + \mathbf{I}_0^+ \cdot \mathbf{M_r} \cdot \Delta\mathbf{I}\right) - K_w \cdot \Delta\mathring{\theta} - \Delta T_l' \tag{27.29}$$

In Eq. 27.28, the term $j\Delta\omega_b\,(\mathbf{L} \cdot \mathbf{I_0})$ is added to include a possible variation of the reference frame speed (e.g. for an instantaneous synchronous reference frame when the supply frequency is varying). If the reference frame were fixed to the steady-state supply frequency, this term would be zero (but the frequency variation would then cause a non-zero voltage variation $\mathbf{\Delta V}$).

27.2.3 Dynamic Model with Pu Time and Speeds

Many studies revert to pu equations. However, in the present textbook absolute values will be used with only a pu time and thus also pu speeds. Pu time will be denoted by $\tau = \omega_n t$, pu frequency or synchronous speed by $\nu = \omega/\omega_n$, pu rotor speed by $\nu_r = \omega_r/\omega_n = (1-s)\nu$ and pu slip frequency by $\nu_s = s\omega/\omega_n = s\nu$.

With these notations, the steady-state equations for a supply frequency ω_0 become:

$$\mathbf{V_0} = \mathbf{R} \cdot \mathbf{I_0} + j\nu_0\,(\omega_n\mathbf{L} \cdot \mathbf{I_0}) - j\nu_{r0}\,(\omega_n\mathbf{M_r} \cdot \mathbf{I_0}) \tag{27.30}$$

$$T = \frac{3}{2}N_p \cdot Im\left(\mathbf{I_0^+} \cdot \mathbf{M_r} \cdot \mathbf{I_0}\right) = T_{l0} \tag{27.31}$$

Or, in vector notation:

$$\underline{V}_{s0} = R_s\underline{I}_{s0} + j\nu_0\omega_n L_s\underline{I}_{s0} + j\nu_0\omega_n M\underline{I}_{r0} \tag{27.32}$$

$$\underline{V}_{r0} = R_r\underline{I}_{r0} + j\nu_{s0}\omega_n L_r\underline{I}_{r0} + j\nu_{s0}\omega_n M\underline{I}_{s0} \tag{27.33}$$

$$T = \frac{3}{2}N_p \cdot Im\left(\underline{I}_{s0} \cdot M \cdot \underline{I}_{r0}^*\right) = T_{l0} \tag{27.34}$$

The linearised equations around the steady state are, introducing the pu Laplace operator $p = d/d(\omega_n t)$:

$$\begin{aligned}\mathbf{\Delta V} = \mathbf{R} \cdot \mathbf{\Delta I} + p\,(\omega_n\mathbf{L} \cdot \mathbf{\Delta I}) + j\nu_{b0}\,(\omega_n\mathbf{L} \cdot \mathbf{\Delta I}) + j\Delta\nu_b\,(\omega_n\mathbf{L} \cdot \mathbf{I_0})\\ - j\nu_{r0}\,(\omega_n\mathbf{M_r} \cdot \Delta\mathbf{I}) - j\Delta\nu_r\,(\omega_n\mathbf{M_r} \cdot \mathbf{I_0})\end{aligned} \tag{27.35}$$

$$(J/N_p)\frac{d^2\Delta\theta}{dt^2} = \frac{3}{2}N_p \cdot Im\left(\Delta\mathbf{I}^+ \cdot \mathbf{M_r} \cdot \mathbf{I}_0 + \mathbf{I}_0^+ \cdot \mathbf{M_r} \cdot \Delta\mathbf{I}\right) - K_w \cdot \Delta\mathring{\theta} - \Delta T_l' \tag{27.36}$$

or, in vector notation:

$$\Delta\underline{V}_s = R_s\Delta\underline{I}_s + p\omega_n L_s\Delta\underline{I}_s + p\omega_n M\Delta\underline{I}_r + j\nu_{b0}\omega_n L_s\Delta\underline{I}_s + j\nu_{b0}\omega_n M\Delta\underline{I}_r + j\Delta\nu_b\omega_n L_s\underline{I}_{s0} + j\Delta\nu_b\omega_n M\underline{I}_{r0} \tag{27.37}$$

$$\Delta\underline{V}_r = R_r\Delta\underline{I}_r + p\omega_n L_r\Delta\underline{I}_r + p\omega_n M\Delta\underline{I}_s + j(\nu_{b0}-\nu_{r0})\omega_n L_r\Delta\underline{I}_r + j(\nu_{b0}-\nu_{r0})\omega_n M\Delta\underline{I}_s + j(\Delta\nu_b - \Delta\nu_r)\omega_n L_r\underline{I}_{r0} + j(\Delta\nu_b - \Delta\nu_r)\omega_n M\underline{I}_{s0} \tag{27.38}$$

$$(J/N_p)\frac{d^2\Delta\theta}{dt^2} = \frac{3}{2}N_p\cdot M\cdot Im\left(\underline{I}_{s0}\Delta\underline{I}_r^* + \underline{I}_{r0}^*\Delta\underline{I}_s\right) - K_w\cdot\omega_n\Delta\nu_r - \Delta T_l' \tag{27.39}$$

Remark: If we use pu, the separate leakages of stator and rotor can be written as $l_{s\sigma} = l_s - l_m$ and $l_{r\sigma} = l_r - l_m$ where $l_m = m$. However, the division of the leakage into the separate leakages depends on the chosen reference values for stator and rotor quantities, in contrast with the total leakages referred to stator and rotor. The latter are $l_{\sigma s} = \sigma l_s$ and $l_{\sigma r} = \sigma l_r$, respectively. In absolute values, separate leakages can also be defined but these will depend on the chosen turns ratio between stator and rotor, which - importantly - cannot be measured.

27.2.4 Approximation for Saturation

The equations above have been derived assuming an unsaturated magnetic circuit. In reality, all electric machines are operated with a saturated magnetic circuit at rated conditions.[5] In traditional machine theory, the steady-state analysis uses the (magnetising field) chord-slope inductances in the operating point (see Fig. 27.2).

This is an acceptable approximation for the constant magnetising condition in steady state. However, for dynamic studies both the chord-slope and the tangent-slope inductances should be used. Indeed, whereas the main flux in the operating point can be written as $\Psi_{m0} = L_{mo}I_{m0}$, for flux variations $\Delta\Psi_m = L_{mt}\Delta I_m$ should be used. However, incorporating both inductances creates a rather complicated machine model (see Ref. [22]). Moreover, in many cases (in particular for induction machines with closed rotor slots) the leakage flux saturation should also be considered, in addition to the main flux saturation. Although some attempts to model both main and leakage flux saturation have been published, some criticism is to be called for as the division of leakage between stator and rotor cannot easily (if at all) be measured.

In the present text, a dynamic model with chord-slope inductances will be used. It turns out that the results of such a dynamic analysis are conservative, as more stable machine behaviour is created if the tangent-slope inductance is also included in the model. Nevertheless, some restrictions apply. When a machine is supplied by

[5]The reason for this is that a relatively high air-gap induction is required to obtain a sufficient power output for a given machine size.

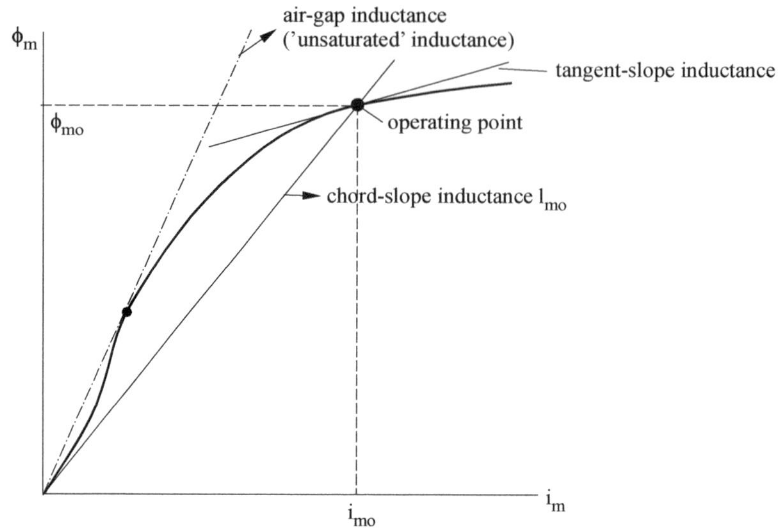

Fig. 27.2 Saturation characteristic

a current source inverter or in self-excitation[6] on a capacitor-resistance bank, the flux level may change considerably and modelling with only the chord-slope inductance is not feasible. Motoring or generating on a grid or a voltage source inverter will typically not lead to any large flux excursions, and modelling with only the chord-slope inductance seems acceptable.

Below, we will limit ourselves to the study of small deviations around a steady state using the approximate model with the chord-slope inductances. For the study of large transients, the most appropriate method is simulation, where the variable saturation can be taken into account in quite a straightforward way.

27.3 Characteristic Dynamic Behaviour of the Induction Machine

27.3.1 Dynamic Model in Real Matrix Form

The dynamic equations for small deviations around a steady state, Eqs. 27.35 and 27.36, can be written in real form as follows:

[6]Another problem with self-excitation is that a model with only the chord-slope inductance will create two degrees of freedom (i.e. two eigenvalues on the imaginary axis) for the self-excitation, which does not exist in real life.

$$\begin{bmatrix} \Delta V_{sx} \\ \Delta V_{sy} \\ \Delta V_{rx} \\ \Delta V_{ry} \\ -\Delta T_l' \end{bmatrix} = \begin{bmatrix} R_s + p\omega_n L_s & -\nu_0\omega_n L_s & p\omega_n M & -\nu_0\omega_n M & 0 \\ \nu_0\omega_n L_s & R_s + p\omega_n L_s & \nu_0\omega_n M & p\omega_n M & 0 \\ p\omega_n M & -\nu_{s0}\omega_n M & R_r + p\omega_n L_r & -\nu_{s0}\omega_n L_r & \omega_n\left(M I_{sy0} + L_r I_{ry0}\right) \\ \nu_{s0}\omega_n M & p\omega_n M & \nu_{s0}\omega_n L_r & R_r + p\omega_n L_r & -\omega_n\left(M I_{sx0} + L_r I_{rx0}\right) \\ \frac{3}{2}N_p M I_{ry0} & -\frac{3}{2}N_p M I_{rx0} & -\frac{3}{2}N_p M I_{sy0} & \frac{3}{2}N_p M I_{sx0} & \frac{J}{N_p}{\omega_n}^2 p + \omega_n K_w \end{bmatrix} \begin{bmatrix} \Delta I_{sx} \\ \Delta I_{sy} \\ \Delta I_{rx} \\ \Delta I_{ry} \\ \Delta\nu_r \end{bmatrix}$$

$$+\,\omega_n \begin{bmatrix} -\left(L_s I_{sy0} + M I_{ry0}\right) \\ +\left(L_s I_{sx0} + M I_{rx0}\right) \\ -\left(M I_{sy0} + L_r I_{ry0}\right) \\ +\left(M I_{sx0} + L_r I_{rx0}\right) \\ 0 \end{bmatrix} \cdot \Delta\nu \qquad (27.40)$$

In this equation, an instantaneously synchronous reference frame is used. This fifth-order model is applicable in both motoring and generating, and for both conventional and power-electronic voltage supply. For current supply, the stator equations can be omitted, and the model becomes third order (the stator equations are only important for the voltages but do not affect the dynamic behaviour of the machine). For self-excitation, also the external relations between stator voltages and currents would be required. However, self-excitation will not be studied because the model does not adequately model main field saturation.

27.3.2 Dimensionless Parameters for Dynamic Analysis

The dynamic model, Eq. 27.40, contains 11 machine parameters: the five electrical parameters (R_s, R_r, L_s, L_r, M), the three mechanical parameters (J, K_w, N_p) and the three parameters that determine the operating point (V_{s0}, ω_0, ω_{r0}). The rotor is assumed to be short-circuited ($V_{r0} = 0$).

The minimum number of (dimensionless) parameters required to describe the dynamic behaviour (open-loop poles and zeros, eigenvalues) is, nevertheless, much smaller: for this fifth-order system with 11 parameters, only six ($= 11 - 5$) parameters are required. A convenient choice is as follows:

- $\nu_0\tau_r$, the product of the rotor short-circuit time constant $\tau_r = \sigma\omega_n L_r/R_r$ and the pu supply frequency $\nu_0 = \omega_0/\omega_n$
- $\nu_0\tau_s$, the product of the stator short-circuit time constant $\tau_s = \sigma\omega_n L_s/R_s$ and the pu supply frequency ν_0 (moreover, $\nu_0\tau_s$ is often replaced by $\alpha = \tau_r/\tau_s \approx r_s/r_r$)
- the magnetic parameter or total leakage coefficient $\sigma = 1 - M^2/L_s L_r$
- the pu slip frequency ν_{s0}, in many cases replaced by $\nu_{s0}\tau_r$
- the electromechanical time constant $\tau_m = \frac{\omega_n\omega_{s0}J}{N_p T_0} = \tau_n\frac{\nu_{s0}}{t_0}$ with $\tau_n = \frac{\omega_n^2 J}{N_p T_n}$ the pu inertia time constant; it is preferably replaced by the ratio τ_r/τ_m
- the dimensionless friction time constant $\tau_w = 1/\varrho_w = \frac{J\omega_n}{N_p K_w}$.

The electrical time constants τ_r and τ_s are in principle constant for a given machine, unless there is important leakage field saturation. However, the magnetic parameter (leakage coefficient) σ depends heavily on the main field saturation level as M as well as the main field parts in L_s and L_r are directly proportional to the chord-slope inductance.

The electromechanical time constant is the ratio of the inertia time constant τ_n and the slope of the line connecting the no-load point ($s = 0$, $T = 0$) and the operating point (s_0, T_0) on the steady-state torque-slip characteristic. In the limit case of no-load operation ($s_0 = 0$), this slope is exactly equal to the slope of the torque-slip characteristic (tangent slope). Then τ_m can be regarded as the time constant of the mechanical system with the steady-state asynchronous torque (characteristic) as the dynamic torque (characteristic). For non-zero steady-state slip values, this interpretation is only approximately true (however, as will be explained in the next chapter, another and exact interpretation valid in all cases can be given). Please note that for constant flux operation ($\approx V_0/\omega_0$ constant), τ_m is essentially constant for a given machine and inertia.

The friction time constant is normally very small (see the next section) and negligible in terms of its effect on the dynamics, except when it represents an important speed-dependent external load (e.g. a ventilator load).

27.3.3 Scaling Laws for the Dynamical Parameters

As is well known, normal induction machines have to obey many constraints (e.g. supply frequency, three-phase) and standards (e.g. minimum ratio of pull-out-torque and rated torque, maximum pu starting current), in addition to economic constraints. As a result, nearly all induction machines end up following some scaling laws (see also Ref. [3]), which relate the machine parameters to the size of the machine (expressed as the pole-pitch τ_p). For example, the pu total leakage inductance can be assumed to be around $x_\sigma \approx 0.2$ (between 0.15 and 0.25 because of minimum pull-out torque and maximum starting current). The pu stator and rotor resistances are almost equal and will decrease with increasing machine size as $r_s \approx r_r \approx 0.005/\tau_p$ (because of economic reasons).

As a result, the dynamic parameters also follow scaling laws, as illustrated in Table 27.1.

These scaling laws indicate the range for the (dynamic) parameters for normal machines. The dynamic analysis can then be restricted to these realistic values in order to describe the dynamic behaviour of most machines. Further, these laws allow us to define an average or typical machine as a function of the power rating and describe its typical dynamic behaviour.

From this table, it is clear that only two dynamic parameters show a significant variation with size, i.e. τ_r/τ_m and τ_r. Moreover, with a variable frequency from e.g. 10–200 Hz, $\nu_0\tau_r$ varies in the range of 1 : 400. With an extra load inertia (up to 10 times the machine inertia for smaller power ratings) τ_r/τ_m will also exhibit a variation in the range of 1 : 200.

Table 27.1 Scaling laws for the dynamic parameters

Parameter	Scaling law	$N_p = 2$; $\tau_p = 0.05\ldots 1m$	$N_p = 3$; $\tau_p = 0.05\ldots 1m$
$\tau_r = \sigma\omega_n L_r/R_r$	$40\tau_p$	$2\ldots 40$	$2\ldots 40$
$\alpha = \tau_r/\tau_s$	$0.5\ldots 2$ (typically 1)	$0.5\ldots 2$ (typically 1)	$0.5\ldots 2$ (typically 1)
σ	$0.02\sqrt{N_p/\tau_p}$	$0.13\ldots 0.028$	$0.16\ldots 0.034$
(ν_{sn})	$0.005\tau_p^{-1}$	$0.1\ldots 0.05$	$0.1\ldots 0.05$
$\nu_{s0}\tau_r$	$0\ldots 0.2$ ($\nu_{sn}\tau_r = 0.2$)	$0\ldots 0.2$	$0\ldots 0.2$
(τ_n)*	$10^3(N_p-1)\tau_p/N_p$	$25\ldots 500$	$30\ldots 650$
(τ_m)*	$5(N_p-1)/N_p$	2.5	3.3
τ_r/τ_m*	$8N_p\tau_p/(N_p-1)$	$0.8\ldots 16$	$0.6\ldots 12$
ϱ_w	$1.6\cdot 10^{-5}(\tau_p + 0.17)\tau_p^{-11/6}$	$(70\ldots 1)10^{-5}$	$(70\ldots 1)10^{-5}$
(P_n)	$10^7\tau_p^{23/6}N_p^2$	400 W … 40 MW	900 W … (90 MW)

*only machine inertia

Of the other parameters, only σ and $\nu_{s0}\tau_r$ show some variation (with power rating for σ and with load for $\nu_{s0}\tau_r$), but as we will see later on, their effect on the dynamic behaviour is somewhat less important. These remaining parameters will be referred to as the secondary parameters.

27.3.4 Block Diagrams and Characteristic Equation

In this section, we will consider a voltage-fed induction machine with short-circuited rotor. External inputs are the stator voltage variations $\Delta V_{sx} = \Delta V_x$ and $\Delta V_{sy} = \Delta V_y$ and the supply frequency variation $\Delta\nu$. The load torque variation $\Delta T_l^{'}$ can be regarded as an external disturbance input. The state variable is the speed variation $\Delta\nu_r$. From Eq. 27.40, the dynamic equation is thus

$$(J/N_p)\omega_n^2(p+\rho_w)\Delta\nu_r + \Delta T_l^{'} = \Delta T = -F_{\nu r}\Delta\nu_r + F_\nu\Delta\nu + F_{\nu x}\Delta V_x + F_{\nu y}\Delta V_y \tag{27.41}$$

corresponding with the block diagram (with internal feedback) in Fig. 27.3.

$F_{\nu r}$, F_ν, $F_{\nu x}$ and $F_{\nu y}$ are the open-loop transfer functions:

$$F_{\nu r}(p) = \frac{-\Delta T}{\Delta\nu_r} = \frac{T_0}{\nu_{s0}}\cdot\frac{N_{\nu r}(p)}{D(p)} \tag{27.42}$$

$$F_\nu(p) = \frac{\Delta T}{\Delta\nu} = \frac{T_0}{\nu_{s0}}\cdot\frac{N_\nu(p)}{D(p)} \tag{27.43}$$

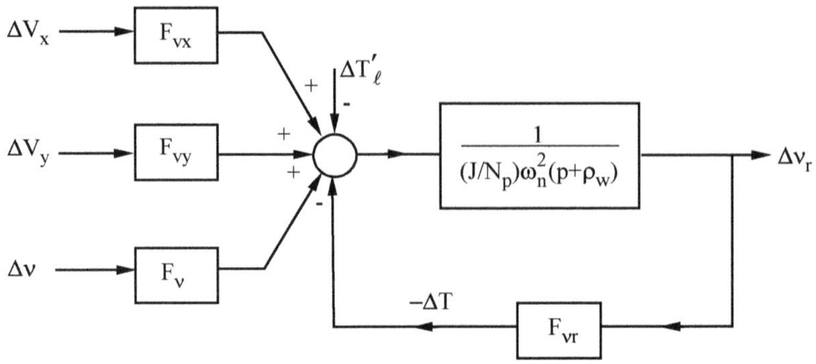

Fig. 27.3 Block diagram for motoring

If we choose the steady-state voltage along the x- or real axis, we get for F_{vx} and F_{vy}

$$F_{vx}(p) = \frac{\Delta T}{\Delta V_x} = \frac{T_0}{\nu_{s0}} \cdot \frac{\nu_0}{V_0} \cdot \frac{N_{vre}(p)}{D(p)} \tag{27.44}$$

$$F_{vy}(p) = \frac{\Delta T}{\Delta V_y} = \frac{T_0}{\nu_{s0}} \cdot \frac{\nu_0}{V_0} \cdot \frac{p\tau_r}{\nu_0 \tau_r} \cdot \frac{N_\nu(p)}{D(p)} \tag{27.45}$$

The polynomials $D(p)$, $F_{\nu r}(p)$, $F_\nu(p)$ and $F_{vx}(p)$ are[7]:

$$D(p) = C(p) \cdot C(p)^* \tag{27.46}$$

with

$$C(p) = (\tau_r p)^2 + (\tau_r p)[(1+\alpha) + j(\nu_0\tau_r)(1+s_0)] + [(\sigma\alpha - s_0(\nu_0\tau_r)^2) + j(1+\alpha s_0)(\nu_0\tau_r)] \tag{27.47}$$

$$N_{\nu r}(p) = (\tau_r p)^3 + (\tau_r p)^2[(1+\sigma)\alpha + 1 - (s_0\nu_0\tau_r)^2] + (\tau_r p)[\sigma\alpha^2 + 2\sigma\alpha - 2\alpha(s_0\nu_0\tau_r)^2 + (\nu_0\tau_r)^2] \\ +\{\sigma^2\alpha^2 + (\nu_0\tau_r)^2 - (s_0\nu_0\tau_r)^2[\alpha^2 + (\nu_0\tau_r)^2]\} \tag{27.48}$$

$$N_\nu(p) = (\tau_r p)^2(\sigma\alpha - s_0(\nu_0\tau_r)^2) + (\tau_r p)[\sigma\alpha(1+\alpha) + (\nu_0\tau_r)^2(1 - \alpha s_0^2 - 2s_0)] \\ +\{\sigma^2\alpha^2 + (\nu_0\tau_r)^2[1 - 2s_0 - s_0^2(\alpha^2 + 2(1-\sigma)\alpha)] - (s_0\nu_0\tau_r)^2(\nu_0\tau_r)^2(1+2s_0)\} \tag{27.49}$$

$$N_{\nu re}(p) = (\tau_r p)^3(1+s_0\alpha) + (\tau_r p)^2\{(1+\alpha)(1+s_0\alpha) + (1-s_0)[(\nu_0\tau_r)^2 s_0 - \sigma\alpha]\} \\ +2s_0(\tau_r p)[(1+\alpha)\sigma\alpha + (1+s_0^2\alpha)(\nu_0\tau_r)^2] + 2s_0\{[\sigma\alpha - (\nu_0\tau_r)^2 s_0]^2 \\ +(\nu_0\tau_r)^2(1+s_0\alpha)^2\} \tag{27.50}$$

---

[7] Please explain why $F_{vy}(p)$ is proportional to $pF_\nu(p)$.

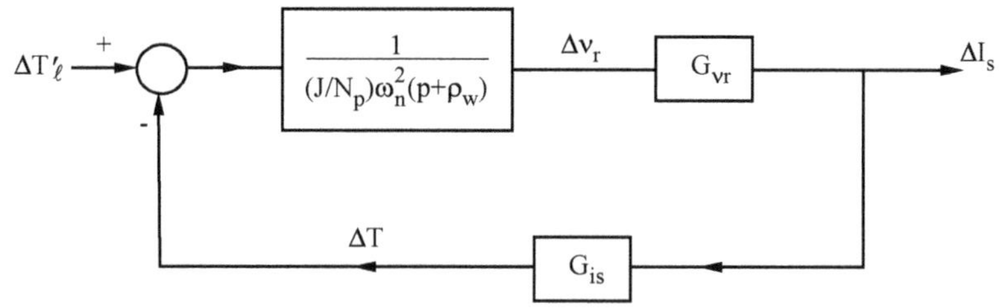

Fig. 27.4 Block diagram for generating

Note that all polynomials are (and can be written as) functions of $\tau_r p$, i.e. the Laplace operator referred to the rotor time constant. Also, all open-loop transfer functions share the same open-loop poles, i.e. the zeros of $D(p)$. All closed-loop transfer functions (with as input frequency or voltage variations and output the speed variation) have as poles the eigenvalues of the system and as open-loop zeros the zeros of the corresponding open-loop transfer function.

The eigenvalues are the zeros of the characteristic equation

$$1 + \frac{F_{\nu r}}{(J/N_p)\omega_n^2(p+\rho_w)} = 0 \tag{27.51}$$

or

$$1 + \frac{\tau_r}{\tau_m} \cdot \frac{N_{\nu r}(p)}{D(p) \cdot (\tau_r p + \tau_r \rho_w)} = 0 \tag{27.52}$$

With p as Laplace operator ($1/\omega_n$ as time reference), the gain of the feedback loop is $1/\tau_r\tau_m$. It turns out to be more convenient to choose the rotor time constant as time reference, i.e. $\lambda = p\tau_r$ as Laplace operator, resulting in τ_r/τ_m as the feedback gain.

The zeros of the open loop are the zeros of the transfer function of speed to electromagnetic torque. The open-loop poles include the poles of this transfer function and the mechanical (inertia) pole.

For generating on a constant voltage supply, we obtain the block diagram in Fig. 27.4; state variable is the current variation and input is the driving (*load*) torque. The voltage variations are assumed to be zero and the current variation therefore also represents the active power variation. $G_{\nu r}$ is the transfer function of speed to current, while G_{is} represents the transfer function of current to electromagnetic torque. As $G_{\nu r} \cdot G_{is} = F_{\nu r}$, the eigenvalues for generating are the same as those for motoring.

27.3.5 Eigenvalue Analysis

27.3.5.1 Open-Loop Poles

The open-loop poles include the mechanical (inertia and friction) pole and the four zeros of $D(p)$. The latter are called the constant-speed poles, as these are the (purely

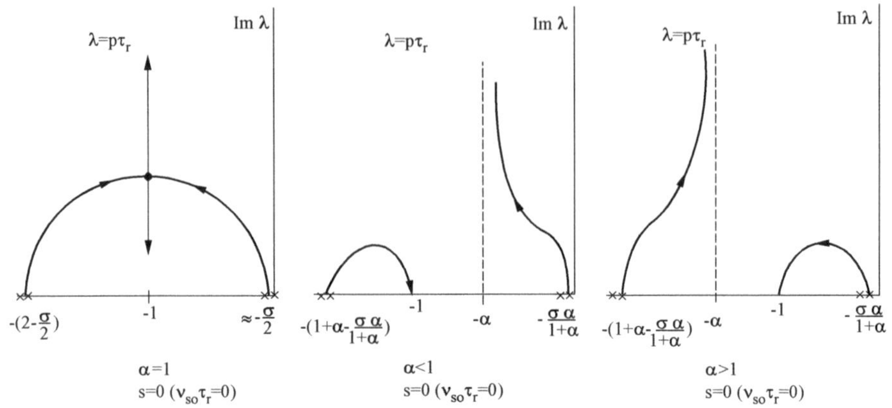

Fig. 27.5 Open-loop poles for $\alpha = 1$, $\alpha < 1$ and $\alpha > 1$

electrical) eigenvalues of the system without speed feedback (for example for an infinite inertia).

To start with, we will analyse these poles for slip zero. Figure 27.5 illustrates the locus of the poles for variable frequency parameter $\nu_0\tau_r$, at zero slip frequency and for three cases: $\alpha = 1$, $\alpha < 1$ and $\alpha > 1$.

For very small $\nu_0\tau_r$, the electrical poles are on or near the real axis. Their real parts correspond with the magnetising field time constants and the short-circuit (or leakage) time constants, respectively.

With an increasing frequency parameter, the imaginary parts increase while the real parts come closer together. If $\alpha = 1$, these four poles come together two by two for $\nu_0\tau_r = 2\sqrt{(1-\sigma)} \approx 2$.

For still higher values of the frequency parameter ($\nu_0\tau_r \gg (1+\alpha)$), the result is two complex conjugate high-frequency poles and two complex conjugate low-frequency poles. The real parts of these two pairs of poles are equal for $\alpha = 1$. The high-frequency imaginary part approximately corresponds with the supply frequency ($\approx \nu_0\tau_r$). Note that for $\alpha \neq 1$ the high-frequency poles have a real part corresponding to $\lambda \approx \alpha$ (the stator time constant), while the real part of the low-frequency poles corresponds to $\lambda \approx 1$ (the rotor time constant).[8]

An acceptable approximation for these constant speed poles is as follows:

- for $\nu_0\tau_r \ll 1 + \alpha$:

$$\lambda \approx -\left(1 + \alpha - \frac{\sigma\alpha}{1+\alpha}\right) \pm j\nu_o\tau_r\frac{\alpha + s_0}{1+\alpha}$$
$$\lambda \approx -\frac{\sigma\alpha}{1+\alpha} \pm j\nu_o\tau_r\frac{1+\alpha s_0}{1+\alpha}$$

[8] Explain this. Keep in mind the chosen (instantaneously) synchronous reference frame.

- for $\nu_0\tau_r \gg 1+\alpha$:

$$\lambda \approx -1 \pm j\nu_0\tau_r \left(s_0 + \frac{(1+\alpha)^2 - 4\alpha\sigma}{4(\nu_0\tau_r)^2(1-s_0)}\right) \approx -1 \pm j\nu_0\tau_r s_0$$
$$\lambda \approx -\alpha \pm j\nu_0\tau_r \left(1 - \frac{(1+\alpha)^2 - 4\alpha\sigma}{4(\nu_0\tau_r)^2(1-s_0)}\right) \approx -\alpha \pm j\nu_0\tau_r$$

The effect of the magnetic parameter σ is particularly important for $\nu_0\tau_r \ll 1+\alpha$, thus for low frequencies. The poles with the smallest damping move up farther into the left half plane for larger σ; this effect can be quite important as the real part of these poles is proportional to σ. The imaginary parts are almost not affected, and neither are the poles with the largest damping. For higher frequencies ($\nu_0\tau_r \gg 1+\alpha$), the effect of σ is limited. The small shift to the left of the high-frequency pole is negligible.

The effect of the slip ($s_0\nu_0\tau_r$) is rather small for normal small slip values. From a detailed analysis, the effect can be summarised as follows.

For low frequencies ($\nu_0\tau_r \ll 1+\alpha$), a larger positive slip results in an increase of the imaginary parts while the difference between the real parts increases slightly. A negative slip, on the other hand, has the inverse effect. For large frequencies ($\nu_0\tau_r \gg 1+\alpha$), the frequency (imaginary part) of the high-frequency pole decreases, while the frequency of the low-frequency pole increases, both by an amount more or less equal to the slip frequency parameter $s_0\nu_0\tau_r$. The real part (damping) is only slightly affected. For intermediate values of the frequency ($\nu_0\tau_r \approx 1+\alpha$), the damping (real part) of the poles is somewhat affected, while the imaginary part is almost not altered.

27.3.5.2 Open-Loop Zeros

The zeros of the open loop are the zeros of the polynomial $N_{\nu r}(p\tau_r) = N_{\nu r}(\lambda)$. To find these zeros, we rewrite the polynomial as follows

$$\left\{\{[\lambda^2 + \lambda(1+\alpha) + \sigma\alpha](\lambda + \sigma\alpha)\} + (\nu_0\tau_r)^2(\lambda+1)\right\} - (\nu_{s0}\tau_r)^2[(\lambda+\alpha)^2 + (\nu_0\tau_r)^2] = 0 \tag{27.53}$$

To find the zeros, we will apply twice the root-locus method, first with $(\nu_0\tau_r)^2$ as gain and next with $-(\nu_{s0}\tau_r)^2$ as gain.

First, we consider zero slip frequency and a variable frequency parameter. The gain is $(\nu_0\tau_r)^2$. This first root locus (cf. the dashed line in Fig. 27.6) has as its starting points $\lambda \approx -\sigma\alpha/(1+\alpha)$, $\lambda \approx -[1+\alpha-\sigma\alpha/(1+\alpha)]$ and $\lambda = -\sigma\alpha$. The end points are $\lambda = -1$ and the asymptote at $\lambda = -(1+\sigma)\alpha/2$.

For the second application of the root-locus method, we consider a constant frequency parameter $\nu_0\tau_r$ and a variable slip frequency parameter $\nu_{s0}\tau_r$. The starting points for the second root locus are points (corresponding with the given value of $\nu_0\tau_r$) on the previous root locus, and the end points are $\lambda = -\alpha \pm j\nu_0\tau_r$ and $\lambda = +\infty$. For a given value of $\nu_0\tau_r$ and variable $\nu_{s0}\tau_r$, results the root locus drawn in full line.

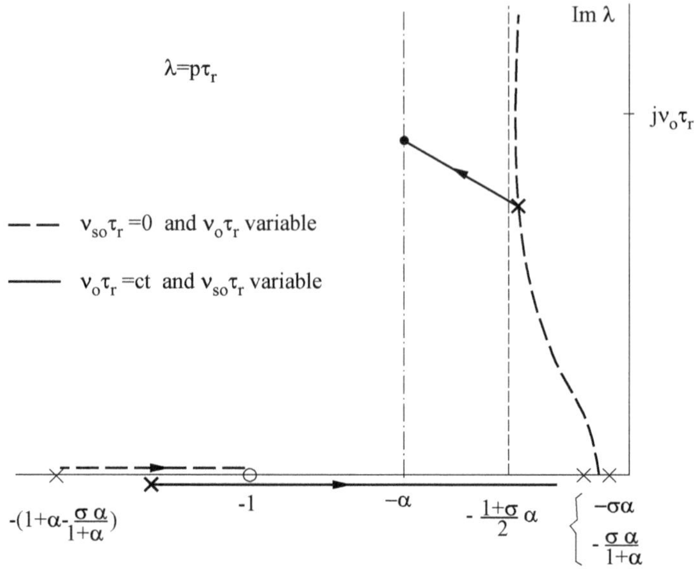

Fig. 27.6 Locus for the open-loop zeros

For sufficiently high frequencies ($\nu_0\tau_r \gg 1+\alpha$) and normal small slip values, an acceptable approximation for the zeros of the open loop is

$$\lambda_1 \approx -1 \tag{27.54}$$

$$\lambda_{2,3} \approx -\alpha\frac{1+\sigma}{2} \pm j\nu_0\tau_r \tag{27.55}$$

For higher slip frequencies (and sufficiently high $\nu_0\tau_r$), only the zero λ_1 should be corrected as

$$\lambda_1 \approx -[1-(\nu_{s0}\tau_r)^2] \tag{27.56}$$

For values of the frequency parameter around $1+\alpha$ (and small slip), an approximation for the open-loop zeros is

$$\begin{aligned}\lambda_1 &\approx -\frac{(\nu_0\tau_r)^2+(1+\alpha)^3}{(\nu_0\tau_r)^2+(1+\alpha)^2} \\ \lambda_{2,3} &\approx -\frac{1}{2}\frac{(\nu_0\tau_r)^2\alpha(1+\sigma)+\sigma\alpha(1+\alpha)(2+\alpha)}{(\nu_0\tau_r)^2+(1+\alpha)^2} \pm j\frac{\nu_0\tau_r}{\sqrt{1+\alpha}}\end{aligned} \tag{27.57}$$

For very low values of the frequency parameter ($\nu_0\tau_r \ll 1+\alpha$), this can be simplified to

$$\begin{aligned}\lambda_1 &\approx -(1+\alpha)\\ \lambda_{2,3} &\approx -\frac{1}{2}\sigma\alpha \cdot \frac{2+\alpha}{1+\alpha} \pm j\frac{\nu_0\tau_r}{\sqrt{1+\alpha}}\end{aligned} \tag{27.58}$$

The complex conjugate zeros are now rather close to the imaginary axis. Note also that their real part is proportional to the magnetic parameter σ.

As is clear from Eqs. 27.55–27.58, the magnetic parameter σ has an important effect on the real part of the complex zeros, especially for low frequencies. Small σ values result in small real parts for these low values of the frequency parameter. For larger frequency parameter values, the effect of σ is negligible. In the next section, it will become clear that the magnetic parameter has a profound effect on the stability behaviour at low supply frequencies.

27.3.5.3 Root Loci for the Eigenvalues

Section 27.3.3 explained that, for conventional induction machines, mainly two parameters exhibit an important variation with machine size (scaling laws) and supply frequency, i.e. the frequency parameter $\nu_0\tau_r$ and the ratio of the rotor time constant to the electromechanical time constant, τ_r/τ_m. Of the other parameters, only σ exhibits any significant effect of machine size (α is for most machines more or less equal to 1 and the slip frequency parameter in rated conditions is more or less equal to 0.2, independently of machine size). On the other hand, from Sects. 27.3.5.1 and 27.3.5.2, it is clear that the open-loop poles and zeros are mainly determined by $\nu_0\tau_r$, and to a minor degree by α, σ and the slip frequency parameter $\nu_{s0}\tau_r$.

As a consequence, machine size and supply frequency can be expected to have an important effect on the dynamic behaviour of induction machines, mainly via the frequency parameter $\nu_0\tau_r$ and the ratio of rotor time constant to electromechanical time constant τ_r/τ_m. The latter parameter is also a convenient choice for the gain of the internal feedback loop, as is clear from the characteristic equation or the block diagrams presented in Sect. 27.3.4. It is important to note that the effect of the operating point (e.g. flux level and slip) is reflected in the slip frequency parameter (slip or thus load torque) and the gain (flux level, inertia).

We will therefore study the eigenvalues by means of the root-locus method with τ_r/τ_m as gain and for a range of values for the other parameters, mainly the frequency parameter $\nu_0\tau_r$ (and in a second instance, the slip frequency parameter $\nu_{s0}\tau_r$, the ratio of rotor to stator time constants α and the magnetic parameter σ).

The root loci exist in mainly two distinct forms (see Fig. 27.7). For larger values of the frequency parameter $\nu_0\tau_r \gg 1+\alpha$, a high-frequency locus results, as is shown in (a) in the figure. For small values of the frequency parameter $\nu_0\tau_r \ll 1+\alpha$, we can observe a low-frequency locus as in (b) in Fig. 27.7. Intermediate locus shapes also exist.

The high-frequency locus (for $\nu_0\tau_r \gg 1+\alpha$) resembles the root locus of an independently excited DC machine. As discussed in Sect. 27.3.5.2, for high $\nu_0\tau_r$ the two

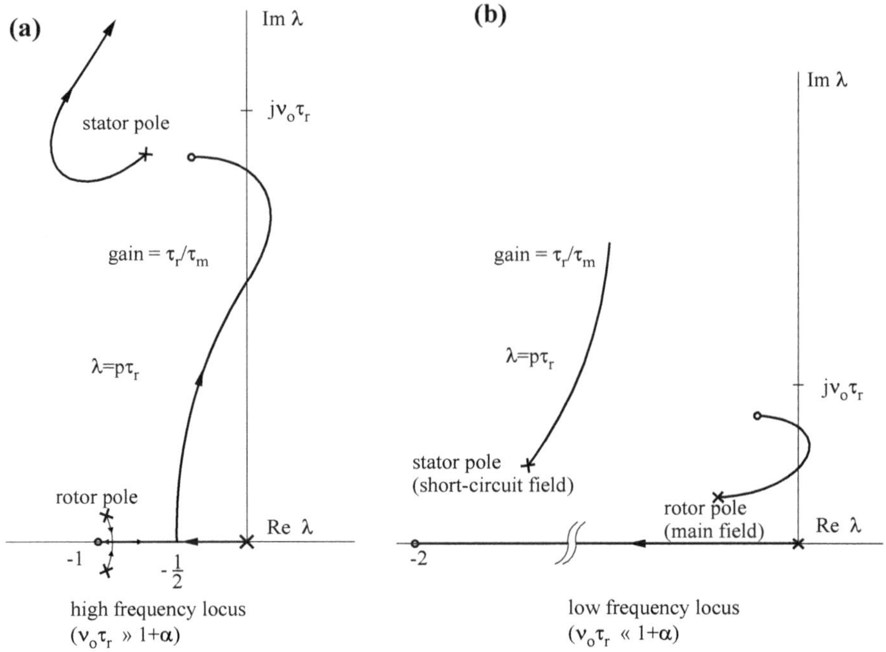

Fig. 27.7 Root loci for the eigenvalues: **a** high frequency; **b** low frequency

pairs of open-loop poles have almost the same damping but a large difference in frequency: $\lambda \approx -\alpha \pm j\nu_0\tau_r$ and $\lambda \approx -1 \pm j\nu_{s0}\tau_r$. As result, there is minor interaction between these high- and low-frequency poles. On the other hand, there is a real zero ($\lambda \approx -1$) of the open loop in the vicinity of the low-frequency poles and a pair of complex conjugate zeros ($\lambda \approx (1+\sigma)\alpha/2 \pm j\nu_0\tau_r$) in the vicinity of the high-frequency poles. As a result, the real zero almost compensates one of the low-frequency poles. The segment on the real axis between the open-loop zero and the inertia pole ($\lambda \approx 0$) belongs to the root locus. Almost exactly in the middle of this segment, there is a break-away point, from which branches depart to the complex conjugate zeros. The high-frequency poles repel these branches so that these branches may move into the right half plane.

Dependent on the gain τ_r/τ_m, three characteristic cases for the dynamic behaviour can be distinguished:

- for very low values of the gain τ_r/τ_m ($\tau_r/\tau_m < 0.1$), the dynamic behaviour is dominated by a real eigenvalue near the origin (on the trajectory starting from the inertia pole $\lambda = \rho_w\tau_r \approx 0$). A good approximation for this eigenvalue is $\lambda = -\tau_r/\tau_m$ or $p = -1/\tau_m$, which corresponds to the quasi-static electromechanical time constant for the inertia and the steady-state load torque characteristic (thus all electrical transients are negligible as these are much faster than the mechanical one).

- for values of the gain τ_r/τ_m approaching 1/4, we subsequently see two real eigenvalues $\lambda \approx -\frac{1}{2} \pm \left[\frac{1}{4} - \frac{\tau_r}{\tau_m}\right]^{1/2}$, two coinciding real eigenvalues $\lambda = -1/2$ for $\tau_r/\tau_m = 1/4$ and then for higher gain, two complex conjugate eigenvalues $\lambda \approx -\frac{1}{2} \pm j\left[\frac{\tau_r}{\tau_m} - \frac{1}{4}\right]^{1/2}$
- for very high values of the gain ($\tau_r/\tau_m \gg 1/4$), the root locus approaches the complex zeros. The real part of the eigenvalues λ is initially equal to -0.5, but for very high gain the root-locus may enter the right half plane, i.e. instability. We will call the uttermost right point on the root locus the *minimal damped point.*

We note that for this high-frequency root locus, the effect of the magnetic parameter σ is rather limited as poles and zeros are almost independent of σ. Only where the root locus comes near to the high-frequency poles and zeros (i.e. where the root locus bends to the right) may a stabilising effect for higher σ-values be observed.

The effect of the rotor-to-stator damping ratio, however, may be quite significant as the real parts of the high-frequency poles and zeros are proportional to α. Low values of α will result in a root locus branch closer to the imaginary axis and possibly bending further into the right half plane. However, normal values for α are around $\alpha = 1$.

The effect of the slip parameter $\nu_{s0}\tau_r$ is very limited for this high-frequency root locus, at least for normal slip values.

From the above discussion, it is tempting to conclude that the high-frequency instability can become a problem as soon as the frequency parameter is sufficiently high. In the next section, we will show that this is only partially true. For very high values of the frequency parameter, the gain values corresponding with possible instability are indeed unrealistically high for practical machines.

The low-frequency locus (for $\nu_0\tau_r \ll 1+\alpha$) is somewhat special as the poles and zeros are quite close together as to frequency (imaginary part). For the poles, we may use the following approximation:

$$\lambda_{1,2} \approx -\left(1+\alpha-\frac{\sigma\alpha}{1+\alpha}\right) \pm j\nu_o\tau_r\frac{\alpha+s_0}{1+\alpha} \approx -\left(1+\alpha-\frac{\sigma\alpha}{1+\alpha}\right) \pm j\nu_0\tau_r\frac{\alpha}{1+\alpha} \tag{27.59}$$

$$\lambda_{3,4} \approx -\frac{\sigma\alpha}{1+\alpha} \pm j\nu_o\tau_r\frac{1+\alpha s_0}{1+\alpha} \approx -\frac{\sigma\alpha}{1+\alpha} \pm j\nu_0\tau_r\frac{1}{1+\alpha}$$

For the zeros, we have already found that (see Eq. 27.58)

$$\lambda_1 \approx -(1+\alpha) \tag{27.60}$$

$$\lambda_{2,3} \approx -\frac{1}{2}\sigma\alpha \cdot \frac{2+\alpha}{1+\alpha} \pm j\frac{\nu_0\tau_r}{\sqrt{1+\alpha}} \tag{27.61}$$

The two pairs of open-loop poles and the complex zeros will now interact quite closely. Moreover, the distance between the real zero and the poles and zeros near to the imaginary axis is rather large, limiting their interaction. The branches of the root locus connecting the poles and the zeros with the smallest real parts will therefore most likely be bent to the right, possibly into the right half plane.

27.3.6 Typical Dynamic Behaviour

27.3.6.1 Parameter Plane

The dynamic properties of induction machines are determined by six dimensionless parameters. However, representing the dynamic properties in a six-dimensional space is not very practical. As discussed in the previous sections, mainly two parameters play a key role in the dynamic properties: the frequency parameter and the gain. Therefore, we will represent the dynamic properties in the parameter plane $(\tau_r/\tau_m, \nu_0\tau_r)$ with the other four parameters as secondary parameters. Each point in this parameter plane will represent a specific dynamic behaviour when associated with values for the other four parameters.

On the other hand, the scaling laws presented in Sect. 27.3.3 have indicated that for normal induction machines, these same two parameters show an important variation with machine size (and operating point, i.e. flux and supply frequency). A *typical* or *average* machine in rated operating conditions can be derived from these scaling laws. Indeed, for $N_p \geq 2$ we derive the following parametric equation in the plane $(\tau_r/\tau_m, \nu_0\tau_r)$:

$$\nu_0\tau_r = 40\tau_p \tag{27.62}$$

$$\tau_r/\tau_m = 8\frac{N_p}{N_p - 1} \cdot \tau_p \tag{27.63}$$

Eliminating τ_p yields the equation of a straight line $\tau_r/\tau_m = \frac{N_p}{5(N_p-1)} \cdot \nu_0\tau_r$ or, on a logarithmic scale,

$$\log(\tau_r/\tau_m) = \log \nu_0\tau_r - \log[5(N_p - 1)/N_p] \tag{27.64}$$

Figure 27.8 depicts this line for rated frequency ($\nu_0 = 1$) and $N_p = 2$. Each point on this line corresponds with a certain value of the pole pitch τ_p or machine size (rated power). Each point also corresponds to scaling law values for the parameter σ. For the parameters α (normally equal to 1) and τ_w/τ_r (negligible in most cases), no annotations are given. The slip parameter $\nu_{s0}\tau_r$ may vary, for example between 0 and $\nu_{sn}\tau_r = 0.2$.

Obviously, actual machines will *not exactly* correspond to this *average* machine line, but the corresponding points will surely approach this line. The secondary parameters for actual machines may also deviate from the average values. In the figure, the arrows indicate how the corresponding point will move when the leakage

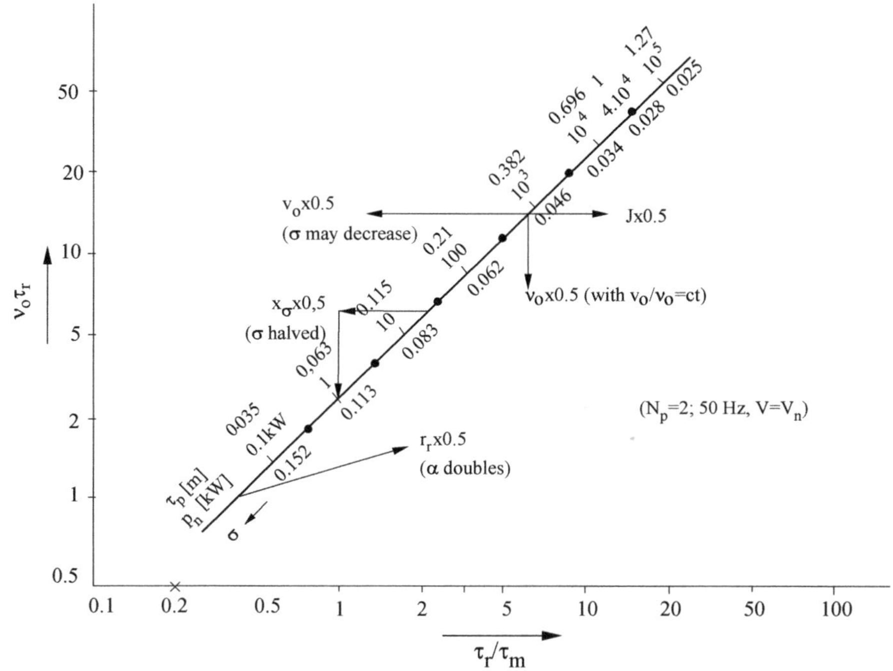

Fig. 27.8 Typical machine line for $N_p = 2$

inductance is halved, or the inertia, the voltage (or flux) or the pu rotor resistance are halved. For variable supply frequency (and the same rated flux level), the line shifts vertically.

27.3.6.2 Stability Boundaries for $\sigma = 0$

An interesting starting point for the dynamic analysis of the induction machine is the ideal case $\sigma = 0$. Note that $\sigma = 0$ does not imply zero leakage, which would correspond to zero rotor and stator short-circuit time constants. Instead, it corresponds to an infinite magnetising inductance, similar to the idealised equivalent circuit without magnetising inductance that is sometimes used for the steady-state analysis.

The advantage of the case $\sigma = 0$ is that analytic solutions are easily obtained. For example, the locus corresponding with marginal stability behaviour (i.e. eigenvalues on the imaginary axis) consists of two intersecting curves:

$$(\nu_0\tau_r)^2 = (1+\alpha)\left[\frac{\tau_r}{\tau_m} - (1+\alpha)\right] \tag{27.65}$$

$$(\nu_0\tau_r)^2 = \frac{\tau_r}{\tau_m} \cdot \frac{(1+\alpha)^2 + (\tau_r/2\tau_m)}{2+\tau_r/\tau_m} \tag{27.66}$$

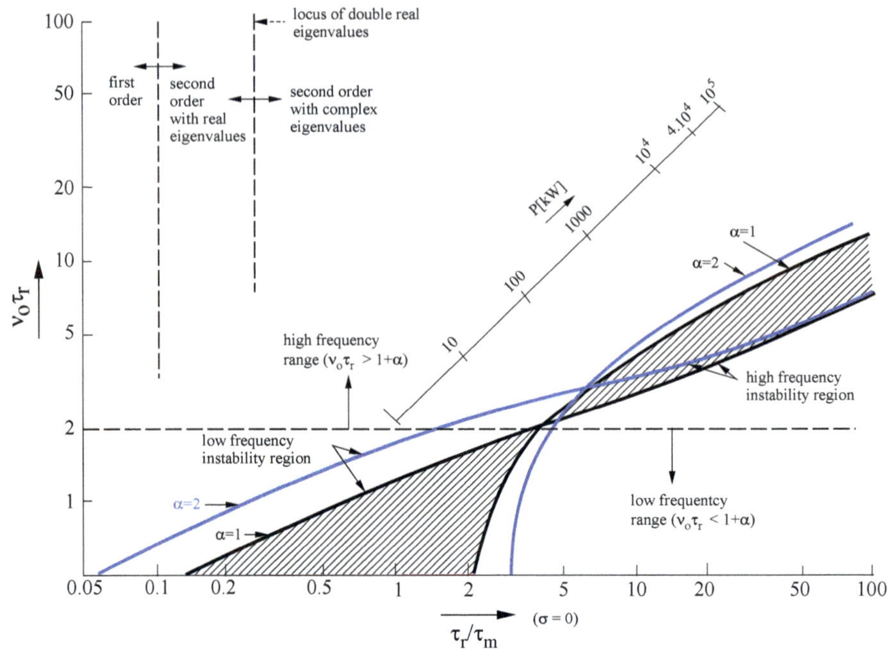

Fig. 27.9 Instability boundaries for $\sigma = 0$ ($\alpha = 1$ and $\alpha = 2$)

These loci (τ_r/τ_m, $\nu_0\tau_r$) are the borderlines between stable and unstable behaviour. The oscillating frequencies (imaginary parts of the eigenvalues) corresponding with these boundaries are

$$(\nu_d\tau_r)^2 = \frac{(\nu_0\tau_r)^2}{1+\alpha} \tag{27.67}$$

$$(\nu_d\tau_r)^2 = \frac{1}{2}\cdot\frac{\tau_r}{\tau_m} \tag{27.68}$$

Figure 27.9 illustrates these boundaries for $\alpha = 1$ and $\alpha = 2$. The hatched region between the boundaries for $\alpha = 1$ corresponds to unstable behaviour (eigenvalues in the right half plane). The region outside the boundaries is stable for $\alpha = 1$. The intersection point of the boundaries is at ($\tau_r/\tau_m = 2(1+\alpha)$, $\nu_0\tau_r = 1+\alpha$) with eigenfrequency $\nu_d\tau_r = \sqrt{(1+\alpha)}$.

The two instability regions demarcated by the boundary curves correspond with the instability parts in the low-frequency and the high-frequency root loci, respectively, and are accordingly called the low- and the high-frequency instability regions. The point $(2(1+\alpha), 1+\alpha)$ where both curves intersect (for $\sigma = 0$) is referred to as the transition point. Note that this value of the frequency parameter corresponds to the break-away point for the open-loop poles.

Figure 27.9 also illustrates the effect of α. Larger α-values cause the instability boundaries (and the transition point) to shift to higher values of $\nu_0\tau_r$ and τ_r/τ_m. As we will see later on for non-zero σ, these larger α-values in fact increase the low-frequency instability region (and slightly decrease the high-frequency instability).

In Fig. 27.9, also the average machine line for rated frequency supply is drawn. As expected, conventional machines are indeed stable for rated frequency operation. As a reduced supply frequency operation causes a vertical shift of the machine line to lower values of the frequency parameter $\nu_0\tau_r$, the figure suggests that at lower frequencies all induction machines may become unstable. This would be a hasty conclusion, as real non-zero values of the magnetic parameter cause an important reduction of the instability regions (see Sect. 27.3.6.4).

Other secondary parameters also affect the stability regions, for example the damping ratio α (see above) or the slip frequency. The latter may have a negative effect on the stability (Sect. 27.3.6.4).

27.3.6.3 Dynamic Behaviour for the Case $\sigma = 0$

In addition to the boundaries between stability and instability, the idealised case of $\sigma = 0$ can also teach us more about the degree of stability. Most important are the dominant eigenvalues, their damping and eigenfrequency.

In the parameter plane, the dominant modes and eigenvalues may be represented as in Fig. 27.10 (for the case $\sigma = 0$). The lines of constant damping may be regarded as constant height contour lines referred to sea level (*i.e. altitude*); a negative value represents a *valley* (stable), a positive value represents a *hill* (unstable). In this representation, the instability regions form a *ridge* from the left lower corner to the right upper corner of the plane.

The red dash-dot line in the figure is *a possible* worst damping line, *here* the locus of the points $(\tau_r/\tau_m, \nu_0\tau_r)$ where for a given $\nu_0\tau_r$ the damping is maximally positive. The points on *this* worst damping line[9] correspond to the uttermost right points on the root loci for given $\nu_0\tau_r$ (see Fig. 27.10). In the ideal case of $\sigma = 0$, all these points are in the right half plane (or, only for the transition point, on the imaginary axis). The transition point is in fact the best damped point for all points on the (or on each) worst damping line and is really a saddle point for all worst damping lines.

In a similar way, the eigenfrequencies (at least for complex dominant eigenvalues) may be represented in the parameter plane (see Fig. 27.11). From both figures, it is again clear that the frequency-parameter value $\nu_0\tau_r = 1 + \alpha$ (or $\nu_0\tau_r = 2$ for $\alpha = 1$) is the real transition between high- and low-frequency behaviour.

As to the instability regions we observe that in the high-frequency range, the (positive) real part of the eigenvalue (*i.e. positive damping*) along the worst damping line

[9]There are an infinite number of worst damping lines, depending on how the concept is defined, for example for constant τ_r/τ_m or for constant $\nu_0\tau_r$. Explain this from a purely mathematical point of view, and think of partial derivatives. In fact, the worst damping line for constant τ_r/τ_m and variable $\nu_0\tau_r$ is somewhat more practical, but the difference between both lines is small.

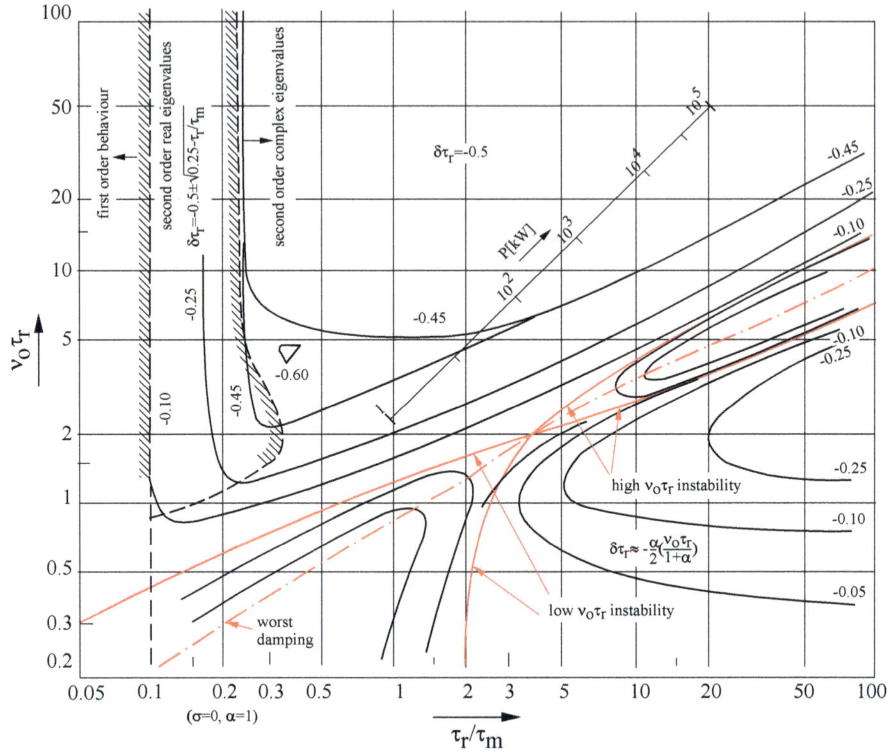

Fig. 27.10 Dynamic behaviour for $\sigma = 0$ and $\alpha = 1$

increases monotonously and quite abruptly with increasing frequency parameter values ($\nu_0\tau_r$). To the right or left of the worst damping line, *positive damping* decreases quite fast (i.e. the high-frequency instability region is steep but quite narrow).

In the low-frequency range, however, the *positive damping* increases much more gently. The positive damping reaches a maximum and then decreases, in the limit ($\nu_0\tau_r \rightarrow 0$) to zero. In other words, the low-frequency instability region is rather weak. As we will see later on, the secondary parameters also have an important effect in this region (in contrast to the high-frequency instability).

In the stable range, particularly above the worst damping line, the damping becomes more and more negative for higher values of the frequency parameter, until an extreme damping of about -0.5 is attained. In this region where the damping is -0.5, the eigenfrequency is $(\tau_r/\tau_m - 0.25)^{1/2}$ (see Sect. 27.3.5.3). For low values of the gain τ_r/τ_m the two dominant eigenvalues become real, and for very low gain values there is only one real dominant eigenvalue, corresponding with the inertia (and quasi-static machine torque). Note that in the stable range far above the transition point, the damping is almost independent of the frequency parameter.

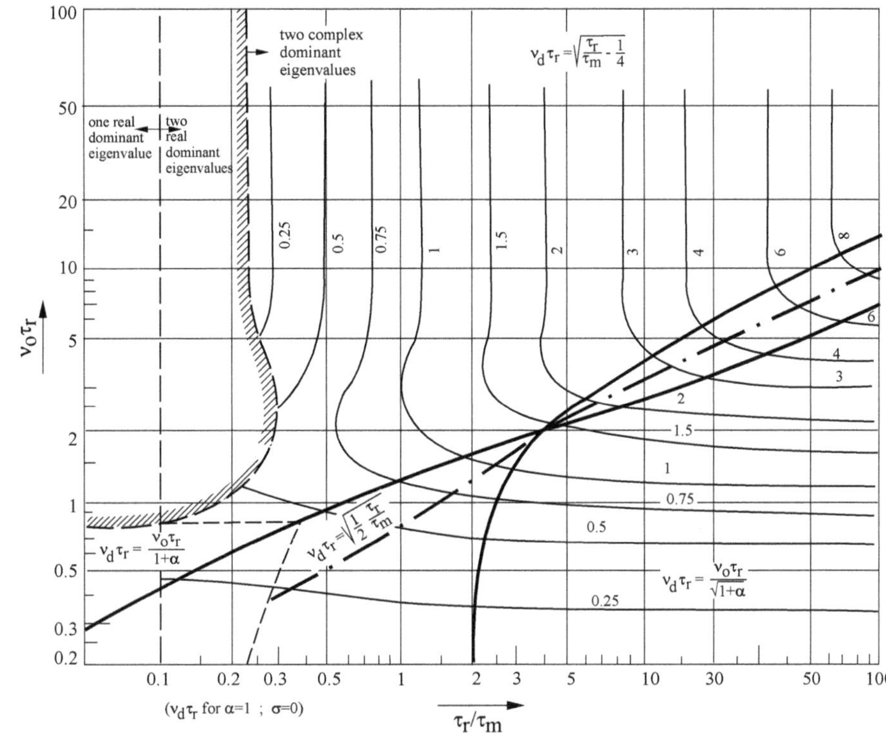

Fig. 27.11 Eigenfrequencies for $\sigma = 0$ and $\alpha = 1$

Below the worst damping line, and particularly in the low-frequency range, the damping changes quite smoothly. For very low $\nu_0\tau_r$ and high τ_r/τ_m the damping becomes independent of τ_r/τ_m and decreases with decreasing $\nu_0\tau_r$. The eigenfrequency then corresponds to the frequency of the open-loop zeros $\nu_d\tau_r = \nu_0\tau_r\sqrt{1+\alpha}$.

From the eigenfrequencies (see Fig. 27.11), similar conclusions as to the difference between the high and low-frequency regions can be deduced. Above the transition point and the worst damping line, the oscillation frequency is mainly dependent on the gain τ_r/τ_m and more or less independent of $\nu_0\tau_r$. Sufficiently below the transition point and the worst damping line, the reverse is true.

27.3.6.4 Effect of the Secondary Parameters

As is the case for the main parameters $\nu_0\tau_r$ and τ_r/τ_m, the effect of the secondary parameters is quite different in the low- and high-frequency regions. In some cases, their effect is completely opposite. This is for example the case for the damping ratio α and the slip parameter $\nu_{s0}\tau_r$. The magnetic parameter σ, on the other hand, has a

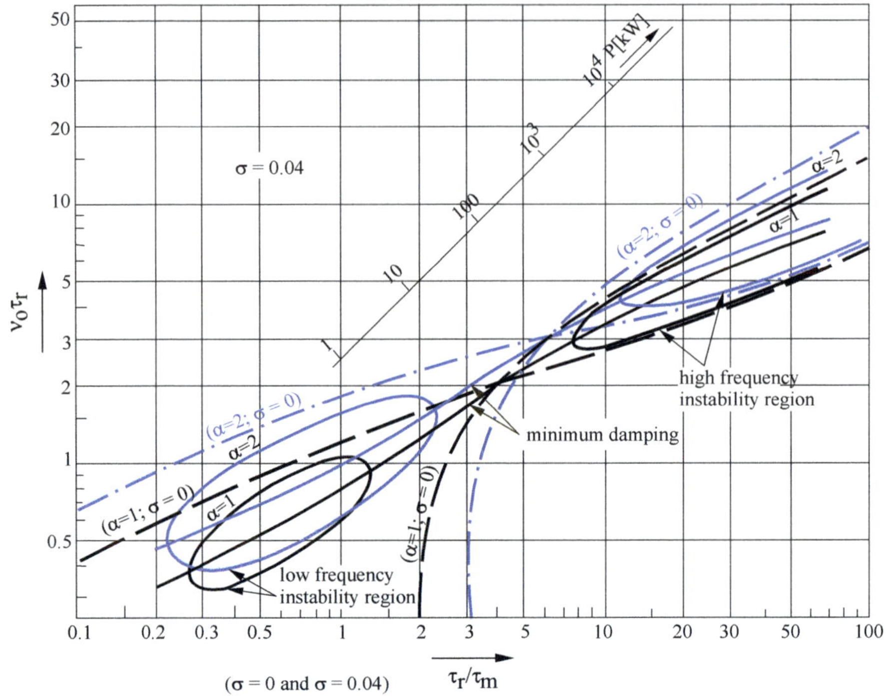

Fig. 27.12 Effect of σ on the stability behaviour

uniform stabilising effect, albeit to a different degree in the low- and high-frequency ranges.

Figure 27.12 shows the instability regions in the parameter plane for $\alpha = 1$ and $\alpha = 2$ and $\sigma = 0.04$. For the sake of comparison, the boundaries for $\sigma = 0$ are also given. The stabilising effect of the magnetic parameter is clear. Even for a low value of $\sigma = 0.04$ the instability regions shrink considerably, especially around the transition point. The two instability regions do not touch any more and around the transition point there is now a band for τ_r/τ_m and $\nu_0\tau_r$ where no instability occurs. A machine of rated power of 100 kW, for example, will never enter instability with decreasing supply frequency (for constant rated flux and machine inertia only). The stabilising effect is, however, strongest in the low-frequency region. The low-frequency instability is indeed much less pronounced than the high-frequency instability (see the previous section). With σ-values different from 0, the low-frequency region becomes closed for low values of $\nu_0\tau_r$, besides shrinking and moving away from the transition point. For $\sigma > 0.055$ and $\alpha = 1$ the low-frequency instability region does not exist any more. The high-frequency instability region also shrinks with higher values of the magnetic parameter but to a much lesser degree.

This may also be illustrated as in Figs. 27.13 and 27.14, which show the damping as a function of τ_r/τ_m along the worst damping line (for $\alpha = 1$ and $\alpha = 2$, respectively).

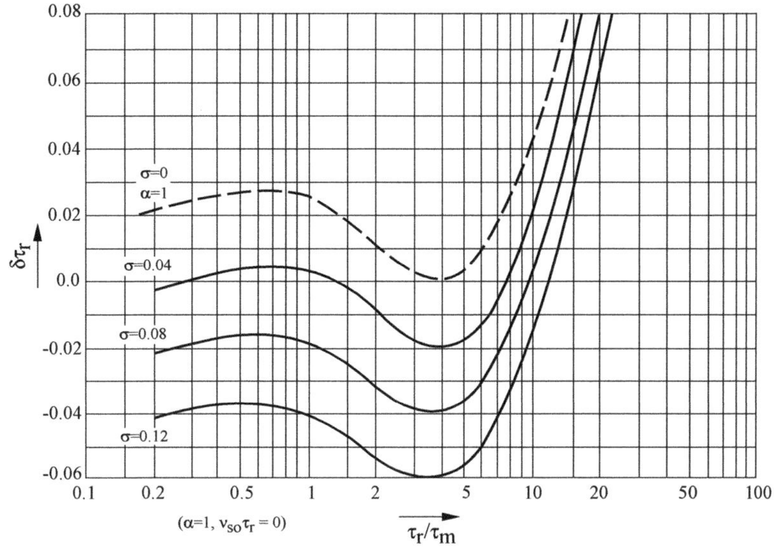

Fig. 27.13 Worst damping for different σ-values and $\alpha = 1$

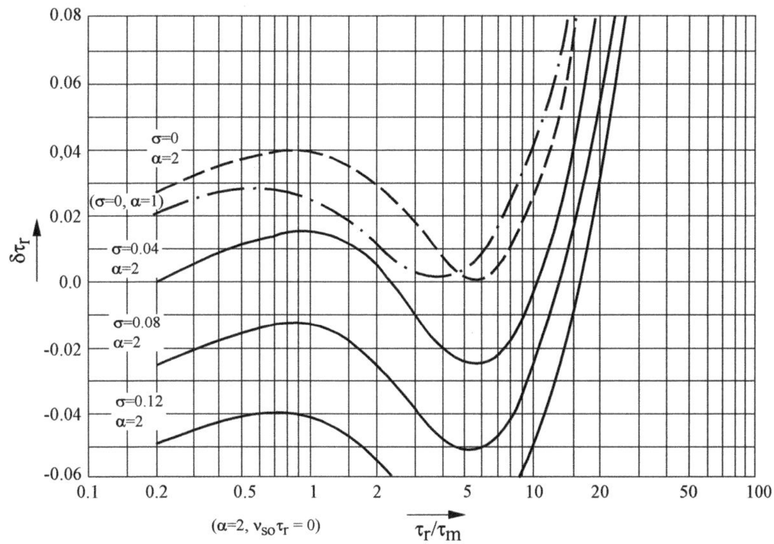

Fig. 27.14 Worst damping for different σ-values and $\alpha = 2$

Whereas a possible instability always remains in the high-frequency region (here large τ_r/τ_m), the instability in the low-frequency region disappears completely for larger σ (positive damping values imply instability, negative values imply stability).

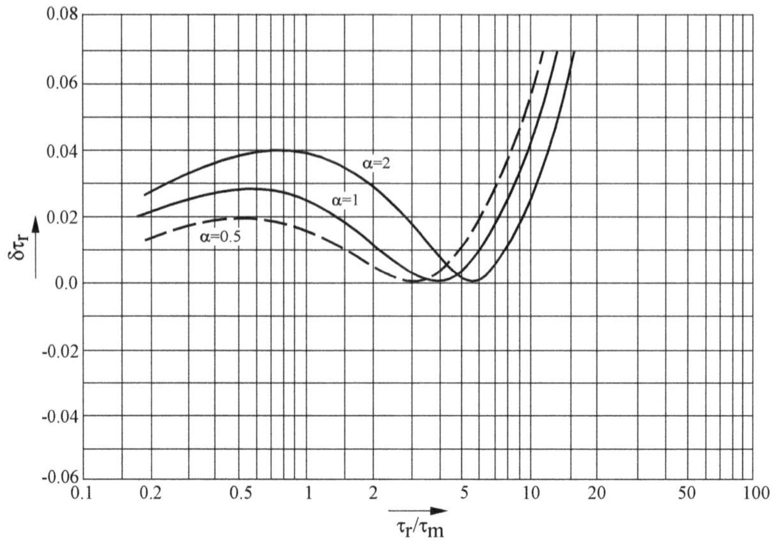

Fig. 27.15 Worst damping for different α-values

The effect of the damping ratio α has already been discussed for $\sigma = 0$. With larger α-values the transition point moves up to higher values of τ_r/τ_m and $\nu_0\tau_r$, and the low-frequency instability region appears to increase in size (Fig. 27.9). A comparison of Figs. 27.13 and 27.14 shows that for non-zero values of σ, with higher α-values the positive damping along the worst damping line increases in the low-frequency range but decreases in the high-frequency range. For example, the low-frequency instability for $\alpha = 1$ is non-existent for values of σ greater than 0.06 but for $\alpha = 2$ and the same σ it will still exist. Also, at the transition point the damping becomes more and more negative (stable) for higher α. Another illustration of the effect of α is given in Fig. 27.15, where for $\sigma = 0$ and three α-values the damping along the worst damping line is depicted.

The effect of the slip or slip frequency is illustrated in Figs. 27.16 and 27.17. The stabilising effect of positive slip is most pronounced in the range of low-frequency parameter values. For positive rated slip frequency $\nu_{s0}\tau_r = 0.2$, the low-frequency instability completely disappears; in the high-frequency range, the positive damping slightly decreases. In contrast, for negative slip values the instability in the low-frequency range increases. In the high-frequency range, stability also improves with negative slip - even more so than for positive slip (see Fig. 27.17).

Although the case $\sigma = 0$ seems to predict possible unstable behaviour for all machine sizes at supply frequency reduction, for non-zero σ-values the instability tendency is much lower. Actual induction machines have non-zero σ-values. The smaller the power rating, the larger the σ-values are. As the stabilising effect of σ is quite important in the range of low values of the frequency parameter and gain (i.e. in the parameter range of these small machines), instability for these small power

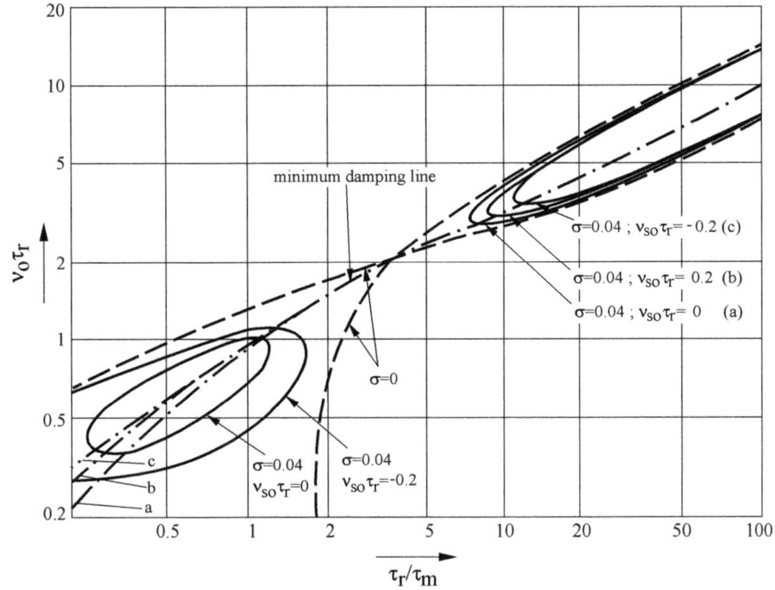

Fig. 27.16 Effect of slip on the stability behaviour

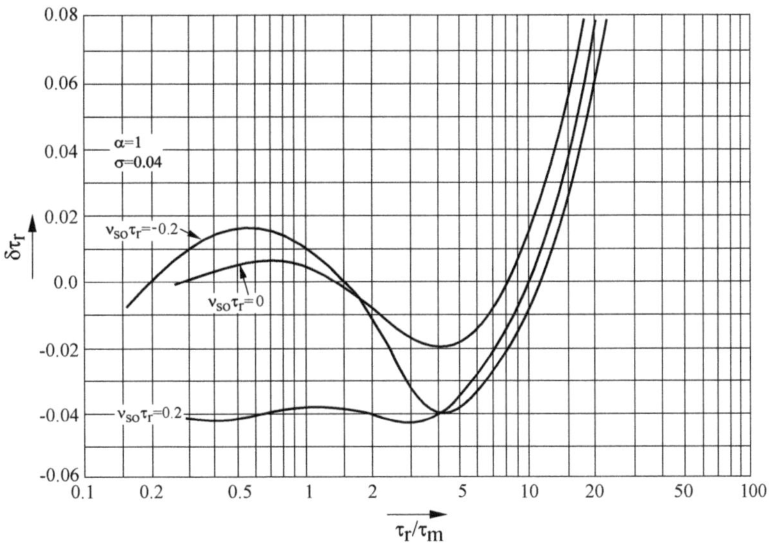

Fig. 27.17 Effect of slip on the worst damping

ratings is very unlikely. Larger machines (with power ratings in excess of 100 kW) may, nevertheless, exhibit instability at frequency reduction. This was first observed in practice in the 1970s and generated this kind of stability studies.[10]

27.4 Conclusions

This chapter began with the derivation of a dynamic model for induction machines. This model presupposes an idealised machine with sinusoidal distributed windings so that mmf harmonics can be ignored. Main field saturation has been modelled in a simplified way, similar to the steady-state analysis model, thus supposing a frozen saturation condition. In terms of supply, the model is very general, however. Both sinusoidal and non-sinusoidal supply can be studied with it.

Next, the model was applied to study the local stability behaviour around a steady-state operating point (with symmetrical sinusoidal supply). For an idealised induction machine with infinite magnetising inductance ($\sigma = 0$), it was found that instability could occur for all power ratings at reduced supply frequency. However, real machines have a finite magnetising inductance and σ-values different from zero. Moreover, as machines with a smaller power rating typically show larger σ-values, the instability tendency for small machines turns out to be non-existent. Yet, machines of large power rating are prone to unstable behaviour (e.g. oscillating, hunting) when operated at lower supply frequencies.

In special cases, small machines may still exhibit unstable behaviour, for example when large resistances are added in the supply lines. Or, with the boom of induction generators for small wind turbines, a more important instance when the machine is in generating mode.

Models with a more accurate modelling of (main) field saturation have quite recently been presented. For example, in [22] main field saturation has been modelled using both chord-slope and tangent-slope inductances. The model has then been applied for a local stability analysis, the results of which show that the simplified model used in this chapter yields somewhat more conservative results than the more accurate model and real machines.

---

[10]On a laboratory setting, it is not possible to demonstrate the instability for large machines as the maximum power rating in most laboratories is too low. An artificial way of demonstrating instability of induction machines is to use machines with a lower power rating and to compensate the stabilising effect of σ by artificially increasing the damping ratio by means of additional line resistances.

Chapter 28
Modelling and Dynamic Behaviour of Synchronous Machines

Abstract Not only induction machines, but synchronous machines as well may exhibit dynamic problems when fed by a variable frequency supply. In this chapter we analyse and represent the dynamic behaviour of synchronous machines in an analogous way as we have done for the induction machine. The results are quite similar. However, for synchronous machines such scaling laws do not exist to the same extent as for induction machines.

28.1 Introduction: Modelling of Rotating Field Machines with Saliency

Compared to induction machines, synchronous machines nearly always exhibit some kind of rotor saliency. In synchronous machines with salient poles and a concentrated DC excitation winding, the saliency is obviously quite pronounced. Yet, smooth-rotor turbo generators (with a distributed DC excitation winding), too, show some saliency. Similar conclusions hold for permanent-magnet synchronous motors. Interior PM machines normally exhibit a more pronounced saliency than surface PM motors, but also in the latter case, saliency is rarely negligible.

The stator of most synchronous machines is similar to that of induction machines, i.e. three-phase with distributed windings (although there are exceptions like small-surface PM motors, also known as BLDC motors).

The usual simplifying assumptions for the steady-state model of synchronous machines are similar to those of induction machines:

- a three-phase (or polyphase) symmetrical machine (both stator and rotor)
- sinusoidal distributed stator windings (no higher space harmonics of current layer, mmf or field)
- neither stator slot-effects nor stator saliency
- a two-phase rotor, possibly with saliency, and with the excitation in the d-axis

J. A. Melkebeek, *Electrical Machines and Drives*, Power Systems,
https://doi.org/10.1007/978-3-319-72730-1_28

- no skin effects
- no saturation (or constant saturation)
- a symmetrical sinusoidal three-phase supply
- steady state only

The rotor may contain saliency effects, and if this is the case, the standard Blondel model applies (see Chap. 5 in Part 1).

However, there is no real need for a restriction to a symmetrical sinusoidal three-phase (or multiphase) supply. If the supply is asymmetrical, this kind of model should be used for each symmetrical component. If the supply voltage contains harmonics, then this kind of model can be used for each harmonic.

For the dynamic model, we will apply mostly the same simplifying assumptions, except of course those unique to the steady state.The supply will not be restricted to be either symmetrical or sinusoidal. To derive the model, we will start from the (dynamical) electrical equations for the three stator phases and for the windings in the two rotor axes, as well as the equation of motion.

Such equations are very apt for direct digital simulation as the number of equations is not very important for simulation.

For an analytical study it is, however, important to reduce the number of equations (and variables). For a wye-connected machine the neutral is almost never connected and the zero-sequence current cannot flow; for a delta-connected machine the zero-sequence voltages are inherently zero. This implies that the zero-sequence equations and components in the stator can be omitted in most or all cases.

To reduce the number of stator equations, the symmetrical components transformation could be used in theory, as for the induction machine. In practice, though, this would initially lead to complex stator equations, while the two-phase rotor equations remain real. Furthermore, while the positive sequence stator component is synchronous with the rotor, the negative sequence component slips with regard to the rotor. Moreover, in case of saliency, the rotor is not two-phase symmetrical (magnetic saliency, different number of windings in the two rotor axes, different windings), and as such space vectors cannot be used. It is therefore more practical to use the Clarke transformation for the stator, while also omitting the zero-sequence components and the corresponding equations.

It will be shown that the resulting equations can be made stationary (i.e. not explicitly dependent on time, whereas the basic equations are time-dependent).

Finally, it is important to stress that neither the model nor the analysis method is restricted to a symmetrical sinusoidal supply. Voltages and currents may in principle contain harmonics and may be asymmetrical.

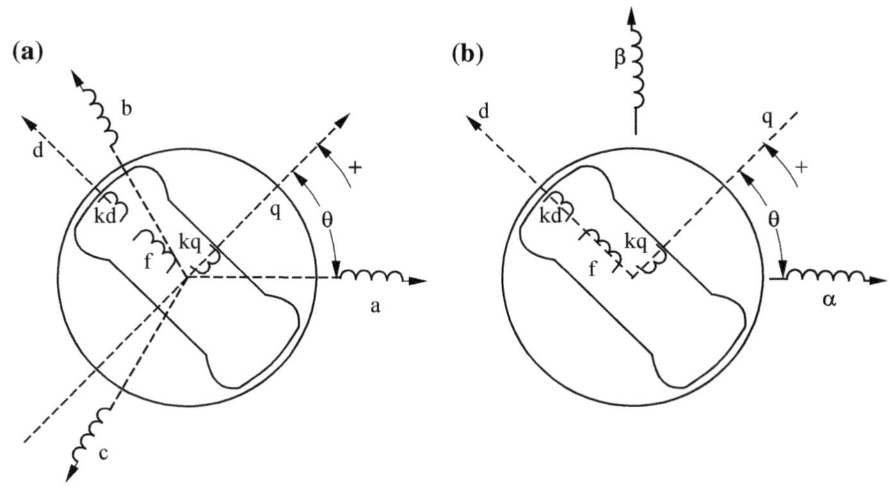

Fig. 28.1 Three-phase synchronous machine and two-phase equivalent

28.2 The Standard Dynamic Model of a Synchronous Machine

28.2.1 *Basic Assumptions and Equations*

Figure 28.1a schematically shows the considered synchronous machine with a three-phase stator and a two-phase rotor, containing, in addition to the field winding f, the two short-circuited damper windings kd and kq.

The differential equations between the instantaneous values of voltages and currents for this machine can be written as

$$\mathbf{V^i} = \mathbf{R^i} \cdot \mathbf{I^i} + \frac{d}{dt}\left(\mathbf{L^i}(\theta) \cdot \mathbf{I^i}\right) \tag{28.1}$$

with

$$\mathbf{V^i} = [\, V_a \; V_b \; V_c \; V_{kq} \; V_{kd} \; V_f]^T$$

$$\mathbf{I^i} = [\, I_a \; I_b \; I_c \; I_{kq} \; I_{kd} \; I_f]^T$$

$\mathbf{R^i}$ and $\mathbf{L^i}$ are the resistance and inductance matrices, respectively:

$$\mathbf{R^i} = diag(R_s, R_s, R_s, R_{kq}, R_{kd}, R_f) \tag{28.2}$$

$$\mathbf{L^i}(\theta) = \begin{bmatrix} \mathbf{L^i_{ss}}(\theta) & \mathbf{L^i_{sr}}(\theta) \\ \mathbf{L^i_{rs}}(\theta) & \mathbf{L^i_{rr}} \end{bmatrix} \tag{28.3}$$

$$\mathbf{L^i_{ss}}(\theta) = \begin{bmatrix} L_s & M_s & M_s \\ M_s & L_s & M_s \\ M_s & M_s & L_s \end{bmatrix} - L_{s2} \begin{bmatrix} \cos 2\theta & \cos 2(\theta - \pi/3) & \cos 2(\theta - 2\pi/3) \\ \cos 2(\theta - \pi/3) & \cos 2(\theta - 2\pi/3) & \cos 2(\theta - 3\pi/3) \\ \cos 2(\theta - 2\pi/3) & \cos 2(\theta - 3\pi/3) & \cos 2(\theta - 4\pi/3) \end{bmatrix} \quad (28.4)$$

$$\mathbf{L^i_{sr}}(\theta) = [\mathbf{L^i_{sr}}(\theta)]^T = \begin{bmatrix} M_{skq}\cos\theta & -M_{skd}\sin\theta & -M_{sf}\sin\theta \\ M_{skq}\cos(\theta - 2\pi/3) & -M_{skd}\sin(\theta - 2\pi/3) & -M_{sf}\sin(\theta - 2\pi/3) \\ M_{skq}\cos(\theta - 4\pi/3) & -M_{skd}\sin(\theta - 4\pi/3) & -M_{sf}\sin(\theta - 4\pi/3) \end{bmatrix} \quad (28.5)$$

$$\mathbf{L^i_{rr}} = \begin{bmatrix} L_{kq} & 0 & 0 \\ 0 & L_{kd} & M_{kdf} \\ 0 & M_{fkd} & L_f \end{bmatrix} \quad (28.6)$$

θ is the instantaneous angle between the stator reference phase and the q-axis of the rotor, i.e. the axis of maximum reluctance (see Fig. 28.1). Voltages and currents are, in principle, the real instantaneous functions of time.

The superscript T indicates matrix transposition. In the absence of saturation L_s, L_{s2}, M_s, M_{skq}, M_{skd}, M_{sf} and L_{kq}, L_{kd}, $M_{fkd} = M_{kdf}$ are constants.

Because of the reluctance difference between the d-axis and q-axis of the rotor, the self-inductance of the stator pulsates with a period of π electrical radians. In the above model, only the fundamental is taken into account and higher harmonics are disregarded. The mutual inductance matrix between stator and rotor contains terms in $\sin\theta$, $\sin(\theta \pm 2\pi/3)$, on the one hand, and terms in $\cos\theta$, $\cos(\theta \pm 2\pi/3)$, on the other hand, because of the two-phase rotor.

To obtain a more practical form of these equations, we will apply some transformations.

In order to obtain the two-axis representation of (b) in Fig. 28.1, we first apply the Clarke transformation to the voltages and currents (a, b, c to $0, \alpha, \beta$) of the stator

$$\mathbf{T_s^{-1}} = \frac{2}{3} \begin{bmatrix} 1/2 & 1/2 & 1/2 \\ 1 & -1/2 & -1/2 \\ 0 & \sqrt{3}/2 & -\sqrt{3}/2 \end{bmatrix} \quad (28.7)$$

After we have omitted the zero-sequence components, this yields

$$\mathbf{V^c} = \mathbf{R^c} \cdot \mathbf{I^c} + \frac{d}{dt}\left(\mathbf{L^c}(\theta) \cdot \mathbf{I^c}]\right) \quad (28.8)$$

with

$$\mathbf{V^c} = [\, V_\alpha \; V_\beta \; V_{kq} \; V_{kd} \; V_f \,]^T \quad (28.9)$$

$$\mathbf{I^c} = [\, I_\alpha \; I_\beta \; I_{kq} \; I_{kd} \; I_f \,]^T \quad (28.10)$$

$$\mathbf{R^c} = diag[\, R_s \; R_s \; R_{kq} \; R_{kd} \; R_f]^T \quad (28.11)$$

$$\mathbf{L^c}(\theta) = \begin{bmatrix} \mathbf{L^c_{ss}}(\theta) & \mathbf{L^c_{sr}}(\theta) \\ \mathbf{L^c_{rs}}(\theta) & \mathbf{L^c_{rr}} \end{bmatrix} \tag{28.12}$$

with

$$\mathbf{L^c_{ss}}(\theta) = \begin{bmatrix} L_s - M_s & 0 \\ 0 & L_s - M_s \end{bmatrix} - \frac{3}{2}L_{s2}\begin{bmatrix} \cos 2\theta & \sin 2\theta \\ \sin 2\theta & -\cos 2\theta \end{bmatrix} \tag{28.13}$$

$$\mathbf{L^c_{sr}}(\theta) = \frac{2}{3}[\mathbf{L^c_{rs}}(\theta)]^T = \begin{bmatrix} M_{skq}\cos\theta & -M_{skd}\sin\theta & -M_{sf}\sin\theta \\ M_{skq}\sin\theta & +M_{skd}\cos\theta & +M_{sf}\cos\theta \end{bmatrix} \tag{28.14}$$

$$\mathbf{L^c_{rr}} = \mathbf{L^i_{rr}} \tag{28.15}$$

Note that the inductance matrix equation 28.12 is not reciprocal. This is the consequence of the non power-invariant Clarke transformation equation 28.7 that we have used (the power-invariant Clarke transformation[1] has the coefficient $\sqrt{\frac{2}{3}}$ instead of $\frac{2}{3}$ before the matrix, and for the zero-sequence components $\frac{1}{\sqrt{2}}$ instead of $\frac{1}{2}$). The non power-invariant form is, nevertheless, most commonly used in machine theory as it preserves the amplitude of currents and voltages (but not the power).

To obtain a stationary machine model, we will transform the α, β stator variables into q, d variables in a reference frame fixed to the rotor (thus with a speed $\omega_r = d\theta/dt$, see (a) in Fig. 28.1):

$$\mathbf{T}_{\cdot}^{-1} = \begin{bmatrix} \cos\theta & \sin\theta \\ -\sin\theta & \cos\theta \end{bmatrix} \tag{28.16}$$

This results in

$$\mathbf{V} = \mathbf{R}\cdot\mathbf{I} + \frac{d}{dt}(\mathbf{L}\cdot\mathbf{I}) - \dot{\theta}[\mathbf{G}\cdot\mathbf{I}] \tag{28.17}$$

with

$$\mathbf{V} = [\, V_q \; V_d \; V_{kq} \; V_{kd} \; V_f \,]^T \tag{28.18}$$

$$\mathbf{I} = [\, I_q \; I_d \; I_{kq} \; I_{kd} \; I_f \,]^T \tag{28.19}$$

$$\mathbf{R} = diag[\, R_s \; R_s \; R_{kq} \; R_{kd} \; R_f \,]^T \tag{28.20}$$

[1] You may also check the Wikipedia pages on the Clarke transformation (which is also called the $\alpha\beta$-transformation).

$$\mathbf{L} = \begin{bmatrix} L_q & 0 & M_{skq} & 0 & 0 \\ 0 & L_d & 0 & M_{skd} & M_{sf} \\ \frac{3}{2}M_{skq} & 0 & L_{kq} & 0 & 0 \\ 0 & \frac{3}{2}M_{skd} & 0 & L_{kd} & M_{fkd} \\ 0 & \frac{3}{2}M_{sf} & 0 & M_{fkd} & L_f \end{bmatrix} \tag{28.21}$$

$$\mathbf{G} = \begin{bmatrix} 0 & L_d & 0 & M_{skd} & M_{sf} \\ -L_q & 0 & -M_{skq} & 0 & 0 \\ 0 & 0 & 0 & 0 & 0 \\ 0 & 0 & 0 & 0 & 0 \\ 0 & 0 & 0 & 0 & 0 \end{bmatrix} \tag{28.22}$$

with $L_d = L_s - M_s + \frac{3}{2}L_{s2}$ and $L_q = L_s - M_s - \frac{3}{2}L_{s2}$ the inductances of the d- and q-windings of the stator; the subscripts s, f, k refer to stator, field and damper windings, respectively.

The instantaneous electromagnetic power is given by

$$P = V_a I_a + V_b I_b + V_c I_c = 3V_0 I_0 + \frac{3}{2}(V_\alpha I_\alpha + V_\beta I_\beta) = 3V_0 I_0 + \frac{3}{2}(V_d I_d + V_q I_q) \tag{28.23}$$

Subtracting the Joule losses and the change in magnetic energy (see Ref. [14] for a proof), we obtain the mechanical power and thus the torque

$$T/N_p = \frac{1}{2}\left\{[\mathbf{I}^{\mathbf{i}}]^T \cdot \frac{d\mathbf{L}^{\mathbf{i}}(\theta)}{d\theta} \cdot \mathbf{I}^{\mathbf{i}}\right\} = \frac{3}{2} \cdot \frac{1}{2} \cdot \{\mathbf{I}^{\mathbf{T}} \cdot [\mathbf{G} + \mathbf{G}^{\mathbf{T}}] \cdot \mathbf{I}\} = \frac{3}{2} \cdot \{\Psi_q I_d - \Psi_d I_q\} \tag{28.24}$$

where Ψ_q and Ψ_d are the fluxes linked with the stator windings

$$\Psi_q = L_q I_q + M_{skq} I_{kq} \tag{28.25}$$

$$\Psi_d = L_d I_d + M_{sf} I_f + M_{skd} I_{kd} \tag{28.26}$$

These fluxes, as well as the fluxes linked with the rotor windings, follow from

$$\mathbf{\Psi} = [\,\Psi_q \ \Psi_d \ \Psi_{kq} \ \Psi_{kd} \ \Psi_f\,]^T = \mathbf{L} \cdot \mathbf{I} \tag{28.27}$$

Each of these fluxes is composed of the main flux (linking all windings in the same axis) and the leakage fluxes. The corresponding inductances may be written as the sum of a main inductance and leakage inductances as follows:

$$L_q = L_{q\sigma} + L_{qm} \tag{28.28}$$

$$L_d = L_{d\sigma} + L_{dm} + L_{dkd\sigma} + L_{df\sigma} \tag{28.29}$$

$$L_{kq} = L_{kq\sigma} + L_{kqm} \tag{28.30}$$

$$L_{kd} = L_{kd\sigma} + L_{kdm} + L_{kdd\sigma} + L_{kdf\sigma} \tag{28.31}$$

$$L_f = L_{f\sigma} + L_{fm} + L_{fkd\sigma} + L_{fd\sigma} \tag{28.32}$$

The main inductances in d- and q-axes refer to the main field or flux in these axes, respectively; thus L_{dm}, L_{kdm}, L_{fm} differ only by their turns ratio. The same holds for L_{qm}, L_{kqm}.

The leakage fluxes in the d-axis (and the corresponding inductances) can be subdivided into leakage stricto sensu ($L_{d\sigma}, L_{kd\sigma}, L_{f\sigma}$) coupled with only one winding, and *mutual* (or *common*) leakages ($L_{dkd\sigma}, L_{df\sigma}, L_{kdf\sigma}$) coupled with exactly two windings.[2] What corresponding leakage fluxes all have in common is that they are not coupled to all three windings and are not considered main flux in this representation. In practice, the common leakage between d-and f-windings is usually rather small and can be disregarded. We will see below that by choosing appropriate turns ratios[3] or reference values, all mutual leakage terms can be made equal to zero.

For the complete electromechanical model, the equation of motion has to be added to the five electrical equations

$$\left(J/N_p\right)\frac{d^2\theta}{dt^2} = T - T_l \tag{28.33}$$

with the torque T given by the expression Eq. 28.24.

Remarks:

1- In the above q, d model, the d-axis is chosen along the rotor flux or excitation axis, while the q-axis is chosen $\pi/2$ electrical radians lagging with respect to the d-axis, i.e. the positive rotation direction is from the q-axis towards the d-axis. This is the preferred choice for modelling a generator in the GRS, as a positive excitation along the d-axis then leads to a positive emf along the q-axis. It may be argued that a choice with the q-axis leading the d-axis is more appropriate for studying motoring, but then two different models have to be dealt with, one for generating and another for motoring. The choice of a d-axis leading the q-axis is in agreement with a 1968 IEEE recommendation (see Ref. [13]).

2- As already mentioned, the α, β- and q, d-models can also be derived by means of the symmetrical components transformation. A space vector representation can also

---

[2]In the literature, this is sometimes called doubly-coupled leakage.

[3]In fact, as for an induction machine or transformer, these turns ratios cannot be measured.

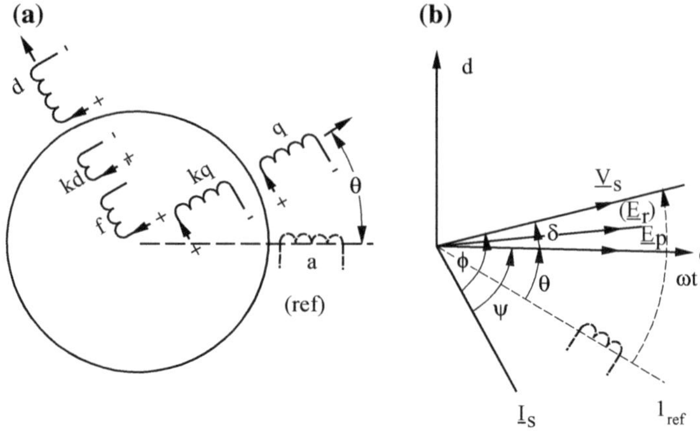

Fig. 28.2 SM d-q-model for steady state

be obtained from the α, β- and q, d-models as follows:

$$\underset{\rightarrow}{V}{}_s^c = V_\alpha + jV_\beta \tag{28.34}$$

$$\underset{\rightarrow}{V}{}_s = V_q + jV_d = \underset{\rightarrow}{V}{}_s^c \exp(-j\theta) \tag{28.35}$$

(where, for sinusoidal supply, space vectors and phasors of voltages and currents are identical, $\underset{\rightarrow}{V}{}_s^c = \underline{V}_s^c$ and $\underset{\rightarrow}{V}{}_s = \underline{V}_s$). Other reference frames than a standstill or a rotor reference can also be used. The problem, however, is that the rotor is not two-axis symmetrical (in terms of saliency, the number of windings and/or the resistances), which automatically excludes a space vector representation of the rotor. In addition, the rotor model has two magnetically uncoupled axes only if a rotor reference frame is used.

3- The steady-state phasor diagram corresponding with the above equations is shown in (b) in Fig. 28.2. The instantaneous angle θ between the stator and rotor reference axes is also the angle between the stator reference axis and the rotor-induced emf $\underset{\rightarrow}{E}{}_p$ (with $\underline{E}_p = \underset{\rightarrow}{E}{}_p$ for sinusoidal steady state): $\theta = \arg(\underset{\rightarrow}{E}{}_p, \underset{\rightarrow}{1}{}_{ref}) = \omega t - \delta$ where ω is the instantaneous supply frequency (ω_0 in steady state) and with the load angle[4] given by $\delta = \arg(\underset{\rightarrow}{V}{}_s, \underset{\rightarrow}{E}{}_p)$. Besides the load angle, also the phase angle φ and the internal angle ψ are used: $\varphi = \arg(\underset{\rightarrow}{V}{}_s, \underset{\rightarrow}{I}{}_s)$ and $\psi = \arg(\underset{\rightarrow}{E}{}_p, \underset{\rightarrow}{I}{}_s)$. In the GRS: $\varphi = \psi - \delta$ because of the GRS definition of δ as $\delta = \arg(\underset{\rightarrow}{E}{}_p, \underset{\rightarrow}{V}{}_s)$; in the URS the relation between φ, ψ and δ is $\varphi = \psi + \delta$ because of the URS definition of δ as $\delta = \arg(\underset{\rightarrow}{V}{}_s, \underset{\rightarrow}{E}{}_p)$.

[4] In the URS, the load angle is positive if the stator voltage leads the emf.

28.2.2 Equations for Sinusoidal Steady State and for Small Deviations Around Steady State

The equations in the previous sections hold for both sinusoidal and non-sinusoidal supply. In the next sections of this chapter, we will study small deviations around a sinusoidal steady state. For a sinusoidal steady state, we have $V_{s0} = V_0$, $V_{d0} = V_0 \sin\delta$, $V_{q0} = V_0 \cos\delta$, $V_f = V_{f0}$, $V_{kq} = V_{kd} = 0$, $\theta_0 = \omega_0 t - \delta_0$, $T_l = T_{l0}$. With all derivatives with respect to time equal to zero in Eq. 28.17, we obtain for the steady-state equations

$$\mathbf{V_0} = \mathbf{R} \cdot \mathbf{I_0} - \omega_0[\mathbf{G} \cdot \mathbf{I_0}] \tag{28.36}$$

$$T_0 = T_{l0} = \frac{3}{2} \cdot N_p \left\{\Psi_{q0} I_{d0} - \Psi_{d0} I_{q0}\right\} \tag{28.37}$$

The equations for small deviations around the steady state are derived by subtracting the steady-state equations from the Eq. 28.17 and omitting terms with second-order variations:

$$\boldsymbol{\Delta V} = \mathbf{R} \cdot \boldsymbol{\Delta I} + \frac{d}{dt}(\mathbf{L} \cdot \boldsymbol{\Delta I}) - \omega_0[\mathbf{G} \cdot \boldsymbol{\Delta I}] - \Delta\omega_r[\mathbf{G} \cdot \mathbf{I_0}] \tag{28.38}$$

$$(J/N_p)\frac{d\omega_r}{dt} = \frac{3}{2} N_p \cdot \left\{\Delta\Psi_q I_{d0} + \Psi_{q0}\Delta I_d - \Delta\Psi_d I_{q0} - \Psi_{d0}\Delta I_q\right\} - K_w \Delta\omega_r - \Delta T_l' \tag{28.39}$$

Note that for the speed and frequency, the following relations hold: $\dot{\theta}_0 = \omega_{r0} = \omega_0$ and $\Delta\omega_r = \Delta\dot{\theta} = \Delta\omega - \Delta\dot{\delta}$

28.2.3 Reciprocity - pu or Absolute Modelling

Due to the non-power-invariant Clarke transformation used for the stator quantities, the Eq. 28.8 (and other equations derived from it) are not reciprocal. Although such a model does not seem physically possible, all results derived based on this model may nevertheless be used, as long as the corresponding non-power-invariant inverse transformations are used to transform back to the original quantities.

Nevertheless, some authors prefer reciprocal equations, which may be obtained by either using a reciprocal transformation or converting the non-reciprocal equations to pu (with appropriate reference values).

The most comprehensive power-invariant transformation (see also Ref. [8]) is as follows. If for the voltages a transformation $\mathbf{V} = \mathbf{T} \cdot \mathbf{V}'$ is used, then the transformation for the currents which automatically guarantees power-invariance is $\mathbf{I}' = \mathbf{T^T} \cdot \mathbf{I}$.

Indeed, $\mathbf{V}^{\mathrm{T}} \cdot \mathbf{I} = \mathbf{V}'^{\mathrm{T}} \cdot \mathbf{T}^{\mathrm{T}} \cdot \mathbf{I} = \mathbf{V}'^{\mathrm{T}} \cdot \mathbf{I}'$ (note that the inverse of the matrix $\mathbf{T}$ is not required to exist).

The Clarke transformation equation 28.7 with $\mathbf{V}' = \mathbf{T}_{\mathbf{c}}^{-1} \cdot \mathbf{V}$ and with the same transformation for the currents $\mathbf{I}' = \mathbf{T}_{\mathbf{c}}^{-1} \cdot \mathbf{I}$ is not power invariant because $\mathbf{T}_{\mathbf{c}}^{\mathrm{T}} \neq \mathbf{T}_{\mathbf{c}}^{-1}$.

A reciprocal form based on the Clarke transformation matrix applies this Clarke matrix to the voltages, $\mathbf{V}' = \mathbf{T}_{\mathbf{c}}^{-1} \cdot \mathbf{V}$, but for the currents it uses $\mathbf{I}' = \mathbf{T}_{\mathbf{c}}^{\mathrm{T}} \cdot \mathbf{I}$ or thus $\mathbf{I} = \left(T_c^{-1}\right)^T \cdot \mathbf{I}'$. It is easy to verify that the coefficients for the positive and negative sequence components in $\left(T_c^{-1}\right)^T$ are those of $\mathbf{T_c}$ multiplied by $\frac{2}{3}$. With this transformation, compared to the original three-phase machine, the voltages are retained but the currents are larger by a factor of $3/2$. Stator resistances and self inductances are smaller by a factor of $2/3$ and all mutual inductances between stator and rotor are equal to the original inductances (M_{skq}, M_{skd}, M_{sf}). The retained voltage amplitudes, the current amplitudes larger by a factor $3/2$ and the resistances and self-inductances smaller by a factor $2/3$ all imply that (1) in the transformed two-phase machine the number of effective stator windings per phase remains equal to that of the original machine, (2) the amplitude of the air-gap mmf remains equal to $\frac{3}{2}(w\xi_1) \cdot \hat{I}$ (meaning that also the flux has the same value), and (3) the section of the stator windings is larger by a factor $3/2$ (i.e. the same amount of copper is used for the windings). As expected, the power remains invariant. For example, the stator joule losses are for the two-phase machine $2 \cdot \frac{2}{3} R_s \cdot \frac{1}{2}(\frac{3}{2}\hat{I})^2 = \frac{3}{2} R_s \hat{I}^2 = 3 R_s I^2$, which is equal to those of the original machine.

In contrast, the non-power-invariant Clarke that we used in the previous sections resulted in a non-reciprocal inductance matrix and a non-power-invariant model. This two-phase machine is therefore not physically realisable[5] (see also Table 28.1).

Another way to obtain an equivalent reciprocal two-phase model is to apply the power-invariant Clarke transformation, but in this case voltage amplitudes are not the same as in the original three-phase machine, and neither are the current amplitudes. As mentioned before, this power-invariant Clarke is rarely used in machine theory.

Although the non-reciprocal equations are useful for studying the machine and its dynamic behaviour, some people prefer a reciprocal model. In that case, either we have to use a reciprocal transformation as discussed above, or we revert to a pu representation where a judicious choice of the reference values results in reciprocal equations. To obtain a reciprocal pu representation starting from the non-reciprocal model derived in Sect. 28.2.1, reference values for the rotor currents have to be chosen which are $3/2$ times larger than those corresponding to the turns ratios between stator and rotor windings. As the voltage levels are retained in both models, the reference values for the voltages differ only by the turns ratios between stator and rotor windings.

In this way, we obtain the pu equations used in this section below.

---

[5]From the equality of the voltages, it follows that the number of effective windings should be the same as in the original machine. In contrast, from the equivalence of the mmf (cf. the inductances of the d- and q-windings compared to the original ones), the new number of effective stator windings should be larger by a factor of $3/2$.

Table 28.1 Three-phase synchronous machine and two-phase equivalents

Three-phase machine	Two-phase non-reciprocal	Two-phase reciprocal
$\hat{V}$	$\hat{V}$	$\hat{V}$
$\hat{I}$	$\hat{I}$	$\frac{3}{2}\hat{I}$
R_s	R_s	$\frac{2}{3}R_s$
$L_s\ [\pm L_{s2}]$	$L_s - M_s \approx \frac{3}{2}L_s\ [\pm L_{s2}]$	$\frac{2}{3}(L_s - M_s) \approx L_s\ [\pm L_{s2}]$
$M_s = -\frac{1}{2}(L_s - L_{s\sigma}) \approx -\frac{1}{2}L_s$	0	0
$M_{sr(d)} = M_{sf}$	$M_{srd} = M_{sf}$	$M_{srd} = M_{sf}$
$M_{rs(d)} = M_{sf}$	$M_{rsd} = \frac{3}{2}M_{sf}$	$M_{srd} = M_{sf}$
$P_j = \frac{3}{2}R_s\hat{I}^2$	$P_j = R_s\hat{I}^2$	$P_j = \frac{3}{2}R_s\hat{I}^2$
m.m.f. $= \frac{3}{2}\hat{I}(w\xi)$	m.m.f. $= \hat{I}(w'\xi')$	m.m.f. $= \frac{3}{2}\hat{I}(w"\xi")$
$\Psi_{phase} = L_s\hat{I}$	$\Psi_{phase} = (L_s - M_s)\hat{I} = \frac{3}{2}L_s\hat{I}$	$\Psi_{phase} = \frac{3}{2}L_s\hat{I}$
$\Psi_{dr,ss} = \frac{3}{2}L_s\hat{I}$	$\Psi_{dr,ss} = \frac{3}{2}L_s\hat{I}$	$\Psi_{dr,ss} = \frac{3}{2}L_s\hat{I}$
$\Psi_{dr,sr} = M_{sf}i_f$	$\Psi_{dr,sr} = M_{sf}i_f$	$\Psi_{dr,sr} = M_{sf}i_f$
$\Psi_{dr,rs} = \frac{3}{2}M_{sf}\hat{I}$	$\Psi_{dr,rs} = \frac{3}{2}M_{sf}\hat{I}$	$\Psi_{dr,rs} = \frac{3}{2}M_{sf}\hat{I}$
	Non-power-invariant, non-reciprocal	Power-invariant; reciprocal
	Equivalence voltages: $w'\xi' = w\xi$	Equivalence voltages: $w"\xi" = w\xi$
	Equivalence mmf: $w'\xi' = \frac{3}{2}w\xi$	Equivalence mmf: $w"\xi" = w\xi$

For the d-axis, the notation l_{dm} has been used for the pu air-gap or main (magnetising) field inductance. Mutual leakage inductances between the d-axis stator winding, on the one hand, and the kd- and f-rotor windings, on the other hand, are absent. This is because we have chosen the current reference values for the rotor windings in the d-axis in accordance with their transformer ratio with respect to the stator d-axis winding (but multiplied by a factor 3/2 to obtain reciprocity). In each of the three d-axis windings, a unit current will result in the same unit air-gap flux in the d-axis and in the same (unit) emf in the stator q-axis winding. A three-phase unit current in the original three-phase stator windings would lead to the same emf. This is well known as the L_{ad}-basis, where L_{ad} stands for the armature reaction inductance. The only remaining mutual leakage[6] is the one between the kd-and f-axes, $l_{fkd\sigma}$.

$$l_d = l_{d\sigma} + l_{dm} \tag{28.40}$$

$$l_f = l_{f\sigma} + m_{fkd} = l_{f\sigma} + l_{fkd\sigma} + l_{dm} \tag{28.41}$$

$$l_{kd} = l_{kd\sigma} + m_{fkd} = l_{kd\sigma} + l_{fkd\sigma} + l_{dm} \tag{28.42}$$

[6]However, by appropriately choosing the turns ratios, we may eliminate this one as well - prove this.

For the q-axis, the notation l_{qm} can be used for the mutual inductance if the actual fundamental transformer ratio between the q- axis windings is used. However, the transformer ratio between the stator and the q-axis rotor windings is not exactly equal to that between the stator and the rotor d-axis windings because of the difference in magnetic permeability for the two axes - referred to the fundamental - (see also Ref. [8]). If, to simplify calculations, the transformer ratio between the d-axis windings is also used for the q-axis, then instead of l_{qm} the notation m_{qkq} should be used[7] for the mutual inductance, as the d-axis transformer ratio is not a realistic representation of the main air-gap field in the q-axis.

$$l_q = l_{q\sigma} + l_{qm} \tag{28.43}$$

$$l_{kq} = l_{kq\sigma} + l_{qm} \tag{28.44}$$

For steady state, with $v_{d0} = v_0 \sin\delta_0$, $v_{q0} = v_0 \cos\delta_0$, $v_f = v_{f0}$, $v_{kq} = v_{kd} = 0$, $t_l = t_{l0}$; $\nu_{r0} = \nu_0$, the pu electrical equations are

$$v_{q0} = v_0 \cos\delta_0 = r_s i_{q0} - \nu_0 \left(l_d i_{d0} + l_{dm} i_{f0}\right) \tag{28.45}$$

$$v_{d0} = v_0 \sin\delta_0 = r_s i_{d0} + \nu_0 l_q i_{q0} \tag{28.46}$$

$$v_{f0} = r_f i_{f0} \tag{28.47}$$

while the torque is given by

$$t_{l0} = t_0 = \left[l_q i_{q0} i_{d0} - \left(l_d i_{d0} + l_{dm} i_{f0}\right) i_{q0}\right] = e_{p0} i_{q0}/\nu_0 + \left(l_q - l_d\right) i_{q0} i_{d0} \tag{28.48}$$

The steady-state currents are, with $e_{p0} = -\nu_0 l_{dm} i_{f0}$ denoting the steady-state emf,

$$i_{q0} = \frac{r_s(v_{q0} - e_{p0}) + \nu_0 l_d v_{d0}}{r_s^2 + \nu_0^2 l_d l_q} \tag{28.49}$$

$$i_{d0} = \frac{r_s v_{d0} - \nu_0 l_q (v_{q0} - e_{p0})}{r_s^2 + \nu_0^2 l_d l_q} \tag{28.50}$$

The steady-state torque is

$$\begin{aligned} t_0 = {} & \frac{l_q v_{d0} e_{p0}}{r_s^2 + \nu_0^2 l_d l_q} + (l_d - l_q)\frac{v_{d0} v_{q0}}{r_s^2 + \nu_0^2 l_d l_q} + (r_s/\nu_0)\frac{(v_{q0} - e_{p0}) e_{p0}}{r_s^2 + \nu_0^2 l_d l_q} \\ & + (l_d - l_q) r_s \frac{\nu_0 l_q (v_{q0} - e_{p0})^2 - \nu_0 l_d v_{d0}^2 - 2 r_s \nu_0 (v_{q0} - e_{p0}) v_{d0}}{r_s^2 + \nu_0^2 l_d l_q} \end{aligned} \tag{28.51}$$

[7] The difference is usually small, however.

or, for zero stator resistance,

$$t_0 = \frac{v_0 e_{p0}}{\nu_0^2 l_d} \sin \delta_0 + \frac{1}{2}(v_0^2/\nu_0^2)(l_q^{-1} - l_d^{-1}) \sin 2\delta_0 \tag{28.52}$$

The pu equations for small deviations around the steady state are (with short-circuited damping windings):

$$\begin{bmatrix} \Delta v \cos \delta_0 \\ \Delta v \sin \delta_0 \\ 0 \\ 0 \\ \Delta v_f \\ -\Delta t_l' \\ \Delta \nu \end{bmatrix} = \begin{bmatrix} r_s + pl_q & -\nu_{r0} l_d & pm_{qkq} & -\nu_{r0} l_{dm} & -\nu_{r0} l_{dm} & -\phi_{d0} & v_{d0} \\ \nu_{r0} l_q & r_s + pl_d & \nu_{r0} m_{qkq} & pl_{dm} & pl_{dm} & \phi_{q0} & -v_{q0} \\ pm_{qkq} & 0 & r_{kq} + pl_{kq} & 0 & 0 & 0 & 0 \\ 0 & pl_{dm} & 0 & r_{kd} + pl_{kd} & pm_{fkd} & 0 & 0 \\ 0 & pl_{dm} & 0 & pm_{fkd} & r_f + pl_f & 0 & 0 \\ \phi_{d0} - i_{do} l_q & -\phi_{q0} + i_{q0} l_d & -i_{d0} l_{qm} & i_{q0} l_{dm} & i_{q0} l_{dm} & \tau_n(p + \varrho_w) & 0 \\ 0 & 0 & 0 & 0 & 0 & 1 & p \end{bmatrix} \cdot \begin{bmatrix} \Delta i_q \\ \Delta i_d \\ \Delta i_{kq} \\ \Delta i_{kd} \\ \Delta i_f \\ \Delta \nu_r \\ \Delta \delta \end{bmatrix} \tag{28.53}$$

in which $p = d/d(\omega_n t)$, $\nu_{r0} = \omega_{r0}/\omega_n = \nu_0 = \omega_0/\omega_n$, $\Delta\nu = \Delta\omega/\omega_n$, $\Delta\nu_r = \Delta\omega_r/\omega_n$, $\phi_{q0} = l_q i_{q0}$, $\phi_{d0} = l_d i_{d0} + l_{dm} i_{f0}$.

In what follows, Eq. 28.53 will be denoted as

$$\mathbf{M} \cdot \boldsymbol{u} = \mathbf{A}(p) \cdot \mathbf{x} \tag{28.54}$$

with

$$\mathbf{u} = \begin{bmatrix} \Delta v \ \Delta v_f \ \Delta t_l' \ \Delta \nu \end{bmatrix}^T \tag{28.55}$$

$$\mathbf{M} = \begin{bmatrix} \cos \delta_0 & 0 & 0 & 0 \\ \sin \delta_0 & 0 & 0 & 0 \\ 0 & 0 & 0 & 0 \\ 0 & 0 & 0 & 0 \\ 0 & 1 & 0 & 0 \\ 0 & 0 & -1 & 0 \\ 0 & 0 & 0 & 1 \end{bmatrix} \tag{28.56}$$

The steady-state equations in absolute values are similar to those in pu. Please keep in mind that the expression for the torque in absolute values needs to be preceded by the factor $(3/2)N_p$.

For the dynamic equations in absolute values for voltages and currents, we obtain

$$\begin{bmatrix} \Delta V \cos \delta_0 \\ \Delta V \sin \delta_0 \\ 0 \\ 0 \\ \Delta V_f \\ -\Delta T_l' \\ \Delta \nu \end{bmatrix} =$$

$$\begin{bmatrix} R_s + p\omega_n L_q & -\nu_{r0}\omega_n L_d & p\omega_n M_{skq} & -\nu_{r0}\omega_n M_{skd} & -\nu_{r0}\omega_n M_{sf} & -\omega_n \Psi_{d0} & V_{d0} \\ \nu_{r0}\omega_n L_q & R_s + p\omega_n L_d & \nu_{r0}\omega_n M_{skq} & p\omega_n M_{skd} & p\omega_n M_{sf} & \omega_n \Psi_{q0} & -V_{q0} \\ \frac{3}{2} p\omega_n M_{skq} & 0 & R_{kq} + p\omega_n L_{kq} & 0 & 0 & 0 & 0 \\ 0 & \frac{3}{2} p\omega_n M_{skd} & 0 & R_{kd} + p\omega_n L_{kd} & p\omega_n M_{fkd} & 0 & 0 \\ 0 & \frac{3}{2} p\omega_n M_{sf} & 0 & p\omega_n M_{fkd} & R_f + p\omega_n L_f & 0 & 0 \\ \frac{3}{2} N_p(\Psi_{d0} - I_{do} L_q) & -\frac{3}{2} N_p(\Psi_{q0} - I_{q0} L_d) & -\frac{3}{2} N_p I_{d0} M_{skq} & \frac{3}{2} N_p I_{q0} M_{skd} & \frac{3}{2} N_p I_{q0} M_{sf} & \frac{J}{N_p} p\omega_n^2 + \omega_n K_w & 0 \\ 0 & 0 & 0 & 0 & 0 & 1 & p \end{bmatrix}$$

$$\begin{bmatrix} \Delta I_q \\ \Delta I_d \\ \Delta I_{kq} \\ \Delta I_{kd} \\ \Delta I_f \\ \Delta \nu_r \\ \Delta \delta \end{bmatrix} \tag{28.57}$$

in which $\Psi_{q0} = L_q I_{q0}$ and $\Psi_{d0} = L_d I_{d0} + M_{sf} I_{f0}$. This equation can also be written in the form of Eq. 28.54.

28.2.4 *Approximation for Saturation in Standard Modelling*

In the above model, the main or magnetising field inductances are assumed to be constant. In reality, however, all (synchronous) machines are designed to work well into saturation, mostly around the knee of the saturation characteristic.

For smooth rotor synchronous machines, the traditional (small-signal) model always uses the chord-slope inductances corresponding with the resulting main flux in the operating point. A more accurate model should use the chord-slope inductances in the operating point for the steady-state flux and the tangent-slope inductances for flux variations, as was the case for the induction machine. In the analysis presented below, we will nevertheless use the standard approach, using only the chord-slope inductance.

For salient pole machines, different approaches are in use, even for the traditional (small-signal) model. The most commonly used model is the Blondel model. Here, first the mmf is split up into its components along the d- and q-axes. The flux components along these two axes are calculated separately and then these fluxes are added vectorially. For the d-axis flux, the (measured) d-axis saturation characteristic is used to calculate the flux and the chord-slope inductances L_{d0} and L_{dm0}. However, for the q-axis, such a measured characteristic is usually not available (although a DC excitation in the q-axis is absent, there are other methods to measure the saturation characteristic[8]). As a consequence, sometimes the d-axis saturation characteristic is used for the q-axis. Another approximation assumes an unsaturated state for the q-axis. More sophisticated methods rely on finite-elements modelling to estimate the permeance and saturation of the different parts of the q-axis flux trajectory (e.g. stator teeth, stator yoke, rotor yoke). Whatever the approximation used for the q-axis, in standard modelling only the chord-slope inductances are considered.

---

[8]These use AC excitation in the q-axis; however, how can stable operation be achieved with only q-axis excitation?.

28.3 Characteristic Dynamic Behaviour of Synchronous Machines

28.3.1 Dynamic Parameters

As was the case with the DC machine and the induction machine, we prefer to study the dynamic behaviour of synchronous machines with a minimal set of dimensionless parameters.

A synchronous machine without excitation winding (reluctance machine) is characterised by 15 parameters: the nine electrical machine parameters (R_s, L_d, L_q, L_{kd}, L_{kq}, M_{skd}, M_{skq}, R_{kd}, R_{kq}), three mechanical parameters (J, N_p, K_w) and the three parameters that define the operating point (V_0, ω_0, δ).

A set of nine ($9 = 15 - 6$) dimensionless parameters is, nevertheless, sufficient to characterise the dynamic behaviour of this sixth-order system. We will choose these in analogy with the parameters chosen for the DC machine and induction machine:

- the magnetic parameters (leakage coefficients) for d- and q-axes:

$$\sigma_d = 1 - \frac{3}{2} M_{skd}^2 / L_d L_{kd} = 1 - l_{dm}^2 / l_d l_{kd} \tag{28.58}$$

$$\sigma_q = 1 - \frac{3}{2} M_{skq}^2 / L_q L_{kq} = 1 - l_{qm}^2 / l_q l_{kq} \tag{28.59}$$

- the dimensionless (short-circuit) stator time constants for d- and q-axes

$$\tau_d = \sigma_d \omega_n L_d / R_s = \sigma_d l_d / r_s \tag{28.60}$$

$$\tau_q = \sigma_q \omega_n L_q / R_s = \sigma_q l_q / r_s \tag{28.61}$$

- the dimensionless (short-circuit) rotor time constants for d- and q-axes

$$\tau_{kd} = \sigma_d \omega_n L_{kd} / R_{kd} = \sigma_d l_{kd} / r_{kd} \tag{28.62}$$

$$\tau_{kq} = \sigma_q \omega_n L_{kq} / R_{kq} = \sigma_q l_{kq} / r_{kq} \tag{28.63}$$

- the dimensionless mechanical damping constant

$$\tau_w = \varrho_w^{-1} = J \omega_n / N_p K_w \tag{28.64}$$

(note that τ_w can also be written in terms of the inertia time constant τ_n: $\tau_w = \tau_n / k_w = (J\omega_n / N_p T_n)/(K_w / T_n)$)

- the dimensionless electromechanical time constant τ_m or the electromechanical eigenfrequency ν_m defined as follows

$$\tau_{kd}\tau_m = \nu_m^{-2} = J\omega_n^2/[N_p \cdot (dT/d\delta)] = \tau_n/(dt/d\delta) \tag{28.65}$$

  where τ_n is the inertia time constant and $dT/d\delta$ (or $dt/d\delta$) is the slope of the steady-state characteristic torque versus load angle
- the load angle δ (δ_0)

The dimensionless supply frequency ν_0 is eliminated by taking it together with the electrical time constants (e.g. $\nu_0\tau_{kd}$), as we have done for the induction machine.

It is sometimes practical to introduce other parameters to illustrate the difference between the d- and q-axes:

- the reluctance ratio

$$\gamma_{dq} = x_d/x_q \tag{28.66}$$

- the damping ratio between d- and kd-windings

$$\alpha_{dkd} = \tau_{kd}/\tau_d \tag{28.67}$$

- the damping ratio between q- and kq-windings

$$\alpha_{qkq} = \tau_{kq}/\tau_q \tag{28.68}$$

- the stator damping ratio

$$\alpha_{dq} = \tau_q/\tau_d \tag{28.69}$$

- the rotor damping ratio

$$\alpha_{kdkq} = \tau_{kq}/\tau_{kd} \tag{28.70}$$

- the leakage ratio

$$\kappa_{dq} = \sigma_d/\sigma_q \tag{28.71}$$

Note that $\alpha_{dkd} \cdot \alpha_{kdkq} = \alpha_{dq} \cdot \alpha_{qkq}$.

For a synchronous machine with excitation winding (and damper winding) in the d-axis, we need seven additional parameters[9] in addition to the three q-axis parameters (τ_q, τ_{kq}, σ_q), the two electromechanical parameters (τ_w, τ_m) and the load angle δ:

[9]There are now 20 original parameters (13 electrical, 3 mechanical and 4 supply and excitation parameters) for the drive while the system order is 7.

- the leakage coefficient $d - kd$:

$$\sigma_{dkd} = 1 - \frac{3}{2} M_{skd}^2 / L_d L_{kd} = 1 - l_{dm}^2 / l_d l_{kd} \tag{28.72}$$

- the leakage coefficient $d - f$

$$\sigma_{df} = 1 - \frac{3}{2} M_{sf}^2 / L_d L_f = 1 - l_{dm}^2 / l_d l_f \tag{28.73}$$

- the leakage coefficient $f - kd$

$$\sigma_{fkd} = 1 - M_{fkd}^2 / L_f L_{kd} = 1 - (l_{dm} + l_{kdf\sigma})^2 / l_f l_{kd} \equiv 1 - m_{fkd}^2 / l_f l_{kd} \tag{28.74}$$

  (if by an appropriate choice of the reference values or turns ratios $l_{kdf\sigma} = 0$, then $m_{fkd} = l_{dm}$)
- the dimensionless stator time constant

$$\tau_d = \sigma_d \omega_n L_d / R_s = \sigma_d l_d / r_s \tag{28.75}$$

  with σ_d the total d-axis leakage coefficient (Eq. 28.78).
- the dimensionless time constant of the field winding

$$\tau_f = \sigma_{df} \omega_n L_f / R_f = \sigma_{df} l_f / r_f \tag{28.76}$$

- the dimensionless time constant of the d-axis damper winding

$$\tau_{kd} = \sigma_{dkd} \omega_n L_{kd} / R_{kd} = \sigma_{dkd} l_{kd} / r_{kd} \tag{28.77}$$

- the excitation parameter

$$e_v = E_{p0} / V_0 = e_{p0} / v_0$$

We note that in the definitions above, four d-axis leakage coefficients have been introduced as there are now three windings in this axis. However, only three of these four leakage coefficients are independent. For example, the total d-axis leakage coefficient σ_d can be expressed in function of the three others as

$$1 - \sigma_d = \frac{(1 - \sigma_{df}) + (1 - \sigma_{dkd}) - 2\sqrt{(1 - \sigma_{df})(1 - \sigma_{dkd})(1 - \sigma_{fkd})}}{\sigma_{fkd}} \tag{28.78}$$

In most cases, σ_d is preferred to σ_{fkd}.

It might be helpful to give a more explicit definition of the meaning of these leakage coefficients. The coefficient σ_{dkd} between the d- and kd-windings is the ratio of the short-circuit (leakage) inductance between these two windings, when measured or viewed from the side of the kd-winding, and the main inductance of the kd-winding, while the f-winding is assumed open, thus: $\sigma_{dkd} L_{kd} / L_{kd}$. Put differently, $\sigma_{dkd} L_{kd}$ represents the remaining field coupled with the kd-winding when the d-winding is shorted and the f-winding is open. These conditions may also be reversed: with the kd-winding shorted and the f-and d-windings open, $\sigma_{dkd} L_d$ is the short-circuit inductance measured from the d-winding, and $\sigma_{dkd} = \sigma_{dkd} L_d / L_d$.

Analogous Definitions Hold for σ_{df} and σ_{fkd}.

The (total) leakage coefficient σ_d, on the other hand, is the ratio of the total short-circuit (leakage) inductance viewed from the d-winding with both kd- and f-windings shorted, and the main inductance of the d-winding, i.e. $\sigma_d L_d / L_d$. Thus, $\sigma_d L_d$ represents the remaining field coupled with the d-winding when the field excited by the d-winding is shorted by both kd- and f-windings.

In addition to the short-circuit time constants τ_d, τ_f, τ_{kd}, τ_q, τ_{kq}, the main field time constants are also sometimes used. These are denoted by τ_{dh}, τ_{fh}, τ_{kdh}, τ_{qh}, τ_{kqh} and are defined in a similar way but with the main inductances instead of the short-circuit inductances (or, with the leakage coefficients replaced by unity).

Although the (short-circuit) time constants and leakage inductances (or corresponding leakage coefficients) are straightforward functions of the winding impedances, in stability studies of grids and synchronous machines other quantities (transient and subtransient time constants and inductances) are quite common. Their definition is as follows:

- the d-axis open-stator transient time constant $\tau'_{d0} = \tau_f / \sigma_{df} = \tau_{fh}$
- the d-axis open-stator subtransient time constant $\tau''_{d0} = \tau_{kd} \cdot (\sigma_{fkd} / \sigma_{dkd})$
- the d-axis transient time constant $\tau'_d = \tau_f$
- the d-axis subtransient time constant $\tau''_d = \tau_{kd} \cdot (\sigma_d \sigma_{fkd} / \sigma_{df} \sigma_{dkd})$
- the q-axis open-stator (sub)transient[10] time constant $\tau''_{q0} = \tau_{kq} / \sigma_q = \tau_{kqh}$
- the q-axis (sub)transient time constant $\tau''_q = \tau_{kq}$
- the d-axis transient inductance $l'_d = l_d \cdot \sigma_{df}$
- the d-axis subtransient inductance $l''_d = l_d \cdot \sigma_d$
- the q-axis (sub)transient inductance $l''_q = l_q \cdot \sigma_q$

The mentioned quantities are idealised and can be derived from the electrical equations derived in the previous section. For example, for all open-stator time constants, the stator resistance has to be assumed to be infinite; for the transient and subtransient time constants the stator resistance is assumed to be zero; in addition, for the subtransient time constant in the d-axis, the field resistance as well needs to be disregarded (zero). For the transient inductance in the d-axis, the damper winding resistance has to be assumed to be infinite (no reaction from the damper winding) and the field resistance has to be put equal to zero. For the transient and subtransient inductances all resistances have to be put equal to zero.

Table 28.2 gives an outline of normal values for these time constants and pu inductances for larger turbo generators. For salient pole machines with massive rotor (without damper windings), the transient and subtransient inductances are somewhat

---

[10] As there is only one rotor winding in the q-axis in our model, it is sometimes referred to as a transient time constant, although the term 'subtransient time constant' is more common.

Table 28.2 Range (and rated values) for time constants and p.u inductances of large synchronous machines

Turbo generators	Two-pole	Two-pole	Four-pole	Four-pole
	Air cooling	Conductor cooling	Air cooling	Conductor cooling
x_d	1.0...1.75 (1.65)	1.5...2.25 (1.85)	1.0...1.75 (1.65)	1.5...2.25 (1.85)
x_q	0.96...1.71 (1.61)	1.46...2.21 (1.8)	0.96...1.71 (1.61)	1.46...2.21 (1.8)
$x_d^{'}$	0.12...0.25 (0.17)	0.2...0.35 (0.28)	0.2...0.3 (0.25)	0.25...0.45 (0.35)
$x_d^{''}$	0.08...0.18 (0.12)	0.15...0.28 (0.22)	0.12...0.20 (0.16)	0.20...0.32 (0.28)
x_2 (inverse field)	$= x_d^{''}$	$= x_d^{''}$	$= x_d^{''}$	$= x_d^{''}$
x_0 (zero sequence)	$(0.1\ldots0.7)x_d^{''}$	$(0.1\ldots0.7)x_d^{''}$	$(0.1\ldots0.7)x_d^{''}$	$(0.1\ldots0.7)x_d^{''}$
x_p (Potier reactance)	0.07...0.17	0.2...0.45	0.12...0.24	0.25...0.45
r_s	0.001...0.007	0.001...0.005	0.001...0.005	0.001...0.005
$T_{d0}^{'}$ (s)	5	5	8	6
$T_d^{'}$ (s)	0.6	0.75	1.0	1.2
$T_d^{''}$ (s)	0.035	0.035	0.035	0.045
T_d (s)	0.13...0.45	0.2...0.55	0.2...0.4	0.25...0.55
τ_i (s)	5...7	5...7	6...8	6...8

different: $x_d^{'} = 0.25\ldots0.5$ and $x_d^{''} = 0.13\ldots0.32$; for salient pole machines with damper windings $x_d^{'} = 0.25\ldots0.5$ and $x_d^{''} = 0.32\ldots0.5$.

As we will see later on, the mentioned (sub)transient time constants are narrowly related to the electrical eigenvalues of a synchronous machine.

28.3.2 Block Diagram and Characteristic Equation

To analyse the dynamic behaviour, we will start from the dynamic equation in the matrix notation (see Sect. 28.2.3)

$$\mathbf{M} \cdot \boldsymbol{u} = \mathbf{A}(p) \cdot \mathbf{x} \tag{28.79}$$

where the matrix $\mathbf{A}(p) = \mathbf{B}p - \mathbf{C}$ can be written in block-matrix form as

$$\mathbf{A}(p) = \begin{bmatrix} \mathbf{A_i}(p) & \mathbf{d} & -\mathbf{c} \\ -\mathbf{m}^T & \left(J\omega_n^2/N_p\right)(p+\varrho_w) & 0 \\ \mathbf{0}^T & 1 & p \end{bmatrix} \tag{28.80}$$

Here, $\mathbf{A_i}(p)$ is the (5×5) matrix of the electrical equations, $\mathbf{d}$, $\mathbf{c}$, $\mathbf{m}$ are (5×1) matrices that can be derived from Eqs. 28.53 or 28.57 (for pu or absolute values, respectively), and $\mathbf{0}$ is a (5×1) zero matrix. As $\det \mathbf{B} \neq 0$, the system equation can also be written as

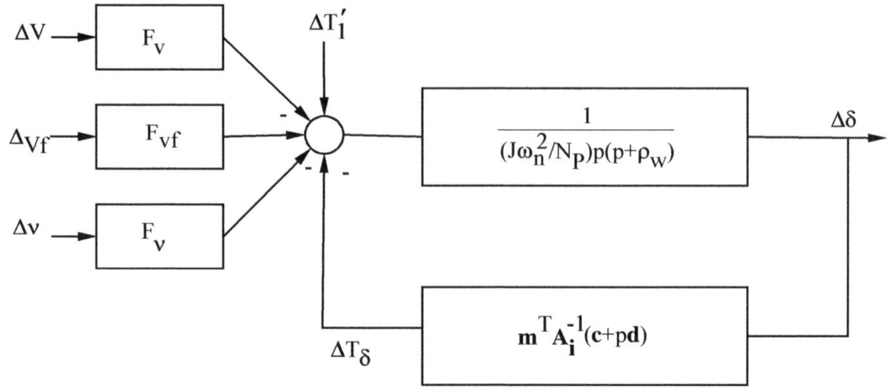

Fig. 28.3 Block diagram for the synchronous machine

$$x = \left(p\mathbf{I} - \mathbf{B}^{-1} \cdot \mathbf{C}\right)^{-1} \cdot \mathbf{B}^{-1} \cdot \mathbf{M} \cdot \mathbf{u} = (p\mathbf{B} - \mathbf{C})^{-1} \cdot \mathbf{M} \cdot \mathbf{u} \tag{28.81}$$

The eigenvalues are the zeros of $G(p) = \det \mathbf{A}(p) = \det(p\mathbf{B} - \mathbf{C})$. Using the block-matrix notation of Eq. 28.80, we obtain

$$G(p) = p \cdot \det \begin{bmatrix} \mathbf{A_i} & \mathbf{d} \\ -\mathbf{m}^T & \left(J\omega_n^2/N_p\right)(p + \varrho_w) \end{bmatrix} - \det \begin{bmatrix} \mathbf{A_i} & -\mathbf{c} \\ -\mathbf{m}^T & 0 \end{bmatrix} \tag{28.82}$$

or

$$G(p) = p\left(J\omega_n^2/N_p\right)(p + \varrho_w)(\det \mathbf{A_i}) + (\det \mathbf{A_i})\,\mathbf{m}^T \cdot \mathbf{A_i}^{-1} \cdot (p\mathbf{d} + \mathbf{c}) \tag{28.83}$$

which may be regarded as corresponding with the feedback control loop in Fig. 28.3 (for motoring).

The open-loop poles comprise the two inertia poles and the zeros of $D(p) = \det \mathbf{A_i}(p)$. The latter may be called the constant-speed poles (cf. the induction machine) as these are the eigenvalues of the electrical system without feedback of speed or load angle. The open-loop zeros are the zeros of (the numerator of) $\mathbf{m}^T \cdot \mathbf{A_i}^{-1} \cdot (p\mathbf{d} + \mathbf{c})$, i.e. the transfer function F_δ from load angle to electromagnetic torque.

Indeed, the torque is given by $\Delta T = \mathbf{m}^T \cdot \boldsymbol{\Delta I}$, while in the absence of external inputs ($\Delta\nu$, ΔV, ΔV_f), $\Delta\mathbf{I} = \mathbf{A_i}^{-1} \cdot (p\mathbf{d} + \mathbf{c})\Delta\delta$.

In fact, the system equation 28.57 or 28.53 may be written as (now with external inputs)

$$-\mathbf{d}\Delta\nu + \mathbf{M_1} \cdot [\,\Delta V \;\; \Delta V_f]^T = \mathbf{A_i}(p) \cdot \Delta\mathbf{I} - (\mathbf{c} + p\mathbf{d})\Delta\delta \tag{28.84}$$

$$\left(J\omega_n^2/N_p\right)(p+\varrho_w)\Delta\nu + \Delta T_l^{'} = \mathbf{m}^T \cdot \boldsymbol{\Delta I} + \left(J\omega_n^2/N_p\right)(p+\varrho_w)p\Delta\delta \quad (28.85)$$

with $\mathbf{M_1} = \mathbf{M}[1,2,3,4,5|1,2]$.

Thus $F_\delta = \frac{\Delta T}{\Delta\delta} = \mathbf{m}^T \cdot \mathbf{A_i}^{-1} \cdot (p\mathbf{d} + \mathbf{c})$.

28.3.3 Gain

For the gain of the feedback loop in Fig. 28.3, different choices are possible. One thing is sure: as the open-loop zeros and poles are completely determined by nearly all dimensionless parameters, except the electromagnetic time constant τ_m (or eigenfrequency ν_m), this gain will (have to) depend on the electromagnetic time constant τ_m. Two possibilities are quite obvious:

- a static gain, proportional to the slope of the stationary characteristic torque versus load angle and inversely proportional to the inertia constant. Mathematically, this can be written as the limit

$$K_s = [\lim_{t\to\infty}(\Delta T/\Delta\delta)]/\left(J\omega_n^2/N_p\right) = [\lim_{p\to 0} F_\delta(p)]/\left(J\omega_n^2/N_p\right) \quad (28.86)$$

or, with $\mathbf{A_i}(p)=p\mathbf{B_i}-\mathbf{C_i}$,

$$K_s = \{\lim_{p\to 0}[\mathbf{m}^T \cdot \mathbf{A_i}^{-1} \cdot (p\mathbf{d} + \mathbf{c})]\}/\left(J\omega_n^2/N_p\right) = -\mathbf{m}^T \cdot \mathbf{C_i}^{-1} \cdot \mathbf{c}/\left(J\omega_n^2/N_p\right) \quad (28.87)$$

- another possibility is the dynamic gain, proportional to the dynamic torque

$$K_d = [\lim_{t\to 0}(\Delta T/\Delta\delta)]/\left(J\omega_n^2/N_p\right) = [\lim_{p\to\infty} F_\delta(p)]/\left(J\omega_n^2/N_p\right) \quad (28.88)$$

or

$$K_d = \{\lim_{p\to\infty}[\mathbf{m}^T \cdot \mathbf{A_i}^{-1} \cdot (p\mathbf{d} + \mathbf{c})]\}/\left(J\omega_n^2/N_p\right) = \mathbf{m}^T \cdot \mathbf{B_i}^{-1} \cdot \mathbf{d}/\left(J\omega_n^2/N_p\right) \quad (28.89)$$

In principle, both definitions can be used. However, we prefer to obtain and use a definition which is analogous to, and an extension of, the gain definition used for the induction machine and the DC machine.

To explore this further, we may rewrite the equation of motion (without external inputs) as follows

$$\left(J\omega_n^2/N_p\right)(p^2+\varrho_w p)\Delta\delta + F_\delta(p)\cdot\Delta\delta = \Delta T_l^{'} \quad (28.90)$$

The transfer function for the closed loop can be written as follows, denoting the transfer function of load angle to torque $F_\delta(p) = N_\delta(p)/D_\delta(p)$,

$$\frac{\Delta\delta}{\Delta T_l'} = \frac{1}{\left(J\omega_n^2/N_p\right)\left(p^2+\varrho_w p\right)+F_\delta} \tag{28.91}$$

or

$$\frac{\Delta\delta}{\Delta T_l'} = \frac{\frac{1}{(J\omega_n^2/N_p)(p^2+\varrho_w p)}}{1+\frac{N_\delta(p)}{(J\omega_n^2/N_p)(p^2+\varrho_w p)D_\delta(p)}} \tag{28.92}$$

On the other hand, the load torque to speed transfer function for the induction machine was, denoting $D_{\nu r}(p\tau_r) = D(p\tau_r)$ and $N_{\nu r}'(p\tau_r) = (T_0/\nu_{s0})N_{\nu r}(p\tau_r)$,

$$\frac{\Delta\nu_r}{\Delta T_l'} = \frac{-\frac{1}{(J\omega_n^2/N_p)(p+\varrho_w)}}{1+\frac{N_{\nu r}'(p\tau_r)}{(J\omega_n^2/N_p)(p+\varrho_w)D_{\nu r}(p\tau_r)}} \tag{28.93}$$

where $F_{\nu r}(p\tau_r) = N_{\nu r}'(p\tau_r)/D_{\nu r}(p\tau_r)$ represents the transfer function of speed to asynchronous machine torque. For the induction machine, we defined the gain τ_r/τ_m as the quotient of the coefficients of the highest degree terms in $\tau_r p$ in the numerator and denominator of

$$\frac{N_{\nu r}'(p\tau_r)}{\left(J\omega_n^2/N_p\right)(p+\varrho_w)\,D_{\nu r}(p\tau_r)} = \frac{\tau_r N_{\nu r}'(p\tau_r)}{\left(J\omega_n^2/N_p\right)(p\tau_r+\varrho_w\tau_r)\,D_{\nu r}(p\tau_r)} \tag{28.94}$$

thus yielding

$$\frac{\tau_r}{\tau_m} = \tau_r\cdot\frac{T_0/\nu_{s0}}{\left(J\omega_n^2/N_p\right)} \tag{28.95}$$

Note, however, that $N_{\nu r}'(p\tau_r)$ and $D_{\nu r}(p\tau_r)$ are polynomials of third and fourth degree, respectively, in $\tau_r p$. As a consequence, the quotient of the coefficients of the highest degree terms in $N_{\nu r}'(p\tau_r)$ and $D_{\nu r}(p\tau_r)$ cannot be a dynamic asynchronous torque, as the limit of $N_{\nu r}'(p\tau_r)/D_{\nu r}(p\tau_r)$ is zero for $p \to \infty$. Nevertheless, we used the quotient of the highest degree terms as gain. The question remains what this quotient actually represents then.

The speed of an induction machine may also be written as the derivative of a *load angle*, similarly to the load angle of a synchronous machine.[11] Thus, with $\nu_r \equiv d\theta/dt$ (θ is the instantaneous rotor angle) and writing θ as $\theta = \omega_0 t - \delta$, we obtain in the Laplace domain $\Delta\nu_r = -p\Delta\delta$.

We may now rewrite the characteristic equation of the induction machine in a form similar to the one for the synchronous machine:

[11] In fact, in a synchronous machine, the rotor excitation current is DC, but the AC rotor current in an induction machine can equally be regarded as an excitation current with slip frequency, inducing an emf in the stator with stator frequency and providing a torque with the flux.

$$\frac{\Delta\delta}{\Delta T_l'} = \frac{\frac{1}{(J\omega_n^2/N_p)p(p+\varrho_w)}}{1+\frac{pN_{\nu r}'(p\tau_r)}{(J\omega_n^2/N_p)p(p+\varrho_w)D_{\nu r}(p\tau_r)}} \tag{28.96}$$

Here, $pN_{\nu r}'(p\tau_r)/D_{\nu r}(p\tau_r)$ is the transfer function of *load angle* to electromagnetic torque, $\Delta T/\Delta\delta$. The limit of $\Delta T/\Delta\delta$ for $p \to \infty$, i.e. the quotient of the highest degree terms in p and equal to $(T_0/\nu_{s0})/\tau_r$, is the dynamic *synchronising* torque.

Now consider the equation of motion corresponding with the characteristic Eq. 27.52:

$$\left(J\omega_n^2/N_p\right)(p^2+\varrho_w p)\Delta\delta + \left(pN_{\nu r}'(p\tau_r)/D_{\nu r}(p\tau_r)\right)\cdot\Delta\delta = \Delta T_l' \tag{28.97}$$

For very large p ($p \to \infty$ thus corresponding to the limit $t \to 0$), we may approximate this as

$$\left(J\omega_n^2/N_p\right)p^2\Delta\delta + ((T_0/\nu_{s0})/\tau_r)\cdot\Delta\delta \approx \Delta T_l'$$

This represents the dynamic equation for $t \to 0$. It follows that $(T_0/\nu_{s0})/(\tau_r \cdot J\omega_n^2/N_p) \equiv 1/\tau_r\tau_m$ has the dimension of the square of a dynamic (or initial) eigenfrequency $\nu_m^2 = (dT/d\delta)_{dyn} \,/\, \left(J\omega_n^2/N_p\right)$. What we called the gain for the induction machine was nothing else than the square of the dynamic eigenfrequency referred to the rotor time constant,

$$\frac{\tau_r}{\tau_m} = \tau_r^2\nu_m^2$$

For the synchronous machine as well,[12] we will therefore use the dynamic gain

$$K_d = \tau_{kd}^2\nu_m^2 = \tau_{kd}^2 \cdot \frac{(dT/d\delta)_{dyn}}{\left(J\omega_n^2/N_p\right)} \tag{28.98}$$

28.3.4 Eigenvalue Analysis of the Synchronous Machine

The eigenvalue analysis of a synchronous machine is obviously possible starting from the 7×7 matrix, Eq. 28.57. Still, a (semi-)analytical method is to be preferred, as it will also allow us to introduce the operational inductances (or more general, impedances) which are very common in machine studies.

The five electrical equations of the synchronous machine are the starting point of this analysis. We rewrite the two stator equations introducing the q- and d-axis flux variations $\Delta\Psi_q = L_q\Delta I_q + M_{skq}\Delta I_{kq}$ and $\Delta\Psi_d = L_d\Delta I_d + M_{skd}\Delta I_{kd} + M_{sf}\Delta I_f$. In these, all rotor current variations can be eliminated and expressed as a function of

[12]Please prove that the gain we used for the DC machine also corresponds to a dynamic gain.

the stator current variations using the three rotor equations. This results in $\Delta\Psi_q = L_q(p)\cdot\Delta I_q$ and $\Delta\Psi_d = L_d(p)\cdot\Delta I_d + F(p)\cdot\Delta V_f$ where

$$L_d(p) = L_d \cdot \frac{p^2\tau_f\tau_{kd}\cdot\frac{\sigma_{kdf}\sigma_d}{\sigma_{dkd}\sigma_{df}} + p\left(\tau_{kd}+\tau_f\right)+1}{p^2\tau_f\tau_{kd}\cdot\frac{\sigma_{kdf}}{\sigma_{dkd}\sigma_{df}} + p\left(\frac{\tau_{kd}}{\sigma_{dkd}}+\frac{\tau_f}{\sigma_{df}}\right)+1} \tag{28.99}$$

$$L_q(p) = L_q\cdot\frac{p\tau_{kq}+1}{p\left(\tau_{kq}/\sigma_q\right)+1} \tag{28.100}$$

$$F(p) = M_{sf}\cdot\frac{p\left(\tau_{kd}/\sigma_{dkd}\right)\left\{\left(1-\sigma_{fkd}\right)\frac{1-\sigma_{dkd}}{1-\sigma_{df}}\right\}^{1/2}+1}{p^2\tau_f\tau_{kd}\cdot\frac{\sigma_{kdf}}{\sigma_{dkd}\sigma_{df}} + p\left(\frac{\tau_{kd}}{\sigma_{dkd}}+\frac{\tau_f}{\sigma_{df}}\right)+1}\cdot R_f^{-1} \tag{28.101}$$

$L_d(p)$ and $L_q(p)$ are called the operational inductances of the d- and q-axes, respectively. These are the transfer functions from current variations to flux variations in these two axes (purely electrical, without load angle variations).

Writing the two stator equations as follows

$$\Delta V\cos\delta_0 = R_s\Delta I_q + p\omega_n\Delta\Psi_q - \nu_{r0}\omega_n\Delta\Psi_d - \psi_{d0}\omega_n(\Delta\nu - p\Delta\delta) + V_{d0}\Delta\delta \tag{28.102}$$

$$\Delta V\sin\delta_0 = R_s\Delta I_d + p\omega_n\Delta\Psi_d + \nu_{r0}\omega_n\Delta\Psi_q + \psi_{q0}\omega_n(\Delta\nu - p\Delta\delta) - V_{q0}\Delta\delta \tag{28.103}$$

we obtain for the current variations (if frequency variations $\Delta\nu$ are not considered)

$$\begin{aligned}\Delta I_q = {} & D(p)^{-1}\cdot\left\{(\Delta V/V_0)\cdot\left[R_sV_{q0} + \omega_nL_d(p)\cdot\left(V_{qo}p + \nu_0V_{do}\right)\right]\right\} + \\ & D(p)^{-1}\cdot\left\{\Delta V_f\cdot\omega_nF(p)\cdot\nu_0R_s + \Delta\delta\cdot\left[\left(V_{q0} + p\omega_n\Psi_{q0}\right)\nu_0\omega_nL_d(p)\right.\right. \\ & \left.\left.-\left(R_s + p\omega_nL_d(p)\right)\left(V_{d0} + p\omega_n\Psi_{do}\right)\right]\right\}\end{aligned} \tag{28.104}$$

$$\begin{aligned}\Delta I_d = {} & D(p)^{-1}\cdot\left\{(\Delta V/V_0)\cdot\left[R_sV_{d0} + \omega_nL_q(p)\cdot\left(V_{do}p - \nu_0V_{qo}\right)\right]\right\} + \\ & D(p)^{-1}\cdot\left\{\Delta V_f\cdot\omega_nF(p)\cdot\left[pR_s + \omega_nL_q(p)\left(p^2+\nu_0^2\right)\right]\right\} + \\ & D(p)^{-1}\cdot\left\{\Delta\delta\cdot\left[\left(V_{d0} + p\omega_n\Psi_{d0}\right)\nu_0\omega_nL_q(p) + \left(R_s + p\omega_nL_q(p)\right)\left(V_{q0} + p\omega_n\Psi_{qo}\right)\right]\right\}\end{aligned} \tag{28.105}$$

with

$$\begin{aligned}D(p) &= \left(R_s + p\omega_nL_d(p)\right)\left(R_s + p\omega_nL_q(p)\right) + \nu_0^2\omega_nL_d^2(p)L_q(p) \\ &= \left(p^2+\nu_0^2\right)\left(\omega_n^2L_d(p)L_q(p)\right) + R_s\omega_np\left(L_d(p)+L_q(p)\right) + R_s^2\end{aligned}$$

The transfer functions with as output the electromagnetic torque and as input the supply voltage, the excitation voltage or the load angle can be derived from

$$\Delta T = \frac{3}{2} N_p \left\{ \left(\Psi_{q0} - I_{q0} L_d(p)\right) \Delta I_d - \left(\Psi_{d0} - I_{d0} L_q(p)\right) \Delta I_q \right\} \tag{28.106}$$

All these transfer functions have the zeros of $D(p)$ as poles. These zeros are also the open-loop poles of the feedback loop of Sect. 28.3.2, i.e. $D(p) = \det A_i(p)$.

Remark: As in most cases $\tau_f \gg \tau_{kd}$ and $\tau_{fh} \gg \tau_{kdh}$, often the operational inductances Eqs. 28.99–28.101 are written as

$$L_d(p) = L_d \cdot \frac{\left(p\tau_d' + 1\right)\left(p\tau_d'' + 1\right)}{\left(p\tau_{d0}' + 1\right)\left(p\tau_{d0}'' + 1\right)} \tag{28.107}$$

$$L_q(p) = L_q \cdot \frac{p\tau_q'' + 1}{p\tau_{q0}'' + 1} \tag{28.108}$$

$$F(p) = \frac{p\tau_{\sigma kd} + 1}{\left(p\tau_{d0}' + 1\right)\left(p\tau_{d0}'' + 1\right)} \cdot M_{sf} \cdot R_f^{-1} \tag{28.109}$$

with the transient and subtransient time constants defined as in Sect. 28.3.1. $\tau_{\sigma kd}$ is the own leakage of the d-axis damper winding only.

28.3.4.1 Open-Loop Poles

The open-loop poles are the two inertia poles and the five zeros of $D(p)$. Approximate expressions for the poles of $D(p)$ depend on the magnitude of the short-circuit time constants.

For larger power ratings, usually $\nu_0\tau_d,\ \nu_0\tau_q,\ \nu_0\tau_{kd},\ \nu_0\tau_{kq},\ \nu_0\tau_{fd} \gg 1$ (e.g. 5…10) and $\tau_f \gg \tau_{kd}$ hold. In that case, the zeros of $D(p)$ can be approximated by $p_{1,2} = \pm j\nu_0$, $p_3 = -\tau_d'^{-1} = -\tau_f^{-1}$, $p_4 = -\tau_d''^{-1} = -\left(\tau_{kd}\frac{\sigma_{kdf}\sigma_d}{\sigma_{dkd}\sigma_{df}}\right)^{-1}$, $p_5 = -\tau_q''^{-1}$, if the stator resistance is negligible.

If the stator resistance is not negligible (but still small, i.e. τ_d and τ_q larger than τ_{kd} and τ_{kq}), then a better approximation for the poles $p_{1,2}$ is $p_{1,2} = -\frac{1}{2}(\tau_d^{-1} + \tau_q^{-1}) \pm j\nu_0$, while the other poles are rather unaffected by the stator resistance.

Typical pu values at rated frequency are, for example, $p_{1,2} \approx -0.02 \pm j$; $p_3 \approx -3 \cdot 10^{-3}$; $p_4 \approx -0.09$; $p_5 \approx -0.09$ which correspond to $\tau_d = \tau_d = 50$, $\tau_f = 350$, $\tau_{kd} = \tau_{kq} = 11$, $\tau_{fh} = \tau_f/\sigma_{df} = 1800$ and $\nu_0 = 1$.

If the conditions $\nu_0\tau_d,\ \nu_0\tau_q,\ \nu_0\tau_{kd},\ \nu_0\tau_{kq},\ \nu_0\tau_{fd} \gg 1$ are not met (e.g. for lower supply frequencies ν_0), then

- the real part of the stator poles becomes $p_{1,2} = -\frac{1}{2}(\tau_{dh}^{-1} + \tau_{qh}^{-1}) \pm j\nu_0$
- the poles corresponding with the subtransient time constants do not change appreciably when the stator resistance is sufficiently small ($\tau_d,\ \tau_q \gg \tau_{kd},\ \tau_{kq}$). For rather small supply frequencies ($\nu_0\tau_{kd},\ \nu_0\tau_{kq} < 1$), these poles may become conjugate complex with as real part their average value $-\frac{1}{2}(\tau_{kd}^{-1} + \tau_{kq}^{-1})$ (and a small

imaginary value). For larger values of the stator resistance, however, the absolute value of their real part increases with an amount approximately equal to the inverse stator time constant, i.e. $p_4 = -\tau_d^{''-1} - \tau_d^{-1}$ and $p_5 = -\tau_q^{''-1} - \tau_q^{-1}$

- The transient pole $p_5 = -\tau_f^{-1}$ is not affected significantly as long as $\tau_f \gg \tau_{kd}$ and $\nu_0 \tau_f \gg 1$.

Note that the behaviour of these open-loop poles is quite similar to the behaviour of the induction machine poles.

28.3.4.2 Open-Loop Zeros

From Eqs. 28.104–28.106, the transfer function from load angle to electromagnetic torque is easily derived

$$
\begin{aligned}
F_\delta(p) = \frac{\Delta T}{\Delta \delta} = \frac{3}{2} N_p \cdot D(p)^{-1} &\{(\Psi_{q0} - I_{q0} L_d(p)) \left[(V_{d0} + p\omega_n \Psi_{d0})\, \nu_0 \omega_n L_q(p)\right. \\
&\left.+ (R_s + p\omega_n L_q(p)) (V_{q0} + p\omega_n \Psi_{qo})\right]\} \\
-\frac{3}{2} N_p \cdot D(p)^{-1} &\{(\Psi_{d0} - I_{d0} L_q(p)) \left[(V_{q0} + p\omega_n \Psi_{q0})\, \nu_0 \omega_n L_d(p)\right. \\
&\left.- (R_s + p\omega_n L_d(p)) (V_{d0} + p\omega_n \Psi_{do})\right]\}
\end{aligned} \tag{28.110}
$$

Substituting $V_{d0} = R_s I_{d0} + \nu_0 \omega_n \Psi_{q0}$ and $V_{q0} = R_s I_{q0} - \nu_0 \omega_n \Psi_{d0}$, we write the numerator of Eq. 28.110 as follows

$$
\begin{aligned}
N(p) = -\frac{3}{2} N_p \cdot &\{[L_q(p) L_d(p) (I_{q0} \Psi_{qo} + I_{do} \Psi_{d0}) \\
&- L_d(p) \Psi_{d0}^2 - L_q(p) \Psi_{q0}^2] (p^2 + \nu_0^2)\, \omega_n^2\} \\
-\frac{3}{2} N_p \cdot R_s &\{[p L_q(p) L_d(p) (I_{q0}^2 + I_{d0}^2) \\
&- p (L_d(p) - L_q(p)) (I_{do} \Psi_{d0} - I_{q0} \Psi_{q0}) - p (\Psi_{q0}^2 + \Psi_{d0}^2)]\, \omega_n\} \\
&- \frac{3}{2} N_p \cdot R_s^2 \{[L_d(p) I_{q0}^2 + L_q(p) I_{d0}^2] - [I_{do} \Psi_{d0} + I_{q0} \Psi_{q0}]\}
\end{aligned} \tag{28.111}
$$

Like the denominator $D(p)$, this numerator is also a fifth degree polynomial in p.

For zero stator resistance, $(p^2 + \nu_0^2)$ is also a factor in the numerator. Thus, in that case the stator modi $p = \pm j\nu_0$ are hidden modes, as these are for the induction machine. The three other zeros are rather complicated functions of the reluctance ratio $\gamma_{dq} = L_d / L_q$, the flux and the excitation.

For small stator resistance and with the flux mainly in the d-axis, two of the remaining zeros are $p = -\tau_f^{-1}$ and $p = -\tau_{kd}^{-1}$. The third zero depends on the reluctance ratio γ_{dq} and the relative excitation $e_{p0} = E_{p0} / V_0$ and varies between $p = -\tau_{kqh}^{-1}(1 - \gamma_{dq}^{-1})$ for $e_{p0} = 0$, $p = -\tau_{kqh}^{-1}$ for $e_{p0} = 1$ and $p = -\tau_{kq}^{-1}$ for

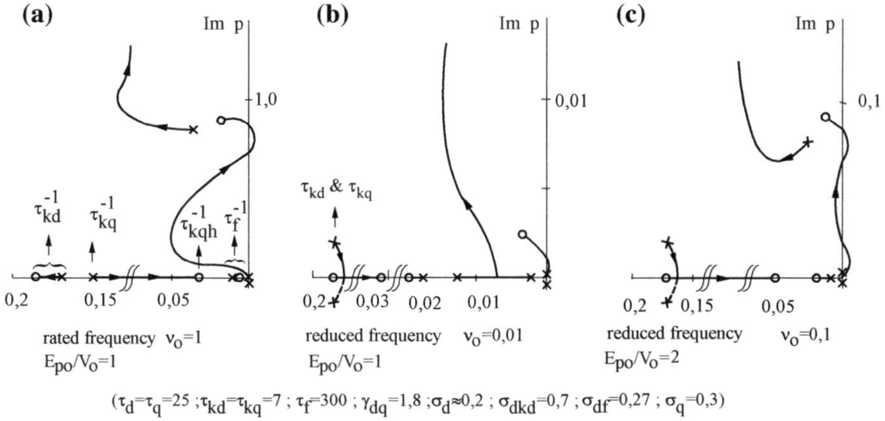

Fig. 28.4 Root loci for rated and reduced frequency

$e_{p0} \to \infty$. For the flux position in the q-axis, one of the zeros is $p = -\tau_{kq}^{-1}$ while one of the other two may lie in the right half plane or in the origin, depending on the reluctance ratio.

For a non-negligible stator resistance, the stator zeros have a real part which is approximately half the real part of the corresponding open-loop poles, i.e. $p = -(1/4)(\tau_d^{-1} + \tau_q^{-1}) \pm j\nu_0$ for large $\nu_0\tau_{kd}$, or $p = -(1/2 \ldots 1/4)(\tau_{dh}^{-1} + \tau_{qh}^{-1}) \pm j\nu_0$ for $\nu_0\tau_{kd} \ll 1$.

28.3.4.3 Root Loci and Eigenvalues

Based on the discussion of poles and zeros above, it can be expected that a *general* dynamic analysis of *the* synchronous machine will be difficult or even impossible. Instead, we will concentrate on the most common root loci shapes and associated dynamic behaviour of synchronous machines.

Figure 28.4a depicts the root locus of a (quite large) synchronous machine with (as usual) low stator resistance, for rated frequency and at no load. Although the branches starting from the inertia poles enter the right half plane around the high-frequency poles and zeros, this is not at all of practical importance as the corresponding gain values are unrealistically high: these values would correspond to a drive inertia 500 times smaller than the machine inertia.

The root locus (b) in Fig. 28.4 could be found for very low supply frequency ($\nu_0 = 0.01$) at no load. The branches starting from the inertia poles will enter the right half plane for very small gains. However, those small gain values correspond to a drive inertia 100 times larger than the machine inertia, which is not realistic.

The root locus (c) corresponds to reduced frequency operation ($\nu_0 = 0.1$) without active load ($\delta_0 = 0$), but over-excited. Here, the root locus enters the right half plane for both low and high gain values. Where the branches enter the right half plane at

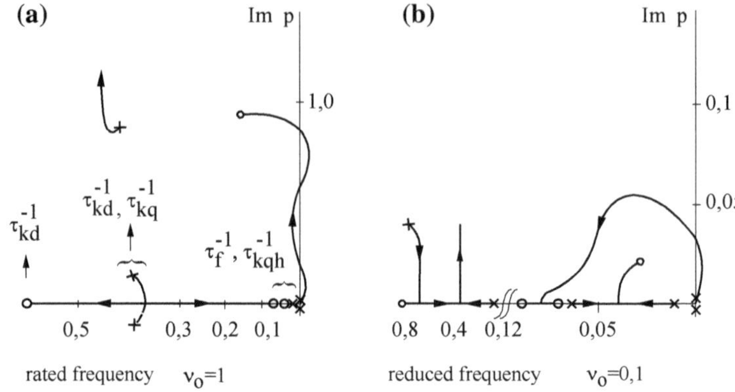

$(\tau_d=2{,}4\ ;\ \tau_q=2\ ;\ \tau_{kd}=\tau_{kq}=3{,}5\ ;\ \tau_f=50\ ;\ \gamma_{dq}=2\ ;\ \sigma_d=0{,}13\ ;\ \sigma_{dkd}=0{,}14\ ;\ \sigma_{df}=0{,}28\ ;\ \sigma_q=0{,}22)$

Fig. 28.5 Root loci for high stator resistance

low gain values (around the inertia poles) is again not of any practical importance. In contrast, the right half plane wandering around the high-frequency poles and zeros now corresponds to gain values for the order of magnitude of the machine inertia and rated flux operation.[13]

For a machine with a rather high stator resistance, on the other hand, we obtain root loci as in Fig. 28.5. For the *high-frequency* root loci (a) at rated frequency, the gain values corresponding with the branches into the right half plane are not of any practical importance. However, for reduced frequency operation ($\nu_0 = 0.1$), see (b) in Fig. 28.5, the gain values corresponding with the right half plane wandering may be realistic, even without over-excitation.

The examples above might be typical but are not general at all. The number of parameters of a synchronous machine (model) is much too high to draw general conclusions for the stability behaviour of synchronous machines. Moreover, the construction variations of synchronous machines are quite huge and not bound by standards as for induction machines (similar scaling laws are non-existent). The main conclusion we may draw is that over-excitation and/or high reluctance ratio may render the synchronous machine more prone to instability, especially at low-frequency operation.

In the following sections of this chapter, we will study the dynamic behaviour of two special types of synchronous machines: the reluctance motor and the symmetrical synchronous machine (i.e. in fact an induction machine with wound rotor).

[13]Over-excitation tends to decrease the stability via its effect on the open-loop zeros; note that even at rated resulting flux, a synchronous machine can be over-excited.

28.3.5 Eigenvalue Analysis of the Reluctance Motor

Reluctance motors exhibit similar properties as induction machines, such as robustness and low cost. In addition, the synchronous operation yields better efficiency, and if a cage is present, asynchronous starting is possible as well.

Reluctance motors usually have modest power ratings, for example less than 10 kW. Older or low-cost reluctance motors often are constructed starting from a similar induction motor design. This most straightforward (and oldest) design is the crenellated rotor. Rotor sheets of an induction machine can be used as a base, by cutting away the gaps in the q-axes to obtain saliency. This has the additional advantage of having a starting cage present if the induction motor rotor sheets already contain the slots for the cage and if these slots are filled. Moreover, the gap can be filled with electrically conducting material, so that the cage is again *complete*.[14] However, the reluctance ratio x_d/x_q that can be obtained in this way is limited to about 2.

The barrier rotor designs, on the other hand, make it possible to increase x_d/x_q beyond 2 without compromising the other requirements (e.g. for sufficiently large main inductances).

The dimensionless parameters for the reluctance machine model have been presented above. For the stator and rotor short-circuit time constants, similar scaling laws hold as for the induction machine, with some slight variations between the d- and $q-$axes and between stator and rotor side: $\tau_d \approx 60\tau_p$; $\tau_q \approx (50 \cdots 60)\tau_p$; $\tau_{kd} \approx (40 \cdots 60)\tau_p$; $\tau_{kq} \approx (50 \cdots 70)\tau_p$. The order of magnitude of the short-circuit inductances is similar to that of the induction machine, i.e. $\sigma_d l_d \approx 0.24$; $\sigma_q l_q \approx 0.2$. Obviously, the main inductances differ from those of an induction machine. For the d-axis inductance, we may write $l_d \approx (10 \cdots 20)\sqrt{\tau_p/N_p}$, thus $1.1 \cdots 2$ for crenellated rotors, and $1.5 \cdots 3$ (or more) for barrier rotors. The reluctance ratio $\gamma_{dq} = l_d/l_q$ depends on the rotor type: $\gamma_{dq} \approx (1 + l_d) \cdots (1 + 2l_d)$ and thus $2 \cdots 5$ for crenellated rotors or $2 \cdots 8$ for barrier rotors (and even higher for multi-segment rotors). The inertia time constant is somewhat smaller than for induction machines, i.e. $\tau_n \approx 300\tau_p/N_p$ which results in $\nu_m^2 \tau_{kd}^2 \approx (0.3 \cdots 0.6)\tau_{kd}$.

The eigenvalue analysis of the reluctance motor can be treated as a simplified version of the analysis for the general synchronous machine, for example as the limit case for $\tau_f \rightarrow \infty$ and $v_f = \Delta v_f = 0$. The operational inductance for the d-axis is now

$$L_d(p) = L_d \cdot \frac{p\tau_{kd} + 1}{p\,(\tau_{kd}/\sigma_d) + 1} \tag{28.112}$$

Figure 28.6 shows the block diagram for the reluctance machine. The numerator and denominator of the transfer function $F_\delta = \Delta T/\Delta\delta$ are both fourth-degree functions of p. The open-loop poles and zeros can be derived directly from the analysis in Sect. 28.3.4. The open-loop poles comprise the two complex conjugate stator poles ($p = -(1/2)(\tau_d^{-1} + \tau_q^{-1}) \pm j\nu_0$) and the two real (at least in most cases, i.e. not

---

[14] I.e. more or less symmetrical in the two axes.

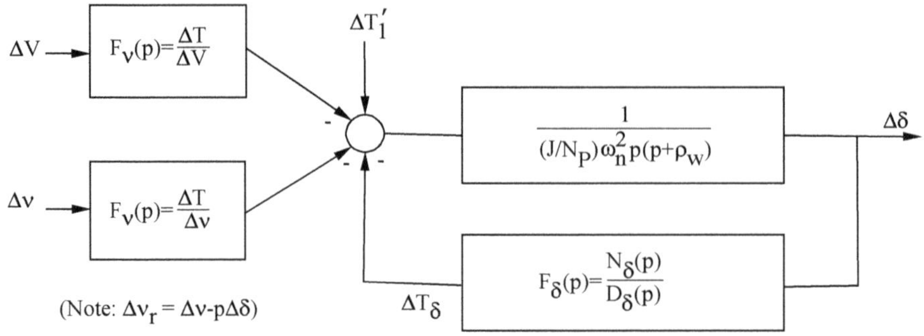

Fig. 28.6 Block diagram for the reluctance machine

for operation at too low frequencies) poles $p = -\tau_{kd}^{-1}$ and $p = -\tau_{kq}^{-1}$. The open-loop zeros comprise the two complex conjugate stator zeros ($p = -(1/4)(\tau_d^{-1} + \tau_q^{-1}) \pm j\nu_0$) and two real zeros which depend on the load condition and reluctance ratio: for no load ($\delta_0 = 0$ and thus also $i_{q0} = 0$), we find $p_3 = -\tau_{kd}^{-1}$ and $p_4 = -\tau_{kqh}^{-1}(1 - \gamma_{dq})$; for $\delta_0 = \pi/2$ and thus also $i_{d0} = 0$ we have an unstable behaviour with $p_3 = +\tau_{kdh}^{-1}(\gamma_{dq} - 1)$ and $p_4 = -\tau_{kq}^{-1}$.

The gain defined above corresponds to the dynamic synchronous torque (which is different from the static[15] synchronous torque). It is also interesting to calculate the limit of (the maximum of) this dynamic synchronous torque for $\gamma_{dq} \to 1$ (and for equal time constants and leakages in d- and q-axes): this limit is equal to the dynamic synchronous torque of the induction machine; or in other words: the gain of the reluctance motor reduces to the gain of the induction motor.

For the root loci of the reluctance motor, we can find fewer variations than for the general synchronous motor. For a reluctance motor of relatively high power rating at rated (or not too low) frequency, we can find a root locus as in Fig. 28.7. Where the high-frequency branch enters the right half plane, the corresponding gain values are much too high to be of any practical importance.

For lower supply frequencies, these gain values decrease and instability may occur in real drives. For still lower frequencies, the root locus may ultimately evolve to a low-frequency shape as (a) in Fig. 28.8. This root locus is quite similar to the low-frequency root locus of an induction machine. The locus (b) corresponds to a machine with a high reluctance ratio, rather high leakages and a larger than normal stator resistance. Despite the equally reduced supply frequency of $\nu_0 = 0.1$, the shape is still similar to a high-frequency root locus, but this is caused by the rather special parameter values. Although the general stability behaviour of the reluctance motor is somewhat similar to that of the induction machine, the larger number of parameters and possible parameter variations and the lack of real scaling laws make it impossible to draw any general conclusions.

---

[15]Calculate both dynamic and static synchronous torques for $\delta_0 = 0$; it will become clear that the static synchronous torque is obviously zero for $\gamma_{dq} = 1$.

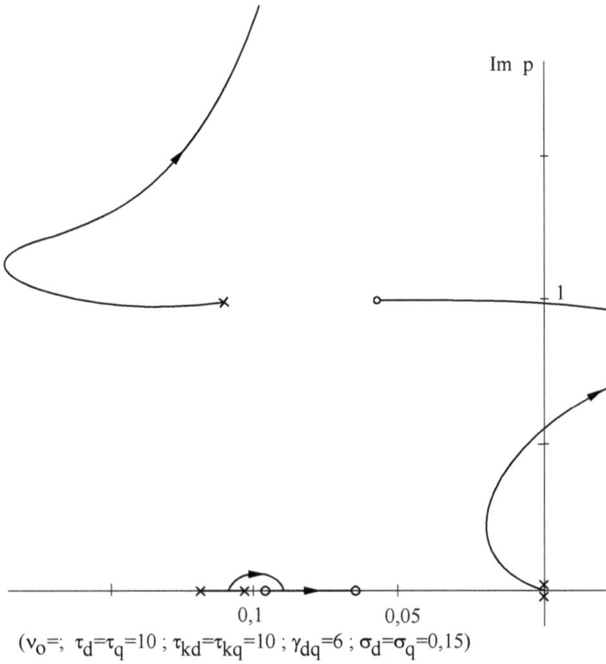

Fig. 28.7 Root loci for the reluctance motor at rated frequency

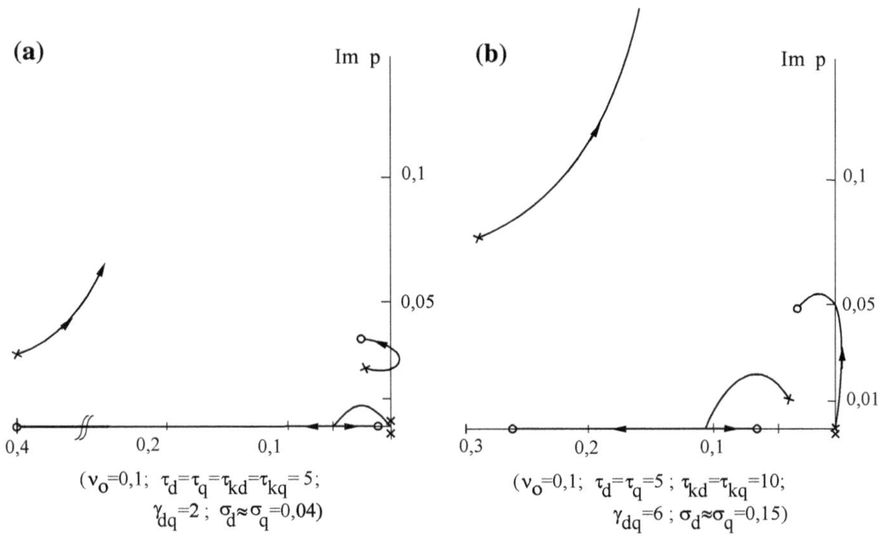

Fig. 28.8 Root loci for the reluctance motor at reduced frequency

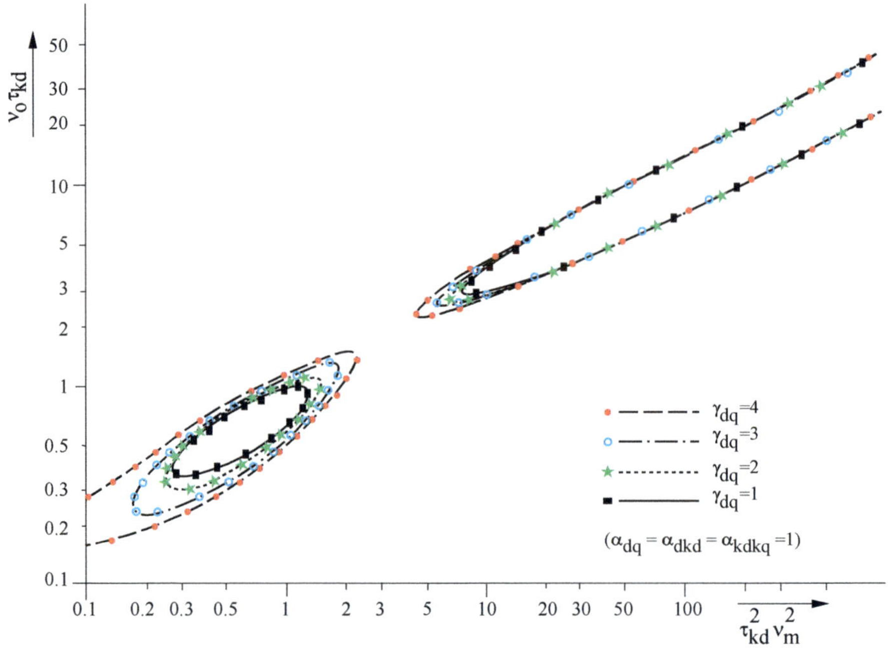

Fig. 28.9 Parameter plane for a reluctance motor with low leakage

Nevertheless, the dynamic properties of a reluctance motor may also be represented in the parameter space (or plane). We will illustrate the behaviour by a few examples. Figure 28.9 shows the instability boundaries for a reluctance machine with low leakage ($\sigma = 0.04$) for different reluctance ratios. Clearly, the larger the reluctance ratio, the larger the instability areas are. Note that the boundary for $\gamma_{dq} = 1$ is the one for an induction machine. For a machine with high reluctance ratio and rather high leakage (see Fig. 28.10) the result is remarkable (compared to the induction machine). A high leakage results in a significant increase of the high-frequency instability region. The instability region may even extend to the whole range of gain and frequency values.

A third illustration shows the effect of the load, together with an average machine line (based on the range of parameters given above) for rated frequency operation. As it is the case for the induction machine, generator operation increases the low-frequency instability region. A second conclusion we may draw from this figure is that stability problems are not likely for rated frequency operation, but that reduced frequency operation may result in unstable behaviour (Fig. 28.11).

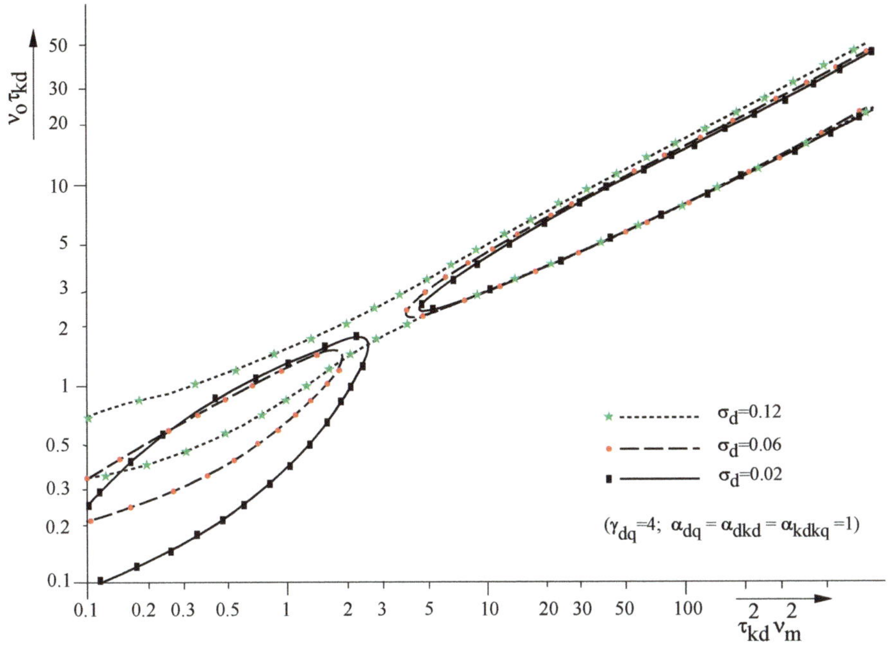

Fig. 28.10 Parameter plane for a reluctance machine with high reluctance

28.3.6 *Eigenvalue Analysis of a Symmetrical Synchronous Machine*

Another interesting topic of study are symmetrical synchronous machines. These are completely symmetrical with respect to the rotor d- and q-axes, both as to saliency and to the windings. Thus, in such a machine, there is only one (and identical) winding in each of the rotor axes. An example could be a wound-rotor induction machine with DC excitation in one of the rotor windings.

For a symmetrical synchronous machine, the dimensionless parameters are the three machine parameters ($\tau_s = \tau_d = \tau_q$; $\tau_r = \tau_f = \tau_{kq}$; $\sigma = \sigma_d = \sigma_q$); the three parameters defining the operating point (the frequency ν_0 which will be taken together with the time constants); the load angle δ_0 and the excitation parameter $e_v = E_{p0}/V_0$; and the two mechanical parameters $\tau_r^2\nu_m^2$ and ρ_w. In other words, the parameters are the same as for the induction machine, with the exception of the slip which is replaced by the load angle and the excitation parameter.

The four open-loop poles are the same as those of the induction motor (see Chap. 27). For frequencies that are not excessively low, we find a fourth zero on the real axis as open-loop zero, in addition to the two complex conjugate zeros $p_{1,2} = -(1/2)\tau_s^{-1} \pm j\nu_0\tau_r$ and the real zero $p_3 = -\tau_r^{-1}$ (see Chap. 27):

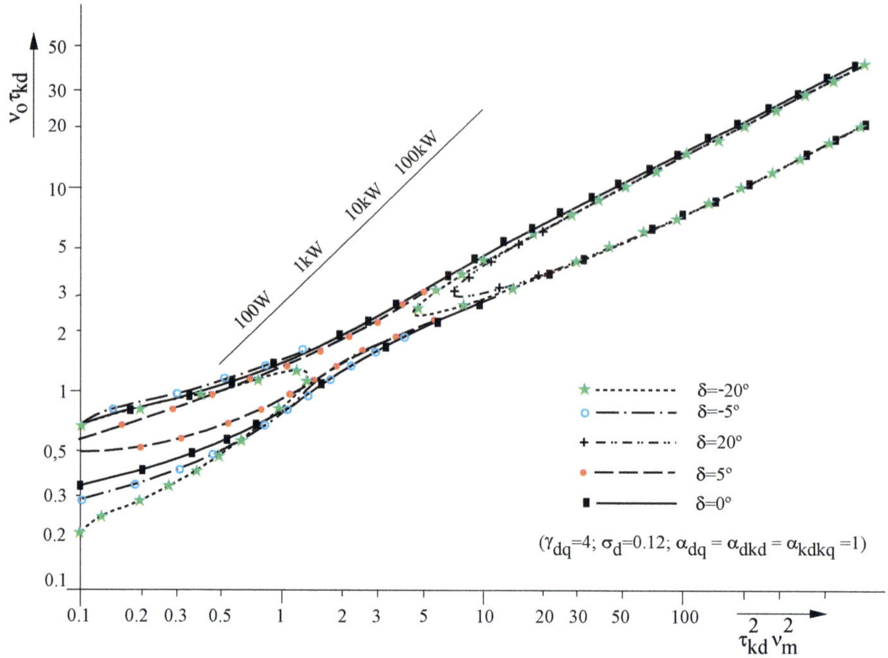

Fig. 28.11 The effect of load on the instability regions in the parameter plane for a reluctance motor

$$p_4 = -\tau_r^{-1} \cdot \frac{1}{1 + \frac{1-\sigma}{\sigma} \cdot \frac{\Psi_o^2}{\Psi_{d0} M_{sf} I_{f0}}} = -\tau_r^{-1} \cdot \frac{1}{1 + \frac{1-\sigma}{\sigma} \cdot \frac{E_o^2}{E_{p0} E_{q0}}} \tag{28.113}$$

The zero p_4 essentially is a function of the excitation. Without excitation ($I_{f0} = 0$ or $E_{p0} = 0$), or also for $\Psi_{d0} = 0$, this zero is in the origin (the machine reduces to an induction machine). For no-load operation ($E_0 = E_{p0} = E_{q0}$), it becomes equal to $-\tau_{rh}^{-1}$, thus with as time constant the inverse main field rotor time constant. For short-circuit operation ($E_0/E_{p0}E_{q0} \to 0$), we get the inverse rotor short-circuit time constant $-\tau_r^{-1}$.

For lower supply frequencies, the complex conjugate zeros shift to the imaginary axis. Their real part is proportional to the inverse stator main field time constant $-\tau_{sh}^{-1}$ (as for the induction machine) and also becomes a function of the excitation, $p_{1,2} = -(1/2)\tau_{sh}^{-1}(2 - E_{p0}/E_{q0}) \pm j\nu_0$. As to the zero p_3, the evolution for lower frequencies is the same as for the induction machine.

The gain is as defined for the general synchronous machine. Note that without excitation this gain is reduced to that of the induction machine.

The root loci appear quite similar to those of an induction machine. There are essentially two types of root loci, one for normal (higher) frequencies and one for low frequencies Fig. 28.12.

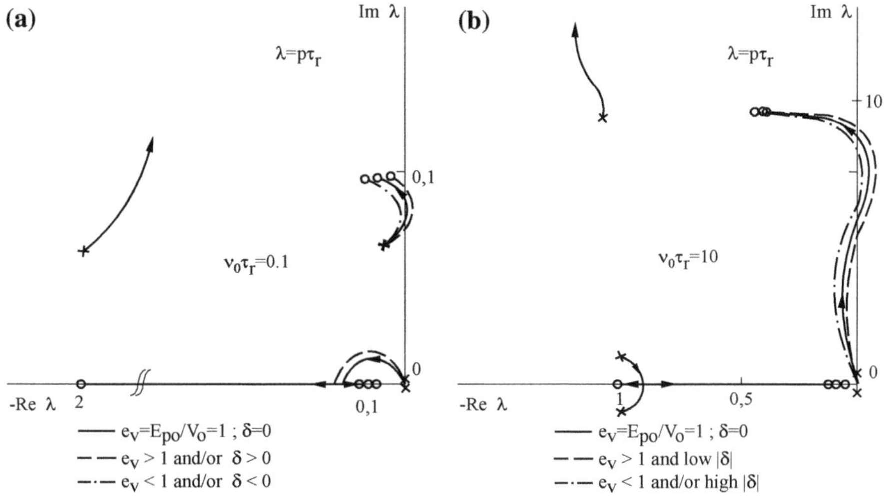

Fig. 28.12 Low- and high-frequency root loci for the symmetrical synchronous machine

In the low-frequency root locus, the poles and zeros around the origin are close to each other, resulting in a branch wandering into the right half plane. A higher excitation shifts the complex conjugate zeros closer to the imaginary axis and therefore results in moving this branch to the right (the shift of the real zero near the origin with variable excitation has little or no effect on this). In a similar way, the effect of the load angle (motoring or generating) can be studied.[16]

The high-frequency root locus, too, seems quite similar to that of the induction machine. Here as well, a higher excitation shifts the branch starting from the inertia poles further to the right (albeit less pronounced than in the low-frequency case). In the high-frequency region, however, higher load angles have a slightly stabilising effect, both for motoring and generating.

In the parameter plane (or space), we see stability boundaries which are more or less similar to those of the induction machine. Figure 28.13 shows these boundaries for a machine with low leakage, for different values of the excitation parameter (at no load). The boundaries for zero excitation ($e_v = 0$) are those of the induction machine for slip zero. Clearly, excitation reduces the stability, in both the low- and high-frequency regions. For very high excitation, the two instability regions even merge into one large region. This effect becomes even more significant for higher stator resistances ($\alpha > 1$).

Figure 28.14 illustrates the effect of load. In the high-frequency range, the effect of load is minimal but in the low-frequency range negative load angles (generating) tend

[16]For the same value of e_v and $|\delta|$, $E_{p0} > E_{q0}$ for $\delta > 0$ and $E_{p0} < E_{q0}$ for $\delta < 0$ from which the distinct shift of the complex zeros can be derived. You may verify this shift by calculating the low-frequency root loci for zero, positive and negative load angles, for example in Matlab.

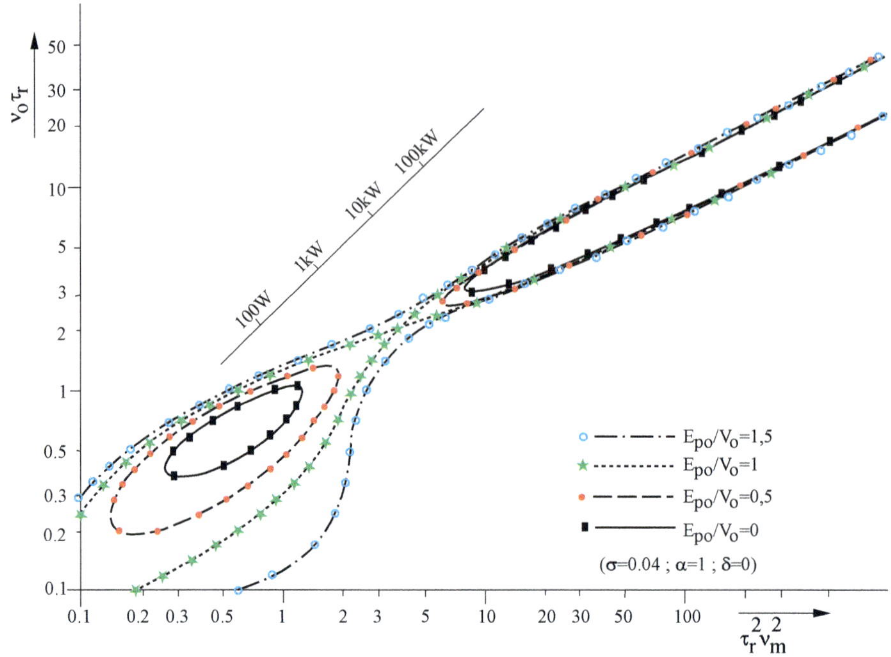

Fig. 28.13 Stability boundaries in the parameter plane for various excitations

to stabilise. Here, very large positive load angles (motoring) result in a significant increase in the low-frequency instability.[17]

28.3.7 *Modelling and Stability for Current Supply*

The models we derived in the previous sections are quite general and can be used for both voltage and current supply. It may be instructive to discuss some general aspects of *ideal* current supply as to dynamic studies.

If the supply is an ideal current source, and if we are not interested in the supply voltages, the stator equations can be omitted from the dynamic equations. For example, the electrical equations of an induction machine or symmetrical synchronous machine become second-order equations rather than fourth-order. In the two remaining rotor equations, the stator currents are imposed and fixed.

If we write down the system equations as a function of dimensionless parameters, it is revealed that these equations and the resulting dynamics are now most conveniently expressed as a function of the main field rotor time constant $\nu_0 \tau_{rh}$, instead of the short-circuit rotor time constant $\nu_0 \tau_r$.

[17] Verify this using a Matlab model.

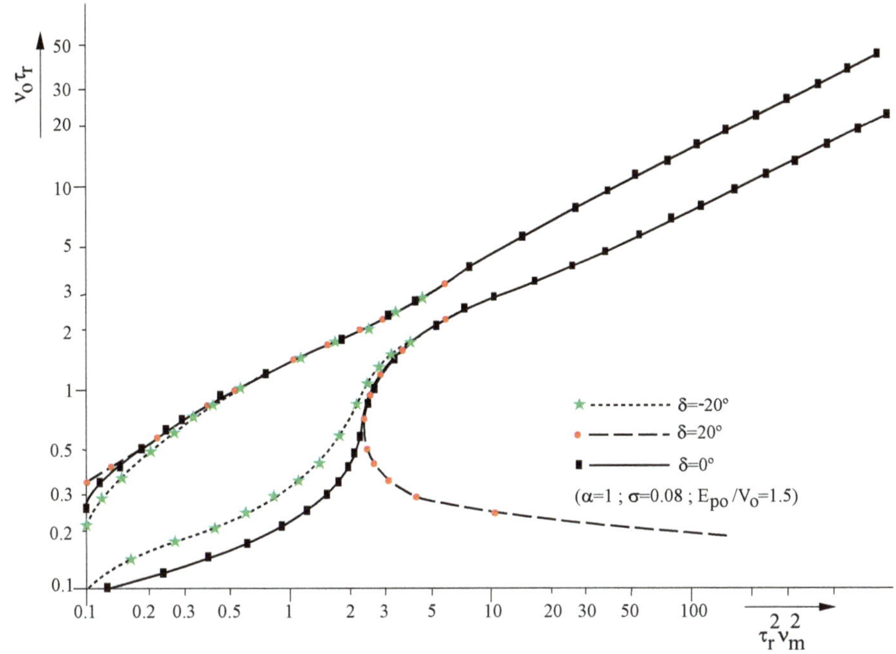

Fig. 28.14 Effect of load on the stability boundaries

28.4 Conclusions and Further Remarks

For the general synchronous machine, the number of parameters and the lack of general scaling laws make it impossible to draw any general conclusions about its stability behaviour (as those derived for the induction machine). However, the techniques presented in this chapter can be used or adapted to study the dynamics of any synchronous machine.

In contrast, for the (maybe somewhat theoretical) case of a symmetrical synchronous machine, i.e. an induction machine with DC excitation in one of the rotor axes, such a general stability analysis is feasible. The results of this analysis show that, especially in the low-frequency range, adding DC excitation to an induction machine increases the instability regions. Whereas the low-frequency instability is of not much practical importance for induction machines (due to the high leakage for smaller power ratings), adding excitation more than annihilates the stabilising effect of high leakage. With high excitation, both low- and high-frequency regions tend to form one large instability region, so that also medium machine sizes (corresponding to the transition zone for induction machines) may exhibit instability at supply frequency reduction.

A similar conclusion can be drawn for reluctance machines. In general, it may be concluded that adding a reluctance effect creates instability regions that are larger than the corresponding regions for induction machines.

From the above, it may be tempting to conclude that adding something that enhances the energetic efficiency reduces the stability behaviour. In fact, such a statement is in line with what Laithwaite [18] once claimed: that the measures which enhance energetic conversion are mostly detrimental to stability.

Chapter 29
Dynamics in Vector Control and Field Orientation

Abstract In Chap. 17 of Part III, vector control and field orientation have already been presented. However, the dynamics have not been analysed. In this chapter we discuss vector control and field orientation more thoroughly, including an analysis of their dynamic behaviour.

29.1 Introduction

For electro-mechanical systems that require four-quadrant operation (even to zero speed) and a fast response, it is essential that the drive has a controllable torque over the whole speed range. In the past, the DC commutator machine was the preferred machine for this kind of application. For constant excitation, the torque of a DC machine is indeed proportional to the current. This controlled current may be provided by a controllable current (DC chopper or controlled rectifier).

Advances in power electronics (and control) allow us to obtain a controllable AC current supply for AC rotating field machines as well. However, to create a similar *ideal* dynamic behaviour as the DC machine, additional control is required. In a DC machine, the (stationary) armature current vector is orthogonal to the flux vector (in the absence of armature reaction, i.e. a compensated machine). In rotating field machines, in contrast, the position of the rotating current vector is not fixed with respect to the rotating field vector. Thus, in addition to the current amplitude, also the relative positions of current and field vectors have to be controlled. The ideal scenario, in which field and current vectors are controlled to be orthogonal, is called *field orientation*. The general case, in which the angle is controlled but not necessarily kept at 90°, is referred to as *vector control*.

29.2 Torque Control of a DC Machine

In a DC machine with the brush axis in the neutral position and without armature reaction (Fig. 29.1), the electromagnetic interaction between the stationary field and armature current layer results in an electromagnetic torque which is proportional to

J. A. Melkebeek, *Electrical Machines and Drives*, Power Systems,
https://doi.org/10.1007/978-3-319-72730-1_29

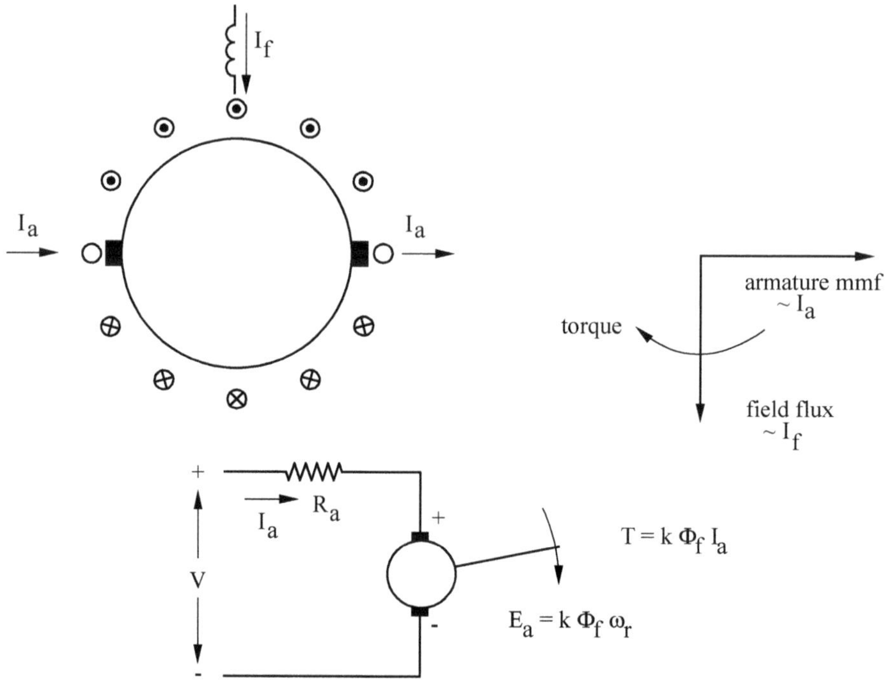

Fig. 29.1 Torque production in the DC machine

the product of main field flux ($\Phi_{fm} = \Phi_f - \Phi_{f\sigma} \approx \Phi_f$) and armature current:

$$T_e = k \cdot \Phi_{fm} \cdot I_a \tag{29.1}$$

Note that the induced emf in the armature is given by $E_a = k \cdot \Phi_{fm} \cdot \Omega_r$, corresponding to the electromagnetic power $T_e \cdot \Omega_r = E_a \cdot I_a$.

The traditional open-loop *speed* control is based on the voltage equation and the expression for the induced emf. From $V_a = E_a - R_a I_a = k \cdot \Phi_{fm} \cdot \Omega_r - R_a I_a \approx k \cdot \Phi_{fm} \cdot \Omega_r$, it follows that for a constant flux the speed will be approximately proportional to the applied armature voltage. The (usually small) speed drop with increasing load due to the armature resistance can be compensated by feedback of the speed, if required.

A completely different control method is based on the *torque* equation. A speed or position control loop can be implemented with, for example, a PID controller with the required torque as output. Using current control, the torque can then be controlled. The speed of the torque control is the same as the speed of the current control. Similar relations holds for the transients in torque and current control.

However, if the flux and current axes are not orthogonal in space (e.g. brush axis shifted from the neutral position), or in case of armature reaction (no compensation windings), then the simple relation between torque and current does not hold any longer. The flux becomes a function of the armature current and additional dynamics for the torque result. In addition, the relation between torque and current becomes dependent on the angle between the two axes. For a given flux and current, the maximum torque is reduced as well.

To realise such an ideal torque control, the following is required:

1. an independent control of the armature current (and if both positive and negative torque values are required the power electronic supply needs to be able to provide positive and negative currents)

2. an independent control of the field flux

3. orthogonality between flux and armature axes (brush axis in the neutral position and absence of armature reaction).

29.3 Vector Control of a Synchronous Machine

29.3.1 Steady State

29.3.1.1 Analogy with the DC Machine

A synchronous machine fed by an ideal current source (for example CSI-fed with ideal DC source) is quite analogous to a current-fed DC machine:

- the CSI is a current supply, with controllable amplitude and phase
- the rotor field is independently controllable
- the position of the rotor field is known from the rotor position.

Figure 29.2 shows a schematic configuration with a CSI as a supply. The CSI can work with load commutation (with an over-excited synchronous machine), or with forced commutation. The rotor position is measured to locate the field axis (d-axis) and the CSI is switched synchronously with the rotor. As a consequence, the stator

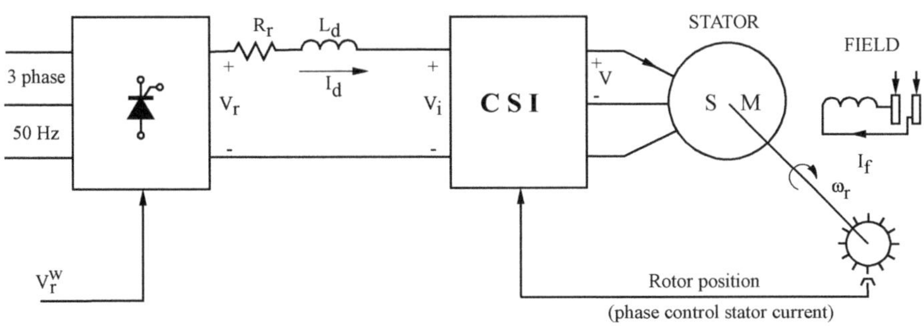

Fig. 29.2 CSI-fed synchronous machine

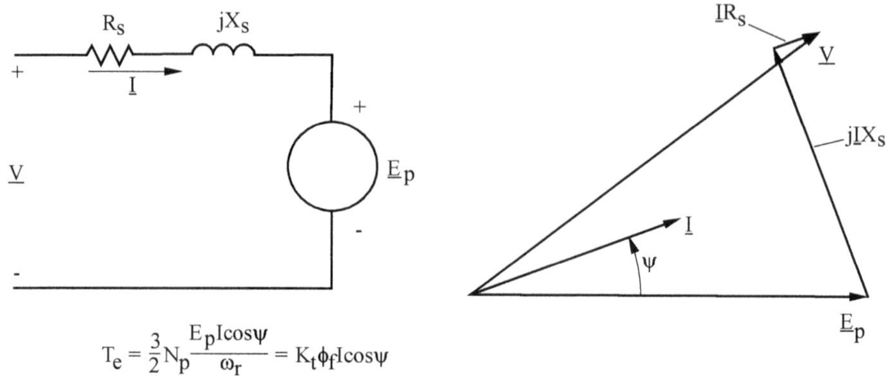

Fig. 29.3 BLAC: equivalent circuit and phasor diagram

frequency is always equal and *follows* the rotor speed. The system may be called *self-synchronous*.

The rotor position information is used to control the angle ϑ between the field axis and the axis of the rotating armature current layer. If this angle is controlled to be 90°, then we get the same situation as for the DC machine. This *ideal* control is called *field orientation*.

To describe the steady state of this *brushless AC machine*[1] (BLAC), we may use the equivalent circuit and the corresponding phasor diagram in Fig. 29.3. The corresponding equations are

$$\underline{E}_p = jK_e \cdot \omega_r \cdot \underline{\Phi}_f = j\omega_r \cdot \underline{\Psi}_f \tag{29.2}$$

with ω_r the rotor speed in electrical radians per second. Note that the magnitude of the emf can be written as for the DC machine: $E_p = K_e \cdot \omega_r \cdot \Phi_f$.

The electromagnetic torque is given by

$$T_e = \frac{3}{2}N_p \cdot (E_p I \cos\psi)/\omega_r = \frac{3}{2}N_p \cdot K_e \cdot I \cdot \Phi_f \cdot \cos\psi = K_t \cdot I \cdot \Phi_f \cdot \cos\psi \tag{29.3}$$

The equations for emf and torque are similar to those of the DC machine. However, the torque depends on the phase angle ψ between the emf and the current (or the angle $\vartheta = \pi/2 + \psi$ between the flux and the current).

The preceding equations show that with a controlled current magnitude and a constant controlled angle between flux (or rotor emf) and current, the synchronous machine may provide torque control (to be used in a drive based on current control). If the CSI is supplied by a DC voltage rather than an ideal DC current, it can be shown, by means of the fundamental CSI model, that the drive behaves as a voltage-

[1]The term brushless AC machine is used when the field profile in space is sinusoidal; for trapezoidal or rectangular field profiles the term brushless DC machine is normally used.

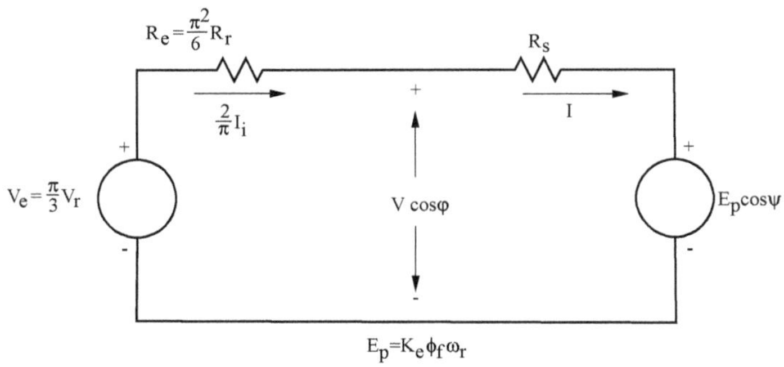

Fig. 29.4 BLAC: reduced equivalent circuit

controlled and independently excited DC machine. Chapter 16 in Part III has already demonstrated that the simplified fundamental equivalent circuit of a voltage-supplied CSI with a synchronous machine as load is as shown in Fig. 29.4. Denoting the rectifier quantities referred to the AC side with the subscript *e,* we obtain

$$V_e = E_p \cos\psi + (R_s + R_e)I \tag{29.4}$$

This equation is quite similar to the armature voltage equation of a DC machine. As $E_p = K_e \cdot \omega_r \cdot \Phi_f$, it shows that a self-synchronising CSI-fed (with DC voltage supply) synchronous machine with a constant internal angle ψ behaves as a separately excited DC machine with a mainly voltage-controlled speed and a slight dependence on the current or the torque if the resistances are small.

29.3.1.2 Torque Control and Choice of ψ

To obtain torque control as in a DC machine, both the current magnitude and its angle ψ with respect to the rotor emf has to be controlled. In other words, the phasor $\underline{I}$ has to be controlled. This may also be formulated as a controlled current space vector $\underrightarrow{I}$ with respect to the flux space vector $\underrightarrow{\Psi}$.

If the internal angle ψ is controlled to remain equal to zero (or, equivalently, the angle ϑ between the flux and current equal to $\pi/2$), then we get an identical situation as in the DC machine. Flux and current magnitudes are independently controlled and the two phasors are orthogonal, because the CSI acts as the electronic commutator in a DC machine. For this ideal case of $\psi = 0$ or $\vartheta = \pi/2$ (called *field orientation*), the ratio T_e/I is at a maximum. As we will see later on, the dynamic behaviour is also optimal in field orientation.

However, if the CSI requires load commutation, field orientation is not possible as then the current should lead the voltage (and will thus lead the rotor emf, see (b) in Fig. 29.5). If the angle ψ (or thus ϑ) is controlled to a fixed value different from the ideal case, the control is called *vector control.*

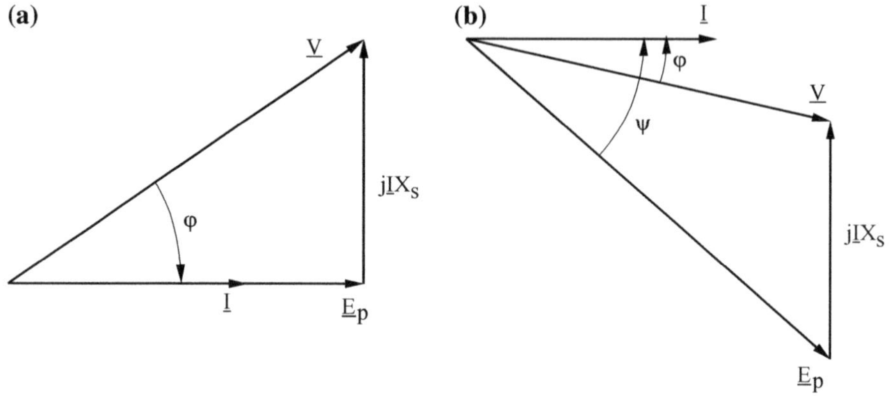

Fig. 29.5 BLAC: zero and large internal angle

29.3.2 Dynamical Analysis

The aim of vector control and field orientation is to obtain a highly dynamic torque control (to be used in a speed or position control loop, for example).

As explained above, this requires an accurate control of the current amplitude and the phase angle of the current vector with respect to the flux vector.

For the dynamic analysis of vector control the $d-q$-axis description with the d-axis fixed to the rotor flux seems the most obvious. The expression for the torque in the $d-q$-model is

$$T_e = \frac{3}{2}N_p \cdot \left(\Psi_q I_d - \Psi_d I_q\right) = \frac{3}{2}N_p \cdot \left(-L_{dm} I_f I_q + (L_q - L_d) I_d I_q - (L_{dm} I_{kd} I_q - L_{qm} I_{kq} I_d)\right) \tag{29.5}$$

In the $d-q$-description, vector control is equivalent to the control of the current components I_d and I_q. From Eq. 29.5, it is therefore clear that even for constant flux and controlled currents I_d and I_q, the torque is not necessarily fully controlled in transients.

In steady state, the voltage equations are

$$V_{q0} = R_s I_{q0} - \omega L_d I_{d0} - \omega L_{dm} I_{f0} \equiv R_s I_{q0} + E_{aq0} + E_{p0} \tag{29.6}$$

$$V_{d0} = R_s I_{d0} + \omega L_q I_{q0} \equiv R_s I_{d0} + E_{d0} \tag{29.7}$$

while the steady-state torque can be written as

$$\begin{aligned} T_{e0} &= \frac{3}{2}N_p \cdot \left(\Psi_{q0} I_{d0} - \Psi_{d0} I_{q0}\right) = \frac{3}{2}N_p \cdot \left(-L_{dm} I_{f0} I_{q0} + (L_q - L_d) I_{d0} I_{q0}\right) \\ &= \frac{3}{2}(N_p/\omega) \cdot \left(E_{p0} I_{q0} + E_{aq0} E_{d0} \left[(\omega L_q)^{-1} - (\omega L_d)^{-1}\right]\right) \end{aligned} \tag{29.8}$$

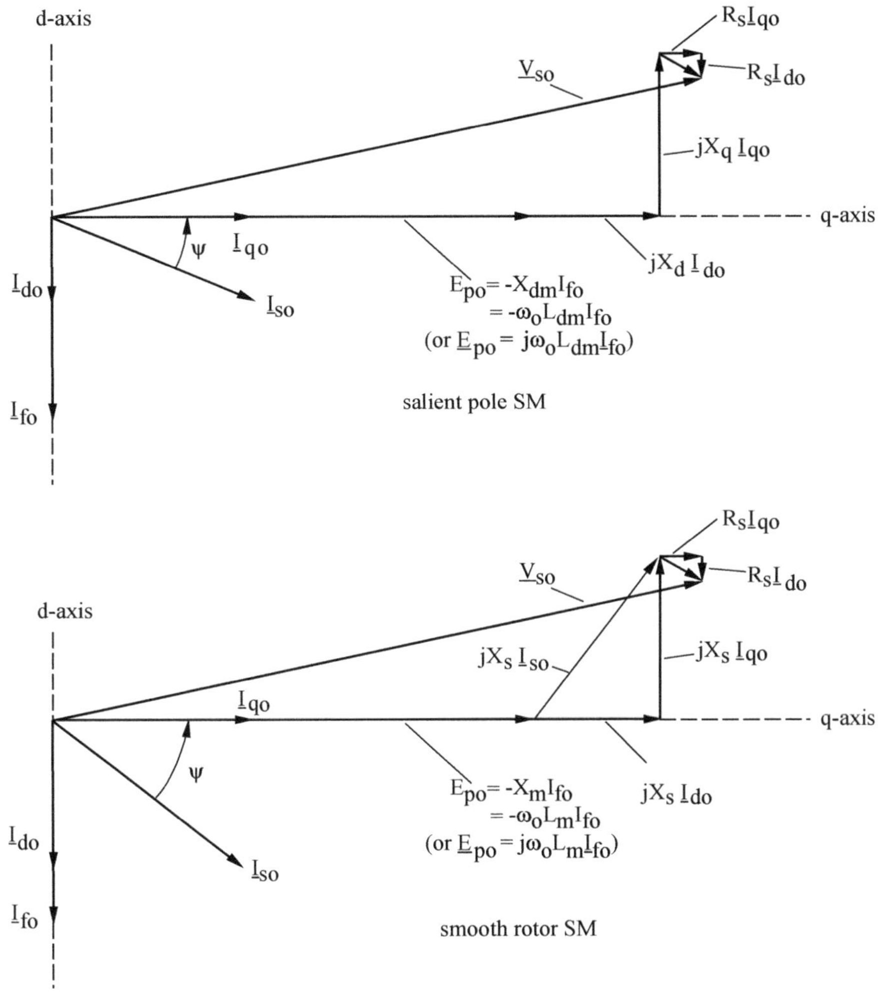

Fig. 29.6 Steady-state vector diagrams

The vector diagrams for steady state are shown in Fig. 29.6 for a salient pole machine and a smooth rotor machine. The angle ψ between the q-axis and the current vector is also the angle between the rotor emf phasor $\underline{E}_{p0}$ and the current phasor $\underline{I}_s$ (space vectors and phasors are identical here and the notations for phasors are used). Note that the angle between the positive d-axis and the current vector is equal to $\vartheta = \pi/2 + \psi$ (the angle between the rotor flux and the current vector is $\vartheta' = \pi/2 - \psi$).

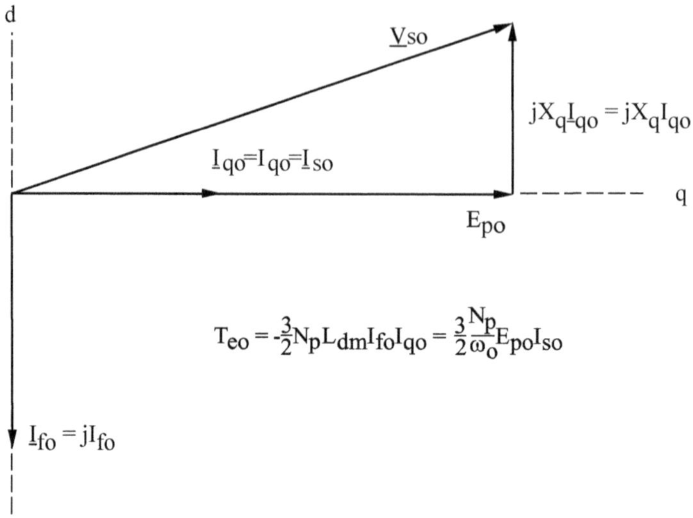

Fig. 29.7 Steady-state vector diagram for field orientation

If the stator current is orthogonal to the rotor field ($\psi = 0$), only a stator q-axis current $I_{q0} = I_{s0}$ is present, and the steady-state voltage and torque equations can be simplified to

$$V_{q0} = R_s I_{q0} - \omega L_{dm} I_{f0} \equiv R_s I_{q0} + E_{p0} \tag{29.9}$$

$$V_{d0} = \omega L_q I_{q0} \equiv E_{d0} \tag{29.10}$$

$$T_{e0} = \frac{3}{2} N_p \cdot \left(-L_{dm} I_{f0} I_{q0}\right) = \frac{3}{2} (N_p/\omega_0) \cdot \left(E_{p0} I_{q0}\right) \tag{29.11}$$

In this case, the reluctance torque is absent, as there is no current in the stator d-axis. The vector diagram is identical to that in Fig. 29.7.

For the dynamical analysis, we will mainly restrict ourselves to the case of field orientation, i.e. without d-axis stator current. Moreover, we will suppose that the orthogonal position of the stator current vector with respect to the rotor flux vector (i.e. stator current in the q-axis) will always be maintained, including during transients. We will also assume that the stator current is the (independent) input variable of the system.

For this case of $I_d = dI_d/dt \equiv 0$, the torque equation 29.5 is simplified to

$$T_e = \frac{3}{2} N_p \cdot \left(-L_{dm} I_f I_q - L_{dm} I_{kd} I_q\right) \tag{29.12}$$

In other words, the reluctance torque component is absent.

The rotor currents are determined by the three rotor equations (with p corresponding with d/dt):

$$R_{kq}I_{kq} + pL_{kq}I_{kq} + pL_{qm}I_q = 0$$

$$R_{kd}I_{kd} + pL_{kd}I_{kd} + p(L_{dm} + L_{fkd\sigma})I_f + 0 = 0 \tag{29.13}$$

$$R_f I_f + pL_f I_f + p(L_{dm} + L_{fkd\sigma})I_{kd} + 0 = V_f$$

For constant field excitation, Eq. 29.13 show that $I_f = V_f/R_f$ and $I_{kd}(p) = \Delta I_{kd}(p) = 0$. Without transient currents in the d-axis, all transients in the torque are absent as well. Although there is a transient current in the q-axis damper winding when the q-axis current changes,

$$I_{kq}(p) = \Delta I_{kq}(p) = \frac{-pL_{qm}}{R_{kq} + pL_{kq}}\Delta I_q(p) \tag{29.14}$$

the torque follows the variations of the q-axis current without any transients. There is thus a complete decoupling of the rotor kd- and f-windings, on the one hand, and the stator winding on the other hand, when the stator current is in the q-axis. This is exactly the same situation as in a DC machine. However, in a DC machine the two current layers are fixed in space, while these are synchronously rotating for the synchronous machine.

In contrast, a variation of the excitation (e.g. in field weakening) results in a transient current in the d-axis damper winding. From Eq. 29.13, we can derive:

$$I_{kd}(p) = \Delta I_{kd}(p) = \frac{-p(L_{dm} + L_{fkd\sigma})}{R_{kd} + pL_{kd}}\Delta I_f(p) \tag{29.15}$$

This transient d-axis damper current results in a torque transient with the same (large) time constant $\tau_{kdh} = L_{kd}/R_{kd}$ (see Eq. 29.5 and Fig. 29.8). In reality, the delay might even be larger as the excitation winding is rather voltage-fed than current-fed. Note that also in a DC machine a variation of the excitation current will result in a torque transient: although there is usually not a real damper winding in a DC machine, the eddy currents in the field yoke of the stator will create a similar damping.

In the general case of vector control ($\psi \neq 0$), the situation is much more complex and the control principles may also differ. For example, the angle ψ may be kept constant and the stator current amplitude may be varied. Or the d-axis current component may be kept constant and the q-axis component of the stator current varied. The latter is frequently used for permanent-magnet motors, with a positive d-axis current causing field weakening.

For example, consider the latter case where the d-axis current is controlled to a fixed value. The non-zero d-axis stator current results in a steady-state reluctance torque. As the d-axis stator current remains constant, the resulting torque is still

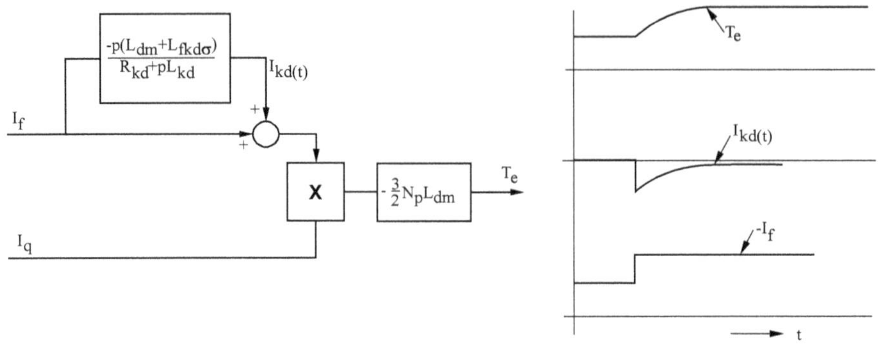

Fig. 29.8 Torque transient for a variation of the field current

proportional to the q-axis current in steady state. However, if the q-axis component is varied, a transient current $I_{kq}(p) = \Delta I_{kq}(p)$ results, which will now cause a transient torque together with the non-zero stator d-axis current.

The situation becomes even more complex if also the d-axis current may change, but this will not be discussed here.

29.3.3 Practical Implementations

In typical implementations, a CSI is utilised (Fig. 29.9). The current magnitude is controlled by the rectifier, while the phase angle is controlled by the switching signals of the inverter.

Advantages include that

- the output is immediately the AC current
- the amplitude is directly controlled by the rectifier while its feedback only requires a DC measurement
- the phase (and frequency) of the output current is directly controlled by the switching signals of the CSI

The disadvantages are that

- the phase control is very fast but the amplitude control of the current is rather slow (due to the large inductance L_{dc})
- the motor current is far from sinusoidal and rather a block wave
- the commutation delay of the inverter results in a current-dependent error that should be compensated and that also depends on the frequency.

When $\psi^w = 0$, the current control is directly a torque control ($I_d \equiv 0$; $I_q = I$) - at least for constant field excitation -. For $\psi^w \neq 0$ this is no longer the case. A desired value for the torque then corresponds to a desired value for the d- and q-axis currents.

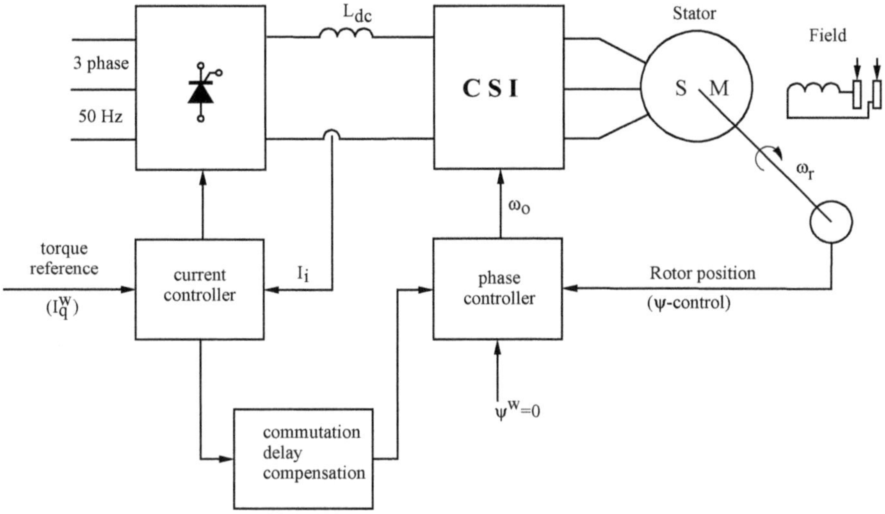

Fig. 29.9 Torque control using a CSI

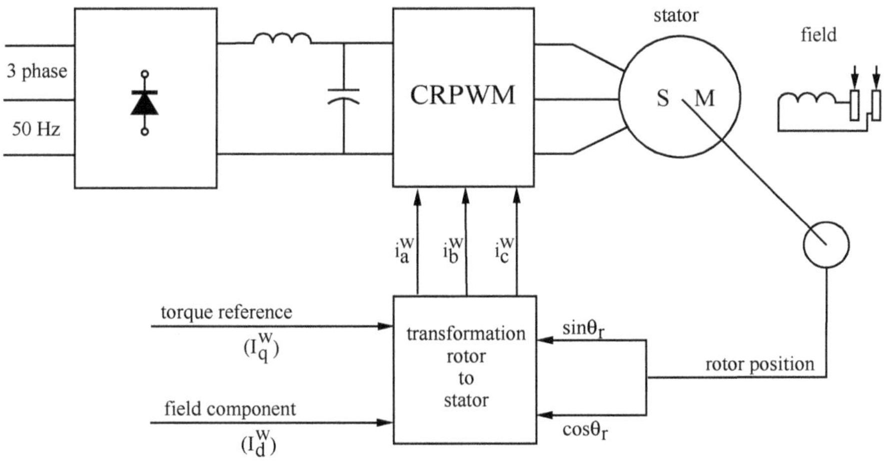

Fig. 29.10 Torque control using a CRPWM

These then have to be transformed into a vector with its amplitude I^w and phase $-\psi$ with respect to the q-axis, using a resolver

$$I_q^w + jI_d^w = \underline{I}_{qd}^w = I^w \cdot \exp(-j\psi^w) \tag{29.16}$$

A more common implementation nowadays makes use of a PWM voltage inverter with current control loop (see Fig. 29.10). The PWM-VSI requires voltage switching signals and therefore a current control loop needs to provide these signals. The current control loop compares the measured AC output current with the desired value. This desired value $I^w(t)$ has to be calculated or synthesised from the desired value of the current magnitude $|I^w|$ and the desired angle ψ^w (or from the values of I_q^w and I_d^w), together with the measured rotor position $\theta_r(t)$. This implies the following transformations:

- the transformation from I_q^w and I_d^w (or $|I^w|$ and ψ^w) to the current vector in the $q-d$ reference frame $\underline{I}_{qd} = I_q^w + jI_d^w$
- the transformation of the current vector in the rotor $q-d$ reference, $\underline{I}_{qd}$, to the stator reference $\underline{I}_{\alpha\beta}(t) = I_\alpha^w + jI_\beta^w$ (using the measured $\theta_r(t)$)
- the transformation from two-phase α, β to three-phase a, b, c

The switching is now quite fast, but the current control often shows dynamical problems. Indeed, this current control is a real-time control in the time domain that regulates both the amplitude and phase of the current (providing the switching signals for the inverter). Both the reference and the feedback signals of the current are AC quantities. This also requires the AC current measurements to be made in real time, with high bandwidth and accuracy.

29.3.4 Vector Control and Field Orientation of Synchronous Machines: Conclusions

The systems described in the previous section (Figs. 29.9 and 29.10) all comply with the basic requirements for field orientation set out in Sect. 29.3.1 (as derived from the DC machine):

- the field winding (or, for PM-motors, the pm-excitation) is the equivalent of the field winding of the DC machine
- the stator ("armature") current amplitude control corresponds to the armature current control of the DC machine
- the phase control of the stator current is the exact electronic equivalent of the mechanical commutator of the DC machine (field orientation for $\psi^w = 0$)

For the PWM-VSI implementation, the latter two items are obviously combined in one controller, but the result is the same.

29.4 Vector Control of the Induction Machine

29.4.1 Introduction

Vector control of the synchronous machine has been derived from the torque control of the independently excited DC machine. For vector control of the induction

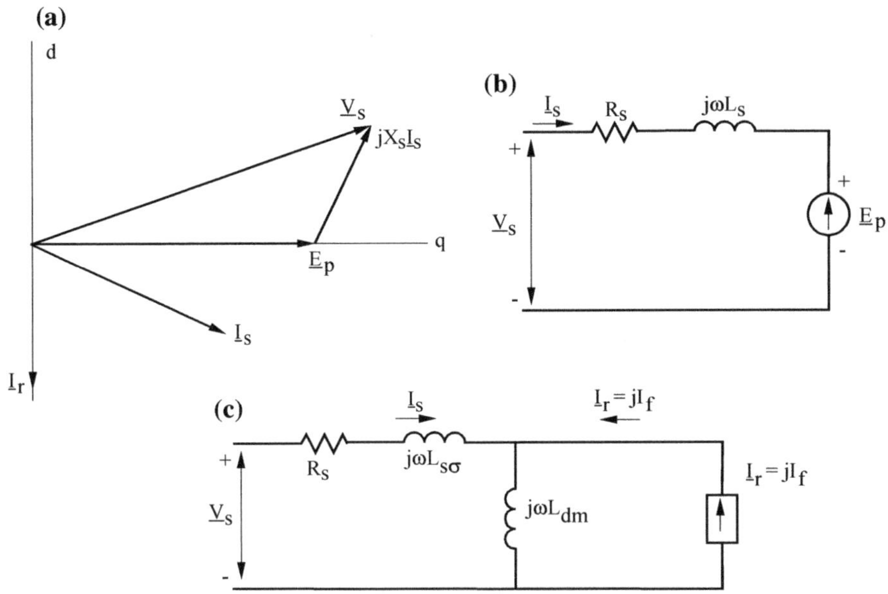

Fig. 29.11 Alternate equivalent circuit for the synchronous machine

machine, we will take the vector control of the synchronous machine as a starting point.

For the smooth-rotor synchronous machine ($L_q = L_d = L_s$), we may rewrite the steady-state equations as follows (we have omitted the subscript "o"):

$$\underline{V}_s = R_s \underline{I}_s + j\omega L_s \underline{I}_s + j\omega L_{dm} \underline{I}_r \equiv R_s \underline{I}_s + j\omega L_s \underline{I}_s + \underline{E}_p \tag{29.17}$$

$$T_e = \frac{3}{2} N_p \cdot \left(\Psi_q I_d - \Psi_d I_q \right) = \frac{3}{2} N_p \cdot \left(-L_{dm} I_f I_q \right) = \frac{3}{2} (N_p/\omega) \cdot Re(\underline{E}_p \cdot \underline{I}_s^*) \tag{29.18}$$

where $\underline{V}_s = V_q + jV_d$, $\underline{I}_s = I_q + jI_d$, $E_p = j\omega L_{dm} \underline{I}_r$ and $\underline{I}_r = jI_f$ (the DC excitation current is replaced by an equivalent complex rotor current referred to the stator).

The usual equivalent circuit with the rotor emf $\underline{E}_p$ (see (b) in Fig. 29.11) can be replaced by an equivalent circuit with a rotor current source $\underline{I}_r$ (see (c) in Fig. 29.11). The current $\underline{I}_r$ is an equivalent AC excitation current as seen from the stator. Together with the stator current $\underline{I}_s$, this equivalent rotor excitation is responsible[2] for the air-gap emf $\underline{E}_r = j\omega L_{dm}(\underline{I}_s + \underline{I}_r)$.

[2]Note that this superposition of mmfs also seems more correct than the traditional one with superposition of emfs, which is not allowed in case of saturation; the air-gap emf is also representative of the main field saturation level in a synchronous machine.

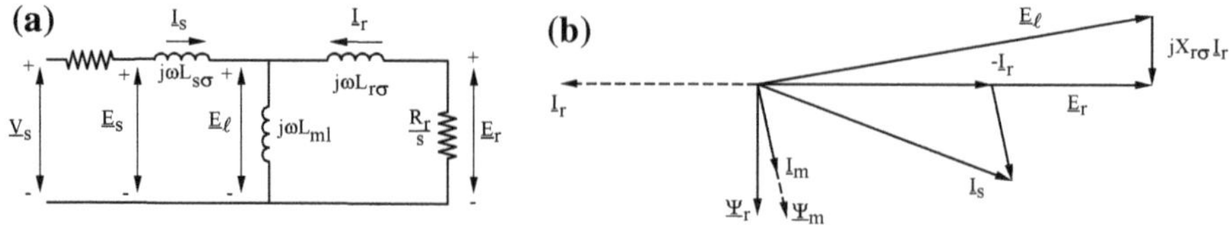

Fig. 29.12 Basic equivalent circuit of the induction machine and phasor diagram

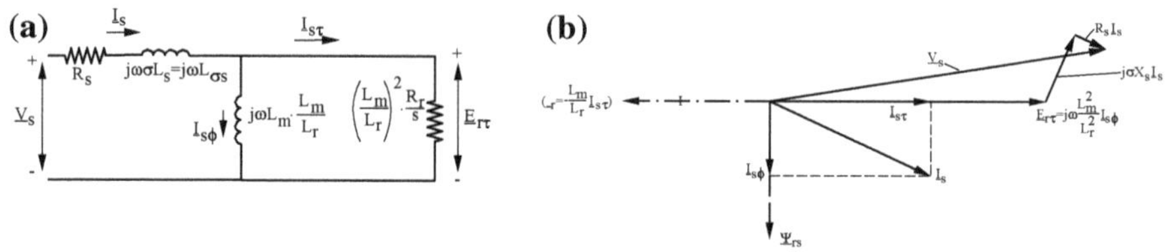

Fig. 29.13 Induction machine equivalent scheme in L

The equivalent circuit (c) clearly bears resemblances with the usual equivalent circuit of the induction machine. We may therefore be tempted to derive a torque control principle for the induction machine from the torque control of the synchronous machine.

Consider the general equivalent circuit and the phasor diagram for the induction machine in Fig. 29.12 (the positive direction of the rotor current has been reversed so as to correspond to the rotor current direction in the circuit of the synchronous machine). The magnetising current for the air-gap field is thus $\underline{I}_m = \underline{I}_s + \underline{I}_r$, with primes to indicate the turns ratio between stator and rotor omitted.

The well-known torque expression for the induction machine can be rewritten as follows:

$$T_e = \frac{3}{2}(N_p/\omega) \cdot |\underline{I}_r|^2 (R_r/s)^2 = \frac{3}{2}(N_p/\omega) \cdot Re(-\underline{E}_r \cdot \underline{I}_r^*) = \frac{3}{2}(N_p/\omega) \cdot Re(-E_r I_r) \tag{29.19}$$

Compared with Eq. 29.3 for the synchronous machine (in which field orientation can be realised based on the field excitation or emf ($\underline{E}_p$) and the *stator* current $\underline{I}_s = I_q$), here the emf is a resulting *rotor* emf and the current is a *rotor* current. If we want to obtain field orientation based on this equation, then both this emf and the rotor current should be controllable by the *stator* current.

As is well known, the general equivalent circuit in T in Fig. 29.12 can be replaced by the completely equivalent circuit in L in (a) in Fig. 29.13. Note that in this circuit,[3] the rotor resistance is in parallel with the rotor inductance.

[3]This circuit may also be derived as the solution of the vectorial equation $\underline{I}_s = \underline{I}_{s\phi} + \underline{I}_{s\tau}$ where the two components are orthogonal, with $\underline{I}_{s\tau}$ in phase with the emf $\underline{E}_r$.

It is easily shown that

$$\underline{I}_{sT} = -\frac{L_r}{L_m}\underline{I}_r \tag{29.20}$$

$$\underline{I}_{s\phi} = \underline{I}_s + \frac{L_r}{L_m}\underline{I}_r \tag{29.21}$$

$$\underline{E}_{rT} = \frac{L_m}{L_r}\underline{E}_r = j\omega\frac{L_m^2}{L_r}\underline{I}_{s\phi} = \left(\frac{R_r}{s}\right)\left(\frac{L_m}{L_r}\right)^2 \cdot \underline{I}_{sT} \tag{29.22}$$

$$\underline{V}_s = R_s\underline{I}_s + j\omega L_{\sigma s}\underline{I}_s + \underline{E}_{rT} \tag{29.23}$$

The two current components $\underline{I}_{sT}$ and $\underline{I}_{s\phi}$ are orthogonal, and their sum is equal to the stator current (see also (b) in Fig. 29.13). The emf $\underline{E}_{rT}$ is the rotor emf $\underline{E}_r = j\omega\underline{\Psi}_r = j\omega(L_m\underline{I}_s + L_r\underline{I}_r) = j\omega L_m\underline{I}_{s\phi}$ transformed to the stator with as *turns ratio* L_m/L_r ($\underline{\Psi}_r$ is indeed the flux coupled with the rotor winding).

29.4.2 Torque Control Based on $\underline{I}_{s\phi}$ and $\underline{I}_{sT}$

29.4.2.1 Steady State

The torque can be rewritten as a function of $\underline{E}_{rT}$ and $\underline{I}_{sT}$:

$$T_e = \left(\frac{3}{2}\right)\left(\frac{N_p}{\omega}\right) Re\left[\underline{E}_{rT} \cdot \underline{I}_{sT}^*\right] = \left(\frac{3}{2}\right)\left(\frac{N_p}{\omega}\right) \cdot [E_{rT} I_{sT}] \tag{29.24}$$

or also

$$T_e = \left(\frac{3}{2}\right) N_p \cdot \left(\frac{L_m^2}{L_r}\right) \cdot Re\left[j\underline{I}_{s\phi} \cdot \underline{I}_{sT}^*\right] = \left(\frac{3}{2}\right) N_p \cdot \left(\frac{L_m^2}{L_r}\right) \cdot \left[-I_{s\phi} I_{sT}\right] \tag{29.25}$$

The torque can thus be controlled by the two orthogonal components of the stator current. The resemblance with the DC machine or synchronous machine is obvious: I_{sT} is the equivalent of the armature current, $I_{s\phi}$ is the equivalent of the field or excitation current.

However, these two current components are not independent: $I_{sT} = I_s \cos\gamma$ and $I_{s\phi} = -I_s \sin\gamma$ (with $\tan\gamma = R_r/s\omega L_r$, see Fig. 29.14). Both are also connected to the slip frequency:

$$\underline{I}_{sT} = j\omega s \cdot \frac{L_r}{R_r} \cdot \underline{I}_{s\phi}$$

This (steady-state) *slip relation* shows that the ratio of both stator current components is strictly related to the slip frequency and, obviously, that the two vectors components

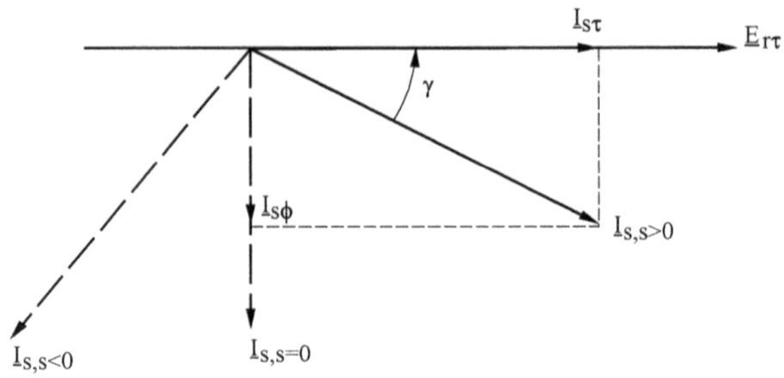

Fig. 29.14 Orientation of $\underline{E}_{r\tau}$, $I_{s\tau}$ and $I_{s\phi}$

are orthogonal. For torque control based on $I_{s\tau}$and $I_{s\phi}$, this also implies that a given torque will correspond to a certain slip frequency. In steady state, this is nothing else but the well-known steady-state torque-slip characteristic. In the next section, we will demonstrate that this also holds for transient conditions, albeit with a modification in case of flux variations.

29.4.2.2 Dynamic Analysis

For the dynamic analysis, we will revert to the usual $q-d$-description. For the time being, the speed ω of this reference frame is deliberate. The equations of the induction machine are the following (see Chap. 27):

$$V_{sq} = R_s I_{sq} + p\Psi_{sq} - \omega\Psi_{sd} \tag{29.26}$$

$$V_{sd} = R_s I_{sd} + p\Psi_{sd} + \omega\Psi_{sq} \tag{29.27}$$

$$0 = R_r I_{rq} + p\Psi_{rq} - (\omega - \omega_r)\Psi_{rd} \tag{29.28}$$

$$0 = R_r I_{rd} + p\Psi_{rd} + (\omega - \omega_r)\Psi_{rq} \tag{29.29}$$

$$T = \left(\frac{3}{2}\right) N_p \cdot \left(\frac{L_m}{L_r}\right) \cdot \left[\Psi_{rq} I_{sd} - \Psi_{rd} I_{sq}\right] \tag{29.30}$$

with p corresponding with d/dt and

$$\Psi_{sq} = L_s I_{sq} + L_m I_{rq} \tag{29.31}$$

$$\Psi_{sd} = L_s I_{sd} + L_m I_{rd} \tag{29.32}$$

$$\Psi_{rq} = L_r I_{rq} + L_m I_{sq} \tag{29.33}$$

$$\Psi_{rd} = L_r I_{rd} + L_m I_{sd} \tag{29.34}$$

In steady state, the voltages, fluxes and currents in these equations are only constants if the speed of the reference system corresponds with the instantaneous supply frequency.

In the previous section we have shown that, in steady state, the torque can be controlled with the two orthogonal components $\underline{I}_{sT}$ and $\underline{I}_{s\phi}$ of the stator current, where $\underline{I}_{s\phi}$ is in phase with the rotor flux. In other words, the stator current phasor is controlled with respect to the rotor flux phasor. However, it is important to keep in mind that this steady-state analysis took place in the time domain, in which voltages, currents and fluxes are phasors in a reference frame synchronous with the steady-state angular supply frequency.

The same steady-state analysis could also have been performed with the space-vector Eq. 27.18. If the speed ω of the reference frame is chosen equal to the steady-state supply angular frequency, then the space-vector description is completely equivalent (or even identical) to the time-phasor description.

What we have discussed in the previous section was nothing else but the normal steady-state operation, only using the equivalent scheme in L and the corresponding equations. It is clear that field orientation is a dynamic control. What we need is a stator current vector that can be controlled in magnitude and position with respect to the rotor flux vector, not only in steady state but also in transients.

To translate this into the $q-d$-representation, we may proceed as follows. We choose the speed of the reference frame a priori equal to the **instantaneous frequency of the stator currents**. The $q-d$-components of these currents are then DC values. Field orientation implies that the flux has a fixed position in this reference frame and, except in the field weakening range, also has a constant magnitude. Vector control in this reference frame implies that the flux has a controllable position.

In analogy with the discussion of field orientation and vector control of the synchronous machine, it seems logical to choose the phase angle of this reference frame so that the flux is along the (negative) d-axis.

Mathematically, the requirement for field orientation can then be translated as

$$\Psi_{rq} \equiv 0$$

In the $q-d$-reference frame, the current and flux vectors are as depicted in Fig. 29.15. This space vector diagram is analogous to the phasor diagram in Fig. 29.14, although the current and flux vectors may now be time variant. Only the d-component of the stator current determines the flux $\Psi_r = \Psi_{rd}$. For constant flux operation, I_{sd} is to be kept constant. Only I_{sq} is then varied (to control the torque). As will be demonstrated below, the torque will then follow the q-axis current without any additional transients.

An alternative way to derive the $q-d$-description of field and vector orientation is to start by choosing the speed of the $q-d$-reference frame equal to the **instantaneous speed of the rotor flux**. For field or vector orientation, the current then needs

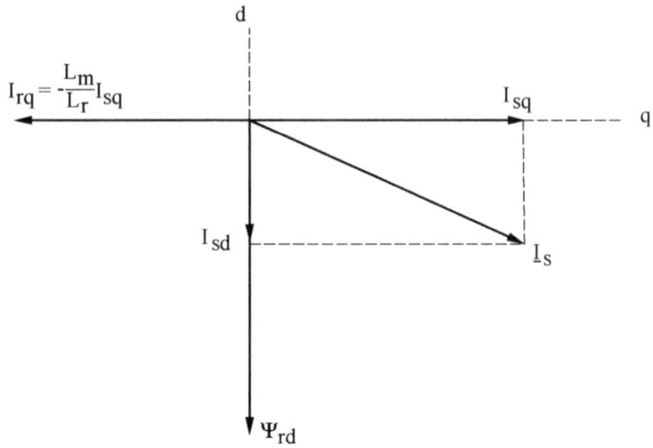

Fig. 29.15 Orientation of $d-q$-currents in a reference frame fixed to the flux

to have a controlled position in this reference frame, i.e. the frequency of the current needs to correspond to the instantaneous speed of the flux.

From the relation Eq. 29.28 $\Psi_{rq} = p\Psi_{rq} = 0$. Substituting in Eqs. 29.28, 29.29 and 29.33 yields

$$I_{sq} = -\frac{L_r}{L_m} I_{rq} \tag{29.35}$$

$$\omega - \omega_r = \omega_s = s \cdot \omega = \frac{R_r I_{rq}}{\psi_{rd}} = -\frac{R_r (L_m/L_r) I_{sq}}{\Psi_{rd}} \tag{29.36}$$

$$p\Psi_{rd} + R_r I_{rd} = 0 \tag{29.37}$$

and for the torque

$$T = \frac{3}{2} N_p \cdot \frac{L_m}{L_r} \cdot \left[-\Psi_{rd} I_{sq} \right] = \frac{3}{2} N_p \cdot \Psi_{rd} I_{rq} \tag{29.38}$$

For constant flux, we observe from Eq. 29.37 that $I_{rd} = 0$. Put differently, there is only a q-axis rotor current, which has to obey the relation with the flux and the slip frequency according to Eq. 29.36. Equation 29.38 shows that the torque will follow the q-axis stator current command without any further transients.

However, if the flux is varied (e.g. in the field weakening range) by a variation of I_{sd}, Eq. 29.29 shows, using $\Psi_{rd} = L_r I_{rd} + L_m I_{sd}$, that there will be a non-zero I_{rd} according to $(R_r + pL_r) \cdot I_{rd} = -pL_m I_{sd}$.

A variation ΔI_{sd} thus leads to a transient $I_{rd} = \Delta I_{rd}$ and a transient $\Delta\Psi_{rd}$ according to

$$\Delta\Psi_{rd} = \Delta I_{sd} \cdot \frac{R_r L_m}{R_r + pL_r} = L_m \Delta I_{sd} \cdot \frac{1}{1 + pT_{rh}} \tag{29.39}$$

with T_{rh} the (relatively large) main rotor field constant. The torque will therefore also exhibit a rather slow[4] transient. The relation between slip frequency and current also becomes time-dependent:

$$\omega_s = \frac{R_r I_{rq}}{\psi_{rd}} = -\frac{(R_r/L_r)I_{sq}}{I_{sd}/(1 + pT_{rh})} \tag{29.40}$$

where $(1 + pT_{rh})^{-1}$ acts on I_{sd}.

Field orientation of the induction machine therefore seems analogous to the field orientation of the synchronous machine. Variations of the torque current component I_{sq} cause a change of the torque proportional to this current variation; variations of the flux current component I_{sd} are accompanied with a transient determined by the field time constant (as the laws of physics prohibit any sudden changes of induction or flux).

Nevertheless, there are some differences:

1- For the synchronous machine, the considered flux is the *(fictitious) rotor flux*, i.e. the flux which *would* exist *if only the rotor excitation* were present. In reality, both rotor and stator mmfs are present and the only real flux is the flux resulting from both stator and rotor mmfs (mmfs may be superposed, not fluxes, as there is actually always saturation). For the induction machine, the considered flux is the *total flux* coupled with the rotor and this is indeed a real flux, resulting from stator and rotor mmfs.

2- Ideal current sources do not exist. In reality, the starting point is always a DC voltage source. As the inductance for the synchronous machine field orientation is the relatively large synchronous reactance, while for the induction machine it is the leakage inductance, the voltage requirements for the inverter are much higher for synchronous machine field orientation.

29.4.3 Implementation of Field Orientation for the Induction Machine

As to the power electronics used, induction machine field orientation schemes may revert to both CSI and PWM-VSI, just as for the synchronous machine. Figure 29.16 illustrates the PWM-VSI and CSI implementations. In the CSI-implementation, the amplitude and phase of the current vector are controlled separately. The PWM-VSI requires a current regulator and a transformation from the $q - d$-reference frame to the stator reference frame.

The basic problem for induction machine field orientation, however, is how the flux angle can be derived from electrical and/or mechanical motor signals. To determine the required position θ_r there are two methods. In the direct method, the flux angle

[4] It is slow, because it is bound by physical laws prohibiting any sudden flux changes.

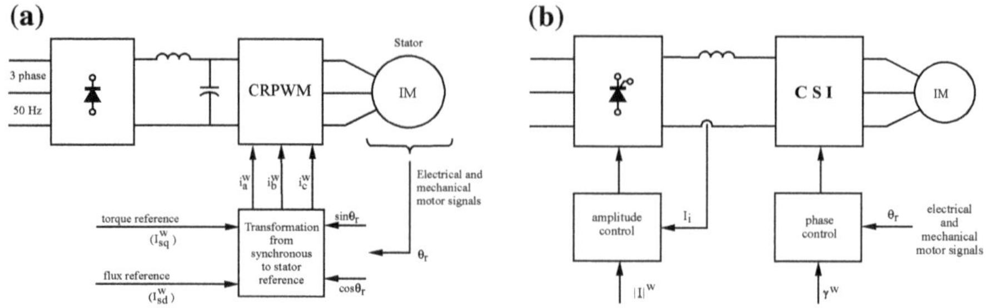

Fig. 29.16 Basic implementations of field orientation using VSI-PWM and CSI

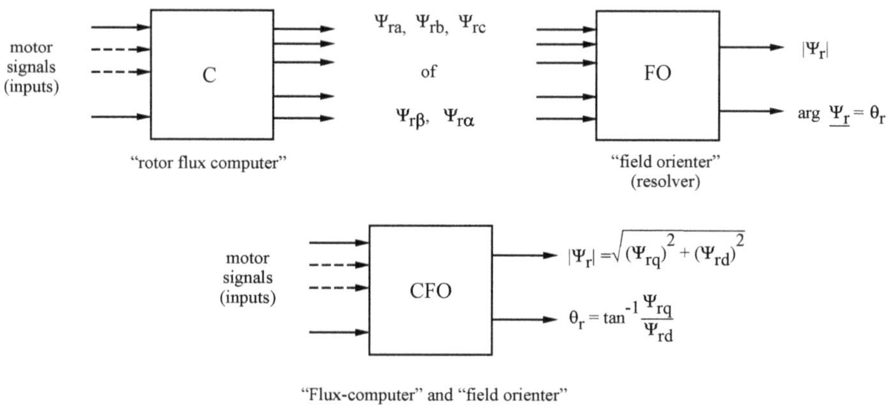

Fig. 29.17 Rotor flux computer and field orienter

is derived from electrical measurements. In the indirect method, the slip relation is used to calculate the angle.

29.4.3.1 Direct Field Orientation

The direct method uses electrical signals to calculate the position of the rotor flux. The principle is shown in Fig. 29.17. In a first step, the *flux computer*, the rotor flux components (Ψ_{ra}, Ψ_{rb}, Ψ_{rc} or $\Psi_{r\alpha}$, $\Psi_{r\beta}$) are calculated. In a second step, a resolver (the *field orienter*) calculates the amplitude and phase angle of the flux vector.

To compute the flux components, the most direct method measures the air-gap flux by means of Hall sensors or flux coils in the air gap (see (a) in Fig. 29.18). Then, to obtain the rotor flux, only a correction for the rotor leakage is required. For this, only two parameters are needed, i.e. $L_{r\sigma}$ and L_r/L_m, which do not vary considerably with saturation.[5] Disadvantages of this method include the requirement to modify the

[5]Except for machines with closed rotor slots where the rotor leakage flux may vary considerably.

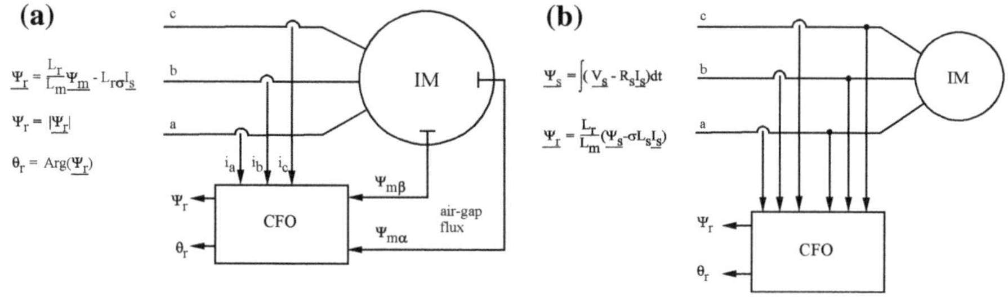

Fig. 29.18 Flux determination

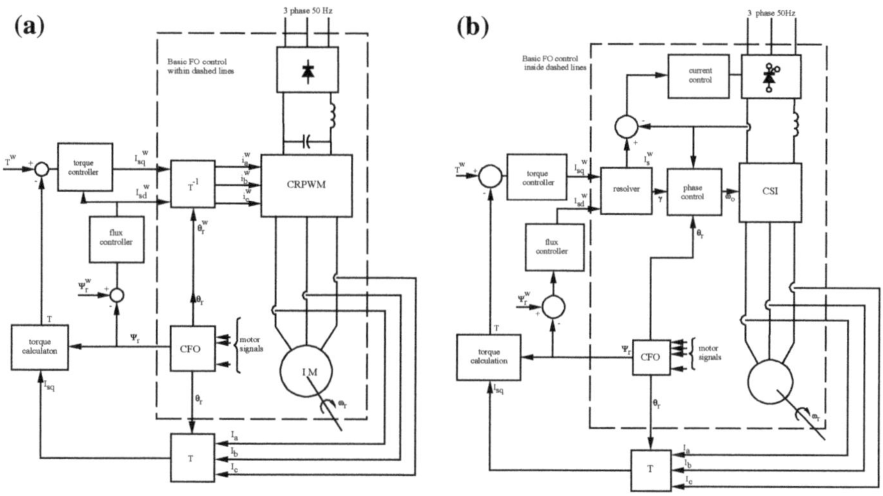

Fig. 29.19 Implementation of direct FO

machine and the poor accuracy of the measurement at low frequencies, particularly for flux coils.

Another, less invasive method, utilises an estimate of the stator flux from voltage and current measurements, as illustrated in (b) in Fig. 29.18. However, to obtain the stator flux a correction using the (temperature-dependent) stator resistance is required. The rotor flux is then obtained with a correction using the total leakage. Here as well, low frequencies pose a problem in terms of accuracy: the integration of low-frequency signals, where the resistive voltage drop has a relatively large impact.

Figure 29.19a illustrates a possible control scheme using a PWM-VSI with current control. The actual field orientation control is inside the dashed lines. The block T^{-1} represents the transformation from the synchronous reference frame to the stator reference frame (using the rotor position signal from the CFO). In the figure, this basic scheme is completed with flux and torque control loops. The flux control loop uses the flux amplitude information from the CFO, while the torque control loop also

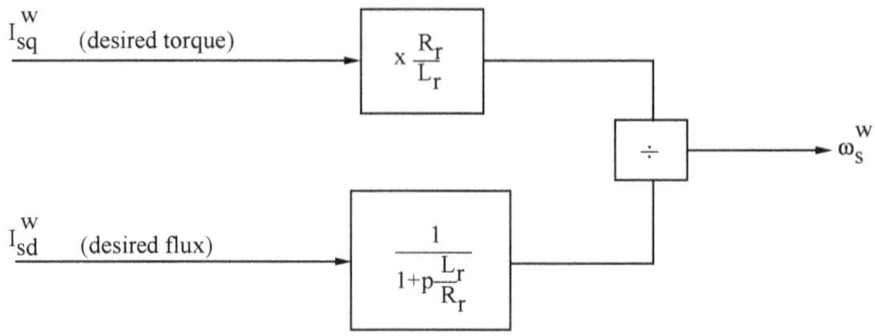

Fig. 29.20 Slip frequency calculator

utilises the q-axis stator current calculated from the measured stator currents and the position information from the CFO.

An implementation of direct field orientation using a CSI is illustrated in (b) in Fig. 29.19. The CSI requires the amplitude and phase signals of the desired current, which are derived in the resolver block. By means of the required phase of the current with respect to the flux and the flux position from the CFO, the switching signals for the inverter are calculated (with a correction for the commutation lag, which depends on the current amplitude). The current amplitude derived from the resolver is used for the rectifier control. The optional flux and torque control loops are similar to those for the PWM-VSI.

29.4.3.2 Indirect Field Orientation

In the indirect methods for field orientation, synchronisation is obtained by means of the slip Eq. 29.40, which is indeed a necessary and sufficient condition for field orientation. From the desired q- and d-axis current components, the required slip frequency is calculated with a *slip calculator* (see Fig. 29.20).

This slip frequency is then combined with the measured rotor speed or angle to obtain the required stator frequency and phase. A disadvantage of the indirect method is that the machine parameters used in the slip calculator are estimated values. The rotor resistance may vary with temperature, while the rotor inductance is dependent on saturation. The danger of incorrect orientation is especially true with varying flux component I_{sd}.

Figure 29.21 shows a scheme using a PWM-VSI. The required phase angle θ_r of the stator current with respect to the rotor flux is calculated from the slip frequency ω_s^w using the slip calculator (after integration to obtain the angle θ_s^w), together with the measured rotor angle[6] θ_{ri}. Usually, digital integration from slip frequency to angle is used to avoid drift and to obtain a higher accuracy.

For the implementation of indirect field orientation with a CSI, Fig. 29.22 shows a possible scheme. On the one hand, the current commands are used to calculate the

[6] Keep in mind that this is not the rotor flux angle but the mechanical rotor angle.

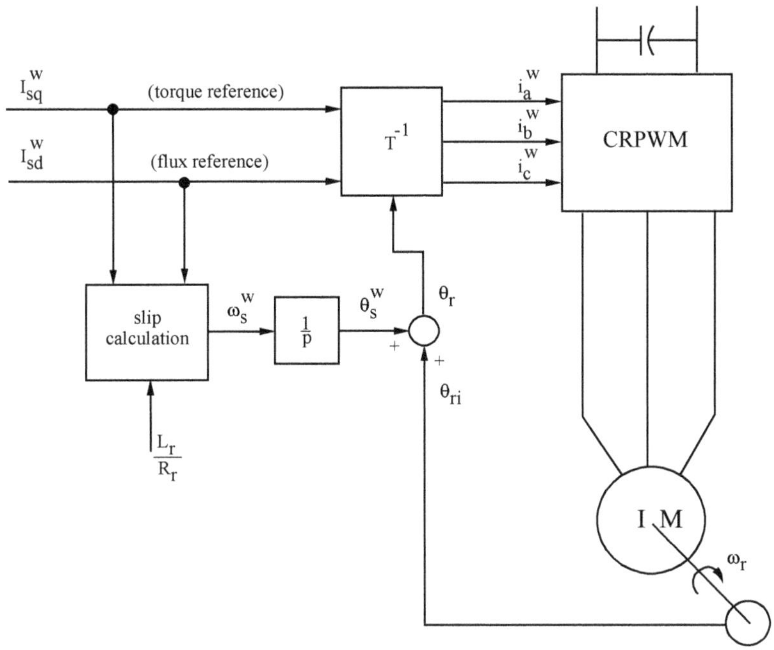

Fig. 29.21 Indirect FO using a CRPWM

required slip frequency ω_s^w and on the other hand, they are transformed by a resolver into amplitude and phase values. The amplitude signal serves as the input signal for the current control of the rectifier, while the phase signal γ is combined with the calculated slip frequency and the measured rotor angular speed ω_r.

However, to take into account instantaneous variations of the commanded currents, a correction $\Delta\gamma$ has to be applied. Figure 29.23 illustrates this necessity. Suppose the torque signal I_{sq} changes from I_{sq1} to I_{sq2} (with I_{sd} unchanged). The phase of $\underline{I}_s$ changes from γ_1 to γ_2, the amplitude from I_{s1} to I_{s2} and the slip frequency from ω_{s1} to ω_{s2}. The new amplitude and slip value will be calculated correctly, but without phase correction the CSI will not take into account the phase change $\Delta\gamma$. The inverter gets the new frequency $\omega_{s2} + \omega_r$, but with the phase γ_1 instead of the frequency $\omega_{s2} + \omega_r$ with the phase γ_2. Without compensation,[7] the new current command would thus correspond to $\underline{I}_s^e$ instead of $\underline{I}_{s2}$. Therefore, without such a compensation the drive would temporarily lose field orientation with a sluggish torque response.

Remarks:

1. In the PWM-VSI implementation, the phase change is correctly taken into account by the integration of the frequency signal, which is subsequently used for the transformation from synchronous to standstill reference.

---

[7]The compensation can also be regarded as an integration constant: the switching angle of the inverter has to change from $\gamma_1 + (\omega_{s1} + \omega_r)t$ to $\gamma_2 + (\omega_{s2} + \omega_r)t$ and not to $\gamma_1 + (\omega_{s2} + \omega_r)t$.

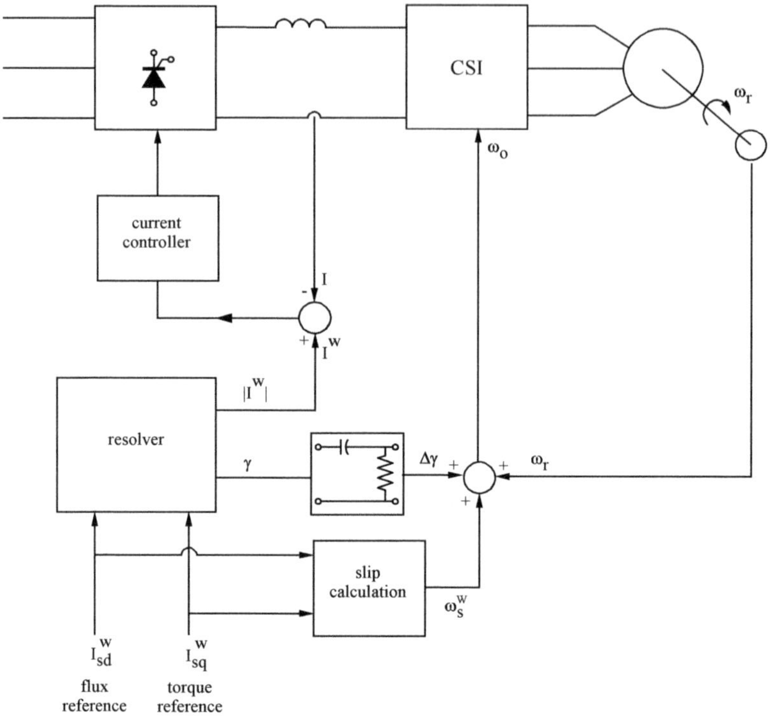

Fig. 29.22 Indirect FO using a CSI

Fig. 29.23 Phase correction for indirect FO with CSI

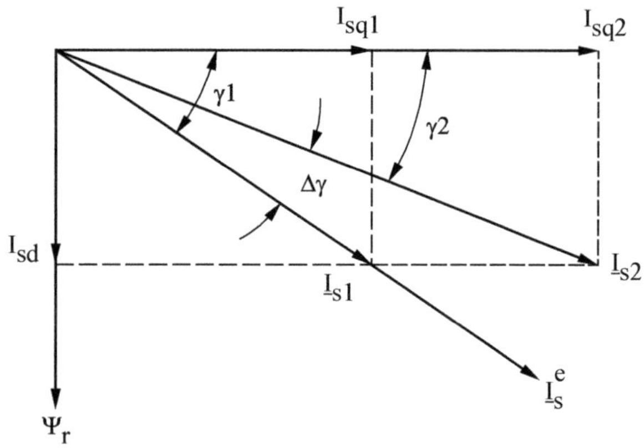

2. In the systems of Figs. 29.21 and 29.22, the desired value of the slip frequency is calculated from the desired values of the q- and d-axis stator currents. For a change of the flux or d-axis component, the calculated slip frequency also remains correct if the value of the rotor time constant L_r/R_r is correct. Yet, the flux will change

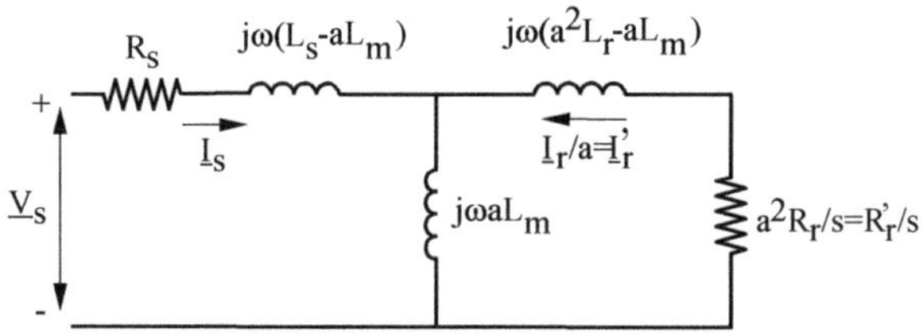

Fig. 29.24 General equivalent scheme

with this same large time constant when the desired value for the stator currents (Fig. 29.21) or the value of the DC current (Fig. 29.22) is calculated from the (new) desired value of I_{sd}. Such implementations are called *uncompensated.*

There are *compensated* systems where the slip frequency is calculated using the value of the flux $\Psi_r = \Psi_{rd}$ (Eq. 29.36). The required d-axis current is calculated using Eq. 29.39 and takes into account the time delay. If the inverter is able to deliver a quite large stator current, the transients will then be limited or short (for flux changes that are not too large, however).

29.4.4 Other Field Orientation Techniques for Induction Machines

The systems described above start from the flux coupled with the rotor (cf. the equivalent scheme with all leakage referred to the stator). This seems the most logical choice as the corresponding two stator current components are completely decoupled. The disadvantage is that this rotor flux is not directly measurable.

However, using the rotor flux for obtaining field orientation is not strictly required. Indeed, the torque can be written in different[8] ways:

$$T = \frac{3}{2}N_p \cdot Re\left[j\underline{\Psi}_r \cdot \underline{I}_r^*\right] = \frac{3}{2}N_p \cdot \frac{L_m}{L_r} \cdot Re\left[j\underline{\Psi}_r \cdot \underline{I}_s^*\right] = \frac{3}{2}N_p \cdot Re\left[j\underline{\Psi}_s \cdot \underline{I}_s^*\right] = \frac{3}{2}N_p \cdot Re\left[j\underline{\Psi}_m \cdot \underline{I}_s^*\right]$$

The last three expressions also correspond to the general equivalent scheme in Fig. 29.24 for different values of the *turns ratio* a ($a = L_m/L_r$, $a = L_s/L_m$ and $a = 1$, respectively).

It is important to note that controlling the flux and current does not necessarily cause field orientation. Field orientation for the induction machine requires the two components of the stator current vector to be able to independently control the flux and torque. This is the case for the torque expression with the rotor flux and the stator current, as explained in the previous sections, but not for the other fluxes.

For example, let us consider the last expression with the air-gap flux. The reference frame is chosen fixed to the flux (with the flux along the d-axis). From the two rotor

[8]Prove this using the flux definitions $\underline{\Psi}_s = L_s\underline{I}_s + L_m\underline{I}_r$, $\underline{\Psi}_r = L_m\underline{I}_s + L_r\underline{I}_r$, $\underline{\Psi}_m = L_m(\underline{I}_s + \underline{I}_r)$ (where the turns ratio has been supposed as equal to 1).

equations, the rotor current components are expressed in terms of the stator current components. This results in two equations,[9] one algebraic and the other a differential equation for the flux:

$$\omega_s = -\frac{(R_r + pL_{r\sigma})I_{sq}}{\frac{L_r}{L_m}\Psi_{md} - L_{r\sigma}I_{sd}}$$

$$p\Psi_{md} = -\frac{R_r}{L_r}\Psi_{md} + \frac{L_m}{L_r}(R_r + pL_{r\sigma})\,I_{sd} + \omega_s\frac{L_{r\sigma}L_m}{L_r}I_{sq}$$

Compared to the equations for rotor flux orientation, both equations are now inherently coupled. For a constant d-axis component of the stator current, the flux will change with a variation of the q-axis current. Therefore, the control will need to calculate the required I_{sd} for a variation of I_{sq} in order to maintain a constant (or desired) flux. Similarly, for the calculation of the slip frequency both I_{sq} and I_{sd} as well as Ψ_{md} come into play. Nevertheless, if the decoupling control is well implemented, a qualitatively equivalent field orientation can be obtained.

Remarks:

1. Other variants involve the implementation of the control algorithms. For example, for the PWM-VSI, the current control is sometimes integrated with the field orientation algorithm and the output of the controller is then a voltage signal for the inverter.
2. There is clearly some relation between field orientation based on the air-gap flux and the standard V/f-control of induction machines. However, in the V/f-control, the orthogonality between the flux and torque current component is not maintained during transients, creating unwanted transient phenomena.
3. The actual magnetising flux in a machine is the combined result of the mmfs of both stator and rotor currents. In this way, the field orientation algorithms for the induction machine are more realistic than the one for the synchronous machine where the (fictitious) *rotor flux* due to only the rotor mmf is considered. For the torque, this is not relevant as long as the d-axis component of the flux is constant (or not affected). For main field saturation, the cross-saturation effect will cause the flux to change for a variation of I_{sq}, just like the armature reaction in an uncompensated DC machine (a compensation winding in the rotor q-axis of a synchronous machine is obviously possible but never or rarely used). The outcome will be an unwanted torque transient.

Although the field orientation algorithms for induction machines are based on the total flux, saturation effects still come into play by the saturation-dependent inductances, required for the calculation of the slip frequency (cf. the indirect method) or the flux calculation or correction (cf. the direct methods).

[9]Prove this, introducing the leakages with $L_s = L_{s\sigma} + L_m$, $L_r = L_{r\sigma} + L_m$.

Chapter 30
Transient Phenomena in Electrical Machines

Abstract While the previous chapters analysed the local stability of electrical machines, here we study a typical case of large transients, i.e. the sudden short-circuit of a synchronous generator. The model used is a simplified constant saturation machine model.

30.1 Introduction

While Chaps. 26, 27 and 28 discussed the stability in the small of electrical machines (i.e. for small deviations around a steady state), the present chapter[1] will focus on large transients. A typical example is the sudden short circuit of a synchronous machine.

In this respect, it is important to point out that an electrical machine is basically a **non-linear** system. Only under simplifying assumptions such as a constant saturation level and a constant speed can a synchronous machine be more or less considered to be linear and allowing analytical solutions.

Nevertheless, even at constant speed, the synchronous machine is essentially a **non-stationary** system. A transformation to a stationary representation is only possible under sinusoidal supply and rather restrictive assumptions (e.g. as to symmetry of the supply). Consider, for example, a two-phase short circuit of phases b and c:

$$I_a = 0;\ V_b = V_c;\ I_b = -I_c \tag{30.1}$$

In the 0, α, β representation, this is equivalent to

$$I_o = 0;\ V_\beta = 0;\ I_\alpha = 0 \tag{30.2}$$

However, in the q, d representation and at constant speed $\Omega_m = \omega/N_p$, we obtain

$$I_q \cos\omega t = I_d \sin\omega t \tag{30.3}$$

[1]This chapter is largely based on [40].

J. A. Melkebeek, *Electrical Machines and Drives*, Power Systems,
https://doi.org/10.1007/978-3-319-72730-1_30

$$V_q \sin \omega t + V_d \cos \omega t = 0 \tag{30.4}$$

i.e. time-dependent relations between the transformed variables.

30.2 Transients in Synchronous Machines at Constant Speed

At constant speed (and disregarding saturation), the synchronous machine model can be considered to be linear. However, it is basically time-dependent (see Chap. 28). The linearity allows us to apply the superposition principle and the theorem of Thévenin for the analysis of transients.

If we limit ourselves to switching phenomena or faults at the AC side, we may study the resulting transients when a sinusoidal but otherwise quite general stator voltage system of the following form is applied at $t = 0$:

$$V_a(t) = V_a \cdot \cos(\omega t + \varphi_a) \cdot \mu(t)$$

$$V_b(t) = V_b \cdot \cos(\omega t + \varphi_b) \cdot \mu(t) \tag{30.5}$$

$$V_c(t) = V_c \cdot \cos(\omega t + \varphi_c) \cdot \mu(t)$$

During the transient, the speed is assumed to be constant and equal to the synchronous speed $\Omega_m = \omega / N_p$ or $\omega_m = \omega$ (i.e. we suppose that the inertia is very large). The excitation voltage V_f is assumed to be constant as well.

The stator voltages in Eq. 30.5 can be split up into a direct, an inverse and a zero-sequence voltage that are applied to the stator at $t = 0$. We will subsequently analyse the transients resulting from the application at $t = 0$ of

1. a positive-sequence voltage with $V_a = V_b = V_c = V$ and $\varphi_b = \varphi_a - 2\pi/3$, $\varphi_c = \varphi_a - 4\pi/3$
2. a negative-sequence voltage with $V_a = V_b = V_c = V$ and $\varphi_b = \varphi_a + 2\pi/3$, $\varphi_c = \varphi_a + 4\pi/3$
3. a zero-sequence voltage with $V_a = V_b = V_c = V$ and $\varphi_a = \varphi_b = \varphi_c$

30.2.1 Direct Transients

In this section, we will analyse the transient when the following positive-sequence voltage is connected at the stator terminals at instant $t = 0$:

$$V_a(t) = V \cdot \cos(\omega t + \delta_1) \cdot \mu(t)$$

$$V_b(t) = V \cdot \cos(\omega t + \delta_1 - 2\pi/3) \cdot \mu(t) \tag{30.6}$$

$$V_c(t) = V \cdot \cos(\omega t + \delta_1 - 4\pi/3) \cdot \mu(t)$$

We introduced the angle δ_1 so as to be able to vary the switching instant ($t = 0$) with regard to the sine period. For the sake of generality, we define the angle between the stator reference axis (or a- axis) and the q- axis (or emf axis) as

$$\theta = \omega t + \delta_2 \tag{30.7}$$

(δ_2 is thus the rotor angle at the switching instant). The load angle is therefore (URS) $\delta = \delta_1 - \delta_2$.

The transformed voltages are

$$V_o(t) = 0$$

$$V_\alpha(t) = V \cdot \cos(\omega t + \delta_1) \cdot \mu(t)$$

$$V_\beta(t) = V \cdot \sin(\omega t + \delta_1) \cdot \mu(t) \tag{30.8}$$

and

$$V_q(t) = V \cdot \cos\delta \cdot \mu(t)$$

$$V_d(t) = V \cdot \sin\delta \cdot \mu(t) \tag{30.9}$$

Transformation to the Laplace domain (with $p = d/dt$, thus in absolute time) yields

$$V_q(p) = \frac{1}{p} \cdot V \cdot \cos\delta$$

$$V_d(p) = \frac{1}{p} \cdot V \cdot \sin\delta \tag{30.10}$$

The relation between the voltages and the currents follows from Chap. 28:

$$\begin{bmatrix} V_q(p) \\ V_d(p) \\ 0 \\ 0 \\ 0 \end{bmatrix} = \begin{bmatrix} R_s + pL_q & -\omega L_d & pM_{skq} & -\omega M_{skd} & -\omega M_{sf} \\ \omega L_q & R_s + pL_d & \omega M_{skq} & pM_{skd} & pM_{sf} \\ \frac{3}{2}pM_{skq} & 0 & R_{kq} + pL_{kq} & 0 & 0 \\ 0 & \frac{3}{2}pM_{skd} & 0 & R_{kd} + pL_{kd} & pM_{fkd} \\ 0 & \frac{3}{2}pM_{sf} & 0 & pM_{fkd} & R_f + pL_f \end{bmatrix} \cdot \begin{bmatrix} I_q(p) \\ I_d(p) \\ I_{kq}(p) \\ I_{kd}(p) \\ I_f(p) \end{bmatrix} \tag{30.11}$$

To calculate the currents, the same method will be applied as in Chap. 28: from the last three equations, the rotor currents are expressed as a function of the stator currents and the stator equations are subsequently written in terms of the operational inductances.

$$V_q(p) = R_s I_q(p) + p\Psi_q(p) - \omega\Psi_d(p) \tag{30.12}$$

$$V_d(p) = R_s I_d(p) + p\Psi_d(p) + \omega\Psi_q(p) \tag{30.13}$$

with

$$\Psi_q(p) = L_q(p) \cdot I_q(p) \tag{30.14}$$

$$\Psi_d(p) = L_d(p) \cdot I_d(p) \tag{30.15}$$

For the stator currents, we find that

$$I_q(p) = D(p)^{-1} \cdot \left[(R_s + pL_d(p)) \cdot V_q(p) + \omega L_d(p) \cdot V_d(p)\right] \tag{30.16}$$

$$I_d(p) = D(p)^{-1} \cdot \left[\left(R_s + pL_q(p)\right) \cdot V_d(p) - \omega L_q(p) \cdot V_q(p)\right] \tag{30.17}$$

with

$$D(p) = (p^2 + \omega^2) \cdot L_q(p) \cdot L_d(p) + R_s[L_q(p) + L_d(p)] + R_s^2 \tag{30.18}$$

The expressions for the operational inductances and the zeros of $D(p)$ have been discussed in Chap. 28. This may be summarised as follows:

$$L_q(p) = L_q \cdot \frac{pT_{kq} + 1}{p\left(T_{kq}/\sigma_q\right) + 1} = L_q \cdot \frac{pT_q^{"} + 1}{pT_{qo}^{"} + 1} \tag{30.19}$$

$$L_d(p) = L_d \cdot \frac{p^2 T_f T_{kd} \cdot \frac{\sigma_{kdf}\sigma_d}{\sigma_{dkd}\sigma_{df}} + p\left(T_{kd} + T_f\right) + 1}{p^2 T_f T_{kd} \cdot \frac{\sigma_{kdf}}{\sigma_{dkd}\sigma_{df}} + p\left(\frac{T_{kd}}{\sigma_{dkd}} + \frac{T_f}{\sigma_{df}}\right) + 1} \approx L_d \cdot \frac{\left(pT_d^{'} + 1\right)\left(pT_d^{"} + 1\right)}{\left(pT_{do}^{'} + 1\right)\left(pT_{do}^{"} + 1\right)} \tag{30.20}$$

Another frequently used expression for the operational inductances is the following[2]:

---

[2]Note that the following notation for $L_d(p)$ as well as the approximation above in terms of the transient and subtransient time constants are only valid if the field time constant is at least 10 times larger than the $d-$axis damper time constant.

$$\frac{1}{L_q(p)} = \frac{1}{L_q} + \left(\frac{1}{L_q^{''}} - \frac{1}{L_q}\right)\frac{pT_q^{''}}{pT_q^{''}+1} \tag{30.21}$$

$$\frac{1}{L_d(p)} = \frac{1}{L_d} + \left(\frac{1}{L_d^{'}} - \frac{1}{L_d}\right)\frac{pT_d^{'}}{pT_d^{'}+1} + \left(\frac{1}{L_d^{''}} - \frac{1}{L_d^{'}}\right)\frac{pT_d^{''}}{pT_d^{''}+1} \tag{30.22}$$

where $L_q^{''} = \sigma_q L_q$; $L_d^{'} = \sigma_{df} L_d$; $L_d^{''} = \sigma_d L_d$. These expressions illustrate quite clearly the behaviour of the operational inductances in the limit cases $t \to 0$ (or $p \to \infty$); $t \to \infty$(or $p \to 0$), or an intermediate value for $L_d(p)$.

As discussed in Chap. 28, for small or zero stator resistance, the zeros of $D(p)$ comprise: the stator poles $p_{1,2} \approx \pm j\omega$; the two $d-$axis rotor poles $p_3 = -T_d^{'-1}$, $p_4 = -T_d^{''-1}$ and the $q-$axis rotor pole $p_5 = -T_q^{''-1}$. For somewhat larger stator resistances, mainly the stator poles change to $p_{1,2} \approx -\frac{1}{2}\left(T_d^{-1} + T_q^{-1}\right) \pm j\omega$, with T_d and T_q the stator time constants for $d-$ and $q-$ axes.

For our purposes (i.e. large machines) we may therefore approximate $D(p)$ as follows:

$$\begin{aligned} D(p) &\approx L_d(p)\cdot L_q(p)\cdot\left[\left(\frac{1}{2}\left(T_d^{-1}+T_q^{-1}\right)+p\right)^2+\omega^2\right] \\ &= L_d(p)\cdot L_q(p)\cdot\left[\left(T_a^{-1}+p\right)^2+\omega^2\right] \end{aligned} \tag{30.23}$$

To find the time expressions for the currents $I_d(t)$ and $I_q(t)$, we have to calculate the residues of $I_d(p)$ and $I_q(p)$ in the zeros of $D(p)$. These time functions will consist of a damped AC term corresponding with the poles $p_{1,2}$ and DC terms corresponding with the poles p_3, p_4 and p_5. To somewhat simplify the solution, we may ignore the resistances in the numerator of Eqs. 30.16 and 30.17. This is essentially the same as disregarding the residue of $I_d(p)$ in the pole p_5 and the residues of $I_q(p)$ in the poles p_3 and p_4, and it leads to

$$I_q(p) = \frac{p\cdot V_q(p)+\omega\cdot V_d(p)}{L_q(p)\cdot\left[\left(T_a^{-1}+p\right)^2+\omega^2\right]} = V\cdot\frac{p\cdot\cos\delta+\omega\cdot\sin\delta}{pL_q(p)\cdot\left[\left(T_a^{-1}+p\right)^2+\omega^2\right]} \tag{30.24}$$

$$I_d(p) = \frac{p\cdot V_d(p)-\omega\cdot V_q(p)}{L_d(p)\cdot\left[\left(T_a^{-1}+p\right)^2+\omega^2\right]} = V\cdot\frac{p\cdot\sin\delta-\omega\cdot\cos\delta}{pL_d(p)\cdot\left[\left(T_a^{-1}+p\right)^2+\omega^2\right]} \tag{30.25}$$

To obtain the residues, it is advantageous to make use of the expressions (30.21) and (30.22) for the operational inductances. It immediately becomes clear that the time function $I_q(t)$ consists of a DC term decreasing with the time constant $T_q^{''}$ and an AC term decreasing with the armature time constant T_a. The time function $I_d(t)$, on the other hand, has two DC terms decreasing with the time constants $T_d^{'}$ and $T_d^{''}$ and an AC term decreasing with the armature time constant T_a.

The initial behaviour (and value) of $I_q(t)$ and $I_d(t)$ can be derived by considering large values of $|p|$ when $L_q(p) \approx L_q^{''}$ and $L_d \approx L_d^{''}$:

$$V \cdot \frac{p \cdot \cos\delta + \omega \cdot \sin\delta}{pL_q^{''} \cdot \left[p^2 + \omega^2\right]} = \frac{V}{X_q^{''}} \cdot \left[\frac{\omega \cdot \cos\delta - p \cdot \sin\delta}{p^2 + \omega^2} + \frac{\sin\delta}{p}\right] \tag{30.26}$$

$$V \cdot \frac{p \cdot \sin\delta - \omega \cdot \cos\delta}{pL_d^{''} \cdot \left[p^2 + \omega^2\right]} = \frac{V}{X_d^{''}} \cdot \left[\frac{\omega \cdot \sin\delta + p \cdot \cos\delta}{p^2 + \omega^2} - \frac{\cos\delta}{p}\right] \tag{30.27}$$

Finally, we obtain

$$I_q(t) = \frac{V}{X_q^{''}} \cdot \sin(\omega t - \delta) \cdot \exp(-T/T_a) + V\sin\delta \cdot \left[\left(\frac{1}{X_q^{''}} - \frac{1}{X_q}\right)\exp(-t/T_q^{''}) + \frac{1}{X_q}\right] \tag{30.28}$$

$$\begin{aligned} I_d(t) = {} & \frac{V}{X_d^{''}} \cdot \cos(\omega t - \delta) \cdot \exp(-T/T_a) \\ & - V\cos\delta \cdot \left[\left(\frac{1}{X_d^{''}} - \frac{1}{X_d^{'}}\right)\exp(-t/T_d^{''}) + \left(\frac{1}{X_d^{'}} - \frac{1}{X_d}\right)\exp(-t/T_d^{'}) + \frac{1}{X_d}\right] \end{aligned} \tag{30.29}$$

Apparently, the damping of the AC terms in $I_q(t)$ and $I_d(t)$ can be attributed to the stator resistance, while the damping of the DC terms results from the rotor time constants. In the $d-$axis, there is first a fast decline with the subtransient time constant of the $d-$axis, with a subsequently slower decrease with the $d-$axis transient time constant. In the $q-$axis, there is only the decrease with the subtransient time constant of the $q-$axis (as we assumed that there would be only one damper in the q- axis). The time span corresponding with the subtransient time constants is called the subtransient interval, and the one corresponding with the transient time constant is the transient interval.

However, the currents $I_q(t)$ and $I_d(t)$ are in a synchronous reference system rotating with the speed ω. The transformation to a standstill reference frame for the AC component in $d-$ and $q-$axes yields for the corresponding current component in phase a:

$$-\frac{V}{2} \cdot \exp(-T/T_a) \cdot \left[\left(\frac{1}{X_d^{''}} + \frac{1}{X_q^{''}}\right) \cdot \sin\delta_1 + \left(\frac{1}{X_d^{''}} - \frac{1}{X_q^{''}}\right) \cdot \sin(2\omega t - \delta + \delta_2)\right] \tag{30.30}$$

The result for phases b and c is analogous (replacing $\delta,\ \delta_1 \rightarrow \delta - 2\pi/3,\ \delta_1 - 2\pi/3$ or $\delta - 4\pi/3,\ \delta_1 - 4\pi/3$, respectively). The phase currents contain a decaying DC component and a small second harmonic. The magnitude of the DC component in a given phase depends on the switching instant. The decaying second harmonic is very small: it is proportional to the difference between the subtransient reactances in $q-$ and $d-$axes.

The decaying DC components in $I_q(t)$ and $I_d(t)$ correspond to decaying AC phase currents with mains frequency. We will refer to I_{qo} and I_{do} as these DC components in $I_q(t)$ and $I_d(t)$. The corresponding AC phase currents components are then

$$I_{qo}\cos(\omega t + \delta_2) - I_{do}\sin(\omega t + \delta_2)$$

$$I_{qo}\cos(\omega t + \delta_2 - 2\pi/3) - I_{do}\sin(\omega t + \delta_2 - 2\pi/3) \tag{30.31}$$

$$I_{qo}\cos(\omega t + \delta_2 - 4\pi/3) - I_{do}\sin(\omega t + \delta_2 - 4\pi/3)$$

The amplitude of these fundamental harmonics is $\left(I_{qo}{}^2 + I_{qo}{}^2\right)^{1/2}$ and decreases exponentially with time. At the start, the decay is very fast, with the subtransient time constants of $q-$ and $d-$axes. After that, the decrease settles down to a much slower pace with the transient time constant of the $d-$axis (the time constant of the excitation winding) - and a transient time constant of the $q-$axis if the machine and its model provide it - until finally the steady-state current is attained. This steady-state value is determined by the reactances X_q and X_d.

To summarise, in the stator currents three components are present: a decaying second harmonic, a damped fundamental harmonic and a decaying DC component. The second harmonic component is typically quite small and decreases with the armature time constant, similar to the DC component. The value of the DC component in a given phase is strongly dependent on the starting instant of the transient. It can be zero in a given phase, but then it will be large in the other phases. The armature time constant is

$$T_a = 2\left(T_d^{-1} + T_q^{-1}\right)^{-1} \tag{30.32}$$

where $T_d = L_d^{"}/R_s$, $T_q = L_q^{"}/R_s$ for machines with damper windings and $T_d = L_d^{'}/R_s$, $T_q = L_q/R_s$ for machines without damper windings.

The damping of the fundamental harmonic is determined by the rotor, i.e. by the subtransient and transient time constants. The initial amplitude of this fundamental harmonic is mainly determined by the subtransient reactances. After the subtransient interval, the remaining amplitude is determined by the transient reactances. The final value of these fundamental harmonic currents is determined by the steady-state rotor reactances.

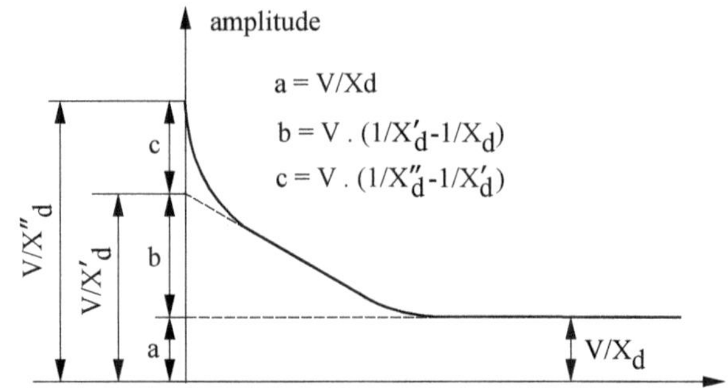

Fig. 30.1 Fundamental harmonic amplitude evolution

Figure 30.1 shows the evolution of the amplitude of the fundamental AC component for the case of $\sin\delta = 0$ (so that only the d-axis determines the value of this current).

As a further illustration, Fig. 30.2 depicts a the currents in phases a, b, c for a symmetrical three-phase short-circuit condition starting from no load. In the figure, $\delta_1 = \pi/2$ is supposed so that no DC current is present in phase a (but in phases b and c two opposite DC components are present, however). Moreover, for this short-circuit condition $\delta = 0$ is supposed so that the subtransient and transient components result from the $d-$axis only. This is usually a good approximation for short circuits in high-voltage AC grids which are mostly inductive.

The currents in the rotor have similar shape as the currents $I_q(t)$ and $I_d(t)$ calculated above. Figure 30.3 shows the currents in the excitation winding and the two damper windings for the same short-circuit phenomenon.

Remarks:

1. For the analysis of a three-phase short circuit as in the example above, the transient voltage to be connected to the stator terminals is a function of both the excitation and the steady-state currents before the short circuit. Consider for example a smooth-rotor machine with a steady-state current I_o before the short circuit. The steady-state excitation emf is $E_o = E_{po}$ and the stator voltage is V_o. The voltage equation before the short circuit is

$$\underline{V}_o = R_s \underline{I}_o + jX\underline{I}_o + \underline{E}_o \tag{30.33}$$

At $t = 0$ the terminal voltage changes to $\underline{V} = \underline{V}_1$. To simplify this analysis, we assume that the machine has no damper windings and that the excitation is by a current source. For $t = 0^+$ we have

$$\underline{V}_1 = R_s \underline{I} + jX\underline{I} + L\frac{d\underline{I}}{dt} + \underline{E}_o \tag{30.34}$$

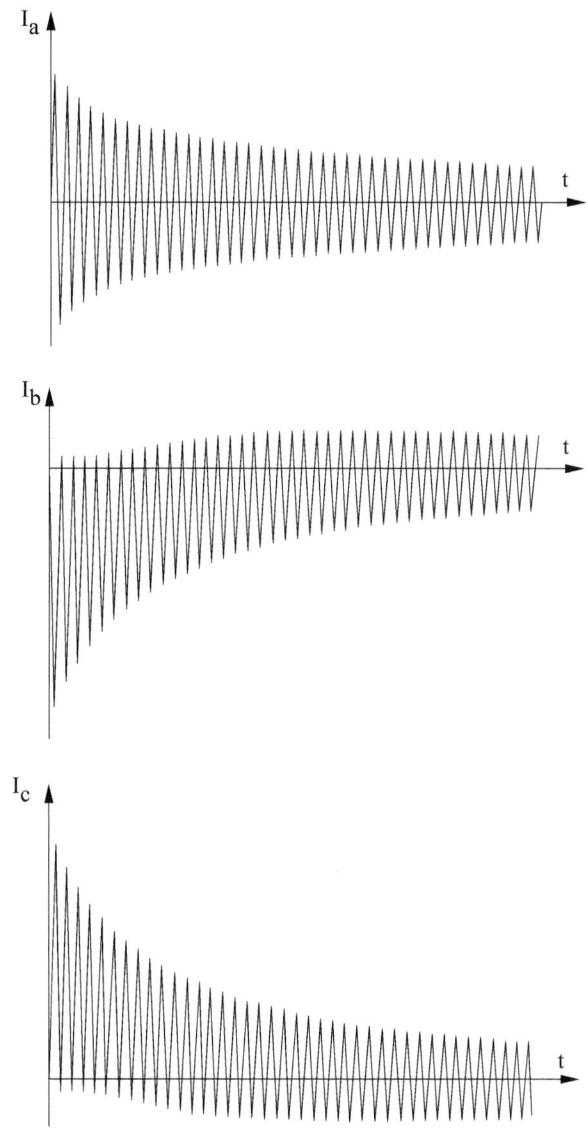

Fig. 30.2 Three-phase AC currents after short circuit

With $\Delta \underline{I} = \underline{I} - \underline{I}_o$, the transient is described by

$$\underline{V} = \underline{V}_1 - \underline{V}_o = R_s \Delta \underline{I} + jX\Delta \underline{I} + L\frac{d\Delta \underline{I}}{dt} \tag{30.35}$$

The right hand side of Eq. 30.35 is the simplified version of Eq. 30.11, for a smooth-rotor machine without damper windings and current supply of the excitation. The left hand side is the voltage to be connected to the terminals to study the transient.

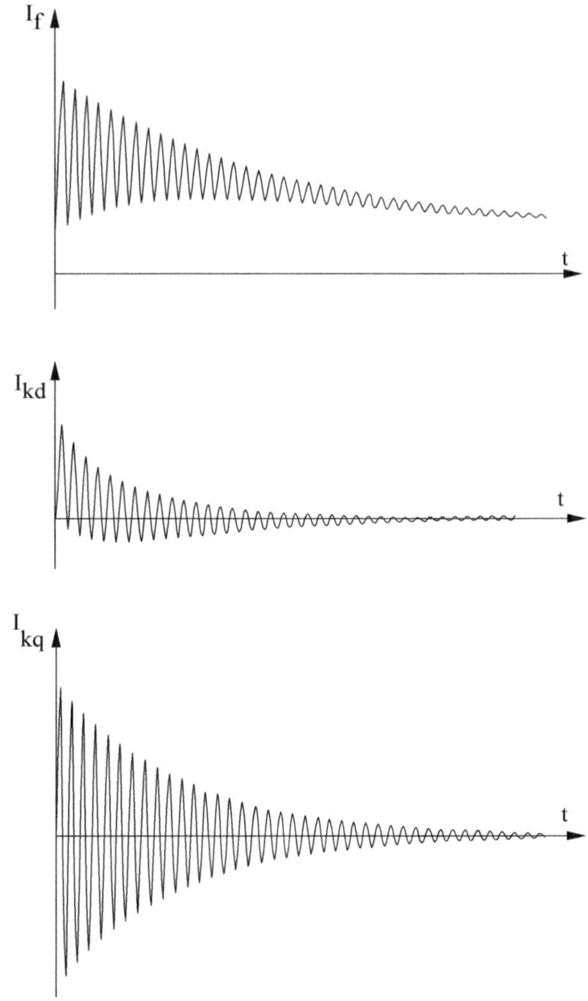

Fig. 30.3 Rotor currents after short circuit

In case of a sudden short circuit, $\underline{V}_1 = 0$ and $\underline{V} = -\underline{V}_o = -R_s\underline{I}_o - jX\underline{I}_o - \underline{E}_o$. This is also the voltage to be considered for Eq. 30.11 in the general case with damper windings and voltage-supplied excitation winding. If the machine was in no load before the short circuit, $\underline{V} = -\underline{V}_o = -\underline{E}_o$. If the machine is carrying current before the short circuit, the voltage to be considered is therefore different for the same excitation voltage.

2. The currents calculated from Eq. 30.11 are the transient currents only. If we are interested in the total current, we need to add the steady-state currents.

In some cases, only the subtransient phenomenon is relevant, i.e. the currents immediately after the short circuit or switching. To simplify the superposition with the steady-state currents, the subtransient reactances may also be used to describe the steady state. In that case, a fictitious rotor emf $\underline{E}_o^{"}$ has to be introduced:

$$\underline{E}_o^{"} = \underline{E}_o + j(X_d - X_d^{"})\underline{I}_{do} + j(X_q - X_q^{"})\underline{I}_{qo} \tag{30.36}$$

This steady state can then also be described by

$$\underline{V}_o = \underline{E}_o^{"} + jX_d^{"}\underline{I}_{do} + jX_q^{"}\underline{I}_{qo} \tag{30.37}$$

If we are only interested in the transient phenomenon, the steady state may be described by the transient reactances $X_q^{'}$ and $X_d^{'}$ by introducing a transient steady-state emf $\underline{E}_o^{'} = \underline{E}_o + j(X_d - X_d^{'})\underline{I}_{do} + j(X_q - X_q^{'})\underline{I}_{qo}$.

The *subtransient state* comes down to the assumption that the currents induced in the damper windings and the excitation winding momentarily keep the flux coupled with them constant. The subtransient emf corresponds with the voltage induced in the stator by this flux. The currents in the stator can only affect the leakage fluxes of these rotor windings with respect to the stator (reactances $X_q^{"}$ and $X_d^{"}$).

Similarly, the *transient state* comes down to the assumption that the currents induced in the excitation winding momentarily keep the flux coupled with it constant. The transient emf corresponds with the voltage induced in the stator by this flux. The currents in the stator can only affect the leakage fluxes of these rotor windings with respect to the stator (reactances $X_q^{'}$ and $X_d^{'}$).

The subtransient representation can be used when only the initial transient phenomena is relevant, i.e. immediately after the switching. This is, for example, useful for analysing the commutation of a rectifier supplied by a synchronous machine. Indeed, the commutation is so fast that the transient currents in damper and excitation windings are barely reduced and, as a consequence, the emf can be considered as constant during the commutation. The reactances involved in the commutation are also the subtransient reactances.

The transient representation can also be used if we are interested in the time span after the subtransient interval but before the steady state.

30.2.2 Zero-Sequence and Negative Sequence Transients

For the zero-sequence equations, we obtain from Eq. 28 in Chap. 28 after the Clarke transformation (28.7):

$$V_o = R_s I_o + \left(L_s + 2M_s\right)\frac{dI_o}{dt} \tag{30.38}$$

or in the Laplace domain

$$V_o(p) = R_s I_o(p) + L_o p I_o(p) \tag{30.39}$$

Apparently, for zero-sequence voltages the machine behaves as a resistor in series with a small inductance. The zero-sequence inductance corresponds with the own leakage of a stator phase (slot leakage, air-gap leakage and end-winding leakage).

To analyse the behaviour for negative-sequence voltages, we consider the following voltage system:

$$V_a(t) = V \cdot \cos(\omega t + \delta_3) \cdot \mu(t)$$

$$V_b(t) = V \cdot \cos(\omega t + \delta_3 + 2\pi/3) \cdot \mu(t) \tag{30.40}$$

$$V_c(t) = V \cdot \cos(\omega t + \delta_3 + 4\pi/3) \cdot \mu(t)$$

This corresponds to

$$V_\alpha(t) = V \cdot \cos(\omega t + \delta_3) \cdot \mu(t)$$

$$V_\beta(t) = -V \cdot \sin(\omega t + \delta_3) \cdot \mu(t) \tag{30.41}$$

and

$$V_q = V \cdot \cos(2\omega t + \delta_3 + \delta_2) \cdot \mu(t)$$

$$V_d = -V \cdot \sin(2\omega t + \delta_3 + \delta_2) \cdot \mu(t) \tag{30.42}$$

Now, the voltages V_q and V_d are not constant but have twice the mains frequency. This is in fact quite obvious as the machine is rotating synchronously in the positive direction while these voltages are rotating in the negative direction at the same speed.

The currents corresponding with Eq. 30.42 can be calculated using Eq. 30.11. We obtain a system of time-variant equations with inputs having twice the mains frequency. Ultimately, this will create a sinusoidal steady-state with twice the mains frequency, which will be analysed next.

In the corresponding equations, the Laplace operator has to be replaced by $2j\omega$ to obtain the steady-state relations (in phasor form). To simplify the calculations, we will disregard the rotor resistances as these are much smaller than the reactances for twice the mains frequency.

From the rotor equations, the rotor currents may be expressed as a function of the stator currents:

$$\underline{I}_{kq} = -\frac{3}{2}\left(M_{skq}/L_{kq}\right)\underline{I}_q \tag{30.43}$$

$$\begin{bmatrix} \underline{I}_f \\ \underline{I}_{kd} \end{bmatrix} = -\frac{3}{2}\begin{bmatrix} L_f & M_{fkd} \\ M_{fkd} & L_{kd} \end{bmatrix}^{-1} \cdot \begin{bmatrix} M_{sf} \\ M_{skd} \end{bmatrix} \cdot \underline{I}_d \tag{30.44}$$

Substitution in the stator equations results in

$$\underline{V}_q = \left(R_s + 2j\omega L_q^{"}\right) \cdot \underline{I}_q - \omega L_d^{"} \underline{I}_d \tag{30.45}$$

$$\underline{V}_d = \omega L_q^{"} \underline{I}_q + \left(R_s + 2j\omega L_d^{"}\right) \cdot \underline{I}_d \tag{30.46}$$

from which

$$\underline{I}_q = \frac{\omega L_d^{"} \underline{V}_d + \left(R_s + 2j\omega L_d^{"}\right) \underline{V}_q}{\left(R_s + 2j\omega L_q^{"}\right)\left(R_s + 2j\omega L_d^{"}\right) + \omega^2 L_q^{"} L_d^{"}} \tag{30.47}$$

$$\underline{I}_d = \frac{-\omega L_q^{"} \underline{V}_q + \left(R_s + 2j\omega L_q^{"}\right) \underline{V}_d}{\left(R_s + 2j\omega L_q^{"}\right)\left(R_s + 2j\omega L_d^{"}\right) + \omega^2 L_q^{"} L_d^{"}} \tag{30.48}$$

with

$$\underline{V}_q = V \cdot \exp j(2\omega t + \delta_2 + \delta_3) \tag{30.49}$$

$$\underline{V}_d = jV \cdot \exp j(2\omega t + \delta_2 + \delta_3) \tag{30.50}$$

To simplify the analysis, we will neglect the stator resistances in Eqs. 30.47 and 30.48, resulting in

$$\underline{I}_q = -\frac{\underline{V}_d}{3\omega L_q^{"}} + \frac{2\underline{V}_q}{3j\omega L_q^{"}} \tag{30.51}$$

$$\underline{I}_d = \frac{\underline{V}_q}{3\omega L_d^{"}} + \frac{2\underline{V}_d}{3j\omega L_d^{"}} \tag{30.52}$$

In the time domain, we then find

$$I_q(t) = \frac{1}{\omega L_q^{"}} V \cdot \sin(2\omega t + \delta_2 + \delta_3) \tag{30.53}$$

$$I_d(t) = \frac{1}{\omega L_d^{"}} V \cdot \cos(2\omega t + \delta_2 + \delta_3) \tag{30.54}$$

The $\alpha,\ \beta-$components are therefore

$$I_\alpha(t) = \frac{1}{\omega L_q^{"}} V \cdot \sin(2\omega t + \delta_2 + \delta_3) \cdot \cos(\omega t + \delta_2) - \frac{1}{\omega L_d^{"}} V \cdot \cos(2\omega t + \delta_2 + \delta_3) \cdot \sin(\omega t + \delta_2) \tag{30.55}$$

$$I_\beta(t) = \frac{1}{\omega L_q^{"}} V \cdot \sin(2\omega t + \delta_2 + \delta_3) \cdot \sin(\omega t + \delta_2) + \frac{1}{\omega L_d^{"}} V \cdot \cos(2\omega t + \delta_2 + \delta_3) \cdot \cos(\omega t + \delta_2) \tag{30.56}$$

If $L_q^{"} = L_d^{"} = L^{"}$, the currents $I_\alpha(t)$ and $I_\beta(t)$ and therefore also the three-phase currents $I_a(t),\ I_b(t),\ I_c(t)$ only contain a first harmonic. The corresponding complex per-phase reactance for the negative sequence is then $\underline{Z}_2 = j\omega L^{"} = jX^{"}$.

However, if $L_q^{"} \neq L_d^{"}$, we can observe that the three-phase currents contain a third harmonic, in addition to the first harmonic.[3] This third harmonic is very small and can usually be ignored.

Nevertheless, for the first harmonic we can observe a somewhat strange behaviour as the impedance turns out to be different for voltage or current supply. For a voltage supply, as in the analysis above, we find for the fundamental harmonic impedance

$$\underline{Z}_2 = j\frac{2X_q^{"}X_d^{"}}{X_q^{"} + X_d^{"}} \tag{30.57}$$

If we consider a current supply, we find for the voltages, in addition to a first harmonic, also a small third harmonic voltage. For the equivalent (and approximate) impedance for the first harmonic, we now find

$$\underline{Z}_2 = j(X_q^{"} + X_d^{"})/2 \tag{30.58}$$

The explanation for this is that, because of the saliency and the associated third harmonic, strictly spoken the machine cannot be represented by a fundamental harmonic impedance.

If the stator resistance is taken into account, the stator resistance R_s should be added to the impedances (30.58) or (30.57).

Conclusion :

For the negative sequence in steady state, the currents $I_q(t)$ and $I_d(t)$ are AC currents (whereas these are DC currents for the positive sequence). The reactances are therefore much more important than the resistances. This is also valid for the initial transient for the negative sequence. As a consequence, we may assume that the machine will enter the negative-sequence steady state almost immediately. With a fair approximation, the machine can be represented by the complex impedances (30.57) or (30.58) for the negative sequence.

[3] What is a more physical explanation for this third harmonic if the machine exhibits saliency?

Chapter 31
Voltage Surge Phenomena in Electrical Machines

Abstract Windings of transformers and electrical machines are often subject to voltage surges, either by switching in the grid or by atmospherical phenomena. The resulting voltage waves may propagate into the windings of transformers and rotating machines, causing voltage stresses in the winding. In this chapter we present two simplified models to calculate these voltage stresses. Although nowadays computer models exist to rather accurately predict these voltage stresses, the simplified models offer valuable insight into the physical causes of the localised voltage stresses.

31.1 Introduction

Windings of transformers and electrical machines are often subject to voltage surges. These are caused either by switching in the grid or by atmospherical phenomena (lightning), resulting in transient voltage waves that propagate in the grid. These waves may also propagate into the windings of transformers and machines. However, because of the high frequencies associated with the steep voltage waves, not only the self and mutual inductances of the windings but also the capacitances between the windings and between windings and earth come into play. Usually, this causes an uneven distribution of the voltage along the winding, resulting in (excessively) large voltage stresses for the entrance coils.

The design of machines and transformers should take this into account by choosing appropriate insulating material for the entrance coils. Other measures are usually also taken to ensure a more even distribution of the voltage across the coils (by modifying the capacitance between the windings and between windings and earth).

Nevertheless, a co-ordination with other measures like lightning conductors, voltage limiters, or protective switches remains necessary.

The most dangerous voltage surges are created by direct strokes of lightning. The resulting induced voltages are of the order of 300–1500 kV, with a dV/dt of $100 \cdots 1500$ kV/μs. In some cases, voltages of several thousands kV are observed. Indirect lighting strokes may generate voltages of $80 \cdots 200$ kV. As long as the rated voltage level of that part of the grid is higher than 30 kV, this will not pose a problem.

J. A. Melkebeek, *Electrical Machines and Drives*, Power Systems,
https://doi.org/10.1007/978-3-319-72730-1_31

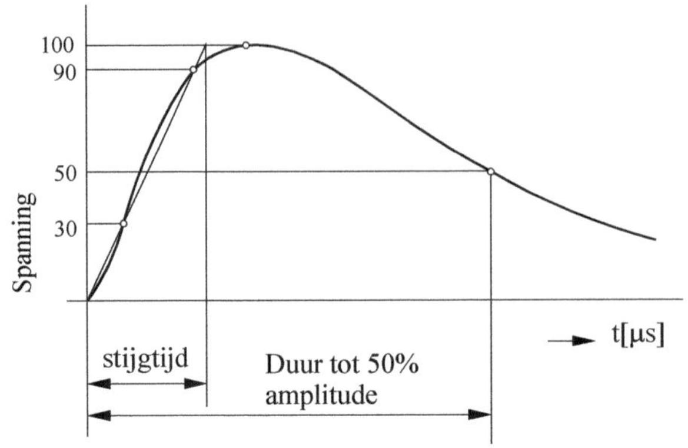

Fig. 31.1 Standardised voltage surge 1, 2/50

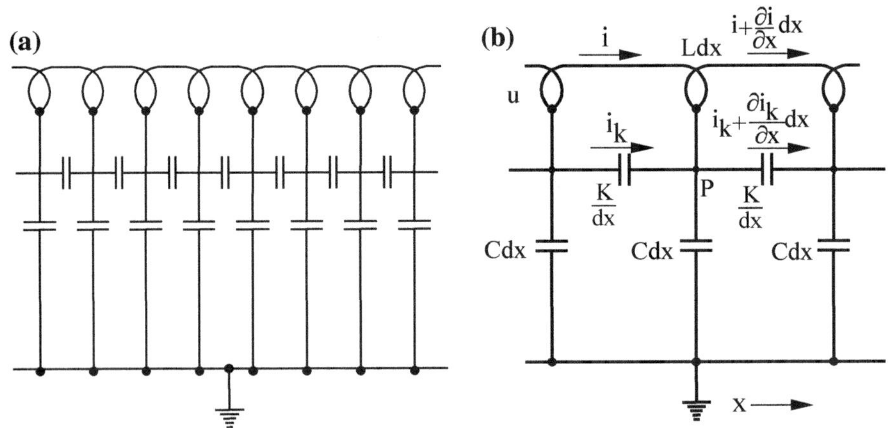

Fig. 31.2 Simplified model of a single coil

The current amplitudes resulting from lightning strokes vary between some hundreds of Ampères to 250 kA. Statistical studies indicate that 97% are lower than 40 kA; only 1% reaches amplitudes higher than 90 kA.

Voltage surges due to switching are much less dangerous. These voltage amplitudes are in fact related to the rated voltage and insulation level of that part of the grid.

To study voltage surge phenomena, standardised reference waves have been introduced. These reference waves are characterised by maximum voltage, polarity, maximum rise time, and duration (up to 50% of the maximum value) (see (a) in Fig. 31.1). A frequently used reference wave is the wave 1, 2/50 (rise time 1 μs, and duration to half the maximum amplitude of 50 μs) standardised by the IEC.

For the analysis (and design of transformers and machines with respect to voltage surges), powerful software tools exist nowadays. These tools should take into account not only the winding inductances but also the capacitances between windings and between windings and earth (see Fig. 31.2). However, these packages are rather complicated and use huge computer time as both the electromagnetic and electrostatic properties are analysed using finite elements. Further, these tools provide little insight into what happens inside the transformer or machine windings. Therefore, we present in the next section some simplified and approximate analytical methods.

31.2 Voltage Surge Waves in a Single-Layer Coil

31.2.1 Simplified Theory Disregarding Mutual Coupling

The winding is represented by the scheme in (b) in Fig. 31.2, where in addition to the inductance of the coils per metre (L, $[H/m]$) also the capacitance (K, $[F \cdot m]$) between turns per metre and the capacitance per metre between turns and earth (C, $[F/m]$) are taken into account.

With the notations in Fig. 31.2, we obtain the following relations for one element:

$$\frac{\partial(i + i_k)}{\partial x} = -C\frac{\partial u}{\partial t} \tag{31.1}$$

$$i_k = -K\frac{\partial^2 u}{\partial x \partial t} \tag{31.2}$$

$$\frac{\partial u}{\partial x} = -L\frac{\partial i}{\partial t} \tag{31.3}$$

Here, x is the distance in axial direction from the first turn of the coil, u is the voltage in x, i is the longitudinal coil current in x and i_k is the capacitive longitudinal current in x. Note that the inductive voltage $L\partial i/\partial t$ only accounts for the magnetic field by the element $L\partial x$ in x (and not the magnetic fields by the other turns).

Eliminating i and i_k from Eqs. 31.1–31.3 yields

$$\frac{\partial^2 u}{\partial x^2} + LK\frac{\partial^4 u}{\partial^2 x \partial^2 t} - LC\frac{\partial^2 u}{\partial t^2} = 0 \tag{31.4}$$

Equation 31.4 describes the transients inside the coil.

We propose a solution of the form

$$u = U \cdot \exp j\omega t \cdot \exp j\alpha x \tag{31.5}$$

Substitution in Eq. 31.4 leads to the condition

$$\alpha^2 - LC\omega^2 - LK\alpha^2\omega^2 = 0 \tag{31.6}$$

or

$$\omega = \frac{\alpha}{\left[LC\left(1 + K\alpha^2/C\right)\right]^{1/2}} \tag{31.7}$$

or else

$$\alpha = \pm\omega\sqrt{\frac{LC}{1 - \omega^2 LK}} \tag{31.8}$$

With each value of α, one value of ω corresponds. When $\alpha \to \infty$, we see that $\omega \to \omega_\infty = 1/\sqrt{LK}$. Clearly, ω_∞ represents the limit frequency for free sinusoidal oscillations in the coil (according to this model).

Since both positive and negative values of α may occur according to Eq. 31.8, the general solution can be represented by

$$u(x,t) = \sum_n \left[a_n \exp(j\alpha_n x) + b_n \exp(-j\alpha_n x)\right] \exp(j\omega_n t) \tag{31.9}$$

with α_n, ω_n satisfying Eq. 31.6. The solution may therefore consist of both standing waves

$$u = U \cdot \sin\alpha x \cdot \cos\omega t \tag{31.10}$$

and travelling waves

$$u = U \cdot \cos(\alpha x \pm \omega t) \tag{31.11}$$

The latter have as linear speed

$$v = \pm\frac{\omega}{\alpha} = \pm\frac{1}{\left[LC\left(1 + K\alpha^2/C\right)\right]^{1/2}} \tag{31.12}$$

which implies that the speed depends on the wave length: the shorter the wave length, the higher the speed. In other words, the coil distorts the waves, i.e. the shape of the waves changes when these waves move farther in the coil.

Let us now study the case of a voltage block wave entering the coil at instant $t = 0$. To do so, we also have to specify the conditions at the other end of the coil. We will study two extreme cases, i.e. with the other end of the coil left open or short-circuited to earth.

First, we will focus on the case with the other end earthed. We suppose the voltage and current in the coil zero for $t \leq 0^-$. At $t = 0^+$, a voltage block wave with amplitude U_o enters the coil, in other words at $t = 0^+$ the voltage at the entry jumps from $u = 0$ to $u = U_o$. Then, currents start to flow, creating a voltage redistribution over the coil and the associated capacitors. This voltage distribution is not instantaneous, as the magnetic field due to the displacement currents in the capacitors opposes the change (see Maxwell's laws). However, as Faccioli [4] has demonstrated, the time span required for this charge redistribution is very small, some nanoseconds, and can be ignored compared to the time required to attain the final state. We will disregard the time required for this redistribution and consider the charge redistribution as instantaneous. This redistribution is therefore the instantaneous result of the voltage jump at $t = 0^+$ to $u = U_o$ to a network of only capacitors (Fig. 31.2 with all inductances set to ∞). The corresponding differential equation describing the initial charge redistribution can be found by expression the charge equilibrium in whatever node P (in x):

$$\left(u_o + \frac{du_o}{dx}dx\right)Cdx + \left(-\frac{du_o}{dx}dx - \frac{d^2u_o}{dx^2}dx\right)\frac{K}{dx} + \frac{du_o}{dx}dx\frac{K}{dx} = 0 \quad (31.13)$$

with $u_o = u(x, t = 0^+)$. Thus

$$\frac{d^2u_o}{dx^2} = \frac{C}{K}u_o \quad (31.14)$$

(this equation can also be derived by putting $L = \infty$ in Eq. 31.4).

For the specific case here with one terminal connected to U_o and the other connected to ground, the boundary conditions are

$$\begin{aligned} x &= 0 \; u_o(0) = U_o \\ x &= l \quad u_o(l) = 0 \end{aligned} \quad (31.15)$$

The solution of the differential equation 31.14 for this case is therefore

$$u_o = U_o\frac{\sinh\gamma(l - x)}{\sinh\gamma l} \quad (31.16)$$

with $\gamma = \sqrt{\frac{C}{K}}$. The (negative) voltage gradient

$$g = -\frac{du_o}{dx} = U_o\frac{\gamma \cdot \cosh\gamma(l - x)}{\sinh\gamma l} \quad (31.17)$$

attains a maximum value for $x = 0$, i.e.

$$g_{max} = U_o\gamma\coth(\gamma l) \quad (31.18)$$

With the average gradient over the length of the coil equal to $g_{av} = U_o/l$, we observe $g_{max}/g_{av} = \gamma l \cdot \coth(\gamma l)$. As in most cases $\gamma l > 3$, $\coth(\gamma l) \approx 1$ and thus $g_{max}/g_{av} \approx \gamma l = l\sqrt{\frac{C}{K}} = \sqrt{\frac{Cl}{K/l}} = \sqrt{\frac{C_t}{K_t}}$ (with K_t and C_t the total series capacitance

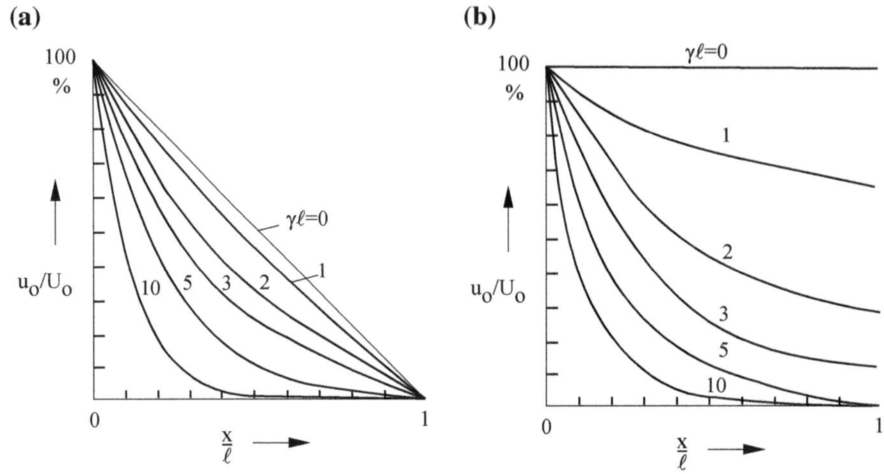

Fig. 31.3 Gradient for **(a)** shorted and **(b)** open coil end

and capacitance to earth, respectively). The greater the capacitance to earth of the coil compared to the series capacitance, the greater the gradient (see (a) in Fig. 31.3).

The physical explanation is that a large coil capacitance to earth results in a large voltage drop for the first turns of the coil as the associated series capacitances have to carry all the current to charge all following capacitances to earth. The voltage drop over the first series capacitor is equal to $-(\partial u/\partial x)dx$. We may therefore define an *equivalent entrance capacitor* C_v for the coil as follows:

$$[i_k]_{x=0} = \frac{\partial}{\partial t}\left[-\frac{\partial u}{\partial x}dx\frac{K}{dx}\right]_{x=0} \doteq C_v\left[\frac{\partial u}{\partial t}\right]_{x=0} \tag{31.19}$$

Integrating equation 31.19 using Eq. 31.16 yields finally

$$C_v = K\gamma \cdot \coth(\gamma l) \approx \sqrt{CK} \tag{31.20}$$

Equation 31.16 represents the initial charge (or voltage) distribution which is the starting point for the free oscillations in the coil. These free oscillations will ultimately end up in the final state (even though all resistances and thus all damping have been ignored).

The differential equation for the final state can be derived from Eq. 31.4 by putting all time derivates equal to zero, i.e.

$$\frac{d^2u_f}{dx^2} = 0 \tag{31.21}$$

With the boundary conditions $u_f(0) = U_o$, $u_f(l) = 0$, the result for the final state is $u_f(x) = U_o(1 - x/l)$. The corresponding current would be $i_f = (U_o/Ll) \cdot t$ but in reality the current will be limited by the resistances, which we have disregarded in our strongly simplified model.

As mentioned, free oscillations (i.e. the solutions of Eq. 31.4) will ensure a transition between the initial charge distribution (i.e. Eq. 31.16) and the final state. The resulting voltage surge wave may be written as follows:

$$u_r(x,t) = u_f(x) + u(x,t) \tag{31.22}$$

$u(x,t)$ represents the free oscillations, i.e. solutions of Eq. 31.4 and of the form of Eq. 31.5. As $u_f(x)$ also satisfies Eq. 31.4, so does $u_r(x,t)$. For $u(x,t)$ we propose the following real form:

$$u(x,t) = \sum_n \left[a_n \cos(\alpha_n x) + b_n \sin(\alpha_n x)\right] \cos(\omega_n t) \tag{31.23}$$

The coefficients $a_n,\ b_n$ result from the boundary conditions:

$$\begin{aligned} &t = 0 \quad u(x,0) = u_o(x) - u_f(x) \\ &x = 0 \qquad\quad u(0,t) = 0 \\ &x = l \qquad\quad u(l,t) = 0 \end{aligned} \tag{31.24}$$

The second condition gives $a_n = 0$. The third condition yields $\alpha_n = n\frac{\pi}{l}$ for $n = 1,\ 2,\ \ldots$. Then, from the first condition, we get

$$\sum_n b_n \sin\left(\frac{n\pi x}{l}\right) = U_o \left[\frac{\sinh \gamma(l-x)}{\sinh \gamma l} - \left(1 - \frac{x}{l}\right)\right] \tag{31.25}$$

Using a Fourier expansion of the right side of this equation, we obtain for the coefficients b_n $(n = 1,\ 2,\ \ldots)$:

$$b_n = \frac{-2}{n\pi} \cdot \frac{\gamma^2 l^2}{n^2\pi^2 + \gamma^2 l^2} \cdot U_o \tag{31.26}$$

Therefore

$$u_r(x,t) = U_o \left\{\left(1 - \frac{x}{l}\right) - 2\sum_n \frac{\gamma^2 l^2}{n^2\pi^2 + \gamma^2 l^2} \cdot \frac{\sin(\frac{n\pi x}{l})}{n\pi} \cos \omega_n t\right\} \tag{31.27}$$

with (see Eq. 31.7):

$$\omega_n = \frac{n\pi/l}{\left[LC + LK\left(n^2\pi^2 + \gamma^2 l^2\right)\right]^{1/2}} \tag{31.28}$$

The case of coil with the other end insulated with respect to earth can be treated in a similar way. For the initial charge distribution, the following boundary conditions hold:

$$\begin{aligned} x &= 0 \quad u_o(0) = U_o \\ x &= l \;\; (\partial u_o/\partial x)_{x=l} = 0 \end{aligned} \tag{31.29}$$

(the condition in $x = l$ results from the zero current in $x = l$). Thus

$$u_o = U_o \frac{\cosh \gamma(l - x)}{\cosh \gamma l} \tag{31.30}$$

The graph (b) in Fig. 31.3 shows $u_o(x)/U_o$ as a function of x/l for some values of γ/l.

For the final voltage distribution, $u_f(x) = U_o$ and $i_f(x) = 0$ hold. As a consequence, the boundary conditions for the free oscillations are

$$\begin{aligned} t &= 0 \;\; u(x, 0) = u_o(x) - U_o \\ x &= 0 \qquad u(0, t) = 0 \\ x &= l \quad\; (\partial u/\partial x)_{x=l} = 0 \end{aligned} \tag{31.31}$$

The third condition yields $\alpha_n = n\frac{\pi}{2l}$ for $n = 1, 3, \ldots$. Then, from the first condition, we get

$$b_n = \frac{-4}{n\pi} \cdot \frac{\gamma^2 l^2}{(n\pi/2)^2 + \gamma^2 l^2} \cdot U_o \tag{31.32}$$

and thus (with $n = 1, 3, \ldots$):

$$u_r(x, t) = U_o \left\{ 1 - 4 \sum_n \frac{\gamma^2 l^2}{(n\pi/2)^2 + \gamma^2 l^2} \cdot \frac{\sin(\frac{n\pi x}{2l})}{n\pi} \cos \omega_n t \right\} \tag{31.33}$$

Figure 31.4 compares the two cases discussed. For an earthed end of the coil the oscillations show an even number of half wave lengths, while for an insulated end of the coil we see an odd number of quarters of wave lengths.

Despite the rather strongly simplifying assumptions, the model (which goes back to 1915, see Ref. [37]) gives a fairly good qualitative idea of the phenomena associated with voltage surges in a coil.

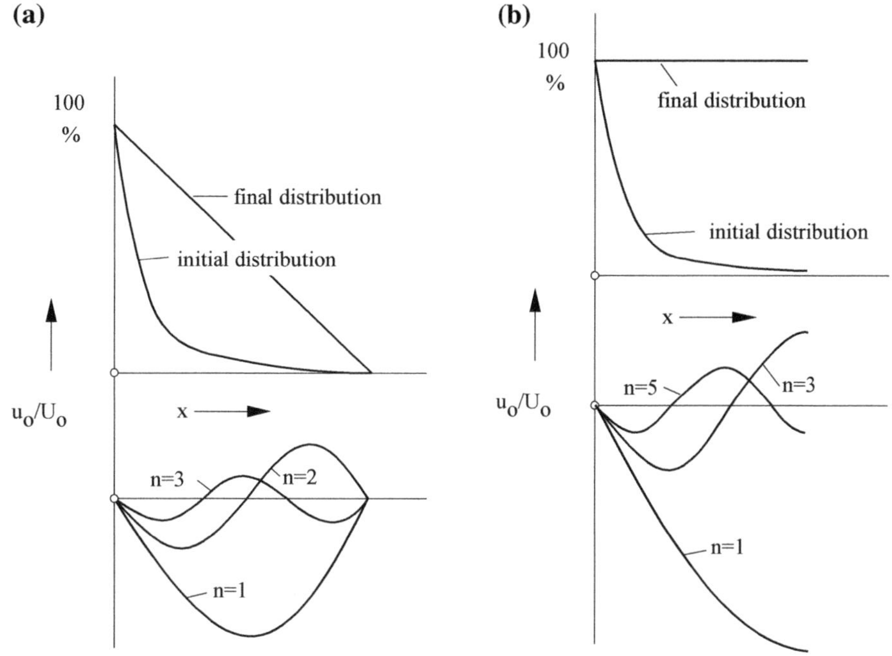

Fig. 31.4 Surge voltage free waves

31.2.2 Effect of the Mutual Coupling

The theory in the previous subsection completely disregards the mutual coupling between the turns of the coil. Although the results describe the actual behaviour quite accurately from a qualitative perspective, quantitatively the results for the free oscillations do not match the actual phenomena very well. For example, a limit frequency (see Eq. 31.8) has never been observed.

An approximate model taking into account the mutual inductive coupling between turns has been developed in Ref. [1]. Consider the schematic representation of a coil with iron core in Fig. 31.5. The current in the coil results in a magnetic field with field lines partially in air (leakage) and partially in the iron core (main field), as depicted in the figure.

Let us call $\varphi(x, t)$ the physical flux (or flux per turn) in x, w the total number of turns and l the axial length of the coil. The flux in x contributes to the voltage according to

$$\frac{\partial u(x,t)}{\partial x} = -\frac{w}{l} \cdot \frac{\partial \varphi(x,t)}{\partial t} \tag{31.34}$$

The flux $\varphi(x, t)$ consists of a main flux φ_o and a leakage flux φ_σ. The field lines of the main flux remain completely in the iron core; this main flux is the same for all

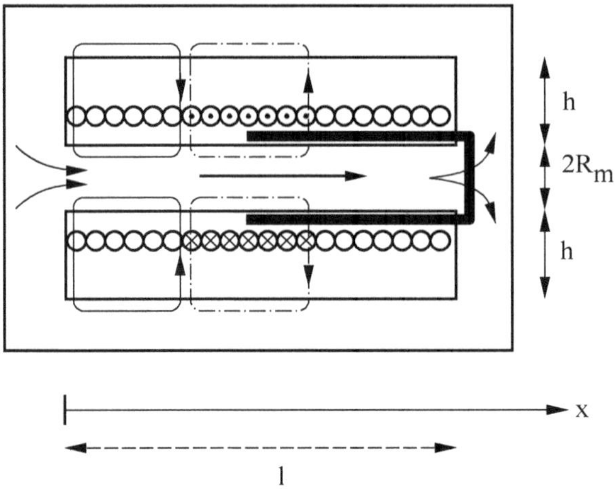

Fig. 31.5 Model with mutual coupling

turns in the coil. For the leakage flux, we may assume that the field lines resemble the dash-dot lines in Fig. 31.5. To calculate the flux $\varphi(x)$ in x, we consider a surface Σ consisting of

- a cylindrical surface from $x = x$ to $x = l$ (with as starting point the turn in $x = x$).
- a basis at $x = l$ completely in the iron yoke

We obtain for the flux

$$\varphi(x) = \iint_{\Sigma} \underline{B} \cdot \underline{n} \cdot d\Sigma = 2\pi R_m \int_x^l B_\sigma(x) \cdot dx + \varphi_o \tag{31.35}$$

with R_m the average radius of the winding (and of the cylinder). To calculate the leakage flux, Ampère's law is applied on the leakage flux line shown in full line in Fig. 31.5:

$$\frac{1}{\mu_o} B_\sigma(x) \cdot h - \frac{1}{\mu_o} B_\sigma(0) \cdot h = \frac{w}{l} \int_0^x i(x) \cdot dx \tag{31.36}$$

(where h is the radial length of a leakage flux line). Substituting Eq. 31.36 in Eq. 31.35 yields

$$\varphi(x) = \varphi_o + 2\pi R_m \int_x^l B_\sigma(0) \cdot dx + 2\pi R_m \int_x^l \left[\int_0^x \frac{w\mu_o}{hl} i(x) \cdot dx \right] \cdot dx \tag{31.37}$$

Differentiating twice with respect to x results in

$$\frac{\partial^2 \varphi(x)}{\partial x^2} = -\mu_o \frac{2\pi R_m w}{hl} \cdot i(x) \tag{31.38}$$

Substituting in Eq. 31.34, we obtain

$$\frac{\partial^3 u}{\partial x^3} = \mu_o \cdot \frac{w^2}{l^2} \cdot \frac{2\pi R_m}{h} \frac{\partial i}{\partial t} \doteq L' \frac{\partial i}{\partial t} \tag{31.39}$$

Equation 31.39 replaces Eq. 31.3. Together with Eqs. 31.1 and 31.2, the differential equation for the voltage in the coil now becomes

$$\frac{1}{L'} \frac{\partial^4 u}{\partial x^4} - K \frac{\partial^4 u}{\partial^2 x \partial^2 t} + C \frac{\partial^2 u}{\partial t^2} = 0 \tag{31.40}$$

Compared to the original differential equation 31.4, the first term undergoes a sign and order change. The reason for this is that now the inductance L' takes into account the mutual leakage flux between the turns, while in the inductance L in the original equation only accounts for the own field of the turn.

We now propose a solution of the form

$$u = U \cdot \exp j\omega t \cdot \exp j\alpha x \tag{31.41}$$

Substitution in Eq. 31.40 leads to the condition

$$\alpha^4 - L' C\omega^2 - L' K\alpha^2\omega^2 = 0 \tag{31.42}$$

or

$$\omega = \frac{\alpha^2}{\left[L' C\left(1 + K\alpha^2/C\right)\right]^{1/2}} \tag{31.43}$$

Compared to the model of Sect. 31.2.1, we find that the angular frequency is somewhat proportional to the square of the space frequency α instead of proportional to the space frequency. Further, there is no limit or critical frequency (ω_∞) above which all oscillations are damped as a function of the distance in the coil (i.e. when $j\alpha$ becomes real).

The remainder of the analysis is similar as in Sect. 31.2.1. For an earthed coil, we also obtain an even number of half waves and for an open end an odd number of quarter waves.

Other important conclusions from the above analysis are the following:

- for an earthed coil the peak voltages may easily attain 150% of the rated voltage of the coil (for switching transients, but much more for a direct lightning stroke)
- for a coil with insulated end the voltages may easily attain 300% of the rated voltage (for switching transients, but much more for a direct lightning stroke).

31.2.3 Discussion of the Models

Experimental research has demonstrated that the model including the mutual leakage is much closer to reality:

- a critical frequency (ω_∞) has never been found experimentally
- the relation between α and ω according to the second model with mutual coupling matches experiments much better than the one according to the first model without mutual coupling
- for higher frequencies in particular, the amplitudes calculated from the first model do not correspond very well with the measured amplitudes (only 50% of the experimental values)

The Blume-Boyajian model is also a simplified representation of reality. In reality, a turn in x_1 and a turn in x_2 are only partially coupled by their leakage fluxes., The greater their distance in the coil, the less they are coupled.

However, we will not discuss more detailed modelling. For more details about modelling surge phenomena, see Ref. [9]. In a real machine, the phenomena are much more complex than in a simple coil (in terms of slots, end windings, and secondary windings, among other things) and, nowadays, finite element software packages are used to study surge phenomena in machines and transformers.

In the next section, we will discuss the most important differences concerning surge phenomena in real machines and transformers compared to a simple coil.

31.3 Surge Phenomena in Real Machines and Transformers

In general, surge phenomena in actual transformers are somewhat similar to those described above for a single coil. However, there are also important differences:

- the presence of a secondary winding, in most cases connected to a secondary grid or load
- for three-phase transformers, the presence of the other phases
- the heterogeneous character of the windings (e.g. disk windings)
- for tap transformers, the effect of unconnected windings
- the effect of the iron core
- etc.

For rotating machines, the voltage surge behaviour is even more distinct from that of a single coil. Indeed, in most cases (if not all) the windings are located in slots. As a result:

- the capacitive coupling between windings in different slots is nearly non-existent but
- the capacitive coupling to earth of the turns in a slot is very large and

- the damping effect of the iron can be very large as the windings in the slots are very near to the iron yoke

Therefore, a first-order model for surge phenomena in an electrical machine (as to the windings in the slots) is more like a transmission line with series inductance L and capacity C to earth (both per metre). The transients in the slot parts of the winding have a linear speed $v = 1/\sqrt{LC}$; the ratio between voltage and current is given by $Z = \sqrt{L/C}$. However, there is a considerable difference between the parts of the windings in the slots and the parts in the overhang. Because of the high series inductance and capacitance to earth for the conductors in the slots, the linear speed of the voltage waves is much smaller for the conductors in the slots ($20 \cdot \exp 6$ m/s versus $200 \cdot \exp 6$ m/s for the conductors in the overhang). The damping due to the iron is also much higher for the conductors in the slots.

An accurate model for the surge phenomena in rotating electrical machines should therefore

- account for the different behaviour of the conductors in the slots and in the overhang
- consider the (longitudinal) resistive components to take into account the eddy current damping; yet, mutual inductive and capacitive coupling between the turns may be disregarded
- take into account the frequency dependence of the resistive and inductive series components

It is, however, not possible to find any analytical solutions for such a complicated model. Instead, numerical models are used and even for these models, approximations are frequently applied.

These models will not be further discussed here. Qualitatively, the models discussed above for a single coil give a good idea of the phenomena in real transformers and rotating machines.

Instead, in the next section we will discuss means to reduce or avoid harmful effects of voltage surges in transformers and rotating machines.

31.4 Protection Against Voltage Surges

To protect transformers and machines against voltage surges, two types of actions are undertaken (in addition to external protection):

1- **mitigating** local high voltage surges by an appropriate (electromagnetic and electrostatic) design of the machine or transformer, so that the voltage surges are redistributed along a larger part of the winding

2- **reducing** the **harmful effects** of voltage surges by increasing the insulation level of the machine or transformer, at least for the first turns

The first kind of measures are mainly applied for transformers; the second kind mostly for rotating machines. Nevertheless, a combination is possible, for example for rotating machines fed at medium voltages, i.e. $V_n > 6kV$.

Below, we will mainly examine the first kind of measures.

For transformers, we will try to make the initial charge distribution as close as possible to the final (linear) distribution so as to reduce the amplitude of the free oscillations. As this initial charge distribution solely depends on the capacitances, a more even distribution is obtained as follows:

1. for coil (cylindrical) windings:

- compensating the capacitances to earth by placing metal shields at the entrance voltage level in close proximity of the windings. These shields compensate the capacitive currents to earth. Basically, they affect the electrostatic fields at the instant of the voltage surge.
- increasing the series capacitances of the windings, either by connecting discrete capacitors in parallel between taps on the windings or by interweaving the turns
- connecting discrete resistances (in parallel with the winding) between the entrance of the coils and well-chosen taps; combinations of resistances and capacitances can also be used.

2. disk windings inherently have a (relatively) larger series capacitance and smaller earth capacitance, except for the first disk. As such, disk windings are more apt for transformer windings with high rated voltage levels. Nevertheless, additional steps may be necessary:

- compensating or reducing the earth capacitance for the input disk (which is usually closest to the core yokes at the top or bottom)
- adding series capacitances in parallel with the winding
- interleaving the turns by transposition inside a disk or, if necessary, with and between adjacent disks

In addition, sometimes

- a tertiary delta-connected winding will be provided, short-circuited on suitable damping resistances (in order to limit the free oscillations), and
- a wye-connected winding will be earthed with a *Peterson* coil

Regarding surge voltages, rotating machines differ from transformers because the voltage waves are propagated rather differently. In addition, rotating machines cannot withstand the same level of voltages because the windings are not submerged in oil and there is a lack of space for additional insulation. Compared to transformers, the available slot space is limited and the fill factor of these slots should be as high as possible.

In practice, a machine is required to withstand the same level of voltages as the minimum industrial voltage test level (imposed by the standards, e.g. 1500 V for a 230 V machine). An alternative way to protect the machine is to limit the steepness of possible voltage fronts by either connecting a capacitor ($>0.1\mu F$) between the terminals and earth, or by connecting the machine to the grid by means of a cable with a suitable capacitance.

For high-voltage machines that are directly connected to the grid (i.e. without a transformer), voltage-limiting devices are applied, sometimes combined with capacitors at both the machine terminals and the connection point of the grid. For wye-connected high-voltage machines, a Peterson coil between the neutral and earth will improve the voltage surge characteristics (compared to an isolated neutral).

Appendix A
Terminal Markings and Markings of Windings

A.1 Markings for Three-Phase Transformers

General rules:

1. The three legs of a three-phase transformer are marked by the capital letters U, V, W.
2. All coils and windings on the same leg are marked by the same letters as their respective legs
   a. Coils that are not intended to be connected in series or parallel get a different numerical prefix (e.g. 1 for the primary, 2 for the secondary: for leg U, 1U and 2U, respectively)
   b. End terminals and taps of a winding are indicated by a numerical suffix. If a winding consists of different coils that are to be connected in series of parallel, the end points are marked by the suffixes (01, 02), (11, 12), ... while the taps get the suffixes (03, 04, 05, ...), (13, 14, 15, ...). The suffixes of a part winding or coil always form an uninterrupted series, except when they are on different coils or part windings, in which case the row is interrupted.
   Examples:
   The example (a) in Fig. A.1 shows a three-phase transformer with three windings (primary, secondary and tertiary) that are not intended to be connected in series or parallel. In (b) a three-phase transformer with taps on the primary winding is illustrated. (c) shows the windings of a three-phase transformer with the primary winding in two parts that are intended to be connected in series or parallel; one of these also has taps.
3. The general rule is that, for all windings on the same leg, the polarity of the end point with the smallest suffix with regard to the polarity of the end point with the largest suffix always is the same; the same rule is valid for the taps.
4. The terminal markings should maintain the same spacial position as the legs. The terminal markings are indicated by the same letter and prefix as the corresponding leg. The suffixes are the same as the suffixes of the end points of the windings

J. A. Melkebeek, *Electrical Machines and Drives*, Power Systems,
https://doi.org/10.1007/978-3-319-72730-1

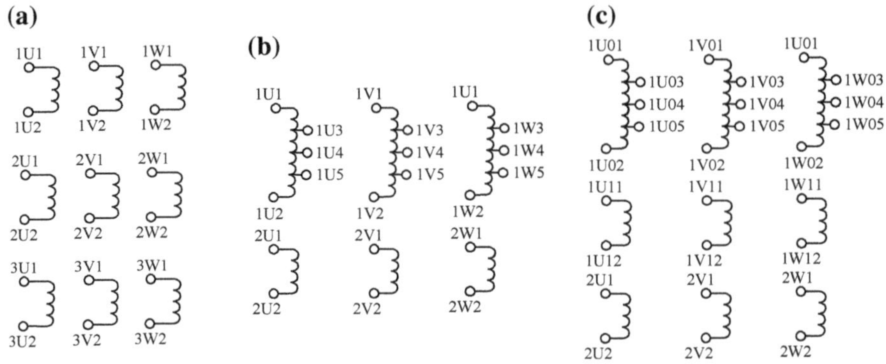

Fig. A.1 Three-phase transformer winding markings

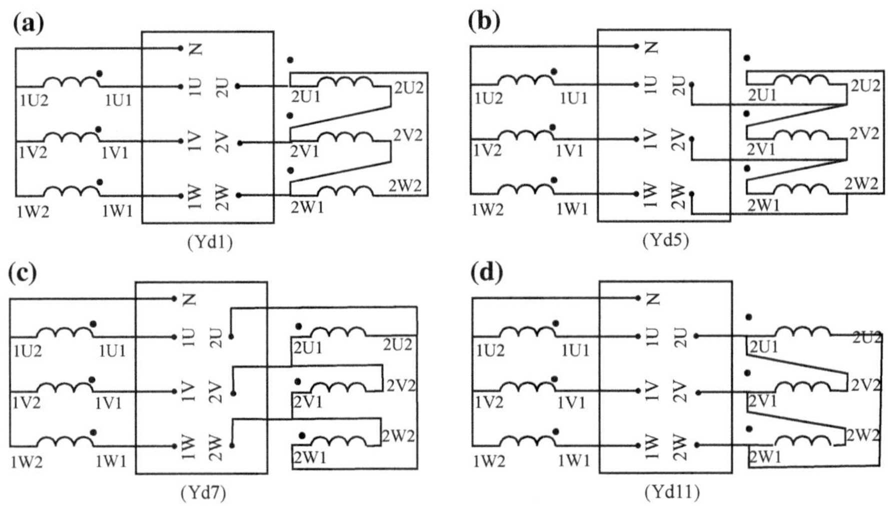

Fig. A.2 Three-phase transformer terminal markings

which they are connected to, see Fig. A.2.
If there is a neutral point, it is indicated by the letter N.

Examples:
A comparison of examples (a) and (b), or (c) and (d) shows that, with unchanged connection of the coils among themselves, there exist two ways to connect the end points to the terminals, resulting in a change of the vector group by 4. The comparison of the examples (a) and (c), or (b) and (d), shows that an interchange of of the end points 1 and 2 (both for the mutual connection of the coils and the connection to the terminals) yields a variation of the vector group of 6. This is also valid for wye connections.

A.2 Markings for Single-Phase Transformers

The markings of single-phase transformers are quite analogous to those of three-phase transformer. As all coils and windings belong to the same leg, only one letter (always the letter U) is used, see Fig. A.3.
Example:

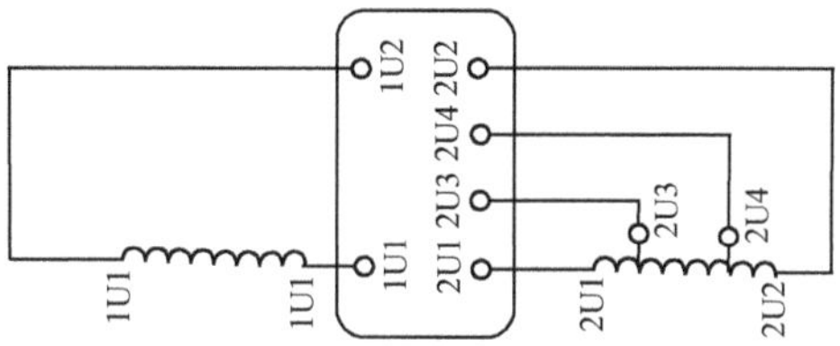

Fig. A.3 Single-phase transformer terminal markings

A.3 Markings and Rotation Direction of Rotating Electrical Machines

A.3.1 General Rules

These general rules are based on the IEC standard: Terminal markings and direction of rotation of rotating electrical machines.

1. The windings are marked by capital letters, e.g. U, V, W . . ..
2. End points and taps of a winding are distinguished by a numerical suffix, e.g. U1, U2.
3. Similar coils of a group of windings are distinguished by a numerical prefix, e.g. 1U, 2U.
4. Prefixes and/or suffixes may be omitted when confusion can be ruled out.
5. For DC commutator machines the letters of the first part of the alphabet are used, e.g. A, B, C. . .. For AC machines the letters belonging to the last part of the alphabet are used, e.g. K, L, M, N, Q, U, V, W, Z.

Remark: The IEC recommendation only refers to external terminals, i.e. those that are available to the user. However, similarly to transformers, these markings may also be used for the internal coils and winding terminals. In Fig. A.4, (a) is for a single stator winding; (b) is for a single stator winding with taps; (c) is for a stator winding in two parts that are to be connected in series or parallel; (d) is for a stator winding in two parts that are not to be connected in series or parallel (except for two-speed machines, see below); (e) is for a single stator winding with neutral point (to be connected or not). If the machine with windings as in (d) is a two-speed machine, then the terminals with the lowest prefix are intended for the lowest speed, and those with the highest suffix for the highest speed.

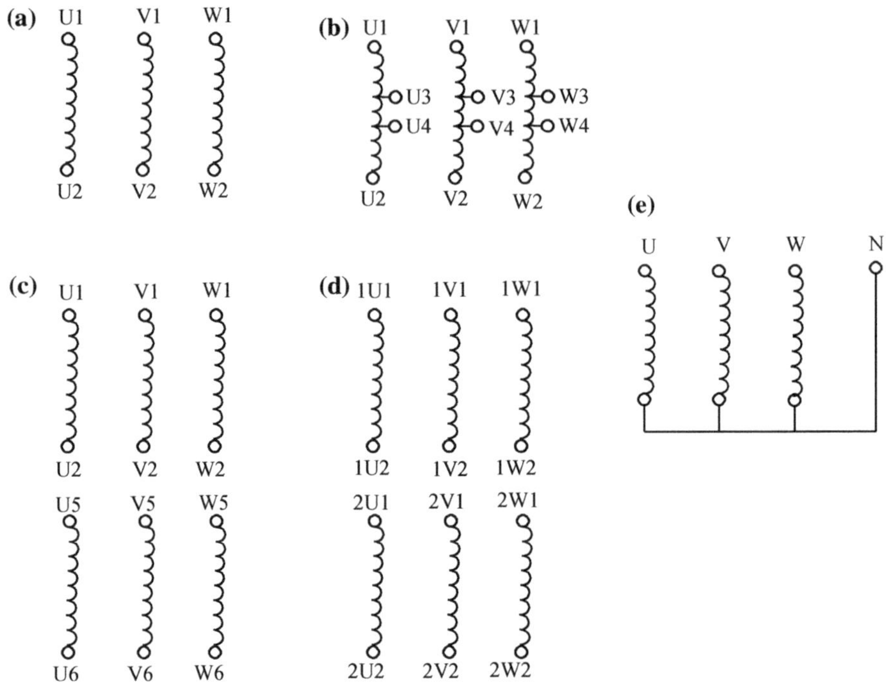

Fig. A.4 Stator winding markings for three-phase machines

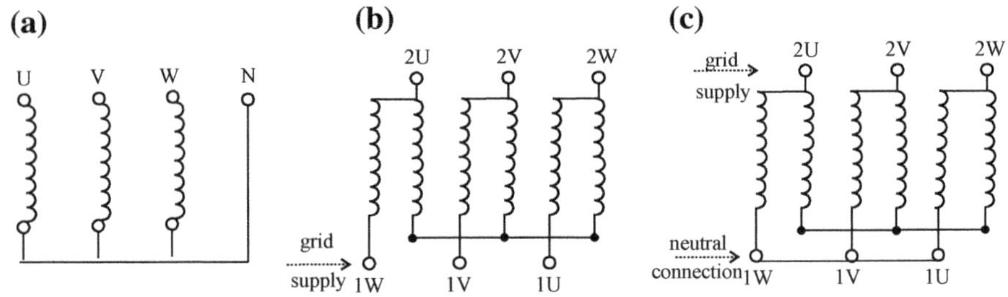

Fig. A.5 Three-phase machine terminal markings: **a** single speed; **b, c** multi-speed

A.3.2 Terminal Markings of Electric Machines

A.3.2.1 Markings for AC Machines (Except AC Commutator Machines)

Three-Phase AC Machines

The terminals are marked as in Fig. A.5, i.e. usually the letters U, V, W, N are used for the stator.

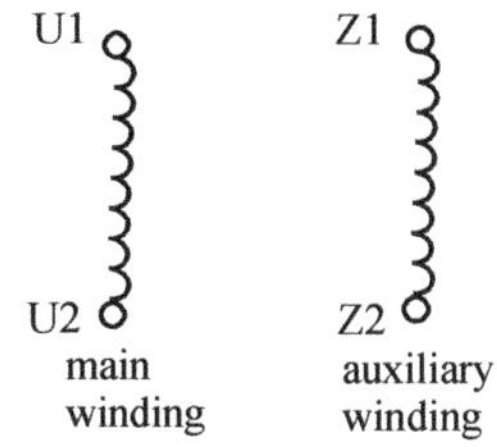

Fig. A.6 Single-phase induction machine terminal markings

In case of multi-speed machines, the terminals with the lowest (highest) prefix of the stator (primary) indicate the lowest (highest) speed. See (b) and (c) in Fig. A.5 (for the lowest and the highest speed, respectively).

For the rotor (secondary), the letters U, V, W, N are replaced by K, L, M, Q, respectively. If the externally supplied winding is on the rotor, this notation is reversed.

Two-Phase AC Machines

The markings are similar to those of three-phase machines, except that the letters W and M are not used.

Single-Phase AC Machines

The markings for single-phase machines are as shown in Fig. A.6.

Excitation Windings of Synchronous Machines

The terminals of a synchronous machine excitation winding are marked as the separate excitation winding of DC machines.

A.3.2.2 DC Commutator Machines

The (winding) terminals of DC commutator machines are marked as in Fig. A.7.
Remarks:

1. the auxiliary (commutation) and compensation windings sometimes consist of two parts, each part to be connected at one side of the armature
2. the series winding may have more than two terminals
3. if the separate excitation consists of two parts (intended for series or parallel connection), the markings are as in Fig. A.8.
4. two (or more) excitation windings result in fluxes in the same sense when their windings are supplied with currents flowing from the terminals with lowest (highest) suffix to those with the highest (lowest) suffix.
5. the fields of the auxiliary (commutation) and compensation windings have the right polarity with respect to each other and to the armature if, in all these windings, the current flows from the terminals with the lowest (highest) suffix to those with the highest (lowest) suffix.

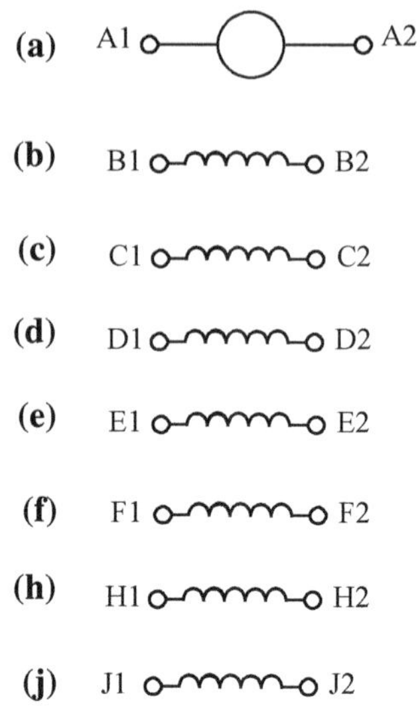

Fig. A.7 DC machine winding and/or terminal markings

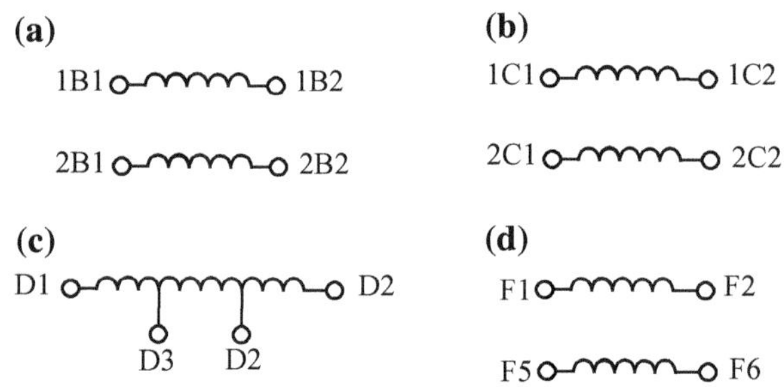

Fig. A.8 DC machine windings in two halves and/or taps

6. the external terminals of the series connection of two coils get the markings of the connected coil extremities.
7. if auxiliary and compensation windings form a single unity, the letter C is used for the markings.
8. when more than one winding is connected to an external terminal, the markings of one or more of the connected extremities are used for the external terminal marking, see Fig. A.9.

A.3.3 Rotation Direction

The rotation direction is assessed by an observer at the shaft end if the machine has only one shaft end, or at the shaft end with the largest diameter if the machine has

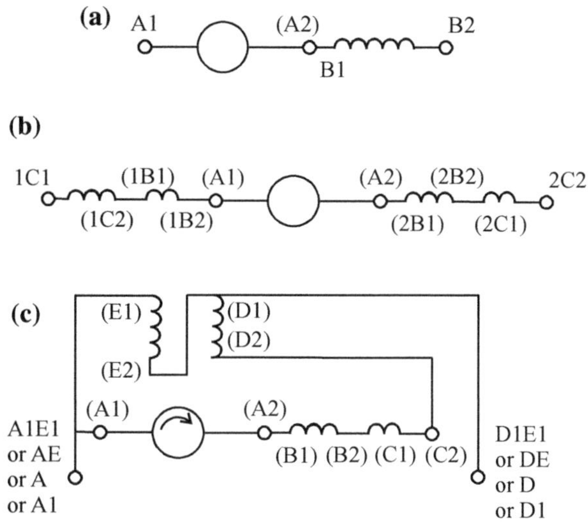

Fig. A.9 Terminal markings for a DC machine

two shaft ends. If the machine has two shaft ends with the same diameter (or no shaft ends at all), then the position of the observer is:

1. at the shaft end opposite to the commutator and/or slip rings if the machine has commutator and/or slip rings only at one end
2. at the shaft end opposite to the commutator if the machine has both commutator and slip rings, each at one side of the armature
3. at the shaft end opposite the ventilator for cage induction motors.

The rotation direction is called right or positive for a clockwise rotation and left or negative for a counterclockwise rotation.

A.3.4 Relation Between the Markings and the Rotation Direction

A.3.4.1 AC Machines

For three-phase machines, the rotation direction is right or positive if the alphabetic order of the markings corresponds to the time order of the voltages.

For single-phase machines, the rotation direction is right or positive for the connections as in Fig. A.10.

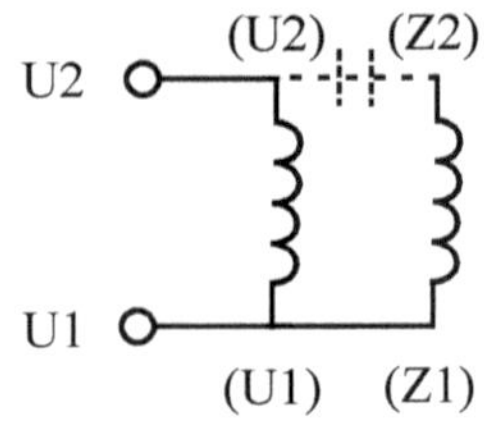

Fig. A.10 Single-phase induction machine: right rotation direction

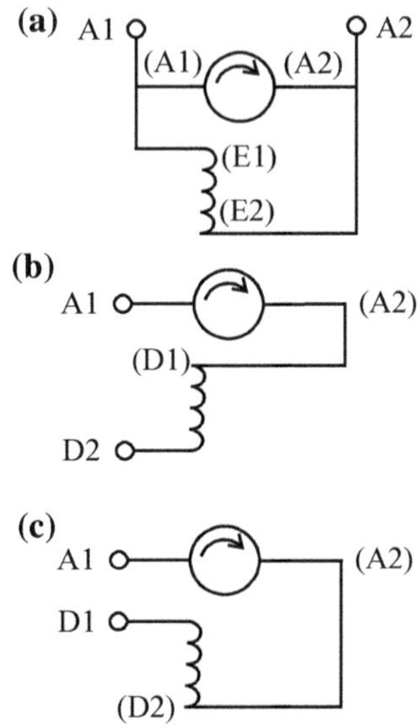

Fig. A.11 DC machine positive rotation direction

A.3.4.2 DC Commutator Machines

The rotation direction is right or positive when the connections are as in Fig. A.11. Remarks:

1. Note that the polarity of the terminals in this figure has no effect on the rotation direction.
2. It is easy to verify that the right rotation direction for motoring corresponds to currents in armature and excitation that flow in the same sense with respect to the suffixes in both windings (lowest to highest, or highest to lowest). For generating, the positive rotation direction corresponds to opposite current directions in armature and excitation with respect to the suffix order.

Appendix B
Static Stability of a Drive

The static stability of a motor with load depends on the slopes of the driving (motor) and load torques at the operating point (the intersection of both). Figure B.1 depicts a stable (a) and an unstable (b) operating point.

Mathematically this becomes clear when the equation of motion is considered (Ω is the state variable)

$$J\frac{d\Omega}{dt} = T_m - T_l \tag{B.1}$$

An equilibrium position (stable or unstable) occurs when $T_m = T_l$. The associated speed is called Ω_0.

Both T_m and T_l are in general (non-linear) functions of the speed. To investigate the local stability of the equilibrium point Ω_0, we have to linearise these torque functions around the equilibrium position and go over to the Laplace domain:

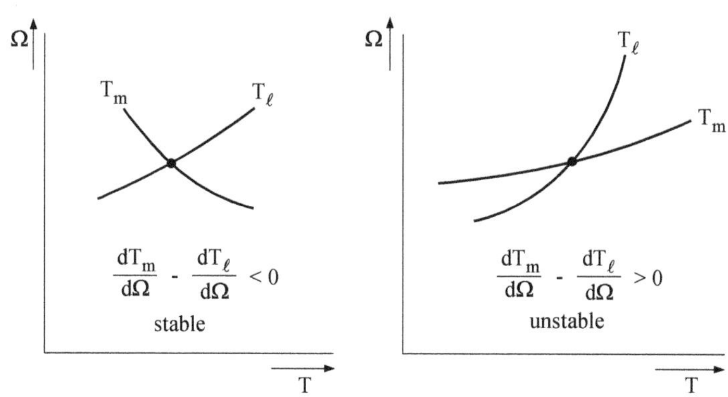

Fig. B.1 Stability of a drive

J. A. Melkebeek, *Electrical Machines and Drives*, Power Systems,
https://doi.org/10.1007/978-3-319-72730-1

$$Jp(\Delta\Omega) = \left[\left(\frac{\partial T_m}{\partial \Omega} \right)_{\Omega_0} - \left(\frac{\partial T_m}{\partial \Omega} \right)_{\Omega_0} \right] \cdot \Delta\Omega \tag{B.2}$$

Clearly the operating point is stable when $\left(\frac{\partial T_m}{\partial \Omega} \right)_{\Omega_0} < \left(\frac{\partial T_m}{\partial \Omega} \right)_{\Omega_0}$ (eigenvalue in the left half plane) and unstable when $\left(\frac{\partial T_m}{\partial \Omega} \right)_{\Omega_0} > \left(\frac{\partial T_m}{\partial \Omega} \right)_{\Omega_0}$ (eigenvalue in the right half plane).

This may also be intuitively deduced from the figure. Consider a deviation $\Delta\Omega$ (e.g. $\Delta\Omega > 0$) from the equilibrium point Ω_0. In case (a) in the figure, the load torque now becomes higher than the driving torque. Thus the speed will decrease and the system will return to the equilibrium point, which is stable. In case (b), the speed will further rise, moving the operating point away from the equilibrium point, which is therefore unstable.

Appendix C
Phasors and Space Vectors

C.1 General: Basic Definitions

Phasors (or time vectors) are commonly used to describe sinusoidal time quantities like voltages or currents. A sinusoidal time quantity (e.g. voltage) $v(t) = \hat{V}\cos(\omega t - \varphi)$ is represented by the complex quantity $\underline{V} = V\exp(-j\varphi)$ where $V = \hat{V}/\sqrt{2}$ is the effective value.[1]

The real time quantity can be regained by multiplying the phasor by $\sqrt{2}\exp(j\omega t)$ and taking the real part[2]: $v(t) = Re[\sqrt{2}\underline{V}\cdot\exp(j\omega t)] = Re[\hat{V}\cdot\exp(j\omega t - j\varphi)] = \hat{V}\cos(\omega t - \varphi)$.

Phasors can be represented in a complex (time) plane as in (a) in Fig. C.1. With a positive angle φ in the above phasor, the voltage phasor is lagging with respect to the real axis, i.e. $v(t)$ is lagging with respect to a voltage $v_0(t) = \hat{V}\cos(\omega t)$.

Space vectors on the other hand are typical for three-phase (or multi-phase) quantities as in rotating field machines. A basic physical example is that of a sinusoidal rotating magnetic field $b(x,t) = \hat{B}\cos(x\pi/\tau_p - \omega t - \psi)$. The space vector representation is $\underset{\rightarrow}{b} = \hat{B}\exp(j\omega t + \psi)$.

To return to the real time representation, multiply the space vector by $\exp(-jx\pi/\tau_p)$ and take the real part: $b(x,t) = Re[\underset{\rightarrow}{b}\cdot\exp(-jx\pi/\tau_p)] = \hat{B}\cos(x\pi/\tau_p - \omega t - \psi)$.

In Fig. C.1, b shows the space vector $\underset{\rightarrow}{b}$ in the complex space plane. In this case, $\underset{\rightarrow}{b}$ is a vector with constant amplitude rotating with constant angular speed in the counter-clockwise direction (usually assumed the positive direction). Remark that a positive angle ψ corresponds to a rotating field that is leading in space with respect to a field $b_0(x,t) = \hat{B}\cos(x\pi/\tau_p - \omega t)$, i.e. at $t = 0$ the field $b(x,t)$ is maximal in

[1]Sometimes amplitude values as well are used for the phasors.

[2]Without taking the real part, we obtain the so-called time phasor $v(t) = \hat{V}\cdot\exp(j\omega t - j\varphi)$.

J. A. Melkebeek, *Electrical Machines and Drives*, Power Systems,
https://doi.org/10.1007/978-3-319-72730-1

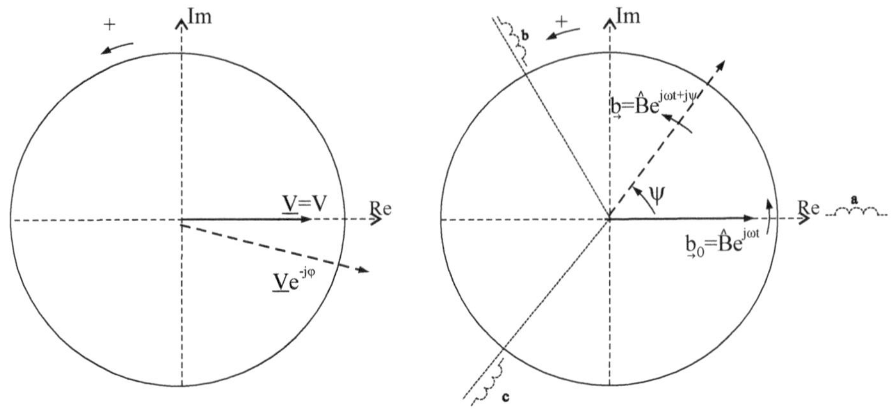

(a): phasors (b): space vectors (at t=0)

Fig. C.1 Phasors and space vectors

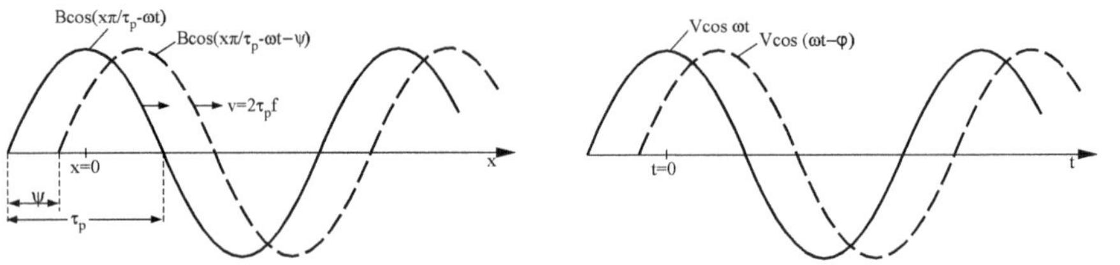

(a): two travelling fields at t=0 (b): two sinusoidal time functions

Fig. C.2 Phasors in time and space vectors in space

$x = \psi\tau_p/\pi$ whereas the field $b_0(x, t)$ is maximal in $x = 0$ at $t = 0$. In the figure, also a schematic representation of the three axes of the stator of a three-phase machine (in two-pole representation) is drawn. When the space vector $\underset{\rightarrow}{b}$ is co-linear with one of the phase axes, the flux in this phase is maximal at that instant (see also Chap. 3). *Remark: the difference between leading and lagging in time and space can also be illustrated as in Fig.* C.2. *Figure* C.2 *a shows the sinusoidal field distributions $b(x, t)$ and $b_0(x, t)$ as a function of the space co-ordinate x. Figure* C.2 *b on the other hand shows the sinusoidal voltages $v(t)$ and $v_0(t)$ versus time.*

C.2 Mathematical Extension

Whereas phasors are limited to pure sinusoidal quantities, space vectors are not. The space vector representation can be used for non-sinusoidal multiphase quantities and are even not limited to multi-phase symmetrical situations. For example a waveform can consist of two (or more) travelling or rotating waves with different angular frequency, however with the same wavelength (or pole pitch for an electrical machine). Also standing waves (with the same wave length) can be included. The space vector will then consist of the superposition of the space vectors for the separate waves.

To derive a more general mathematical approach, we start with considering three-phase purely sinusoidal quantities, e.g. the stator currents of a three-phase machine: $i_a(t) = \hat{I}_a \cos(\omega t - \varphi_a)$, $i_b(t) = \hat{I}_b \cos(\omega t - \varphi_b)$, $i_c(t) = \hat{I}_c \cos(\omega t - \varphi_c)$. Their phasor and time phasor representations are (using amplitude values): $\underline{I}_a = \hat{I}_a \exp(-j\varphi_a)$, $\underline{I}_b = \hat{I}_b \exp(-j\varphi_b)$, $\underline{I}_c = \hat{I}_c \exp(-j\varphi_c)$ and $\underline{i}_a(t) = \hat{I}_a \exp(j\omega t - j\varphi_a)$, $\underline{i}_b(t) = \hat{I}_b \exp(j\omega t - j\varphi_b)$, $\underline{i}_c(t) = \hat{I}_c \exp(j\omega t - j\varphi_c)$ respectively.

The currents considered are not necessarily 3-phase symmetrical, as indicated by the general amplitudes and phase angles. To study the resulting current layers or mmfs in the machine the symmetrical components transform can be used: $[\,\underline{I}_0\ \underline{I}_p\ \underline{I}_n]^T = \mathbf{T_s^{-1}}[\,\underline{I}_a\ \underline{I}_b\ \underline{I}_c]^T$ with

$$\mathbf{T}_s^{-1} = \frac{1}{3}\begin{bmatrix} 1 & 1 & 1 \\ 1 & a & a^2 \\ 1 & a^2 & a \end{bmatrix}$$

and $a = \exp(j2\pi/3)$. Usually, symmetrical components are applied to the phasor representation, but they can as well be used for the time phasor representation (the only difference being the time exponential for both input and output of the transform).

Applying the resulting positive, negative and zero-sequence components to the machine equations yields the corresponding current layer or mmf components. As the zero-sequence components are in most cases not important for machines, we will concentrate on the positive and negative sequence components, $\underline{I}_p$, $\underline{I}_n$ and $\underline{i}_p(t)$, $\underline{i}_n(t)$ for phasors and time phasors, respectively. For this sinusoidal currents $\underline{i}_p(t) = \underline{I}_p \exp(j\omega t)$, $\underline{i}_n(t) = \underline{I}_n \exp(j\omega t)$.

The space vector is now defined as $\underset{\rightarrow}{i}(t) = \underline{i}_p(t) + \underline{i}_n^*(t)$. If the original three-phase currents are symmetrical with positive sequence, then $\underline{i}_n(t) = \underline{I}_n = 0$ and $\underset{\rightarrow}{i}(t) = \underline{I}_p(t) = \underline{I}_p \exp(j\omega t)$. This is a vector with constant amplitude and speed rotating in the positive (counter-clockwise) direction in the space plane. Conversely, if the original currents are symmetrical with negative sequence then $\underline{i}_p(t) = \underline{I}_p = 0$ and $\underset{\rightarrow}{i}(t) = \underline{I}_n^*(t) = \underline{I}_n^* \exp(-j\omega t)$, which is a vector rotating clockwise, i.e. in the negative direction.

If the original currents are asymmetrical, i.e. contain both positive and negative sequence components (with the same frequency), the resulting space vector is the superposition of two counter-rotating waves. The result is then an elliptic wave,[3] see Fig. C.3.

If the original currents contain multiple frequency components, the resulting space vector is a superposition of the space vectors for each frequency. Therefore, the space

[3]This elliptical wave can also be regarded as the superposition of a standing wave and a rotating wave; prove this.

Fig. C.3 Elliptic wave

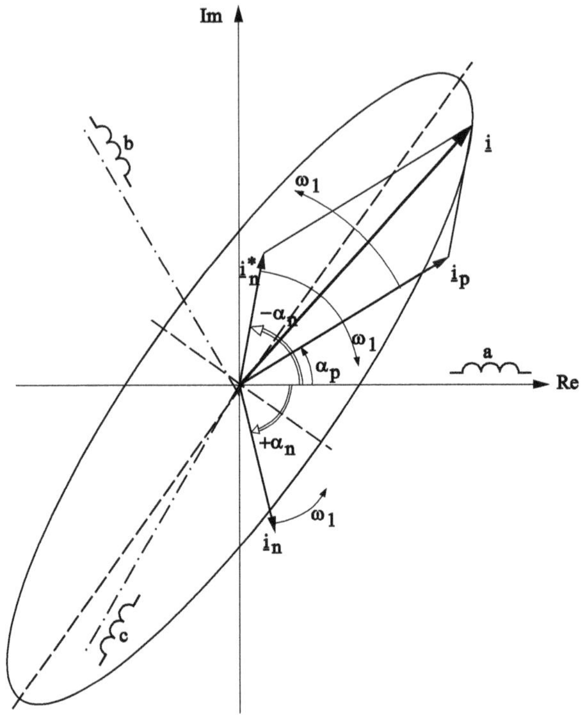

vector approach can also be used for power-electronic waveforms (by considering the Fourier expansion of the waveform).

From the foregoing, it can easily be shown that a more general mathematical definition of the space vector is as follows, using the real time quantities:

$$\underset{\rightarrow}{i} = \frac{2}{3}\left(i_a(t) + a \cdot i_b(t) + a^2 \cdot i_c(t)\right)$$

Using this expression, the space vector for a power-electronic waveform can quite easily be derived.

Physically, the notion of a space vector speaks for itself for current layers, mmfs or induction waves in an electrical machine. The space vector for a current layer ($\underset{\rightarrow}{a}$), mmf ($\underset{\rightarrow}{f}$) or induction ($\underset{\rightarrow}{b}$) always points in the direction of the maximal conductor current, mmf or induction, respectively, at that instant.[4] For voltages to the contrary, a physical interpretation is less clear. A somewhat artificial interpretation (for an electrical machine) is as follows. Suppose a layer of conductors (in the slots or in the air gap) at the circumference of the stator (or rotor). The emfs induced in these

[4]Note that if the space vector $\underset{\rightarrow}{i}$ for a current i is depicted, by convention this vector is aligned along the axis of the maximal current at that instant.

separate conductors by a rotating magnetic field form a rotating *voltage wave* in phase with the rotating magnetic field.

Nevertheless, space vectors (in the mathematical sense and definition), are frequently used for poly-phase systems, even without reference to an electrical machine.

References

1. L.F. Blume, A. Boyajian, Abnormal voltages within transformers. Trans. AIEE 577 (1919)
2. T. Bödefeld, H. Sequenz, *Elektrische Machinen* (Springer, Berlin, 1971)
3. H. de Jong, *A.C. Motor Design* (Clarendon Press, Oxford, 1976)
4. Faccioli, General Electric Review (1914), p. 749
5. A.J. Forsyth, High-frequency load-resonant DC-DC converters. Power Eng. J. (1999)
6. J.F. Gieras, M. Wing, *Permanent Motor Technology* (Marcel Dekker Inc., New York, 2002) (ISBN 0-8247-0739-7)
7. B.S. Guru, H.R. Hiziroglu, *Electric Machinery and Transformers*, 3rd edn. (Oxford University Press, Oxford, 2001)
8. M.R. Harris, P.J. Lawrenson, J.M. Stephenson, *Per-Unit Systems with Special Reference to Electrical Machines* (Cambridge University Press, Cambridge, 1970)
9. B. Heller, A. Veverka, *Surge Phenomena in Electrical Machines* (London Iliffe Books Ltd., London, 1968)
10. J. Hindmarsh, *Electrical Machines and their Applications*, 4th edn. (Pergamon Press, Oxford, 1991)
11. J. Hindmarsh, *Electrical Machines and Drives: Worked Examples* (Pergamon Press, Oxford, 1991)
12. D.G. Holmes, T.A. Lipo, *Pulse Width Modulation for Power Converters* (IEEE Press, Wiley-Interscience, 2003) (ISBN 0-471-20814-0)
13. IEEE Committee Report, Recommended phasor diagram for synchronous machines. Trans. IEEE **PAS-88**(11), 1593–1610 (1968)
14. C.V. Jones, *The Unified Theory of Electrical Machines* (Butterworths, London, 1967)
15. T. Kenja, A. Sugawara, *Stepping Motors and their Microprocessor Controls* (Clarendon Press, Oxford, 1994) (ISNB 0-19-859386-6)
16. D. Kinzer, S. Oliver, Monolithic HV GaN power ICs. IEEE Power Electron. Mag. 14–21 (2016)
17. R. Krishnan, *Switched Reluctance Motor Drives* (CRS Press, Boca Raton, 2001) (ISBN 0-8493-0838-0)
18. E.R. Laithwaite, The goodness of a machine. Proc. IEE **112**(3), 531–538 (1965)
19. W. Leonhard, *Control of Electric Drives*, 3rd edn. (Springer, Berlin, 2001) (ISBN 3-540-41820-2)
20. G. McPherson, R.D. Laramore, *Electrical Machines and Transformers* (Wiley, New York, 1990)
21. E. Levi, *Polyphase Motors: A Direct Approach to their Design* (Wiley, New York, 1984)
22. J.A. Melkebeek, Dynamisch gedrag van verzadigde gelijkstroommachines, induktiemachines en synkrone machines bij konventionele en vermogenselektronische voeding, UGent (1986)

J. A. Melkebeek, *Electrical Machines and Drives*, Power Systems,
https://doi.org/10.1007/978-3-319-72730-1

23. J.A.A. Melkebeek, D.W. Novotny, The influence of saturation on induction machine drive dynamics. Trans. IEEE **IA–19**(5), 671–681 (1983)
24. J.A. Melkebeek, J.L. Willems, Reciprocity relations for the mutual inductances between orthogonal axis windings in saturated salient pole machines. IEEE Trans. Ind. Appl. **26**(1), 107–114 (1990)
25. J. Melkebeek, Electrical drives (course text in Dutch: Elektrische Aandrijftechniek), Ghent University, 2016
26. J. Melkebeek, Controlled Electrical Drives, course text, Ghent University, 2016
27. J. Melkebeek, Dynamics of Electrical Machines and Drives, course text, Ghent University, 2016
28. T.J.E. Miller, *Switched Reluctance Motors and their Control*, Oxford Science Publications (Monographs in Electrical and Electronic Engineering) (Magna Physics Publishing, 1993). (ISBN 0-19-859387-2)
29. Murphy, Turnbull, *Power Electronic Control of AC Motors* (Pergamon Press, Oxford)
30. D.W. Novotny, Equivalent circuit steady state analysis of inverter driven electric machines. Department of ECE, University of Wisconsin-Madison
31. D.W. Novotny, T.A. Lipo, *Vector Control and Dynamics of AC Drives, Monographs in Electrical and Electronic Engineering* (Clarendon Press, Oxford, 1996). (ISBN 0-19-856439-2)
32. M.H. Rashid, *Power Electronics*, 3rd edn. (Pearson Education International, London)
33. R. Richter, *Elektrische Machinen* (Springer, Berlin, 1950)
34. P.C. Sen, *Principles of Electric Machines and Power Electronics*, 2nd edn. (Wiley, New York, 1996)
35. P. Sergeant, De synchrone machine, Chapter 5 from course text (in Dutch), Ghent University, 2016
36. R. Singh, S. Sundaresan, Fulfilling the promise of high-temperature operation with silicon carbide devices. IEEE Power Electron. Mag. 27–35 (2015)
37. K.W. Wagner, Das Eindringen einer elektromagnetischen Welle in eine Spule mit Windungskapazität, Elektrotechnik u. Masch.-Bau (1915) p. 89
38. T. Wildi, *Electrical Machines, Drives and Power Systems* (Pearson Education International, London, 2002)
39. J.L. Willems, *Stability Theory of Dynamical Systems* (Nelson, London, 1970)
40. J.L. Willems, *Elektrische Netten, course text* (in Dutch), Ghent University, Story, 1978
41. D. Zorbas, *Electric Machines* (West Publishing Co., Eagan, 1989)

GPSR Compliance
The European Union's (EU) General Product Safety Regulation (GPSR) is a set of rules that requires consumer products to be safe and our obligations to ensure this.

If you have any concerns about our products, you can contact us on

ProductSafety@springernature.com

In case Publisher is established outside the EU, the EU authorized representative is:

Springer Nature Customer Service Center GmbH
Europaplatz 3
69115 Heidelberg, Germany

www.ingramcontent.com/pod-product-compliance
Ingram Content Group UK Ltd.
Pitfield, Milton Keynes, MK11 3LW, UK
UKHW022218051026
15624UKWH00001B/219
9783319727295